Whither Turbulence and Big Data in the 21st Century?

Andrew Pollard • Luciano Castillo
Luminita Danaila • Mark Glauser
Editors

Whither Turbulence and Big Data in the 21st Century?

Springer

Editors
Andrew Pollard
Department of Mechanical and Materials Engineering
Queen's University
Kingston, ON, Canada

Luminita Danaila
Department of Mechanical Engineering
University of Rouen
Mont-Saint-Aignan, France

Luciano Castillo
Department of Mechanical Engineering
Texas Tech University
Lubbock, TX, USA

Mark Glauser
College of Engineering and Computer Science
Syracuse University
Syracuse, NY, USA

ISBN 978-3-319-41215-3 ISBN 978-3-319-41217-7 (eBook)
DOI 10.1007/978-3-319-41217-7

Library of Congress Control Number: 2016945165

Printed on acid-free paper

This Springer imprint is published by Springer Nature
The registered company is Springer International Publishing AG Switzerland

Bill George and John Lumley shared a passion for turbulence but had other passionate interests too.

John Lumley standing in front of a restored 1951 Mk. VI Bentley, August 4, 2007, which he did in his "spare" time. John took great pride and pleasure as he acted as a chauffeur for weddings of close friends. Photo taken by Shelley Blackler (the groom's mother and dear friend of the Lumleys). Photo supplied by John's daughter, Jennifer. John wrote a book Still Life with Cars: An Automotive Memoir, McFarland & Co., Jefferson, North Carolina and London, 2005 (ISBN 0-7864-2053-7).

Bill George aboard "Wings," the 42-foot sailboat that he and his wife sailed to Europe from America in 1995 and have lived aboard for extended periods since in Sweden and in various research locations throughout America and Europe. (Photo in Calais, FR, 2009, by Abofazl Shiri, one of many of Bill's Ph.D. students who wrote all or part of their dissertations aboard)

Preface

It is often said that advances in most fields of endeavor result from "standing on the shoulders of giants," and this meeting is no exception. In 1989, John Lumley, who needs no introduction to readers and researchers interested in turbulence, brought together leading thinkers and doers in turbulence to discuss the then-current controversies in the subject as well as to consider the role of public policy (and therefore funding) decisions that help to steer the field in either a direct way or through decisions which have unintended consequences. The meeting was international in scope and attendance and there resulted from this meeting a volume entitled "Whither Turbulence? Turbulence at the Crossroads." The present volume summarises the findings and presentations of another meeting that considered the broad question of "Whither Turbulence" in the context of the ubiquitous network of computers and networks. John Lumley was invited to and indeed enthusiastically supported it: "I am honored....you have my blessing, for what it is worth." This simple statement is a testament to his kind demeanor. Unfortunately, John could not attend due to illness and sadly he passed away in late May 2015.

In the intervening years or so between Lumley's volume and the present one, much has happened and new giants have emerged in this, the oft-said "last unsolved problem in classical physics." A significant disrupter to and leader in our field is Professor W. K. "Bill" George who was also Lumley's student. From George, there have emerged many academic children and now grandchildren, each of whom continues to provide leadership and impact on the field. Given his 5-decade long career, the meeting, details of which are provided within these pages, was dedicated to Bill on the occasion of his 70th birthday.

In 25 years, the world of research in turbulence has changed to where computation and simulation has grown to become the third leg of the scientific stool. In fact, with the web/Internet, commodity computing, high-performance computing, and significant advances in experimental tools, especially particle image velocimetry, it could be said that what was a dream in 1989, say active control of turbulence, is now becoming a reality because the three legs of the stool (theory, simulation, and experiment) have each advanced, and Bill has been leading the charge on at least two of those.

However, it remains to be seen what state we will be in 2040. The meeting in Cargese began 50 years after Gordon Moore predicted the future of the semiconductor: a doubling of computer processing speeds every two years. A look at Bill Reynolds' paper in the Lumley volume (Fig. 1, p. 342) suggests similar growth such that peta-flops are now reachable (as at 2008), with exascale computing on the near horizon (expected by 2020). One can imagine even further ubiquitous computational infrastructure and new and even more exciting methods, algorithms, and most importantly, ideas. But a significant issue now is data and this will continue to grow. In 1989, again with reference to Reynolds' paper, computer memory sizes were of order gigabytes, while in 2015, terabyte drives are ubiquitous and cheap. An example of drivers for increased data storage and bandwidth is the square kilometer array (radio telescope) that will produce about 30 exabytes of data per month, which will require a doubling of the current Internet traffic bandwidth, worldwide! One can imagine that while the turbulence community will continue to push the Reynolds number envelope, it will be in combination with other physicochemical processes (e.g., high-Schmidt-number turbulent mass transfer) over the full spectrum of scales (nano- to full scale, including planetary scale).

Kingston, ON, Canada — Andrew Pollard
Lubbock, TX, USA — Luciano Castillo
Mont-Saint-Aignan, France — Luminita Danaila
Syracuse, NY, USA — Mark Glauser

Acknowledgments

The organizers of the Whither Turbulence meeting would like to extend their thanks to the supporters of the meeting. These are:

The National Science Foundation
Queen's University at Kingston
US Office of Naval Research
Global Office of Naval Research
CNRS
Institut P'
ERCOFTAC
CORIA
Association Française de Mécanique

The meeting organization was ably assisted by Gabrielle Whan of Queen's University and Linda Whitehead at Texas Tech University.

The organizers also wish to acknowledge the following individuals for their assistance during the meeting, which ensured that any difficulty was resolved with notable aplomb:

Christophe Latailleur, CORIA
Amandine Cyr, CORIA
The Cargese staff: Dominique Donzella, Nathalie Giudicelli, Pierre-Eric Grossi, and Brigitte Cassegrain

Finally, we wish to thank Springer-Verlag for enthusiastically supporting the publication of this volume, especially Michael Luby and Brinda Megasyamalan.

Overview of Volume

The meeting itself followed a traditional format held over a 5-day period in the delightfully salubrious environment of the Institut d'Etudes Scientifiques de Cargèse, Corsica. The meeting was divided into 15 sessions and two extensive discussion sessions. These were:

Turbulence, Then, Now and Future; Turbulence Control; Turbulent Boundary Layers; Simulations and Fluid Dynamics; Turbulent Structures and Jets; Turbulent Boundary Layers over Rough Surfaces; High Reynolds Numbers; Atmospheric Flows and Theory; Turbulence Theory; Turbulence and Renewable Energy; Wall-Bounded Flows; Large Data; Complex and Industrial Flows; Turbulence; and Simulations and Experiments.

The two discussion sessions were devoted to Challenges in Turbulence in the Twenty-First Century: What Problems We Should Focus On in the Next 20 Years? and Big Data: Opportunities for Collaborations and Dealing with Large Databases, which benefitted from an extensive video conversation between Profs. Hacker and Dyke at Purdue University and the delegates at the meeting. These two discussion sessions were recorded and the transcription of them is included here.

The volume begins with two papers that provide some perspective on the field of turbulence: Christos Vassilicos' delightful article "From Tennekes and Lumley to Townsend and to George: A Slow March to Freedom" and Bill George's "A 50-Year Retrospective and the Future," wherein he once again challenges the community to think differently. Thereafter, we have assembled the remaining contributions according to themes. These are Turbulent Boundary Layers, Jets, Environmental and Wind Energy, Data Manipulation, and General Topics.

The remaining 22 contributions reflect the diversity of the topics considered and extensively discussed at the meeting. These original research contributions are arranged as noted above, and each author has endeavored to place into context the link between their contribution and Big Data.

The Turbulent Boundary Layer section contains four papers. Marusic et al. consider the response of a TBL to different tripping conditions under "fixed and carefully quantified initial conditions" and to determine "under what conditions the effects of upstream trip and other initial conditions no longer play a role in defining the state of the boundary layer." They find that after sufficient development length they reach a "converged state." From their perspective, the data set obtained from such broad spatial and temporal resolution experiments requires a community-type decision on what data should be retained for future use. Soria et al. present DNS of zero pressure gradient and adverse pressure gradient TBLs starting from the same initial condition argue the existence of a self-similar flow regardless of the pressure gradient. Their perspective on Big Data is for our community to recognize that many of the tools required have been and continue to be developed by computer scientists and recommend the need for broader interaction with them. Doosttalab et al. consider transitionally rough TBL using DNS and find that Townsend's hypothesis, where the characteristics of a turbulent flow are independent of surface roughness beyond about 5 roughness heights, is invalid. They note that the Big Data issue revolves around having access to adequate HPC resources. Shahab et al. perform a quadrant analysis of a TBL that is perturbed by a shock and find that the higher-order correlations inform further those data obtained from lower-order statistical data. They introduce the idea of the 4 Vs of Big Data, volume, velocity, value, and variety, where velocity refers to the timeliness of the data.

The section on Jets begins with deepening fundamental insight on equilibrium similarity and scale-by-scale energy budgets in the near to intermediate field of a round free jet. They comment on data analytics and provide a useful excursion into high-performance computing, networks, and data archiving. Tinney et al. focus on rocket nozzle plumes and noise generation to focus on the "link between the sources of most intense vibro-acoustic loads that form during ignition ...." They speculate that in the future there will be a need for "strong synergy between both experimental and numerical disciplines that leverages carefully designed measurements with robust, yet simple, computational models with built-in analytics." Voivenel et al. consider variable viscosity jets and determine that "the presence of a strong viscosity discontinuity across the jet edge results in an increase in both the scalar spread rate and the turbulent fluctuations." At this stage of their work, the Big Data issue seems to be manageable with current resources. Hodzic et al. performed stereoscopic PIV in a round jet to test the robustness of Lumley's projection approach. They noted the sensitivity of the results to spatial resolution. They argue that proper orthogonal decomposition is a useful tool to filter and therefore compress the size of a database to encapsulate the essential information contained in a larger database.

In the Environmental and Wind Energy section, we begin with Armenio's consideration of spatial scales: from the laboratory to the real world. He focuses on physics of turbulence at laboratory scale which then informs the real-world environmental fluid mechanics issues. Hangan et al. introduce a novel wind engineering facility called the WINDEEE dome, which is "a hexagonal chamber of 25 m in diameter surrounded by a 'return circuit' of the same hexagonal shape of 40 m in diameter with the aim to create a wide variety of wind systems (e.g., tornadoes, downburst, all kind of gusts and currents, shear winds and boundary layers, etc.) at large scales and Reynolds numbers." Clearly, the data deluge from real-world simulations and experiments as considered by Armenio and Hangan et al.

is challenging! Martin et al. consider supervisory control and data acquisition (SCADA) data gathered over an 18-month period from a 67-wind-turbine farm. From these data, they identified four wake effects and provided insight into wind farm performance that is richer than simple power curve analysis. It is clear that the time and spatial scales involved produce huge "Big Data" issues that the SCADA approach may alleviate. Ali et al. delve more deeply into wind turbine wakes through probing with hot wires of the wake of a wind turbine array in a wind tunnel.

Big Data suggests volumes of data rather than necessarily the other 3 Vs introduced by Shahab et al. In this section which is devoted to data manipulation, various compression and filtering approaches are introduced so that inherently each author recognizes the Big Data issue and have decided to address it from a different perspective. Magstadt et al. consider various jet flows that produce pressure, velocity, acoustic, and other data and "different levels of granularity, density (or sparseness)/distribution (uniform, checkered, lattice, random, etc.) and span in space and time to develop a holistic systems-level understanding". They apply Big Data analyses/modeling tools (the right filters) to identify patterns and predictive models rather than just a posteriori trends, statistics and distributions. Bai et al. consider the use of machine learning where the "overarching goal is to reduce the burden of data acquisition and processing." Buchave and Velte recognize the limits of data obtained in the temporal domain, and they propose a method to eliminate the need to impose Taylor's frozen eddy hypothesis and find a way to obtain spatial gradients from both hot wire and laser Doppler signals.

The papers to be found in the General Aspects of Turbulence are listed in alphabetical order. Barros et al. consider blowing as a control strategy and deduce a 30 % recovery in base pressure for a simplified blunt vehicle. Fureby explores the state-of-the-art large eddy simulation in the context of large-scale engineering flows with reliance on DNS and a variety of other thermophysical models. He provides an extensive bibliography culminating in petascale DNS of combustion. Grinstein et al. consider under-resolved velocity fields (and initial conditions) as a prelude to accurate predictions and quantifiable uncertainty in turbulent material mixing and relate these ideas to Bill George's ideas on initial conditions. They conclude that "Ensemble averaging over a suitably complete set of realizations covering the relevant IC (initial condition, ed.) variability is a data reduction strategy of choice". Meldi and Sagaut consider homogeneous, isotropic turbulence using an eddy-damped quasi-normal Markovian (EDQNM) model and provide insight into the time evolution of $C_{\varepsilon.}$ Orlandi et al. return to a DNS of the minimal flow unit (channel flow) to explore why an inertial range forms and to further inform the separate effects of large- and small-scale structures. Rahbari and Scalo consider compressible channel flow (DNS) with a general impedance boundary condition to effect flow control. They introduce the idea of "Small Data" and "Big Data" and emphasize the need for both. Tardu rounds out the individual papers by addressing the important problem of dissipation and particularly how the "palm" statistics ("palm statistics are the statistics of a given quantity under the condition that another stochastic process crosses a fixed level") that help inform mean dissipation conditioned by level crossings of the spanwise velocity seems independent of y (essentially the buffer layer region, ed.) using a DNS of channel flow at $Re_\tau = 1100$. He argues that for computational turbulence scientists should agree on common data formats given the multi-terabyte databases being generated.

The transcription of the 90-minute Discussion Session 1 provides a broad exploration of Whither Turbulence, and the reader will note many different threads. Of course, given the 60+ attendees, you can imagine there are at least 61+ arguments to consider! It became clear that cross-laboratory collaboration is an important theme that will enable large-scale experiments and computations to take place in a collaborative manner. There are many internationally competitive and unique experimental facilities across the world that are willing to share access; of course funding issues remain, including transportation and subsistence costs for enabling this mobility.

Discussion Session 2 considers Big Data. This session, which was two hours long, included a video link with Purdue University and Professor Tom Hacker of the Department of Computer and Information Technology and Professor Shirley Dyke of Mechanical and Civil Engineering. They are connected to the National Earthquake Engineering Simulation (http://nees.org) and provided a basis for the extensive discussions on Big Data, cyberinfrastructure, and collaborative environments. The perspective they brought to the general discussion triggered many parallel ideas on data archiving, transmission, and sharing. As a result of these discussions, Prof. Hacker and Dyke were invited to provide a written contribution to this volume. Furthermore, the editors invited Profs. Meneveau and Marusic (of Johns Hopkins and Melbourne, respectively), the former being unable to join the meeting directly; Dr. Sillero and Prof. Jimenez of Universidad Politecnica de Madrid, who could not participate directly; and Prof. Menon and Ranjan of Georgia Tech and Dr. Oefelein of Sandia National Labs, who were approached after the meeting to contribute their ideas and perspectives on Big Data. As the reader will appreciate, there are manifold interpretations of what Big Data means. The emergence of big databases for turbulence research will continue to evolve, and the community will continue to find efficacious ways to both create and interface with them. The invited contributors address the Big Data issue from very different perspectives, and with the other perspectives mentioned, we hope that they will stimulate the turbulence and fluid dynamics community to work with the broader Big Data communities and agencies before they force us into a position that is not in our best interest.

Queen's University, Kingston, ON, Canada — Andrew Pollard
Texas Tech University, Lubbock, TX, USA — Luciano Castillo
University of Rouen, Mont-Saint-Aignan, France — Luminita Danaila
Syracuse University, Syracuse, NY, USA — Mark Glauser

Contents

Part I
Historical Perspectives

Chapter 1
From Tennekes and Lumley to Townsend and to George: A Slow March to Freedom

J.C. Vassilicos

1.1 Introduction

The spring of 2015 has been a landmark for the turbulence research community. In April, Bill George turned 70 and a conference was held in Cargese, Corsica, in his honour. A little more than a month later, on May 30th, John Lumley passed away. John was a central reference figure in turbulence research in the second half of the twentieth century and also the PhD supervisor of Bill George. The spring of 2015 has therefore brought some pause for thought and reflection on the ups and downs, attempts, failures and achievements, vigorous debates, agreements and disagreements on various aspects of the turbulence problem(s) over the past 50 years.

I know Bill George far better than I have known John Lumley. The last time I talked to John Lumley was in the summer of 2000 at the European Turbulence Conference in Barcelona. He came up to me when we were heading towards a social function organised by the conference and said: 'do you mind if I stick with you, I do not know anyone here any longer'! I knew instantly that I would never forget this comment. Perhaps it meant that the European turbulence community did not interact much with the American one at the time? But it surely also signalled the advent of a new generation and with it, perhaps, a shift of emphasis on what is worth researching. The other two landmarks that come to my mind whenever I think of this episode are June 2nd 1986 when Stan Corrsin passed away, and March 1989 when the meeting 'Whither Turbulence? Turbulence at the Crossroads' was held in Cornell, having been organised by John Lumley. I have often had the sense that a way was lost in turbulence research after these two dates'. Turbulence at

J.C. Vassilicos (✉)
Turbulence, Mixing and Flow Control Group, Department of Aeronautics, Imperial College London, London SW7 2BY, UK
e-mail: j.c.vassilicos@imperial.ac.uk

A. Pollard et al. (eds.), *Whither Turbulence and Big Data in the 21st Century?*,
DOI 10.1007/978-3-319-41217-7_1

the Crossroads' was perhaps a prescient warning. A decade later one of the central figures in turbulence did not recognise his own research community.

Bill George often recounts how, at a turbulence workshop in Monte Verita, Switzerland, in 1998, John Lumley approached him, and referring to Chap. 4 'Boundary-free shear flows' of the classic 1972 textbook 'A first course on Turbulence' (that John had co-authored with H. Tennekes), told him: 'I now teach it your way'. This is a good example of the vigorous debates, agreements and disagreements that gripped the turbulence community through the time which spanned the careers of these two men and I have chosen to concentrate my contribution to this book on boundary-free shear flows. This book is a memento of the Cargese April 2015 meeting which celebrated the achievements of Bill George but which also intentionally echoed John Lumley's 1989 meeting through its title 'Whither turbulence and Big Data in the 21st Century'; I guess, perhaps trying to find our way again.

1.2 Self-preserving Turbulent Wake: Townsend [13] and Tennekes and Lumley [12]

The particular boundary-free turbulent shear flow which I discuss here for the purpose of concrete illustration is the self-preserving turbulent wake. Usually one assumes that a planar or axisymmetric turbulent wake becomes self-preserving/self-similar at some distance downstream where the details of the wake-producing obstacle have lost their influence on the flow. The problem is to know how the turbulence wake's mean velocity deficit and mean width vary with streamwise distance.

Following Townsend [13], Tennekes and Lumley [12] present a solution to this problem which is based on the Reynolds-averaged streamwise momentum equation. Concentrating attention on the axisymmetric case to more concretely ground thoughts, the Reynolds-averaged streamwise momentum equation is

$$U_\infty \frac{\partial}{\partial x}(U_\infty - U) = -\frac{1}{r}\frac{\partial}{\partial r} r\langle u_x u_r\rangle \tag{1.1}$$

where U_∞ is the uniform free stream velocity and U is the streamwise mean flow velocity at streamwise distance x from the obstacle and radial distance r from the streamwise centreline. The Reynolds shear stress is $\langle u_x u_r\rangle$ in usual understandable notation.

The self-preserving solutions considered by Townsend [13] and Tennekes and Lumley [12] are of the form

$$U_\infty - U = u_0(x) f[r/\delta(x)] \tag{1.2}$$

and

$$\langle u_x u_r \rangle = u_0^2 g[r/\delta(x)]. \tag{1.3}$$

Note the strong assumption, later relaxed by Townsend [14] and George [2], that the Reynolds shear stress scales with the square of the centreline mean velocity deficit u_0. With this assumption, the problem to find the x-dependencies of $u_0(x)$ and the wake width $\delta(x)$ is simply solved by noting that (1.1) admits solutions of the form (1.2) and (1.3) under the conditions

$$\frac{d}{dx}\delta(x) \sim \frac{u_0}{U_\infty} \text{ and } u_0\delta^2 \sim U_\infty\theta^2 = \text{Const}$$

where θ is the conserved momentum thickness. There are therefore two conditions for two unknowns, u_0 and δ, leading to the solution (where x_0 is a virtual origin)

$$u_0/U_\infty \sim \left(\frac{x - x_0}{\theta}\right)^{-2/3} \tag{1.4}$$

and

$$\delta/\theta \sim \left(\frac{x - x_0}{\theta}\right)^{1/3} \tag{1.5}$$

which can be found in many turbulence textbooks, in particular Townsend [13, 14], Tennekes and Lumley [12], Mathieu and Scott [6] and Pope [9].

1.3 Townsend [14]

With Townsend [14] starts the 'slow march to freedom' mentioned in the title of this contribution. In the 1976 revision of his 1956 book, Townsend relaxed the self-preservation ansatz of the Reynolds shear stress and replaced (1.3) with

$$\langle u_x u_r \rangle = R_0 g[r/\delta(x)]. \tag{1.6}$$

where R_0 is an extra free parameter and not necessarily proportional to u_0^2. This extra freedom leads to the two conditions

$$\frac{d}{dx}\delta(x) \sim \frac{R_0}{U_\infty u_0} \text{ and } u_0\delta^2 \sim U_\infty\theta^2 = \text{Const}$$

which are necessary for (1.2) and (1.6) to be consistent with (1.1). These two conditions now involve three, rather than two, unknowns, namely u_0, δ and R_0 and are therefore not conclusive by themselves. This is effectively a closure problem, as is so typically the case in turbulence, and Townsend [14] introduced the idea of using the turbulent kinetic energy equation and the Taylor–Kolmogorov turbulence dissipation scaling to close this problem. However he still over-restricted the problem by assuming that the turbulent kinetic energy's self-preservation form is

$$K = R_0 k[r/\delta(x)]. \tag{1.7}$$

The result of Townsend's [14] approach was identical to Townsend's [13] and Tennekes and Lumley's [12] equations (1.4) and (1.5). Allowing more freedom to the problem led to the same x-dependencies of the wake's mean flow velocity deficit and width. If this was not enough, Townsend's [14] approach returned $R_0 \sim u_0^2$ by itself. The extra freedom Townsend [14] gave was not appreciated by the problem which naturally returned to the same conclusions and even assumptions of Townsend [13] and Tennekes and Lumley [12]. The one extra gain of Townsend's [14] approach was the scaling $K = R_0 k[r/\delta(x)] \sim u_0^2 k[r/\delta(x)]$.

1.4 George [2]

George [2] took an extra, crucial, step in this slow march to freedom. Like Townsend [14] he left R_0 unconstrained but unlike Townsend [14] he assumed

$$K = K_0 k[r/\delta(x)]. \tag{1.8}$$

and did not constrain K_0 to be proportional to R_0. He left K_0 as a free parameter.

George's [2] approach is based on the momentum equation (1.1) and associated self-preservation forms (1.2) and (1.6) as well as the turbulence kinetic equation

$$U_\infty \frac{\partial}{\partial x} K = -\langle u_x u_r \rangle \frac{\partial}{\partial r} U + T - \epsilon \tag{1.9}$$

and associated self-preservation forms (1.8),

$$T = T_0 t[r/\delta(x)]. \tag{1.10}$$

for the transport and pressure terms and

$$\epsilon = D_0 e[r/\delta(x)]. \tag{1.11}$$

for the turbulence dissipation rate. This is similar to Townsend [14] except that Townsend used a slightly different form of the turbulent kinetic energy equation and took K_0 to be proportional to R_0.

George's [2] theory leads to five conditions for six unknowns, namely u_0, δ, R_0, K_0, T_o and D_0 and, again, requires closure to provide answers. This closure is achieved as in Townsend [14] by using the Taylor–Kolmogorov scaling

$$D_0 \sim K_0^{3/2}/\delta. \tag{1.12}$$

Readers who are novices in turbulence research might be surprised to read at this stage that all this extra freedom given by George [2] is in some sense inconsequential because the results of George's [2] theory are (1.4) and (1.5) again and therefore identical to those of Townsend [13], Tennekes and Lumley [12] and Townsend [14]. Furthermore, Townsend [14] and George [2] both predict $K_0 \sim u_0^2$. It would seem that however much freedom you allow into the problem, the predictions are always the same. Even so, John Lumley felt it necessary to tell Bill George in 1998 that 'I now teach it your way', meaning in the way of this section rather than Sect. 1.2.

There is in fact one important difference between George [2] and the other theories. Whereas for Townsend [13], Tennekes and Lumley [12] and Townsend [14] $R_0 \sim u_0^2$, George [2] predicts $R_0 \sim U_\infty u_0 \frac{d}{dx}\delta$. The scalings (1.4) and (1.5) imply that $R_0 \sim u_0^2$ and $R_0 \sim U_\infty u_0 \frac{d}{dx}\delta$ return the same x-dependence for R_0. However, if the x-scalings of u_0 and δ were different, as can happen if the dissipation's scalings differ from (1.12), then it would be possible to distinguish between $R_0 \sim u_0^2$ and $R_0 \sim U_\infty u_0 \frac{d}{dx}\delta$.

Dissipation scalings different from (1.12) have indeed been discovered recently in various turbulent flows, and in the axisymmetric turbulent wake in particular (see [1, 3, 8, 15]). This has created an opportunity for the distinction between George [2] and the other theories of self-preserving axisymmetric turbulent wakes [12–14] to become manifest and meaningful.

1.5 Turbulence Dissipation Scalings

The Taylor–Kolmogorov dissipation law first proposed by Taylor [11] and then given a theoretical underpinning by Kolmogorov [5] is

$$D_0 \sim K_0^{3/2}/L \tag{1.13}$$

where L is an integral length-scale. Kolmogorov's [5] theoretical justification of (1.13) is given in the framework of his equilibrium cascade theory where the interscale energy flux at length-scales comparable to L scales as $K_0^{3/2}/L$ and balances the turbulence dissipation at the smallest, viscosity-dominated, length-scales (see [15]). Townsend [13, 14] presents arguments in support of $L \sim \delta$ which allows (1.13) to be written as (1.12), the form which provides the closure leading to the wake laws (1.4) and (1.5).

It is intriguing that there is another way to obtain (1.12), and this is via the strong self-preservation theorem proved by Johansson et al. [4] in the Appendix of their paper. These authors showed that if one uses all individual component Reynolds stress equations (neglecting viscous terms for high enough Reynolds number) instead of the single turbulent kinetic energy equation used by Townsend [14] and George [2] and if one assumes that every single term in each one of these equations is self-preserving (strong self-preservation), then (1.12) follows directly. There is no need to assume the validity of the dissipation scaling (1.12) in this strong self-preservation scenario, it just follows.

Conversely, the strong self-preservation theorem implies that if (1.12) is violated then there have to be some terms in some of the individual component Reynolds stress equations which are not self-preserving. Recent wind tunnel experiments of high Reynolds number axisymmetric turbulent wakes [1, 8] have shown that in a region extending in the streamwise direction between about ten and at least fifty times the size L_B of the wake-generating obstacle, the dissipation scalings are given by

$$D_0 \sim U_\infty L_B K_0/\delta^2 \tag{1.14}$$

which is very different from (1.12). This new dissipation law characterises non-equilibrium interscale energy exchanges because one can still expect the interscale energy flux at length-scales comparable to L to scale as $K_0^{3/2}/L$ (see arguments based on the Karman–Howarth equation in Vassilicos [15]), in which case the turbulence dissipation at small scales clearly does not balance this large-scale flux. It is remarkable that forms equivalent to (1.14) seem to hold in a variety of turbulent flows (see [3, 15]), suggesting some universality to non-equilibrium turbulence.

Replacing the dissipation relation (1.12) with the non-equilibrium relation (1.14) in the theory of George [2] leads to

$$u_0/U_\infty = C_1 \left(\frac{x-x_0}{\theta}\right)^{-1} \theta/L_B \tag{1.15}$$

and

$$\delta(x)/L_B = C_2 \left(\frac{x-x_0}{\theta}\right)^{1/2} (L_B/\theta)^{1/2}. \tag{1.16}$$

Extensive hot wire anemometry measurements have confirmed these two scalings in a streamwise region between $x \approx 5L_B$ and at least $x = 50L_B$ [1, 7, 8]. The use of systematically different plates with same surface area A has made it clear that the dimensionless constants C_1 and C_2 depend on the geometrical details of the wake generator [1, 7, 8].

Given that the wake scalings in this region are different from (1.4) and (1.5), this is the ideal ground where to test the George [2] approach against those of Townsend [13, 14] and Tennekes and Lumley [12]. The direct numerical simulations (DNS) of Dairay et al. [1] gives clear support for George's

$$R_0 \sim U_\infty u_0 \frac{d}{dx}\delta(x) \tag{1.17}$$

and clearly invalidates $R_0 \sim u_0^2$ which is an assumption for Townsend [13] and Tennekes and Lumley [12] and a prediction for Townsend [14]. George's [2] footsteps in this slow march to freedom have therefore been indispensable, at least for the region where the non-equilibrium dissipation law rules and (1.15) and (1.16) hold too. The DNS of Redford et al. [10] suggest that (1.4) and (1.5) actually hold very much further downstream where the local Reynolds number has dropped to much lower values and, presumably, the dissipation scalings have changed from (1.14) to (1.12).

1.6 $K_0 \sim u_0^2$? The March to Freedom is Not Over

As mentioned in Sect. 1.2, the justification often given for self-preservation of flow profiles is that the details of the wake-producing obstacle lose their influence on the flow at some distance far enough downstream. It is clear from the self-preservation theorem of Johansson et al. [4] that there cannot be strong self-preservation in the region where (1.14) holds instead of (1.12). Recent wind tunnel experiments and DNS by Nedic et al. [7] and Dairay et al. [1] have shown that profiles such as those of the mean flow velocity, the Reynolds shear stress, the turbulent kinetic energy and the turbulence dissipation are self-preserving in this region whereas other flow profiles are not. Their data also invalidate $K_0 \sim u_0^2$ in this region and strongly suggest

$$K_0 \sim R_0 \tag{1.18}$$

instead. This is important because both theories which can make a prediction on K_0, those of Townsend [14] and George [2], predict $K_0 \sim u_0^2$.

The failure of $K_0 \sim u_0^2$ points to a failure of (1.9) in the context of George's [2] approach because approximating the turbulence production by $-\langle u_x u_r \rangle \frac{\partial}{\partial r} U$ is essential for obtaining $K_0 \sim u_0^2$. The DNS of Dairay et al. [1] shows that the turbulence production is dominated by normal stress terms on and around the centreline and that these normal stress terms are not negligible off centreline either. It also shows that the profiles of these normal stresses are not self-preserving at least till $x = 100L_B$ which is consistent with the strong self-preservation theorem and the failure of the Taylor–Kolmogorov dissipation law in this region.

There is therefore a need to grant new freedoms by not constraining all terms in the energy equation to be self-preserving. Dairay et al. [1] have therefore been led to propose a new approach which is based on the momentum equation (1.1) and associated self-preservation forms (1.2) and (1.6) as well as the turbulence kinetic equation

$$U_\infty \frac{\partial}{\partial x} K = P + T - \epsilon \tag{1.19}$$

where only K and ϵ are self-preserving but not P and T (though it of course follows from (1.19) that $P + T$ is self-preserving). The self-preservation forms of K and ϵ are again given by (1.8) and (1.11), respectively, and the problem to determine the a priori independent quantities u_0, R_0, δ, K_0 and D_0 requires more information to be closed. The information is enough as it is to give (1.17), very much like in George [2], but not more. If one assumes that the turbulence is out of two-point equilibrium and obeys (1.14), then this extra piece of information allows (1.15) and (1.16) to follow, in agreement with experimental and numerical data. Hence we have a consistent theoretical framework in agreement with observations, at least for the region where the turbulent dissipation scales as (1.14). More details and discussion can be found in Dairay et al. [1].

1.7 Conclusion

This slow march to freedom has taken about 60 years and does not seem to have fully ended yet. The important progress brought about by the seminal contributions of Townsend [14] and George [2] has led us to new questions. For example, given that the dimensionless constants C_1 and C_2 in (1.15) and (1.16) depend on the details of the wake-generating body, thereby indicating a clear dependence of the flow on initial conditions, why is it that some flow profiles are self-preserving in the region where (1.15) and (1.16) hold? What is the root cause of self-preservation?

Secondly, what happens at the far downstream point where the wake laws (1.15) and (1.16) cease to hold and the traditional wake laws (1.4) and (1.5) take over as the DNS of Redford et al. [10] would suggest? Are there similar transitions from one scaling to another in other boundary-free turbulent shear flows? And what does such a transition imply for self-preservation?

And finally, what is the reason for (1.18) and what are the cascade physics behind the non-equilibrium dissipation law (1.14)? How do these physics change to give rise to the expected scaling (1.12) at some point downstream?

Big data sets generated by experimental measurements and DNS have helped make progress beyond Townsend [14] and George [2] and will definitely play an important role in answering these new questions. But most important of all, one conclusion of the 2015 meeting 'Whither turbulence and Big Data in the 21st Century' must surely be that one must ask the right questions in the first place.

Acknowledgements This may be the right place to record my deepest gratitude to Myriam Scheel Larsen for her very careful, caring and conscientious teaching. She has been a link between Bill George and myself decades before we knew it. There must have been days 35 or so years ago in Denmark, when she would come to school to teach introductory physics and chemistry to my class in the morning and then have dinner in the evening with Bill George who was already Professor and who was visiting her husband, Professor Poul Scheel Larsen.

References

1. T. Dairay, M. Obligado, J.C. Vassilicos, Non-equilibrium scaling laws in axisymmetric turbulent wakes. J. Fluid Mech. **781**, 166–195 (2015)
2. W.K. George, The self-preservation of turbulent flows and its relation to initial conditions and coherent structures, in *Advances in Turbulence*, ed. by W.K. George, R. Arndt (Hemisphere Publishing Corp., New York, 1989), pp. 39–73

3. S. Goto, J.C. Vassilicos, Energy dissipation and flux laws for unsteady turbulence. Phys. Lett. A **379**(16–17), 1144–1148 (2015)
4. P.B.V. Johansson, W.K. George, M. Gourlay, Equilibrium similarity, effects of initial conditions and local Reynolds number on the axisymmetric wake. Phys. Fluids **15**(3), 603–617 (2003)
5. A.N. Kolmogorov, Dissipation of energy in locally isotropic turbulence. Dokl. Akad. Nauk. SSSR **32**, 16–18 (1941)
6. J. Mathieu, J. Scott, *An Introduction to Turbulent Flow* (Cambridge University Press, Cambridge, 2000)
7. J. Nedic, J.C. Vassilicos, B. Ganapathisubramani, Axisymmetric turbulent wakes with new non-equilibrium similarity scalings. Phys. Rev. Lett. **111**(14), 144503 (2013)
8. M. Obligado, T. Dairay, J.C. Vassilicos, Non-equilibrium scalings of turbulent wakes. Phys. Rev. Fluids (2016, to appear)
9. S.B. Pope, *Turbulent Flows* (Cambridge University Press, Cambridge, 2000)
10. J.A. Redford, I.P. Castro, G.N. Coleman, On the universality of turbulent axisymmetric wakes. J. Fluid Mech. **710**, 419–452 (2012)
11. G.I. Taylor, Statistical theory of turbulence. Proc. R. Soc. Lond. A **151**, 421–444 (1935)
12. H. Tennekes, J.L. Lumley, *A First Course in Turbulence* (MIT Press, Cambridge, 1972)
13. A.A. Townsend, *The Structure of Turbulent Shear Flow* (Cambridge University Press, Cambridge, 1956)
14. A.A. Townsend, *The Structure of Turbulent Shear Flow* (Cambridge University Press, Cambridge, 1976)
15. J.C. Vassilicos, Dissipation in turbulent flows. Ann. Rev. Fluid Mech. **47**, 95–114 (2015)

Chapter 2
A 50-Year Retrospective and the Future

William K. George

2.1 Big Data

Experimental turbulence research has always been about "Big Data"—and usually never enough of it. Part of the reason has been because of the need to use statistical measures. Data records measured in thousands of time integral scales are necessary to make even the simplest estimators converge, sometimes even tens and hundreds of thousands of integral scales in length for probability density functions and correlations at large lags. As a general rule the time (or length of record required) for a given statistical error is proportional to the rms fluctuations of the statistical quantity being estimated divided by the square root of the number of effectively independent realizations of it. Note that the variance of *the quantity being measured* is not the same as the variance of the underlying process. For example, if a second moment is to be measured its variance is $\langle [u^2 - \langle u^2 \rangle]^2 \rangle$, which for a Gaussian process is 3 $[\langle u^2 \rangle]^2$, and can be much larger for non-Gaussian processes which are common in turbulence. The pre-multiplying factor for simple powers of the variance increases rapidly with the order of the moment, so demands on data length can increase very rapidly (v. [15, 29, 30], or appendices of my turbulence notes available at www.turbulence-online.com). The same is true for attempts to measure events of decreasing probability (like the tails of a pdf), since the lower the probability of it being observed, the more "statistically independent" data that must be acquired to measure it. Fractional statistical error, or variability, is the rms fluctuation of the quantity being measured divided by its average or expected value, or the variability of the quantity desired itself. So the higher the variability of the process, the more independent samples are required to estimate it. Quantities with zero mean will always have infinite variabilities, but finite errors. Many a student has thrown away

W.K. George (✉)
Department of Aeronautics, Imperial College of Science, Technology and Medicine, London, UK
e-mail: georgewilliamk@gmail.com

A. Pollard et al. (eds.), *Whither Turbulence and Big Data in the 21st Century?*,
DOI 10.1007/978-3-319-41217-7_2

excellent statistical estimates as "too noisy" because they failed to realize that the quantity they were estimating was zero. Modern DNS is now encountering these same difficulties that have plagued experimentalists for decades: as Reynolds numbers get higher, simulations must be much longer to examine the rare events which have become of increasing interest.

So whether experiment or numerical simulation, how to store the data for even a short while is a major issue. Long term storage is even more problematic, so often it is easier (at least for DNS) to contemplate doing the simulation again instead of storing the data. Or alternatively, only storing certain "starting points." Or even more efficient, storing the snapshot POD coefficients at these starting points. Given the difficulty to get an experiment right in the first place and the cost of maintaining the facilities, this is not an option for most experimentalists. But neither is simply storing the data a viable option either, since the march of technology will soon make it unreadable, or simply make it useless to anyone but those who actually recorded it. This is of course a colossal waste, and any efforts (like this meeting) to address ways to preserve data for posterity are laudable.

Sadly I have little to offer as a solution to this problem. In spite of my best efforts, the rush of universities to clear disks and tapes, the urgency of students to get on with their careers, and the simple press of academic survival has meant that some really wonderful data has simply disappeared, or cannot be recovered for use even if preserved. Even preserving a simple email address or website after retirement or change of location seems to be too much of a stretch for modern university administrations. Only it seems to me if there is a national repository and a contractual commitment to use it might there be the possibility of preserving it.

So while I do not have any panacea for the problem of how to keep "big data" once we have acquired it, I do have considerable experience in generating it. The experiments of me and my students almost always stretched the bounds of what was possible at the time we did them. Since that is likely to be true for all turbulence experiments in the future (whether laboratory or computer-generated) the amount of data collected will undoubtedly increase. In the absence of a general solution to the storage problem, it therefore becomes more and more crucial to make sure we use it properly when we have it. And that is the point of this paper.

2.2 What Should We Be Looking For?

There is no point it seems to me to be carrying out large simulations and experiments if we don't have some idea what we are looking for in the first place. The last thing turbulence needs is more random numbers, more disks of poorly generated data, and more poorly thought-out applications of existing or yet-to-be-developed technology. Or for that matter, more pictures of random intertwined vortices. What exactly are we trying to find? How will we know when we've found it?

2.2.1 We Have No "Exact" Solutions

The single biggest problem of turbulence research is that we really have no exact solutions. Similarity solutions for averaged quantities (especially multipoint) should have been particularly useful (e.g., [33, 35, 40]). These have, however, been quite controversial. In large part I believe because most of these early similarity solutions were done incorrectly, usually by making wrong assumptions at the first step or overconstraining the solutions. I have written a great deal about this starting in the late 1980s [10, 12, 14]. Of course these "wrong solutions" did not agree with most experiments. And the few that appeared to were discredited as better experiments were performed. So a mythology was born and took root which argued that such solutions never applied to turbulence. And alternative ideas took over—like that turbulence could never be described by a single length scale, which meant that pretty much any simulation or experiment could be justified if it collapsed in a local scaling over a range of scales. So our field has become cluttered with experiments which claim to be one thing—like jets or decaying turbulence or wakes—but really are flows largely determined by initial and spatially confined boundary conditions.

In the absence of believable analytical solutions, or at least some way to decompose turbulence, there is no way to sort out boundary and initial condition effects from actual dynamical processes. We do have such tools for homogenous flows, even though we seldom use them as such. And (as we shall see below), we have them for several other classes of turbulent flows as well. But first let's look at a field which closely resembles our own, so we can see what we are missing.

2.2.2 Examples from Wave Research of What We Are Missing in Turbulence

My very first research experience was using a small internal wave tank in the lab of Owen Phillips at the Johns Hopkins University. In many ways this set a pattern for much of my career which was to follow. Phillips was at heart an applied mathematician with extraordinary abilities and physical insight. He was also a mesmerizing lecturer, and it was the simple elegance of the sophomore mechanics course he taught which led me to work in his lab in the first place.[1] Phillips' experimental work, mostly on wind waves and non-linear interactions, was pretty much left to his Ph.D. students and undergrads like me.

The original goal of my study was to examine how internal waves at a saltwater/freshwater interface broke and propagated when subjected to a current—which we generated by towing a contraction through the tank. Sinusoidal waves indeed

[1]A fraternity brother and physics student, Ben Wegbreit, who was already working for Phillips, made the suggestion to me that I trade my job in the library sorting books for one in the lab, and for that I have been forever grateful.

steepened, broke, then reformed and propagated as symmetrical disturbances. What we were seeing were multiple solutions to the governing equations which very much resembled the Schrödinger equation. For a budding young physicist the excitement was very real. Parallel experiments at Cambridge in the UK pretty much scooped our focus on the second and higher modes, so nothing on our work was published. But thanks to an elegant analysis by Phillips, and hundreds of meters of 8 mm movie film of aluminum-flake particle paths, we did manage to provide a rather nice analysis and experimental confirmation of how the waves were steepened by the current [32].

It was this synthesis of phenomena we could clearly see, elegant analysis of well-posed equations, and massive amounts of data (for that era at least) which really captured me. I fell in love with waves, and it very much influenced my professional (and non-professional) life thereafter. It was the very need of such theory for turbulence which attracted me to it. And in an interesting way to my linking up with John Lumley.

What is nice about surface water waves in particular is that you can easily see and generate them. Throw a stone into a pond, pull a boat through the water, or just blow on a surface, and all sorts of interesting mathematics take real form before your eyes. They often organize themselves into recognizable forms, and they can be analyzed by equations which are for the most part tractable. The reason is that wave non-linearities are relatively weak—at least compared to turbulence. For surface waves, these non-linearities enter at fourth and fifth order in expansions about the wave slope (amplitude divided by length), so waves can travel great distances without significant modification.

This weak non-linearity implies that any coherent features of surface waves are maintained or destroyed almost entirely by the phase speeds and bandwidth of the Fourier components that are present. The more monochromatic the wave, the farther it can propagate, even if different wavelengths are traveling at different phase velocities. In fact the "visibility" (or lifetime) of a group of deep water waves is entirely determined by the bandwidth of the disturbances comprising it. The reason is that for deep water waves the speed is proportional to the square root of the wavelength. (A deep water wave is one whose length is much less than the water depth.) This is why it is impossible to track a deep water wave on the open ocean for very long by watching it—it just disappears (or comes "unglued") as the different components comprising it propagate through. Shallow water waves whose length is greater than the depth by contrast can be followed indefinitely since all phases propagate at the same speed which is proportional to the square root of the depth of the water. This is why we can follow the surf in at the beach, and why the energy piles up as the depth is reduced and the wave slows down.

Now all of the above is well known, and has been for a long time. Coherent features, clearly visible, and their dynamics completely explained (at least until they start to break) as the action, interaction, and superposition of eigensolutions of the governing equations. The structures themselves are what we see. But it is the behavior of the underlying eigensolutions in the governing equations which enables us to understand the dynamics.

Now what does this all have to do with turbulence? And big data?

2.2.3 How Can We Decompose Turbulence?

Everyone who has ever looked at turbulence sees structures. Artists painted them centuries ago. Children and adults alike are fascinated by them. We see them in moving rivers because of the disturbances they generate. Clouds track them in our skies, even the simple processes of stirring our food and drinks present them to us. But what are they? The last five decades have seen a massive effort to find and quantify turbulence structures. And while these have cured us of any illusions about whether an average flow really exists, they have contributed little to our understanding of turbulence. We still don't know how to write equations for them, much less predict them. We suspect they are important but really can't prove it. We are sure they are a necessary part of turbulence and many believe they must be accounted for if turbulence is to be controlled. But we have made almost no progress in proving the first nor moving forward with the latter.

Many have argued that these structures are vortical, and that we simply need to concentrate on vortex dynamics to understand and predict them. Indeed DNS and PIV coupled with clever vortex recognition algorithms show tantalizing strings of vorticity—usually quite concentrated in strings or sheets with lots of empty space around them. Whether these are related to the large scale coherent features we see is debatable, but the role of vorticity is not! Even so, recognizing these concentrations of vorticity has not been particularly helpful. Thanks to the Biot–Savart law, each vortex feels the velocity of all the other vortices in the field. So the problem is quite complex, even without the complications of viscous effects (like "cut and connect," vorticity diffusion, etc.). Like the water waves above if we had just taken pictures of them, simply studying pictures of what we see in turbulence and labeling it vortical has not led us to a methodology for predicting anything about them.

Our modern quest for coherent structures really started with the Schlieren pictures of a mixing layer [3] and the low Reynolds number near-wall dye boundary layer studies of Kline and co-workers [26]. But even before that Townsend and his co-workers (v. [35]) had noticed that correlation functions seemed to go to zero at large separations and time lags slower than they might have guessed. And they postulated that these were a result of "large eddies," which provided spatial coherence over large distances, but relatively little energy. No dynamic role for these was suggested, so they were quite different than the role postulated for coherent structures in the early 1970s by many (e.g., [21, 22]).

It was in trying to understand and contribute to the original Townsend idea that John Lumley [28] made what I believe was his most important contribution to turbulence. Published in an obscure Russian proceedings in 1967, I first became aware of it scarcely a year later. I was taking my first turbulence course, and was chasing down a paper on the buoyant subrange by Phillips who had presented it at the same meeting. And there in the same volume was Lumley's paper, seemingly the answer to the questions I had been posing above to myself about the differences between waves and turbulence. Little did I know that in a short while we would be working together, first with Lumley as my mentor and thesis advisor, then as my colleague. In order not to interrupt the narrative of this paper, I have placed in Appendix 1 my own mostly personal history of Lumley's idea and my involvement with it. It might be of special interest to anyone struggling to be heard in the hostile world that sometimes turbulence becomes for new ideas.

2.3 Lumley's Great Idea

What Lumley had proposed was an objective way to decompose a turbulent flow—or for that matter, any random process. His stated goal was quite modest—to find an objective way to identify Townsend's "big eddy." What he found was a way to decompose almost any flow into an infinite set of eigenfunctions. What he did NOT find was a unique way to identify any one of them as a large eddy or coherent structure.

It is not clear when Lumley himself recognized this, but most likely it was when Mark Glauser (then my Ph.D. student) and I showed him the results from our round jet mixing layer experiments in the early 1980s (see Figs. 2.1 and 2.2). What was clear to us from our very first results was that our "Lumley decomposition" had led us to a set of eigenfunctions which turned on and off, different eigenfunctions describing whatever structure was there at different points in its life cycle. Unfortunately (for us) a journal editor pretty effectively prevented us from publishing our results archivally. It is with some satisfaction that our meeting papers have exceeded 100 citations anyway (e.g., [16]). And probably not unrelated that Holmes et al. [20] included an entire chapter on our early results.

So what exactly did Lumley propose? This: imagine a random vector field of space and time. If you like pictures, think of time-resolved holographic PIV images

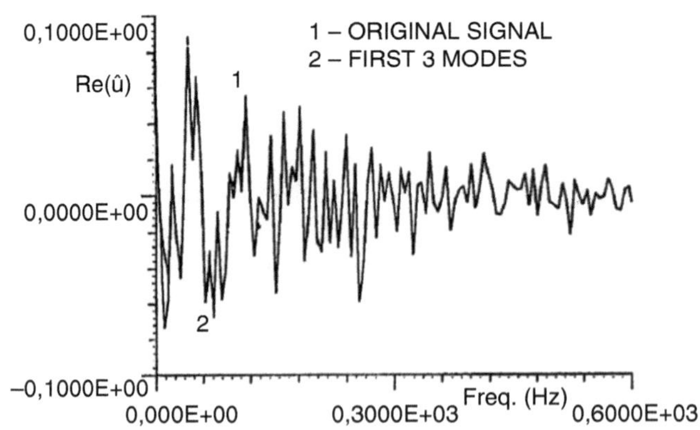

Fig. 2.1 Reconstruction of Fourier transform of instantaneous streamwise fluctuating velocity in center of axisymmetric jet mixing layer at $x/D = 3$ with only three POD modes. POD performed across mixing layer using seven probes. From Leib et al. [17, 27]

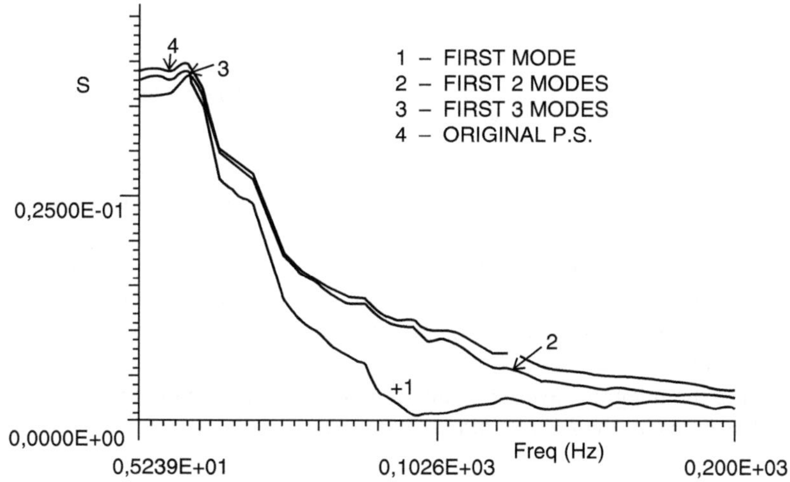

Fig. 2.2 Reconstruction of streamwise fluctuating velocity spectrum in center of axisymmetric jet mixing layer at $x/D = 3$ with only three POD modes. POD performed across mixing layer using seven probes. From Leib et al. [17, 27]

of an entire field in which the vectors are changing with time at every point and different in every realization. Now imagine another vector field that is not random, but is doing its best to track the random one. How could one chose the deterministic field so that it optimally (in some statistical sense) follows the random one?

Now here it is the brilliant part of what Lumley did. And by doing so managed to formulate the problem mathematically so that the solutions could be related to *existing* mathematics—and the Navier–Stokes equations. Lumley proposed to maximize the mean square inner product *in two senses*. First, the usual scalar product of two vectors; but then second, the inner product of two fields in the sense of projections in Riemann space. This was in my mind the great moment (and Lumley's great talent)—taking an idea, and turning it into mathematics. Not even new mathematics, but very old established mathematics *and* statistics. This was new—very new! That he initially misinterpreted what he did (as did most others) is irrelevant. A great idea was born, and a whole new way of thinking about turbulence had begun. Thanks to my brief background in waves, I recognized this immediately, even at this very early point in my career and while still a student. It was therefore a great shock for me to discover subsequently how controversial Lumley's approach had become—even ridiculed (see Appendix 1).

So how does this great idea work? Here it is. If $u_i(\mathbf{x}, t)$ represents the random field of space and time, and $\phi_i(\mathbf{x}, t)$ the deterministic one which is to optimally track it in a mean square sense, then we need to maximize the square of their "double inner product," $\langle|\alpha|^2\rangle$; i.e.,

$$\langle|u_i(\mathbf{x}, t)\ \phi_i(\mathbf{x}, t)|^2\rangle = \langle|\alpha|^2\rangle \tag{2.1}$$

where

$$u_i(\mathbf{x},t)\ \phi_i(\mathbf{x},t) \equiv \int \cdots \int_{\text{all space, time}} u_i(\mathbf{x},t)\ \phi_i(\mathbf{x},t) d\mathbf{x}dt \tag{2.2}$$

defines the inner product in *both* senses. This was the brilliant part! For want of a better terminology I will call this the "*Lumley integral*" or the "*Lumley projection*."

From here on the rest is easy, and could have been performed by even the most mediocre mathematician. From all the hostility the results generated in some quarters (then and even now), and the misuse of them since, it must be presumed that many turbulence researchers qualified (at least then) as less than mediocre mathematicians.

Squaring the integrals, averaging and maximizing the variation of it yields what I have chosen to call the *Lumley Integral Equation*:

$$\int \cdots \int_{\text{all space, time}} R_{i,j}(\mathbf{x},\mathbf{x}',t,t')\ \phi_j(\mathbf{x}',t') d\mathbf{x}'dt' \ = \ \lambda\ \phi_i(\mathbf{x},t) \tag{2.3}$$

where the kernel, $R_{i,j}(\mathbf{x},\mathbf{x}',t,t') \ = \ \langle u_i(\mathbf{x},t)u_j(\mathbf{x}',t')\rangle$ is the two-point two-time correlation (or two-point two-time Reynolds stress tensor). Equation (2.3) is sometimes erroneously called the *POD integral*. It most certainly is NOT! There are circumstances under which it reduces to a proper orthogonal decomposition. As we shall be reminded below, more often it does not!

What is important is that the integral converges and does so independent of the boundaries used in practice to truncate its estimation (e.g., field of view, size of tunnel, computational domain, etc.). In fact, it is not at all obvious whether the integral even exists in many turbulent flows; for example, flows which are statistically homogenous and/or stationary. Both of these types of flows are of necessity of infinite extent, and hence have infinite energy. Not surprisingly, these flows behave very differently than flows which are of "finite total energy," meaning that either they are bounded in space or die off rapidly enough in all directions (and time).

Only when the finite total energy property is satisfied in ALL directions AND time does the integral equation yield what can properly be referred to as POD solutions. Failure to recognize this has been the source of much confusion over the years since Lumley first derived it. I have discussed these various conditions in detail in several papers [11, 13], and in several places in my turbulence notes (available on www.turbulence-online.com). And of course pretty much all of the information is there quite cryptically in Lumley's original paper and even more obscurely in his *Stochastic Tools in Turbulence* [29]. In the succeeding sections, I shall review what Lumley told us, and then some of what we have learned since.

2.4 It is All About the Total Energy in the Flow

If the integral of $R_{i,i}(\mathbf{x},\mathbf{x},t,t) = \langle u_i(\mathbf{x},t)u_i(\mathbf{x},t)\rangle$ itself converges when integrated over all directions and time, then the flow is said to be of *finite total energy*; i.e.,

$$\int_{\text{all time}} \int\int\int_{\text{all space}} \langle u_i(\mathbf{x},t)u_i(\mathbf{x},t)\rangle d\mathbf{x}dt < \infty. \tag{2.4}$$

Note that this "finiteness" does not imply that the field itself need be finite, only that the total energy in it is finite.

"Finiteness" is, of course, automatically satisfied if the field is of finite extent or time, so the results of it can be applied to all experiments and computer simulations. But this "finite energy by truncation" can be quite artificial and can yield solutions which have nothing to do with the flow and everything to do with the boundary conditions imposed upon it. Obviously it makes a difference whether we are truncating just the integral by our "field of view" or actually creating the flow in a finite space. Note that a similar phenomenon occurs with spectral analysis when the window is too small relative to the flow integral scale—the resulting spectrum looks more like the Fourier transform of the window than the flow. In optics and signal analysis usually we can evaluate the effect of windows since we know the eigenfunctions. In turbulence we do not—at least without Lumley. All of our experiments and DNS are always of finite total energy, even when we are trying to model flows which are not. So we have much to be concerned about. Since these "eigenfunctions" in principle vary from flow to flow, no general conclusions are possible about the validity of our results without first finding them. Hence what I believe should be the primary goal of our "big data"—making it possible to find the appropriate basis functions for the flow of interest.

2.4.1 *Flows of Finite Total Energy*

If the energy integral converges, then the solutions are indeed what are referred to as the "classical proper orthogonal decomposition solutions" (or "classical POD"). When most people refer to the POD (or the Lumley decomposition), these are the kinds of solutions they think they are referring to. There can be denumerably infinite eigensolutions, they are orthogonal, and proper (meaning their eigenvalues can be arranged so the lowest order one has the most energy, the next the second most energy, etc.). The sum of the eigenvalues is equal to the total energy in the field; i.e.,

$$\int \cdots \int_{\text{all space time}} R_{i,i}(\mathbf{x},\mathbf{x},t,t)d\mathbf{x}dt = \Sigma_{n=1}^{\infty}\lambda_n \tag{2.5}$$

Finally the two-point two-time Reynolds stress tensor, the original kernel, can be reconstructed like this:

$$\langle u_i(\mathbf{x}, t)\; u_j(\mathbf{x}', t')\rangle = \Sigma_{n=1}^{\infty}\; \lambda_n\; \phi_i^{(n)}(\mathbf{x}, t)\; \phi_j^{(n)}(\mathbf{x}', t') \tag{2.6}$$

Most importantly, any random realization of the field can be represented using appropriate "random" coefficients determined by projecting the entire field onto the eigenfunctions; i.e.,

$$u_i(\mathbf{x}, t) = \Sigma_{n=1}^{\infty}\; a_n\; \phi_i^{(n)}(\mathbf{x}, t) \tag{2.7}$$

where the random coefficients, a_n, are determined by:

$$a_n = \int\int\int\int_{\text{all space}} \phi_i^{(n)}(\mathbf{x}, t)\; u_i(\mathbf{x}, t) d\mathbf{x} dt \tag{2.8}$$

The eigenvalues are given by:

$$\lambda_n = \langle a_n a_m \rangle\; \delta_{mn} \tag{2.9}$$

since the coefficients are uncorrelated.

Thus each realization of the entire instantaneous field can be reproduced from its POD decomposition. But to determine the coefficients requires the entire field at once. This is where the really big data problem enters. The actual determination of the eigenfunctions needs only statistics, which in turn requires only simultaneous measurements at two space-time points. So pairs of points can be considered one pair at a time. But to reconstruct the instantaneous field we need the entire space-time field at once. It should be noted that while this POD representation can in fact describe any existing field, there may also be other types of solutions (non-orthogonal) needed to get to any field from a transient condition. This is no different than the transient versus steady-state solutions of every other differential equation.

It is a common misconception that there are only empirical solutions to Eq. (2.3). A simple analytical example proves otherwise. Assume the two-point correlation for a one-dimensional scalar field to be given by $R(x, x') = R(0, 0)\; \exp[-(x^2 + x'^2)/2L^2]$. Substitution into the Lumley integral yields

$$\int_{-\infty}^{\infty} R(0, 0)\; e^{-[x^2+x'^2]/2L^2}\; \phi(x')\; dx' = \lambda\; \phi(x) \tag{2.10}$$

But the kernel can be separated to yield:

$$R(0, 0)\; e^{-x^2/2L^2} \left\{ \int_{-\infty}^{\infty} e^{-x'^2/2L^2}\; \phi(x')\; dx' \right\} = \lambda\; \phi(x) \tag{2.11}$$

The integral in brackets is just a number proportional to L. So clearly $\phi(x) \propto L\; e^{-x^2/2L^2}$ where the factor of proportionality must be chosen to make the integral of $\phi^*(x)$ $\phi(x)$ unity. Clearly in this case there is only a single eigenfunction, and the corresponding eigenvalue contains 100 % of the energy. It should not be at all surprising that a flow might have eigenfunctions that resemble the correlation function, since it might be simply a random superposition of them sprinkled in space.

Unfortunately very few flows are of finite total energy, and in fact none of the "ideal" ones we realize in the laboratory (since most are statistically stationary and many are assumed to be "homogeneous" in one or more directions). A number of investigators, however, not realizing this, have produced data (usually using the snapshot POD) which showed the "POD-modes." And many of them noticed the domain dependence of their data. The reason for this is obvious (at least in hindsight): the finite domain they could see from their PIV (or DNS) was an arbitrary truncation of the energy to make it finite, so indeed it was the domain that determined the eigenfunctions, not the flow. This was *not* what Lumley had in mind! But it is *exactly* the problem we have in all of turbulence—and one of the main points of this paper. In the absence of an exact solution, or at least basis functions, how can we know the degree to which our computational or experimental domains are determining or affecting our solutions?

An example of a flow that truly fits the POD solution might be a mass of turbulence without boundaries which is decaying. But even such a mass (like the sun) is probably spinning. So this imposes a preferred axis with consequent azimuthal homogeneity and periodicities—which like the periodic box turbulence moves it into a different mode of solution—at least in the homogenous periodic direction. And this makes the solutions and coefficients (POD though they may be) dependent on the azimuthal mode number. And since they will be complex, phase is also important. This immediately almost guarantees that any actual "coherent features" or structures will be transient, as they will be comprised of different modes at different times. In other words, they will "evolve"; meaning sometimes you will see them, then other times you won't. Most maddenly they will seemingly vanish before your eyes!

2.4.2 Homogenous Fields of Infinite Energy

In spite of their highly idealized nature, statistically stationary and homogeneous flows are those that have received the most attention theoretically. In particular, homogeneous turbulence which is either forced or decaying, and homogeneous shear flow turbulence. In the former the energy is usually assumed (or derived from similarity considerations) to be decaying either exponentially or as a power law. And in the later to be increasing exponentially in time. Since they are all assumed homogeneous they are of infinite energy in the spatial directions quite independent of the time dimension. Few (if any) have considered such flows in the context of the Lumley integral above. But without understanding the implications of the Lumley integral on these solutions, flows with partial homogeneities cannot be understood.

And it is that lack of understanding which I believe has held our field back for the past half-century. We have looked for things that were not there; and when we have seen things, we often did not understand them.

Most of the flows we create in the laboratory are presumed to be excellent approximations to stationary random processes. And for decades we have attempted to study them by using various grids in wind and water tunnels. While the flows we investigate are of course of finite total energy (since we have truncated them by our experimental boundaries or record lengths), it is important to recognize that our goal is to measure flows that can be interpreted theoretically. Hence, the short-comings of our attempts to create them notwithstanding, the "real" flows are of infinite extent in all directions and time. Otherwise the statistics cannot be independent of origin, a necessary condition for both homogeneity and stationarity. And as a consequence, most "real flows" of interest are of infinite energy, no matter which direction or time is used to examine it.

Since homogeneity and stationarity imply that the integrals are infinite and the flows of infinite energy, then the solutions will be entirely determined by the boundaries we arbitrarily impose on them— *unless there is some property of the kernel itself which makes the integral converge*. Solutions which depend only on how the field is truncated are pretty useless. CFD people are very conscious of these problems, since they are always trying to make the computational domain as small as possible. In my experience experimentalists are generally not nearly as concerned as they should be, and for sure not as concerned as the integral equation above demands they must be. Not unexpectedly, a number of experimenters through the years have noticed this field dependence of their "raw" application of the POD solution of Lumley's integral (or its snapshot version). Few though recognized the reasons.

2.4.3 A One-Dimensional Field of Infinite Energy

That the Lumley integral can converge even with infinite total energy can best be illustrated by a simple one-dimensional example. Consider a stationary random process with the two-time correlation function, $R(t, t') = B(\tau)$ where $\tau = t' - t$. If the limits of integration are infinite, then the Lumley integral equation reduces to:

$$\int_{-\infty}^{\infty} R(t, t')\phi(t')dt' = \int_{-\infty}^{\infty} B(\tau)\phi(t+\tau)d\tau = \lambda\, \phi(t) \tag{2.12}$$

But this can be rearranged as follows (since the integral is only over τ):

$$\int_{-\infty}^{\infty} B(\tau) \left\{ \frac{\phi(t+\tau)}{\phi(t)} \right\} d\tau = \lambda \tag{2.13}$$

There is no time-dependence left on the right-hand side, so clearly the solution must have the property that makes the bracketed term t-independent! Only an exponential function has this property, and only if the argument is pure imaginary does it not either blow up or die out.

Clearly this implies that the eigenfunctions for a stationary random process are harmonic functions (sines and cosines or complex exponentials). The "eigenvalues" are just their spectral energy content. And as before we determine the coefficients of the eigenfunctions by projecting the field upon them. But these are just Fourier transforms in the continuous radial frequency variable ω; i.e.,

$$\hat{u}(\omega) = \frac{1}{2\pi} \int_{-\infty}^{\infty} e^{-i\omega t} u(t) dt \tag{2.14}$$

where the 2π is placed before the integral so that the spectrum defined below integrates to the energy.

So statistical stationarity implies Fourier transforms in time. And of course *in the generalized sense* for the instantaneous random fields that are being represented. Statistical stationarity also implies the Fourier coefficients at different frequencies are uncorrelated, so:

$$\langle \hat{u}^*(\omega)\hat{u}(\omega')\rangle = F(\omega)\delta(\omega' - \omega) \tag{2.15}$$

where $F(\omega)$ is the "spectrum" defined as just the Fourier transform (in the ordinary sense) of the two-time correlation, $B(\tau)$; i.e.,

$$F(\omega) = \frac{1}{2\pi} \int_{-\infty}^{\infty} e^{-i\omega\tau} B(\tau) d\tau. \tag{2.16}$$

2.4.4 A Homogeneous and Stationary Random Field

So what does this imply if the field is homogeneous and stationary? Then the optimal representation is also harmonic functions and the coefficients are Fourier transforms *in the sense of generalized functions*, but of *four* dimensions; i.e.,

$$\hat{u}_i(\mathbf{k}, \omega) = \left[\frac{1}{2\pi}\right]^4 \int_{-\infty}^{\infty} e^{-i[\mathbf{k}\cdot\mathbf{x}+\omega\, t]}\; u_i(\mathbf{x}, t)\; d\mathbf{x}dt. \tag{2.17}$$

where the $\hat{u}_i(\mathbf{k}, \omega)$ are the Fourier coefficients, which are functions of both spatial wavenumber, $\mathbf{k}$, and radial frequency, ω, both defined over infinite domains. These Fourier "coefficients" tell how much of each Fourier "eigenfunction" is present in the field.

The field can of course be reconstructed using them and the eigenfunctions as:

$$u_i(\mathbf{x},t) = \int_{-\infty}^{\infty} e^{+i[\mathbf{k}\cdot\mathbf{x}+\omega\, t]}\; \hat{u}_i(\mathbf{k},\omega)\; d\mathbf{k}d\omega. \tag{2.18}$$

Homogeneity and stationarity both dictate that the Fourier coefficients in non-overlapping bands are uncorrelated; i.e.,

$$\langle \hat{u}_i^*(\mathbf{k},\omega)\hat{u}_j(\mathbf{k}',\omega')\rangle \;=\; F_{i,j}(\mathbf{k},\omega)\;\delta(\mathbf{k}'-\mathbf{k},\omega'-\omega) \tag{2.19}$$

where $F_{i,j}(\mathbf{k},\omega)$ is the four-dimensional cross-spectral tensor which itself is the four-dimensional Fourier transform (in the ordinary sense) of the two-point two-time cross-correlation (or Reynolds stress tensor) given by:

$$B_{i,j}(\mathbf{r},\tau) = \langle\; u_i(\mathbf{x},t)\; u_j(\mathbf{x}+\mathbf{r},\; t+\tau)\rangle \tag{2.20}$$

$F_{i,j}(\mathbf{k},\omega)$ and $B_{i,j}(\mathbf{r},\tau)$ are well known to be a four-dimensional Fourier transform pair; i.e.,

$$F_{i,j}(\mathbf{k},\omega) = \frac{1}{(2\pi)^4}\int\int\int\int_{-\infty}^{\infty} e^{-i[\mathbf{k}\cdot\mathbf{x}+\omega\,\tau]}B_{i,j}(\mathbf{r},\tau)d\mathbf{r}d\tau \tag{2.21}$$

and

$$B_{i,j}(\mathbf{r},\tau) = \int\int\int\int_{-\infty}^{\infty} e^{+i[\mathbf{k}\cdot\mathbf{x}+\omega\,\tau]}\; F_{i,j}(\mathbf{k},\omega)d\mathbf{k}d\omega \tag{2.22}$$

where $\mathbf{r} = \mathbf{x}' - \mathbf{x}$ and $\tau = t' - t$ are the spatial separation and time lag, respectively. This is, of course, just the classical Wiener–Khintchine theorem stating that in homogeneous and stationary random fields the Fourier coefficients at different wavenumbers and frequencies are uncorrelated.

So these are exactly like the Fourier decompositions of the random water waves (or acoustic or optical waves) described in Sect. 2.2.2 above. And it gives a rationale for books on homogeneous turbulence which seem to have almost arbitrarily chosen to represent homogeneous fields by continuous Fourier modes.

Now there are a several things we can surmise immediately if a homogeneous turbulent field of uniform density were to be decomposed into these optimal Fourier modes. First, consider what happens after transforming in four dimensions the $u_j\,\partial u_i/\partial x_j$ term in the instantaneous velocity fluctuation equation. Since the Fourier transform of a product is simply the convolution of its Fourier transforms, then the four-dimensional transform is

$$\begin{aligned}
&\frac{1}{(2\pi)^4}\int_{\mathbf{x}}\int_t e^{-i[\mathbf{k}\cdot\mathbf{x}+\omega t]}\left[u_j(\mathbf{x},t)\;\frac{\partial u_i(\mathbf{x},t)}{\partial x_j}\right]\; d\mathbf{x}dt \qquad (2.23)\\
&= \int_{\mathbf{k}'}\int_{\omega'}\left[-i\mathbf{k}_j'\hat{u}_j(\mathbf{k}',\omega')\hat{u}_i(\mathbf{k}-\mathbf{k}',\omega-\omega')\right]\; d\mathbf{k}'d\omega
\end{aligned}$$

where all integrals are over the infinite domain. Thus *all* the non-linear interactions are among triads of the four-dimensional wavenumber–frequency combinations $(\mathbf{k}, \omega)$, $(\mathbf{k}', \omega')$; and $(\mathbf{k}' - \mathbf{k}, \omega' - \omega)$.

That there are interactions among triads of wavenumbers of Fourier decomposed homogeneous fields has been well known since the 1940s at least. But that frequency also interacts as triads of frequencies does not seem to have been previously noticed. *What this also means is that unless there is some mechanism to make the phases of these superimposed eigenfunctions lock together, then a stationary random field can have no definitive lasting (or coherent) structure.* The reason is that the phases will cancel each other out as the eigenfunctions of different wavenumbers and frequencies move and evolve. Or said another way, analysis of a field with time-dependent Fourier coefficients (as is common), say $\hat{v}_i(\mathbf{k}, t)$, will be very different from a complete four-dimensional analysis using $\hat{u}_i(\mathbf{k}, \omega)$ since any complex phase frequency information in the latter will be smeared out in time in the former. This could have important implications for our studies using forced turbulence since it is both homogeneous and stationary (and in fact the only flow we know that is). I know of no attempts to look at four-dimensional triadic interactions, but they could be very interesting as noted below.

Turbulence in this regard is very different than a homogeneous field of surface water waves, for example, where the dispersion relation between wavenumber and frequency means different wavenumbers propagate at different speeds. To the best of my knowledge no such general relations for turbulence exist. Maybe structures can form and persist anyway. But without including frequency in the analysis along with wavenumbers we will never really be able to tell. For example, one question which might be asked is: when we see these vortical structures in DNS of forced turbulence, how do they evolve and how long do they persist? And what Fourier wavenumber–frequency combinations make them up? This we know from wave theory: if these vortices are spatially compact, they must have a very wide spectrum of contributing wavenumbers. But if they persist for long times (say relative to the Kolmogorov or Taylor microtimes), then they must have very narrow bandwidth in the frequency domain. Persistence time will be inversely proportional to frequency bandwidth, just as spatial localization corresponds to a broad spectral content in wavenumber space. The wider the bandwidth, the less localized the disturbance is in either space or time. So the whole idea of associating small scales with high wavenumbers and frequencies is intrinsically wrong. If one is localized in space, then a broad band of wavenumbers must be in play. If it persists for a long time, however, the spectral band in frequency must be quite narrow. (Note that in a real flow, "frequency" becomes a bit garbled by advection, so frequency is probably best thought of in a moving frame. Or by removing $\mathbf{k} \cdot \mathbf{U}$ from it where $\mathbf{U}$ is the mean velocity; i.e., $\omega - \mathbf{k} \cdot \mathbf{U}$.)

2.4.5 Homogeneous Periodic Flows

Homogeneous periodic flows are a special case as was noted by Lumley [28]. (See also [11, 13], the latter of which shows how the above relate to them.) The eigenfunctions can similarly be shown to be Fourier modes, but this time at frequencies which are integral multiples of the fundamental (or inverse of the period). This is of course just classical Fourier series.

Note that it is commonly assumed that when we model homogenous flows as periodic, then "real" turbulence must also be periodic. This is false. There is a fundamental difference between a truly periodic flow, and one which we have only imagined as such. The difference is whether the flow has been "windowed" or not. Periodic flows are not "windowed." But flows artificially treated as periodic are. The differences can be quite important. For example, it is quite common to analyze finite records of time series of statistically stationary flows by assuming them to be periodic. This is incorrect, and inverse transformation leads to estimates of the correlations that are wrong unless the window is included.

2.4.6 Mixed Flows

Flows that we commonly encounter can be a mixture of all the possibilities above: inhomogeneous and of finite energy in one-direction, homogeneous in one (or more) directions, perhaps periodic in one and stationary in time. This is in fact pretty close to flow in a turbulent pipe, and fairly typical of many axisymmetric flows we generate in the laboratory. Axisymmetric jets and wakes have received particular attention (e.g., [4, 8, 9, 16, 23–25, 34, 36, 41]). Even high Reynolds number turbulent boundary layers have been investigated using Lumley's approach [37].

The easiest way to address these flows is to carry out the Fourier decompositions first, usually on the instantaneous velocities; then carry out the Lumley integral in the inhomogeneous and non-periodic directions using the cross-spectral tensors computed from them. Finally determine the random POD coefficients by projecting the resulting eigenfunctions onto realizations of the random Fourier coefficients. For example, imagine the flow to be homogeneous in x, periodic in an azimuthal coordinate θ, stationary in time t, and inhomogeneous in a transverse coordinate y. The corresponding Lumley projection is

$$\int_{\text{finite energy direction}} F_{\alpha,\beta}(y, y'; k, \omega, m)\psi_\beta(y'; k, \omega, m)dy' = \lambda(k, \omega, m)\ \psi_\alpha(y; k, \omega, m) \tag{2.24}$$

where the subscripts α, β indicate the appropriate components in the various directions (streamwise, azimuthal, and transverse). Only the triple transform of the instantaneous velocity can be constructed by summing up its POD modes, all of which are functions of not only y, but also the wavenumber, k, the frequency, ω, and the periodic mode number, m; i.e.,

$$\hat{u}_\alpha(y; k, \omega, m) = \sum_{n=1}^{\infty} a_n(k, \omega, m)\ \psi_\alpha^{(n)}(y; k, \omega, m) \tag{2.25}$$

The velocity itself (or some portion) of it can only be obtained by inverse transformation over k, ω and inverting the Fourier series for each harmonic component m. Since these Fourier coefficients are complex, this can yield results that look nothing like the individual eigenfunctions. And reconstructions and inferences from partial decompositions may look totally different.

It should be easy to see from this example why simplifying the decomposition to make the random coefficients a function of time only most likely will not succeed, since the various Fourier contributions from k, ω, m to the complex random coefficient $a_n(k, \omega, m)$ can change very much which POD modes contribute to the sum as various Fourier modes turn off or on, or propagate through. In general, homogeneities, stationarity, and periodicities tend to "destroy" the very coherent features that they have built up. This is very much like the disappearing deep water wave groups described earlier. It is probably more correct to think of coherent features in turbulence as "*groups*" rather than structures, "groups" whose lifetime and visibility depend on the frequency bandwidth of the Fourier components comprising them. Or they may act more like solitons and actually propagate as fixed form solutions of the equations. In the absence of suitable eigenfunctions to study turbulence we simply can never know.

Unfortunately it has taken us (me in particular) decades to realize the source of the problem with "partial dimension" decompositions—which in fact describes all our experiments to date (and those of others as well). While those applications of Lumley's methodology using partial-dimension decompositions have given us considerable insight into many flows, they have not produced for us the eigenfunctions we sought. The reason is that no partial decomposition can avoid the phase-scrambling that comes from ignoring even one dimension. This is even more of a problem with the snapshot decomposition as noted in Appendix 2. The good news is that we have now learned enough to know how to carry out a complete four-dimensional Lumley decomposition, especially using the multipoint similarity methodologies described in the next section. And experimental and computational capabilities are now large enough to be able to support it. So my challenge to the generations behind me: do it!

2.5 Extensions Beyond Lumley

There is little in the preceding section which was not in Lumley's original paper, or at least could not be inferred from it. But part of my life-long quest has been to extend the same kind of analysis to flows for which there was no obvious solution to the Lumley integral, nor any reason to believe such solutions might exist. All of these extensions involve similarity solutions to the actual governing equations, equations which were only tangential to Lumley's decompositions. But for the extensions described briefly below, it is similarity solutions to the actual two-point two-time equations which provide the key to solving the integral, and indeed to whether or not it is solvable at all. Amazingly, it turns out that in these cases at least, the integral not only has solutions, they are analytical. So the real challenge for the future for these flows becomes not to find empirical eigenfunctions from knowing the flow, but to use the analytical ones to calculate it. Two examples are described in the following sections, the second for the first time.

2.5.1 Partially Inhomogeneous Flows of Infinite Extent and Infinite Energy

Since this example has already been partially published elsewhere, I shall keep this discussion brief. But I would call attention to the paper on the fully developed jet by Velte et al. [39] in this volume which discusses some recent extensions of this work and summarizes it nicely.

Inhomogeneous flows of infinite extent were not a case considered by Lumley, but it was always a major interest for me and my students because of our interest in turbulent jets exhausting into a quiescent environment. And it should be a major concern for anyone worried about most turbulent shear flows. In the ideal case the jet in the streamwise direction continues forever. If the ideal flow can be assumed to statistically stationary (as opposed to a starting jet), then momentum conservation ensures that total energy associated with that direction is infinite. So the Lumley integral can converge ONLY if some property of the kernel ensures that it does. But what is that property?

The answer is: two-point similarity of the averaged equations [6, 7]. What we found was that two-point similarity of the axisymmetric far jet allowed us to transform the streamwise coordinate logarithmically. As a result the kernel of Eq. (2.3) could be rewritten as $R(x, x') = U_s(x)U_s(x')B(\xi' - \xi)$ where $\xi = \ln x/D$ where D can be any convenient dimension (like jet exit diameter). The result of this is that the transformed flow is now homogeneous in the transformed streamwise direction. And as noted above, a direct consequence is that the eigenfunctions are now Fourier transform modes in the logarithmic coordinate, ξ. *In other words, it was NOT necessary to solve the Lumley Integral Equations empirically, the solutions were analytical!* Note that these results are not approximations, they represent an exact solution of the instantaneous equations.

That real jets behaved this way were confirmed by the extensive experiments of Bettina Frohnapfel (reported in her master's thesis) and in Ewing et al. [7]. Subsequently we were able to confirm these results in the studies at the Danish Technical University by Maja Wänström [41, 42] using both cross-plane and streamwise plane

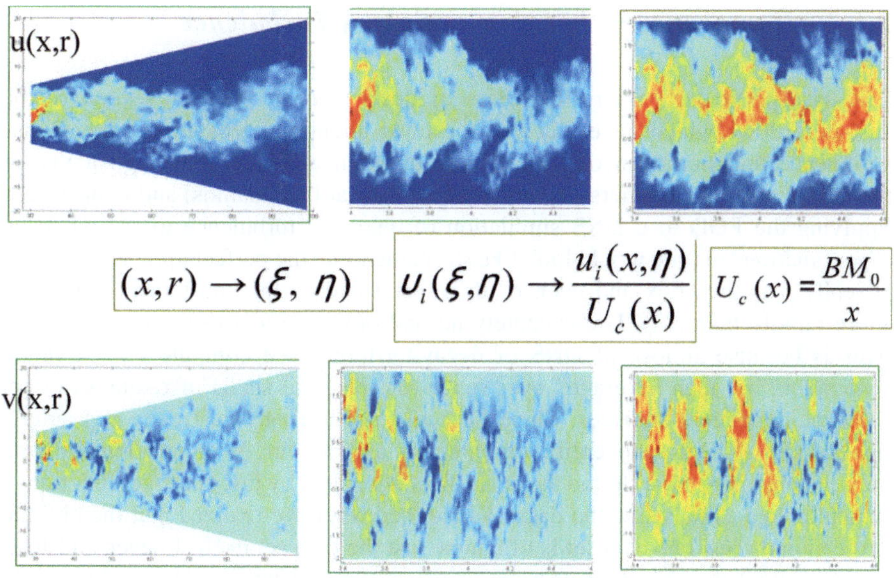

Fig. 2.3 Composite images of streamwise and radial velocities showing homogenization of field by mapping from physical to similarity coordinates. From Wänström et al. [42]

PIV. (As noted above, Clara Velte and Azur Hodzic are continuing these studies at the Danish Technical University.) But more importantly we were able to extend them to the *instantaneous field*. The thesis of Wänström reports these studies in detail. But Fig. 2.3 summarizes the important points. The actual jet measurements from 30 to 100 diameters were transformed using only the local downstream half-width and centerline velocity into a flow which was homogenous in the streamwise direction. And by stretching the streamwise coordinate logarithmically, it was possible to analytically determine the eigenfunctions to be Fourier modes in the logarithmically stretched coordinate. The great streamwise extent of the flow was necessary to avoid windowing when Fourier transforming in the stretched streamwise variable.

Unfortunately we were not able to measure all directions simultaneously. So we ran into exactly the problem that quite surreptitiously complicates most attempts to use various decompositions (or to determine turbulence structures). Even though we had three (of four) dimensions, the missing fourth dimension meant the eigenfunctions were contaminated by the complex phases of the missing dimension. So even though the results were quite robust and repeatable, each partial decomposition gave different profiles for the actual radial dependence of the eigenfunctions. In the light of our discussions above, this is exactly what we should have expected. Efforts are now underway by Velte and Azur and co-workers to use the similarity results and log stretching to perform a DNS of a jet of sufficient length ($x/D > 100$) to allow Fourier analysis in the streamwise direction and decompose all four dimensions simultaneously. Someday, hopefully soon, we will see what the full four-dimensional eigenfunctions really look like.

2.5.2 Homogeneous, but Non-stationary Turbulence

This is a particularly interesting class of flows since it includes several of the turbulent flows which have received the most attention: decaying homogeneous turbulence and homogeneous shear flow turbulence. Some years ago a master student of mine at Chalmers (Adam Wachtor, now at Los Alamos) and I considered applying the POD to a DNS simulation of decaying turbulence to see what the eigenfunctions in time would look like. It seemed like the perfect application, since it really was in a box and even of finite total energy in time (since it decayed more rapidly than t^{-1}). Unfortunately he had only one realization to work with. This is because almost all DNS of decaying turbulence compute only a single realization, but then "average" over space (or spherical shells of radius k) to get the time-dependent statistics. Since he could not average we (really he) simply applied the POD to the entire time record and to our great puzzlement (and subsequent amusement) found the entire flow described by just one eigenfunction—which looked exactly like the data. We laughed when we realized that the optimal projection on just one data set is of course just the data set itself. I never told John Lumley about this, but I suspect we were not the first nor the last to do this. In fact a few months later a Danish Ph.D. student showed me exactly the same result applied to the wake of a wind generator. Obviously statistics are an important part of the whole Lumley projection idea. Clearly application of Lumley's methodology can advance only if we have enough realizations to perform reasonable statistics. Unfortunately, except for "forced turbulence," this is seldom the case for DNS. And while experiments often have well-converged statistics, they often have a very limited spatial field, thus either arbitrarily truncating (or windowing) the flow, or limited resolution and not resolving the important scales.

But let's think ahead and presume that someday we will have enough statistical information at sufficient spatial resolution. Here is what I think the lucky person who does this will find—probably from a DNS with many integral scales in the field of view.

First since the three space dimensions are presumed to be homogeneous, then the spatial eigenfunctions will be Fourier modes. Moreover, since the scales grow in time, the energy moves to lower wavenumbers with time. If we further assume the George similarity theory for decaying turbulence to be correct [12, 14], then we know there are single length scale similarity solutions for which the Taylor microscale is the best choice of scaling parameter, and the integral scale is proportional to it. Since for long times at least our length scales grow relative to the computational box (or move to lower wavenumbers in a fixed wavenumber domain), our resolution will improve at the smallest scales in time, but the large scales will try to grow out of the box. The result will be more energy at not just the lowest wavenumbers, but also at the highest wavenumbers because of the triadic interactions (between two high and one low wavenumber). The net result will be to "artificially" increase the dissipation—at least relative to the flow we were trying to create. So to avoid this, what we really need is a simulation which rescales dynamically so the box grows in time with the turbulence length scales. This actually seems quite straightforward, but surprisingly does not seem to have been done. So let's imagine that we have done it. How do we apply Lumley's projection to this field?

If $F_{i,j}(\mathbf{k}, t)$ is the two-time three-dimensional cross-spectral Reynolds stress tensor from the result of decomposing in space first, Lumley's projection integral leaves us with the following integral equation for the time-dependence:

$$\int_{t_o}^{\infty} F_{i,j}(\mathbf{k}, t, t')\, \phi_j(\mathbf{k}, t')dt' = \lambda(\mathbf{k})\, \phi_i(\mathbf{k}, t) \tag{2.26}$$

where the wavenumber part of the three-dimensional cross-spectral tensor is given by the three-dimensional spatial transform of the two-point two-time correlation tensor; i.e.,

$$F_{i,j}(\mathbf{k}, t, t') = \frac{1}{(2\pi)^3} \int\int\int_{\text{all space}} e^{-i\mathbf{k}\cdot\mathbf{r}} \langle u_i(\mathbf{x}, t)u_j(\mathbf{x} + \mathbf{r}, t')\rangle\, d\mathbf{r} \tag{2.27}$$

where $\mathbf{r} = \mathbf{x}' - \mathbf{x}$ is the separation vector in the homogeneous directions. Clearly we need the two-point two-time cross-correlation, $\langle u_i(\mathbf{x}, t)\; u_j(\mathbf{x}', t')\rangle$, or its three-dimensional spatial transform. In spite of all the experiments and DNS of decaying turbulence I am not aware that this has ever been computed.

But there are a few facts which can make our job easier. First, since the flow is homogeneous at every time even though its scales grow in time, we can define a growing coordinate system using $\boldsymbol{\eta} = \mathbf{x}/\delta(t)$ where $\mathbf{x}$ is our space coordinate at time, t, and $\delta(t)$ is our length scale, and $\boldsymbol{\eta}' = \mathbf{y}/\delta(t')$ where $\mathbf{y}$ is our coordinate system at time t' and $\delta(t')$ is the length scale at t'. (Note that $\delta(t)$ can be either the Taylor microscale or the physical integral scale, since they are proportional. But the former is usually more accurately determined.) By examining the two-point two-time Reynolds stress equations (which I have been working on with Clay Byers and Marcus Hultmark of Princeton University) it is possible to show that they admit to an exact similarity solution for all times and spatial points of the following type:

$$\langle u_i(\mathbf{x}, t)\; u_j(\mathbf{x}', t')\rangle = R_s(t, t')\, f_{i,j}(\boldsymbol{\eta}, \boldsymbol{\eta}', t, t') \tag{2.28}$$

By substituting into the two-point two-time Reynolds stress equations, the function $R_s(t, t')$ can further be shown by equilibrium similarity arguments to decompose into two functions, one of t only, the other of t' only, and they are the same functions; i.e., $R_s(t, t') = U_s(t)\; U_s(t')$. It is easy to show that this reduces to the two-point, single time result of [12] if $U_s(t)^2$ is identified as the turbulence kinetic energy per unit mass.

Second, it turns out that the turbulence is *also homogeneous in the scaled space coordinates*. Now we already know that homogeneity in any coordinate (scaled or unscaled) implies that the Fourier modes are the solutions to the Lumley integral in these directions. So our two-point two-time Reynolds stress tensor in scaled space will behave very much as it does in regular space with one important difference—in scaled space wavenumbers, say $\boldsymbol{\kappa}$, the scaled spectra will be fixed. There will be no evolution as no energy moves to lower wavenumbers. The energy will, of course, be moving to lower wavenumbers in physical space wavenumbers, but not in their scaled space counterparts. The computational advantages of using this coordinate system for any simulation should be obvious, and completely eliminate some of the concerns expressed earlier about resolution and box-size by making sure that the scales of turbulence do not change relative to them. This should help immensely in sorting out how things really vary in time.

Third, it also turns out that time should be measured (and discretized) logarithmically. This should have been obvious to us from the single point similarity equations, but for some reason was not. It leaps out at us inescapably from the two-point two-time similarity Reynolds stress equations in similarity variables. So it means we really should be looking at everything from the perspective of things evolving in *logarithmic time*, say $\tau = \ln t$ and $\tau' = \ln t'$.

But it gets even better yet. It is straightforward to show that in fact in these new logarithmically stretched time-coordinates *the scaled statistics are statistically stationary*. To anyone who has been reading and understanding the essential points of this paper, the implications for the Lumley integral for this flow are obvious—and striking! The solutions in logarithmic time are also Fourier modes—Fourier modes of dimensionless frequency and logarithmically stretched time. In other words the solutions are known analytically!

2.5.3 But What If Turbulence Decays as t^{-1}, or More Slowly?

One of the oldest points of discussion in turbulence is about whether homogeneous decaying turbulence should decay as t^{-1}. Many theories suggest it should. All experiments show it decays more rapidly. The reason for the latter is now obvious in light of the previous section: there is no way to do an experiment or DNS which does not have a finite total energy in the field. Even if allowing for the finiteness in the spatial directions, there is no way to feed an infinite amount of energy in. So the energy must decay more rapidly than t^{-1}. So our "theoretical" flow is never obtainable.

In spite of our inability to generate it, there is something even more interesting that we can learn from the theoretical possibility of an infinite energy flow, a BIG BANG TURBULENCE if you like. Let's suppose that the integral of Eq. (2.26) is indeed infinite. Can the Lumley projection still apply? Amazingly the answer is yes—*if the two-time kernel can be shown to be statistically stationary in an appropriately stretched coordinate system*. But this is exactly what we just discovered in the preceding section.

What this means in practical terms is that the solutions in a truly infinite time, infinite energy domain (the flow we think we are analyzing) are fundamentally different solutions than the ones we can realize in the lab or the computer. Thus any attempt to verify infinite domain theory by experiment is doomed to failure—at least unless the differences are recognized and accounted for. Some aspects will be testable, others will not. The point is that the eigenfunctions describing them are fundamentally different.

Before leaving this point, let me note the analogy to standard Fourier analysis. We have long known that windowing affects our ability to do Fourier analysis. The same is true in our fluid experiments but with one difference. We are not really windowing an infinite solution, the flow we have generated inside our finite domains is truly different. It is only at best an approximation to those we analyze theoretically with assumptions which make analysis possible. The great challenge is to understand how the two relate. When is the experimental data telling us the theory is wrong? And when is the theory telling us the experiment is not a valid test? Lumley's Projection seems to have provided a way.

2.6 Conclusions: Similarity and the Lumley Projection Integral

The importance of symmetries for POD has long been recognized (c.f. Holmes et al. [20]). But the interrelation of similarity theory and the Lumley integral for inhomogeneous flows of infinite extent are only just beginning to be explored. The relation of two-point two-time similarity and the Lumley integral has been presented here for the first time. And it raises several interesting questions.

First, it is clear that in flows which are not of *finite total energy*, unless there are symmetries or some type of similarity (of the type exploited above), then there is no possibility of representing (or decomposing) the flow in the manner suggested by Lumley. It is only the additional information that the statistical properties bring to bear that makes the integral solvable.

Second, is it possible that Lumley's optimal projection integral is really more than that? Is it possible that the optimization criterion Lumley has applied, also implies that similarity solutions exist for these flows. Clearly the inverse is true. Two-point two-time similarity in at least the few cases we have considered yields analytical solutions to the Lumley integral equation. In both cases (decaying turbulence and the axisymmetric jet), the optimal bases are Fourier modes in a logarithmically stretched coordinate system. But are solutions of the Lumley type a necessary feature of all flows? And in turn, do they imply similarity in some? If so, this would be a very powerful result indeed, and substantially enhance our understanding of turbulence.

Acknowledgements I want to express my gratitude to the organizers of this conference and those who participated in it. A number of participants, especially Clay Byers and Azur Hodzic, provided helpful comments about the manuscript which were very much appreciated.

One has to be incredibly lucky to have been able to enjoy working in turbulence for five decades. And even more lucky to be honored by his fellow travelers and former students. I am humbled, and truly grateful to all those who have been a part of my life's journey to this point.

Appendix 1: A Brief History of Lumley's Projection and My Involvement with It

History of science is always difficult, since the published literature often does not correspond to the way things happened, and especially since order of publication is often unrelated to the actual chronology. The "first results" are sometimes not published at all, or published much later. As a result, newcomers frequently are misled about who did what and when. This is especially true when an area of research is suppressed, as was the case with Lumley's early work on this subject as well as my own work with my students. Since I joined this area of research just about the time Lumley was losing his enthusiasm for it, and we played some small role in both keeping it alive and re-inspiring him, the following account might be of some value to those struggling in their own isolated corners of turbulence.

With just a few exceptions, Lumley's ideas beginning in the early 1960s were not received with great enthusiasm by the turbulence community. Nils Busch, a colleague and friend of Lumley's in Meteorology at Penn State returned to his native Denmark to set up the Meteorology group at the Danish National Lab (RISOE) and carried Lumley's ideas with him, most notably the Ph.D. thesis of Eric Lundtang Petersen (of Wind Atlas fame) who applied it to turbulent gusts. And Rex Reed at U. Missouri at Rolla (in Switzerland at the time) also saw the advantages, and for years tried to carry out experiments to obtain enough information to apply them. The group at Poitiers under the leadership of Jean-Paul Bonnet embraced it somewhat later, and collaborated extensively with Glauser starting in the 1980s (e.g., [5, 38]). So this particular respect for Lumley's work was very much a part of the reason that the U. Poitiers gave him an honorary doctorate. But most others in the world, and especially the USA, ignored this part of Lumley's work. In part, this was because they didn't understand it. But that alone cannot explain why a few were so openly very hostile to his idea.

Most of this hostility I believe was based on misinformation, in part a consequence of the early applications of Lumley and his students themselves. In particular, Bakewell [2] carried out experiments in the viscous sublayer of the glycerine tunnel at Penn State built especially to apply the decomposition to near-wall turbulence, and Payne [31] applied it to existing measurements of a turbulent plane wake using Grant's measurements [19]. Both had the tremendous disadvantage of working with very little data. Bakewell's near-wall measurements were taken with a single hot-film probe along a single line perpendicular to the wall, with only a single component of velocity and only out to $y^+ = 40$. By using a series of "tricks" to fill in some of the missing component and cross-stream data, they inferred that the near-wall structure might be counter-rotating vortices, and created the schematic seen in many publications. Payne had even less data to work with, but managed to produce a pair of counter-rotating structures which spanned the wake. Unfortunately both of these efforts, instead of stimulating more work, just increased the criticism and the cynicism.

I don't have any accounts of the Moscow meeting where Lumley presented his now famous paper, other than the paper which got my attention as a first year grad student. But I do know a bit about what happened later—mostly from Lumley himself. Lumley's problems with the fluids community's lack of acceptance, or even interest, began with his presentation at the 1967 APS meeting in Hawaii. He felt humiliated by the comments after his presentation, especially by the public criticism from Otto Laporte, and often cited this bad experience as one of his reasons for avoiding such meetings as much as possible—at least in his earlier years. Ironically Laporte did not criticize the ideas themselves, but instead castigated (Lumley's words) him for using dots ("·") instead of symbols for the independent variables $(\mathbf{x}, t)$, even though these were quite commonly used by mathematicians.

But the bigger problem with acceptance came from the emerging coherent structures community, and from Lumley's own misunderstanding of what he had done. The coherent structure people led by the Cal Tech and Stanford groups were observing very active events, but Lumley was still thinking like Townsend's big and mostly passive eddies. The attacks at meetings were fierce, so that by the time I arrived on the scene in the late 1960s Lumley had already lost interest (at least in the fight). His unwillingness to respond to unsolicited attacks often left me in the 1970s as his lone defender at APS and coherent structure meetings, even when he was present. He often thanked me for my efforts to defend his ideas in public, but clearly was disheartened by the need to do so.

By contrast with Townsend's "passive large eddies," and probably thanks to my wave background, I had never thought of Lumley's decomposition as passive. From the very first I saw the dynamic possibilities, and went to Penn State to work with Lumley—in part because of my enthusiasm for working on the decomposition with him. (The other reason was to avoid being sent to Vietnam. Thanks to being hired at Penn State on a Navy contract I was able to avoid a war I did not approve of. And still finish my dissertation at Hopkins, but under Lumley's supervision, thanks to Stan Corrsin's intervention on my behalf.)

Once at Penn State, and somewhat to my disappointment, I instead ended up taking over the non-Newtonian drag-reduction experiments which were underway. And this in turn led to my early work with polymer drag reduction and the LDA. But immediately upon finishing my Ph.D. I set about to reactivate the glycerine tunnel to look for dynamic near-wall events. My very first proposal was to the GHR program of ONR and it was funded in 1972 with Lumley as co-PI. The idea was to include *time* in the measurements as well as multiple velocity components in multiple planes so we could see how things changed in time, not just space. Unfortunately I left for Buffalo before the facility modifications were finished, and the work was ultimately taken over by Siegfried Herzog. Shortly thereafter Lumley moved to Cornell, leaving "my" experiment and Herzog behind to finish it. (Siggy eventually wrote his dissertation at Cornell about 10 years later.) As per normal the experiments had proven more difficult than we had hoped, mostly because of probe and data storage issues. Herzog did complete them, but by this time the DNS efforts of Moin and Kim had caught up, in part because of our interaction with them. Our combined efforts proved important in laying the groundwork for the dynamic systems work of Aubrey et al. [1] later.

At Buffalo starting in 1974 I set out to do the same type of experiments that I had started at Penn State, but in the axisymmetric jet mixing layer. With a modest grant from NSF and some support from AFOSR in a collaborative program with Roger Arndt at Minnesota and Hassan Nagib at IIT, my students and I began the series of experiments which continue until today. Given the acceptance and widespread use of POD-based techniques today, it is hard in hindsight to imagine the hostility we faced at every step of the way. While I had gotten used to the negative proposal reviews, I really was quite surprised at how ready opponents were to make their disdain obvious in public. It was never clear, at least early on, whether the opposition was to Lumley's idea, or to Lumley himself with me as surrogate for their attacks.

The turning point for me came at the 1976 APS meeting in Eugene, OR where in an invited talk, Hans Liepman of Cal Tech went out of his way to trash Lumley's ideas. His specific comment after a brief tirade: "I've never seen any structures just sitting there." There was really no opportunity to question him in the plenary session, but seeing him with a small group at the coffee break which followed I tried to engage him about his comment (quite gently—since I was only 31 at the time). I suggested he was being unfair and clearly lacked understanding of what Lumley had actually done. And I tried to explain briefly why the time-dependence was really all there. Hans was not in a mood to listen and pretty much exploded in my face. By this time a rather larger group of 20 or so had gathered around us, probably smelling my blood. At the most intense moment I felt an arm pushing me aside and Bill Reynolds of Stanford stepped between us. He tapped Hans on the stomach with this program and pointed over his shoulder with his thumb at me and said with a big smile: "The kids's right, Hans!". Liepman said not a word, turned on his heel and left.[2] Needless to say, Bill Reynolds was my hero after that. And while I got my reputation as "controversial," not many took cheap shots in public after that at either me or Lumley or his decomposition.

The real breakthrough in our collective thinking about Lumley's decomposition actually came with the work of Mark Glauser and Stewart Lieb, both Ph.D. students of mine in the early 1980s. My colleague Andres Soom at Buffalo had several MS students design for us a special computer controlled rig to make the measurement

[2]Seven years later I went to Liepman's 70th birthday celebration at Cal Tech, a marvelous event celebrating his life and career. We had not spoken since the meeting in Oregon. When he expressed his surprise at seeing me there, I explained that many of the things I thought I had learned from Corrsin and Lumley actually started with him. And I wanted to personally thank him. He seemed quite appreciative. Indeed he was the bridge between turbulence and the classical physics of Europe. But I never understood his problem with Lumley, nor to the best of my knowledge did Lumley.

program possible—a real novelty in the late 1970s. It was constructed by Scott Woodward, who for many years afterward was an important part of my life and lab. Mark copied an idea from Hassan Nagib of making rakes of probes instead of individual ones. So we had finally both space and time information simultaneously. Figures 2.1 and 2.2 show the results of this decomposition. It was clear to us from the moment we saw the first reconstructed velocity traces that the Lumley integral had produced almost exactly the most dynamic events. And it had done so with only three eigenfunctions.

We showed these very plots to John Lumley during a break in a meeting at Cornell in the summer of 1983 and presented them about the same time [17, 27]. Looking at the plots together was truly the moment I think that John first realized the meaning and potential for what he had done. And it is also this moment I think that his interest in POD was reborn, but recast this time in the context of dynamic systems. Our paths also diverged—no longer was his decomposition about coherent structures or large eddies. He, with the collaboration of Phil Holmes and students Aubrey and Podvin and Bergooz, went for how to model these dynamic events. A whole new field of dynamic systems approach to turbulence was born—about which much has been written (c.f. [1, 18, 20]).

My group, by contrast, continued our quest to find the eigenfunctions and learn how the flow itself put them together. And that quest continues until this day. Not as much as been written, but enough to merit some discussion of its theoretical underpinnings. That is what this article was about—understanding what Lumley's integral truly implies about (and demands of) the underlying flows, whether experimental, computational, or theoretical.

Appendix 2: The Problem with the "Snapshot POD"

The so-called Snatpshot POD was introduced by Sirovich in the early 1980s and has been used extensively since for a variety of purposes. It has been extremely popular for the dynamics system attempts to understand and control turbulence, especially since the pioneering study of Aubrey et al. [1] and the book by Holmes et al. [20]. It is easily implemented if one has many 'snapshots' of data like those commonly produced by PIV. But it is a mistake to confuse it with what has come to be known as the 'classical POD', and it has only a superficial connection to the Lumley decomposition discussed above.

The primary problem with it for real turbulent flows can be demonstrated quite easily. Understanding why there is a problem is a bit more subtle. To simplify things, consider a field of only the spatial variable, x, and time, t. The snapshot POD basically replaces the instantaneous velocity with the following expansion:

$$u(x,t) = \sum_{n=1}^{N} a_n(t)\, \phi^n(x) \tag{2.29}$$

Like the classical POD, the eigenfunctions, $\phi^n(x)$, are still orthogonal and the random coefficients, $a_n(t)$, are uncorrelated at different times. The two-point two-time correlation can therefore readily be computed as:

$$\langle u(x,t)u(x',t')\rangle = \sum_{n=1}^{N} \langle a_n^*(t)a_n(t')\rangle \; \phi^{n*}(x)\phi^n(x') \tag{2.30}$$

The connection to the classical POD (first noted by Sirovich in the early 1980s) comes about by replacing the classical POD integral with its finite difference approximation over space. In most applications this number is quite low so the matrices involved are quite manageable. But when using all the data available from DNS or PIV, these spatial arrays can be very large—typically 10^6 or more. It might be argued, why not just take a smaller number of points—say a subset of those available. Unfortunately this causes serious aliasing, exactly like what happens if a time series is sampled too slowly. Now this can be overcome by spatially filtering the instantaneous data, but this can also be quite computationally intensive.

So enter the snapshot POD. By assuming the flow to be stationary, time averaging can be recognized to also be a summation, not over space but over snapshots. And by comparison the number of snapshots can be quite manageable, thousands instead of millions. Now comes Sirovich's clever trick: interchange the order of summation so the time 'average' is outside the double summation and solve that eigenvalue problem instead. Presto, the classical POD and snapshot appear to have produced exactly the same result! So where is the problem?

Implicit in the derivation of the snapshot POD is the assumption of statistical stationarity, hence $\langle a_n(t)a_n(t')\rangle = F^n(t'-t)$ only. So letting $\tau = t' - t$, we can rewrite Eq. (2.30) as:

$$\langle u(x,t)u(x',t')\rangle = \sum_{n=1}^{N} F^n(\tau)\, R^n(x,x') \tag{2.31}$$

where $R^n(x,x') = \phi^{n*}(x)\phi^n(x')$, There is nothing in principle wrong with this except for the fact that I know of no turbulence which behaves this way. It would be a very rare flow indeed were the turbulence scales uncoupled from the temporal evolution of the flow.

This problem with the snapshot was first pointed out by me and Mark Glauser in 1986 in an APS/DFD presentation, but ignored and even disputed (see note by Aubrey et al. [1]). The failure of near-wall models and other dynamic system models to advance beyond relatively simple or separated flows, I believe can still be largely attributed to this underlying deficiency.

The underlying rationale for the snapshot POD lies in the clever interchanging of order of summation and the numerical approximations to both the Lumley integral and the finite sum arithmetic used to estimate an averaged value. The basic problem lies in the fact that Lumley's optimization applied to a field which is inhomogeneous in space but stationary in time implies that the time-modes are Fourier modes in frequency. And this means all of the spatial eigenfunctions are functions of both space and frequency, not separate functions of space and time. This was one of the most important points of the body of this paper. Similarly, if the field has some directions which are homogeneous and/or periodic, then the eigenfunctions in these directions are Fourier modes, and so in the other directions they depend on frequency and mode number as well as space. Moreover, stationarity implies complex coefficients so that the same eigenfunctions can be used with different phases between them. The snapshot POD cannot reflect this since it is missing information, and simply mixes them.

Now my negativity about the snapshot POD should not be interpreted to mean that I think it cannot be useful. It can be very useful—just not in the sense of Lumley's projection. Sometimes the periodic spatial pieces can be sorted as noted in the work of my students and co-workers and those of Glauser, Tinney, and co-workers. But in general, the time–frequency problem does not appear to be tractable, and this complicates attempts to use it for understanding dynamics. On the other hand, the snapshot POD can be a VERY useful way to sort and reduce data sets. One example previously mentioned was suggested to me by J. Freund who stored snapshot POD coefficients to provide a starting field for large scale CFD computations of a compressible mixing layer well into the run. Another example is from the work of Wänström et al. [41, 42] who used the snapshot POD results to filter and reconstruct the cross-correlation *before* applying the classical POD to the snapshot results.

References

1. N. Aubrey, P. Holmes, J. Lumley, E. Stone, The dynamics of coherent structures in the wall region of a turbulent boundary layer. J. Fluid Mech. **192**, 115–173 (1988)
2. H. Bakewell, Viscous sublayer and adjacent wall region in a turbulent pipe flow. Ph.D. thesis, Department of Aerospace Engineering, Pennsylvania State University, 1966
3. G.L. Brown, A. Roshko, On density effects and large structure in turbulent mixing layers. J. Fluid Mech. **64**, 775–816 (1974)
4. J.H. Citriniti, W.K. George, Reconstruction of the global velocity field in the axisymmetric mixing layer utilizing the proper orthogonal decomposition. J. Fluid Mech. **418**, 137–166 (2000)
5. J. Delville, L. Ukeiley, L. Cordier, J.P. Bonnet, M. Glauser, Examination of the large scale structure in a turbulent mixing layer part 1 : proper orthogonal decomposition. J. Fluid Mech. **390**, 91–122 (1999)
6. D. Ewing, On multi-point similarity solutions in turbulent free-shear flows. Ph.D. thesis, Department of Mechanical and Aerospace Engineering, State University of New York at Buffalo, 1995
7. D. Ewing, B. Frohnapfel, W.K. George, J.M. Pedersen, J. Westerweel, Two-point similarity in the round jet. J. Fluid Mech. **577**, 309–330 (2007)
8. S. Gamard, W.K. George, D. Jung, S. Woodward, Application of a "slice" proper orthogonal decomposition to the far field of an axisymmetric turbulent jet. Phys. Fluids **14**(7), 2515–2522 (2002)

9. S. Gamard, D. Jung, W.K. George, Downstream evolution of the most energetic modes in a turbulent axisymmetric jet at high Reynolds number. Part 2. The far-field region. J. Fluid Mech. **514**, 205–230 (2004)
10. W. George, The self-preservation of turbulent flows and its relation to initial conditions and coherent structures, in *Advances in Turbulence*, ed. by W. George, R. Arndt (Hemisphere (now Francis and Bacon), NY, 1989), pp. 1–41
11. W.K. George, Insight into the dynamics of coherent structures from a proper orthogonal decomposition, in *The Structure of Near Wall Turbulence, Proceedings of the 1988 Symposium on Near Wall Turbulence*, Dubrovnik, ed. by S.K. Robinson et al. (Taylor and Francis, New York, 1990), pp. 168–180
12. W.K. George, The decay of homogeneous isotropic turbulence. Phys. Fluids A **4**(7), 1492–1509 (1992)
13. W.K. George, Some thoughts on similarity, the POD, and finite boundaries, in *Fundamental Problematic Issues in Turbulence*, ed. by W.K.A. Gyr, A. Tsinober. Trends in Mathematics (Birkhäuser, Basel, 1999), pp. 117–128
14. W.K. George, Asymptotic effect of initial and upstream conditions on turbulence. ASME J. Fluids Eng. **134**(6), 1–27 (2012)
15. W.K. George, P.D. Beuther, J.L. Lumley, Processing of random signals, in *Proceedings of the Marseille-Baltimore Dynamic Flow Conference on Dynamic Measurements in Unsteady Flow* (Dantec, Skovlunde, 1978)
16. M.N. Glauser, W.K. George, An orthogonal decomposition of the axisymmetric jet mixing layer utilizing cross-wire measurements, in *Proceedings of the Sixth Symposium on Turbulent Shear Flow*, Toulouse (1987), pp. 10.1.1–10.1.6
17. M.N. Glauser, S.J. Leib, W.K. George, An application of Lumley's orthogonal decomposition to the axisymmetric jet mixing layer, in *Bulletin of American Physical Society, Division of Fluid Dynamics Annual Meeting*, Houston, TX (1983)
18. M.N. Glauser, X. Zheng, C. Doering, The dynamics of organized structures in the axisymmetric jet mixing layer, in *Turbulence and Coherent Structures. From Turbulence 89: Organized Structures and Turbulence in Fluid Mechanics*, Grenoble, Sept. 1989, vol. 2, ed. by M. Lesieur, O. Metais (Kluwer Academic Publisher, Dordrecht, 1991), pp. 253–265
19. H.L. Grant, The large eddies of turbulent motion. J. Fluid Mech. **4**, 149–190 (1958)
20. P. Holmes, J.L. Lumley, G. Berkooz, *Turbulence, Coherent Structures, Dynamical Systems and Symmetry* (Cambridge University Press, Cambridge, 1996)
21. A.K.M.F. Hussain, Coherent structures-reality and myth. Phys. Fluids **26**(10), 2816–2850 (1983)
22. A.K.M.F. Hussain, Coherent structures and turbulence. J. Fluid Mech. **173**, 303 (1986)
23. P.B.V. Johansson, W. George, The far downstream evolution of the high-Reynolds number axisymmetric wake behind a disk: Part 2. Slice proper orthogonal decomposition. J. Fluid Mech. **555**, 387–408 (2006)
24. D. Jung, S. Gamard, W.K. George, S.H. Woodward, Downstream evolution of the most energetic POD modes in the mixing layer of a high Reynolds number axisymmetric jet, in *Turbulent Mixing and Combustion. Proceedings of the IUTAM Symposium*, Kingston, June 3–6, 2001, ed. by A. Pollard, S. Candel (Kluwer Academic Publisher, Dordrecht, 2002), pp. 23–32.
25. D. Jung, S. Gamard, W.K. George, Downstream evolution of the most energetic modes in a turbulent axisymmetric jet at high Reynolds number. Part 1. The near field region. J. Fluid Mech. **514**, 173–204 (2004)
26. S.J. Kline, W.C. Reynolds , F.A. Schraub, P.W. Rundstatler, The structure of turbulent boundary layers. J. Fluid Mech. **30**, 741–773 (1967)
27. S. Leib, M. Glauser, W. George, An application of Lumley's orthogonal decomposition to the axisymmetric turbulent jet mixing layer, in *Proceedings of the 9th Rolla Symposium on Turbulence in Fluids* (University of Missouri-Rolla, Rolla, MI, 1984)
28. J.L. Lumley, The structure of inhomogeneous turbulent flows, in *Atmospheric Turbulence and Radio Wave Propagation*, Nauka, Moscow, ed. by A.M. Yaglom, V.I. Tatarsky (1967)

29. J.L. Lumley, *Stochastic Tools in Turbulence* (Academic, New York, NY, 1970)
30. J.L. Lumley, H. Panofsky, *The Structure of Atmospheric Turbulence* (Interscience, Hoboken, NJ, 1964)
31. F. Payne, Large eddy structure of a turbulent wake. Ph.D. thesis, Department of Aerospace Engineering, Pennsylvania State University, 1966
32. O.M. Phillips, W.K. George, R.P. Mied, A note on the interaction between internal gravity waves and currents. J. Deep Sea Res. **15**, 267–273 (1968)
33. H. Tennekes, J.L. Lumley, *A First Course in Turbulence* (MIT Press, Cambridge, MA, 1972)
34. C. Tinney, M.N. Glauser, L.S. Ukeiley, Low dimensional characteristics of a transonic jet. Part 1. Proper orthogonal decomposition. J. Fluid Mech. **615**, 107–141 (2008)
35. A.A. Townsend, *The Structure of Turbulent Shear Flow* (Cambridge University Press, Cambridge, 1956)
36. M. Tutkun, P.B.V. Johansson, W.K. George, Three-component vectorial proper orthogonal decomposition of axisymmetric wake behind a disk. AIAA J. **46**, 1118–1134 (2008)
37. M. Tutkun, W.K. George, M. Stanislas, J. Delville, J.M. Foucaut, S. Coudert, Two-point correlations and POD analysis of the wallturb experiment using the hot-wire rake database, in *Progress in Wall Turbulence: Understanding and Modeling*. ERCOFTAC Series, vol. 14, ed. by J.J.M. Stanislas, I. Marusic (Springer, New York, 2009), pp. 95–102
38. L. Ukeiley, L. Cordier, R. Manceau, J. Delville, M. Glauser, J. Bonnet, Examination of large-scale structures in a turbulent plane mixing layer. Part 2. Dynamical systems model. J. Fluid Mech. **441**, 67–108 (2001)
39. C.M. Velte, A. Hodzic, K. Meyer, POD mode robustness for the turbulent jet sampled with PIV, in *Whither Turbulence and Big Data 2015*, ed. by A. Pollard et al. (Springer, Berlin, 2016)
40. T. von Karman, L. Howarth, On the statistical theory of isotropic turbulence. Proc. R. Soc. Lond. A **164**, 192 (1938)
41. M. Wänström, Spatial decomposition of a fully-developed turbulent round jet sampled with particle image velocimetry. Ph.D. thesis, Department of Applied Mechanics, Chalmers Tech. U, Gothenburg, 2009
42. M. Wänström, W.K. George, K.E. Meyer, Streamwise and radial decomposition of a turbulent axisymmetric jet, in *Progress in Turbulence and Wind Energy IV, 2010 ITI Conference*, Bertinoro, IT (Springer, Berlin, 2012), pp. 147–150

Part II
Turbulent Boundary Layers

Chapter 3
Study of the Streamwise Evolution of Turbulent Boundary Layers to High Reynolds Numbers

I. Marusic, K.A. Chauhan, V. Kulandaivelu, and N. Hutchins

We consider the streamwise evolution of zero-pressure-gradient (ZPG) turbulent boundary layers developing on the smooth floor of the Melbourne wind tunnel. The flat plate extends over 27 m, and three different tripping devices are used to set the initial conditions. The first trip consists of standard sandpaper, and the second and third trips consist of the addition of threaded rods of diameter 6 and 10 mm, respectively, that lead to "over-tripped" conditions. Fixed Reynolds number per metre U_∞/ν, where $U_\infty \approx 20$ m/s is the free-stream velocity and ν is kinematic viscosity, is maintained with a well-established ZPG for all tripping configurations. As the boundary layer evolves along the length of the tunnel floor, the mean velocity profiles are found to approach a constant wake parameter Π (as suggested by Coles [7]) for all the three tripping configurations after a sufficient distance downstream of the tripping devices. The broadband turbulence intensities and higher-order moments are found to show variations up the same streamwise distance, here corresponding to O(2000) trip-height lengths downstream of the trips. Further downstream the boundary layers appear to be independent of initial upstream condition and reach converged states. The discussion is aided by computations based on a modified approach originally described by Perry et al. [17], where given an initial upstream mean velocity profile, mean flow parameters are computed for different streamwise stations. The results are shown to compare well with the experimental results.

I. Marusic (✉) • V. Kulandaivelu • N. Hutchins
Department of Mechanical Engineering, University of Melbourne, Parkville, VIC 3010, Australia
e-mail: imarusic@unimelb.edu.au

K.A. Chauhan
School of Civil Engineering, University of Sydney, Sydney, NSW 2006, Australia

A. Pollard et al. (eds.), *Whither Turbulence and Big Data in the 21st Century?*,
DOI 10.1007/978-3-319-41217-7_3

3.1 Introduction

A long-standing topic in the study of wall-bounded turbulent flows is to understand their scaling behaviour. This is required for both theoretical and practical reasons, and invariably the basis for evaluating a proposed scaling law relies on comparisons with empirical data. In the case of zero-pressure-gradient (ZPG) boundary layers these comparisons are conducted over a range of Reynolds numbers, as this is the assumed sole non-dimensional parameter that specifies the state of the boundary layer. However, comparisons of experimental data from different studies rely on an assumption that scaled results (that is, statistical measurements made non-dimensional with the appropriate scaling parameters) are equivalent provided the local Reynolds numbers are equivalent. Figure 3.1 shows schematically a simple example of this for a typical comparison in the same wind tunnel. The Reynolds number can be matched by varying a combination of U_∞, the free-stream velocity, and x, the streamwise distance from the trip. When making such comparisons the local Reynolds number can be either the friction Reynolds number (also known as the Karman number) $Re_\tau = \delta U_\tau/\nu$, or the Reynolds numbers based on momentum or displacement thicknesses, $Re_\theta = U_\infty\theta/\nu$, $Re_{\delta*} = U_\infty\delta^*/\nu$, respectively. Here, δ is the boundary layer thickness, U_τ is the skin-friction velocity, ν is the kinematic viscosity of the fluid, θ is the momentum thickness, and δ^* is the displacement thickness. Therefore, when comparing one experiment to another it is implicitly assumed that there exists a one-to-one correspondence between any of these local Reynolds numbers. The aim of this paper is to understand under what conditions these assumptions are valid, and under what conditions the effects of upstream trip and other initial conditions no longer play a role in defining the state of the boundary layer.

To someone new to the field it may seem strange as to why this is still a question given that these flows have been studied over many decades. The issue

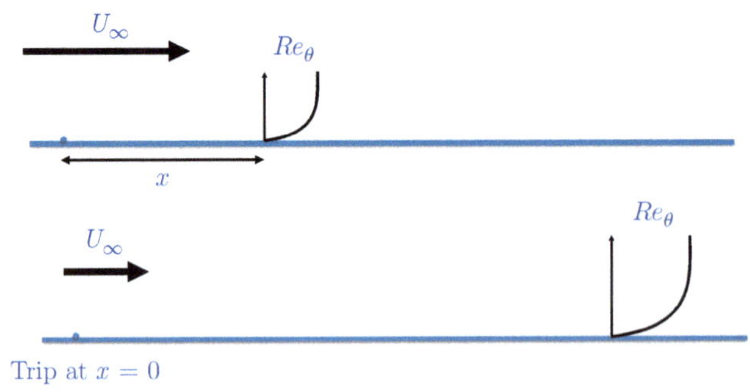

Fig. 3.1 A schematic of two experiments. The local Reynolds number, Re_θ, is the same for both cases but the top case uses a higher velocity and shorter streamwise distance from the trip to achieve the required Reynolds number

certainly is not new, but unfortunately although researchers have long recognised the importance of initial conditions in boundary layers, experimental challenges have remained and there have only been a small number of studies that have documented the streamwise evolution of boundary layers from fixed and carefully quantified initial conditions. One of the most rigorous studies performed in this area was by Erm and Joubert [8] who investigated the effect of various tripping conditions on turbulent boundary layers for Reynolds numbers between $715 \leq Re_\theta \leq 2810$. Erm and Joubert [8] proposed a technique for obtaining correctly stimulated turbulent boundary layers for a particular tripping device by changing the dimension of the trip iteratively until the measured [6] wake factor Π agrees with the [7] curve of Π versus Re_θ. For ZPG flows, then, all experiments should fall on one curve for Π versus a local Reynolds number. However, an extensive compilation of experiments by Chauhan and Nagib [3], reproduced here in Fig. 3.2, shows that there exists considerable scatter in the available data, with a wide range of Π values reported for a given value of Re_{δ^*}. Such differences have led to some investigators, such as [1, 9] and [2] to produce new scaling laws where the initial conditions persistent for all time for boundary layers. While there are a number of reasons for the observed differences between different studies, we contend that the major differences in ZPG flows can be explained due to the insufficient evolution lengths in boundary layers and this is focus of the presented results. It is also noted that the trend of the curve shown in Fig. 3.2 where Π varies with Re_{δ^*} has been described by Coles [7] as a

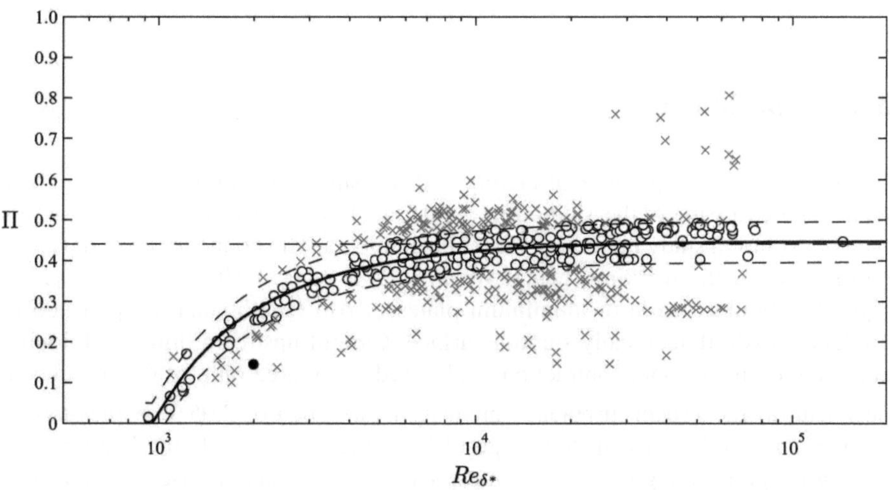

Fig. 3.2 Compilation of experimental results for Coles wake factor versus Re_{δ^*} from [5]. A total of 508 data points from a large number of experiments are shown of which 235 (*circles*) nominally agree with the *solid line*, which is obtained from an integration of the composite profile of [5]. *Gray crosses* represent profiles that have Π values beyond ± 0.05 of the *solid line*

low-Reynolds number effect. Here, we will show that a more appropriate description is one based on an evolution effect resulting from the initial conditions set-up by the trip and/or the initial inflow conditions.

3.1.1 Implications for Big Data

Presently it is not feasible to store all of the data that is generated in direct numerical simulations of evolving high Reynolds number wall-bounded flows, or to measure the full spatial domain of the flow in time-resolved experiments. The size of such datasets would simply be too onerous. Rather, decisions need to be made as to what given locations measurements are made in experiments, and what time-steps are kept in numerical simulations in order to recover the statistical quantities that are of interest. Therefore, there are obvious implications for the quantity of stored data that is required to properly document the evolution history of boundary layers. In the following we will study boundary layers with comparisons of statistics for flows that have evolved from three types of tripping conditions. By making detailed measurements at multiple streamwise locations for each flow we can determine over what streamwise distance the transients effects of the trips persist.

It is noted that the present paper should be seen as a supplement to the full paper of [12]. Some of the main results are repeated here but additional results are presented in order to present a more complete report.

3.2 Experiments

Full details of the experimental facility and measurements are given in [12]. The main features are use of the Melbourne wind tunnel, which was purpose built for this type of evolution study up to high Reynolds numbers. The boundary layer develops on the floor of the working section which is 27 m long, 1.89 m wide and 0.92 m high. The floor is made of aluminium plates of 6 m lengths that are polished to produce a hydrodynamically smooth surface. Careful upstream flow conditioning and a three-dimensional contraction, with a reduction area ratio of 6.2, produce a flow with a free-stream turbulence intensity ($\sqrt{\overline{u^2}}/U_\infty$) of 0.05 % at the start of the working section, and in the range of 0.15–0.2 % at $x \approx$ 18 m. The working section is operated slightly above atmospheric pressure, and a series of adjustable slots and panels on the ceiling of the working section enables precise control of the streamwise pressure gradient. For all experiments presented here the coefficient of pressure C_p along the entire working section is constant to within ± 0.87 % (for all tripping configurations). For all cases, measurements were conducted with a nominal free-stream velocity of 20 m/s, with a reference Reynolds number per metre of $U_\infty/\nu = 1.295 \times 10^6$ m^{-1}.

The long development length of the working section produces thick boundary layers, which enables us to attain good spatial resolution with a conventional hot-wire probe design. In terms of spatial resolution, since the unit Reynolds number is matched everywhere, the only variation in the viscous scaled wire-length, $l^+ = lU_\tau/\nu$, occurs due to the weak variation in U_τ along the development length of the facility. At $x = 1.6$, the 0.54 mm long sensing element yields $l^+ = 25.6$, falling to 22.3 at $x = 18$ m. Hence, the probes can be considered to be nominally matched in terms of viscous lengths ($l^+ = 24 \pm 2$, and precise values are given in Table 3.1). Another important feature of the experiments was the use of a special calibration procedure to reduce drift. The hot-wire probes were statically calibrated against a Pitot-static tube positioned along the centreline of the tunnel in the undisturbed free-stream, and to account for calibration drift during the experiments, the probe was periodically traversed to the free-stream within the boundary layer profile measurement. At this free-stream excursion, the mean voltage measured by the hot-wire and the mean velocity measured by the Pitot-static tube provide an additional *re-calibration* point at various intervals during the boundary layer traverse. Effectively, this means that for every six measurements during the boundary layer traverse (which consisted of between 33 and 50 logarithmically spaced measurement stations), the probe is re-calibrated. This procedure leads to a considerable reduction in the scatter between repeat experiments and is described in detail by Talluru et al. [18].

Three flow cases were studied corresponding to three different tripping configurations. All trips were introduced at the inlet to the working section (at $x = 0$ for the axis system used in this paper). The initial set of measurements was performed with the 'standard' tripping configuration, which consisted of a strip of P40 grit sandpaper, of length 154 mm in the streamwise direction, and was chosen based on the criteria outlined by Erm and Joubert [8]. The next two tripping configurations were deliberately chosen to over-stimulate the boundary layers. In these configurations, 6 and 10 mm diameter threaded rods were added at $x = 0$. We refer to the sandpaper and threaded rod configurations throughout as SP40, TR06 and TR10. Table 3.1 presents a summary of the main experimental parameters for each of the three cases. Here, $U_\infty^+ = U_\infty/U_\tau$ and in all cases U_τ, δ and Π are obtained by using the composite profile fit of [5] using the log-law constants $\kappa = 0.384$ and $B = 4.17$.

Given the limited cross-sectional area of the working section ($1.89 \times 0.92\,\mathrm{m}^2$), compared to the very long length (27 m), careful attention was given to ensure that the boundary layers were nominally two-dimensional in the mean for the streamwise stations considered here. This is discussed in detail in [12], and Fig. 3.3 shows the results of [11] who conducted a spanwise survey of free-stream velocity at $U_\infty \approx 20$ m/s at $x = 10.5$ m over a spanwise distance of 0.8 m either side of the tunnel centre-line, and found the variation to be less than $\pm 0.35\%$ with no distinguishable slope in the velocity variation across the width of the boundary layer (corresponding to over $\Delta y \approx 8\delta$ at this streamwise station). The wind tunnel also contains corner fillets throughout the facility to minimise the effect of secondary flows in the corners of the wind tunnel. Comparisons of turbulence statistics at $x = 21$ m, with and

Table 3.1 Experimental parameters for the SP40, TR06 and TR10 configurations, respectively

Station	x (m)	Re_τ	Re_{δ^*}	Re_θ	δ (mm)	Π	U_∞^+	l^+
SP40: Sandpaper (regular) trip								
1	**1.60**	2800	9800	7200	57.5	0.49	27.37	26.1
2	2.65	3600	12,500	9300	75.2	0.48	27.97	25.5
3	3.75	4300	14,900	11,200	92.2	0.46	28.36	25.3
4	**4.75**	5100	16,800	12,700	108	0.41	28.54	25.3
5	6.30	6000	20,600	15,600	132	0.46	29.21	24.5
6	**7.50**	6700	22,600	17,200	150	0.43	29.36	24.1
7	10.00	8400	28,400	21,700	188	0.44	29.97	24.1
8	**12.80**	10,500	33,700	26,100	237	0.38	30.25	23.9
9	17.50	13,000	42,800	33,300	304	0.40	30.95	23.1
10	18.90	13,400	45,000	34,900	319	0.43	31.15	22.7
TR06: 6 mm threaded rod								
1	**1.60**	3700	10,700	8100	75.6	0.25	26.88	26.1
2	2.65	4900	13,500	10,400	101	0.21	27.40	25.8
4	**4.75**	6100	18,400	14,200	130	0.30	28.46	24.9
5	6.30	6800	22,100	16,900	148	0.39	29.20	24.3
6	**7.50**	7600	24,500	18,800	167	0.38	29.42	24.1
8	**12.80**	10,700	35,100	27,100	244	0.40	30.41	23.2
–	15.50	12,300	40,000	31,100	284	0.38	30.71	23.0
–	22.10	16,300	53,500	41,800	385	0.40	31.51	22.4
TR10: 10 mm threaded rod								
1	**1.60**	8000	13,900	11,000	161	–	26.75	26.8
4	**4.75**	8700	20,400	16,100	184	0.07	28.13	25.5
6	**7.50**	9300	26,200	20,500	203	0.24	29.19	24.6
8	**12.80**	11,800	36,700	28,600	269	0.34	30.36	23.6
–	15.50	13,800	41,700	32,700	317	0.30	30.59	23.1
10	18.90	15,400	48,600	38,100	361	0.36	31.13	23.0

Bold notation for streamwise distance x indicates that it is common measurement station for all three tripping configurations. For all cases, measurements were conducted with a nominal free-stream velocity of 20 m/s, with a reference Reynolds number per metre of $U_\infty/\nu = 1.295 \times 10^6\,\text{m}^{-1}$

without corner fillets in the working section revealed no discernible differences, providing further confidence that the boundary layers at all streamwise locations reported here are nominally two-dimensional in the mean.

Another important issue that should be clarified for this type of study is the appropriate Reynolds number for comparison based on streamwise distance. It has been proposed that a virtual origin, x_0, needs to be accounted for. That is, $Re_{(x-x_0)}$ should be used instead of Re_x. We consider this in Fig. 3.4 where we show boundary layer thickness versus streamwise distance for the three cases. It is noted that the boundary layers are abruptly tripped for the TR06 and TR10

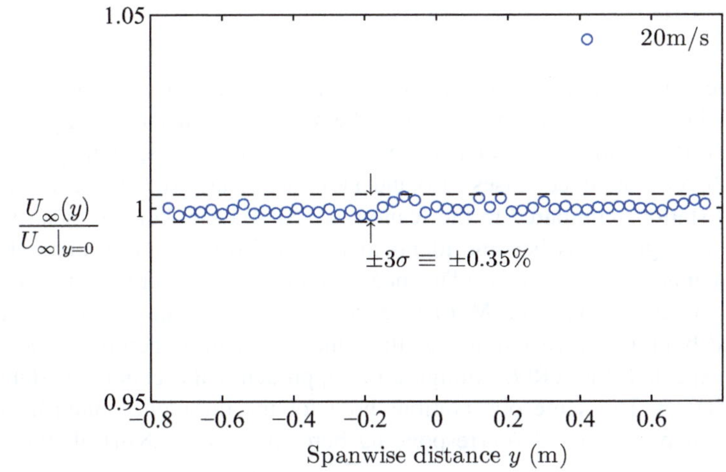

Fig. 3.3 The relative spanwise variation of the free-stream velocity $U_\infty = 20$ m/s at $x = 10.4$ m downstream of the trip. The wall-normal location is $z = 0.5$ m

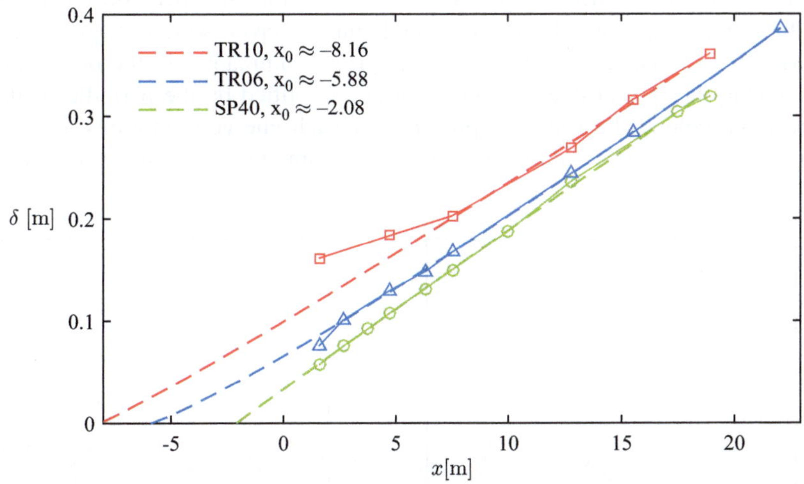

Fig. 3.4 Boundary layer thickness versus streamwise distance for the three tripping conditions

configurations and hence the δ values for these two cases are significantly higher than the SP40 configuration. The low Reynolds number development of these boundary layers is not self-similar (see Marusic et al. [12]) and therefore the virtual origin calculated in Fig. 3.4 is not relevant. Here the objective is to examine the downstream influence of a particular tripping configuration on the flow measured at a particular streamwise distance on the floor. Hence, a streamwise variable x that is relative to the location where the transition trigger is placed is appropriate. This allows for a direct comparison of profile measurements at same the streamwise location as will be shown in the next section.

3.3 Results

In this section we compare the streamwise velocity statistics for the three cases. Figures 3.5a and 3.6a, respectively, show the normalised mean velocity and turbulence intensity profiles for four matched stations, thus corresponding to matched Re_x. The 'over-stimulated' cases TR6 and TR10 are clearly seen to be influenced at the initial stations compared to the 'canonical' SP40 case. For the mean velocity, the three configurations become identical further downstream as marked by the collapse in inner-scaled profiles. (The mean velocity-deficit profiles also agree when shown in outer scaling—see Marusic et al. [12].) The rate at which the over-stimulated boundary layers return to the canonical state depends on the type of trip. As expected, the TR06 configuration approaches the canonical state faster than the TR10. The same observations hold for the turbulence intensity results, where again by Station 8 (corresponding here to $x = 12.8$ m) all the profiles nominally agree. The differences in the $\overline{u^2}^+$ profiles are seen at low Reynolds number, particularly in the outer region of the boundary layer, with the larger trip (TR10) resulting in larger deviations from canonical behaviour. Comparisons of spectra (shown in Marusic et al. [12]) reveal that the over-stimulated trips introduce large-scale disturbances into the boundary layer, which are prominent at low-Re. These large-scale disturbances reside predominantly in the outer part of the boundary layer, and likely originate or are amplified by the periodic shedding of the wake behind the rod. The presence of such energetic motions due to the abrupt tripping of the boundary layer manifests as an outer peak in the spectrogram

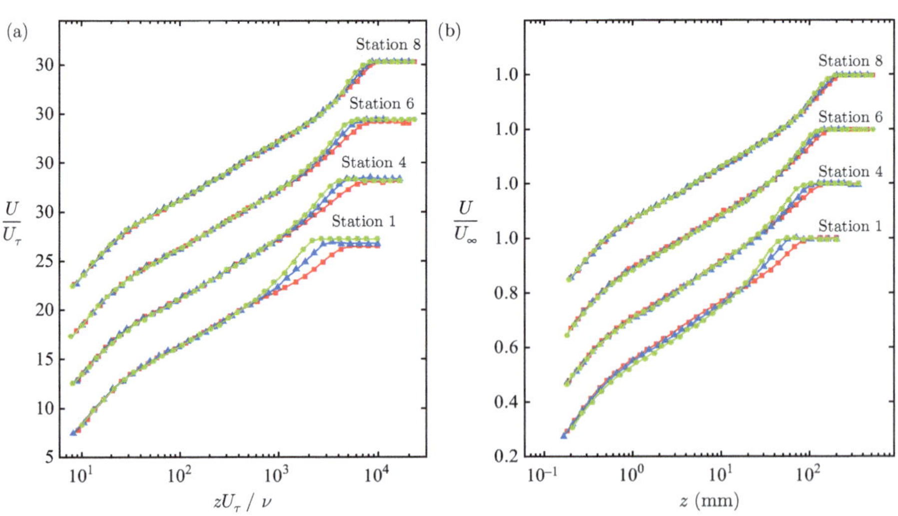

Fig. 3.5 Comparison of mean velocity profiles for the three trips at four streamwise locations. *Filled circle*, SP40; *filled triangle*, TR06; *filled square*, TR10. Note the vertical shift in profiles. (**a**) With inner scaling; (**b**) where mean velocity is scaled with free-stream velocity

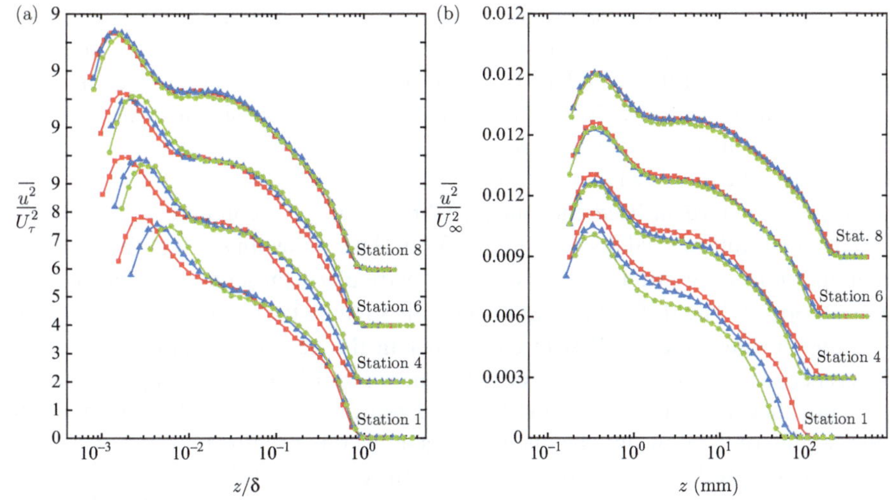

Fig. 3.6 Comparison of streamwise turbulence intensity profiles for the three trips at four streamwise locations. *Filled circle*, SP40; *filled triangle*, TR06; *filled square*, TR10. Note the vertical shift in profiles. (**a**) With outer scaling; (**b**) scaled with free-stream velocity

at low Re_x while at the same Re_x such large-scales are absent in the SP40 case. This artificial outer peak is different to the naturally occurring outer spectral peak that occurs at high Re in canonical ZPG flows. The remnants of the 'over-tripped' conditions are seen to persist at least until Station 8, at which position the non-canonical boundary layers (TR06 and TR10) exhibit a weak memory of their initial conditions only for the large-scales $O(10\delta)$.

One concern when making the above comparisons is the accuracy of determining U_τ for the over-tripped configurations, especially at low Reynolds numbers. In the absence of direct measurement of U_τ we rely on a composite velocity formulation that describes the mean velocity from the wall to the outer edge of the boundary layer. Chauhan et al. [4] compared U_τ estimates from the composite profile formulation with direct measurements and found an agreement within $\pm 2\,\%$ for data at $Re_{\delta*} < 10{,}000$. Also, Chauhan et al. [5] found that fitted U_∞^+ values are accurate within $\pm 1.5\,\%$ even for low Reynolds number data, i.e. $Re_{\delta*} < 10{,}000$ when compared with U_∞^+ obtained from the integral skin-friction relation of [14] which is calibrated using oil-film interferometry data. Hence using the composite fit approach to determine U_τ at the initial stations seems reasonable. Further, the fitting algorithm considers equal weighting for all measured data. Therefore any non-canonical characteristics of the flow that are present in the mean velocity (in the inner, outer or both regions) will be accounted for in the fitting procedure. This results in the non-equilibrium influences to appear in the form of variations in U_τ, Π or δ compared to the canonical case of SP40. Non-equilibrium influences in ZPG flows were also illustrated using the parameters obtained from the composite velocity profile by Chauhan et al. [5]. To further check this, we plot U/U_∞ versus z in Fig. 3.5b. In this form, Fig. 3.5b is free of any fitted parameters (U_τ or δ) and provides a direct comparison of momentum at a fixed x location between the three trip configurations. At the initial station we observe clear differences in the mean velocity throughout the boundary layer. Further downstream these differences are only seen in the outer region of the profiles. In the dimensional form, Fig. 3.5b clearly shows that the near-wall region of the over-stimulated flow recovers the earliest to match the canonical mean flow behaviour of the SP40 configuration. Similarly, we have utilised the outer scaling of U_∞^2 to compare $\overline{u^2}$ versus the dimensional wall-normal distance in Fig. 3.6b. Here we again see that the three trip configurations have different levels of $\overline{u^2}$ at particular z throughout the layer at the initial stations. The differences diminish with streamwise distance and by Station 8 the $\overline{u^2}$ profiles agree well with each other in the near-wall region while slight differences are observed in the outer region near $z \approx 15$ cm. This comparison again indicates that the transition by the threaded rod most significantly influences the outer region of the flow and the disturbances in the outer region persist downstream while the near-wall region recovers earlier to the canonical behaviour.

Higher-order statistics were also considered. Figures 3.7 and 3.8 show the comparisons, respectively, for even and odd higher-order moments of u up to tenth order. The even moments are presented in the form $(\overline{u^{2p}})^{1/p}$ following the work of [13] who showed that even moments represented in this way have a logarithmic behaviour with distance from the wall in the log region of fully developed ZPG flows, and this is seen to be the case for the SP40 profiles. Comparison between the profiles in Figs. 3.7 and 3.8 and Fig. 3.5 indicate that the recovery length required for the statistics to become independent of the trip is not dependent on the order of the statistic (at least not up to tenth order), with all statistics nominally agreeing by Station 8. This suggests that while the larger trips introduce additional length scales into the flow, these perturbations and interactions relax as the boundary layer evolves downstream, and once they have decayed to the point of no longer influencing the mean velocity profiles, their effect also appears to be negligible for the higher-order statistics. This finding implies that in order to determine if a flow has sufficiently recovered from a trip and reached a canonical ZPG boundary layer state, only information about the evolution of the mean velocity profile is required. This is particularly advantageous as a reliable computation scheme can be developed for mean flow evolution, and this is considered in the following section.

3.4 Computing Boundary Layer Evolution

Here we compute the streamwise evolution of the mean flow parameters for a ZPG boundary layer using the computational scheme outlined in [12]. The approach uses the hypothesis that the mean velocity-defect profile is uniquely described by a two parameter family. In addition, a relation between the mean flow and shear-stress parameters is required to close the system of equations. The equations that govern the streamwise evolution of a turbulent boundary layer can be found after considerable algebra by using the momentum integral and differential equations, the

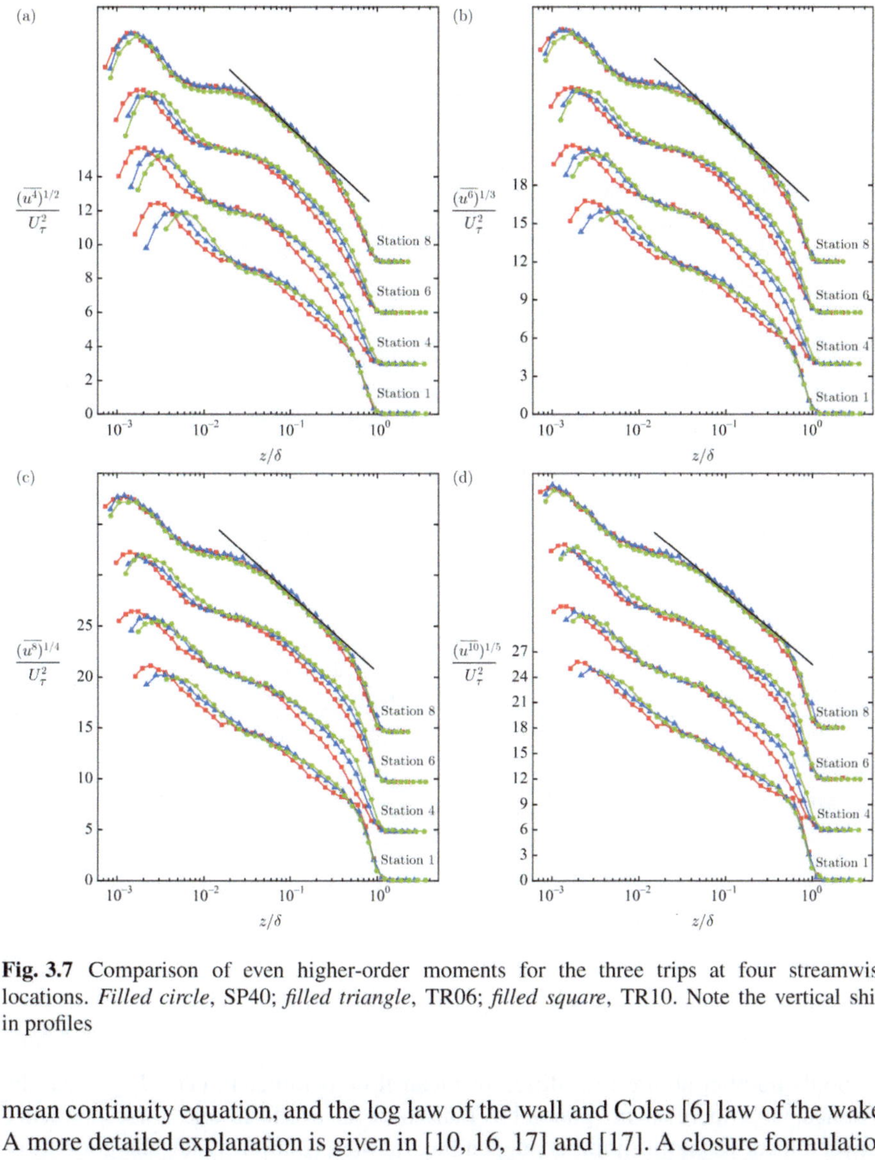

Fig. 3.7 Comparison of even higher-order moments for the three trips at four streamwise locations. *Filled circle*, SP40; *filled triangle*, TR06; *filled square*, TR10. Note the vertical shift in profiles

mean continuity equation, and the log law of the wall and Coles [6] law of the wake. A more detailed explanation is given in [10, 16, 17] and [17]. A closure formulation is required and this is based on the limited empirical data available in the literature where the streamwise evolution of the boundary layer is fully documented.

The results of the computational scheme for the SP40, TR6 and TR10 cases are shown in Fig. 3.9 together with the data presented in [12] in addition to the data of [15] in the NDF facility at IIT. Overall the scheme is seen to agree well with all the experimental data. Here, $H = \delta^*/\theta$ is the shape factor and $C_f = 2(U_\tau/U_\infty)^2$ is the friction coefficient. It can be seen that through the local parameters such as the Reynolds number Re_x is matched (for the sake of

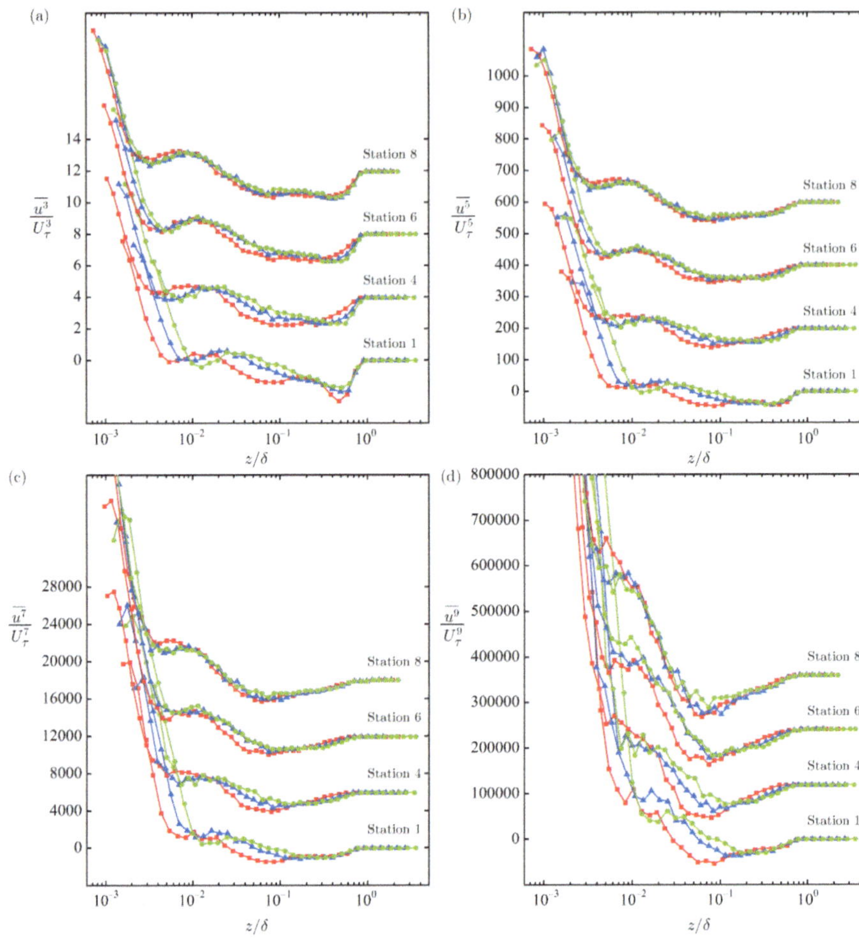

Fig. 3.8 Comparison of odd higher-order moments for the three trips at four streamwise locations. *Filled circle*, SP40; *filled triangle*, TR06; *filled square*, TR10. Note the vertical shift in profiles

Reynolds number similarity), different mean flow parameters (Π, H, C_f) can be obtained, and there is no guarantee that there is a one-to-one correspondence between local Reynolds numbers depending on the evolution state. The good fit of the computational scheme gives some confidence in evaluating the state of evolution for boundary layers.

All ZPG boundary layers are then expected to evolve to an equivalent form where Reynolds number similarity holds, provided the Reynolds number is sufficiently high, and the larger the trip is above the ideal trip size, the longer the boundary layer will need to recover. However, it is not clear a priori how long this will take and how far downstream of the trip will occur. The computational approach used here can provide this information and appears to be a valuable tool towards determining whether a given ZPG boundary layer is of a canonical form.

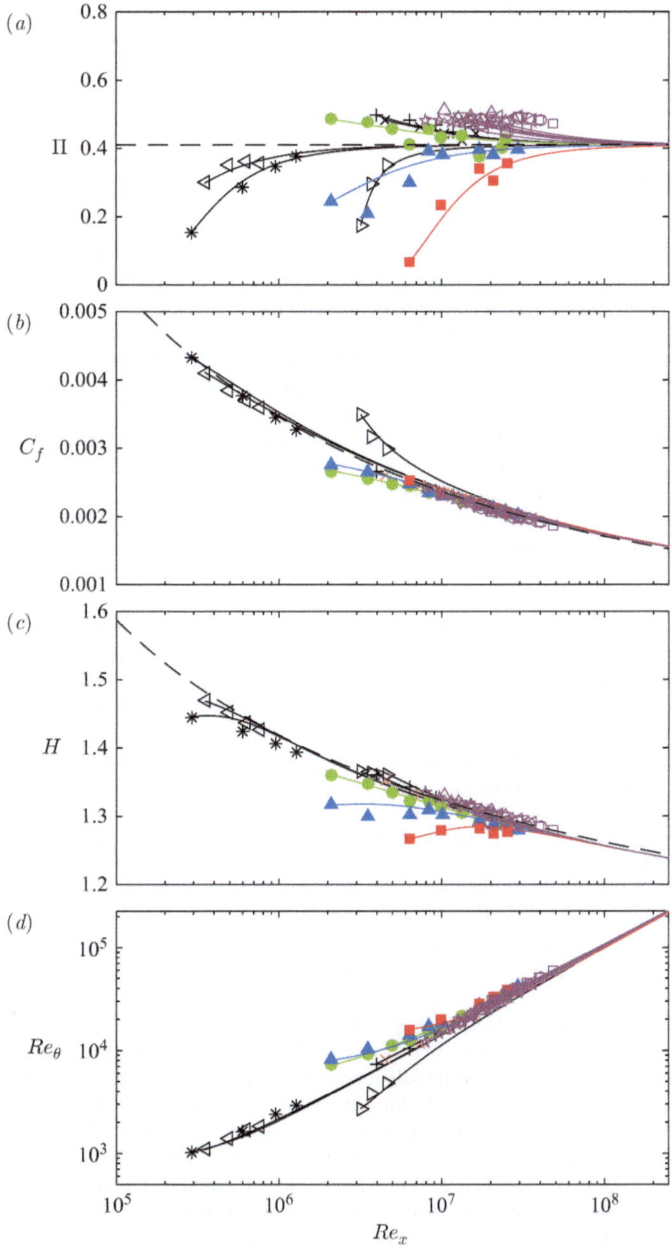

Fig. 3.9 Comparison of experimental boundary layer parameters with computed evolution. (**a**) Π versus Re_x. *Dashed line* is the asymptotic limit of Π according to the definition of [5]; $\Pi \rightarrow 0.42$. (**b**) C_f versus Re_x. *Dashed line* is Schlichting's fit, $C_f = [2\log_{10}(Re_x) - 0.65]^{-7/3}$. (**c**) H versus Re_x. *Dashed line* is $H = (1 - 7.1/\sqrt{2/C_f})^{-1}$. (**d**) Re_θ versus Re_x. *Purple coloured* symbols are the data of [15]. A description of the other datasets and further description of the *dashed curves* is given in [12]

Acknowledgements The authors gratefully acknowledge the Australian Research Council for the financial support of this project. We also thank Prof. Hassan Nagib for sharing the NDF data taken by Chris Christophorou.

References

1. L. Castillo, T.G. Johansson, The effects of the upstream conditions on a low Reynolds number turbulent boundary layer with zero pressure gradient. J. Turbul. **3**(31), 1–19 (2002)
2. L. Castillo, D. Walker, Effect of upstream conditions on the outer flow of turbulent boundary layers. AIAA J. **40**(7), 1292–1299 (2002)
3. K.A. Chauhan, H.M. Nagib, On the development of wall-bounded turbulent flows, in *IUTAM Symposium on Computational Physics and New Perspectives in Turbulence* (Springer, New York, 2008), pp. 183–189
4. K. Chauhan, H. Nagib, P. Monkewitz, On the composite logarithmic profile in zero pressure gradient turbulent boundary layers, in *Proceedings of the 45th AIAA Aerospace Sciences Meeting and Exhibit, Reno, NV*, Paper No. AIAA, vol. 532, 2007, pp. 1–18
5. K.A. Chauhan, H.M. Nagib, P.A. Monkewitz, Criteria for assessing experiments in zero pressure gradient boundary layers. Fluid Dyn. Res. **41**, 021404 (2009)
6. D.E. Coles, The law of the wake in the turbulent boundary layer. J. Fluid Mech. **1**, 191–226 (1956)
7. D.E. Coles, The turbulent boundary layer in a compressible fluid. Appendix A: a manual of experimental boundary-layer practice for low-speed flow. Tech. Rep. R-403-PR. USAF The Rand Corporation, 1962
8. L.P. Erm, P.N. Joubert, Low-Reynolds-number turbulent boundary layer. J. Fluid Mech. **230**, 1–44 (1991)
9. G. Johansson, L. Castillo, LDA measurements in turbulent boundary layers with zero pressure gradient, in *Proceedings of the 2nd International Symposium Turbulent Shear Flow Phenomena*, Stockholm, 2001
10. M.B. Jones, I. Marusic, A.E. Perry, Evolution and structure of sink-flow turbulent boundary layers. J. Fluid Mech. **428**, 1–27 (2001)
11. V. Kulandaivelu, Evolution of zero pressure gradient turbulent boundary layers from different initial conditions. Ph.D. thesis, University of Melbourne, 2012
12. I. Marusic, K. Chauhan, V. Kulandaivelu, N. Hutchins (2015) Evolution of zero-pressure-gradient boundary layers from different tripping conditions. J. Fluid Mech., **783**, 379–411
13. C. Meneveau, I. Marusic, Generalized logarithmic law for high-order moments in turbulent boundary layers. J. Fluid Mech. **719**, R1 (2013)
14. P.A. Monkewitz, K.A. Chauhan, H.M. Nagib, Self-contained high Reynolds-number asymptotics for zero-pressure-gradient turbulent boundary layers. Phys. Fluids **19**, 115101 (2007)
15. H.M. Nagib, C. Christophorou, P.A. Monkewitz, High Reynolds number turbulent boundary layers subjected to various pressure-gradient conditions, in *IUTAM Symposium on One Hundred Years of Boundary Layer Research* (Springer, New York, 2006), pp. 383–394
16. A.E. Perry, I. Marusic, J.D. Li, Wall turbulence closure based on classical similarity laws and the attached eddy hypothesis. Phys. Fluids **6**(2), 1024–1035 (1994)
17. A.E. Perry, I. Marusic, M.B. Jones, On the streamwise evolution of turbulent boundary layers in arbitrary pressure gradients. J. Fluid Mech. **461**, 61–91 (2002)
18. K.M. Talluru, V. Kulandaivelu, N. Hutchins, I. Marusic, A calibration technique to correct sensor drift issues in hot-wire anemometry. Meas. Sci. Technol. **25**(10), 105304 (2014)

Chapter 4
Towards the Direct Numerical Simulation of a Self-similar Adverse Pressure Gradient Turbulent Boundary Layer Flow

J. Soria, V. Kitsios, C. Atkinson, J.A. Sillero, G. Borrell, A.G. Gungar, and J. Jimenez

4.1 Introduction

Adverse pressure gradient turbulent boundary layer (APG-TBL) flow is in practical applications the norm rather than the exception and includes applications such as the flow over aircraft wings, wind turbine blades, road vehicles, ships and flow in turbo-machinery and diffusers. The accurate prediction of TBL separation, a direct consequence of APG-TBL flow, remains a significant challenge for engineering

J. Soria (✉)
Laboratory For Turbulence Research in Aerospace and Combustion, Department of Mechanical and Aerospace Engineering, Monash University (Clayton Campus), Melbourne, VIC 3800, Australia

Department of Aeronautical Engineering, King Abdulaziz University, Jeddah 21589, Kingdom of Saudi Arabia
e-mail: julio.soria@monash.edu

V. Kitsios • C. Atkinson
Laboratory For Turbulence Research in Aerospace and Combustion, Department of Mechanical and Aerospace Engineering, Monash University (Clayton Campus), Melbourne, VIC 3800, Australia
e-mail: vassili.kitsios@monash.edu; callum.atkinson@monash.edu

J.A. Sillero • G. Borrell
School of Aeronautics, Universidad Politécnica de Madrid, E-28040 Madrid, Spain
e-mail: sillero@torroja.dmt.upm.es; guillem@torroja.dmt.upm.es

A.G. Gungar
Department of Astronautical Engineering, ITU Faculty of Aeronautics and Astronautics, Maslak, 34469 Istanbul, Turkey
e-mail: ayse.gungor@itu.edu.tr

J. Jimenez
School of Aeronautics, Universidad Politécnica de Madrid, E-28040 Madrid, Spain
e-mail: jimenez@torroja.dmt.upm.es

A. Pollard et al. (eds.), *Whither Turbulence and Big Data in the 21st Century?*,
DOI 10.1007/978-3-319-41217-7_4

design. One of the complexities of the practical applications of APG-TBL is that the pressure gradient is constantly changing in the streamwise direction, as for example, in the large eddy simulation of Kitsios et al. [6].

In the present study our aim is to focus our attention on the canonical flow configuration of a self-similar TBL subjected to an APG such that the TBL is at the verge of separation, akin to the configuration in the experimental study of Skåre and Krogstad [15]. A self-similar APG-TBL is defined as having a constant ratio of friction velocity, u_τ, to free-stream streamwise velocity, U_∞, and also a constant ratio of pressure velocity, U_P to U_∞ [10].

Previous direct numerical simulations (DNS) of APG-TBL flows include non-self-similar separated flow [3] and self-similar APG-TBL cases at relatively low Reynolds numbers [16] and [8]. In the present study we undertake DNS of a self-similar APG-TBL developing on a flat surface using the TBL DNS code developed by Simens et al. [14] and Borrell et al. [1]. The code has been modified in the wall-normal far-field to implement the APG boundary condition (BC). The maximum momentum based Reynolds number in the present simulations is $Re_{\delta_2} = 6700$.

4.2 Self-similar Analysis of APG-TBL

The self-similar analysis as it pertains to APG-TBl is briefly reviewed and the necessary conditions for self-similar APG-TBL to exist are presented. These conditions will be tested using the DNS data to investigate which of them are satisfied. We assume the classical Reynolds decomposition of the dependent variables,

$$\begin{aligned} U_i(\underline{x}, t) &= \overline{u}_i(\underline{x}) + u_i(\underline{x}, t) \\ P(\underline{x}, t) &= \overline{p}(\underline{x}) + p(\underline{x}, t), \end{aligned} \tag{4.1}$$

where $\underline{x} = (x, y, z)$, the subscript i refers to the streamwise, x, wall-normal, y and spanwise, z, directions, with respective instantaneous velocity components in these directions of $U = U_1$, $V = U_2$ and $W = U_3$. The overlined dependent variables denote the ensemble average of the dependent variables. The associated fluctuating components are denoted by (u, v, w). Assuming that the mean flow is two-dimensional, i.e.,

$$\begin{aligned} \overline{u}_i(\underline{x}) &= \overline{u}_i(x, y) \\ \overline{p}(\underline{x}) &= \overline{p}(x, y), \end{aligned} \tag{4.2}$$

then the governing equations for the mean flow are the mean two-dimensional boundary layer equations,

$$\begin{aligned} \partial_x \overline{u} + \partial_y \overline{v} &= 0 \\ \overline{u}\, \partial_x \overline{u} + \overline{v}\, \partial_y \overline{u} &= -\tfrac{1}{\rho}\, \partial_x P_e + \partial_x \overline{v^2} - \partial_y \overline{u^2} - \partial_y \overline{uv} + \nu\, \partial_{y,y} \overline{u}, \end{aligned} \tag{4.3}$$

where $P_e(x)$ is the streamwise mean pressure distribution in the free-stream, which is related to the streamwise free-stream velocity $U_e(x)$ via Bernoulli's equation. ∂_x, ∂_y and $\partial_{y,y}$ represent partial differentiation with respect to the indicated subscripted independent variable.

The self-similar analysis of the mean governing equations follows the approach initially proposed by Townsend [17, 18] and with the less restrictive assumptions suggested by George and Castillo [2]. Assuming the ansatz for the dependent variables,

$$\begin{aligned} \overline{u}(x, y) &= U_e(x) + U_0(x) f(\eta) \\ \overline{uv}(x, y) &= -R_{uv}(x)\, r_{uv}(\eta) \\ \overline{u^2}(x, y) &= R_{uu}(x)\, r_{uu}(\eta) \\ \overline{v^2}(x, y) &= R_{vv}(x)\, r_{vv}(\eta) \end{aligned} \tag{4.4}$$

where $U_0(x)$ is a local velocity scale yet to be determined and $\eta \equiv \frac{y}{L_0(x)}$, where $L_o(x)$ is a local length scale also yet to be determined. In (4.4) $f(\eta)$ represents the self-similar solution, if it exists, while $r_{uv}(\eta)$, $r_{uu}(\eta)$, $r_{vv}(\eta)$ represent the self-similar functions that describe the relevant Reynolds stresses in (4.3).

The displacement thickness is defined as

$$\delta_1 \equiv \int_0^\infty \left(1 - \frac{\overline{u}(x, y)}{U_e(x)}\right) dy. \tag{4.5}$$

Substituting the ansatz yields

$$\delta_1 = \left(-\frac{U_0\, L_0}{U_e}\right) \int_0^\infty f(\eta)\, d\eta. \tag{4.6}$$

Without any loss of generality one can define

$$L_0(x) \equiv \frac{U_0\, \delta_1}{U_e} \quad \Rightarrow \quad \int_0^\infty f(\eta)\, d\eta = -1. \tag{4.7}$$

the momentum thickness is defined as,

$$\delta_2 \equiv \int_0^\infty \frac{\overline{u}(x, y)}{U_e(x)} \left(1 - \frac{\overline{u}(x, y)}{U_e(x)}\right) dy, \tag{4.8}$$

which can be shown to be given by,

$$\delta_2 = \delta_1(x)\left[1 + \frac{U_0(x)}{U_e(x)}\int_0^{\infty}\{f(\eta)\}^2\, d\eta\right], \tag{4.9}$$

and yields the following relationship for the shape factor:

$$H \equiv \frac{\delta_1(x)}{\delta_2(x)} = \frac{1}{\left[1 + \frac{U_0(x)}{U_e(x)}\int_0^{\infty}\{f(\eta)\}^2\, d\eta\right]}. \tag{4.10}$$

Substitution of the ansatz (4.4) into (4.3) and after some algebra yields the following mean boundary layer x-momentum equation form:

$$\begin{aligned}&\partial_x(U_oU_e)\, f + \left(\tfrac{U_o\,U_e}{\partial_x\delta_1} - U_o\,\partial_x U_e\right)\eta f' + \tfrac{1}{2}\,\partial_x\left(U_o{}^2\right) f^2 \\ &-U_o{}^2\,\partial_x\left(\text{Log}\left[\delta_1\, U_e\right]\right) F f' \\ &= \tfrac{R_{vv}}{\partial_x\delta_1}\,\eta\, r'_{vv} - \tfrac{R_{uu}}{\partial_x\delta_1}\,\eta\, r'_{uu} + \tfrac{R_{uv}}{\delta_1^*}\, r'_{uv} + \partial_x R_{vv}\, r_{vv} - \partial_x R_{uu}\, r_{uu} + \tfrac{\nu\, U_o}{\delta_1^{*2}} f''\end{aligned} \tag{4.11}$$

where

$$\begin{aligned}\delta_1^*(x) &\equiv \tfrac{\delta_1(x)U_e(x)}{U_o(x)} \\ \delta_1'(x) &\equiv \left\{\partial_x\left(\text{Log}\left[\tfrac{U_o(x)}{\delta_1(x)\,U_e(x)}\right]\right)\right\}^{-1}.\end{aligned} \tag{4.12}$$

Existence of a self-similar solution $f(\eta)$ requires that all x-dependent coefficients are proportional to each other such that their ratios are independent of x, i.e. the following relationships must be simultaneously satisfied:

$$\begin{aligned}\tfrac{1}{2}\,\partial_x\left(U_o{}^2\right) &= c_1\,\partial_x\left(U_o\, U_e\right) \\ \left(\tfrac{U_o\,U_e}{\delta_1'} - U_o\,\partial_x U_e\right) &= c_2\,\partial_x\left(U_o\, U_e\right) \\ U_o^2\,\partial_x\left(\text{Log}\left[\delta_1\, U_e\right]\right) &= c_3\,\partial_x\left(U_o\, U_e\right) \\ \tfrac{R_{vv}}{\delta_1'} &= c_4\,\partial_x\left(U_o\, U_e\right) \\ \tfrac{R_{uu}}{\delta_1'} &= c_5\,\partial_x\left(U_o\, U_e\right) \\ \tfrac{R_{uv}}{\delta_1^*} &= c_6\,\partial_x\left(U_o\, U_e\right) \\ \partial_x R_{vv} &= c_7\,\partial_x\left(U_o\, U_e\right) \\ \partial_x R_{uu} &= c_8\,\partial_x\left(U_o\, U_e\right) \\ \tfrac{\nu\, U_o}{\delta_1^{*2}} &= c_9\,\partial_x\left(U_o\, U_e\right)\end{aligned} \tag{4.13}$$

where $c_1, c_2, \ldots, c_9$ are constants. The first of the conditions in (4.13) yields

$$U_o(x) = K\ U_e(x), \text{ where } K \text{ is a constant,} \tag{4.14}$$

This implies that the local velocity scale, $U_o(x)$, is proportional to the local streamwise free-stream velocity, $U_e(x)$. Using (4.14) permits the definition of Λ as proposed by George and Castillo [2]:

$$\begin{aligned} \Lambda &= -\frac{\partial_x(\text{Log}[U_o])}{\partial_x(\text{Log}[\delta_1])} = -\frac{\partial_x(\text{Log}[U_e])}{\partial_x(\text{Log}[\delta_1])} \\ &= \frac{\partial_x P_e}{\rho\, U_e^2\, \partial_x(\text{Log}[\delta_1])} = \left(\frac{U_p}{U_e}\right)^2 \frac{1}{\partial_x \delta_1} \end{aligned} \tag{4.15}$$

where $U_p = \sqrt{\frac{\delta_1\, \partial_x P_e}{\rho}}$ is known as the pressure velocity [10]. This allows (4.11) to be written in the following form:

$$\begin{aligned} &\tfrac{2}{K}\,\Lambda f + \Lambda f^2 - \tfrac{1}{K}(\Lambda - 1)\,\eta f' - (\Lambda - 1)\,F f' \\ &= C_{vv}\,\eta\, r'_{vv} - C_{uu}\,\eta\, r'_{uu} - K\, C_{uv}\, r'_{uv} - D_{vv}\, r_{vv} + D_{uu}\, r_{uu} - K^2\, C_\nu\, f'' \end{aligned} \tag{4.16}$$

where

$$\begin{aligned} C_{uu} &= \frac{R_{uu}}{U_o^2} \\ C_{vv} &= \frac{R_{vv}}{U_o^2} \\ C_{uv} &= \frac{R_{uv}}{U_o^2\, \partial_x \delta_1} \\ D_{uu} &= \frac{\partial_x R_{uu}}{U_o^2\, \partial_x(\text{Log}[\delta_1])} \\ D_{vv} &= \frac{\partial_x R_{vv}}{U_o^2\, \partial_x(\text{Log}[\delta_1])} \\ C_\nu &= \frac{-2\,\nu}{U_o\, \partial_x \delta_1^2} \end{aligned} \tag{4.17}$$

must all be independent of x for a self-similar solution $f(\eta)$ to exist. Since Λ is a constant, (4.15) can be integrated to yield the following power-law relationship between the streamwise free-stream velocity and the displacement thickness with the exponent given by $-\Lambda$

$$\begin{aligned} -\Lambda\, \partial_x\, (\text{Log}\,[\delta_1]) &= \partial_x\, (\text{Log}\,[U_e]) \\ \Longrightarrow U_e(x) \quad &= A\, \delta_1^{-\Lambda}. \end{aligned} \tag{4.18}$$

This specific relationship is also a consequence of a self-similar flow. Furthermore, it can be shown that if Λ, C_{uu} and C_{vv} are independent of x, then D_{uu} and D_{vv} must also be independent of x. Therefore, only five conditions need to be independent of x to ensure that a self-similar flow exists, i.e. Λ, C_{uv}, C_{uu}, C_{vv} and C_ν must be independent of x. These conditions can be tested in the APG-TBL and ZPG-TBL DNS to establish if a self-similar solution describing the mean flow can exist.

4.3 Direct Numerical Simulation Details

The DNS code solves the Navier–Stokes equations in a three-dimensional rectangular volume, with constant density, ρ, and kinematic viscosity, ν. A fractional-step method [4, 11] is used to solve the governing equations for the velocity and pressure (P) fields. Fourier decomposition is used in the periodic spanwise direction, with compact finite differences [9] in the aperiodic wall-normal and streamwise directions. The equations are stepped forward in time using a modified three sub-step Runge–Kutta scheme [14].

The code utilises MPI and openMP parallelisation. For each MPI process the physical domain is decomposed into streamwise regions containing all spanwise and wall-normal points. The physical subdomain is further decomposed into wall-normal planes for each openMP thread [1]. All spatial derivatives in the spanwise and wall-normal direction can then be calculated with no MPI message passing. To calculate the streamwise derivatives the data is rearranged into streamwise oriented lines [1].

The boundary conditions of the original ZPG-TBL version of the DNS code are as follows. The bottom surface is a flat plate with a no-slip (zero velocity) BC. The spanwise boundaries are periodic. A downstream streamwise normal plane is copied and mapped to the inlet BC [13] after being appropriately scaled to account for the ZPG-TBL growth. At the wall-normal far-field boundary the spanwise vorticity is zero, and the wall-normal velocity is given by

$$V_{\mathrm{ZPG}}(x) = \frac{d\delta_1(x)}{dx} U_{\infty\,\mathrm{ZPG}}, \tag{4.19}$$

where $U_{\infty\,\mathrm{ZPG}}$ is the constant free-stream streamwise velocity, and δ_1 is the displacement thickness [12]. Note that wall-normal far-field wall-normal velocity represents a suction velocity.

The numerical details of the ZPG-TBL and APG-TBL simulations are summarised in Table 4.1. The number of grid points in each of the three flow directions is given by (N_x, N_y, N_z). The extents of the computational domain in the respective directions (L_x, L_y, L_z) are also presented, along with the constant grid spacings in the streamwise (Δx) and spanwise directions (Δz) and the wall-normal grid spacing at the far-field boundary (Δy_∞) and at the wall (Δy_{wall}). Note the APG simulation has a larger wall-normal domain (L_y) and more points in this direction (N_y). This is due to the fact that the APG-TBL expands more quickly in the streamwise direction than the ZPG-TBL. In both simulations the Courant number was set to unity. The length scales in Table 4.1 are non-dimensionalised by $\delta_{99,I}$ as viscous scaling is inappropriate for the APG-TBL flow.

Table 4.1 Numerical details of the APG-TBL and ZPG-TBL DNS

	$(L_x, L_y, L_z)/\delta_{99,I}$	$N_x \times N_y \times N_z$	$(\Delta x, \Delta y_{\text{wall}}, \Delta y_\infty, \Delta z)/\delta_{99,I}$
ZPG	$(850, 40, 71)$	$8193 \times 315 \times 1362$	$(0.10, 0.0027, 0.18, 0.052)$
APG	$(850, 74, 71)$	$8193 \times 500 \times 1362$	$(0.10, 0.0027, 0.18, 0.052)$

4.3.1 Adverse Pressure Gradient Boundary Condition

In order to implement the desired APG on the turbulent boundary layer development an appropriate wall-normal far-field streamwise wall-normal velocity distribution needs to be specified. Since the interest of the present study is the consideration of an APG-TBL at the point of incipient separation, we refer to the work of Mellor and Gibson [10] who found that the free-stream streamwise velocity distribution for incipient separation scales like $U_{\infty\,\text{APG}}(x) \propto x^m$ where $m = -0.23$ and ($u_\tau \to 0$). The wall-normal far-field streamwise distribution of the wall-normal (suction) velocity $V_{\infty\,\text{APG}}(x)$ can be deduced from $U_{\infty\,\text{APG}}(x)$ using a boundary layer streamfunction solution in the wall-normal far-field region as

$$V_{\infty\,\text{APG}}(x) = -\frac{dU_{\infty\,\text{APG}}(x)}{dx}\left[y_{\text{BC}} - \delta_1(x)\right] + \frac{d\delta_1(x)}{dx}\,U_{\infty\,\text{APG}}(x)\,, \qquad (4.20)$$

where y_{BC} is the wall-normal position of the far-field boundary [10]. Note that for the case of a constant streamwise velocity, i.e. the ZPG-TBL, (4.20) becomes equivalent to (4.19).

The structure of the complete wall-normal far-field BC, $V_\infty(x)$ shown in Fig. 4.1, is as follows. In the APG-TBL DNS, to allow the rescaling necessary for the inlet BC an initial domain of a ZPG-TBL is simulated up until the streamwise position $x_s = 100\delta_{99,I}$, which is located after the recycling plane, by applying $V_{\infty\,\text{ZPG}}(x)$ as defined by (4.19). Note that $\delta_{99,I}$ is the boundary layer thickness at the inlet. Downstream of the position $x_f = 140\delta_{99,I}$ the wall-normal velocity $V_{\infty\,\text{APG}}(x)$ given by (4.20) is applied at the wall-normal far-field boundary. This results in the desired free-stream deceleration and hence, expansion of the turbulent boundary layer. From x_s to x_f the velocity $V_{\infty\,\text{APG}}(x)$ is gradually introduced using a smoothing function. Finally towards the outflow of the computational domain the wall-normal velocity in the wall-normal far-field transitions from suction ($V_\infty(x) > 0$) at $x_o = 760\delta_{99,I}$ to blowing ($V_\infty(x) < 0$) to reduce the number of instantaneous reversed flow events at the outflow boundary conditions and thus maintain the numerical stability of the DNS. The streamwise functional form of the ZPG and APG wall-normal far-field boundary conditions, $V_\infty(x)/U_{\infty,I}$, is shown in Fig. 4.2, where $U_{\infty,I}$ is the free-stream streamwise velocity at the inlet.

This wall-normal far-field BC was first implemented and tested using two-dimensional Reynolds averaged Navier–Stokes (RANS) simulations, yielding the appropriate self-similar velocity profiles and subsequently implemented in the DNS code.

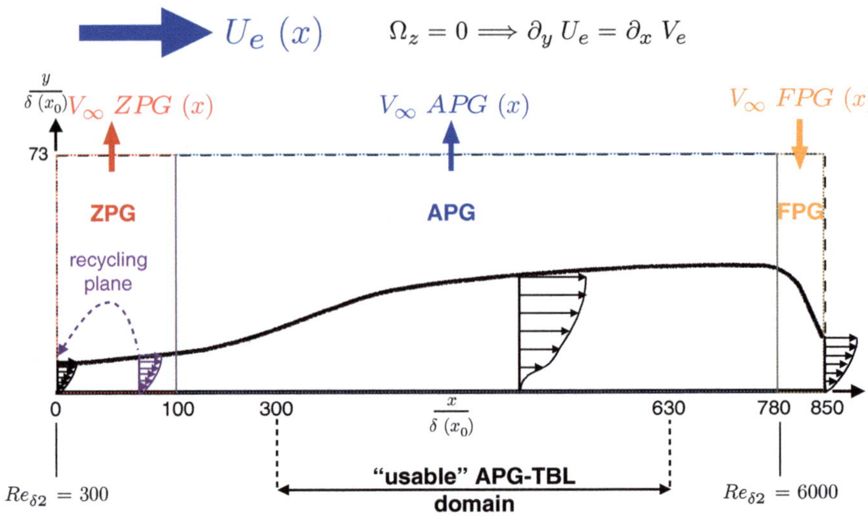

Fig. 4.1 Schematic of the streamwise wall-normal domain and pertinent parameters of the DNS

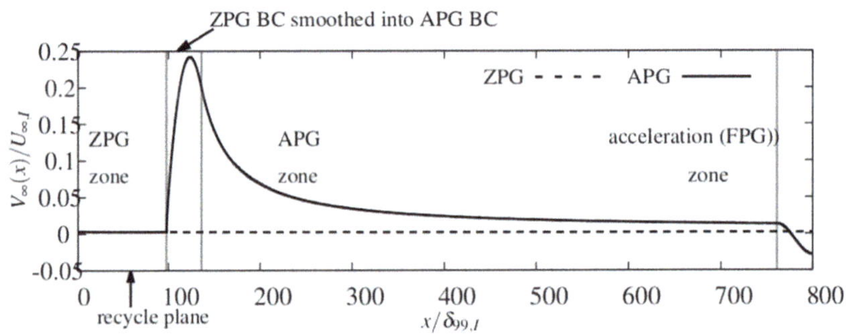

Fig. 4.2 Far-field wall-normal velocity boundary condition in the adverse pressure gradient (*solid line*) and zero pressure gradient (*dashed line*) direct numerical simulation

4.4 Results

4.4.1 Streamwise Variation of Pertinent Parameters

The streamwise variation of Λ, C_{uv}, C_{uu}, C_{vv} and C_ν for the APG-TBL and ZPG-TBL DNS is shown in Fig. 4.3 for the domain $300 < x/\delta(x_0) < 630$. In this domain of the DNS, Λ, C_{uv}, C_{uu} and C_{vv} are approximately constant for the APG-TBL, C_ν has a slight streamwise decay. In contrast the ZPG-TBL shows a significant streamwise variation in C_{uv} and C_ν. Given these results a self-similar flow for the APG-TBL is only to be expected where viscous effects are negligible, that is, in the

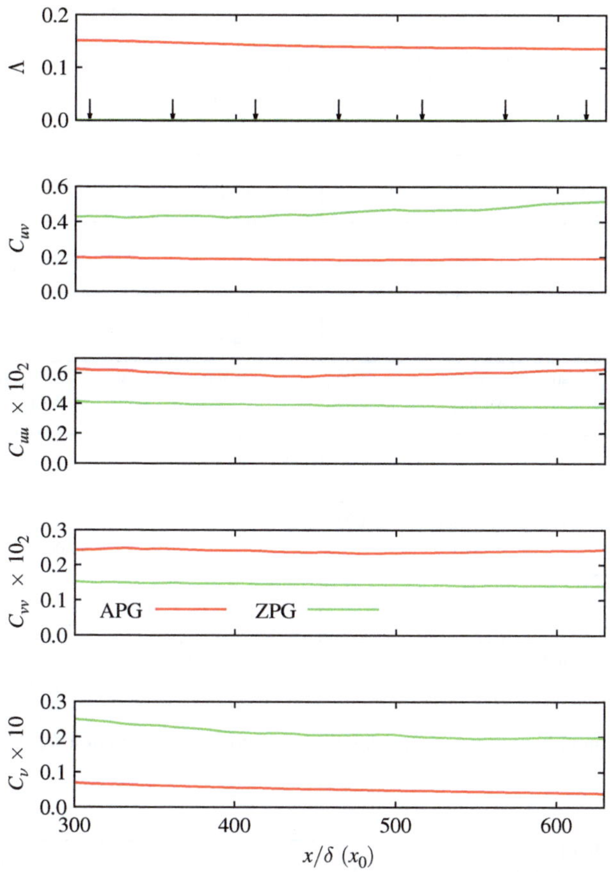

Fig. 4.3 Streamwise variation of Λ, C_{uv}, C_{uu}, C_{vv} and C_ν for the APG-TBL and ZPG-TBL DNS

outer layer of the APG-TBL. The arrows along the x-axis in the Λ figure of Fig. 4.3 correspond to the streamwise locations where the wall-normal profiles for the data shown in Sect. 4.4.2 are sampled.

For a self-similar flow (4.10) implies that $H =$ constant which is shown to be the case for the APG-TBL in the corresponding streamwise variation of H in Fig. 4.4. The various boundary layer thicknesses are found to vary linearly with streamwise directions, as shown in Fig. 4.4, which is consistent with the constant shape factor. Figure 4.5 shows the variation of the free-stream velocity, $U_e(x)$, with displacement thickness, $\delta_1(x)$ for the APG-TBL and ZPG-TBL. For a self-similar flow according to (4.18), the relationship should be a power law with an exponent equal to $-\Lambda$, which as the data in Fig. 4.5 indicates, is approximately satisfied for the APG-TBL DNS.

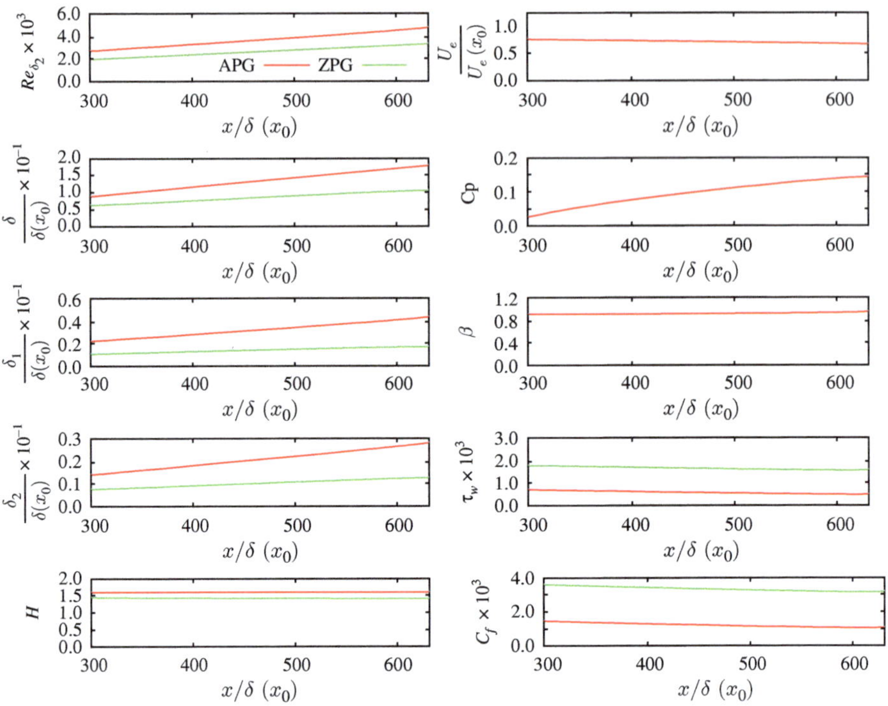

Fig. 4.4 Variation of boundary layer parameters with streamwise direction for the APG-TBL and ZPG-TBL DNS

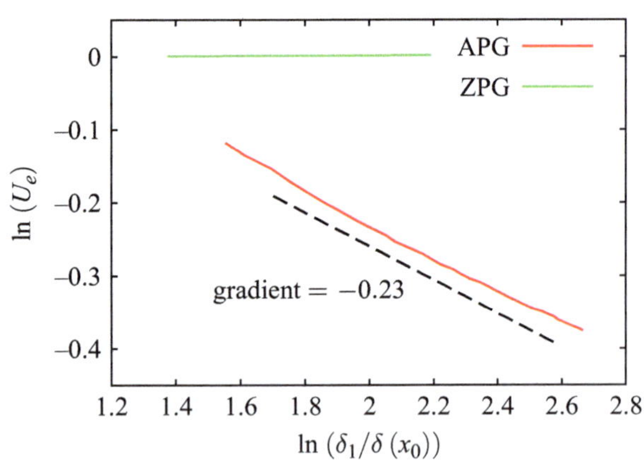

Fig. 4.5 Variation of free-stream velocity $U_e(x)$ in with displacement thickness $\delta_1(x)$ for the APG-TBL and ZPG-TBL DNS

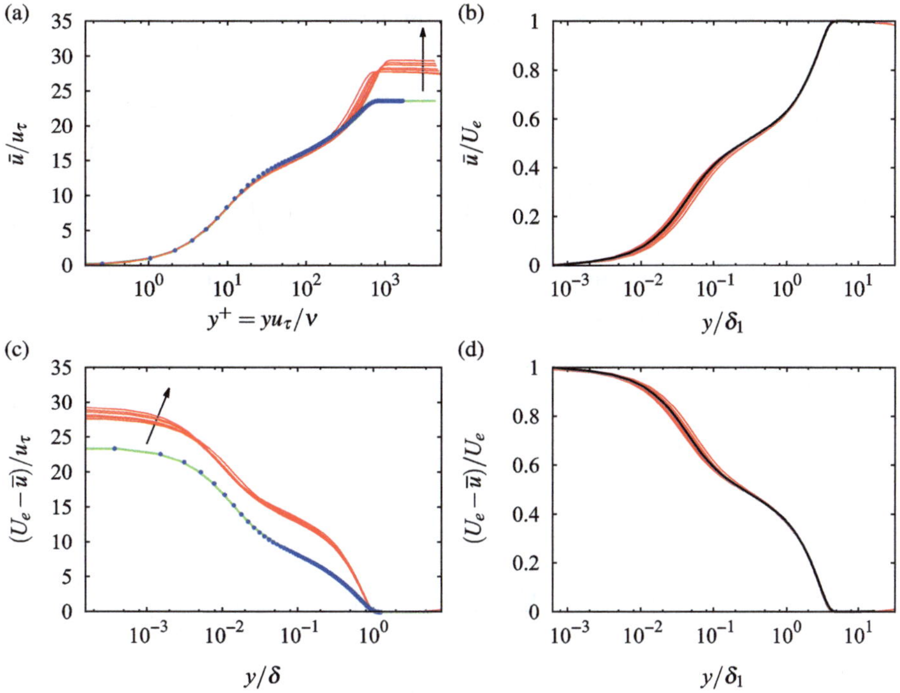

Fig. 4.6 Streamwise mean velocity profiles, *red solid line* APG-TBL, *green solid line* ZPG-TBL, *blue filled circle* [5]. (**a**) and (**c**) scaled using viscous units, (**b**) and (**d**) scaled using outer scaling using δ_1 and U_e with (**c**) and (**d**) plotted in velocity defect form. Note that the *black solid line* in (**b**) and (**d**) represents the average of all profiles shown at the sampling locations indicated in Fig. 4.3 by the *arrows*

4.4.2 Wall-Normal Velocity Statistics Profiles

The mean streamwise velocity profiles for the APG-TBL and ZPG-TBL are shown in Fig. 4.6. Figure 4.6a shows that the mean velocity profile scales with viscous units for the entire profile of the ZPG-TBL, while for the APG-TBL this in not the case. The APG-TBL does not scale in defect form either when the velocity is scaled with the viscous velocity u_τ. However, as Fig. 4.6b, d show, when the APG-TBL mean velocity is scaled with δ_1 and U_e as indicated by the self-similar analysis of Sect. 4.2, then the scaled mean velocity profiles collapse across the entire layer including in the viscous layer close to the wall.

Figure 4.7a, c, d shows the tangential Reynolds stress profiles for the APG-TBL and ZPG-TBL scaled in viscous units, which clearly shows the lack of scaling using u_τ and viscosity for the APG-TBL. Figure 4.7b, d and f shows the tangential Reynolds stress profiles scaled by the variables δ_1 and U_e required from the self-similar analysis, showing much better collapse of the scaled profiles across the entire APG-TBL layer.

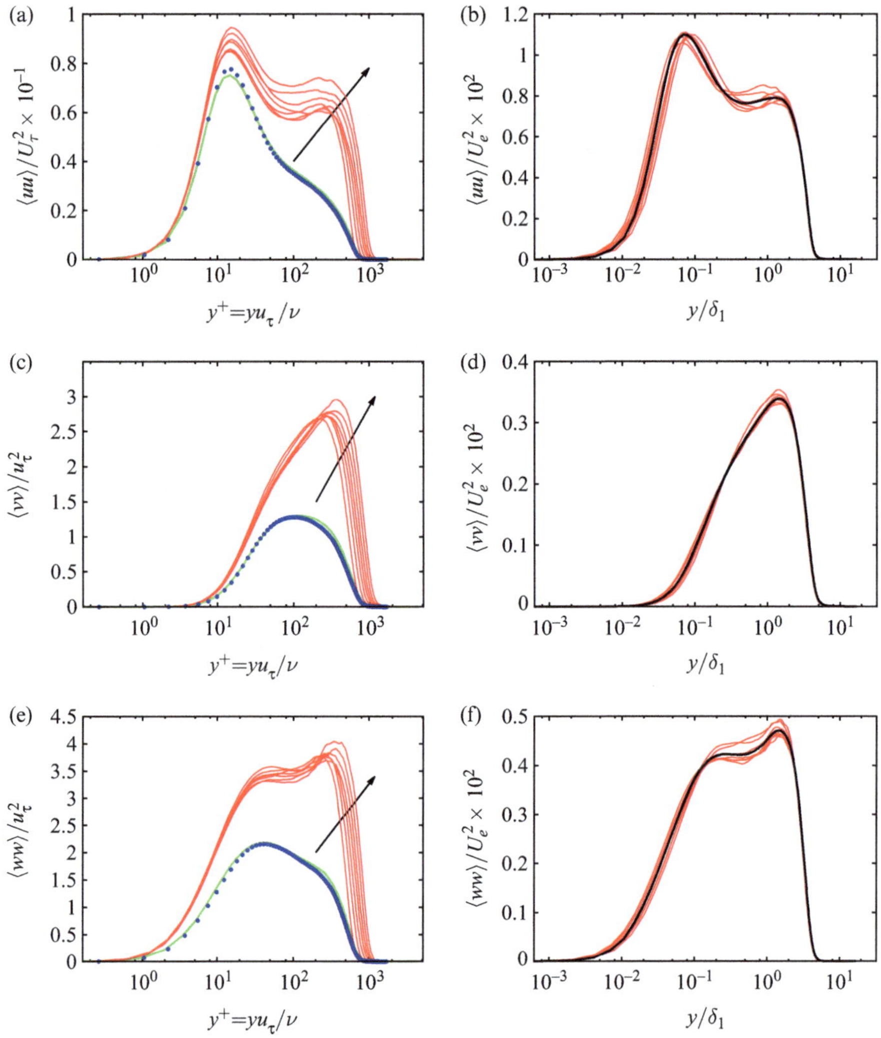

Fig. 4.7 Tangential Reynolds stress profiles *red solid line* APG-TBL, *green solid line* ZPG-TBL, *blue filled circle* [5]. (**a**), (**c**) and (**e**) scaled using viscous units; (**b**), (**d**) and (**f**) scaled using outer scaling using δ_1 and U_e. Note that the *black solid line* in (**b**), (**d**) and (**f**) represents the average of all profiles shown at the sampling locations indicated in Fig. 4.3 by the *arrows*

There are further significant differences between the tangential Reynolds stress profiles of the ZPG-TBL and the APG-TBL. The streamwise Reynolds stress profile has a secondary maximum in the outer layer $y > \delta_1$ in the APG-TBL, which is not present in the ZPG-TBL. The maximum in the wall-normal Reynolds stress profile of the APG-TBL is also located at $y > \delta_1$ in the APG-TBL and is much narrower

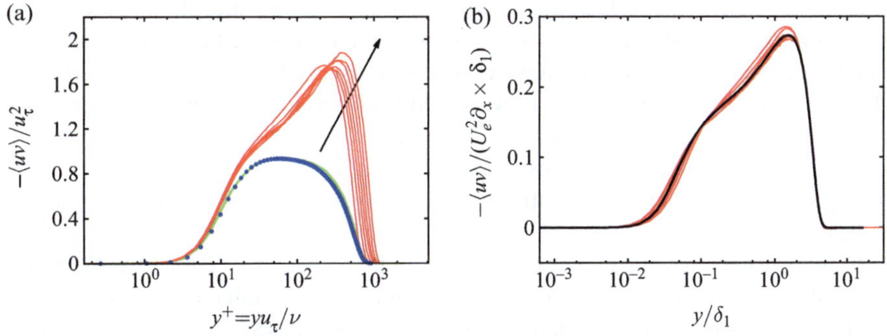

Fig. 4.8 Shear Reynolds stress profiles *red solid line* APG-TBL, *green solid line* ZPG-TBL, *blue filled circle* [5]. (**a**) Scaled using viscous units; (**b**) scaled using outer scaling using δ_1 and U_e. Note that the *black solid line* in (**b**) represents the average of all profiles shown at the sampling locations indicated in Fig. 4.3 by the *arrows*

than the much broader peak in the ZPG-TBL that is located in the inner viscous layer around $y^+ \approx 100$.

The spanwise Reynolds stress profile also has a much larger maximum at $y > \delta_1$ in the APG-TBL, with a small inner maximum near $y \approx 0.1\,\delta_1$. The location of this inner maximum coincides approximately with the location of the maximum in the ZPG-TBL spanwise Reynolds stress profile, as shown in Fig. 4.7e.

The shear Reynolds stress profiles for the APG-TBL and the ZPG-TBL DNS are shown in Fig. 4.8. As in the case with the tangential Reynolds stress profiles, there is no collapse of the APG-TBL data when scaled with viscous variables, as shown in Fig. 4.8a. However when the APG-TBL shear Reynolds stress profiles are scaled with δ_1 and U_e, then the collapse is much better, as shown in Fig. 4.8b. There is also quite a difference in the shear Reynolds stress distribution between the ZPG-TBL and the APG-TBL. The former has a broad maximum spanning $20 < y^+ < 120$, whereas the latter has a maximum in the outer layer at $y \approx 1.5\,\delta_1$. In the APG-TBL there is also a secondary smaller maximum or plateau in the inner layer at $y \approx 0.1\,\delta_1$, with an inflection point located in the profile between these two locations.

4.5 Discussion and Concluding Remarks

The appropriate wall-normal far-field (free-stream) boundary conditions have been developed to simulate a self-similar APG-TBL flow. DNS of a ZPG-TBL and a self-similar APG-TBL, with the latter not yet at the verge of separations, has been carried out. Both simulations were started from the same mean inflow ZPG-TBL conditions. The ZPG-TBL DNS reached a Reynolds number $Re_{\delta_2} \approx 4107$, which is usable up to $Re_{\delta_2} \approx 4000$, while the APG-TBL reached a Reynolds $Re_{\delta_2} \approx 6700$, which is usable up to $Re_{\delta_2} \approx 4820$.

Analysis of the DNS database of the APG-TBL shows that the coefficients in the appropriately scaled mean x-momentum boundary layer equations are approximately independent of the streamwise direction, except for the viscous coefficient. This suggests that a self-similar flow, scaled with the local length scale $\delta_1(x)$ and the local velocity scale $U_e(x)$, exists, except possibly in the viscous sublayer. The mean velocity and Reynolds stress profiles scaled by these variables, which show quite good collapse across the entire boundary layer of the APG-TBL, support this notion.

It is pertinent to make a few remarks with respect to "BIG DATA" as the DNS of the APG-TBL presented in this paper clearly falls into this category. Staring with some details regarding the APG-TBL DNS: each restart point in time of the DNS requires four files for the fluid velocity vector field and the pressure, amounting to approximately 90 GB of storage for each instant in time of the simulation. Given that it is desirable to store at least 1000 or possibly more of these statistically independent fields for turbulence structure analysis and to determine the statistical characteristics of the turbulence structures, it is self-evident that the storage requirements for one simulation very quickly reach requirements in the 100s of TB.

In addition to these storage requirements, the other point that needs to be noted is that this simulation, which is quite typical, was conducted non-locally from where the researchers are located, i.e. in this case the majority of the simulation was conducted in Germany using the SuperMUC Petascale System in Munich, while the researchers setting up the simulation and monitoring it were located in Melbourne, Australia. Furthermore, the DNS fields could not be stored in Germany and had to be transferred using the internet to Melbourne, Australia. In this instance this was done using standard *scp* or *rsync* commands. However, better and more parallel methods of international big data movement need to developed and implemented, in particular, if this data is to be shared with national and the international research community as the demand for open access sharing increases. One possible approach is to use BitTorrent peer-to-peer file sharing technology, an example of this approach is BioTorrents [7].

One observation is the necessity to collaborate with computer system scientists on these issues of BIG DATA access around the world and how to design robust and efficient web-portals to allow for BIG DATA access and distributed BIG DATA post-processing and visualisation. It seems that many tools for these operations have already been developed by computer scientists, but what seems to be missing is the implementation within our discipline, due to a lack of awareness and/or knowledge of these on our behalf and a lack of awareness and/or knowledge of our requirements by computer scientists. The only way for us to progress on these issues is to make linkages and discuss our BIG DATA issues with the computer scientists—"BIG DATA" is a multi-disciplinary problem.

Acknowledgements The authors would like to acknowledge the support of this research by the Australian Research Council and European Research Council. The computational resources, which are gratefully acknowledged, were provided by the Australian National Computational Infrastructure supported by the Australian government and the Pawsey Supercomputer Centre supported by the Western Australian and Australian governments and PRACE who is supported by the European Union. Julio Soria gratefully acknowledges the support of an Australian Research Council Discovery Outstanding Researcher Award fellowship.

References

1. G. Borrell, J. Sillero, J. Jiménez, A code for direct numerical simulation of turbulent boundary layers at high Reynolds numbers in BG/P supercomputers. Comput. Fluids **80**, 37–43 (2013)
2. W.K. George, L. Castillo, Boundary layers with pressure gradient: another look at the equilibrium boundary layer, in *Near Wall Turbulent Flows*, ed. by RMC So, C.G. Speziale, B.E. Launder (Elsevier, Amsterdam, 1993)
3. A. Gungor, M. Simens, J. Jiménez, Direct numerical simulation of wake-perturbed separated boundary layers. J. Turbul. **134**, 061024 (2012)
4. F. Harrow, J. Welch, Numerical calculation of time-dependent viscous incompressible flow of fluid with free surface. Phys. Fluids **8**(12), 2182–2189 (1965)
5. J. Jiménez, S. Hoyas, M. Simens, Y. Mizuno, Turbulent boundary layers and channels at moderate Reynolds numbers. J. Fluid Mech. **657**(22), 335–360 (2010)
6. V. Kitsios, L. Cordier, J.P., Bonnet, A. Ooi, J. Soria, On the coherent structures and stability properties of a leading edge separated airfoil with turbulent recirculation. J. Fluid Mech. **683**, 395–416 (2011)
7. M. Langille, J.A. Elsen, Biotorrents: a file sharing service for scientific data. PLoS ONE **5**(4), e10071 (2010). doi:10.1371/journal.pone.0010,071
8. J.H. Lee, J. Sung, Effects of an adverse pressure gradient on a turbulent boundary layer. Int. J. Heat Fluid Flow **29**, 568–578 (2008)
9. S. Lele, Compact finite difference schemes with spectral-like resolution. J. Comput. Phys. **103**, 16–42 (1992)
10. G. Mellor, D. Gibson, Equilibrium turbulent boundary layers. J. Fluid Mech. **24**, 225–253 (1966)
11. J. Perot, An analysis of the fractional step method. J. Comput. Phys. **108**, 51–58 (1993)
12. J. Sillero, High Reynolds number turbulent boundary layers. Ph.D. thesis, Universidad Politénica de Madrid, 2014
13. J. Sillero, J. Jiménez, R. Moser, One-point statistics for turbulent wall-bounded at Reynolds numbers up to $\delta^+ \approx 2000$. Phys. Fluids **25**, 105102 (2013)
14. M.P. Simens, J. Jiménez, S. Hoyas, Y. Mizuno, A high-resolution code for turbulent boundary layers. J. Comput. Phys. **228**, 4128–4231 (2009)
15. P. Skåre, P.A. Krogstad, A turbulent equilibrium boundary layer near separation. J. Fluid Mech. **272**, 319–348 (1994)
16. M. Skote, D.S. Henningson, R.A.W.M. Henkes, Direct numerical simulation of self-similar turbulent boundary layers in adverse pressure gradients. Flow Turbul. Combust. **60**(1), 47–85 (1998)
17. A.A. Townsend, The structure of the turbulent boundary layer. Math. Proc. Camb. Philos. Soc. **47**(02), 375–395 (1951)
18. A. Townsend, *The Structure of Turbulent Shear Flow* (Cambridge University Press, New York, 1976)

Chapter 5
Analysis of Velocity Structures in a Transitionally Rough Turbulent Boundary Layer

Ali Doosttalab, Suranga Dharmarathne, Murat Tutkun, Ronald Adrian, and Luciano Castillo

5.1 Introduction

Studies on rough turbulent boundary layers have shown that the surface roughness alters the velocity field near the wall, leading to an increased skin friction. This effect is in particular felt in the roughness sub-layer ($3k - 5k$, where k denotes the characteristic roughness height) and may extend across the boundary layer for $k/\delta < 50$ as mentioned in [9] (note that δ represents the boundary layer thickness). Knowledge of the effects of roughness on turbulent structures is vital for modelling purposes and flow control strategies which can then reduce skin friction drag in turbulent boundary layers by modifying the structures of turbulence.

Flores et al. [5] performed direct numerical simulation (DNS) of a moderately rough channel flow and found that the attached vortical structures were independent of the details of the wall. This was attributed to memory of the structures as the structures "forget their origin" quickly. Wu and Christensen [15] experimentally studied the impact of realistic roughness on the spatial structures of wall turbulence at moderate Reynolds numbers and reported shortening of streamwise velocity coherent structures. Their proper orthogonal decomposition (POD) analysis revealed that the structures with large streamwise dimension are more sensitive

A. Doosttalab (✉) • S. Dharmarathne • L. Castillo
Department of Mechanical Engineering, Texas Tech University, Lubbock, TX 79409, USA
e-mail: ali.doosttalab@ttu.edu

M. Tutkun
Department of Process and Fluid Flow Technology, IFE, Kjeller, Norway

Department of Mathematics, University of Oslo, Blindern, 0316 Oslo, Norway

R. Adrian
School for Engineering of Matter, Transport and Energy, Arizona State University, P.O. Box 876106, Tempe, AZ 85287-6106, USA

A. Pollard et al. (eds.), *Whither Turbulence and Big Data in the 21st Century?*,
DOI 10.1007/978-3-319-41217-7_5

to surface roughness when compared to the structures which have shorter extent in streamwise direction. Lee et al. [10] conducted DNS of turbulent flows over a smooth wall, and a rough wall with either two-dimensional or three-dimensional roughness elements. The results showed that the packet of coherent structures was more inclined over the rough surface geometry compared to the smooth one.

Townsend [13] proposed the Reynolds number similarity hypothesis, which states that at sufficiently high Reynolds number and outside the roughness sub-layer ($3k - 5k$), the characteristics of turbulent flow are independent of surface roughness. Doosttalab et al. [4] showed that the Townsend's wall similarity hypothesis is valid for single-point statistics for both velocity and thermal fields for $k/\delta > 71$. However, validity of the Townsend's hypothesis for multi-point statistics and structures of the velocity field still needs to be addressed in more depth.

The objective of the present study is to investigate: (1) the influence of surface roughness on the structure of turbulent velocity field and (2) validity of the Townsend's hypothesis for the multi-point statistics. To accomplish this objective we use visualizations of spatial two-point correlations and compare turbulence structures over smooth and rough surfaces. The paper is organized as follows: Sect. 5.2 describes the basis of the dynamic multi-scale approach (DMA) employed in the simulations, Sect. 5.3 describes the roughness modelling and description of boundary conditions, Sect. 5.4 discusses the velocity field results, Sect. 5.5 addresses the big data issue and Sect. 5.6 presents the concluding observations.

5.2 The Dynamic Multi-Scale Approach (DMA) for Spatially Evolving Flows

The current DNS employs the DMA of [1] for prescribing turbulent inlet boundary conditions. DMA is based on the classical recycling/rescaling method developed by Lund et al. [11], but uses only a single domain instead of two domains. In order to achieve a statistically steady flow, computational domain consisted of large box with a development region. The computational domain often had to be very long in the streamwise direction to incorporate transition and only a small section of the domain contained the fully turbulent flow. The original recycling/rescaling method was developed to eliminate the developing region of the computational domain, therefore reducing the computational costs of simulating spatially evolving boundary layer flow. By introducing a recycling plane into the computational domain, the flow at the recycling plane is rescaled and re-injected at the inlet plane of the domain as explained in [3]. By allowing the flow to follow this recycling, rescaling and re-injection process, it is found that after some time, the flow reaches a statistically steady state. Thus, the recycling method provides certain advantages since it is possible to simulate turbulent boundary layer using a smaller, less

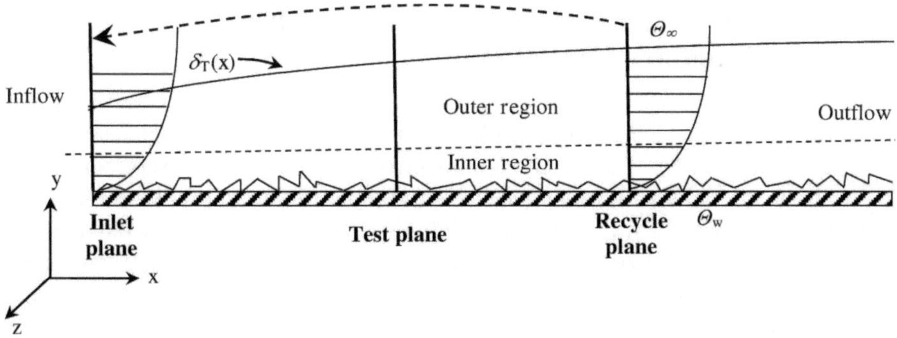

Fig. 5.1 Schematic of the rough thermal boundary layer with different regions and planes: inlet, test and recycle stations

Table 5.1 Velocity scaling functions

	Inner		Outer	
Variable	Classical	Equilibrium similarity	Classical	Equilibrium similarity
$\overline{U}, u', w'$	u_τ	u_τ	u_τ	U_∞
V	U_∞	u_τ	U_∞	$U_\infty d\delta/dx$
v'	u_τ	u_τ	u_τ	$U_\infty d\delta/dx$
$\langle u'v' \rangle$	u_τ^2	u_τ^2	u_τ^2	$U_\infty^2 d\delta/dx$

computationally demanding domain compared to other techniques. By introducing a "test-plane" between the recycling and the inlet planes, as shown in Fig. 5.1, the required flow parameters at the inlet plane (e.g. the inlet friction velocity) are extrapolated based on the solution downstream [1]. This improvement to the original rescaling/recycling method avoids the need to use some empirical correlations, which are required in the method of [11]. A complete detail of DMA can be found in [1]. DMA is slightly modified in our case to incorporate surface roughness. The modifications follow the implementation of [3].

One of the most important aspects of DMA is the selection of appropriate scaling law to be used in the re-scaling process (i.e. different scaling in inner and outer layers as shown in Table 5.1). In this investigation, we used the velocity scaling law based on the equilibrium similarity approach developed in [6]. It should, however, be noted that the proposed method herein is generic to allow any other scaling law to be easily implemented. Table 5.1 depicts the classical scaling parameters and the equilibrium similarity scaling parameters which are implemented in this study while generating the turbulent inflow.

5.3 Surface Roughness Modelling, Flow Solver and Boundary Conditions

Cardillo [3] modified the original DMA to incorporate the surface roughness into the computational domain. Direct simulations of transport equations were performed by using parallel hierarchic adaptive stabilized transient analysis (PHASTA) code. PHASTA is based on finite element method with a streamline upwind Petrov–Galerkin (SUPG) stabilization [7, 14]. PHASTA code was also used in [3] where the surface roughness and the velocity field data were validated using the experimental data documented in [2].

For the velocity field, the no-slip condition was imposed at the wall and displaced to the height of the roughness element at the corresponding node for the rough case; meanwhile at the free stream location, the streamwise velocity is given by U_∞ and zero derivatives of the normal and spanwise velocities are prescribed. At the outflow plane, the pressure was prescribed (Dirichlet condition) and at the inlet station, the time-dependent inflow conditions for the velocity and temperature fields are generated based on DMA. Periodic boundary conditions were set in the spanwise direction for the instantaneous velocity and temperature. In order to have good performance for turbulent simulations, structured meshes with hexahedral elements were employed. In the wall-normal direction, non-uniform mesh sizes were used with the mesh properties indicated in Table 5.2. Differences between the zero-pressure-gradient (ZPG) smooth and ZPG rough simulations were the spatio-temporal resolution (i.e. Δx^+, Δy^+, Δz^+ and Δt^+), the sample time and δ^+_{inl}, where inl indicates the inlet plane. Both cases were run on 128 processors (N_{procs}), as shown in Table 5.2.

Table 5.2 Proposed DNS cases and domain parameters for smooth and rough ZPG simulations

Parameters	ZPG smooth	ZPG rough
Re_θ	1940–2300	2077–2439
L_x	$10\delta_{\mathrm{inl}}$	$10\delta_{\mathrm{inl}}$
L_y	$3\delta_{\mathrm{inl}}$	$3\delta_{\mathrm{inl}}$
L_z	$1.57\delta_{\mathrm{inl}}$	$1.57\delta_{\mathrm{inl}}$
$N_x\ N_y\ N_z$	400 × 150 × 125	400 × 150 × 125
Δx^+	23	24.6
$\Delta y^+_{\min}$, $\Delta y^+_{\max}$	0.5, 25	0.545, 26.74
Δz^+	11	11.77
Δt^+	0.44	0.50
$T_{\mathrm{sample}} \frac{u_\tau^2}{\nu}$	2640	6500
δ^+_{inl}	980	1048
N_{procs}	128	128

5.4 Numerical Results

5.4.1 Displacement Height

In order to make the rough and smooth surface statistics comparable, the origin of the profiles was displaced in the wall-normal direction. This origin displacement appears in the profile diagrams as a reference height for vertical coordinates, and data is shifted by the displacement height. The idea behind this approach is to satisfy the wall conditions. Figure 5.2 shows two criteria that could be used to prescribe the origin for the rough surface. The first criterion is based on the location where the mean streamwise velocity, $\overline{U}^+$(over bar represents averaging over both time and spanwise direction), is zero, and the second one is the point where $\overline{u'v'}^+$ is zero. Because these conditions are not simultaneously satisfied at the same location, the wall-normal coordinate where $\overline{U}^+$ is zero (i.e. to satisfy the no-slip condition) is considered as the reference height, and thus, given by $y^+ = \varepsilon^+ = 6.312$. The negative mean streamwise velocity near the rough surface can be explained by the recirculating flow between roughness elements or the alignment of vortex axes primarily in the spanwise direction, causing the negative mean velocity near the wall over the rough surface.

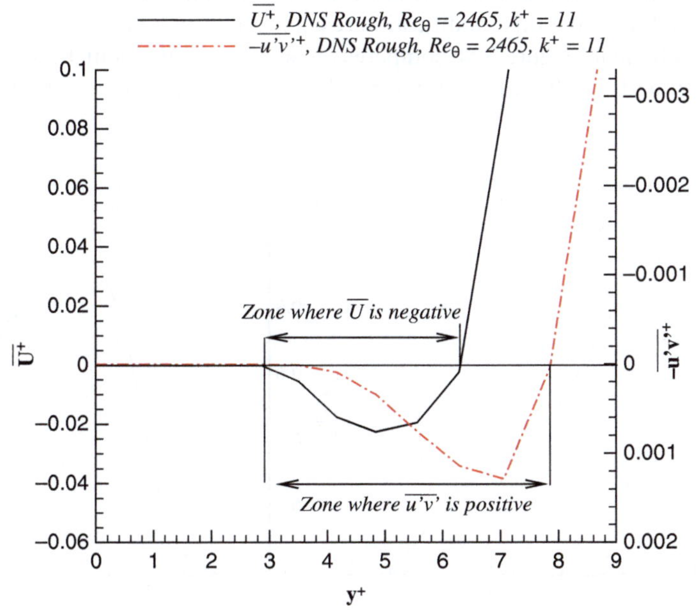

Fig. 5.2 Mean velocity and Reynolds shear stress profiles near the rough wall

Given the moderate Reynolds number (as compared to typical high Reynolds numbers of experiments), an inertial sub-layer does not exist in these simulations. On the other hand, a meso-layer region exists. Thus, the roughness elements in the current study occupy most of the viscous sub-layer. This eventually leads to significant alteration in the inner region of boundary layer up to $y^+ \approx 11$.

5.4.2 Single-Point Statistics

Figure 5.3a shows the normalized mean streamwise velocity deficit profiles scaled using classical outer units for both smooth and rough cases. It can be observed that there is an upward shift in the deficit form profile for the rough case which extends well into the outer region when compared to the profile of the smooth case. This implies a flow retardation in near-wall region due to the surface roughness elements. Reverse flow region between roughness elements causes the flow to lose momentum inside the roughness canopy. Figure 5.3b depicts the Reynolds shear stress profiles for the smooth and the rough surface. In a similar fashion, the classical outer units are used for scaling the profiles. The differences which can be observed between these profiles are caused by the change of friction velocity due to surface roughness.

In Fig. 5.4a, streamwise variation of skin friction coefficient is shown. The rough surface increases the skin friction coefficient by about 23 %. The question which needs to be addressed here is how surface roughness modifies velocity structures near the wall and how this modification results further in a drastic increase in skin friction coefficient? Figure 5.4b compares r.m.s. of vorticity fluctuation components for smooth and rough cases, normalized by ν/u_τ^2 in outer units. The two shown

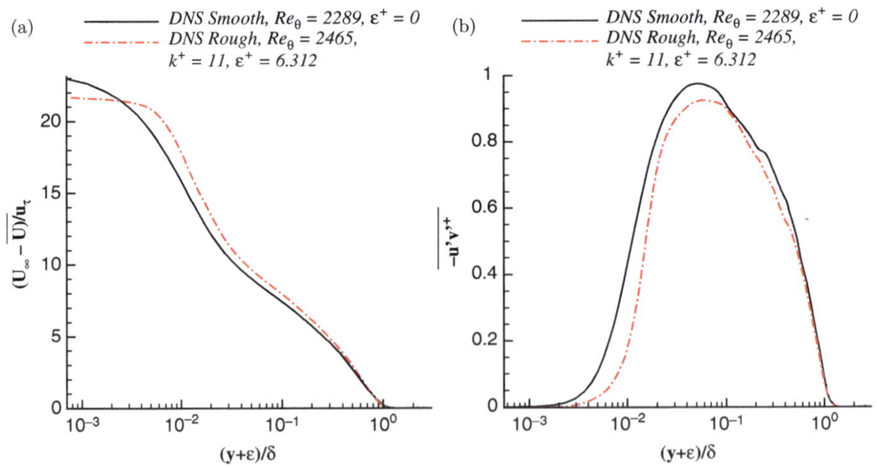

Fig. 5.3 (**a**) Mean velocity deficit profiles, (**b**) Reynolds shear stress profiles

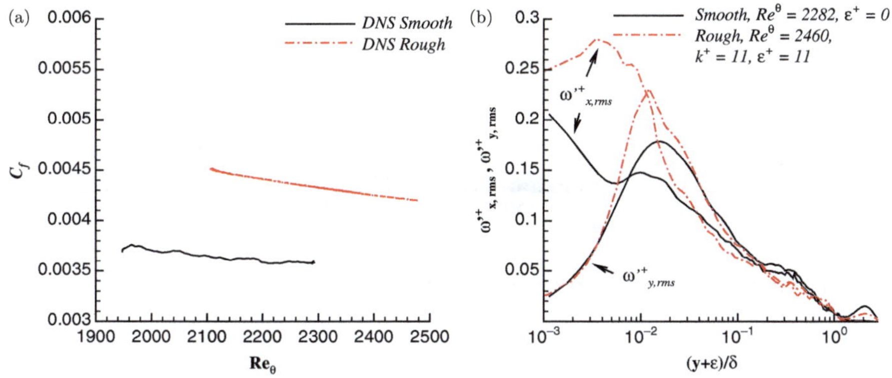

Fig. 5.4 (**a**) Streamwise variation of skin friction coefficient (C_f), (**b**) r.m.s. of vorticity fluctuations for x and y components

components are nearly identical for smooth and rough cases away from the wall for $(y + \varepsilon)/\delta > 0.03$, showing the validity of Townsend's hypothesis for single-point statistics in velocity field.

5.4.3 Two-Point Cross-Correlations

In this section turbulent structures are studied for the smooth and the rough case by means of two-point correlations. Two-point spatial correlation of any two mean zero variables α' and β' is defined as

$$\rho_{\alpha'\beta'}(x_0, x, y_0, y, \delta z) = \frac{R_{\alpha'\beta'}(x_0, x, y_0, y, \delta z)}{\sigma_{\alpha'}(x_0, y_0)\sigma_{\beta'}(x, y)}. \tag{5.1}$$

Here, (x_0, y_0, z_0) is the reference location where the correlation is based on and (x, y, z) is another point in space. The flow is assumed homogeneous only in the spanwise direction. Hence, the correlation depends only on the separation in this direction, $\delta z(= z - z_0)$ regardless of the spanwise coordinate of the reference location. $R_{\alpha'\beta'}(x_0, x, y_0, y, \delta z)$ is the covariance of α' and β' which is the inverse Fourier transform of $\hat{R}_{\alpha'\beta'}(x_0, x, y_0, y, k_z)$. Covariance is computed in Fourier space and it is given by

$$\hat{R}_{\alpha'\beta'}(x_0, x, y_0, y, k_z) = \left\langle \hat{\alpha}'(x_0, y_0, k_z)\hat{\beta'}^*(x, y, k_z) \right\rangle. \tag{5.2}$$

Here, k_z is spanwise wavenumber, $\hat{\alpha}'$ is Fourier transform of α' and $\hat{\beta'}^*$ is the complex conjugate of $\hat{\beta}'$. Two-dimensional sections of $\rho_{u'u'}$, $\rho_{u'v'}$ and $\rho_{v'v'}$ at three different wall-normal locations are described in this section.

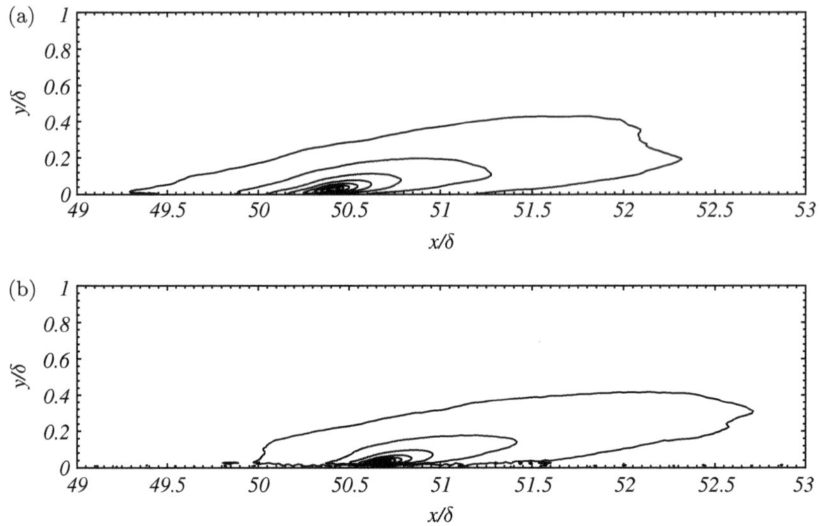

Fig. 5.5 Streamwise section (xy) of correlations of $\overline{u'u'}$ at $(y+\varepsilon)^+ \approx 20$, the outerline represents $C_{uu} = 0.1$ and increments are set as 0.1 (**a**) the smooth case, (**b**) the rough case

Figure 5.5a, b presents the $\rho_{u'u'}$ correlations at $(y + \varepsilon)^+ \approx 20$ for the smooth case and the rough case. The length of correlations is slightly shorter for the rough case ($\approx 2.7\delta$) than that of the smooth case ($\approx 3\delta$). However, the shape of structures is very similar to each other with similar inclination angle from the wall.

In Fig. 5.6, $\rho_{u'v'}$ correlations at $(y + \varepsilon)^+ \approx 20$ are shown for the both cases. A more pronounced reduction in the extent of correlations is observed in the rough case which can be due to stronger effect of surface roughness on wall-normal fluctuations in this region. The streamwise length of the correlations for the rough case ($\approx 0.3\delta$) is less than half of the smooth case ($\approx 0.65\delta$). Also in the wall-normal direction the correlated area is much thinner in the rough case compared to the smooth case. Some noise is observed for the rough case close to the wall which is the effect of proximity to the irregular rough surface. It can be concluded that the surface roughness has similar effect on shorter structures as well as longer structures.

Figure 5.7a, b shows $\rho_{v'v'}$ at $(y + \varepsilon)^+ \approx 20$ for the smooth and the rough cases. Similar to $\rho_{u'v'}$, the rough case indicates shorter structures, e.g. the correlation length in the streamwise direction reduces from $\sim 0.35\delta$ to $\sim 0.2\delta$.

Figure 5.8 depicts $\rho_{u'u'}$ correlations at $(y + \varepsilon)^+ \approx 80$ in the log-layer. As it can be seen, effect of the surface roughness on streamwise velocity fluctuations in this region is minimal. The length of the $\rho_{u'u'}$ for the smooth case is approximately 2.5δ, which is roughly the same as 2.5δ for the rough case. Also, it can be observed that the inclination of the contour lines is slightly more towards the wall in the rough case.

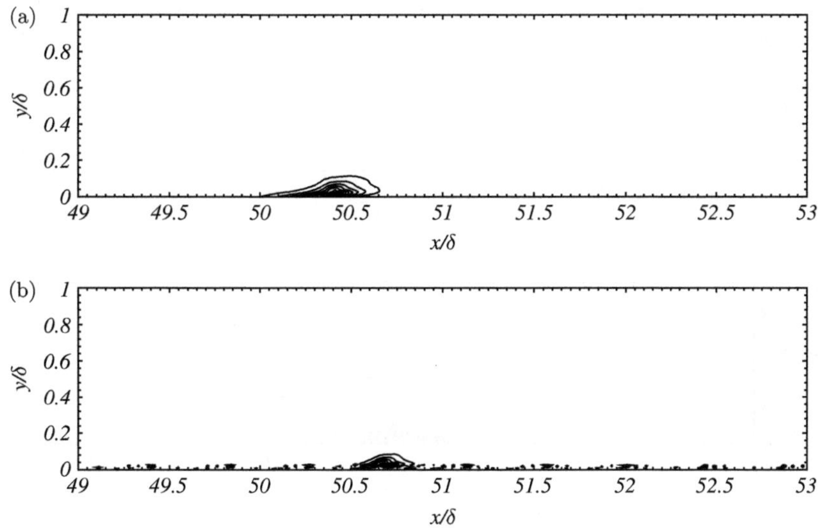

Fig. 5.6 Streamwise section (xy) of correlations of $\overline{u'v'}$ at $(y+\varepsilon)^+ \approx 20$, the outerline represents $C_{uv} = -0.1$ and decrements are set as -0.1 (**a**) the smooth case, (**b**) the rough case

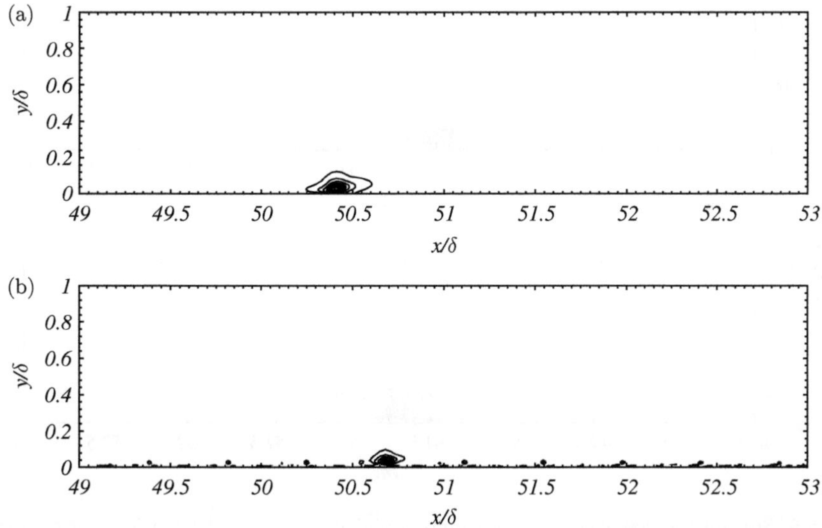

Fig. 5.7 Streamwise section (xy) of correlations of $\overline{v'v'}$ at $(y+\varepsilon)^+ \approx 20$, the outerline represents $C_{vv} = 0.1$ and increments are set as 0.1 (**a**) the smooth case, (**b**) the rough case

As we move away from the wall, size of $\rho_{u'v'}$ at $(y+\varepsilon)^+ \approx 80$, shown in Fig. 5.9, becomes smaller when compared to the ones observed near-wall region. The streamwise extent of these correlations reduced from $\sim 1.1\delta$ for the smooth case to $\sim 0.8\delta$ for the rough case.

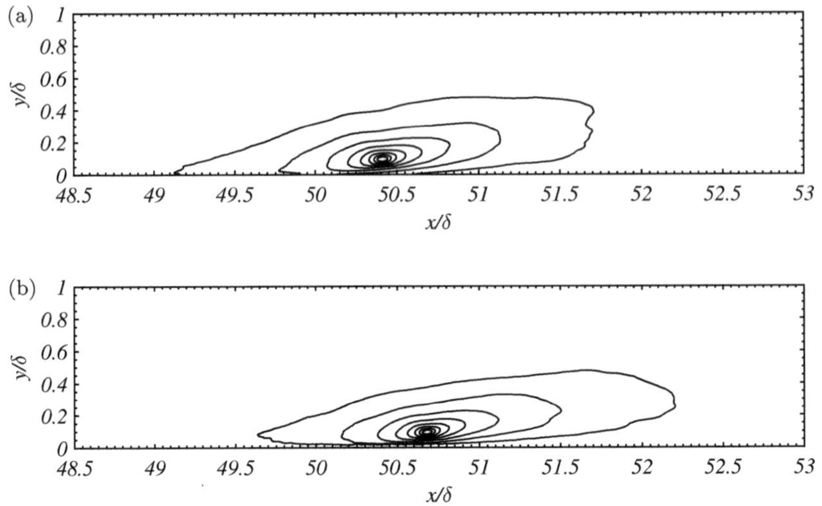

Fig. 5.8 Streamwise section (xy) of correlations of $\overline{u'u'}$ at $(y+\varepsilon)^+ \approx 80$, the outerline represents $C_{uu} = 0.15$ and increments are set as 0.75/8 (**a**) the smooth case, (**b**) the rough case

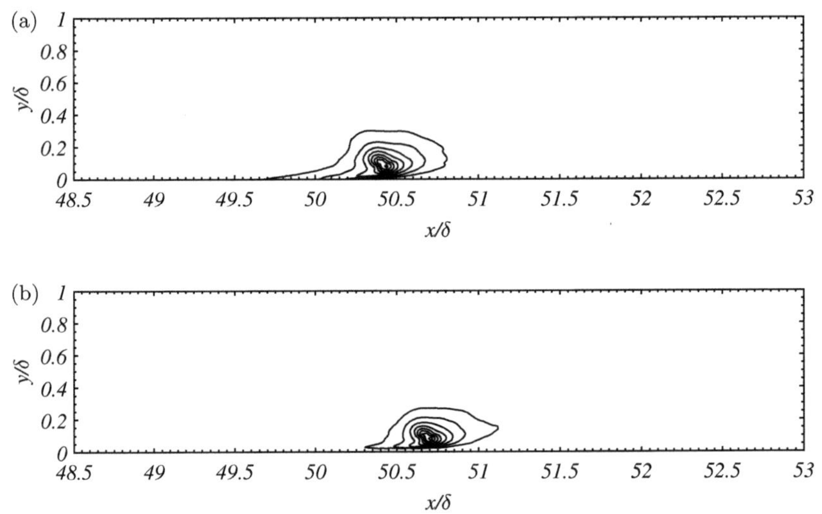

Fig. 5.9 Streamwise section (xy) of correlations of $\overline{u'v'}$ at $(y+\varepsilon)^+ \approx 80$, the outerline represents $C_{uv} = -0.1$ and decrements are set as -0.1 (**a**) the smooth case, (**b**) the rough case

$\rho_{v'v'}$ at $(y + \varepsilon)^+ \approx 80$ are shown in Fig. 5.10. In this figure it can be seen that the correlations are very similar in shape, but the streamwise extent is much shorter for the rough case ($\approx 0.3\delta$) compared to the smooth case ($\approx 0.45\delta$). From these results it is evident that the effect of surface roughness on the structures is still felt at $(y + \varepsilon)^+ \approx 80$, which indeed is in the log-layer.

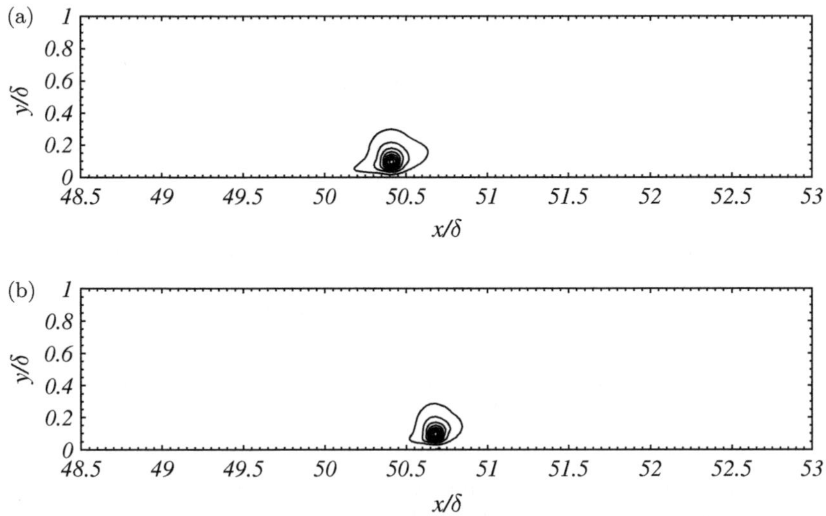

Fig. 5.10 Streamwise section (xy) of correlations of $\overline{v'v'}$ at $(y + \varepsilon)^+ \approx 80$, the outerline represents $C_{vv} = 0.1$ and increments are set as 0.1 (**a**) the smooth case, (**b**) the rough case

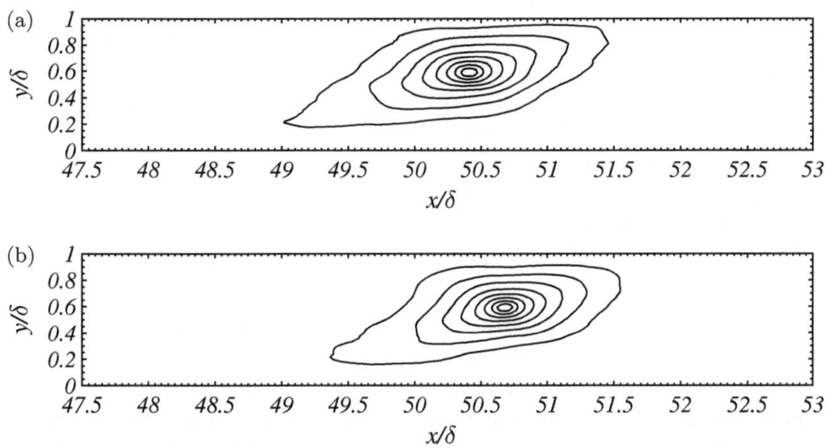

Fig. 5.11 Streamwise section (xy) of correlations of $\overline{u'u'}$ at $(y + \varepsilon)/\delta \approx 0.6$, the outerline represents $C_{uu} = 0.15$ and increments are set as 0.75/8 (**a**) the smooth case, (**b**) the rough case

Figure 5.11 shows contours of $\rho_{u'u'}$ at $(y + \varepsilon)/\delta \approx 0.6$, which is inside the so-called outer layer of the turbulent boundary layer. The length of the correlations in the streamwise direction is about 2.5δ for the smooth case and it is about 2.2δ for the rough case. This is approximately a 12 % reduction in length. However, they are very similar in shape, hence it can be inferred that the effect of rough wall slightly manifests itself even in the outer region.

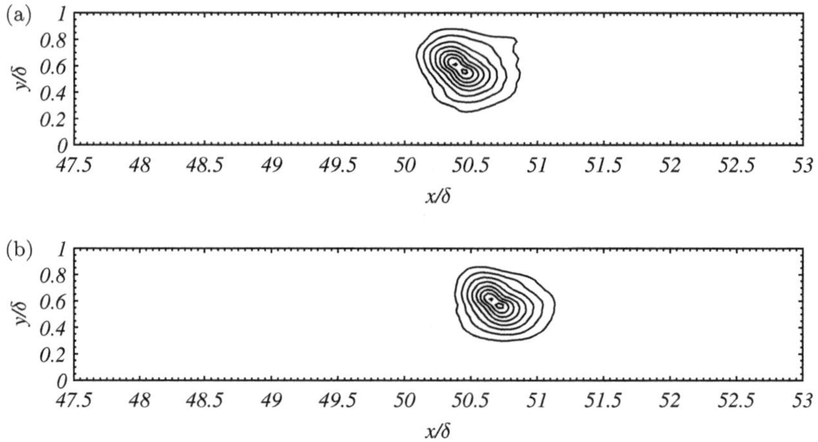

Fig. 5.12 Streamwise section (*xy*) of correlations of $\overline{u'v'}$ at $(y + \varepsilon)/\delta \approx 0.6$, the outerline represents $C_{uv} = -0.15$ and decrements are set as $-0.75/8$ (**a**) the smooth case, (**b**) the rough case

Figure 5.12 depicts $\rho_{u'v'}$ correlations at $(y + \varepsilon)/\delta \approx 0.6$. It is observed that both smooth and rough cases produce similar results in terms of both size ($\approx 0.75\delta$), and shape. We can conclude that the surface roughness effect is only limited to the outer layer structures producing $\overline{u'u'}$ correlations. In Fig. 5.13 the $\rho_{v'v'}$ correlations at $(y+\varepsilon)/\delta \approx 0.6$ can be seen. Similar to the $\overline{u'v'}$ correlations the difference in shape and sizes of $\overline{v'v'}$ correlations is negligible in the outer layer and the roughness effect on the structures producing wall-normal–wall-normal correlations cannot be felt in this region of boundary layer.

This perhaps indicates the so-called structures defined as inactive motion by [12]. Even though these type of motions produce significant streamwise kinetic energy, and display long streamwise correlations, they do not contribute to the Reynolds shear stress because of lack of correlation between the streamwise and wall-normal components. If the outer layer is populated by these type of structures at this moderate Reynolds number, the loss of streamwise momentum caused by the surface roughness only appears in the $u'u'$ correlations, while not effecting both $u'v'$ and $v'v'$ correlations.

Present results imply that the effect of roughness on velocity structures persists well into the outer layer. However, [4] found that Townsend's hypothesis is valid for high-order single-point statistics of the velocity field. That means that turbulent characteristics are similar above the roughness layer (that is, $3k - 5k$ away from the wall) for the rough and smooth cases in the single-point statistics. But when one takes closer inspection into multi-point statistics it can be seen that the structures are affected above the roughness layer in the log-layer and also into the outer layer of the boundary layer, especially for the structures of streamwise velocity fluctuations ($\rho_{u'u'}$).

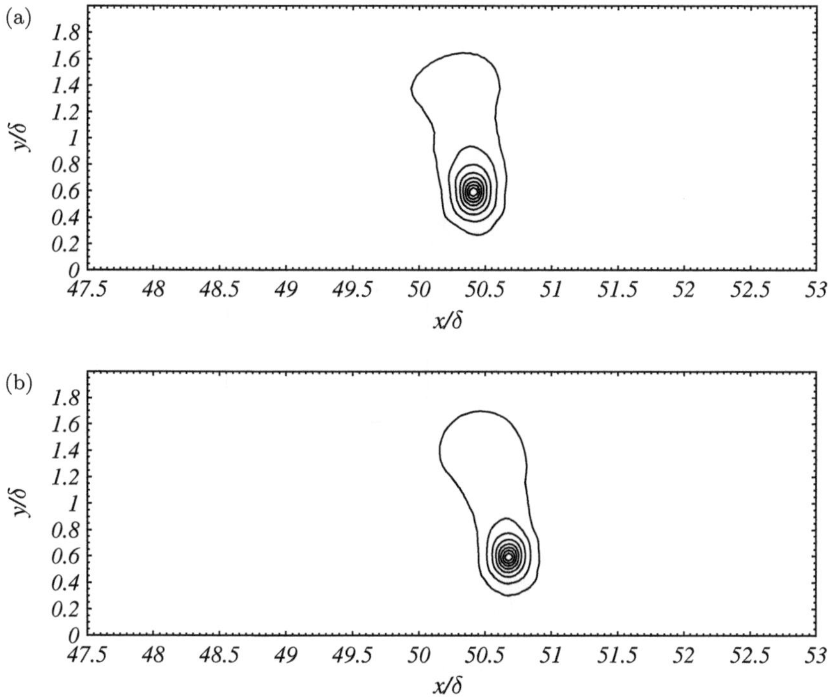

Fig. 5.13 Streamwise section (*xy*) of correlations of $\overline{v'v'}$ at $(y + \varepsilon)/\delta \approx 0.6$, the outerline represents $C_{vv} = 0.15$ and increments are set as 0.75/8 (**a**) the smooth case, (**b**) the rough case

In Fig. 5.14, a comparison of the λ_2 [8] coloured with y^+ value is shown for the smooth and rough cases. It can be seen that surface roughness significantly promotes the production of vortical structures near the wall and into the log-layer. Although vortical structures are more prevalent over the rough wall, their lengths are shorter compared to the smooth case. This increase in the number of smaller vortical structures close to the surface can explain higher skin friction of the rough surface boundary layer.

5.5 Big Data Issue

In this article several aspects of big data were obvious during the simulation and post-processing: (1) Most DNS studies in turbulent boundary layers are limited to relatively low Reynolds numbers due to the demanding nature of these simulations. Even moderate Reynolds number (Re_θ of 2400) simulation like the one we carried out needs large computation resources. In the end of the simulation, there were approximately 5 TB of raw data generated. The major challenge was to allocate

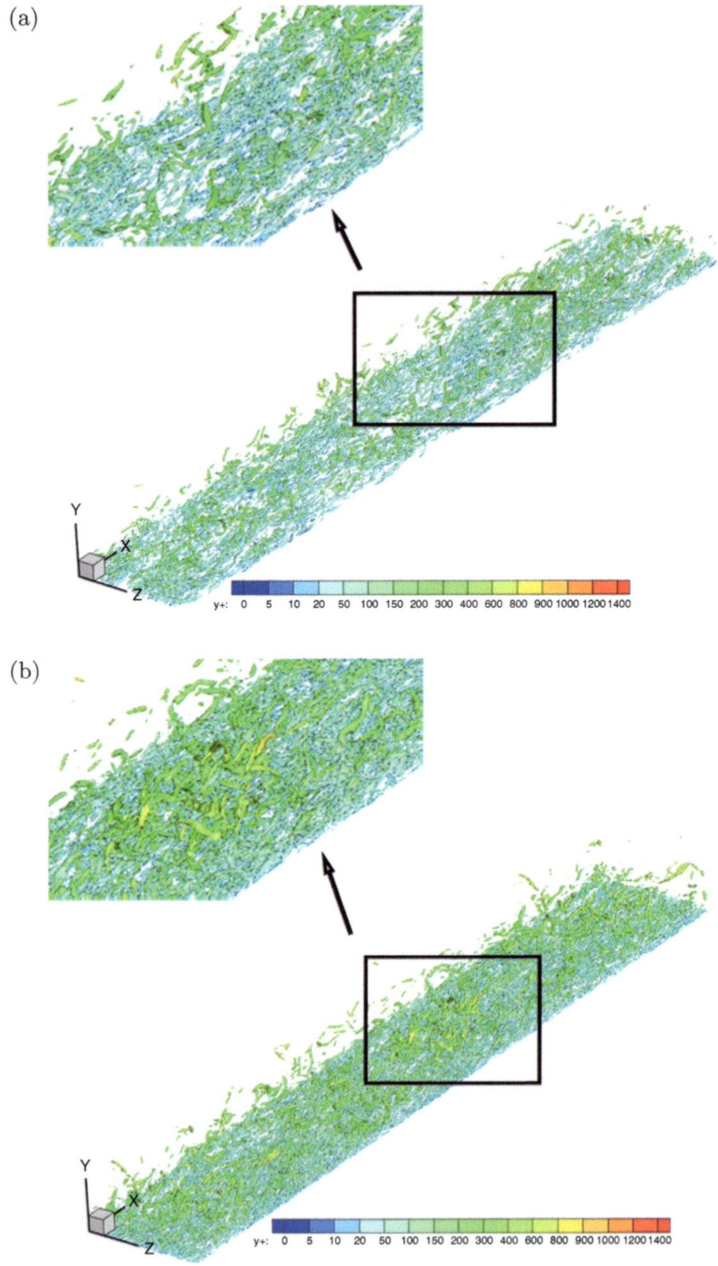

Fig. 5.14 Identification of vortical structures with λ_2-criterion (**a**) the smooth case, (**b**) the rough case

enough space for data creation, storage and transfer between different mediums. (2) As the Reynolds number increases, the turbulence cascade widens, and smaller scales become even smaller. This needs further refinement in grid resolution, in particular near the wall. Roughness complicates the system one step further. A realistic simulation which solves the Navier–Stokes equation in fine mesh is quite challenging and special care has to be taken. In our simulations it was necessary to have $400 \times 125 \times 150$ grid points, which led to about 0.5 GB per time step in memory space. (3) High quality DNS simulations always need large spatial domain in order to remove boundary effects from the simulation. The large domain is necessary to capture the correct physical mechanism and the largest scales of motions which forms the beginning of turbulence cascade. The challenge is also to accommodate realistic turbulent inlet conditions for the velocity and thermal fields. (4) In addition, there comes extensive post-processing of the raw data and computational resources needed for detailed analysis.

5.6 Conclusions

Two-point statistics of a ZPG turbulent boundary layer flowing over a transitionally rough surface (24-grit sandpaper) with k^+ of 11 ($\delta/k \approx 71$) and Reynolds numbers based on momentum thickness of $Re_\theta \sim 2400$ is studied using DNS. The velocity structures of the developing boundary layer were studied by means of two-point correlations at different wall-normal locations (i.e. inner, logarithmic and outer layers). The structures are compared for the rough and the smooth cases to evaluate the effect of surface roughness on these structures. It was shown that the Townsend's wall similarity hypothesis does hold for single-point statistics but not for the turbulent structures above the roughness layer ($3k - 5k$). The longer structures ($\rho_{u'u'}$) seem to be the most affected by the surface roughness in the outer region, whereas the shorter $\overline{u'v'}$ and $\overline{v'v'}$ structures are the most affected close to the rough surface. It was observed that the angle of inclination of $\rho_{u'u'}$ is not affected close to the wall at $(y + \varepsilon)^+ \approx 20$ due to wall roughness, but as we move away from the wall into the log-layer at $(y + \varepsilon)^+ \approx 80$, the $\rho_{u'u'}$ is slightly more inclined towards the wall for the rough case. In the outer region the angle of inclination and geometry of $\rho_{v'v'}$ and $\rho_{u'v'}$ are very similar in both the smooth and rough cases and a difference in structure size is only evident for $\rho_{u'u'}$ correlations. Furthermore, in presented single-point statistics it was shown that the skin friction is increased about 23 % in the rough case compared to the smooth case which can be attributed to the modification of turbulent structures near the wall which is consistent with vortical structure identification with λ_2-criterion. It was also shown that the correlations are not as much affected close to the wall (at $(y + \varepsilon)^+ \approx 20$) region compared to the log-layer, and the modification of structures that leads to increase of skin friction seems to be confined in the region very close to the surface and lays in the complex interactions of surface roughness elements and the flow.

Acknowledgements This research was supported by the grants from NSF-CBET No. 1512393, NSF No. 21P407 B56589 200 and ONR No. 21C211 B56589 200. Dr. Tutkun's work is partially financed by the research project DOMT—Developments in Optical Measurement Technologies funded by the Research Council of Norway with project number 231491 under the FRINATEK program.

References

1. G. Araya, L. Castillo, C. Meneveau, K. Jansen, A dynamic multi-scale approach for turbulent inflow boundary conditions in spatially developing flows. J. Fluid Mech. **670**, 581–605 (2011)
2. B. Brzek, R.B. Cal, G. Johansson, L. Castillo, Transitionally rough zero pressure gradient turbulent boundary layers. Exp. Fluids **44**(1), 115–124 (2008)
3. J. Cardillo, Y. Chen, G. Araya, J. Newman, K. Jansen, L. Castillo, DNS of a turbulent boundary layer with surface roughness. J. Fluid Mech. **729**, 603–637 (2013)
4. A. Doosttalab, G. Araya, J. Newman, R.J. Adrian, K. Jansen, L. Castillo, Effect of small roughness elements on thermal statistics of a turbulent boundary layer at moderate Reynolds number. J. Fluid Mech. **787**, 84–115 (2016)
5. O. Flores, J. Jimenez, J.C. Del Alamo, Vorticity organization in the outer layer of turbulent channels with disturbed walls. J. Fluid Mech. **591**, 145–154 (2007)
6. W.K. George, L. Castillo, Zero-pressure-gradient turbulent boundary layer. Appl. Mech. Rev. **50**, 689–729 (1997)
7. K.E. Jansen, A stabilized finite element method for computing turbulence. Comput. Methods Appl. Mech. Eng. **174**(3–4), 299–317 (1999)
8. J. Jeong, F. Hussain, On the identification of a vortex. J. Fluid Mech. **285**, 69–94 (1995)
9. J. Jiménez, Turbulent flows over rough walls. Annu. Rev. Fluid Mech. **36**(1), 173–196 (2004). arXiv: http://dx.doi.org/10.1146/annurev.fluid.36.050802.122103
10. J. Lee, H.J. Sung, P. Krogstad, Direct numerical simulation of the turbulent boundary layer over a cube-roughened wall. J. Fluid Mech. **669**, 397–431 (2011)
11. T.S. Lund, X. Wu, K.D. Squires, Generation of turbulent inflow data for spatially-developing boundary layer simulations. J. Comput. Phys. **140**(2), 233–258 (1998)
12. A.A. Townsend, Equilibrium layers and wall turbulence. J. Fluid Mech. **11**, 97–120 (1961)
13. A.A Townsend, *The Structure of Turbulent Shear Flow* (Cambridge University Press, New York, 1976)
14. C.H. Whiting, K.E. Jansen, A stabilized finite element method for the incompressible Navier–Stokes equations using a hierarchical basis. Int. J. Numer. Methods Fluids **35**(1), 93–116 (2001)
15. Y. Wu, K.T. Christensen, Spatial structure of a turbulent boundary layer with irregular surface roughness. J. Fluid Mech. **655**, 380–418 (2010)

Chapter 6
Streamwise Relaxation of a Shock Perturbed Turbulent Boundary Layer

M.F. Shahab, G. Lehnasch, and T.B. Gatski

6.1 Introduction

For more than four decades direct numerical simulations (DNS) have been performed on an ever increasing range of compressible flow fields. Such large scale simulations, that rapidly generate massive volumes of data of various types and that need to be stored in a very low density format, require big data technologies and architectures. These technologies and architectures are designed to economically and efficiently extract detailed information from large volumes of data. In engineering applications, such as the interaction of shock waves with turbulent flows, this new enhanced data acquisition and management facilitates the data analysis and provides new information about the diversity of dynamic interactions occurring in such flows. These interactions can be two- or three-dimensional and can include, for example, separation shocks in front of compression

M.F. Shahab
Department of Fluid Flow, Heat Transfer and Combustion, Institute Pprime, Université de Poitiers, ENSMA, CNRS, Futuroscope Chasseneuil Cedex, France

Department of Mechanical Engineering and Aeronautics, City University, London EC1V 0HB, UK

Department of Mechanical Engineering, DHA Suffa University, Karachi 75500, Pakistan

G. Lehnasch (✉)
Department of Fluid Flow, Heat Transfer and Combustion, Institute Pprime, Université de Poitiers, ENSMA, CNRS, Futuroscope Chasseneuil Cedex, France
e-mail: guillaume.lehnasch@isae-ensma.fr

T.B. Gatski
Department of Fluid Flow, Heat Transfer and Combustion, Institute Pprime, Université de Poitiers, ENSMA, CNRS, Futuroscope Chasseneuil Cedex, France

Center for Coastal Physical Oceanography and Ocean, Earth and Atmospheric Sciences, Old Dominion University, Norfolk, VA 23529, USA

A. Pollard et al. (eds.), *Whither Turbulence and Big Data in the 21st Century?*,
DOI 10.1007/978-3-319-41217-7_6

corners, and impinging-reflecting oblique shocks, or around swept wedges and blunt fins. All of these have characterizing features that distinguish their dynamics. The interest in this study is inherently constrained to impinging-reflecting oblique shocks on a plane boundary layer flow. Nevertheless, even here the literature is vast and numerous experimental and numerical studies have taken place over the last decade [1, 7, 8, 13, 15, 20, among others]. Due to the complex interactions inherent in the phenomenon, a complete understanding of the dynamics remains elusive, however.

The interaction between the turbulence and a shock wave results in a modification of both fields. For the turbulence, the characteristic velocity, time and length scales change considerably, and for the shock wave, it can exhibit substantial unsteadiness and deformation. The extent to which each of these modifications occur depends on a wide variety of parameters such as the strength, orientation, location, and shape of the shock wave, as well as the flow geometry and boundary conditions. The state of turbulence and compressibility of the incoming (upstream) flow are additional parameters that will affect the interaction. In this numerical study the incident shock induces an incipient separation at the surface. For such cases, the upstream influence of the disturbance, initiated by the penetration of the incident shock, is not felt over too large a streamwise extent, and in addition, the impact on shock unsteadiness is also reduced. Nevertheless, downstream of the (weak) reattachment, the turbulent field has now passed through a series of complex dynamic processes both inside and outside of the boundary layer and undergoes a recovery process. The effect of the interaction on the recovering boundary layer is less pronounced, and the mean and turbulent properties of the boundary layer may take $\mathscr{O}(10)$ undisturbed boundary layer thicknesses to fully recover.

DNS of spatially evolving turbulent boundary layers at a free-stream Mach number of 2.25 with an impinging shock wave (flow deflection angle of 8°) have been performed with an adiabatic wall, zero heat flux condition, and an isothermal cold-wall (with respect to the adiabatic wall temperature) condition. The corresponding reference flow no-shock simulations have been analyzed previously [16]. The complex dynamic interactions induced by the impinging shocks suggest a strong effect on the wall-normal behavior of the turbulence structures, and on the downstream relaxation toward a new (undistorted) developed turbulent boundary layer.

In order to probe the structural dynamics associated with such a shock impinging flow, a variety of statistical quantities are examined with emphasis on the streamwise evolution downstream of the shock interaction zone. Included in this analysis are the turbulent velocity second-moments and the associated turbulent anisotropy fields and invariant maps, turbulent stress quadrant analysis coupled with skewness and flatness factor distributions. All these quantities are compared between an adiabatic thermal wall condition and an isothermal condition involving a cold wall (relative to adiabatic).

6.2 Problem Formulation

The equations solved in the DNS are the non-dimensional compressible Navier–Stokes equations comprising the density ρ, momentum $\rho u_i = (\rho u, \rho v, \rho w)$, and total energy ρE conservation equations, and coupled with the equation of state for a perfect gas, $p = \rho T$. The conservation equations are given by

$$\frac{\partial \rho}{\partial t} + \frac{\partial}{\partial x_j}(\rho u_j) = 0 \tag{6.1}$$

$$\frac{\partial(\rho u_i)}{\partial t} + \frac{\partial(u_i \rho u_j)}{\partial x_j} = -\frac{\partial p}{\partial x_i} + \frac{M_\infty \sqrt{\gamma}}{Re_\infty} \frac{\partial \sigma_{ij}}{\partial x_j} \tag{6.2}$$

$$\frac{\partial(\rho E)}{\partial t} + \frac{\partial(u_j \rho H)}{\partial x_j} = \frac{M_\infty \sqrt{\gamma}}{Re_\infty} \left\{ \frac{\partial}{\partial x_j} \left[u_i \sigma_{ij} - \left(\frac{\gamma}{\gamma - 1} \right) q_j \right] \right\}, \tag{6.3}$$

where the viscous stress σ_{ij}, total energy and enthalpy H (total temperature, T_t), and heat flux q_j are given by

$$\sigma_{ij} = 2\mu \left(S_{ij} - \frac{\delta_{ij}}{3} S_{kk} \right), \qquad S_{ij} = \frac{1}{2} \left(\frac{\partial u_i}{\partial x_j} + \frac{\partial u_j}{\partial x_i} \right) \tag{6.4a}$$

$$E = \frac{T}{\gamma - 1} + \frac{u_i u_i}{2}, \qquad H = \frac{\gamma}{\gamma - 1} T_t = \frac{\gamma T}{\gamma - 1} + \frac{u_i u_i}{2} \tag{6.4b}$$

$$q_j = -\frac{\mu}{\mathrm{Pr}} \frac{\partial T}{\partial x_j}. \tag{6.4c}$$

with the molecular Prandtl number Pr assigned a value of 0.72 throughout. The non-dimensional scaling is based on a characteristic velocity scale given by the free-stream speed of sound, c_∞, divided by $\sqrt{\gamma}$ ($\gamma = 1.4$ is the constant ratio of specific heats for a perfect gas), unit length scale, and free-stream thermodynamic variables. The Mach number M_∞ and Reynolds number Re_∞ are based on free-stream values, with $M_\infty = 2.25$, a free-stream unit Reynolds number of 25×10^6/m.

For the adiabatic case, the simulations were performed with the (Neumann) condition of no heat flux at the wall $q_w = 0$. The free-stream temperature was $T^*_\infty = 170\,\mathrm{K}$ (* quantities are dimensional) and the corresponding (recovery) temperature at the wall was $T^*_w = 323\,\mathrm{K}$ which varied little along the plate. This corresponded to a recovery factor of $r \approx 0.89$ which is consistent with the value obtained from $r = \mathrm{Pr}^{1/3}$. The friction Mach number M_τ ($= u_\tau / c_w$, with $c_w = \sqrt{\gamma T_w}$ and $u_\tau = \sqrt{\tau_w / \rho_w}$) is 0.076. For the isothermal (cold) wall case, the free-stream temperature was once again 170 K and the wall temperature was set at $T^*_w = 230\,\mathrm{K}$ which corresponded to a wall temperature ratio $T^*_w / T^*_{aw} = 0.67$ ($T^*_{aw} = T^*_\infty[1 + (\gamma + 1)M^2_\infty/2]$); whereas, the wall temperature ratio based on the

Table 6.1 Parameter specification

(a) Physical parameters

Wall condition	M_∞	$Re_\infty/m.$	Re_θ	H	T_∞	B_q	M_τ	T_w
Adiabatic	2.25	25×10^6	3390	3.40	170 K	0.0	0.077	323 K
Isothermal	2.25	25×10^6	3775	2.65	170 K	-0.017	0.079	230 K

(b) Numerical parameters

Wall condition	N_x	N_y	N_z	Δx^+	Δy_w^+	Δz^+
Adiabatic	2650	111	255	13.90	0.76	6.50
Isothermal	2650	111	255	22.05	1.19	10.30

Table 6.2 Diagnostic plane positions: x_i is the impingement position of incident shock wave (inviscid reflection point—see Fig. 6.1), X_{pos} is the distance from the upstream plate origin, and δ_ω^o is the boundary layer thickness at reference station $P01$

$x/\delta_\omega^o = (x_{pos} - x_i)/\delta_\omega^o$	
Plane	Adiabatic/isothermal
$P01$	−2.9
$P02$	0.4
$P03$	3.0
$P04$	5.6
$P05$	8.2
$P06$	16.7

recovery temperature $T_r^* = 323\,\text{K}(= T_\infty^*[1 + r(\gamma + 1)M_\infty^2/2])$ was 0.71. Finally, for the isothermal case the heat flux parameter, $B_q = q_w/\rho_w u_\tau T_w$, is −0.017 with friction temperature $-B_q T_w = 0.023$, and a friction Mach number of 0.079. The physical and numerical parameters are summarized in Table 6.1.

Statistics are obtained at six different streamwise positions, with one position upstream of the shock-system and the remaining in the relaxation region downstream of the incident shock impingement point (see Table 6.2). At position $P01$, $Re_x = 5.2 \times 10^6$ with a momentum thickness Reynolds number, Re_θ (based on free-stream values), of 3390 for the adiabatic case and 3775 for the isothermal case. Due to the significant distortion of the mean streamwise velocity, a vorticity measure is chosen for the boundary layer thickness, δ_ω. The vorticity measure is based on the invariant $\sqrt{2\text{II}_W}$ ($2\text{II}_W = -W_{ij}W_{ji}$, and W_{ij} is the skew-symmetric part of the velocity gradient tensor). This allows for a consistent boundary layer measure at all the streamwise positions examined. The threshold vorticity level (2.5) was chosen so that the vorticity thickness ($\delta_\omega^o = 1.95 \times 10^{-3}$ m) at position $P01$ coincided with the boundary layer thickness based on 0.99 of the free-stream velocity. Nevertheless, other boundary layer thickness measures have also been used [6, 14] and are based on the spanwise vorticity itself.

At position $P01$, the Reynolds number based on δ_ω is $Re_{\delta_\omega} = 4.86 \times 10^4$. At each streamwise position, statistics are gathered using 2500 flow-field sets extending over a total time period of $65\delta_\omega^o/U_\infty$. The Reynolds averages are obtained through a combination of spatial averaging in the (homogeneous) spanwise, z, direction followed by an average in time. The Reynolds averaged quantities are designated by an overbar, and the fluctuating part by a single $'$.

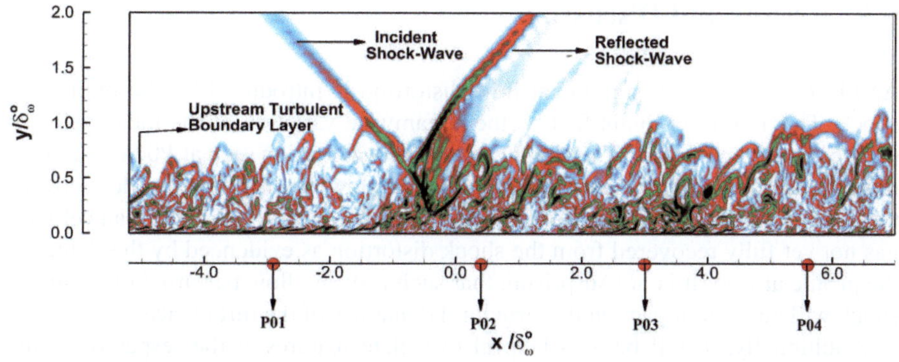

Fig. 6.1 Instantaneous numerical schlieren plot for the present direct numerical simulation (DNS) along with the relative location of the diagnostic planes. Location of stations *P05* and *P06* that are not shown is given in Table 6.2. Contours represent instantaneous density gradient

A numerical schlieren snapshot of the flow field for the adiabatic case is presented in Fig. 6.1 in order to illustrate the enhancement of turbulent structures and the change of flow dynamics that occur as a consequence of interaction with the shock-system, and also to indicate the positions of the diagnostic planes used in the present analysis.

A mixed weighted-ENO compact-difference numerical algorithm is used for the discretization of the three-dimensional Navier–Stokes equations. The inviscid flux function uses a seventh-order weighted-ENO reconstruction of the characteristic inviscid fluxes. The viscous derivative and flux function are determined using a fourth-order compact-difference scheme, and the integration in time is performed by means of the four-step fourth-order Runge–Kutta algorithm.

The computational domain in the streamwise direction is partitioned into three zones: laminar-transition zone ($N_{x_1} = 596$), turbulent zone ($N_{x_2} = 1900$), and buffer (sponge) zone ($N_{x_3} = 154$). The grid spacing in the laminar-transition zone is relatively coarser than that in the developing turbulence regime, and in the buffer zone the grid is progressively coarsened to dampen out any numerical oscillations. The grid spacing in the wall-normal direction through the boundary layer is variable and is based on a geometric stretching algorithm given by $y_j = y_{j-1} + \Delta y_w r^{j-2}$ for $j = 2, n_y$, where $\Delta y_w = y_2 - y_1$ ($y_1 = y_w = 0$) is set so that $\Delta y_w^+ \simeq 1$ and r is the stretching parameter whose value is slightly greater than 1 ($r \simeq 1.053$, $y_w = 2.675 \times 10^{-6}$ m). For the isothermal case, the grid stretching ratio in the wall-normal direction is greater than the ratio for the adiabatic case ($r \simeq 1.055$, $y_w = 2.032 \times 10^{-6}$ m), since numerical validation tests found that the grid had to be more refined for the isothermal case near the wall. Details of the numerical solutions for the no-shock adiabatic and isothermal cases can be found in [16]. For the spanwise direction, the grid spacing is uniform.

6.3 Results and Discussion

As Fig. 6.1 suggests, a significant flow distortion is introduced by the impinging shock. This is further quantified by the streamwise variation of the mean velocity profiles shown in Fig. 6.2. The significant velocity distortion seen at *P*02 is followed in both the adiabatic and isothermal cases by a slow downstream profile recovery toward an undistorted profile. Nevertheless, even after $16\delta_\omega^o$ (*P*06) the mean velocity has not yet fully recovered from the shock distortion as evidenced by the defect in the profile at *P*06. It is not surprising that such a strong flow perturbation from the shock will have an impact on the structural dynamics of the turbulence.

Additionally, it will be also helpful to isolate features at the respective wall-normal locations of the evolving boundary layer. For this purpose, the van Driest transformed mean velocity profile, $\overline{u}_c^+ = \int_0^{\overline{u}^+} (\overline{\rho}/\overline{\rho}_w)^{1/2}\, d\overline{u}^+$, will be used as a guide in correlating boundary layer location at the different streamwise positions. In Fig. 6.3a, the van Driest profiles at positions *P*01 and *P*06 are shown. At position *P*01, the log-layers are clearly visible and extend approximately from $y^+ = 40$ to $y^+ = 250$. For the isothermal case, the boundary layer has thickened relative to the adiabatic case with the outer layer wake occurring farther from the wall. Nevertheless, for both the adiabatic and isothermal cases the wake functions and wake parameters (peak of wake function distribution) are similar. At position *P*06, a log-layer has not yet been fully established within the length of the computational domain but, when compared to Fig. 6.2, the van Driest velocity appears to be closer to an equilibrium form than that suggested by the mean velocity profiles in Fig. 6.2. Both the turbulent second-moments and higher-order correlations can provide some quantification of the effect on turbulent structure that occurs due to flow distortion introduced by the impinging shock. In the sequel, the structural changes in the turbulence field and the subsequent streamwise relaxation back to an undistorted state will be investigated.

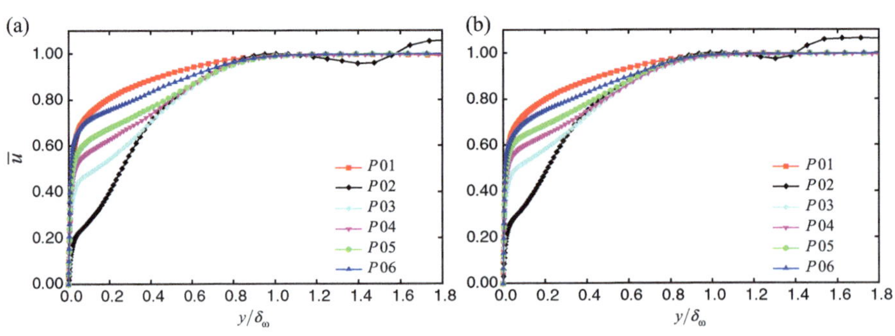

Fig. 6.2 Mean velocity profiles at positions *P*01–*P*06. (**a**) Adiabatic. (**b**) Isothermal

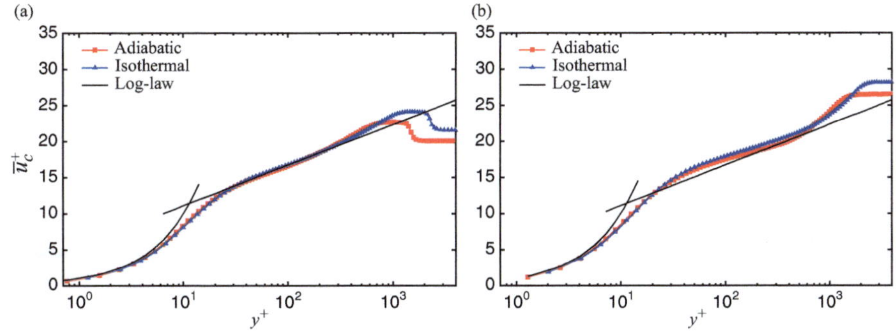

Fig. 6.3 van Driest mean velocity profile at positions *P*01 and *P*06. Log-law line has slope 1/0.41 and intercept 5.5. (**a**) Position *P*01. (**b**) Position *P*06

6.3.1 Turbulent Reynolds Stress and Anisotropy

The boundary layer distributions of the normalized Reynolds stress components at the different streamwise positions for both the adiabatic and isothermal cases are shown in Fig. 6.4 with the normalization velocity U_e which is the mean streamwise velocity at wall-normal distance $y = \delta_\omega$.

6.3.1.1 Reynolds Stress Components

For both the adiabatic and isothermal cases, there is a significant alteration of the Reynolds stress component profiles after the impingement point at position *P*02. The streamwise component $\overline{u'^2}$ displays an amplitude increase across the boundary layer at position *P*02. This increase is highlighted by the significant shift in peak amplitude location away from the wall relative to the unperturbed state. Slight differences occur between the adiabatic and cold-wall cases shown, but the overall trends are similar. As Fig. 6.4 shows, the $\overline{u'^2}$ component subsequently relaxes from the distorted condition at *P*02, but is unable to recover to its distribution at *P*01 even at the far downstream position *P*06. Nevertheless, a more rapid relaxation in the isothermal case is observed.

In contrast to the streamwise component, the wall-normal and spanwise components, $\overline{v'^2}$ and $\overline{w'^2}$, respectively, are characterized at position *P*01 with a distribution across the boundary layer that rises to a maximum value as distance from the wall increases and slowly decreases as the boundary layer edge is approached. At position *P*02, the shock perturbation has significantly increased both the maximum amplitude as well as the amplitude of both components well away from the wall. Similar to the streamwise component, the location of the maximum $\overline{v'^2}$ and $\overline{w'^2}$ has moved away from the wall. Although all three normal Reynolds stress components showed an increase in maximum amplitude at position *P*02 relative to position

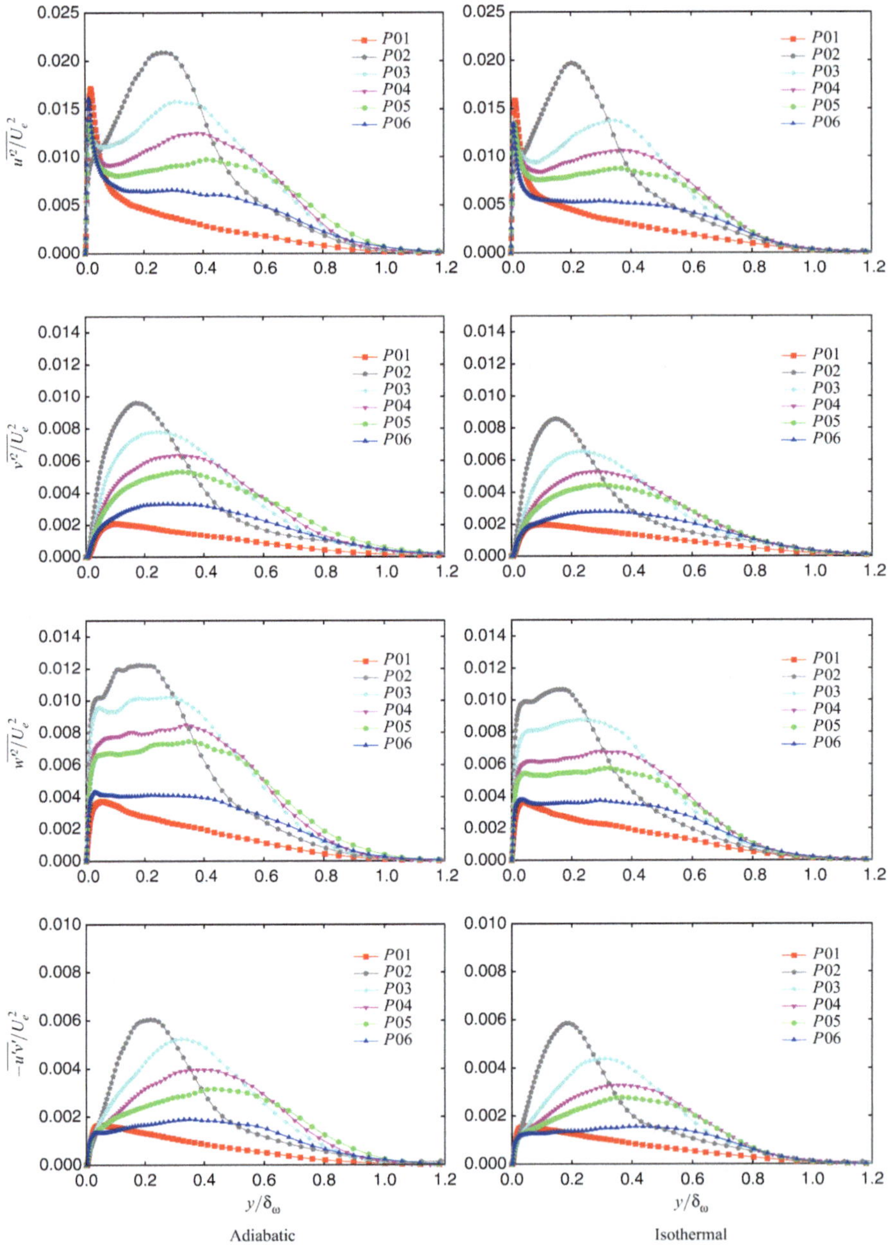

Fig. 6.4 Turbulent Reynolds stress distributions across boundary layer for adiabatic and isothermal wall conditions

$P01$, the increases varied. For the adiabatic case, the wall-normal and spanwise components were increased by a factor of approximately 2.5 and 3, respectively; whereas, for the streamwise component it was only approximately 1.3. A similar trend was also observed in the isothermal wall case but the increases in all components were slightly less.

For the shear stress distribution, the qualitative and quantitative description of the streamwise effect of the shock perturbation follows that for the wall-normal and spanwise normal stress components. Note that the increase in maximum amplitude for this component is similar to that for the wall-normal component. As with all the other Reynolds stress components, the turbulent velocity field (at the level of statistical second-moments) has not relaxed back at position $P06$ to the pre-distorted state. While the inner layer region in closer proximity to the wall has more fully recovered, the outer layer region appears to be a non-negligible distance away from full recovery.

The streamwise and wall-normal variations of the Reynolds stress components just discussed are consistent with the mean streamwise velocity profiles shown in Fig. 6.2 for both the adiabatic and isothermal cases. In the mean velocity figure the profiles in close proximity to the wall for both cases tend to collapse to their undistorted state, but farther away from the wall the relaxation is much slower. A relative quantitative comparison of the extent to which the mean velocity profiles in Fig. 6.2 and the Reynolds stress profiles in Fig. 6.4 have relaxed back at position $P06$ to position $P01$ is difficult. However, both mean velocity and Reynolds stress profiles clearly suggest that a much longer recovery length is needed for both the adiabatic and isothermal cases. In contrast, as pointed out previously, the van Driest profiles in Fig. 6.3 may be less conclusive and may suggest a more complete recovery than the profiles in Fig. 6.2. This can be attributed to the fact that the van Driest field is obtained from an integration involving the square root of the density (or temperature) field and which subsequently minimizes the differences that would otherwise appear. Although Fig. 6.4 showed that the increase of all the Reynolds stress components was less in the isothermal case than the adiabatic, the reason for this inhibited response is not clear at this point and what connection it has with augmented downstream recovery is, as yet, unanswered.

6.3.1.2 Reynolds Stress Invariants

As just shown, the turbulent velocity second-moments are significantly distorted in the interaction zone and subsequently relax with increasing distance downstream of the shock/boundary layer interaction. A further quantification of the flow distortion is provided by the Reynolds stress anisotropy and the associated anisotropy (Lumley) invariant maps [10, 12, 17]. Figures 6.5 and 6.6 show the anisotropy invariant mappings for the adiabatic and isothermal cases, respectively, at the six streamwise positions ($P01$–$P06$) upstream and downstream of the impingement point. In the figures, the invariant definitions used are II_b $(= -b_{ij}b_{ji}/2 = -b_{ii}^2/2)$

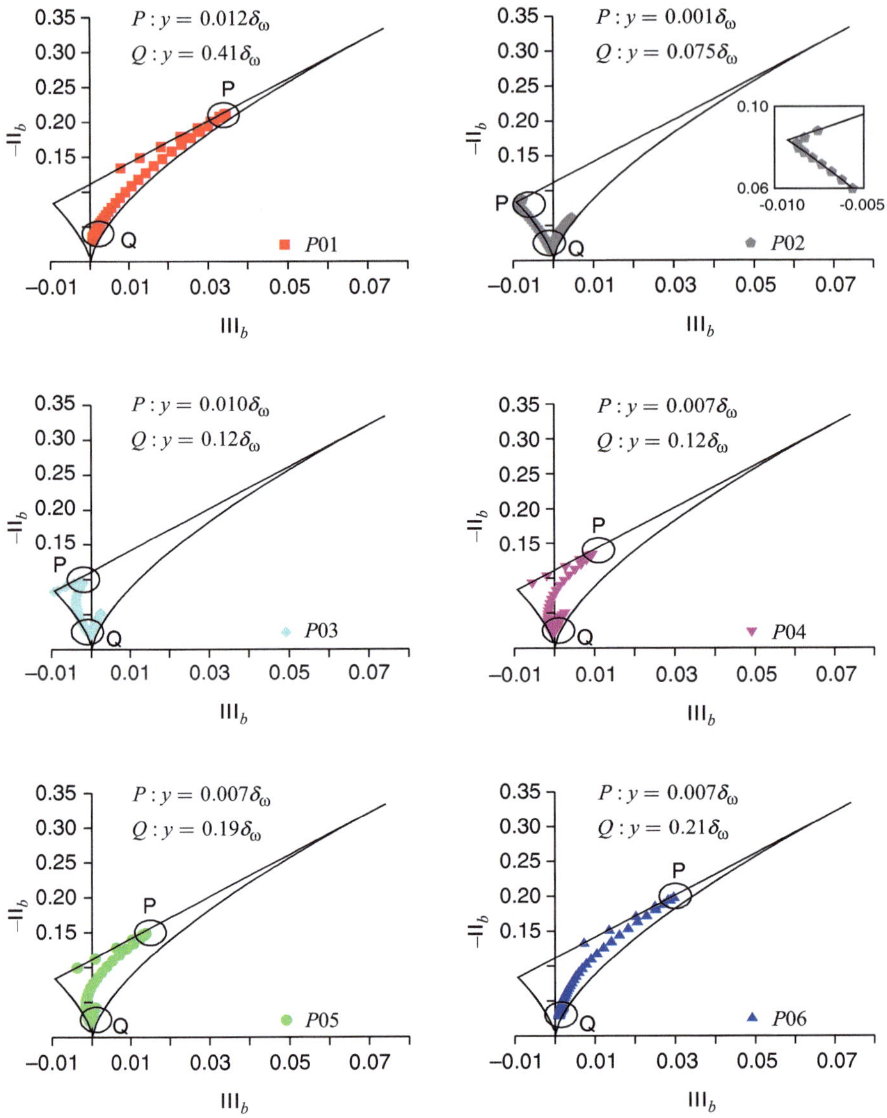

Fig. 6.5 Reynolds stress anisotropy map for adiabatic case

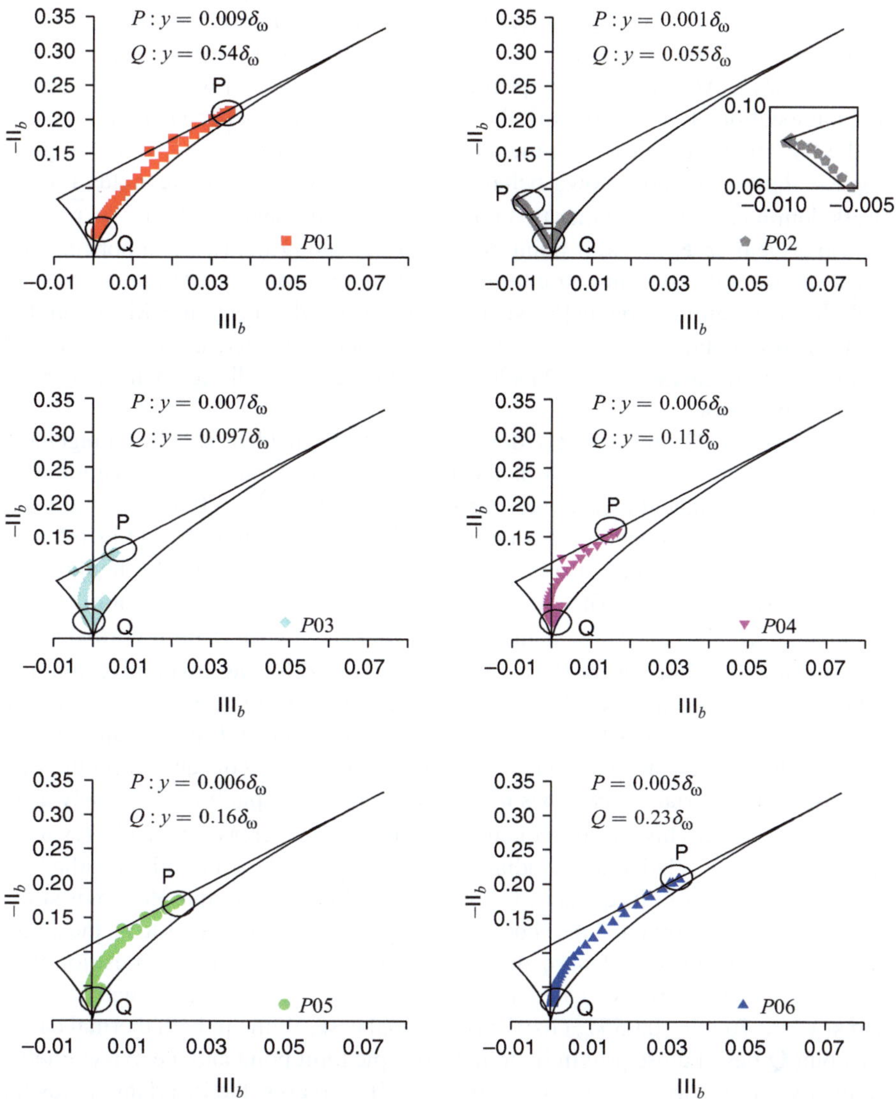

Fig. 6.6 Reynolds stress anisotropy map for isothermal case

and III_b $(= b_{ij}b_{jk}b_{ki}/3 = b_{ii}^3/3)$, with b_{ij} given by $(\overline{u'_i u'_j}/\overline{u'_k u'_k}) - \delta_{ij}/3$. For each case, two locations are marked: **P** marks the location on the invariant curve where $-\mathrm{II}_b$ is maximum and **Q** marks the location on the invariant curve where $-\mathrm{II}_b$ is minimum.

For both the adiabatic and isothermal cases, the upstream invariant map (position *P*01) is similar to the distribution usually found for a boundary layer flow. The point **P** is located at $y^+ \approx 8$ $(y = 0.012\delta_\omega)$ and the point **Q** is located at $y^+ \approx 300$

($y = 0.41\delta_\omega$) for the adiabatic case. In the isothermal case, the corresponding locations are for point **P** at $y^+ \approx 10$ ($y = 0.009\delta_\omega$) and point **Q** at $y^+ \approx 650$ ($y = 0.54\delta_\omega$). At this upstream position $P01$, a comparison with Fig. 6.3 shows that the locations **P** and **Q** delimit the length between the buffer layer and the end of the log-layer region for both the adiabatic and isothermal cases. In this region, Fig. 6.4 shows the well-known behavior where the streamwise Reynolds normal stress dominates over the other two normal stress components with $\overline{w'^2} > \overline{v'^2}$. For the isothermal case, the wall point clearly moves along the two-component limit line away from the axisymmetric limit toward the one-component limit. Consistent with the observation made by [5] where an increase in the mean flow Mach number shifts the wall points more toward the one-component limit, here a decrement in wall temperature increases the local Mach number in the near wall region and generates the same behavior.

At position $P02$, immediately downstream of the incident shock impingement location, a significant deviation of the invariant curve from that at position $P01$ occurs. For the adiabatic case (see Fig. 6.5), point **P** at $y^+ \approx 0.8$ ($y = 0.001\delta_\omega$) lies on the two-component limit boundary (see inset at position $P02$ in Fig. 6.5). As the distance from the wall increases, the invariant curve migrates to the adjacent axisymmetric boundary. This axisymmetric limit represents the action of two dominant normal stress components ($\overline{u'^2}$ and $\overline{w'^2}$) and a weaker third normal stress component ($\overline{v'^2}$). For the isothermal case (see Fig. 6.6), the qualitative trend is similar to the adiabatic case for position $P02$; however, there is a change in the quantitative values. Point **P**, while still on the two-component limit boundary, lies closer to the adjacent axisymmetric boundary. In this case, as distance from the wall increases, the invariant curve does not approach the isotropic limit as closely as in the adiabatic case. This is a consequence of a more rapid increase in the $\overline{v'^2}$ Reynolds stress component relative to the adiabatic case. Nevertheless, in both the adiabatic and isothermal cases this enhanced suppression of the wall-normal Reynolds stress immediately downstream of the shock impingement location appears to be the dominant influence on the turbulent second-moment statistics. In contrast to position $P01$, point **Q** at position $P02$ is located at $y^+ \approx 55$ ($y = 0.075\delta_\omega$) for the adiabatic case, and at $y^+ \approx 70$ ($y = 0.055\delta_\omega$) for the isothermal case. While in both thermal cases, the point **Q** lies in close proximity to the isotropic limit point and the axisymmetric boundary intersection, as distance from the wall increases, the invariant curves for each case assume a form consistent with the outer limits of the undistorted invariant curve of position $P01$. Recall that at position $P01$, a large part of the boundary layer thickness was delimited by points **P** and **Q**. The shock impingement has significantly altered the distance within the boundary layer delimited by **P** and **Q**; although, this has now provided a quantitative estimate of the region where the turbulence is most susceptible to the effects of the impinging shock.

The remaining streamwise positions in Figs. 6.5 and 6.6 exhibit a slow streamwise reconstruction of the turbulent stress anisotropy field that originates near the wall and spreads away from the wall with increasing streamwise distance. Once again, a comparison with the invariant maps at position $P06$ and those at position $P01$ suggests that the isothermal case more rapidly recovered from the shock perturbation.

6.3.2 Quadrant Analysis

Quadrant analysis [11] is one technique that is commonly used to detect and to quantify the occurrence of different turbulent events (e.g., ejections and sweeps) in a turbulent boundary layer. The technique involves the splitting of the u'-v' velocity plane into four quadrants according to the signs of the velocity fluctuations (u' and v'). The events associated with the instantaneous shear stresses, $(u'v')_i$, can then be detected by identifying the stresses with their corresponding quadrant based on the signs of fluctuations. On the basis of this conditional sampling, the averages in each quadrant are estimated by

$$\overline{(u'v')}_i = \frac{1}{N}\sum_{j=1}^{N} S_i \times (u'v')_j, \tag{6.5}$$

where $S_i = 1$ is the point $(u'v')_j$ in the ith quadrant, $S_i = 0$ otherwise, and N is the number of data samples at each point of consideration. The turbulent events relative to each u'-v' quadrant are defined as follows: the first quadrant, where $u' > 0$ and $v' > 0$, contains the events associated with the outward motion of high speed fluid; the second quadrant, where $u' < 0$ and $v' > 0$, contains the events that are associated with the ejections (or bursts) of low speed fluid away from the wall; the third quadrant, where $u' < 0$ and $v' < 0$, represents the events corresponding to the inward motion of low speed fluid; and the fourth quadrant, where $u' > 0$ and $v' < 0$, represents the events that characterize the sweeping of high speed fluid toward the wall.

Figures 6.7 and 6.8 show the downstream evolution of the contributions of the conditionally averaged Reynolds shear stresses $\overline{(u'v')}_i$ across the boundary layer for the adiabatic and isothermal cases. The distributions are normalized with respect to the total turbulent shear stress $\overline{u'v'}$ to highlight the relative contribution of each quadrant. Upstream of the shock interaction zone ($P01$) the trends are in qualitative agreement with the results obtained in incompressible and compressible (undisturbed) wall-bounded flows [4, 9, 21]. There is little difference between the adiabatic and isothermal cases at the upstream station $P01$. In both thermal cases, the ejection and sweep events are the dominant contributors to the total turbulent shear. In the sublayer and buffer layer ($y^+ \lesssim 10$, $y \approx 0.02\delta_\omega$) the sweep events dominate since it is in this region where fluid being brought toward the wall dominates over that being ejected away from the wall. In contrast, in the equilibrium log-layer, the opposite occurs with the ejection events dominating over the sweep events. In this log-layer region, the combination ejection and sweep events account for about 130 % of the shear stress with the 1- and 3-quadrants (equally) accounting for −30 %.

Downstream of the impingement area ($P02$–$P06$), a notable change in the ejection and sweep contributions is observed in Figs. 6.7 and 6.8. At position $P02$, the most significant alteration of the quadrant distributions for both thermal cases is

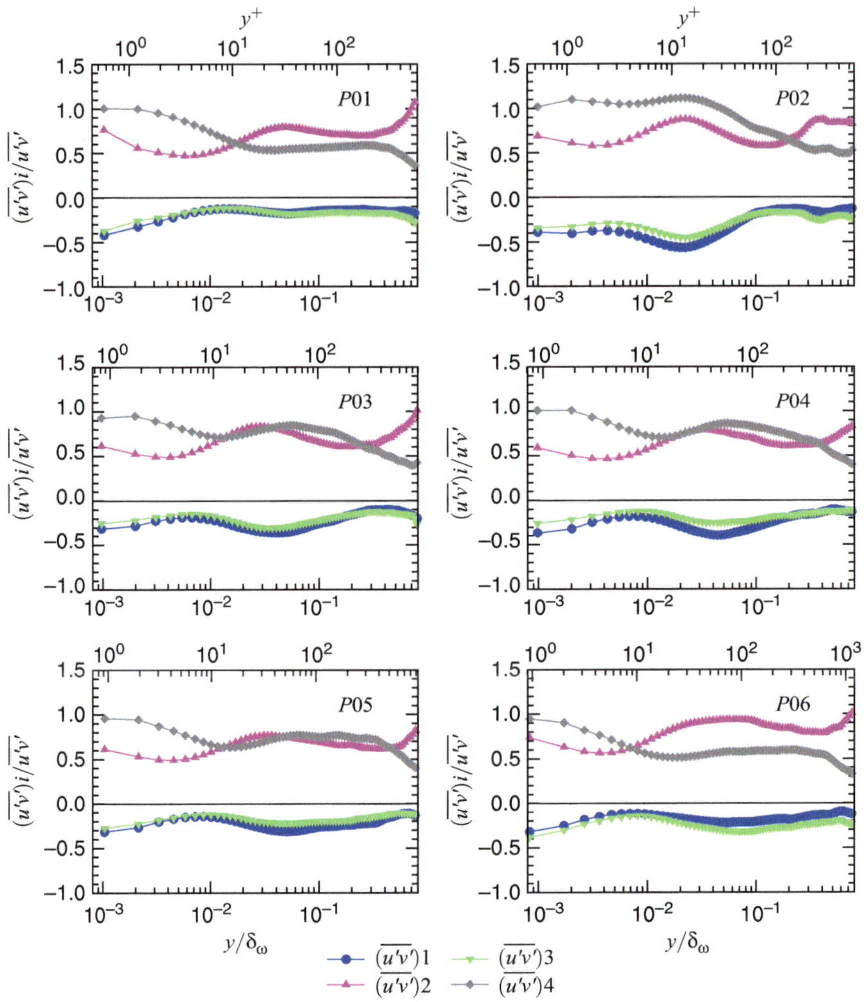

Fig. 6.7 Normalized quadrant distribution across boundary layer for adiabatic wall case

the significant increase in wall-normal distance where the ejection events dominate over the sweep events [see 3]. This general trend is to be expected since the location *P*02 is located downstream of the expansion waves that result from the separation/reattachment process and associated with the convex trajectories of the turbulent structures. These waves, coupled with the appearance of a separation bubble, enhance the sweeping motion over a larger portion of the inner layer region. The cross-over location now occurs at $y^+ \approx 100$ in both the adiabatic and isothermal cases; however, in the isothermal case the ejection event is only slightly larger than the sweep event from $y^+ \approx 25$ outward. Overall, the region where sweeps remain dominant over the ejections is comparatively less in the

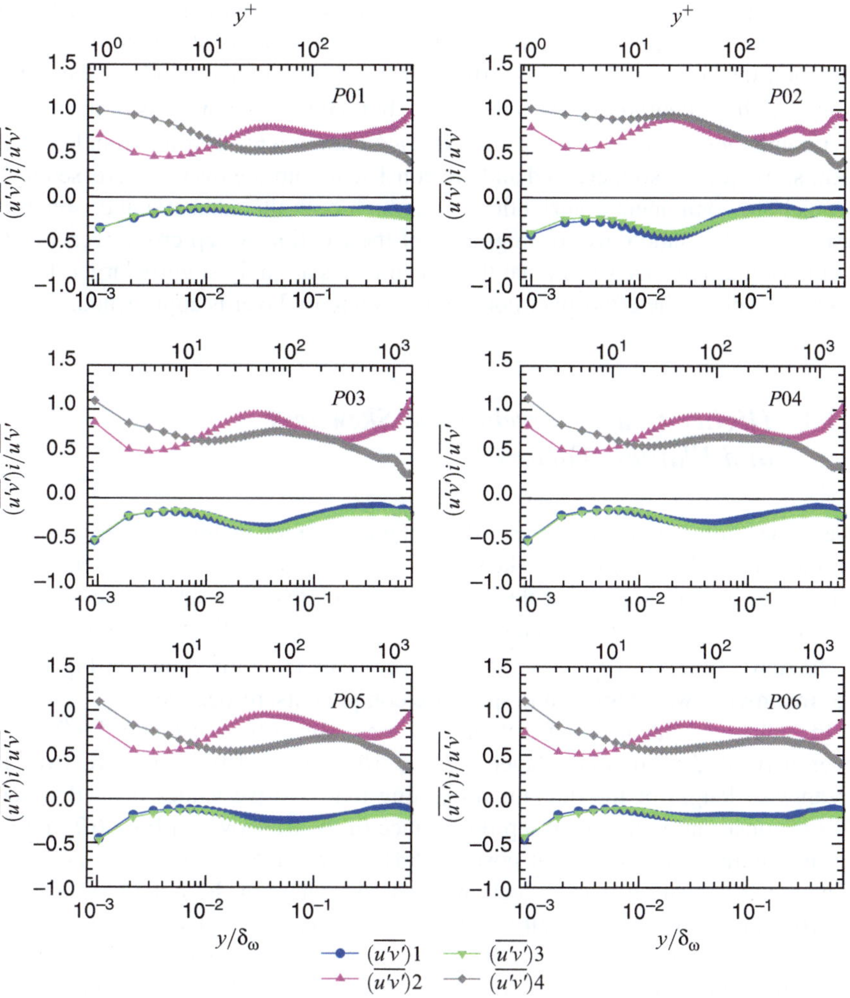

Fig. 6.8 Normalized quadrant distribution across boundary layer for isothermal cold-wall case

isothermal case relative to the adiabatic case further affirming that the relaxation process is relatively quick in the isothermal case. As noted, this outward migration of cross-over location is due primarily to an increase in the contributions of the sweep events with little change to ejection events. The overall balance with the total stress being achieved by an increase in the magnitudes of the 1- and 3-quadrant events. Both the outward migration of the ejection and sweep cross-over location and intensification of the sweep event can be further corroborated with the velocity second-moment distributions presented in Fig. 6.4. The flow distortion produced by the shock relaxes with increasing distance downstream and the results obtained at the station *P*06 show a set of conditional Reynolds stress distributions similar to that of the undisturbed boundary layer (cf. Figs. 6.7 and 6.8).

A characteristic feature of the Reynolds stress distributions shown in Fig. 6.4 was the shift in peak amplitude for the various Reynolds stress components at position $P02$ to $y \approx 0.2\delta_\omega$ ($y^+ \approx 10^2$). Additionally, the magnitude of the normal Reynolds stress components was increased, but with the magnitudes of the $\overline{v'^2}$ (and $\overline{w'^2}$) component increased more than the $\overline{u'^2}$ component. The magnitudes of the shear stress level also increased and are consistent with the overall increased levels of quadrant component contributions. Although both the adiabatic and isothermal cases show the same outward migration of the ejection/sweep cross-over, Fig. 6.7 for the isothermal case shows that the sweep levels are only slightly larger than the ejection levels as the cross-over point in the boundary layer is approached.

6.3.3 Higher-Order Correlations: Skewness and Flatness Factors

The higher-order moments are useful supplements in analyzing the structure of the turbulent field. Recall that the skewness is associated with the asymmetry of the tails of the *pdf* function and the flatness is a relative measure of the weight in the tails of the distribution. The downstream evolution of the third-order velocity fluctuation moments $S(u_i') = \overline{u_i'^3}/(\overline{u_i'^2})^{3/2}$ ($u_i' = u', v', w'$) is shown in Fig. 6.9 for the streamwise, wall-normal and spanwise components, respectively.

In both the adiabatic and isothermal cases, the skewness factor $S(u')$ in the inner part of the boundary layer ($y/\delta_\omega \lesssim 10^{-2}$, $y^+ \lesssim 10$) is positive over the streamwise length of the domain suggesting that positive values of u' dominate. This is consistent with the sweep dominance ($u' > 0$) shown in Figs. 6.7 and 6.8. Correspondingly, the switch at position $P01$ to negative skewness values $u' < 0$ occurs at the location where ejection events dominate in the boundary layer (cf. Figs. 6.7 and 6.8). Additionally, [18] and [19] have found that in a turbulent boundary layer the streamwise skewness factor changes sign at a location in a boundary layer where Reynolds shear stress and the turbulent intensities are maximum. This characterizing feature is also found here from an examination of the second-moment statistics shown in Fig. 6.4.

In contrast, the skewness factor $S(v')$ at position $P01$ shows a predominance for positive values, and which slightly differs with incompressible channel flow [9] at low Re_τ where the wall-normal skewness was negative over the narrow range $6 < y^+ < 30$. These $S(v') > 0$ values imply, when coupled with the distributions for $S(u')$, a dominance of first quadrant (outward interaction) events. However, analysis of the joint *pdf* of the shear stress $u'v'$ shown in Fig. 6.10 reveals that the negative shear stress is a consequence of the correlation between large positive u' and negative v' fluctuations. (Note that since large amplitude velocity fluctuation levels are less likely to occur, the associated *pdf* is necessarily smaller.) As the distance from the wall increases $y/\delta_\omega \gtrsim 10^{-2}$, ejection events dominate requiring correlation between large $u' < 0$ and $v' > 0$ fluctuations. This behavior is corroborated by

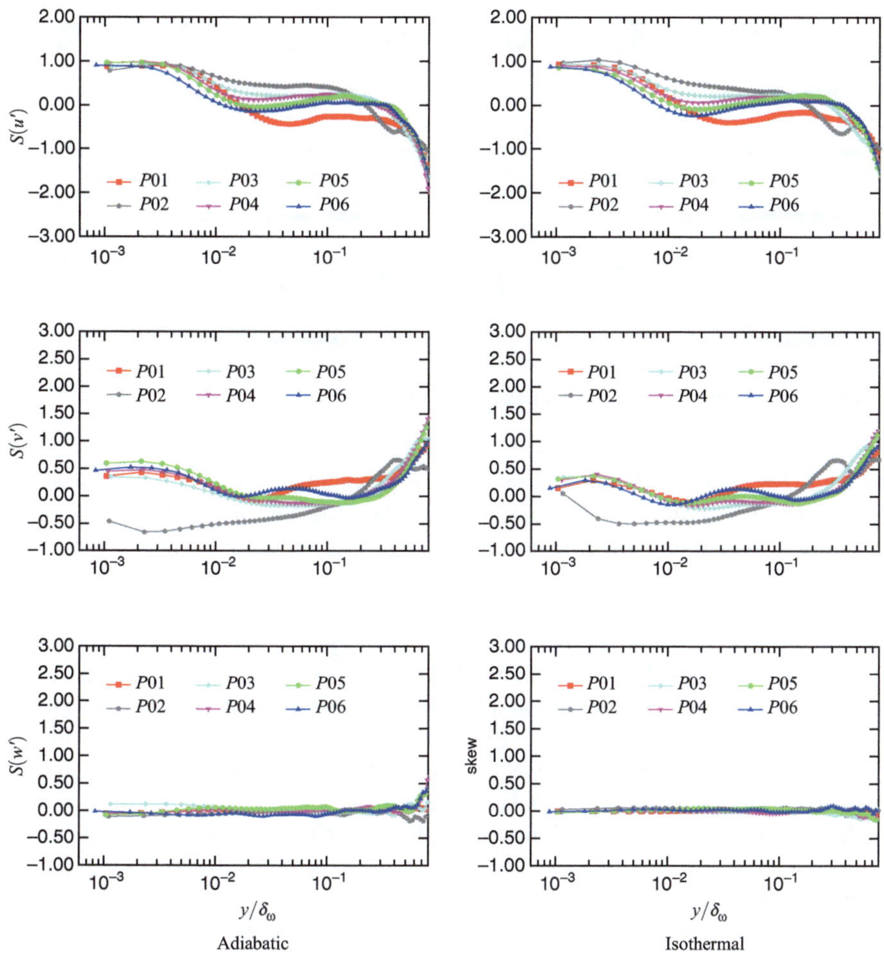

Fig. 6.9 Skewness factor distributions within boundary layer at positions *P*01–*P*06

the $S(v')$ skewness factor in Fig. 6.9 which is positive in the boundary layer region ($y/\delta_\omega \gtrsim 10^{-2}$), and also reaffirmed from the joint *pdf* $u'v'$ in Fig. 6.10.

At position *P*02, closest to the shock impingement point, the skewness factors $S(u')$ and $S(v')$ are significantly displaced away from the wall relative to the upstream *P*01 behavior and to a lesser extent relative to the remaining downstream positions (especially for $S(v')$). The same qualitative behavior is shown for the isothermal case. The streamwise skewness factor $S(u')$ now remains positive in the region $y/\delta_\omega \lesssim 2 \times 10^{-1}$ ($y^+ \lesssim 1.5 \times 10^2$) and the wall-normal skewness factor $S(v')$ remains negative over the same boundary layer range. This is consistent with the quadrant analysis which showed that sweep events dominated the dynamics immediately downstream of the shock/boundary layer interaction point. In addition,

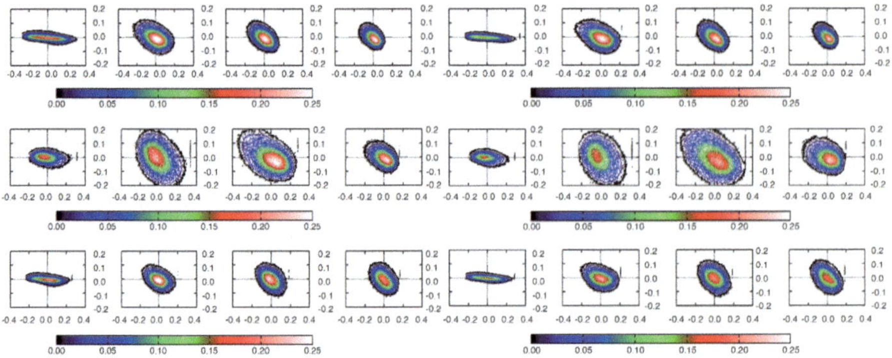

Fig. 6.10 Quadrant distribution of joint *pdf* shear stress $u'v'$

the joint *pdf* in Fig. 6.10 also shows a clear probabilistic bias to fourth quadrant events. The proximity of the shock impingement location to position *P*02 shows that this strong influx of fluid toward the wall is induced by the influence of the shock penetration into the boundary layer. In addition to the influence on the second and fourth quadrant events, a general increase in first and third quadrant, outward and inward motions are also evident although to a much less extent than the ejections and sweeps. In the remaining downstream positions (*P*03–*P*06) an almost monotonic relaxation of the $S(u')$ and $S(v')$ factors is observed. Nevertheless, while position *P*06 has the same qualitative behavior as position *P*01, even after almost 17 boundary layer thicknesses the exact details of the third-moment statistics are not the same as the pre-distorted boundary layer.

Although not of dynamic significance, the adiabatic and isothermal skewness distributions for the spanwise component w' are significantly smaller than the skewness factors for the other components. While not completely negligible, particularly near the end of the log-layer region, the relatively small magnitudes confirm the homogeneity assumption upstream of the shock impingement, and suggest that downstream of the interaction, fluctuations induced by the unsteady shock motion in the spanwise direction may not influence the dynamics over a large extent of the boundary layer.

Up to this point, the analysis of the second- and third-moments of velocity fluctuations has produced a cohesive description of the composite motions active within the boundary layer. For the third-moments, some information about *pdf* of the instantaneous turbulence process was also obtained. While the relative asymmetry of the *pdf* was quantified by $S(u_i')$, the deviation from the Gaussian behavior can be featured through the flatness factor $F(u_i') = \overline{u_i'^4}/(\overline{u_i'^2})^2$ $(u_i' = u', v')$. Figure 6.11 shows the downstream evolution of $F(u_i')$ for both the adiabatic and isothermal cases. In the sublayer region, all the flatness factor values exceed the Gaussian value consistent with previous studies [e.g., 18, 19]. As suggested by [18], this is due to the dominance of the sweeping motion in this region where high speed fluid is brought toward the wall. In the range $8\times10^{-3} \gtrsim y/\delta_\omega \lesssim 4\times10^{-2}$ $(3 \gtrsim y^+ \lesssim 40)$,

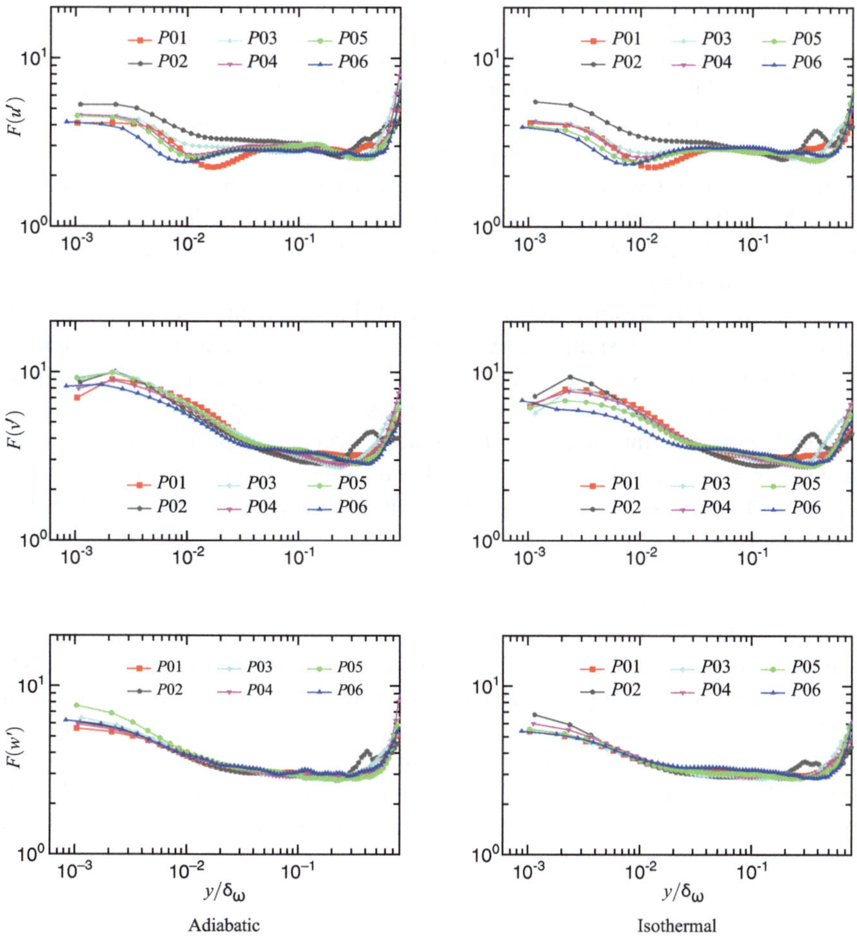

Fig. 6.11 Flatness factor distributions within boundary layer at positions *P*01–*P*06

at the upstream position *P*01, the $F(u')$ flatness factor is below 3 and reaches a minimum in close proximity where the streamwise skewness factor $S(u')$ changes sign (see Fig. 6.9). Farther from the wall, $F(u')$ rises to near the Gaussian value. In this same region, both the $F(v')$ and $F(w')$ flatness factors decrease toward the Gaussian value. As distance from the wall is increased further, the flatness factors increase as expected due to an increased intermittency of the flow. As shown, the same qualitative behavior holds for both the adiabatic and isothermal cases. These trends are qualitatively similar to those found in the incompressible [e.g., 9] and the compressible [3] regimes. In the log-layer region, near Gaussian values are expected and this is confirmed at position *P*01 where the log-layer spans the range $4 \times 10^{-2} \gtrsim y/\delta_\omega \lesssim 4 \times 10^{-1}$ ($30 \gtrsim y^+ \lesssim 200$). At *P*01, high values of the streamwise flatness factor $F(u')$ (relative to the Gaussian value) are found in close proximity of the wall for $y/\delta_\omega \lesssim 8 \times 10^{-3}$ ($y^+ \lesssim 6$) for both the adiabatic and isothermal cases. Correspondingly, within the same boundary layer range, the wall-normal and spanwise flatness factors, $F(v')$ and $F(w')$, also exhibit high values with $F(v')$ greater than either $F(u')$ or $F(w')$. In this sublayer region, the increased influence of viscous effects reduces both the magnitude and scale (spatial and temporal) of the velocity fluctuations which results in the higher flatness values. The larger values of the $F(v')$ factor suggest a more significant reduction which reflects the strong influence of the wall. Once again, the flatness factors are unaffected by a change in the thermal boundary condition.

At position $P02$ downstream in the shock impingement interaction zone, the flatness factor $F(u')$ is significantly affected in both the adiabatic and isothermal cases. For the adiabatic case, the $F(v')$ and $F(w')$ factors are only minimally influenced; however, in the isothermal case there is an increase in the sublayer region for both factors. The increase in the $F(u')$ factor over a large portion of the boundary layer suggests a reduction of streamwise fluctuation levels relative to the undistorted state at $P01$. This is consistent with the general thickening of the boundary layer downstream of the shock impingement that results in lower streamwise velocity values over a portion of the boundary layer relative to the upstream, undistorted values. Both the adiabatic and isothermal $F(u')$ factors show a similar behavior across the boundary layer; however, the $F(v')$ and $F(w')$ factors in the isothermal case are increased relative to their adiabatic counterparts. Although the increase is not significant, it does suggest an attenuation of these components relative to the adiabatic case.

With increasing distance downstream, positions $P03$–$P06$, a relaxation process ensues; although, as in the previous results that have been analyzed, the relaxation process and flow structure reconstruction have not been fully completed even at the far downstream position $P06$.

6.4 Big Data: The 4Vs

It is natural to ask how Big Data differs from other terms such as massive data or very large data. While all such terms share the common qualitative feature of large capacity, the Big Data description has come to be associated with some quantitative features or measures. An excellent recent text by [2] provides a detailed discussion of the features characterizing Big Data—termed the 4Vs. These four Big Data characteristics are: (1) *volume*, (2) *velocity*, (3) *value*, and (4) *variety*. (As noted previously, in the context of quantifying Big Data, the term *velocity* is here associated with timeliness of data collection and processing.)

It is clear that the fluid flow simulations and analysis described here meet the *velocity* and *value* characteristics because such simulations provide important new physical insight into the flow dynamics that previously had not been available using either experimental or theoretical methodologies. For the *variety* characteristic, the dataset content here is based on a numerical format using a fixed Eulerian frame with a structured mesh. Other dataset content are readily available using algorithms based on unstructured meshes or grid free formulations based on Lagrangian frame approaches. For this study, the volume of data generated for each case (adiabatic, isothermal; with, without shock) was 7.5 TB or 30 TB in total. While large, the Big Data concept of *volume* is often associated datasets having petabyte, exabyte, or larger.

Nevertheless, while such simulations have proven value, they are still limited by the cost and availability of computer resources. In practice, extracting statistical correlations from full databases containing several terabytes of data is still prohibitive from a cost and time standpoint. As in the database analysis described here, it is common to save only $y - z$ plane slices and extract the various correlations needed (such as though in an energy or stress budget analysis). Such slices are "thick" in that a sufficient number of adjacent planes are needed and saved to complete a high-order derivative stencil that is often required. Nevertheless, significant savings in cost and time are achieved through such a selective plane procedure. In each of the simulations (adiabatic, isothermal; with, without shock) of the current study, the required storage is reduced to 0.5 TB or 2 TB in total. However, even with this significant saving in database storage, the computations in each case still required one million hours of the IBM Blue Gene/Q supercomputer using between 256 and 1024 cores in parallel.

Simulations of the type described here can be expanded to improve the resolution and therefore the detail of the flow dynamics analysis. While this can significantly increase the database size, what is of more *value* is the accumulation of datasets similar to the one described here, but over a much broader parameter range. For example, relevant parameters for such flow fields include Reynolds number, wall heat flux conditions, inflow conditions, wall surface variation and curvature, etc. Accumulation, ordering, and effective processing of all these datasets in total would allow for a more in-depth perspective of flow dynamics over a broad range of conditions. Such datasets would clearly benefit from efficiencies in Big Data acquisition, storage, management, and analysis advances.

6.5 Summary

In this study, an analysis of the turbulent velocity field statistics in the vicinity of, and downstream of, a shock impingement on a supersonic boundary layer has been presented. Both the velocity second-moments and higher-order correlations have been used to quantify the effects on the turbulence and turbulent structure due to the impinging shock with emphasis on the streamwise evolution and relaxation back to an undisturbed (equilibrium) state. This has included the effect of two different thermal wall conditions, adiabatic and isothermal (cold wall), on the dynamic evolution.

For all the Reynolds stress components, there is an increase in peak magnitude and shift in peak location due to the impingement immediately downstream of the mean reattachment point of the incipient separation bubble. With increasing distance downstream away from the impingement point, the peak magnitude relaxes although the relaxation is still not back to upstream levels even after 17 (before interaction) boundary layer thicknesses. Similar trends were observed for the isothermal (cold wall) case but the enhanced Reynolds stress component levels were not as large as the adiabatic case.

The evolution of the Reynolds stress anisotropy invariant maps at various downstream wall-normal planes showed a marked departure from the usual equilibrium boundary layer features. At the location immediately downstream of the mean reattachment point, in the vicinity of the wall, the streamwise and spanwise stress components dominated, with the wall-normal component significantly attenuated. Downstream relaxation was most rapid near the wall, while with increasing distance from the wall the large scale structures were slower to respond. In the isothermal case, the qualitative trend was similar, but the distortion as represented by the distribution of invariants along the axisymmetric boundary was weaker.

A quadrant analysis of the distorted boundary layer flow showed that sweeps and ejections dominated the flow dynamics and that the effect of the shock impingement significantly moved the cross-over point between sweeps and ejections away from the wall in both adiabatic and isothermal conditions. This was linked to the expansion waves formed in the vicinity of the mean reattachment point. While first and third (outward and inward interactions, respectively) quadrants were affected, their overall magnitude and qualitative behavior were not significantly affected by the shock impingement.

The higher-order correlation skewness and flatness factors provided further confirmation of the dynamic interactions deduced from the lower-order statistical analysis. These factors were found to be consistent with those for incompressible separated flows as well as compressible shock impingement flows. For the skewness factor, the wall-normal factor was most significantly affected by the shock distortion, and for the flatness factor the streamwise component was the most affected.

In the context of Big Data such simulations optimally form just one dataset contained within a much larger volume of data containing datasets covering a broad range of parameters as well as enhanced resolution. Such data volumes will undoubtedly be of petabyte and exabyte sizes requiring all the enhanced processing and data management tools applied to Big Data.

MFS acknowledges the support of the Higher Education Commission Pakistan. This work was performed using HPC resources from GENCI-[CCRT/CINES/IDRIS] (Grant 2011—[x2011090912]) under the support of the Agence Nationale de la Recherche [SPICEX project (ANR-07-CIS7-009)]. TBG acknowledges the support of the DFG SFB-TR40 Project and the TRR40 Summer Program at Technische Universität München. It is regrettable that our colleague and friend Pierre Comte was unable to see the completion of this work. Pierre's interests were broad and his enthusiasm unbounded. He always had insightful questions, often subtle and almost always requiring some extra thought.

References

1. P. Bookey, C. Wyckham, A. Smits, M.P. Martin, New experimental data of STBLI at DNS/LES accessible Reynolds numbers. Paper No. 2005–0309, AIAA (2005)
2. M. Chen, S. Mao, Y. Zhang, V.C.M. Leung, *Big Data: Related Technologies, Challenges and Future Prospects*. Springer Briefs in Computer Science (Springer, New York, 2014)
3. J. Deleuze, Structure d'une couche limite turbulente soumise à une onde de choc incidente. PhD thesis, Université de la Mediterranee Aix - Marseille (1995)
4. J. Deleuze, N. Audiffren, M. Elena, Quadrant analysis in a heated-wall supersonic boundary layer. Phys. Fluids **6**, 4031–4041 (1994)
5. B. Frohnapfel, P. Lammers, J. Jovanović, F. Durst, Interpretation of the mechanism associated with turbulent drag reduction in terms of anisotropy invariants. J. Fluids Mech. **577**, 457–466 (2007)
6. E. Garnier, P. Sagaut, M. Deville, Large eddy simulation of shock/boundary-layer interaction. AIAA J. **40**, 1935–1944 (2002)
7. E. Garnier, P. Sagaut, M. Deville, Large eddy simulation of shock/homogeneous turbulence interaction. Comput. Fluids **31**, 245–268 (2002)
8. R.A. Humble, Unsteady flow organization of a shock wave/boundary layer interaction. Ph.D. thesis, Delft University of Technology, 2008
9. J. Kim, P. Moin, R.D. Moser, Turbulence statistics in fully developed channel flow at low Reynolds number. J. Fluids Mech. **177**, 133–166 (1987)
10. P.A. Krogstad, L.E. Torbergsen, Invariant analysis of turbulent pipe flow. Flow Turbul. Combust. **64**, 161–181 (2000)
11. S.S. Lu, W.W. Willmarth, Measurements of the structure of the Reynolds stress in a turbulent boundary layer. J. Fluids Mech. **60**, 481–511 (1973)
12. J.L. Lumley, Computational modeling of turbulent flows. Adv. Appl. Mech. **18**, 123–176 (1978)
13. S. Piponniau, J.P. Dussauge, J.F. Debiève, P. Dupont, A simple model for low-frequency unsteadiness in shock-induced separation. J. Fluids Mech. **629**, 87–108 (2009)
14. S. Pirozzoli, M. Bernardini, F. Grasso, Direct numerical simulation of transonic shock/boundary layer interaction under conditions of incipient separation. J. Fluids Mech. **657**, 361–393 (2010)
15. S. Priebe, M. Wu, M.P. Martin, Direct numerical simulation of a reflected-shock-wave/turbulent-boundary-layer interaction. AIAA J. **47**, 1173–1185 (2009)
16. M.F. Shahab, G. Lehnasch, T.B. Gatski, P. Comte, Statistical characteristics of an isothermal, supersonic developing boundary layer flow from DNS data. Flow Turbul. Combust. **86**, 369–397 (2011)
17. A.J. Simonsen, P.A. Krogstad, Turbulent stress invariant analysis: clarification of existing terminology. Phys. Fluids **17**, 088103-1–088103-4 (2005)
18. R.L. Simpson, Y.T. Chew, B.G. Shivaprasad, The structure of a separating turbulent boundary layer. Part 2. Higher-order turbulence results. J. Fluids Mech. **113**, 53–73 (1981)
19. P.E. Skåre, P.Å. Krogstad, A turbulent equilibrium boundary layer near separation. J. Fluids Mech. **272**, 319–348 (1994)
20. E. Touber, N.D. Sanadham, Large-eddy simulation of low-frequency unsteadiness in a turbulent shock-induced separation bubble. Theor. Comput. Fluid Dyn. **23**, 79–107 (2009)
21. J.M. Wallace, H. Eckelmann, R.S. Brodkey, The wall region in turbulent shear flow. J. Fluids Mech. **54**, 39–48 (1972)

Part III
Jets

Chapter 7
Turbulence and Data Analytics in the Twenty-First Century: The Round Free Jet

P. Lavoie, A. Pollard, and H. Sadeghi

7.1 Introduction

This paper concerns turbulent, round free jets. It is also dedicated to Bill George on the occasion of his 70th birthday. Turbulent jets have been a focus for Bill over most of his career; and indeed, they have been the focus of his academic fathers and grandfathers too: von Kármán/Liepmann begat Corrsin, whose Ph.D. thesis at Caltech was on jets; Corrsin begat Lumley whose celebrated work on jets with Panchapakesan continues to be quoted; and Lumley begat Bill. With his many students, Bill has applied theory and experimental expertise to jets, including the use of proper orthogonal decomposition, laser interrogation versus stationary and flying hot wires, 138 hot wire array and not to mention providing significant theoretical insight. Given the academic genealogy, and the time and effort spent on this flow, one would think it should be a 'solved problem'. As will become evident, there remain some significant gaps in our understanding of turbulent jets, which we

---

Data analytics (DA) is the science of examining raw data with the purpose of enabling conclusions to be drawn about that information.

P. Lavoie (✉)
Institute for Aerospace Studies, University of Toronto, Toronto, ON, Canada
e-mail: lavoie@utias.utoronto.ca

A. Pollard
Department of Mechanical and Materials Engineering, Queen's University at Kingston, Kingston, ON, Canada

H. Sadeghi
Institute for Aerospace Studies, University of Toronto, Toronto, ON, Canada

Department of Mechanical and Materials Engineering, Queen's University at Kingston, Kingston, ON, Canada

A. Pollard et al. (eds.), *Whither Turbulence and Big Data in the 21st Century?*,
DOI 10.1007/978-3-319-41217-7_7

believe can only be filled by judicious interplay between theory, experiment and simulation. Moreover, and keeping with the theme of the volume, the opportunities for data analytics and 'big data' will be central to advancing our understanding of this class of free turbulent shear flows, to which some attention will be paid at the end of the paper.

7.2 Some Background on Round Free Jets

Turbulent jets are one canonical flow, turbulent boundary layers and wakes are two others. While the boundary layer provides opportunities for continuous injection of vorticity into the flow from the zero-slip wall condition, the jet and wake differ, of course, in that one is in a continual state of dissipation while the other eventually loses the large-scale remnants of the vorticity generating body that created it. It is because the jet continually dissipates the energy injected into it from its source and because the Navier–Stokes equations (and the kinematic property, conservation of mass) must be obeyed, that questions remain as to the magnitude and relative importance of each of the terms in the N–S equations during the evolution and ultimate dissipation of the jet energy. The source conditions, as we now know from [18], play a defining role in this process, at least as far as self-preservation is concerned.

Ball et al. [4] reviewed the literature associated with a round turbulent jet from 1926 using as points of reference: characterisation, entrainment, effects of initial conditions and similarity, length scales, spectral analysis and coherent structures. In all these areas, George has made significant contributions through both novel theory and experiments. We are reminded by a passage from Lumley [20, p.1] 'After all, there is a saying in German, that a professor is a man who thinks otherwise', which we think characterises Bill! Theoretical understanding begins with the N–S equations; similarity in the sense of George [18, 19] is a cornerstone of some aspects of turbulent flow, follows from the thin shear layer approximations to N–S and the introduction of scaling parameters (without assuming their functional dependence).

The majority of research in the near and intermediate fields (NIF) of turbulent jets may be divided into two research thrusts: the study of the influence of the flow at the vicinity of the jet origin (commonly termed initial conditions); and, the control of large-scale structures that appear mostly in the near field. The effect of the initial (exit) conditions on the downstream development of a jet flow is significant. The exit conditions of a jet are conventionally defined to be the exit Reynolds number, the exit profile characteristics (mean velocity and turbulence intensity), nozzle-exit geometry and, in the case of non-circular jets, its aspect ratio. It is well known that exit conditions affect higher-order statistics in a round jet, e.g. [1, 16–18, 21, 34, 43]. From another perspective, the near-field development of a round jet is strongly characterised by a regular vortex structure (organised transitional flow) that results from the initial instability (see, e.g., [10, 44]). This structure is amplified and subsequently forms large eddies.

These large eddies are unstable and break down to form smaller structures. The control of the fluid in the near exit region that employ different methods to change the initial mean profile (passive methods) or excitation at a particular frequency (active methods) has been the subject of different studies that purport to investigate the effects of large-scale coherent structures on the flow development (e.g. [24, 30–32, 42]). An effective way to change the initial conditions in turbulent jet flows is to use passive objects near the exit. Most recently, Sadeghi and Pollard [36] studied the effect of placing a thin square ring inside the shear layer of a round free jet. It was found that the stable vortex pairing of the shear layer mode completely disappeared when the ring was introduced into the shear layer (see Figs. 15–17 in their paper).

7.3 Theoretical Framework

The theories developed for round jets tend to focus on its far field since the flow may be assumed there to be homogeneous and isotropic, at least on the centreline. The spectral energy equation, or the Karman–Howarth equation, which is an alternate, inherently assumes local isotropy. George and Hussein [22] addressed departure from local isotropy and suggested local axisymmetric turbulence as an alternate form and noted 'The conventional view of turbulence [27, 28] - that the turbulent fluctuations will tend to be more isotropic the smaller their scale - can be reconciled with the findings here if it is argued that it is only the larger scales which contribute to the departures from isotropy'. Thus, in the case of laboratory-scale experiments, and in regions not too far from the jet exit, it is probably intuitively obvious that when attention is removed from the jet centreline, the large-scale motions should increasing become influential. Furthermore, this region is complicated through the interaction between mean shear and small- and large-scale intermittency. An intriguing question is: how do these motions influence the smaller scales and the assumptions about dissipation mechanisms?

Our understanding of the round turbulent jet, equilibrium similarity and scale-by-scale energy budget, in particular, has been recently improved in a series of papers by Burattini et al. [7, 8] and Thiesset et al. [41]. Burattini et al. [7, 8] focussed on the far field (say $x/D \geq 30$), and along the jet centreline. Equilibrium similarity was considered along the jet axis in the context of the axial velocity structure functions and spectra. They noted that since the Taylor Reynolds number, Re_λ, is constant along the jet axis, the longitudinal velocity energy structure functions and the spectra collapse when normalised by either Kolmogorov or integral length scales. Their scale-by-scale energy budget data indicated that for scale separations beyond λ, the scale-by-scale budget did not completely balance. Thus, inhomogeneities that persist in the intermediate field (say $10 \leq x/D \leq 30$) disappear somewhat in the far field but such that the 'relationship between second- and third-order structure functions is an asymptotic result, unlikely to be satisfied in turbulent flows encountered in the laboratory at moderate values of Reynolds number', [7].

It must be remarked that the scale-by-scale energy budget of the jet assumes that pressure plays a negligible role, which is probably appropriate when consideration is only given to the centreline behaviour. Furthermore, as noted by Thiesset et al. [41], the transport equation for the mean kinetic energy dissipation 'imposes a fundamental constraint on the balance between S (e.g. skewness of the longitudinal velocity derivative) and G (e.g. the destruction coefficient of enstrophy) that depends on the type of large-scale forcing and may thus differ from flow to flow'. From a different perspective, George [18] also argued that 'differences in the far field could be due to differences in the large-scale structures generated in the near-field that propagate into the fully developed region', [14].

7.4 Effects of Passive Control on Scale-by-Scale Energy Budget

Recently, Sadeghi et al. [38] focussed attention on the near-to-intermediate field, $10 \leq x/D \leq 25$. They determined that the normalised third-order structure function can be uniquely determined when the normalised second-order structure function, the power-law exponent of $\langle q^2 \rangle$ and the decay rate constants of $\langle u^2 \rangle$ and $\langle v^2 \rangle$ are available and if the axial mean velocity decay is known, a power-law decay of $\langle q^2 \rangle \propto (x - x_0)^m$ results (jet centreline). In the case of the far field, see [40, 41], $m = -2$; however, in the near to intermediate field $m \neq -2$, rather it is slightly larger (of order $m \sim -1.85$), which suggests incomplete similarity over the axial range considered (the second-order structure functions were found to collapse approximately at each streamwise location, but the self-similarity was partial for the third-order structure functions). Moreover, the scale-by-scale budget displayed similar imbalances to those of Burattini et al. [7] but with subtle differences. These differences occur at scale separations, r, above and below λ. When the axial velocity spectra are considered, again there are subtle differences in their collapse for $10 < x/D < 20$ which are directly attributable to what forms are used to determine $\langle \epsilon \rangle$, λ and L_u, where the latter variable is the integral length scale. While it may be argued that perhaps full equilibrium similarity has yet to be achieved in the intermediate field, it does pose an interesting question about what causes the departure. The same question can be posed if the focus is off-axis, say $0 < r/R_{1/2} < 1$. Indeed it is these two questions where attention is focussed here.

The concept of similarity, or self-preservation, which assumes the flow scales with single velocity and length scales, has been an important analytical tool in turbulence research. In this work, an equilibrium similarity analysis, first introduced by George [19], has been developed for the transport equation of the second-order energy structure function of $\langle (\delta q)^2 \rangle$ along the centreline of a round turbulent jet [12], viz.

$$-\langle(\delta u)(\delta q)^2\rangle + 2\nu\frac{\mathrm{d}}{\mathrm{d}r}\langle(\delta q)^2\rangle - \frac{U_c}{r^2}\int_0^r s^2\frac{\partial}{\partial x}\langle(\delta q)^2\rangle \mathrm{d}s$$
$$-2\frac{\partial U_c}{\partial x}\frac{1}{r^2}\int_0^r s^2(\langle(\delta u)^2\rangle - \langle(\delta v)^2\rangle)\mathrm{d}s = \frac{4}{3}\langle\epsilon\rangle r\,. \tag{7.1}$$

where $\langle(\delta q)^2\rangle$ $(= \langle(\delta u)^2\rangle + \langle(\delta v)^2\rangle + \langle(\delta w)^2\rangle)$ is the total turbulent energy structure function. Here, U_c is the local mean streamwise velocity along the centreline and s is a dummy separation variable. The equilibrium similarity forms of the second- and third-order structure functions of u, v and q are required to obtain the similarity form of (7.1). Following the same procedure as [12, 19, 38], we can show that these are given by

$$f(r/\lambda) = \langle(\delta q)^2\rangle/\langle q^2\rangle, \tag{7.2}$$

$$e(r/\lambda) = \langle(\delta u)^2\rangle/\langle u^2\rangle, \tag{7.3}$$

$$h(r/\lambda) = \langle(\delta v)^2\rangle/\langle v^2\rangle \tag{7.4}$$

and

$$g(r/\lambda) = -\langle(\delta u)(\delta q)^2\rangle/(3^{-1/2}Re_\lambda^{-1}\langle q^2\rangle^{3/2}), \tag{7.5}$$

respectively. Here, g is the normalised third-order structure function, and f, e and h are the normalised second-order structure functions. Assuming axisymmetry, $\langle q^2\rangle = \langle u^2\rangle + 2\langle v^2\rangle$. The accuracy of this assumption has been confirmed both on and off the centreline of round jets by Hussein et al. [25]. The general definitions of Taylor microscale and Taylor microscale Reynolds number are

$$\lambda^2 = 5\nu\frac{\langle q^2\rangle}{\langle\epsilon\rangle} \tag{7.6}$$

and

$$Re_\lambda = \frac{\langle q^2\rangle^{1/2}\lambda}{3^{1/2}\,\nu}, \tag{7.7}$$

respectively, [2, 8]. One possible equilibrium similarity solution of (7.1) is a power-law of the form

$$\langle q^2\rangle = A(x - x_0)^m, \tag{7.8}$$

where x_0 is the virtual origin, m is the power-law exponent and A is a constant of proportionality. The same power-law behaviour is also suggested for $\langle u^2\rangle$ and $\langle v^2\rangle$ as

$$\langle u^2\rangle = A_1(x - x_0)^m \tag{7.9}$$

and

$$\langle v^2 \rangle = A_2(x - x_0)^m. \tag{7.10}$$

The virtual origin follows from the variation of the mean velocity along the centreline, viz.

$$U = C/(x - x_0), \tag{7.11}$$

where C is a constant. For the region near the axisymmetric jet centreline, Burattini et al. [7] approximated the kinetic energy budget equation as

$$\langle \epsilon \rangle_{\mathrm{ls},q} = -\frac{U}{2}\frac{\mathrm{d}\langle q^2 \rangle}{\mathrm{d}x} - (\langle u^2 \rangle - \langle v^2 \rangle)\frac{\mathrm{d}U}{\mathrm{d}x}. \tag{7.12}$$

The subscript $_{\mathrm{ls},q}$ indicates that the dissipation is evaluated from the large-scale quantities via the energy budget of $\langle q^2 \rangle$. Assuming axisymmetry, and introducing (7.8)–(7.11) into (7.12), we obtain

$$\langle \epsilon \rangle_{\mathrm{ls},q} = C\left[\frac{-(A_1 + 2A_2)m}{2} + (A_1 - A_2)\right](x - x_0)^{m-2}. \tag{7.13}$$

The similarity form of (7.1) follows after substituting (7.2)–(7.5) and (7.13) into (7.1), viz.

$$\begin{aligned} g + 2\frac{\mathrm{d}f}{\mathrm{d}(r/\lambda)} + 10(c_1 + 2c_2)\frac{\Gamma_1}{(r/\lambda)^2} - 10m(c_1 + 2c_2)\frac{\Gamma_2}{(r/\lambda)^2} \\ +20c_1\frac{\Gamma_3}{(r/\lambda)^2} - 20c_2\frac{\Gamma_4}{(r/\lambda)^2} = \frac{20}{3}(r/\lambda), \end{aligned} \tag{7.14}$$

where Γ_1, Γ_2, Γ_3 and Γ_4 are given

$$\Gamma_1 = \int_0^{r/\lambda}\left(\frac{s}{\lambda}\right)^3 \frac{\mathrm{d}f}{\mathrm{d}(r/\lambda)}\,\mathrm{d}\left(\frac{s}{\lambda}\right), \quad \Gamma_2 = \int_0^{r/\lambda}\left(\frac{s}{\lambda}\right)^2 f\,\mathrm{d}\left(\frac{s}{\lambda}\right),$$

$$\Gamma_3 = \int_0^{r/\lambda}\left(\frac{s}{\lambda}\right)^2 e\,\mathrm{d}\left(\frac{s}{\lambda}\right), \quad \Gamma_4 = \int_0^{r/\lambda}\left(\frac{s}{\lambda}\right)^2 h\,\mathrm{d}\left(\frac{s}{\lambda}\right).$$

Here,

$$c_1 = \frac{A_1}{-Am + 2(A_1 - A_2)} \tag{7.15}$$

and

$$c_2 = \frac{A_2}{-Am + 2(A_1 - A_2)}. \tag{7.16}$$

Dividing by $(20/3)r/\lambda$, (7.14) can be rewritten symbolically as

$$A^* + B^* + D^* + P^* = C^*, \tag{7.17}$$

where A^* is the turbulent advection term (the first term in (7.14)), B^* is the diffusion term (the second term in (7.14)), D^* is the inhomogeneous decay term along the streamwise direction x (the sum of third and forth terms in (7.14)), P^* is the production term (the sum of fifth and sixth terms in (7.14)) and C^* is the balance of all other terms.

The experiments were conducted to verify the accuracy of this new equation and investigate the effect of introducing a passive ring into the jet shear layer on each term of the transport equation. An air jet was generated using a fan mounted on anti-vibration pads. The air then exits a settling chamber via a round duct to the inlet of a smoothly contracting axisymmetric nozzle with exit diameter $D = 73.6$ mm. A wire ring, with square cross-section, of side $h = 1.5$ mm, and outer diameter $D_{\text{wire}} = 71.6$ mm was placed at a stand-off distance (to the ring leading edge) downstream of the jet nozzle exit plane $x/D = 0.03$. The ring was supported by three prongs (1.5 mm square and length 2.2 mm) located at 120° intervals. More details about the current experimental setup can be found in [36, 37]. The experiments were carried out at an exit Reynolds number of $Re_D = 50{,}000$, where Re_D is calculated based on the jet exit mean velocity (U_j=10.65 m/s) and the nozzle-exit diameter. The measurements were performed for $10 \leq x/D \leq 20$, where x is the downstream location, along the jet centreline. Measurements of the turbulence statistics were obtained using a stationary cross-wire probe. The wires were made of 2.5 μm diameter tungsten wire with a 0.5 mm sensing length. The cross-wire was calibrated using a look-up table, with calibration angles within the range ±40, in intervals of 10°. The signals were low-pass filtered at a cut-off frequency f_c, which was selected based on the onset of electronic noise and close to the Kolmogorov frequency, $f_k \equiv U/2\pi\eta$, where $\eta \equiv \nu^{3/4}/\langle\epsilon\rangle^{1/4}$. A few basic parameters at $x/D = 15$ along the jet centreline for both modified and unmodified jets are listed in Table 7.1.

First, in order to illustrate the validity of (7.14), the term $g(r/\lambda)$ is calculated from this Eq. (7.14) using the corresponding power-law exponents m and the decay rates A_1 and A_2 at $x/D = 15$ (identified as g_c) and compared with the measured profile of $g(r/\lambda)$ (denoted by g_m) in Fig. 7.1a for both modified and unmodified jets. A relatively good agreement (within ±12 %, similar to results from grid turbulence

Table 7.1 A few basic parameters at $x/D = 15$ along the jet centreline for both modified and unmodified jets

Unmodified jet (no ring)					Modified jet (with ring)				
x/D	$\langle\epsilon\rangle_{\text{hom}}$ (m^2 s^{-3})	λ_{hom} (mm)	Re_λ	η (mm)	x/D	$\langle\epsilon\rangle_{\text{hom}}$ (m^2 s^{-3})	λ_{hom} (mm)	Re_λ	η (mm)
15	9.30	4.35	241	0.15	15	13.7	4.06	254	0.13

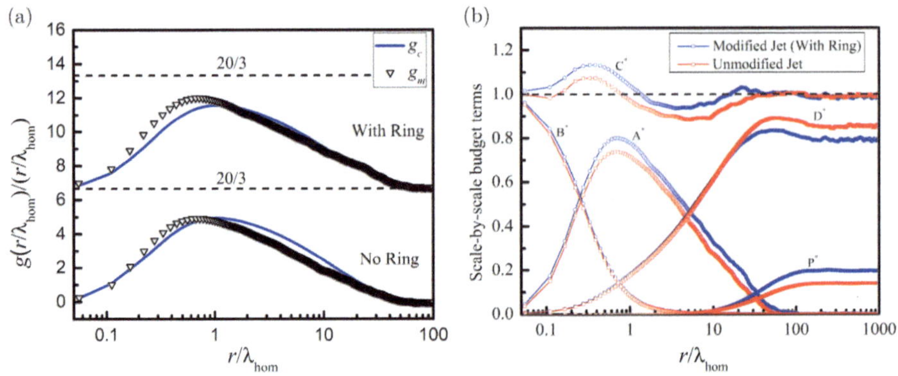

Fig. 7.1 (**a**) Comparison between measured (*triangles*) and calculated (*solid line*) distributions of g divided by r/λ_{hom}(at $x/D = 15$); note for the 'with the ring' case have been shifted successively (offset 20/3) with respect to the lower one. *Dashed line* is 20/3. (**b**) Scale-by-scale budget terms with and without ring, see Eq. (7.17)

experiments) is found between g_m and g_c for both cases. Note that the normalised third-order structure functions are divided using r/λ so that their maximum peaks can be compared with the onset of the inertial range. It can be observed that the asymptotic value of 20/3, which represents the onset of the inertial range for a high Reynolds number, is significantly higher than the maximum measured and calculated g. Sadeghi et al. [37] showed that a proper inertial range is unlikely to be established along the jet axis unless a very high Reynolds number of $Re_\lambda = 10^4$ can be reached. Equation (7.17), which is a normalised form of (7.14), is used as a scale-by-scale budget to quantify the effect of initial conditions on all scales of a round free jet flow on its centreline. The scale-by-scale budget terms, measured at $x/D = 15$ on the centreline and $Re_D = 50{,}000$, are given in Fig. 7.1b. This figure demonstrates that (7.17) is adequately satisfied by the experimental data ($0.88 \leq C^* \leq 1.12$) for nearly all scales of the jet flows on the centreline. For both jets, at small separations, the diffusion term B^* dominates, while at large separations, the decay term D^* and the production term B^* are the dominant terms. The advection term A^* goes to zero at both very small and large separations, while its maximum is located at nearly the same location for both modified and unmodified jets ($r \simeq \lambda$). Direct comparison of each term for the modified and unmodified jet indicates that B^* is nearly unaffected by the use of the ring. The impact of the initial condition can be observed mainly on the large scales for P^* and D^*. The magnitude of A^* is higher for the modified jet, which can be related to the increase in turbulent Reynolds number.

7.5 Limitations on Current Experimental Approaches

Characterisation of the small-scale motions in turbulent flows has been one of the challenging problems of turbulence research, especially in turbulent shear flows such as a jet. One of the primary concerns has been the determination of the mean-square gradients of the fluctuating velocity field to estimate the rate of dissipation of turbulence kinetic energy, viz.

$$\langle \epsilon \rangle = \frac{1}{2}\nu \left\langle \left(\frac{\partial u_i}{\partial x_j} + \frac{\partial u_j}{\partial x_i} \right)^2 \right\rangle . \tag{7.18}$$

Twelve separate derivative correlations must be determined before the dissipation can be truly calculated. Since it is difficult to calculate all velocity derivatives accurately from experimental measurements, it has been common to use alternative assumptions to estimate the dissipation rate. The most well-known assumption is local isotropy. The dissipation for isotropic turbulence reduces to

$$\langle \epsilon \rangle_{\text{iso}} = 15\nu \left\langle \left(\frac{\partial u}{\partial x} \right)^2 \right\rangle . \tag{7.19}$$

This is the simplest definition for the dissipation. Using homogeneity and incompressibility, another estimate for the dissipation can be obtained as

$$\langle \epsilon \rangle_{\text{hom}} = \nu \left\langle \left(\frac{\partial u_i}{\partial x_j} \right)^2 \right\rangle . \tag{7.20}$$

By assuming axisymmetry, a relatively crude approximation for this equation can be obtained as (e.g. [7, 8, 11])

$$\langle \epsilon \rangle_{\text{hom}} = 3\nu \left[\left\langle \left(\frac{\partial u}{\partial x} \right)^2 \right\rangle + 2 \left\langle \left(\frac{\partial v}{\partial x} \right)^2 \right\rangle \right] . \tag{7.21}$$

However, the assumptions of local isotropy and homogeneity for small scales are suspect. While it is difficult to evaluate ‘true’ dissipation in a turbulent jet experimentally, numerical studies are the best tools to verify this problem. Until recently most numerical studies have used large eddy simulations (LES) or other models to approximate the turbulence in jet flows (e.g. [5]). Perhaps we can expect that direct numerical simulations (DNS) (where no modeling is used) can give the best answer to this problem in the near future.

7.6 Big Data, Data Analytics and Turbulence

In 1989 the community was learning how to do DNS, which was just emerging to what has become the gold standard simulation method for turbulence. Petersen et al. [33] projected that LES over an aircraft would require petaword memory and exaflop computer speeds; and, given the danger of projecting too far into the future, they restricted their projections for the development of increased capability to a decade beyond ca. 1990. It is prudent to take a similar cautious approach today, although there is evidence that we are on track to reach these memory and flops within about 10 years, i.e., 2025, and perhaps sooner, see [3]. Clearly, zetaflops (10^{21}) as well as novel algorithm development will be necessary to perform DNS calculations of high Re flows that include significant additional physico-chemical processes. For example, exploring fundamental questions about structure functions in high Schmidt number mass transfer, see, for example, Borgas et al. [6] where the resolved scales become proportional to the Batchelor scale $\eta_B = \eta_K / \sqrt{(Sc)}$ or premixed combustion phenomena using complex molecule fuels such as biofuels, see the Sandia Combustion Research Facility and its Center for Exascale Simulation of Combustion in Turbulence (Exact) and Shalf et al. [39].

In many fields of enquiry, data are becoming the currency for discovery. For the turbulence community, it is useful to consider the efforts taken by other groups as they generate and are learning how to distribute massive amounts of data. As in 2015, the square kilometer array, initial discussions for its formation began in 1993, is an 11 country consortium that will use radio astronomy to explore the non-visible spectrum of the universe. It will eventually consist of over 350 dishes located in both South Africa and Australia with its main headquarters at the University of Manchester Jodrell Bank site (see http://www.skatelescope.org). It will eventually generate data at a rate of 10^{15} *bytes per second*, which will require computers with 100 petaflop capability for data processing interconnected with 80 Tb/s optical bandwidth. It should be fully operational by 2030. Furthermore, and more generally, it is estimated by CISCO that annual internet traffic will blossom to about 2 zettabytes 10^{21} per year by 2019, [9], which will require sustained terabit bandwidth between 'important' links, such as the recently announced NSF Pacific Research Platform that will connect most California universities at 100 gigabits per second. These examples are mentioned to draw attention to an issue that must be faced by the 'turbulence' community: coordination of data, archiving and providing a stewarded window to effectively interface with these databases via multi-investigator and multi-institutional collaborations: it will not happen overnight, but initial efforts have begun, see, for example, the Registry of Research Data Repositories, http://www.re3data.org that is international in scope.

Compare the SKA example to current sizes of DNS databases for the highest Re number channel flow performed to date. For a 'simple' channel flow at $Re_\tau = 5400$, using 250×10^9 control volumes, and 10^8 time steps produced a datafile of order 10^{20} [29]. When placed in the context of the research life cycle of data (see http://www.data-archive.ac.uk/create-manage/life-cycle) funding agencies are increasing their expectation that data will be archived and made widely available for use by others; and this means the raw data, not their post-processed results. Thus 'data analytics' and their life cycle will become the norm beyond 2015. Important, too, is 'who pays' for this? At the moment, and by example, the John Hopkins group, the Institute for Data Intensive Engineering and Science (http://idies.jhu.edu/), relies on both internal and external funding sources to pay for computing and storage hardware, which must be continuously refreshed and its size increased as the number of cases (canonical flows in the first instance) expands. It is not clear if our community will be eventually well served and appropriately by, for example, http://www.re3data.org referred to above. As Sillero and Jimenez argue in this volume, funding agencies are increasingly emphasising the need to store data generated through publicly funded research. It may be that without community input, these agencies could dictate a format that is not suitable for, say, our community, which would then require additional efforts and structures to be put in place further eroding research time itself.

Indeed, the unseen and often unappreciated contributions are the efforts expended by academics and their teams whose sole purpose is to enable access, but which does not enhance their reputation. Thus a 'reputational economy' must emerge where such activities increase a person's reputation because they enabled others to succeed. For example, in the turbulence community, the databases maintained by Meneveau's and Jimenez's groups, for example, have provided data that have enabled other research groups around the world to probe, manipulate, ask new questions of those data and have to date generated over 50 peer-reviewed publications. Therefore, 'big data' also means mining data for what was otherwise hidden, since at the time of their generation, the originators were asking different questions. The Chair of the recently formed USA Dept. of Energy 'Exascale Computing Initiative', Dan Reed, noted: 'Keep in mind in that data intensive and computationally intensive workflows both matter and in fact most of the time they are the same thing. They are intertwined pretty deeply [and] draw on the same ecosystems of hardware and software. Both matter. That drives as a corollary some focus on a new generation of analysis tools and libraries that will be needed to interpret that data'. (see http://www.hpcwire.com/2015/07/28/doe-exascale-plan-gets-support-with-caveats/).[1]

In the context of this volume and this paper in particular, and as a challenge for future investigations, a DNS of a round jet at $Re = 10{,}000$ and 30,000, which straddles the mixing transition Reynolds number of $\geq 20{,}000$, (see [13, 16]); Fellouah and Pollard [15] will generate a dataset of order 10^{20} velocity and pressure values. This assumes a grid structure consisting $10^7 \times 10^3 \times 10^3$ in the axial, radial and azimuthal directions, based on a mean Kolmogorov length scale of order 10^{-4} m throughout the domain and approximately 200 flow through times (after reaching

[1]At the time of writing (July/August 2015) the USA has just announced a National Strategic Computing Initiative to broadly invest in and deploy exascale computing, that would include turbulence simulations at a variety of scales, i.e. combustion simulations, aircraft design and atmospheric/climate change.

statistically steady state) using a longitudinal integral scale of 50 mm, see [38]. This grid size aligns with the value obtained from $N_{xyz} \approx 0.1Re_{\lambda}^{9/2}$, see [35], assuming Re_{λ} remains constant, which has been found to be correct from numerous studies, see, for example, [41]. A less brute force approach to constructing the integration domain will lower this estimate, but the data so generated is far too large to be offered to the community directly at this point in our evolution in communication bandwidths. Thus, options for accessing those data must be found, perhaps utilising the example of Meneveau's group at Johns Hopkins, see [23, 26].

It is anticipated the Johns Hopkins model may indeed become more the norm as the continual huge investments required to build, run (both the infrastructure and technical support) and maintain a standalone high performance computing centre that will adequately serve all simulation communities become increasingly competitive. In the case of the DNS of a jet, current theoretical understanding is focussed on the far field, where isotropic assumptions work rather well. But, moving closer to the nozzle, and off-axis where large-scale inhomogeneities exist and may even be tailored through either passive or active control, will mean dispensing with Kolmogorov's ideas, and even perhaps the notion of axisymmetric turbulence [22] to estimate dissipation. At what location does Taylor's frozen eddy hypothesis break down? What secrets will emerge as simulations go beyond the mixing transition? Answers to these and other yet unimagined questions will be obtained only when both experimental and simulation databases (and all conditions relevant to their formation) become generally available to and maybe probed either locally or remotely by the whole community.

Acknowledgements This work forms part of HS Ph.D. thesis. The Big Data sections draw upon one of us (AP) who has long term interest in establishing HPC (see, e.g., HPCVL.org, which he was the original P.I.) and sharing of data from both simulations and experiments. Funding from NSERC, Queen's University and the University of Toronto is gratefully acknowledged.

References

1. A.A. Abdel-Rahman, W. Chakroun, S.F. Al-Fahed, LDA measurements in the turbulent round jet. Mech. Res. Commun. **24**, 277 (1997)
2. R.A. Antonia, R.J. Smalley, T. Zhou, F. Anselmet, L. Danaila, Similarity of energy structure functions in decaying homogeneous isotropic turbulence. J. Fluid Mech. **487**, 245–269 (2003)
3. S. Ashby, P. Beckman, J. Chen, P. Colella, B. Collins, D. Crawford, J. Dongarra, D. Kothe, R. Lusk, P. Messina, T. Mezzacappa, P. Moin, M. Norman, R. Rosner, V. Sarkar, A. Siegel, F. Streitz, A. White, M. Wright, The opportunities and challenges of exascale computing. Technical report, US Dept. of Energy (2010)
4. C.G. Ball, H. Fellouah, A. Pollard, The flow field of a turbulent round jet, progress in aerospace sciences. Prog. Aerosp. Sci. **50**, 1–26 (2012)
5. C. Bogey, C. Bailly, Turbulence and energy budget in a self-preserving round jet: direct evaluation using large eddy simulation. J. Fluid Mech. **627**, 29–160 (2009)
6. M.S. Borgas, B.L. Sawforda, S. Xu, D.A. Donzis, P.K. Yeung, High Schmidt number scalars in turbulence: structure functions and Lagrangian theory. Phys. Fluids **16**, 3888–3899 (2004)

7. P. Burattini, R.A. Antonia, L. Danaila, Scale-by-scale energy budget on the axis of a turbulent round jet. J. Turbul. **6**, 1–11 (2005)
8. P. Burattini, R.A. Antonia, L. Danaila, Similarity in the far field of a turbulent round jet. Phys. Fluids **17**, 025101 (2005)
9. Cisco, The zettabyte era: trends and analysis. White Paper (2015), http://www.cisco.com/c/en/us/solutions/collateral/service-provider/visual-networking-index-vni/VNI_Hyperconnectivity_WP.pdf
10. S.C. Crow, F.H. Champagne, Orderly structure in jet turbulence. J. Fluid Mech. **48**, 547–591 (1971)
11. L. Danaila, F. Anselmet, R.A. Antonia, An overview of the effect of large-scale inhomogeneities on small-scale turbulence. Phys. Fluids **14**, 2475 (2002)
12. L. Danaila, R.A. Antonia, P. Burattini, Progress in studying small-scale turbulence using 'exact' two-point equations. New J. Phys. **6**, 1–28 (2004)
13. P.E. Dimotakis, The mixing transition in turbulent flows. J. Fluid Mech. **409**, 69–98 (2000)
14. D. Ewing, B. Frohnapefl, W.K. George, J.M. Pedersen, J. Westerweel, Two-point similarity in the round jet. J. Fluid Mech. **577**, 309–330 (2007)
15. H. Fellouah, A. Pollard, The velocity spectra and turbulence length scale distributions in the near to intermediate regions of a round free turbulent jet. Phys. Fluids **21**, 115101–115101 (2009)
16. H. Fellouah, C.G. Ball, A. Pollard, Reynolds number effects within the development region of a turbulent round free jet. Int. J. Heat Mass Transf. **52**, 3943–3954 (2009)
17. E. Ferdman, M.V. Otugen, S. Kim, Effect of initial velocity profile on the development of the round jet. J. Propul. Power **4**, 676–686 (2000)
18. W.K. George, The self-preservation of turbulent flows and its relation to initial conditions and coherent structures, in *Advances in Turbulence*, ed. by W.K. George, R. Arndt (Springer, Berlin, 1989)
19. W.K. George, The decay of homogeneous isotropic turbulence. Phys. Fluids **4**(7), 1492–1509 (1992)
20. W.K. George, R. Arndt, *Advances in Turbulence* (Springer, New York, 1989)
21. W.K. George, L. Davidson, Role of initial conditions in establishing asymptotic flow behavior. AIAA J. **42**, 438–446 (2004)
22. W.K. George, H.J. Hussein, Locally axisymmetric turbulence. J. Fluid Mech. **233**, 1–23 (1991)
23. J. Graham, K. Kanov, X.I.A Yang, M. Lee, N. Malaya, C.C. Lalescu, R. Burns, G. Eyink, A. Szalay, R.D. Moser, C. Meneveau, A Web services accessible database of turbulent channel flow and its use for testing a new integral wall model for LES. J. turbul. **17**(2), 179–213 (2016)
24. A. Hussain, K. Zaman, The preferred mode of the axisymmetric jet. J. Fluid Mech. **110**, 39–71 (1981)
25. H.J. Hussein, S. Capp, W.K. George, Velocity measurements in a high-Reynolds number, momentum-conserving, axisymmetric, turbulent jet. J. Fluid Mech. **258**, 31–75 (1994)
26. K. Kanov, R. Burns, C. Lalescu, G. Eyink, The Johns Hopkins turbulence databases, an open simulation laboratory for turbulence research. Comput. Sci. Eng. **17**(10), 10–17 (2015)
27. A.N. Kolmogorov, The local structure of turbulence in incompressible viscous fluids for very large Reynolds numbers. Dokl. Akad. Nauk SSSR **30**(4), 301–305 (1941)
28. A.N. Kolmogorov, Dissipation of energy in the locally isotropic turbulence. Dokl. Akad. Nauk SSSR **32**(1), 16–18 (1941)
29. M. Lee, R. Ulerich, N. Malaya, R.D. Moser, Experiences from leadership computing in simulations of turbulent fluid flows. Comput. Sci. Eng. **16**, 24–31 (2014)
30. S. McIlwain, T. Holme, S. Waterman, A. Pollard, Effects of one, two and quadruple tabs on the mixing and entrainment in the near field of round jets, in *IUTAM Symposium on Turbulent Mixing and Combustion* (Kluwer Academic Press, Dordrecht, NL, 2002)
31. R. Parker, S. Rajagopalan, R.A. Antonia, Control of an axisymmetric jet using a passive ring. Exp. Thermal Fluid Sci. **27**, 545–552 (2003)
32. R.A. Petersen, M.M. Samet, On the preferred mode of jet instability. J. Fluid Mech. **194**, 153–173 (1988)

33. V.L. Petersen, J. Kim, T.L. Holst, G.S. Deiwert, D.M. Cooper, A.B. Watson, F.R. Bailey, Supercomputing requirements for selected disciplines important to aerospace. IEEE **77**, 1038–1055 (1989)
34. W.R. Quinn, Upstream nozzle shaping effects on near field flow in round turbulent free jets. Eur. J. Mech. B Fluids **25**, 279–301 (2006)
35. W.C. Reynolds, The potential and limitations of direct and large eddy simulation, in *Whither Turbulence? Turbulence at the Crossroads Proceedings*. Technical report (Springer, New York, 1990)
36. H. Sadeghi, A. Pollard, Effects of passive control rings positioned in the shear layer and potential core of a turbulent round jet. Phys. Fluids **24**, 115103 (2012)
37. H. Sadeghi, P. Lavoie, A. Pollard, The effect of Reynolds number on the scaling range along the centreline of a round turbulent jet. J. Turbul. **15**, 335–349 (2014)
38. H. Sadeghi, P. Lavoie, A. Pollard, Equilibrium similarity solution of the turbulent transport equation along the centreline of a round jet. J. Fluid Mech. **772**, 740–755 (2015)
39. J. Shalf, D. Quinlan, C.J. Rethinking, Hardware-software codesign for exascale systems. IEEE Comput. **44**, 22–30 (2011)
40. H. Tennekes, J.L. Lumley, *A First Course in Turbulence* (The MIT Press, Cambridge, 1972)
41. F. Thiesset, R.A. Antonia, L. Djenidi, Consequences of self-preservation on the axis of a turbulent round jet. J. Fluid Mech. **748**, R2 (2015)
42. C. Tong, Z. Warhaft, Turbulence suppression in a jet by means of a fine ring. Phys. Fluids **6**, 328–333 (1994)
43. G. Xu, R.A. Antonia, Effect of different initial conditions on a turbulent round free jet. Exp. Fluids **33**, 677–683 (2002)
44. A.J. Yule, Large scale structure in the mixing layer of a round jet. J. Fluid Mech. **89**, 413–432 (1978)

Chapter 8
The Sound-Field Produced by Clustered Rockets During Start-Up

C.E. Tinney, A. Canchero, R. Rojo, G. Mack, N.E. Murray, and J.H. Ruf

8.1 Introduction

Space vehicles are subject to violent unsteady loads caused by the intense sound pressure levels and transient side-loads produced during rocket engine ignition, as well as unsteady aerodynamic loads during flight. These loads are known to cause electronic and mechanical component failures, to increase the likelihood of fatigue failure on lightweight exterior structures such as aerodynamic fins and thermal protection systems, as well as produce adverse environmental conditions for vehicle occupants and payload. At launch, vibro-acoustic loads form during the ignition of both the solid- and liquid-propellant engines, but are more prolonged in the latter due to the longer time (up to 3–5 s) required for the turbine to produce the hydrodynamic head needed for the nozzle and combustion chamber to flow full. During flight, however, aerodynamic loads caused by unsteady compressions and expansions of the surrounding gas produce shock-wave boundary layer interactions (SWBLI), acoustic/pressure resonance, and friction heating on the vehicle surface. In order to withstand these fluid/thermal and acoustic loads, the robustness of

C.E. Tinney (✉)
Applied Research Laboratories, The University of Texas at Austin, 10000 Burnet Rd., Austin, TX 78713, USA
e-mail: cetinney@utexas.edu

A. Canchero • R. Rojo
Center for Aeromechanics Research, The University of Texas at Austin, Austin, TX, USA

G. Mack
Institute for Flight Propulsion, The Technische Universität München, München, Germany

N.E. Murray
National Center for Physical Acoustics, The University of Mississippi, University, MS, USA

J.H. Ruf
Fluid Dynamics Branch, NASA Marshall Space Flight Center, Huntsville, AL, USA

A. Pollard et al. (eds.), *Whither Turbulence and Big Data in the 21st Century?*,
DOI 10.1007/978-3-319-41217-7_8

the vehicles supporting structure is increased all the while reducing the allowable weight of its payload. This weight penalty increases both the number of launches required to complete a mission and the overall operating cost of the launch.

Previous efforts at The University of Texas at Austin, in collaboration with NASA Marshall Space Flight Center, have focused on single nozzle configurations. This encompassed measurements of the effect of shock foot asymmetry on lateral accelerations [3], an analysis of the effect of free-shock separated (FSS) or restricted-shock separated (RSS) flow on the far-field sound [6], the influence of startup transients on SWBLI relative to fixed flow analysis [2], methods for predicting full-scale launch loads using laboratory scale measurements [12], as well as the effect of pressure gradient on SWBLI [4].

The current work focuses on developing an understanding of the flow mechanisms responsible for generating the vibro-acoustic loads that act on the base of a space launch vehicle during engine ignition. The effort is a natural extension to our understanding of this complex problem where multiple nozzles (in this case, two) are operated in close vicinity to one another and at identical pressure ratios. This includes additional factors related to vibro-acoustic loads not accounted for in the single nozzle configuration. Numerous efforts to understand these flow states, their lateral and thermal loads, as well as the acoustic waveforms that they generate, has been a topic of considerable interest as is evidenced by the work of Nave and Coffey [17], McInerny [13], Frey and Hagemann [8], Nguyen et al. [18], Verma and Haidn [22], Ruf et al. [21], Baars and Tinney [2], and Donald et al. [6] and the references therein.

8.2 Overview of Data Collection and Modeling Methods

8.2.1 Facility and Instrumentation

All measurements were acquired in the anechoic chamber and high speed flow facility at The University of Texas at Austin; see Baars and Tinney [2] and Donald et al. [6] for a description of this facility. An illustration of the full setup is provided in Fig. 8.2 and comprised a cluster of rocket nozzles mounted to the jet rig located in the center of an open jet wind tunnel and anechoic chamber. For the current configuration, a two-rocket cluster was utilized and comprised only the upper and lower nozzles identified to the right of the image in Fig. 8.1. Unheated air (T_o = 293 K) is used as the working fluid with atmospheric temperature (T_∞) and relative humidity (%RH) in the anechoic chamber being 290 K and 70 %, respectively. Two geometrically identical TOP contour nozzles comprising an exit-to-throat area ratio of $A_j/A^* = 30.29$ were used as the test hardware. This particular contour was designed to ensure both FSS and RSS flow states during start-up as well as an end-effects-regime [17] at a nozzle pressure ratio (NPR; ratio of plenum to atmospheric pressure) between 35 and 40 [21]. The principal focus of this discussion will be on

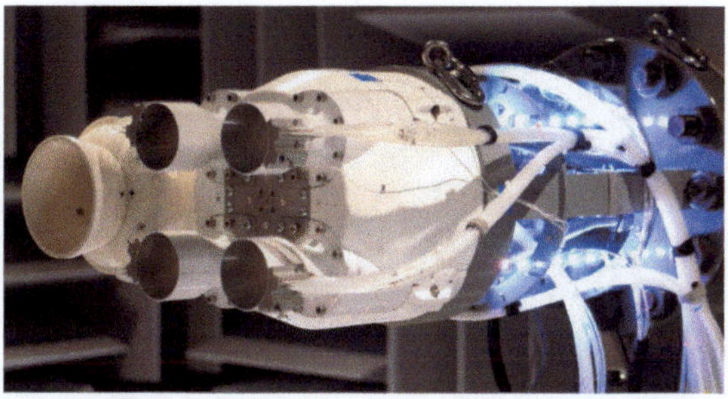

Fig. 8.1 Instrumented rocket cluster installed on the jet rig at UT Austin. Center to center spacing between the upper and lower nozzles is $1.9273D_j$. Nozzles comprise a throat radius of $r^* = 6.35$ mm and an exit diameter of $D_j = 69.89$ mm. The design Mach number of 5.24 is achieved at NPR $\sim$ 700

transient start-ups of the rocket cluster. Albeit, data from two steady runs at NPR 25 and 37 will be presented in an effort to provide a statistical description of the sound-field using a new diagnostic tool called acoustic imaging. For this particular nozzle contour, FSS$\rightarrow$RSS transition occurs at NPR 24, so the steady runs are for RSS flow only.

The experimental component of the study encompassed an eighth-inch microphone, two sixteen-channel static wall pressure sensing arrays, and a shadowgraphy system. The first of these, the eighth-inch microphone, comprised a 3.2 mm (eighth-inch) diameter pressure field microphone (G.R.A.S. type 46DD capsule with G.R.A.S. type 26CB preamplifier) mounted flush with the surface of the base plate located behind the nozzle cluster. Static wall pressures inside each nozzle were captured using two Scanivalve DSA3218 gas pressure scanners in order to reveal the state (FSS, RSS) and location of the shock foot pattern for a given NPR. This utilized a total of 20 ports uniformly distributed in the axial direction along the upper nozzle with constant axial spacing of $\Delta x/r^* = 0.4$ spanning between $x/r^* = 4.5$ and 12.1. The nozzle exit plane is located at $x/r^* = 12.5$ from the nozzle throat. As for the third instrument, this comprised a retroreflective shadowgraphy system that was oriented to visualize the plume and acoustic field of the exhausting flow. The setup was similar to the one described by Hargather and Settles [9] and encompassed a retroreflective screen, a digital camera (PCO.Edge sCMOS), a xenon arc lamp point source (flash synchronization), and a 45° rod mirror. An illustration of the shadowgraphy system, relative to the rocket cluster and chamber, is shown in Fig. 8.2. All instruments were synchronized using a standard 5 V trigger pulse.

8.2.2 *RANS Model*

As the shadowgraphy system provides visuals of the flow outside of the nozzle, a steady Reynolds-averaged Navier–Stokes (RANS) model of the same nozzle

Fig. 8.2 Optical arrangement of the shadowgraphy system relative to the nozzle test stand in the anechoic chamber and high speed flow test facility at The University of Texas at Austin

contour and gas properties (unheated air) was computed to visualize the internal flow and shock foot pattern. This RANS model was constructed using the Loci-Chem code [11] with an axisymmetric, time-accurate analysis, and the Menter baseline turbulence model [14]. This turbulence model encompasses a blending of the $\kappa - \epsilon$ and $\kappa - \omega$ models as well as a compressibility correction. The results are displayed here as the magnitude of the density gradient $|\nabla \rho|$. Wall pressure profiles from the simulation have been shown to compare well with the measurements performed in the UT-Austin facility [4, 6].

8.3 Non-stationary Start-Up

Non-stationary start-up of these nozzles is first considered as it encompasses all of the significant flow events of interest. RANS model snapshots and shadowgraphy images are provided in Figs. 8.3, 8.4 and 8.5, while microphone data is shown in Fig. 8.6 followed by static wall pressure contours in Fig. 8.7. Where microphone data is concerned, a time–frequency analysis is performed using the Morlet wavelet transform in order to view the spectral content as function of time (or NPR). Details concerning time–frequency analysis are provided elsewhere [1, 7] with applications to jet and rocket noise studies being found in the work of Baars and Tinney [2] and Donald et al. [6].

Beginning with the microphone data, the raw acoustic waveform is shown at the bottom of Fig. 8.6 along with a window averaged overall sound pressure level (OASPL; dB, re $20\mu\text{Pa}\sqrt{\text{Hz}}$). The OASPL gives a sense of the average pressure

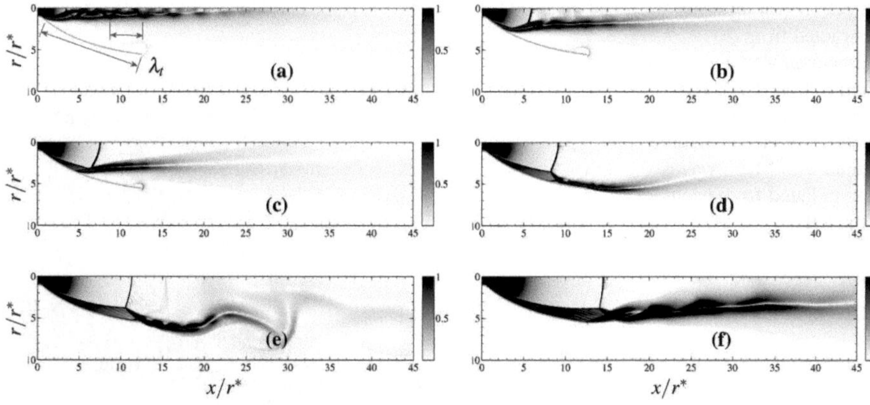

Fig. 8.3 RANS model of the TOP nozzle during start-up (plotted here using the magnitude of the density gradient ($|\nabla\rho|$). (**a**) Free-shock separated flow with supersonic core at NPR 6. (**b**) Free-shock separated flow with subsonic core at NPR 15. (**c**) Free-shock separated flow prior to FSS→RSS transition at NPR 20. (**d**) Restricted shock separated flow after FSS→RSS transition at NPR 25. (**e**) End-effects-regime at NPR 37. (**f**) Full flowing nozzle at NPR 70

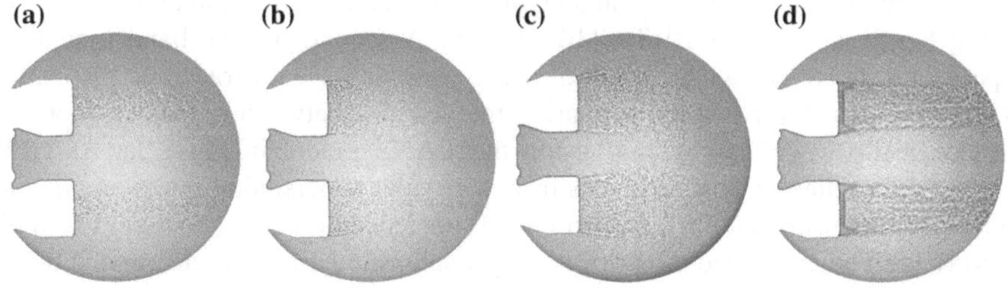

Fig. 8.4 Shadowgraphy images of the instantaneous rocket plumes during start-up at (**a**) NPR 15, (**b**) NPR 23, (**c**) NPR 37 and (**d**) NPR 70

levels that occur at the base of this two-nozzle configuration. The static wall pressure illustrated in Fig. 8.7 demonstrates well-known features corresponding to various operating states of a TOP nozzle. That is, during start-up, FSS→RSS transition is shown to occur around NPR 23 for the two-nozzle configuration, followed by an end-effects-regime at NPR 37. A more thorough explanation of these wall pressure signatures is provided by Baars and Tinney [2] and Mack [12] and the references therein.

At low nozzle pressure ratios (NPR 6 in Fig. 8.3a), a throat shock, Mach disk, and Mach stem immediately form due to the non-ideal contour of the nozzle. The Mach disk is confirmed by the pressure rise across the normal shock as well as the merging of shocks cells along the centerline of the nozzle downstream [8]. A separation shock also forms due to the overexpanded state of the flow and resides near the nozzle throat at this NPR. Shock cells residing on the nozzle centerline comprise spacings that vary from $\lambda_b/r^* = 4.56$ upstream to $\lambda_b/r^* = 19.6$

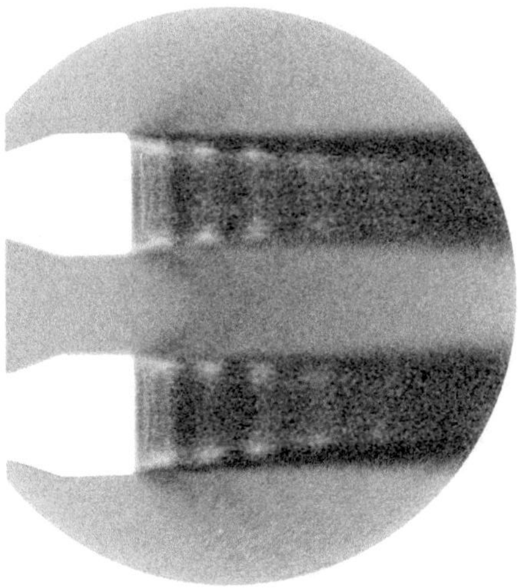

Fig. 8.5 Mean intensity constructed from 20 shadowgraphy images in the vicinity of NPR 70. The image reveals shock cell structures in the supersonic annulus with Mach waves being radiated from each shock cell

downstream, which eventually fade and give way to a shock free supersonic plume. Wavelet spectra at this NPR reveal a broadband, high energy hump around 6.5 kHz in Fig. 8.6 as a result of BBSN [10, 19, 20]. A second narrow band tone with a peak at 1.2 kHz is seen in Fig. 8.6, and is associated with transonic resonance (TR) [23]. This is verified by measuring the distance between the shock foot and the nozzle lip ($\lambda_t/r^* = 12.54$) using the RANS model in Fig. 8.3a. As NPR increases, the internal flow advances from a circular supersonic flow to an annular supersonic flow surrounding an open-ended subsonic core, as is shown in Fig. 8.3b at NPR 15. The combined effect of having a subsonic core flow surrounded by an annular supersonic flow results in a relatively higher acoustic impedance, which is most noticeable at shallow angles where turbulence mixing noise is greatest [6]. Spherically shaped wavefronts emanating from the shear layer regions are captured by the shadowgraphy image in Fig. 8.4a. Further increases in NPR results in a larger diameter supersonic annulus, as shown in Fig. 8.3c. The thinness and orientation of the supersonic annulus computed by the RANS at NPR 25 in Fig. 8.3d is complementary in shape to the shadowgraphy images at similar NPRs in Fig. 8.4b.

A second series of high amplitude broadband shock tones are encountered and are shown to begin around NPR 35 in Fig. 8.6. Closer inspection reveals how the first series of strong BBSN tones (at low NPRs) are connected to this second series (at high NPRs). This is not surprising as the shock pattern persists throughout the duration of the start-up. By following the BBSN tone in Fig. 8.6, along with the RANS model, a richer understanding of the shock cell behavior during start-up can be gleaned. The end-effects-regime is eventually encountered at higher NPRs and is caused by the intermittent breathing of the last trapped annular separation bubble with the ambient [17, 18]. This is identified in the static wall pressure (Fig. 8.7) by the intersection of the last dark contour region with the nozzle lip and occurs at NPR 37 for this nozzle contour. An instantaneous snapshot from the time-accurate RANS solution is illustrated in Fig. 8.3e.

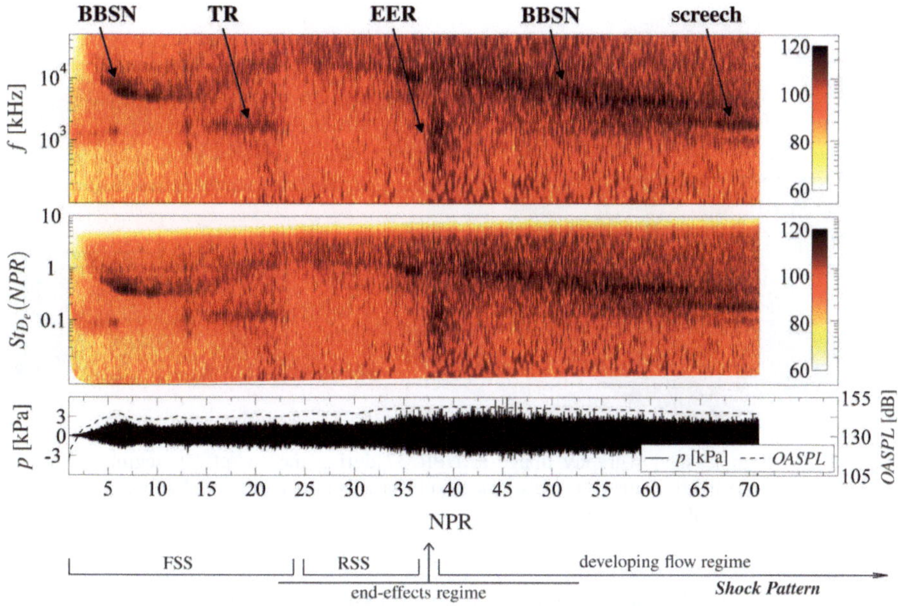

Fig. 8.6 Morlet wavelet power spectra of acoustic pressure measured behind the nozzles during start-up expressed as $10\ \log_{10}(E(f,t)/\text{ref})$, [dB, re $20\ \mu\text{Pa}\sqrt{\text{Hz}}$] identify broadband shock associated noise (BBSN), transonic resonance tones (TR), the end-effects-regime (EER) and screech. Upper and lower contours are plotted with frequency [Hz] and Strouhal number on the ordinate axis, respectively. The raw pressure signal is displayed at the bottom [kPa] with a moving window averaged OASPL

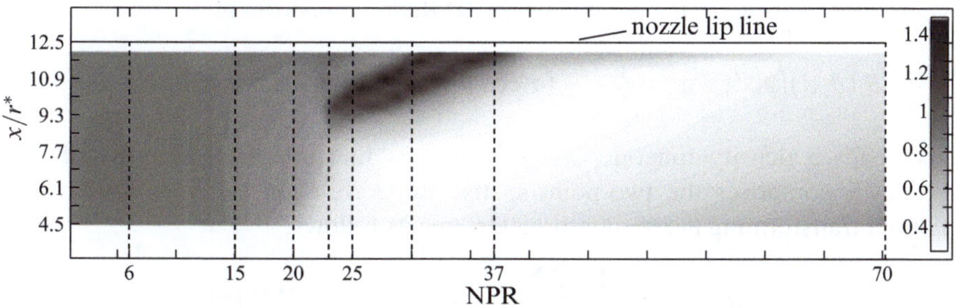

Fig. 8.7 Static wall pressure (P_w/P_∞) during start-up of the upper nozzle

8.4 Stationary Nozzle Operations

We focus our attention now on the shadowgraphy images acquired during stationary operations of the nozzles at NPR 25 and 37. The second of these corresponds to the end-effects-regime where intermittent breathing of the last separation bubble causes the exiting flow to expand and contract violently. Glancing at Fig. 8.8a, two

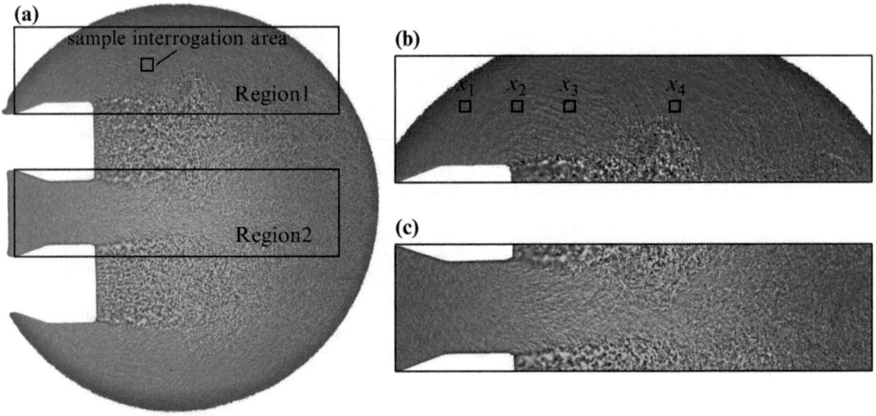

Fig. 8.8 (**a**) Sample shadowgraphy image at NPR 37 during the end-effects-regime with two regions of interest being defined for subsequent analysis. Detailed view of (**b**) Region 1 and (**c**) Region 2 from a sample raw image

regions are outlined which are then scrutinized using Radon transforms to quantify the direction of the wavefronts passing through these regions. By combining Radon transforms with the shadowgraphy images, insightful information concerning the speed and location of the sound sources responsible for generating these acoustic wavefronts are revealed. The use of Radon transforms to analyze shadowgraphy images of acoustic wavefronts was first performed by Murray [15] and Murray and Lyons [16] while studying Mach waves emitted from a heated supersonic jet. The general expression for the parallel beam 2D Radon transform is

$$R_i(\vartheta, \alpha)[\varphi(x', y')] = \int_{-\infty}^{\infty} \varphi_i(\alpha \cos \vartheta - \delta \sin \vartheta, \alpha \sin \vartheta + \delta \cos \vartheta) \mathrm{d}\delta, \quad (8.1)$$

and maps a density function $\varphi_i(x', y')$ into its line integrals. The spatial field, $\varphi_i(x', y')$, comprises the two-point spatial autocorrelation and is constructed by Fourier transforming the ith interrogation area as follows:

$$\phi_i(k_x, k_y) = \frac{1}{2\pi} \int_{-\infty}^{\infty} \int_{-\infty}^{\infty} \phi_i(x, y) e^{-i(k_x x + k_y y)} \mathrm{d}x\mathrm{d}y. \quad (8.2)$$

Since $\varphi_i(k_x, k_y) = \phi_i(k_x, k_y)\phi_i^*(k_x, k_y)$, where ϕ_i^* is the complex conjugate of ϕ_i, then the average wavenumber spectra associated with the spatial wavefronts, as they pass through an interrogation area at a given time, is realized. A sample interrogation area is shown in Fig. 8.9a alongside its resultant wavenumber spectra in Fig. 8.9b. The inverse Fourier transform of Eq. (8.2) yields the two-point spatial autocorrelation,

$$\varphi_i(x', y') = \int_{-\infty}^{\infty} \int_{-\infty}^{\infty} \varphi_i(k_x, k_y) e^{i(k_x x' + k_y y')} \mathrm{d}k_x \mathrm{d}k_y, \quad (8.3)$$

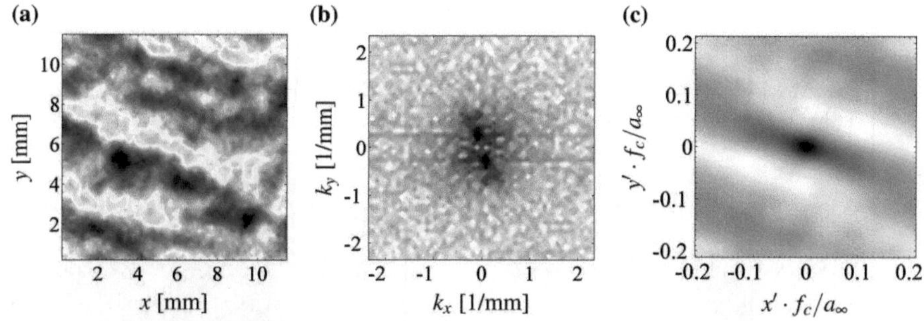

Fig. 8.9 (**a**) Sample interrogation area from a raw image. (**b**) Corresponding double Fourier transformation. (**c**) Resultant two-point spatial autocorrelation

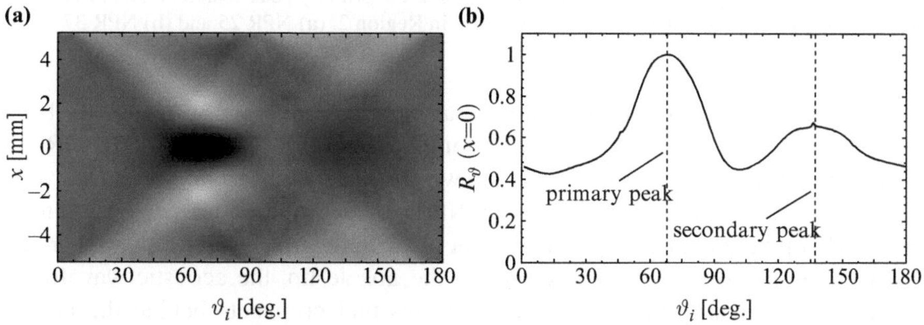

Fig. 8.10 (**a**) Radon transform of the sample image. (**b**) Central distribution of the Radon transform ($x = 0$)

which is shown in Fig. 8.9c using the sample image. After having computed $\varphi_i(x', y')$ from Eq. (8.3), the Radon transform is performed by solving Eq. (8.1). The analysis is not restricted to shadowgraphy measurements and can be easily applied to schlieren, or other wavefront imaging techniques.

Upon analyzing Eq. (8.1) with the sample autocorrelation in Fig. 8.9c, a spatial mapping forms for a range of search angles over which the transform is performed. The Radon transform yields the projection of the spatial autocorrelation onto a line normal to the search angle, ϑ_{ij} for a given interrogation area. The radial line is swept through a range of search angles between $0° \leq \vartheta \leq 180°$, as shown in Fig. 8.10. In doing so, several peaks are revealed. The first primary peak ($j = 1$) identifies the angle of the most energetic wavefront passing through the interrogation area, while the second peak ($j = 2$) identifies the angle of the second most energetic wavefront, and so on. A sensitivity study conducted by Canchero [5] showed how an interrogation area of 55×55 pixels is adequate for this analysis.[1]

[1]The interested reader is referred to the work of Canchero [5] for a more thorough discussion of the application of Radon transforms to these shadowgraphy images.

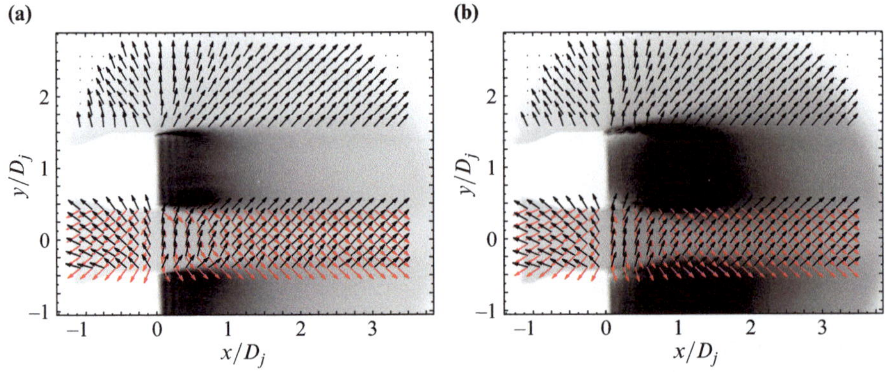

Fig. 8.11 Mean vector field of acoustic wavefronts of the primary peak search angle in Region 1, and the primary and secondary peak search angles in Region 2. (**a**) NPR 25 and (**b**) NPR 37

Having analyzed all interrogation areas in Regions 1 and 2 using all 1190 samples, the statistical properties are compared for the NPR 25 and NPR 37 conditions. The mean vector field is shown in Fig. 8.11 overlaid the standard of deviation of the intensity field for a given NPR. As expected, vectors downstream of the nozzle lip propagate at angles that coincide with the Mach wave radiation angle. At steep angles to the flow and closer to the nozzle lip, the acoustic wavefronts propagate spherically. In Region 2, wavefronts that propagate behind the nozzle reflect off the outer surface of the adjacent nozzle while intersecting with oncoming acoustics waves from the same nozzle flow. As for the primary and secondary peak search angles in Region 2, it is not surprising to see a crisscross pattern corresponding to sound waves propagating from the upper and lower shear layers, respectively. The robustness of the analysis methodology is demonstrated by the sensibility of the overlapping sound wave patterns captured in Region 2.

Where the shadowgraphy images are concerned, both flows exhibit shock cell patterns corresponding to the supersonic annulus along the nozzle lip line with the larger cell spacing occurring at NPR 37 in Fig. 8.11b as expected. Likewise the wider and thicker shear layer at NPR 37 is attributed to the flapping behavior that occurs during the end-effects-regime. The effect of this flapping behavior on the acoustic field (increased sound levels over a broader range of frequencies) is not easily revealed in these vector maps, and so a closer inspection is considered with Fig. 8.12, which comprises contours of the mean of the primary and secondary peak search angles.

Beginning with the average peak search angle in Region 1 (Fig. 8.12), the effect of NPR appears only in the downstream regions and is attributed to the faster subsonic core flow in the higher NPR case. Comparing the mean contour associated with the primary peak search angle in Region 1 to that in Region 2, and for a given NPR, a lack of symmetry is revealed and is attributed to vectoring of the flow and plume due to the lower local ambient pressure that forms in the interaction zone between the nozzles. The effect is the same for both NPRs.

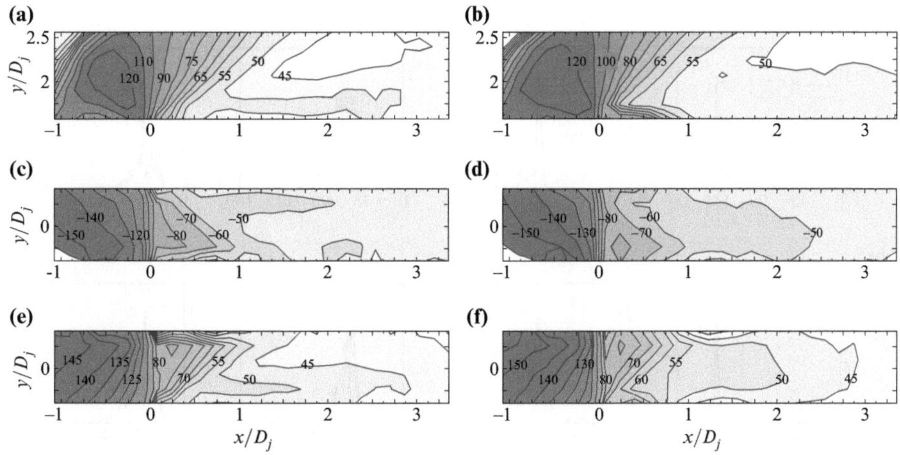

Fig. 8.12 Mean primary peak search angle in Region 1 at (**a**) NPR 25 and (**b**) 37. Mean primary peak search angle in Region 2 at (**c**) NPR 25 and (**d**) 37. Mean secondary peak search angle in Region 2 at (**e**) NPR 25 and (**f**) 37

The probability density function (p.d.f) of the primary search angle for a range of interrogation areas in Region 1 covering shallow, sideline, and steep angle observers is shown in Fig. 8.13 with each distribution appearing to be centered around a common angle. The locations where data is acquired are shown in Fig. 8.8b and are normalized using the maximum value. A generalized extreme value distribution (g.e.v.) of each data set is also shown and demonstrates the deviations of each distribution from a normal Gaussian. Shifts in the primary search angle between x_1 and x_4 complement the mean vector field shown in Fig. 8.12a. Likewise, multiple peaks at x_1 and x_2, not observed at x_3 and x_4, are attributed to difficulties in imaging the relatively weaker wavefronts in this region of the acoustic field. Generally speaking, Fig. 8.13a, b appear quite similar, which suggests that the differences in the wavefronts produced during the end-effects-regime, relative to other nozzle operating conditions, are subtle.

8.5 Future Outlook

Upon celebrating this very special occasion of William K George's 70th birthday with a technical symposium, and an open forum discussion that followed, it is natural that we reflect on past efforts in order to project the future needs and anticipated outcomes of the scientific community where turbulence and big data issues are concerned.

Large data computing has become a quintessential element in the deployment of current and future high speed propulsive technologies. Much of this is attributed to

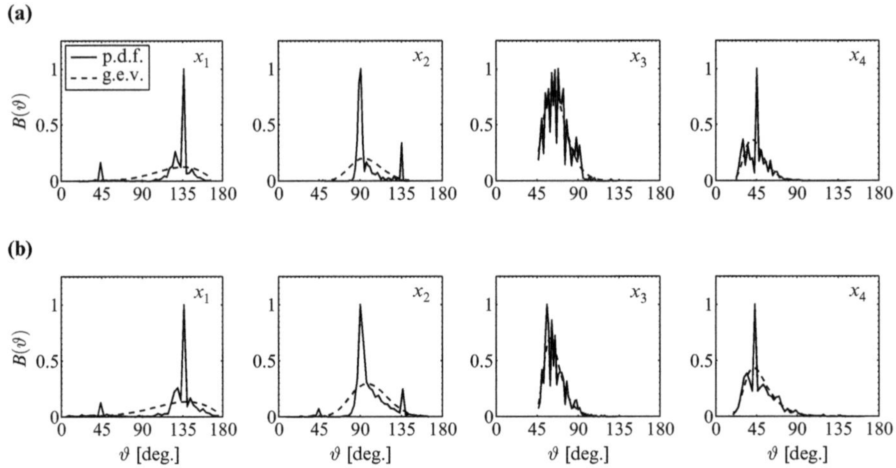

Fig. 8.13 p.d.f. of the primary search angle for interrogation areas x_1 through x_4 identified in Fig. 8.8. (**a**) NPR 25 and (**b**) NPR 37

the rapid development in computing power, which has replaced many of the theoretical branches in fluid mechanics with run of the mill simulations. RANS models are undoubtedly the industrial norm, given its simplicity and robustness. Albeit, time-accurate solutions are becoming more readily available, such as LES and even DNS, with the inclusion of chemically reacting agents, multi-species flows, and fluid structure coupling now being realized. Where laboratory diagnostics are concerned, the continued progression in the development of non-intrusive, volumetric, and time-resolved measurement apparatus have succeeded the single-point measurement tool that once monopolized the technical discussion floor. Likewise, with the advent of more affordable and sophisticated additive manufacturing techniques, experiments are being conducted on realistically complex hardware that were once the cornerstone of the numerical disciplines; the gamut of syndicate project scopes is being rebalanced. On the contrary, we are becoming victims of our own success as the enormity of the data from both disciplines is cumbersome and is plaguing our ability to produce practical and timely engineering solutions. The future success will be revealed by a strong synergy between both experimental and numerical disciplines that leverages carefully designed measurements with robust, yet simple, computational models with built-in analytics.

Acknowledgements Funding for this study was graciously provided by the NASA Engineering and Safety Center.

References

1. P.S. Addison, *The Illustrated Wavelet Transform Handbook* (Taylor & Francis Group, New York, 2002)
2. W.J. Baars, C.E. Tinney, Transient wall pressures in an overexpanded and large area ratio nozzle. Exp. Fluids **54**(1468), 1–17 (2013)
3. W.J. Baars, C.E. Tinney, J.H. Ruf, A.M. Brown, D.M. McDaniels, Wall pressure unsteadiness and side loads in overexpanded rocket nozzles. AIAA J. **50**(1), 61–73 (2012)
4. W.J. Baars, C.E. Tinney, J.H. Ruf, Non-stationary shock motion unsteadiness in an axisymmetric geometry with pressure gradient. Exp. Fluids **56**(92), 1–18 (2015)
5. A. Canchero, Retroreflective shadowgraph of clustered rocket nozzles. MS thesis, The University of Texas at Austin, Austin, TX, 2015
6. B.W. Donald, W.J. Baars, C.E. Tinney, J.H. Ruf, Sound produced by large area-ratio nozzles during fixed and transient operations. AIAA J. **52**(7), 1474–1485 (2014)
7. M. Farge, Wavelet transforms and their application to turbulence. Ann. Rev. Fluid Mech. **24**, 395–458 (1992)
8. M. Frey, G. Hagemann, Restricted shock separation in rocket nozzles. J. Prop. Power **16**(3), 478–484 (2000)
9. M.J. Hargather, G.S. Settles, Retroreflective shadowgraph technique for large scale flow visualization. Appl. Opt. **48**(22), 4449–4457 (2009)
10. M. Harper-Bourne, M.J. Fisher, The noise from shock waves in supersonic jets, in *Proceedings of AGARD Conference on Noise Mechanics*, AGARD-CP-131-11 (1973), pp. 1–13
11. E. Luke, P. Cinella, Numerical simulations of mixtures of fluids using upwind algorithms. Comput. Fluids **36**(10), 1547–1566 (2007)
12. G. Mack, Scaling laws and acoustic characteristics derived from experimental investigations of TOP-nozzle configurations. MS thesis, The University of Texas at Austin; Prof. C. E. Tinney as advisor, Austin, USA, The Technische Universität München; Prof. O. J. Haidn as advisor, München (2014)
13. S.A. McInerny, Launch vehicle acoustics part 1: overall levels and spectral characterization. J. Aircr. **33**(3), 511–517 (1996)
14. F.R. Menter, Two-equation eddy-viscosity turbulence models for engineering applications. AIAA J. **32**(8), 1598–1605 (1994)
15. N.E. Murray, Detection and analysis of shock-like waves emitted by heated supersonic jets using shadowgraph flow visualization. J. Acoust. Soc. Am. **136**(4), 2101 (2014)
16. N.E. Murray, G.W. Lyons, On the convection velocity of source events related to supersonic jet crackle. J. Fluid Mech. 793, 477–503 (2016)
17. L.H. Nave, G.H. Coffey, Sea level side loads in high-area-ratio rocket engines. AIAA Paper 1973–1284 (1973)
18. A.T. Nguyen, H. Deniau, S. Girard, T. Alziary de Roquefort, Unsteadiness of flow separation and end-effects regime in a thrust-optimized contour rocket nozzle. Flow Turbul. Combust. **71**(1–4), 161–181 (2003)
19. T.D. Norum, J.M. Seiner, Broadband shock noise from supersonic jets. AIAA J. **20**(1), 68–73 (1982)
20. H.S. Ribner, Acoustic flux from shock-turbulence interaction. J. Fluid Mech. **35**(2), 299–310 (1969)
21. J.H. Ruf, D.M. McDaniels, A.M. Brown, Nozzle side load testing and analysis at Marshall space flight center. AIAA Paper 2009-4856 (2009)
22. S.B. Verma, O. Haidn, Study of restricted shock separation phenomena in a thrust optimized parabolic nozzle. J. Prop. Power **25**(5), 1046–1057 (2009)
23. K.B.M.Q. Zaman, M.D. Dahl, T.J. Bencic, C.Y. Loh, Investigation of a 'transonic resonance' with convergent-divergent nozzles. J. Fluid Mech. **463**, 313–343 (2002)

Chapter 9
Variable Viscosity Jets: Entrainment and Mixing Process

L. Voivenel, E. Varea, L. Danaila, B. Renou, and M. Cazalens

9.1 Introduction

The theory of Kolmogorov [13] premises that at infinitely large Reynolds numbers, the statistical properties of the small scales should be universally determined by ν and $\overline{\epsilon}$, the kinematic viscosity and the mean energy dissipation rate, respectively. Implicit to this theory is that viscosity, considered as one independent parameter of the flow, is a 'small scale' quantity and thus should not affect large scale mixing. This is one possible explanation for why most studies focus on homogeneous fluids (same density and viscosity), or on variable-density flows [2, 18]. Nonetheless, many flows deal with real fluids, for which both density and viscosity fluctuate in space and time.

One of the first studies devoted to effects of viscosity was that of Campbell and Turner [8]. In order to determine the composition of a magmatic layer, they studied the injection of a fluid in a more viscous one (whose kinematic viscosities are, respectively, ν_l and ν_h, subscripts 'l' and 'h' stand for 'low' and 'high', respectively), for several ratios $R_\upsilon = \frac{\nu_h}{\nu_l}$ spread from 1 to 400. Campbell and Turner [8] observed a very different behaviour for the two extreme cases. Indeed, mixing does not occur at all for the $R_\upsilon = 400$ case. This phenomenon is due to a competition between the destabilizing inertial forces and the stabilizing viscous ones at the interface. Thus, the study of Campbell and Turner highlights that the large scale mixing is in fact, greatly viscosity-dependent and that Variable-Viscosity-Flow (hereafter referred to as VVF) should be carefully studied. Indeed, this kind of

L. Voivenel (✉) • L. Danaila • E. Varea • B. Renou
CORIA-UMR 6614, Normandie Université, CNRS, INSA et Université de Rouen, Avenue de l'Université, 76800 St Etienne du Rouvray, Rouen, France
e-mail: voivenel@coria.fr

M. Cazalens
SAFRAN R&T Center, Paris, France

A. Pollard et al. (eds.), *Whither Turbulence and Big Data in the 21st Century?*,
DOI 10.1007/978-3-319-41217-7_9

flow is frequent in industrial applications. Differences in the morphology of the VVFs were acknowledged, e.g. [7], but generally for large values of R_v. Particular attention has been given to the instabilities born at the interface of two variable-viscosity fluids [9–11, 17, 20]. To cite one example, combustion processes involve fluids with different physical and chemical properties (e.g. fuel and oxidizer).

Numerous questions, however, remain without clear answers. Some of them are fundamental, such as those dealing with the rate of entrainment and the associated phenomenology [1], or the exact expression of the mean energy dissipation rate [14, 19] which appears to be of great importance for flame stabilization and quenching [16]. Hence, experimental, numerical and theoretical efforts are to be devoted to traditional aerodynamic configurations (gaseous flow and relatively high Reynolds number).

The present study aims at furthering our understanding of VVFs, with a particular view on the very near field and turbulence generation. The roadmap of the paper is as follows. Section 9.2 details the experimental facility, whereas in Sect. 9.3 the optical diagnostics are presented. The fourth and fifth sections aim at developing results on the dynamic and scalar fields in VVF versus CVF, based on the same momentum and Reynolds number, respectively. Finally some conclusions drawn for the present study are provided.

9.2 Experimental Set-Up

The effects of viscosity variations are quantified by comparing the following cases:

- Constant-Viscosity Flow (CVF), which is the baseline case. A nitrogen jet issues in a coflow of nitrogen. The viscosity ratio of the two fluids is $R_v = 1$.
- Variable Viscosity Flow (VVF). A propane jet issues in a coflow of nitrogen. The latter is 3.5 times more viscous than the propane, so that $R_v = 3.5$. The density ratio is very nearly equal to 1.

The comparison between the two cases is based on the *same initial condition*, i.e. the same initial jet momentum, therefore the same injection velocity which is $U_{\text{inj}} = 1.45$ m/s. To remove any ambiguity regarding the role of the Reynolds number, the comparison between CVF and VVF at the same initial Reynolds numbers will be done in Sect. 9.5.

The flow facility is a round jet of diameter $D = 30$ mm surrounded by a (slight) coflow. Jet and coflow are enclosed in order to get well-defined boundary conditions allowing future accurate comparison with numerical simulations, Fig. 9.1. The coflow diameter, $D_{\text{cof}} = 800$ mm, is sufficiently large to restrain the wall influence on the main jet while isolating it from the exterior environment.

The main jet issues from a contraction designed to ensure a *top-hat* velocity profile at the nozzle exit. To achieve this objective, the two key parameters to be chosen are:

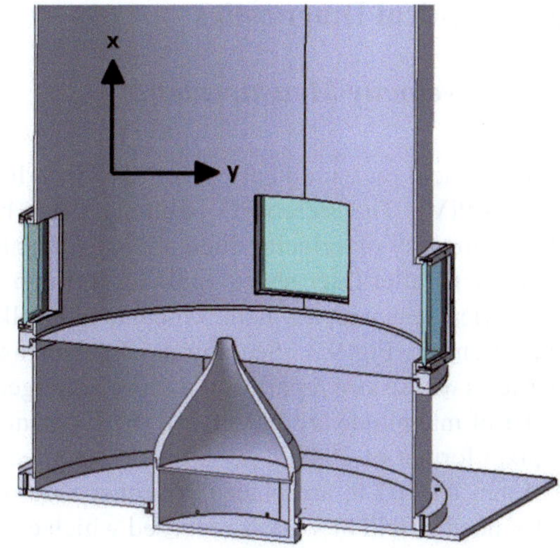

Fig. 9.1 Sketch of the experimental facility. Nozzle, confinement and optical accesses

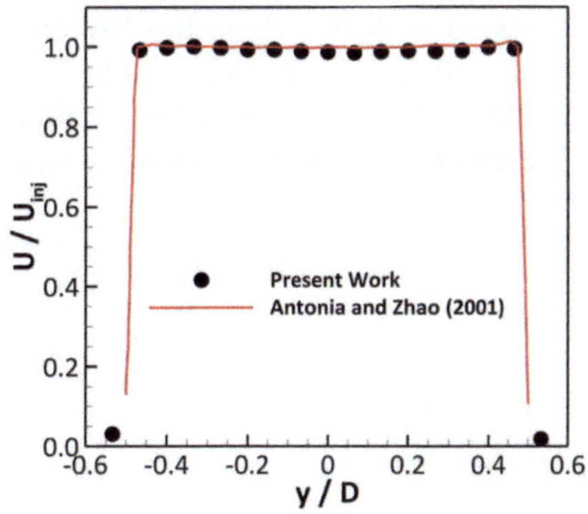

Fig. 9.2 Comparison of the velocity profile at the nozzle exit obtained in the current study and in Antonia and Zhao work

- the contraction ratio $C_R = \frac{D_{\text{in}}^2}{D_{\text{out}}^2}$, where D_{in} and D_{out} are the initial diameter of the contraction and the diameter at the contraction exit, respectively,
- the length on in-diameter ratio $\frac{L}{D_{\text{in}}}$, where L is the length of the contraction.

We have chosen to use the same values as Antonia and Zhao [3], i.e. $C_R = 87$ and $\frac{L}{D_{\text{in}}} \approx 1$. These parameters have then been used in the correlations provided by Bell and Mehta [4] to design contraction walls. The velocity profiles measured at the immediate vicinity of the nozzle exit by hot wire anemometry are consistent with those obtained by Antonia and Zhao [3], Fig. 9.2. The initial turbulence intensity is 1 %.

The flow-rate is controlled by a Bronkhorst Coriolis Mass Flow Controller (model SNB13201070A/s) for the main jet and Bronkhorst Thermal Mass Flow Controller (model SNM4209650B) for the coflow. Both mass flow controllers have been calibrated or checked using an in-house calibration bench.

9.3 Optical Diagnostics

9.3.1 Velocity Measurements

Velocity field measurements were performed by stereo-Particle Image Velocimetry (stereo-PIV). The stereo-PIV technique has been chosen because of the three-dimensionality of velocity fluctuations. A Quantel Ultra Twin laser at 532 nm was used. A parallel laser sheet—passing through the jet center—is obtained using a cylindrical lens with a −40 mm focal length followed by a spherical lens of 500 mm focal length, Fig. 9.3. Seeding is done using Di-Ethyl-Hexyl-Sebacat (DEHS) particles whose size repartition is more homogeneous than that of vegetable oil (size order of magnitude around 1 μm), [6]. Two Imager ProX cameras (LaVision) with a pixel format of 2048 × 2048 pixels, coupled with two visible objectives Nikkor 105 mm and f/2.8, are placed on either side of the laser sheet at a 45° angle. A 60 × 60 mm field of view is recorded which corresponds to a magnification ratio of 35 pixels/mm. Each camera records particle images that are independently post-processed with the algorithm 'Adaptive PIV'—provided by the Dantec software *Dynamic Studio* (3.4 release). Then, using a previously performed calibration, the *n*th 2D field from camera #1 is combined with the corresponding field from camera #2, creating a single 2 dimensions-3 components (2D-3C) velocity field.

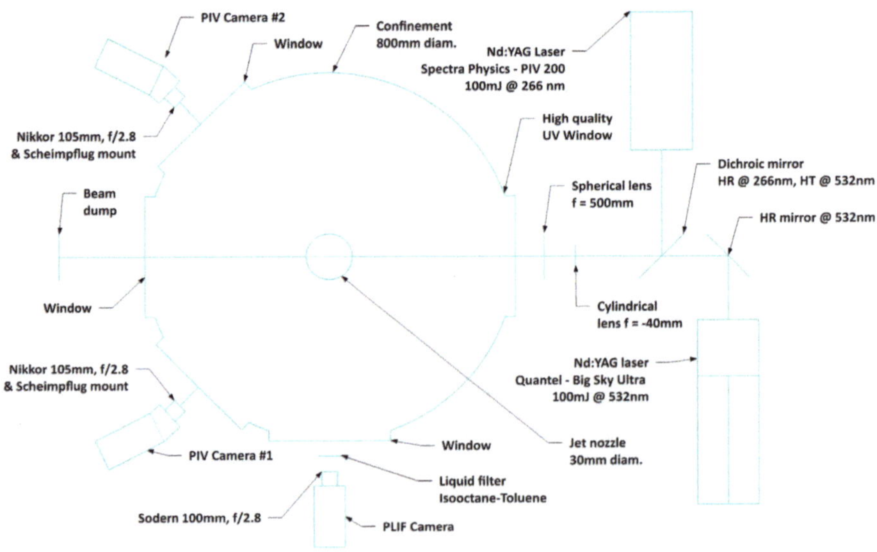

Fig. 9.3 Schematic diagram of the jet experiment

9.3.2 Scalar Field Measurements

Whilst the velocity field measurements are standard, the main experimental difficulty rests upon the scalar field measurements. Usually, acetone molecules are used as a tracer to perform Planar Laser Induced Fluorescence (PLIF) measurements. However, in order to obtain a sufficient signal-to-noise ratio, a great amount of tracer has to be used, leading to a modification of the seeded fluid properties. The focus of this work is on the effects of viscosity, therefore acetone is not the best choice here. Thus, an alternative molecule allowing a better signal-to-noise ratio while conserving the studied fluid properties was used. In addition to the previously discussed restrictions, the tracer must satisfy several other criteria:

- absorption wavelength has to be compatible with highly energetic laser at our disposal ($\lambda = 266$ nm),
- fluorescence spectrum must be shifted from the excitation wavelength
- evaporation properties must allow mixing with a gas.

The chosen tracer was anisole. To avoid ignition of the mixture when VVF cases are performed (propane is injected), the jet issues into nitrogen and not into an air coflow. Therefore, the strong quenching of anisole with O_2 is not an issue in our case. Moreover, the quantum yield, the ratio of photons absorbed to photons emitted through fluorescence, is very high which allows to inject a small quantity of anisole into the jet. The physical properties of the jet are therefore not altered.

To validate this technique, the linearity of the PLIF signal with the laser energy and the tracer concentration was tested. Another critical point is the proximity of the anisole absorption and emission bands. To eliminate the Mie signal at 266 nm coming from the DEHS particles as well as the stray light due to reflection at 266 nm, a liquid filter composed of iso-octane (spectroscopically neutral) and toluene as suggested in [15] was used. Indeed, the toluene absorbs predominantly at 266 nm and dimly from 270 nm, which is the beginning of the anisole fluorescence signal [5].

The tracer particles are excited using an Nd:YAG laser (Spectra Physics) with a fourth-harmonic generating crystal that produces a Q-switched laser output in the UV ($\lambda = 266$ nm, 100 mJ). A dichroic mirror is used to optically combine the PIV laser beam with LIF laser beam. The fluorescence signal is collected by an intensified CCD (ICCD) camera coupled to a UV Cerco 100 mm, f/2.8 lens. The ICCD camera is a Roper Scientific PIMAX 4 (16 bits) manufactured by Princeton Instruments with a pixel format of 1024×1024 pixels. The exposure time is set to 500 ns which is a compromise between fluorescence signal collection and the increase in noise level.

9.3.3 Big Data

With the emergence of new sensors for cameras, the recorded images can easily rise resolution of 5 Mega pixel. Moreover, the repetition rate is enhanced and a 5 Hz acquisition frequency is achieved. Therefore, a huge amount of data is accessible which makes it possible to obtain fully converged statistical results. However, the post-processing as well as the recording techniques needs to be designed and optimized in order to minimize the lag (during recording) and the overall post-processing time.

In the present study, a set of 3000 images is recorded for both the dynamic and the concentration fields. The issue concerning the acquisition is solved by using super-computers (Dell Z-800 Workstation) still affordable with enhanced access to memory (RAID 0). Two super-computers are used to record the dynamic field and the scalar fields.

As soon as the velocity field is concerned, two cameras of 5 M pixel each are used. Within 10 min of experiments, a set of two times 3000 images is stored on the hard drives which correspond to memory of approximately 94 Go. For the scalar field, the corresponding 3000 images are recorded using a Mega pixel camera which correspond to 12 Go of memory on the second workstation. Therefore, each data set represents a total amount of raw data of 108 Go. The question of classical storage of data on hard drives is addressed. To facilitate the access to the files, the data sets are uploaded on the Centre de Ressources Informatiques de Haute-Normandie (CRIHAN) storage servers where Tera bites of memory are available. Therefore, thanks to Giga bites ethernet connexion, an easy access to data is possible. In order to post-process the recorded data specific parallelized routines must be applied. This is facilitate since images are matrix and available toolbox in Matlab R2105 already exists. However, the post-processing time on a third workstation is around 1 week for both the scalar and the dynamic fields. A total amount of 92 Go of post-processed images is obtained. To optimize and decrease the calculation time, we are thinking on developing in-house post-processing routines which would be run on the CRIHAN servers where linux platform is in use. Therefore, storage and post-processing would be done on super-computer from the CRIHAN organization.

In the following sections, CVF and VVF cases will be compared on the basis of same jet momentum—Sect. 9.4—and same Reynolds number—Sect. 9.5—respectively.

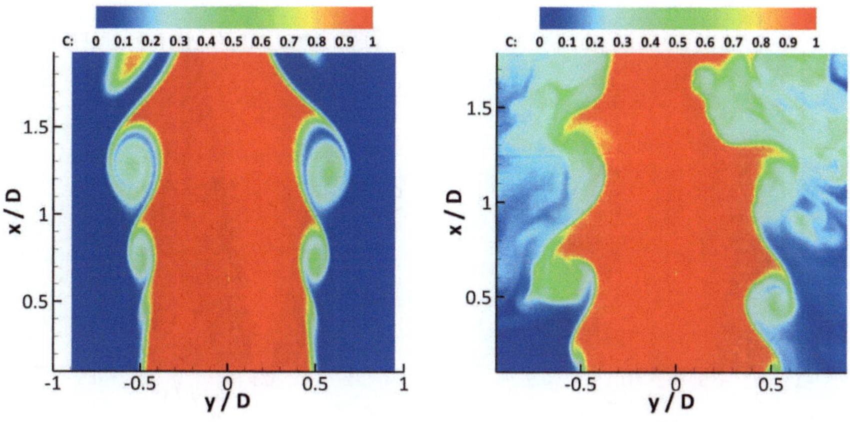

Fig. 9.4 Instantaneous images of mixing in N_2/N_2 jet and Propane/N_2 jet. (**a**) CVF. (**b**) VVF

9.4 Comparison of CVF and VVF Based on the Same Initial Jet Momentum

9.4.1 Phenomenology

Figure 9.4 illustrates instantaneous images of the scalar distribution. Here C is the propane concentration—the mixture fraction—normalized such that $C = 1$ in the propane core jet and $C = 0$ in the N_2 coflow. A careful analysis of the scalar mixing provides a qualitative way to compare the two flows. Whilst the CVF exhibits classical Kelvin–Helmholtz vortices, Fig. 9.4a, the VVF, Fig. 9.4b, only provides a hint of the large scale, lateral engulfment of the ambient fluid, together with mixing at scales distributed over a much wider range.

Planar distributions of the mean and RMS (root mean squared) of the scalar are represented for the very near field of the flow, spanning between 0 and 2 jet diameters, in Fig. 9.5. Several observations may be made.

- The CVF potential core, Fig. 9.5a, image left-half-side, is wider than that of the VVF, Fig. 9.5a, image right-half-side, which suggests a better mixing for the latter. This statement is supported by the presence of propane in the full field of view for VVF. This is in contrast with the N_2/N_2 jet, where the core jet fluid (seeded N_2) is completely absent on the image edges.
- The largest RMS values are not located at the same axial locations: for the CVF flow, the largest values of the scalar RMS are located at 2 D, whereas for the VVF, the maxima are distributed much closer to the nozzle, between $0.5\,D - 1\,D$.

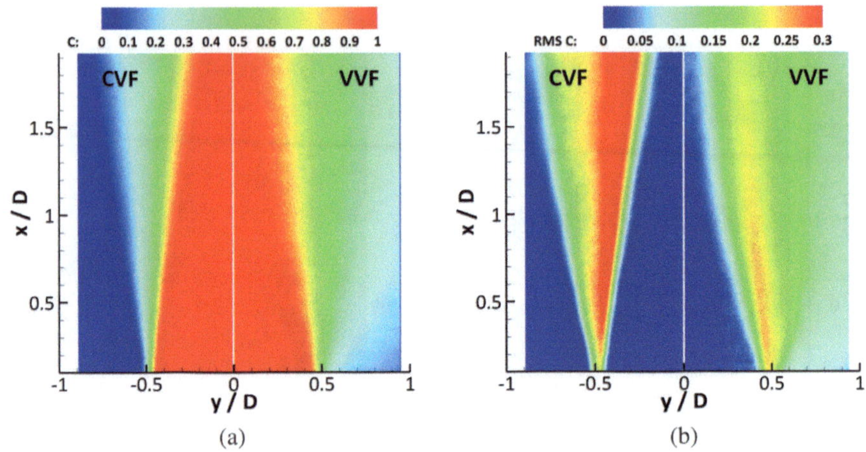

Fig. 9.5 Planar distributions of the (**a**) scalar mean and (**b**) RMS in CVF (N_2/N_2 jet), image *left-half-side*, and VVF (Propane/N_2 jet), image *right-half-side*

The latter observation is to be understood in connection with the instantaneous images. The intense fluctuations are strongly correlated with the presence of the large structure—Kelvin–Helmholtz. For the VVF case, while at $y/D = 1$ engulfment only occurs, the mixing exhibits smaller and smaller scales at $y/D = 2$.

As far as the CVF is concerned, only large scale mixing occurs, thus explaining that larger fluctuations are observed, compared to the VVF case. This observation is strengthened by the study of the velocity field and more particularly of the mean lateral fluctuations (not shown here), whose evolution is similar to that of the scalar. Indeed, at a downstream position of one diameter, the lateral fluctuations are more intense in VVF than in CVF.

Moreover, there is a stronger decrease of the axial mean velocity in VVF, starting at the very early stage of injection, Fig. 9.6, indicating an increased entrainment of the ambient fluid into the jet fluid and an accelerated trend towards self-similarity.

Intense values of the axial velocity fluctuations, Fig. 9.7, as well as a faster trend towards isotropy (here quantified through the ratio $u_{\mathrm{RMS}}/v_{\mathrm{RMS}}$, Fig. 9.8, u and v being the axial and radial velocities, respectively) in VVF than in the baseline case (CVF) are observed.

9.4.2 Analysis

The birth of the turbulent fluctuations most likely results from a combination of four factors:

(1) Kelvin–Helmholtz instabilities;
(2) Wake instabilities behind the injector lip;

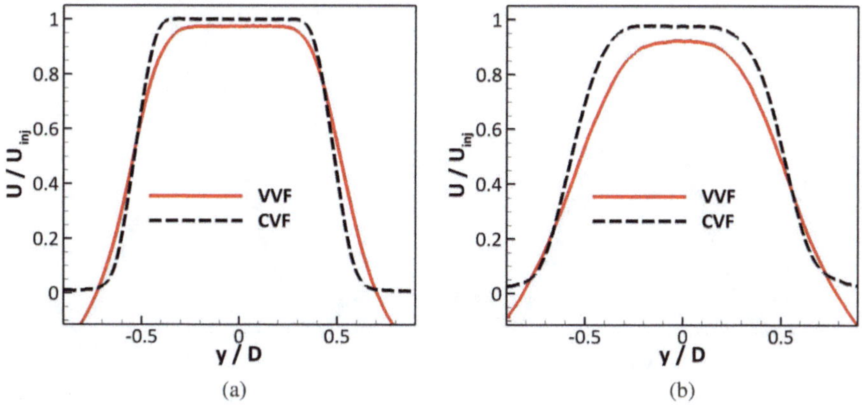

Fig. 9.6 Mean axial velocity normalized with respect to the injection velocity, for both CVF and VVF, at two axial locations: (**a**) $y = 1\ D$ and (**b**) $y = 2\ D$

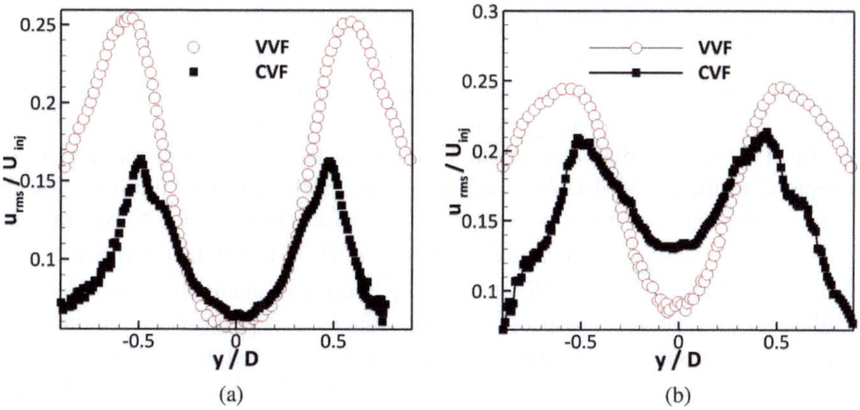

Fig. 9.7 Radial RMS normalized with respect to the injection velocity, for both CVF and VVF, at two axial locations: (**a**) $y = 1\ D$ and (**b**) $y = 2\ D$

(3) Interface instabilities due to density gradients;
(4) Interface instabilities due to viscosity jumps [12, 20].

Points (1) and (2) are characteristic of jet flows, constant-viscosity or not, thus, they cannot be responsible for such different behaviours. As far as the density effects are concerned, the studied configuration is that of a *heavy jet* (heavy fluid injected in a lighter one). Yet, according to Amielh et al. [2], density stratification for heavy jets results in mixing inhibition. The opposite is observed here, so the density effects are not responsible for the observed behaviour. Mixing enhancement is subdued to the occurrence of the four different types of instabilities at the jet edges.

Experimentally, the local viscosity is linked to the local concentration. A two species mixing law is applied. Hence, since the scalar field modifies the velocity

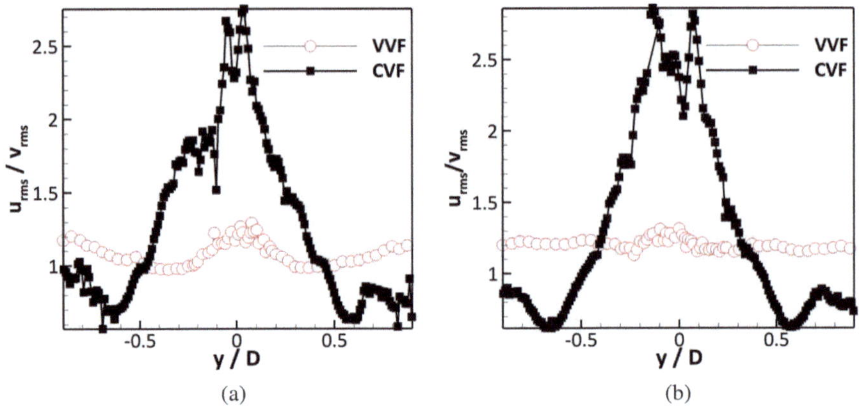

Fig. 9.8 Ratio u_{RMS}/v_{RMS} for VVF and CVF, at two axial locations: (**a**) $y = 1\,D$ and (**b**) $y = 2\,D$

field—active scalar—both fields (dynamic and scalar) are coupled and must be studied together. By computing the joint probability function density (PDF) of the **mean** concentration and **mean** velocity, a complete mean flow mapping is obtained. Figure 9.9 reports the joint PDF of the mean concentration and mean axial velocity at 0.5 and 1.5 diameter, top and bottom, respectively. Right and left images correspond to the cases of CVF and VVF, respectively. To avoid ambiguity in the analysis, only statistics from the left part of the jet are presented. The two maxima located at ($\overline{C} = 1, \frac{\overline{U}}{U_{\text{inj}}} = 1$) and ($\overline{C} = 0, \frac{\overline{U}}{U_{\text{inj}}} = 0$) in Fig. 9.9a, c indicate a clear bimodal distribution for the CVF case. These two maxima are characteristic of the jet fluid and of the coflow, respectively. Thus, the lack of points between these two extremes confirms that the mixing is still at a very early stage in the N_2/N_2 jet.

In the propane case, at $y = 0.5\,D$, Fig. 9.9b, the bimodal distribution is attenuated compared to the CVF case, Fig. 9.9a. Once again, this is consistent with the observation of a more advanced mixing in the VVF case. It is interesting to note that, if the maximum located at ($\overline{C} = 1, \frac{\overline{U}}{U_{\text{inj}}} = 1$) is still present in the propane jet, it is not the case anymore for the second extremum ($\overline{C} = 0, \frac{\overline{U}}{U_{\text{inj}}} = 0$). The former is shifted to a negative axial velocity ($\frac{\overline{U}}{U_{\text{inj}}} = -0.2$) and a more important value of $\overline{C}$ (e.g. $\overline{C} = 0.2$). These values can be explained by the presence of a recirculation zone which bring propane into the coflow.

The mean radial velocity and mean concentration joint PDF's allow us to highlight the processes at play—entrainment or jet expansion—at the considered downstream locations. Figure 9.10 shows the joint PDF of the mean concentration and mean radial velocity at 0.5 and 1.5 diameter, top and bottom, respectively. Right and left images correspond to the cases of CVF and VVF, respectively.

Looking at the CVF case, at $y = 0.5\,D$, the existence of the bimodal distribution previously observed is confirmed, Fig. 9.10a. The few points ($\overline{C} = 0, \frac{\overline{V}}{U_{\text{inj}}} > 0$) indicate the birth of the Kelvin–Helmholtz instabilities in the coflow. At downstream

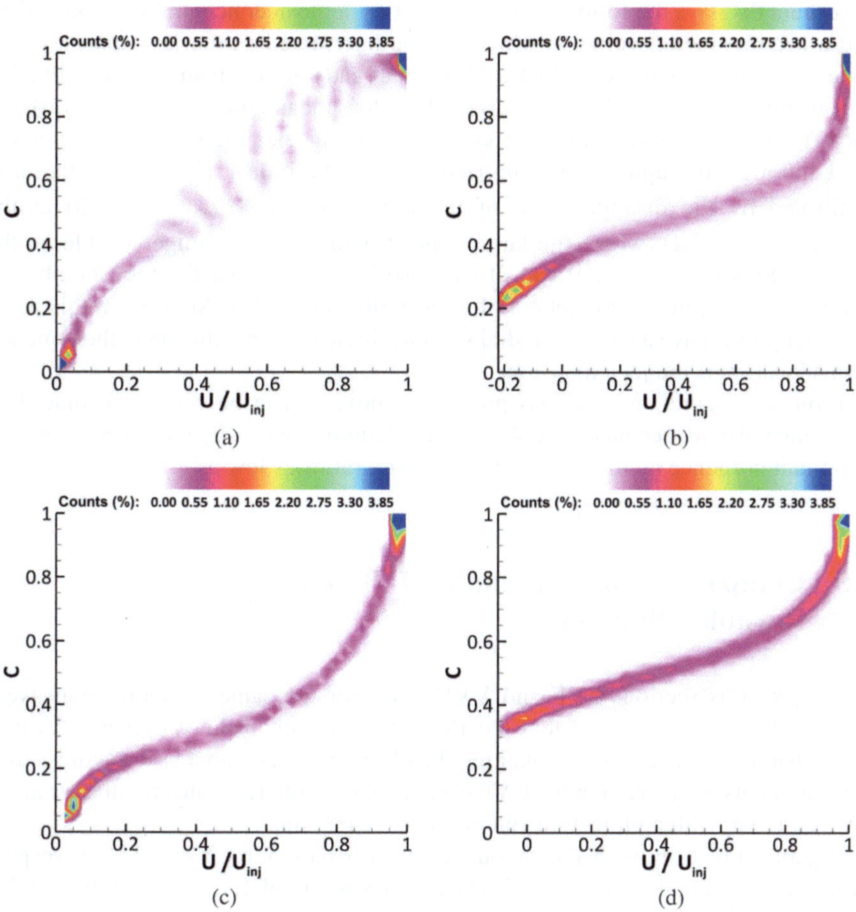

Fig. 9.9 Joint probability density functions of the mean concentration and mean axial velocity ($\overline{C}$, $\overline{U}$) for CVF (*left*) and VVF (*right*), at two axial locations: $y = 0.5\ D$ (*top*) and $y = 1.5\ D$ (*bottom*). (**a**) CVF at $y = 0.5\ D$, (**b**) VVF at $y = 0.5\ D$, (**c**) CVF at $y = 1.5\ D$, (**d**) VVF at $y = 1.5\,D$

location $y = 1.5\ D$, the positive radial velocities are now associated with a lower concentration, $\overline{C} \approx 0.1$, Fig. 9.10c. This can be interpreted as the beginning of the large scale mixing—ensured by the Kelvin–Helmholtz instabilities—between jet and coflow. Moreover, the start of the jet expansion is observed through the points $(\overline{C} = 1\ ,\ \frac{\overline{V}}{U_{\text{inj}}} < 0)$.

The phenomenology identified for the VVF case is again very different from that observed in the N_2 jet. At $y = 0.5\ D$, three particular zones may be distinguished, Fig. 9.10b. The first one, whose meaning is the most easily explained, is the maximum located at $(\overline{C} = 1,\ \frac{\overline{V}}{U_{\text{inj}}} = 0)$. It corresponds to the jet core where the velocity vectors are only oriented along the axial direction. The second local maximum displays the following characteristics: a mean concentration around 0.55 and a negative radial velocity, which can be interpreted as the jet expansion in a zone where the mixing with the host fluid is already at an advanced stage. Finally, the last remarkable area is characterized by a low mean concentration, $\overline{C} \approx 0.2$— and a positive radial velocity. This combination indicates the presence of ambient fluid inflows—through the Kelvin–Helmholtz instabilities. This interpretation is confirmed by the disappearance of the points ($\overline{C} \approx 0.2$, $\frac{\overline{V}}{U_{inj}} < 0$) at the location $y = 1.5\,D$, where the large scale structures are no longer visible on the instantaneous images, Fig. 9.10d. It can also be noticed that the total number of points corresponding to the jet core has been divided by 1.5. Apart from this zone, the other points have a negative radial velocity, indicating that the main phenomenon at this location is the jet expansion.

From the results and analysis presented above, it can be concluded that clear experimental evidence has been obtained to claim that viscosity stratification has an important influence on turbulence, for viscosity ratios as low as 3.5.

9.5 Comparison of VVF and CVF Based on the Same Reynolds Number

In the previous section, CVF and VVF cases for the same jet momentum were compared. It can be argued that the Reynolds number in the N_2 jet is 3.5 times lower than propane jet, thus explaining the observed discrepancies. To address this, measurements were taken in the CVF case with the same Reynolds number as in the VVF case, i.e. with an injection velocity $U_{inj} = 4.37$ m/s.

Figure 9.11a, b show instantaneous scalar distribution in CVF and VVF, respectively. Once again, the topology of the two cases is completely different. If the CVF case is indeed more turbulent than in the previous experiments, Fig. 9.4, it still does not present the large range of scales exhibited by the VVF.

The map of the scalar mean is presented in Fig. 9.12a. It is observed that a shift occurs in the virtual origin of the N_2/N_2 jet, compared to the constant-viscosity case detailed in the previous section. As far as the jet angle is concerned, it is smaller in the CVF case than in the VVF case, indicating a less advanced mixing.

The maxima of the RMS of the longitudinal fluctuations are once again correlated with the presence of large structures. This is particularly visible when attention is focused on the top of the image in Fig. 9.12b (from $y = 1.2\,D$ up to $y = 1.9\,D$). Indeed, this is the largest zone of intense fluctuations. Confronting with the instantaneous image, Fig. 9.11a, it also corresponds to the location of the largest structures. Thus, even if the CVF topology differs slightly from that of the previous section (Sect. 9.4, same jet momentum), up to this point the observations previously made still hold.

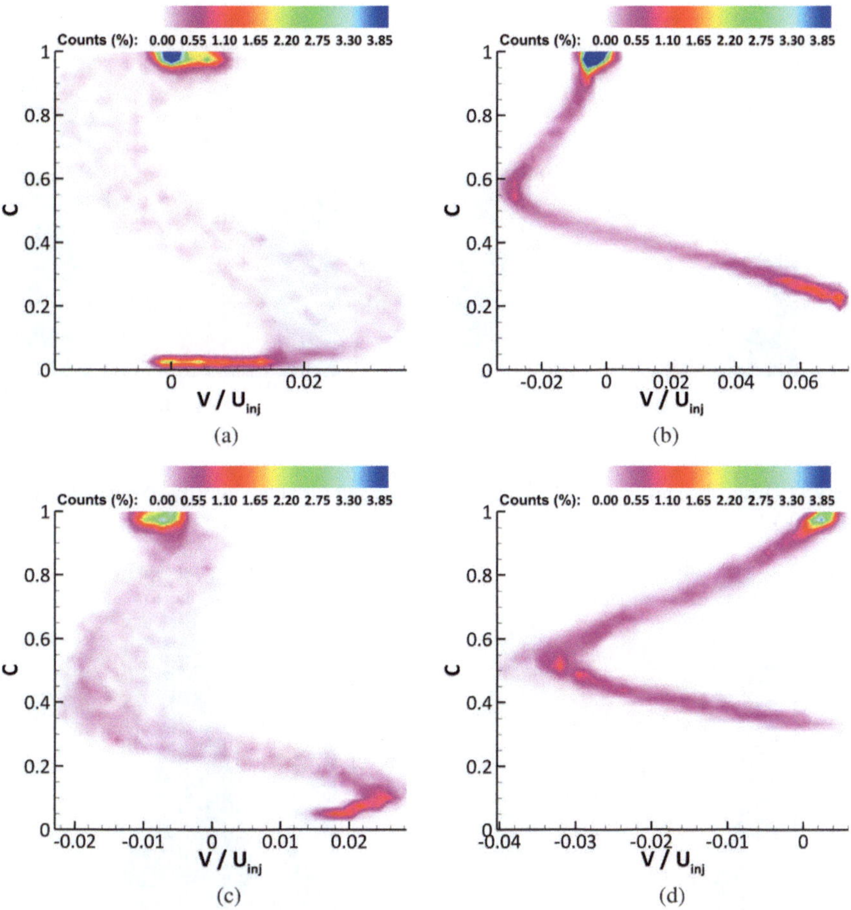

Fig. 9.10 Joint probability density functions of the mean concentration and mean radial velocity ($\overline{C}$, $\frac{\overline{V}}{U_{\text{inj}}}$) for CVF (*left*) and VVF (*right*), at two axial locations: $y = 0.5\ D$ (*top*) and $y = 1.5\ D$ (*bottom*). (**a**) CVF at $y = 0.5\ D$, (**b**) VVF at $y = 0.5\ D$, (**c**) CVF at $y = 1.5\ D$, (**d**) VVF at $y = 1.5\ D$

Profiles of mean axial velocity are reported in Fig. 9.13. For a given axial location, they present a more advanced decrease in the variable-viscosity case than in the N_2/N_2 jet. Similarly, Fig. 9.14 shows the longitudinal fluctuations are stronger in the VVF case and seem to have started their decrease contrary to those in the CVF which still increase with the axial location.

To conclude, this section illustrates that the discrepancies are still present, even if a little less pronounced, when comparing VVF and CVF configurations with the same Reynolds number (i.e. different injection velocities).

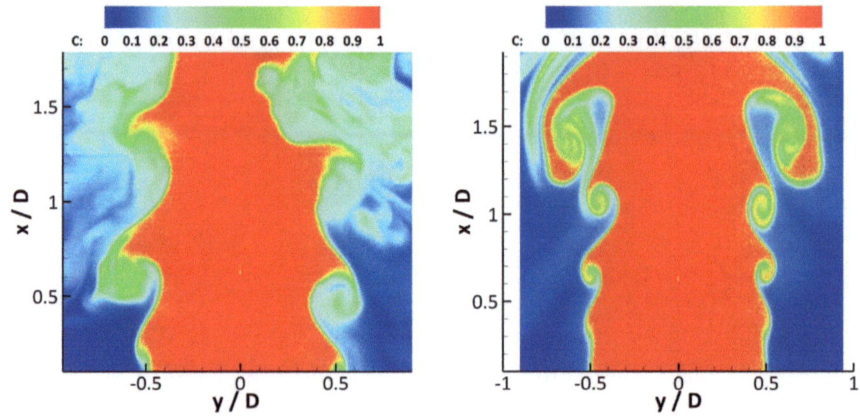

(a) (b)

Fig. 9.11 Instantaneous images of mixing in N_2/N_2 jet and variable-viscosity (Propane/N_2) jet. (**a**) CVF. (**b**) VVF

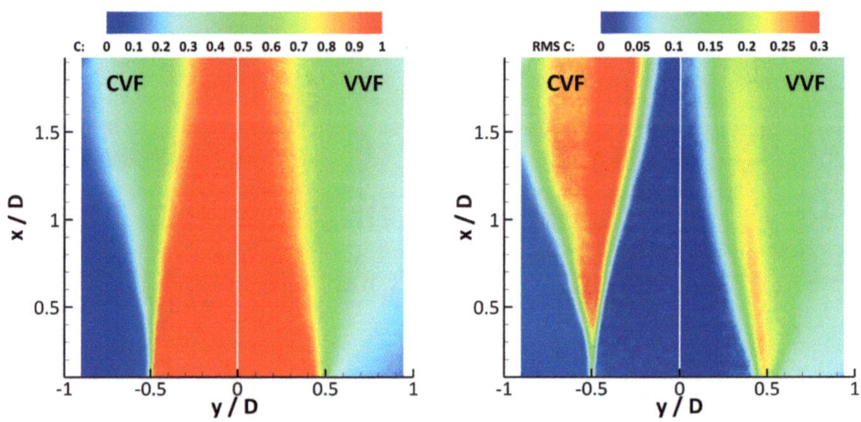

(a) (b)

Fig. 9.12 Planar distributions of the (**a**) scalar mean and (**b**) RMS in CVF (N_2/N_2 jet), image *left-half-side*, and VVF (Propane/N_2 jet), image *right-half-side*

9.6 Conclusion

With respect to the classical constant-viscosity jet, the variable-viscosity jet of a fluid issuing into a more viscous ambient fluid exhibits in the very near field:

- enhanced entrainment
- more important turbulent fluctuations.

We explain these phenomena by stating that if the different 'steps' of the turbulence are the same by nature (birth, growth, decrease and death), their duration is shorter

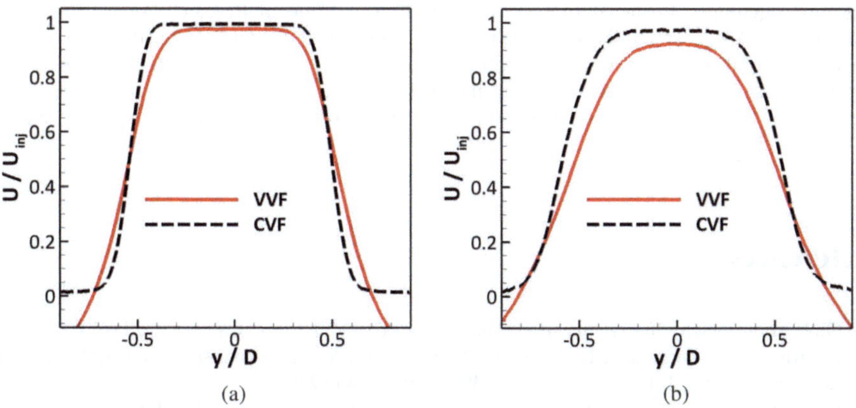

Fig. 9.13 Mean axial velocity normalized with respect to the injection velocity, for both CVF and VVF, at two axial locations: (**a**) $y = 1\ D$ and (**b**) $y = 2\ D$

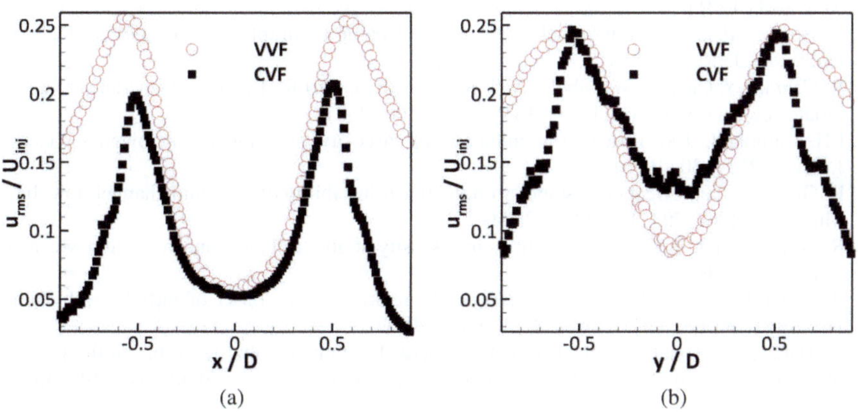

Fig. 9.14 Radial RMS normalized with respect to the injection velocity, for both CVF and VVF, at two axial locations: (**a**) $y = 1\ D$ and (**b**) $y = 2\ D$

in VVF than in CVF. Moreover, processes like fluctuation production are more intense in flow with variable viscosities. It means that, even when viscosity gradients disappear (far from the injection where the mixing is achieved), they have already significantly modified the flow dynamics. Thus, its final state will be different from a flow which has not be subjected to viscosity effects, even if their initial conditions—Re or jet momentum—are identical. The general message of this contribution is that whereas the viscosity itself indeed acts at the level of smallest scales, flows with viscosity variations at a large scale (such as jets issuing in different environment) are characterized by effects of viscosity variations at any scale, including the largest. A simple visualization of the scalar dispersion allows us to observe a significant disparity between VVF and CVF behaviours, leading us to state that the viscosity affects the topology and the dynamics of the whole flow at all scales.

Acknowledgements Financial support from the French National Research Agency under the project 'MUVAR' is gratefully acknowledged. We thank Dr. G. Godard, A. Vandel and C. Gobin for their helpful assistance in the experimental set-up. The authors thank H. Cavalier for his expertise in safety procedures concerning flammable mixtures. A first version of the paper was presented at the 10th Pacific Symposium in Flow Visualization and Image Processing, Naples 15–18 June 2015 as paper 240.

References

1. G.N. Abramovich, *The Theory of Turbulent Jets* (The MIT Press, Cambridge, 1963)
2. M. Amielh, T. Djeridane, F. Anselmet, L. Fulachier, Velocity near-field of variable density turbulent jets. Int. J. Heat Mass Transf. **39**, 2149–2164 (1996)
3. R.A. Antonia, Q. Zhao, Effects of initial conditions on a circular jet. Exp. Fluids **31**, 319–323 (2001)
4. J.H. Bell, R.D. Mehta, Contraction design for small low-speed wind tunnel. TR **84**, 1–37 (1988)
5. I. Berlman, *Handbook of Fluorescence Spectra of Aromatic Molecules*, 2nd edn. (Academic, New York, 1971)
6. G. Boutin, Mélange et micro-mélange dans un réacteur à multiples jets cisaillés. Ph.D. thesis, Université de Rouen, 2010
7. R. Cabra, J.Y. Chen, R.W. Dibble, A.N. Karpetis, R.S. Barlow, Lifted methane-air jet flame in vitiated co-flow. Combust. Flame **143**(4), 491–506 (2005)
8. I.H. Campbell, J.S. Turner, The influence of viscosity on fountains in magma chamber. J. Petrol. **27**, 1–30 (1986)
9. R. Govindarajan, Effect of miscibility on the linear instability of two-fluid channel flow. Int. J. Multiphase Flow **30**, 1177–1192 (2004)
10. R. Govindarajan, K. Sahu, Instabilities in viscosity-stratified flow. Ann. Rev. Fluid Mech. **46**, 331–353 (2014)
11. R. Govindarajan, V. L' vov, I. Procaccia, Retardation of the onset of turbulence by minor viscosity contrasts. Phys. Rev. L **87**, 174501 (2001)
12. A. Harang, O. Thual, P. Brancher, T. Bonometti, Kelvinhelmholtz instability in the presence of variable viscosity for mudflow resuspension in estuaries. Environ. Fluid Mech. **14**, 743–769 (2014)
13. A.N. Kolmogorov, The local structure of turbulence in incompressible viscous fluids for very large Reynolds numbers. Dokl. Akad. Nauk SSSR **30**(4), 301–305 (1941)
14. K. Lee, S. Girimaji, J. Kerimo, Validity of Taylor's dissipation-viscosity independence postulate in variable-viscosity turbulent fluid mixtures. Phys. Rev. Lett. **101**(7), 074501 (2008)
15. N. Pasquier, B. Lecordier, A. Cessou, Investigation of flame propagation through a stratified mixture by simultaneous piv/lif measurements, in *Proceedings of the Combustion Institute* (2007)
16. N. Peters, F. Williams, Lift-off characteristics of turbulent jet diffusion flames. AIAA J. **21**, 3 (1983)
17. A. Pinarbasi, A. Liakopoulos, The effect of variable viscosity on the interfacial stability of two-layer Poiseuille flow. Phys. Fluids **7**(6), 1318–1324 (1995)
18. W.M. Pitts, Effects of global density ratio on the centerline mixing behavior of axisymmetric turbulent jets. Exp. Fluids **11**, 125–134 (1991)
19. B. Talbot, L. Danaila, B. Renou, Variable-viscosity mixing in the very near field of a round jet. Phys. Scr. **T155**, 014006 (2013)
20. C.S. Yih, Instability due to viscosity stratification. J. Fluids Mech. **27**(2), 337–352 (1967)

Chapter 10
POD Mode Robustness for the Turbulent Jet Sampled with PIV

A. Hodžić, K.E. Meyer, and C.M. Velte

10.1 POD and Big Data

An important challenge in the description and simulation of turbulence is the large amount of information that is needed to describe even relatively simple flows in detail. The frequent disagreement between Reynolds averaged Navier–Stokes-based simulations and experiments is well known. Albeit, direct numerical simulations and in certain cases large eddy simulations tend to agree fairly well with experiments, their practical implementation introduces the problem of data storage. The experimentalist, however, experiences the same problem, using high speed particle image velocimetry (PIV) systems and even high speed volumetric PIV systems providing fully three dimensional velocity fields. Another challenge is how do we verify simulations against experiments and ensure that we indeed have simulated the same flow that we have measured?

POD is a useful tool allowing us to extract statistical data related to dynamic flow structures and can thereby not only be used to gain understanding of the dynamics of the flow, but can also be used as a tool to validate the dynamical aspect of simulations to experiments. Finally, the POD can be used to filter down large batches of data to encapsulate only the most essential information of the flow and thereby compressing the data amount to a mere fraction of its original size.

A. Hodžić (✉) • K.E. Meyer • C.M. Velte
Department of Mechanical Engineering, Technical University of Denmark, Building 403, 2800 Kgs. Lyngby, Denmark
e-mail: azuhod@mek.dtu.dk; cmve@dtu.dk; kem@mek.dtu.dk

A. Pollard et al. (eds.), *Whither Turbulence and Big Data in the 21st Century?*,
DOI 10.1007/978-3-319-41217-7_10

10.2 Introduction

The turbulent free axisymmetric jet constitutes a unique opportunity for the study of turbulence due to its homogeneous streamwise direction recovered from the logarithmic mapping of the velocity field [4, 26].

The technical and practical aspects and limitations of the study of Wänström [26], not least regarding domain size coverage and spatial resolution of the PIV measurements, have left open the question of the robustness of the results with respect to the domain. With the equipment at hand, [26] managed only to cover the jet to within a similarity radius of approximately $\eta = 2.0$ for the streamwise and crossplane. Similar attempts to test the robustness of the POD analysis using extensive (138) hot-wire rakes can be found in [7] and also in the work of [18]. Going beyond this limit naturally requires evaluation for confidence in the robustness of the results, not least since the POD analysis assumes a domain of finite energy for the analysis to hold [8].

The POD was applied in the crossplane of the turbulent jet by Glauser et al. [12] using an array of hot-wire probes, and later by Holmes et al. [16], Citriniti and George [2], Gamard et al. [6], Jung et al. [19] and Gamard et al. [7]. Based on equilibrium similarity theory [5, 9, 10] and the two-point similarity solutions deduced by Ewing et al. [4], a spatial decomposition of the turbulent jet was performed in both the streamwise and crossplane in two separate experiments by Wänström [26] using PIV sampled data. Recently, measurements have been carried out with 16 MPix cameras [15], covering the full $\eta = 2.5$ domain, as compared to the 4 MPix cameras used by Wänström [26], covering only the $\eta = 2.0$ domain. Although the light sensitivity of the 16 MPix cameras is poorer than for the 4 MPix cameras and therefore display an impaired signal to noise ratio, a 200 mJ laser was used in the current work compared to the 120 mJ YAG laser in [26].

In the current work, streamwise measurements from [15] and stereoscopic crossplane measurements of the jet at $40D$ downstream of the nozzle exit were analyzed with the aim to test the effect of radial coverage and radial grid resolutions on the modes. 11,000 and 1000 realizations were obtained from the streamwise and crossplane, respectively, using two 16 Mpix cameras at $Re_D \approx 20{,}000$ for the two independent experiments. The *Lumley approach* was applied to decompose the flow, i.e., a Fourier decomposition was performed in the azimuthal and streamwise directions and a proper orthogonal decomposition was applied along the (inhomogeneous) radial direction where the Fourier coefficients comprised the kernel.

10.3 Experimental Approach

Two separate experiments were performed in the investigation of the jet. (1) Two-component PIV measurement in the streamwise plane and (2) stereoscopic PIV measurements in the crossplane.

10.3.1 Facility and Equipment

The jet facility consists of a settling chamber with baffles and screens to dampen out significant spatial and temporal variations. The flow is then directed through a contracting axisymmetric plexiglas nozzle following a fifth order polynomial with an inlet diameter of 32 mm that connects the settling box with the ambient through a nozzle exit diameter of $D = 10$ mm. A fan connected to the jet box with a flexible tube is driving the flow. Both jet and ambient are seeded to provide as homogeneous a seeding distribution in the measured flow as possible. The seeding particles for the streamwise experiment were generated from liquid Glycerine using an atomizing nozzle built in-house, providing tracers with a typical size of about 2–3 μm while DEHS with a typical particle diameter of 1–2 μm was used as seeding in the crossplane experiment. The enclosure consisted of a tent of dimensions $2.5 \times 3.0 \times 10.0\,\text{m}^3$, yielding a cross-sectional area of $2.5 \times 3.0\,\text{m}^2$ that resulted in a momentum conservation of 100.0 % at $x/D = 30$ and 98.4 % at $x/D = 100$, where x is the downstream distance from the jet exit [17]. Sketches of the experimental setups are seen in Fig. 10.1.

The PIV apparatus included a 200 mJ ND:YAG 532 nm laser with dual cavities and two Dantec FlowSenseEO 16 MPix cameras (4872×3248 pix with a pixel pitch of 7.4 μm) with 60 mm Nikon lenses with an aperture of $f^{\#}2.8$ and $f^{\#}4.0$ for the streamwise and crossplane experiments, respectively. In order to produce a light sheet following the downstream development of the jet, the laser was positioned below the jet nozzle where an optical mirror was placed. The cameras were used to extend the total field of view for the streamwise experiment while they were used for stereoscopic measurements in the crossplane experiment at a 45° angle with respect to the lasersheet (see Fig. 10.1).

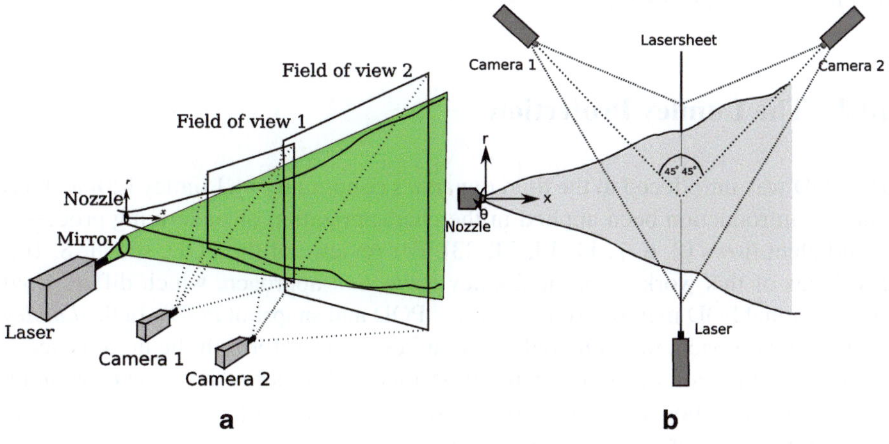

Fig. 10.1 Experimental setups for (**a**) the streamwise and (**b**) the crossplane experiments. In the streamwise experiment two-component PIV was performed while stereo PIV was used for the crossplane configuration

Table 10.1 Grid specifications for the streamwise and crossplane experiments

Experiment	Coordinate	No. points	Step size	Unit
Streamwise	$\Delta\xi$	600	0.0023	–
	$\Delta\eta$	201	0.0250	–
Crossplane	Δr	121	0.79	mm
	$\Delta\eta$	121	0.025	–
	$\Delta\theta$	201	1.8	Deg

10.3.2 Data Processing and Grid

Preliminary measurements in the streamwise plane were performed and estimated the integral time scale to 0.059 s from the local jet-half width velocity at *Field of view 2* seen in Fig. 10.1a. The sampling frequency was set to 1 Hz and a time between pulses of 150 μs for the streamwise experiment while a sampling frequency of 2 Hz and a time between pulses of 80 μs was used for the crossplane experiment. An ensemble of $N = 11{,}000$ and $N = 1000$ samples were acquired for the streamwise and crossplane, respectively, yielding a variability $\epsilon_N \propto \sqrt{N} = \mathcal{O}(10^{-2})$ and $\mathcal{O}(10^{-1})$ for the respective data sets.

The particle displacements were obtained from the particle images with afast Fourier transform (FFT)-based interrogation scheme. A correlation method was used with 32 × 32 pix interrogation areas with 50 % overlap and window shifting. Subpixel interpolation was performed in order to be able to resolve particle displacements below 1 pixel.

For the streamwise measurements, the obtained Cartesian grid density was approximately $\Delta x \times \Delta y = 1.69 \times 1.69$ mm and 2.65×2.65 mm for cameras 1 and 2, respectively, while for the crossplane measurements the Cartesian grid density was approximately $\Delta y \times \Delta z = 1.58 \times 1.58$ mm and was interpolated onto a polar grid (r, θ). The grid specifications are summarized in Table 10.1.

10.4 The Lumley Projection

The POD was introduced to the fluid dynamics community by Lumley [20] and has since its introduction been applied in the characterization of underlying processes of turbulent flows [2, 6, 7, 11–14, 21, 23]. For reviews of the POD, see [1, 8, 16]. The focus of this work is on the *Lumley projection* approach, which differs from the traditional POD methods (e.g., classical POD and snapshot POD). In the *Lumley Projection* approach, the POD reduces to a Fourier transform in the homogeneous streamwise direction of the jet, while it becomes a Fourier series expansion along the azimuthal homogeneous periodic θ-coordinate, [24–26]. The streamwise and radial similarity coordinates, (ξ, η), are defined as

Table 10.2 Optimized scaling parameters, obtained from the optimization scheme represented by (10.9)

Experiment	Camera	Description	Symbol	Value	Unit
Streamwise	1	Centerline decay rate	B_u	5.76	–
		Virtual origin	x_0	3.7D	–
		Spreading constant	A	0.093	–
	2	Centerline decay rate	B_u	5.68	–
		Virtual origin	x_0	2.6D	–
		Spreading constant	A	0.093	–
Crossplane	–	Centerline decay rate	B_u	5.97	–
		Virtual origin	x_0	6.0D	–
		Spreading constant	A	0.093	–

$$\xi = \ln\left((x - x_0)/D\right) \tag{10.1}$$

$$\eta = r/\delta \tag{10.2}$$

$$\delta = A\,(x - x_0) \tag{10.3}$$

where these are obtained from equilibrium similarity [4] where x_0 is the virtual origin, D is the nozzle diameter, δ is the jet-half width, and A is the jet-spreading rate. The scaling parameters obtained from the two cameras, x_0, A, and B_u, are given in Table 10.2. The Fourier coefficients for the streamwise and azimuthal directions, $\hat{u}_i$ and $\hat{v}_i$, respectively, are obtained from an FFT algorithm from which the two-point energy density cross spectra are constructed

$$\Phi_{i,j}^{(1)}\left(\kappa,\eta,\eta'\right) = \frac{2\pi}{L_\xi}\left\langle \hat{u}_i^*\left(\kappa,\eta,t\right)\hat{u}_i\left(\kappa,\eta',t\right)\right\rangle \tag{10.4}$$

$$\Phi_{i,j}^{(3)}\left(x,\eta,\eta',m\right) = \left\langle \hat{v}_i^{(m)*}\left(x,\eta,t\right)\hat{v}_j^{(m)}\left(x,\eta',t\right)\right\rangle \tag{10.5}$$

where the superscripts (10.1) and (10.3) denote the streamwise and azimuthal direction along which the Fourier coefficients were estimated and m is the azimuthal mode number. L_ξ is the length of the domain in the streamwise direction, "$*$" designates the complex conjugate, κ is the dimensionless wave number, t is time, and the angle brackets represent ensemble averaging. The Fredholm equation of the second kind takes on the following shape for this system:

$$\int_0^{|\eta|} \Phi_{i,j}\phi_j\eta' d\eta' = \lambda\phi_i \tag{10.6}$$

where the upper bound of the integral is truncated at various $|\eta|$. The following substitution was performed in order to ensure a Hermitian symmetric kernel:

$$\widetilde{\Phi}_{i,j}^{(\gamma)} = \left(\eta\eta'\right)^{\frac{1}{2}} \Phi_{i,j}^{(\gamma)} \tag{10.7}$$

$$\widetilde{\phi}_i = \eta^{\frac{1}{2}} \phi_i \tag{10.8}$$

where $\gamma = 1, 3$ for the streamwise and crossplane experiment, respectively. For the robustness analysis of the modal energies, the integral eigenvalue problem, (10.6), was estimated from the full cross-correlation tensor $\widetilde{\Phi}_{i,j}$ while for the robustness tests of the eigenfunctions, (10.6) was solved separately for $i = j$ allowing a distinction between the streamwise, radial, and azimuthal modes to be made.

10.5 Experimental Results

Using a non-linear optimization scheme, a curve fit was applied on the mean streamwise velocity component in order to estimate the scaling parameters, B_u, x_0, and A objectively for both cameras. Minimizing the deviations in a least-squares sense between the measured streamwise mean velocity profile, U_x, and a modelled profile expressed by a sech-function, the scheme determined the sought scaling parameters is formulated as

$$\left| U_x - U_c \operatorname{sech}^2\left(c\frac{r - \alpha\,(x - x_0)}{\delta\,(x)} \right) \right|_{\min} \tag{10.9}$$

$$\left| U_x - U_c \operatorname{sech}^2\left(c\frac{\sqrt{(y - y_0)^2 + (z - z_0)^2}}{\delta\,(x)} \right) \right|_{\min} \tag{10.10}$$

where $U_c = B\sqrt{M_0}/\,(x - x_0)$ is the modelled centerline velocity and $B = 2B_u/\sqrt{\pi}$, M_0 is the momentum at the nozzle, $c = \operatorname{asech}(1/\sqrt{2})$, α is the in-plane angle of rotation of the coordinate system, and y_0 and z_0 are the jet center position in the crossplane. The scaling parameters for both experiments feature in Table 10.2 and show close resemblance. For the analysis of the streamwise experiment, an average value of the scaling parameters obtained from the cameras 1 and 2 was used corresponding to $B_u = 5.72$, $x_0 = 3.1D$, and $A = 0.093$. These parameters can be compared to the parameters from [26] who obtained $B_u = 5.76$, $x_0 = 2.4D$, and $A = 0.093$ and $B_u = 5.8$, $x_0 = 4.0D$, and $A = 0.093$ from the laser doppler anemometry (LDA) measurements in [17]. Finally, the radial mean velocity profiles can be tested against those obtained from continuity from the following expression:

$$\frac{U_r}{U_c} = \frac{d\delta}{dx}\left(\eta f - \frac{1}{\eta}\int_0^{\eta} f\tilde{\eta}\,d\tilde{\eta} \right) \tag{10.11}$$

where $f = U_x/U_c$ is the mean streamwise velocity normalized by the centerline velocity.

10.5.1 Single-Point Statistics

10.5.1.1 Streamwise Plane

The first order statistical moments measured in the streamwise plane are shown in Fig. 10.2 at different streamwise distances from the nozzle. The profiles are seen for cameras 1 and 2 in the left and right column, respectively, and are compared to hot-wire (HW) and LDA data from [17]. Expressed in similarity coordinates, the profiles collapse at all downstream locations and agree well with the HW (at least around the center of the jet where the turbulence intensities are sufficiently low for the HW to provide reliable results, [17]) and LDA profiles. In Fig. 10.2c, d the mean radial velocity profiles are shown. Although these are two orders of magnitude smaller than the streamwise components, they seem to collapse relatively well with the profile obtained from continuity, deviating mostly at the radial extremities. The small values of the measured velocities are of the order of the inherent errors when sampled with PIV, as the radial displacement of seeding particles is very small during the time between pulses. For these measurements, this time was set to 150 μ s, corresponding to a particle displacement of 0.3 Pix at $x = 72$D, indicating that pixel locking could have a noticeable effect on the results. The slight differences between the collapsed profiles and the theoretical profiles in Fig. 10.2c, d are likely to primarily be a result of these factors and the reduction of laser sheet intensity when moving downstream.

The second moments measured in the streamwise plane are seen in Fig. 10.3. Figure 10.3(a)–(b), 10.3(c)–(d), and 10.3(e)–(f) show the streamwise-, radial- and shear-stresses, respectively, at various streamwise locations. The most noticeable difference between the profiles from the cameras is the degree to which the profiles collapse. The profiles from camera 2 collapse to a higher degree than those from camera 1. From similarity analysis, it is possible to show that the jet width increases linearly with downstream distance from the nozzle [3, 26], meaning that the scales increase relative to the (fixed size) interrogation areas with downstream distance from the nozzle, reducing the effective spatial filtering. It is therefore not surprising to see that the second order moments display less scatter for camera 2 than for camera 1.

The PIV data shows in general lower normal stress values, especially around the centerline, compared to the data from [17]. This is most likely due to the much lower Reynolds number (20,000) in the current experiments [15] than those from [17] which are $\mathcal{O}\left(10^5\right)$. A second contributory factor arises from the magnification of self-noise in the normal stresses resulting from the second power working on the moments. This is contrasted by the shear stresses in Fig. 10.3e, f, where the results are closer to the LDA data as the noise from the two components is uncorrelated and

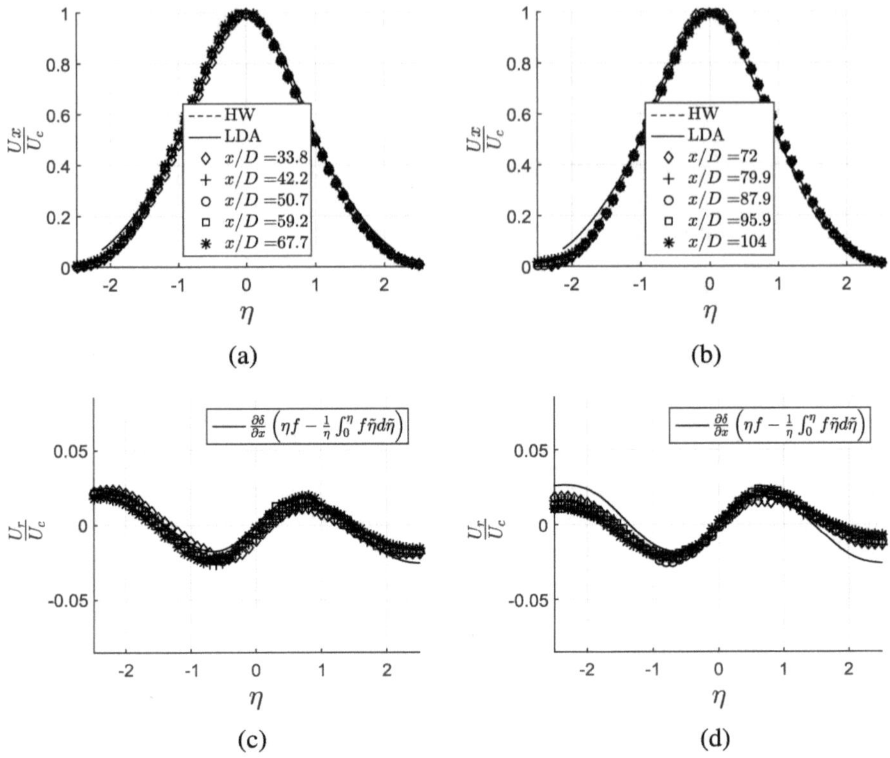

Fig. 10.2 First order moments for the streamwise plane at various downstream locations in similarity coordinates with mean profiles and hot-wire (HW) and LDA models from [17]. The mean streamwise velocites for cameras 1 and 2 are shown in (**a**) and (**b**), while the mean radial velocities are shown in (**c**) and (**d**) together with the profiles obtained from (10.11)

therefore cancel out by averaging. The random noise becomes perfectly correlated when the random numbers are multiplied by themselves such as happens for the normal stresses. For shear stresses, the correlation is only performed between uncorrelated random numbers and thus average out with increasing statistical basis.

10.5.1.2 Crossplane

The single-point statistics from the crossplane measurements are shown in Fig. 10.4 together with hot-wire and LDA curve fits from [17]. The data is taken from six radial "slices" with a $\frac{\pi}{3}$ rad spacing along the azimuthal direction. The results agree fairly well with the LDA measurements, again taking into consideration that these were obtained from a much higher Re-number flow. Note that LDA measurements are more reliable than hot-wire measurements due to their ability to distinguish

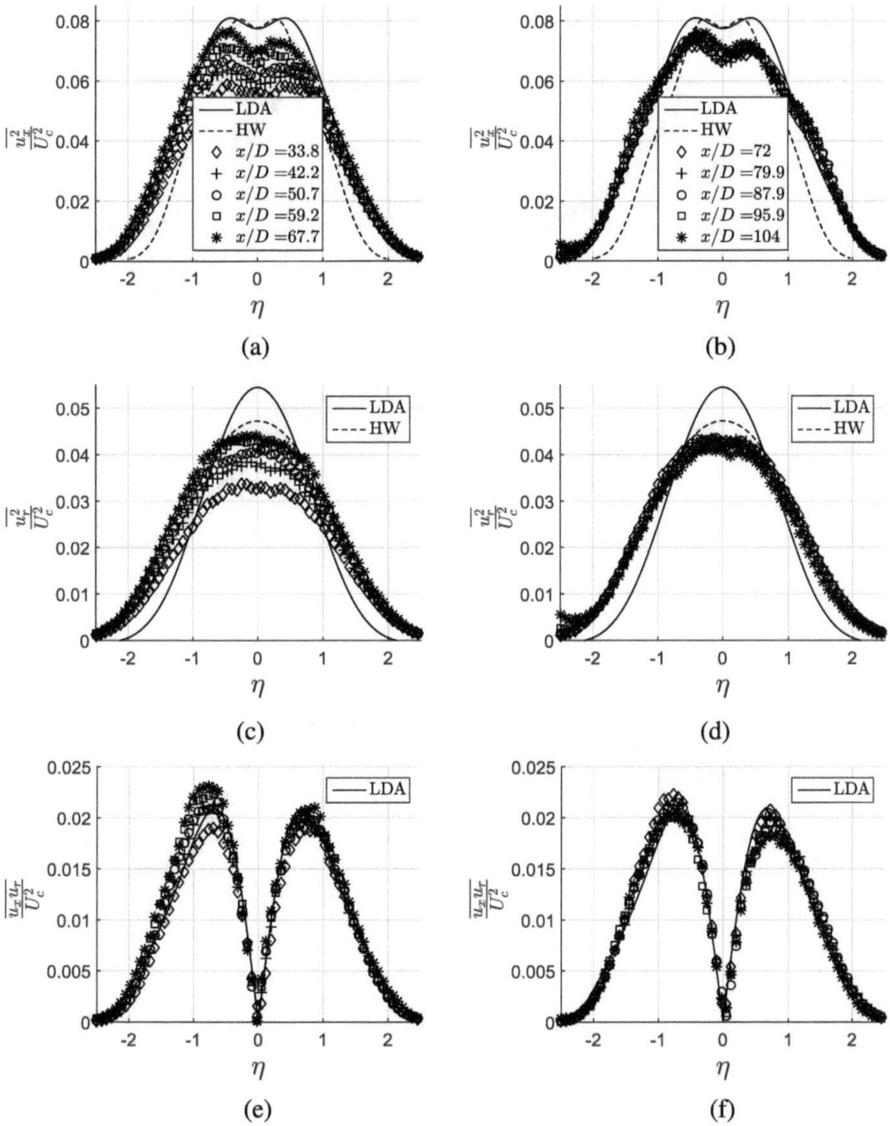

Fig. 10.3 Second order moments for the stremwise plane at various streamwise coordinates together with HW and LDA curve fits from [17]. (**a**)–(**b**), (**c**)–(**d**), and (**e**)–(**f**) show the streamwise-, radial and shear-stresses

flow directions, particularly in the high turbulence intensity regions, e.g., the off-axis regions of the jet. The mean radial velocity profile in Fig. 10.4c is seen to follow (10.11) relatively well, baring in mind that these velocities are two orders of magnitude smaller than U_x. The mean azimuthal velocity component is plotted in Fig. 10.4e to ensure that the resulting swirl is zero, as is seen to be the case from

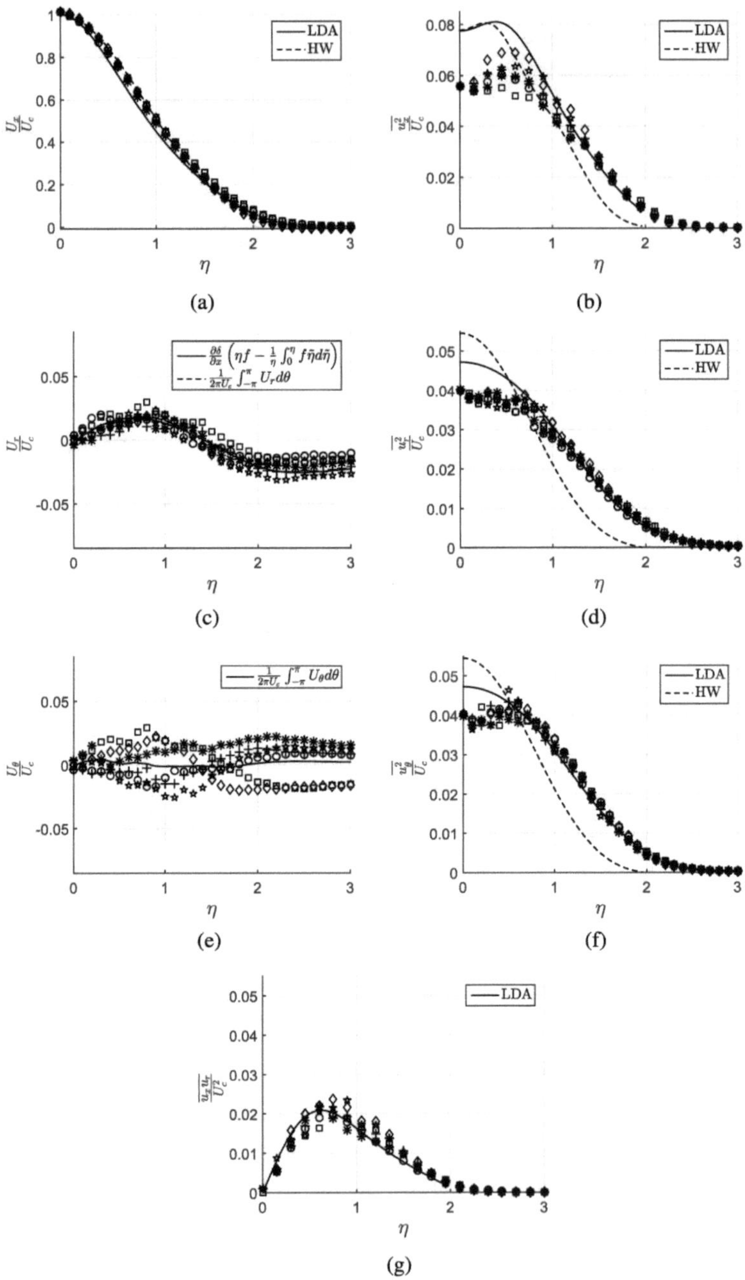

Fig. 10.4 First- and second order moments at various azimuthal locations in similarity coordinates with hot-wire (HW) and LDA curve fits from [17]. (**a**), (**c**) and (**e**) show the mean streamwise-, radial- and azimuthal velocities. (**b**), (**d**) and (**f**) show the streamwise-, radial and azimuthal stresses, while (**g**) shows the shear stresses. In (**c**) the azimuthally averaged profile and the radial profile obtained from (10.11) are also shown. In (**e**) the azimuthally averaged profile is also plotted

the azimuthally averaged profile represented by the full line. The normal stresses shown in Fig. 10.4b, d, f are inconsistent with the LDA data for $|\eta| \leq 0.75$. Similar tendencies were observed in [26] and [15] for PIV measurements in the streamwise plane using the same jet facility and in the hot-wire measurements from [22] at $Re_D = 11{,}000$. The lower values in these experiments compared to those in [17] are possibly due to the large difference in Reynolds numbers.

10.6 Modal Robustness

10.6.1 Streamwise Plane

Since the radial extent and resolution have an effect on the POD modes [7, 24], these parameters were varied systematically using high resolution 16 M cameras in order to quantify the consequences. Two analyses were performed: (1) For a constant grid resolution the radial extent was varied and (2) For a constant radial extent, the radial grid resolution was varied.

Figure 10.5a shows the relative energy distribution of mode numbers for various radial extents while a constant grid resolution of $\Delta\eta = 0.025$ is held. The wavenumber dependency has been integrated out and the energies are normalized by the turbulence kinetic energy. As seen in Fig. 10.5a, a large fraction of the energy resides in the first couple of modes and the dominance of the first mode is greatly enhanced by the shortening of the radial coverage. The percent-wise energy differences (relative to the energy for $|\eta| = 2.5$) are shown in Fig. 10.5c. These are supplemented by Table 10.3 summarizing the maximum and minimum energy differences of the first twenty modes in response to the radial extent and radial resolution variations. $n = 1$ and $n \geq 5$ project large energy variations when the radial extent is reduced. Modes $n = 1{:}3$ show an increase in relative energy while the large percent-wise decrease in energy for higher n indicates that low-energetic modes are affected to a higher degree, by the domain extent, than the most energetic modes. The results indicate a redistribution of energy from low to high mode numbers.

Analogously, results of the grid resolution analysis for a constant $|\eta| = 2.5$ are seen in Fig. 10.5b, d. Sensitivity with respect to the grid resolution manifests itself as an energy increase for $n \leq 10$. Contrary to the results from the radial extent variations, here the first mode is affected to a much lower degree.

The eigenfunction response to radial extent reductions, for $\Delta\eta = 0.025$, is seen in Fig. 10.6. Only the real parts of the modes are shown since the imaginary parts are negligible. The first three POD modes of the streamwise and radial components are seen in the left and right columns, respectively. As the radial extent is reduced, two effects are seen, namely an amplitude increase and a shift of the radial position of the nodes and anti-nodes towards the jet center. It should be noted that the

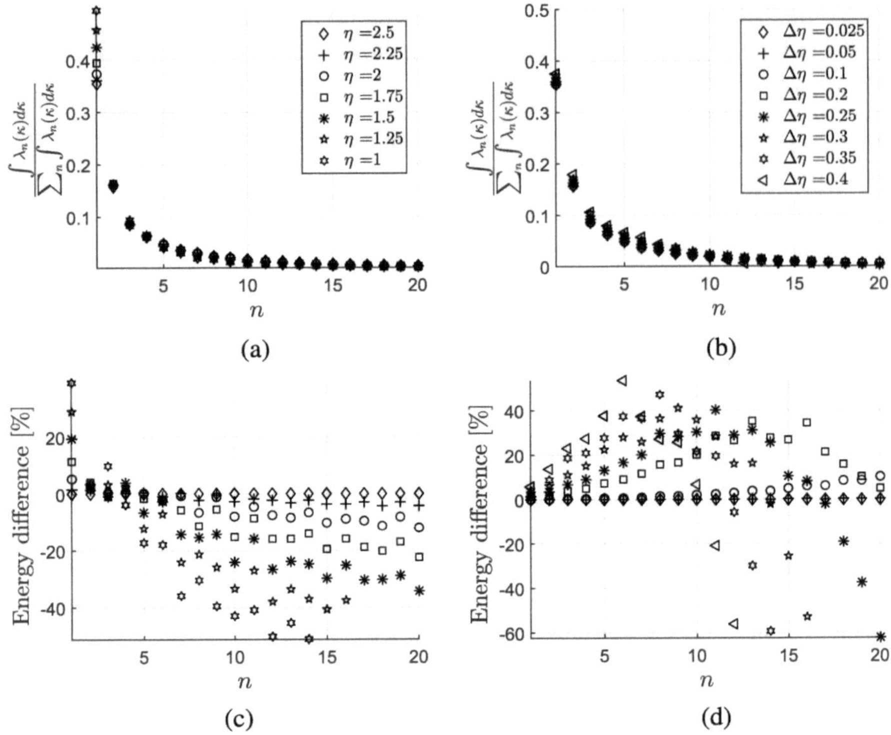

Fig. 10.5 Normalized energies of modes for the streamwise plane for various radial extents and radial resolutions. The *left column* (**a**), (**c**) shows the modal variation with radial extent for a constant grid resolution $\Delta\eta = 0.025$ while the *right column* (**b**), (**d**) demonstrates the corresponding variations for various radial grid resolutions at a fixed radial extent $|\eta| = 2.5$

eigenfunctions exert high robustness against radial extent for $|\eta| \geq 1.75$, as no significant variations to the eigenfunctions are seen for the first three mode numbers.

In Fig. 10.7 the radial resolution is varied while the radial extent is kept constant at $|\eta| = 2.5$. Here it is seen that the streamwise and radial modes experience an increase in amplitudes, but contradictory to the effects of the radial extent variations, the nodes remain unaffected.

10.7 Modal Robustness: Crossplane

From the crossplane SPIV measurements the radial extent variation ranged over $|\eta| \in [1 : 3]$ with a step size of 0.25 jet-half widths, while the spatial resolution was again kept at a constant $\Delta\eta = 0.025$. The resulting eigenvalue variations for the first 20 Fourier mode numbers, m, are shown in Fig. 10.8a and c for $n = 1$ and $n = 2$. The energies were normalized by the total energies of the respective fields.

Table 10.3 Percent-wise energy differences for the streamwise plane for various radial extents and resolutions, relative to the largest radial extent, $|\eta| = 2.5$, and radial resolution, $\Delta\eta = 0.025$

Mode		Radial extent, $\lvert\eta\rvert$						Radial resolution, $\Delta\eta$						
		2.25	2.0	1.75	1.5	1.25	1.0	0.05	0.1	0.2	0.25	0.3	0.35	0.4
$n \leq 20$	Max	1.8	5.4	11.4	19.5	29.1	39.3	0.3	10.3	35.2	40.1	41.2	47.3	53.5
	Min	−4.3	−11.9	−22.4	−34.4	−40.6	−51.2	0.0	0.0	1.1	−62.4	−52.8	−59.2	−56.2

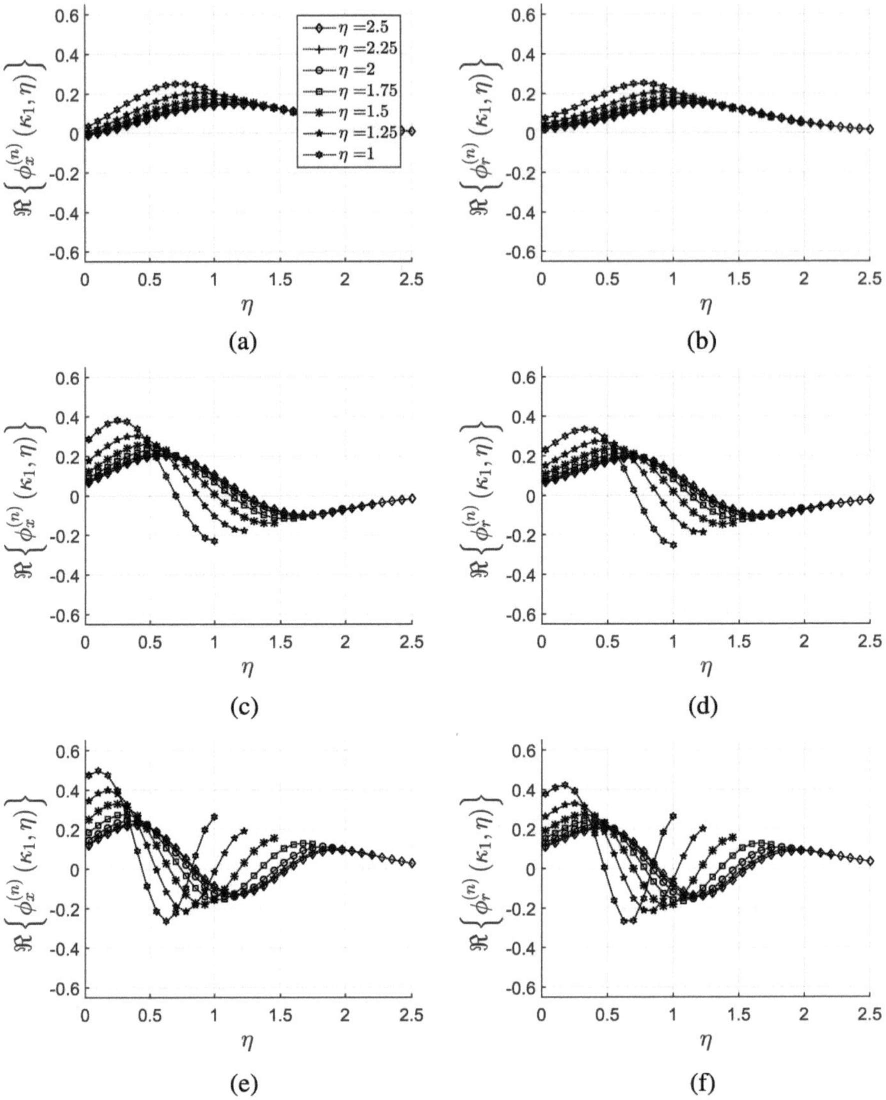

Fig. 10.6 Real parts of the first three POD modes for the streamwise plane for various radial extents. In the *left* and *right columns* the streamwise and radial components are seen, respectively. The legend in (**a**) is global. (**a, b**) $n = 1$. (**c, d**) $n = 2$. (**e, f**) $n = 3$

The results show that $m = 2$ is the dominant azimuthal mode for $n = 1$, as observed in [7], but only for $|\eta| \geq 1.25$. For the extreme case of $\eta = 1$ the energy peak shifts from $m = 2$ to $m = 1$. The energy differences for $n = 1$ and $n = 2$ are quantified in Fig. 10.8e and g as the percent-wise differences relative to the case of $|\eta| = 3.0$ and $\Delta\eta = 0.025$. Table 10.4 provides an overview of the maximum

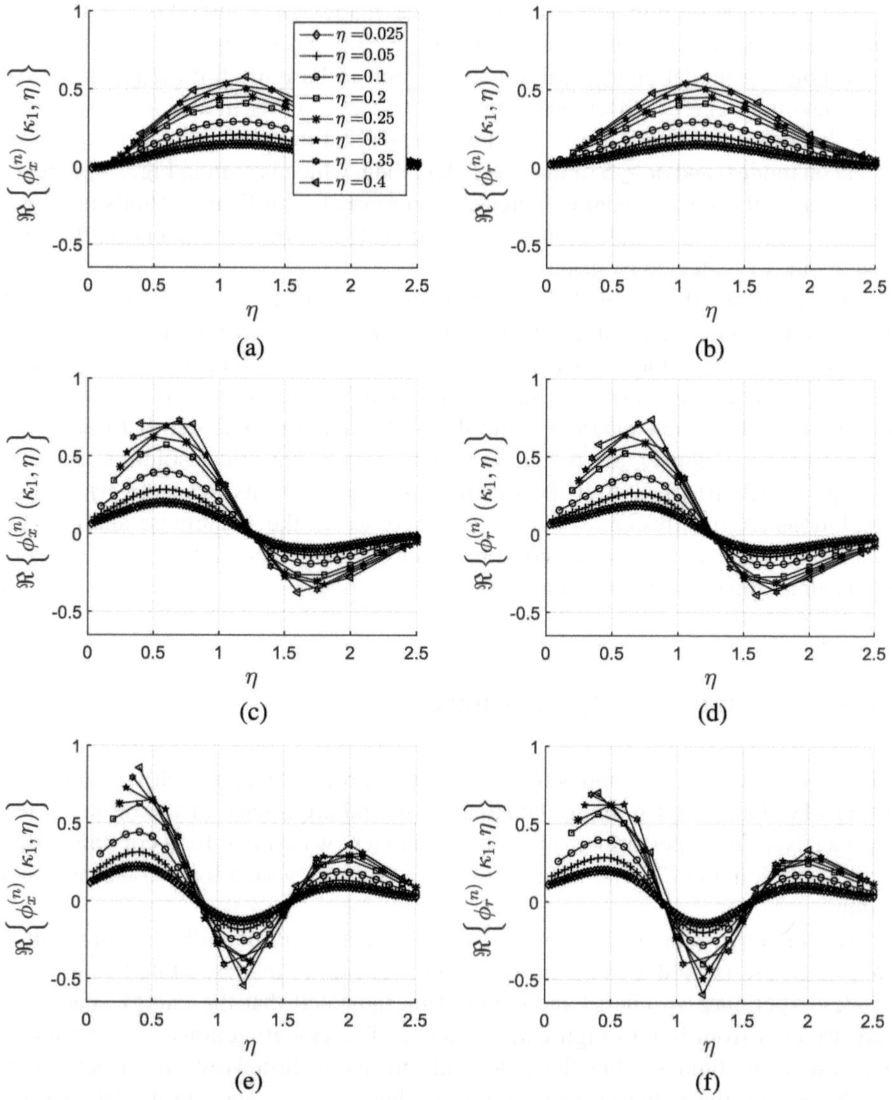

Fig. 10.7 Real parts of the first three POD modes for the streamwise plane for various grid densities. In the *left* and *right columns* the streamwise and radial components are seen, respectively. The legend in (**a**) is global. (**a, b**) $n = 1$. (**c, d**) $n = 2$. (**e, f**) $n = 3$

and minimum differences. It is clear that $m = 0$, 1, and 2 are mostly affected. The data shows that the energy differences are below 3.3 % for $|\eta| \geq 2.25$ for the first two POD modes and are consistently above 5.6 % percent for $|\eta| \geq 2.0$, while the maximum deviations for $n = 1$ and $n = 2$ are above 7.5 % for $|\eta| < 2.0$.

The radial resolution was varied over the range $\Delta\eta \in [0.025 : 0.4]$ while the radial extent was kept constant at $|\eta| = 3.0$. From Fig. 10.8b, d, f, h the corresponding modal energy variations are seen to be quite limited for $m \leq 10$. The maximum energy difference across the first 20 Fourier modes for $n = 1$ and $n = 2$ is below 0.3 % for $\Delta\eta \leq 0.1$. For $\Delta\eta \geq 0.2$, the radial grid has the greatest effect on mode 0 and $m \geq 5$ of the first POD mode, which is seen in Fig. 10.8f and h. The energy differences seem to increase proportionally for Fourier numbers $m \geq 5$ for $n = 1$ and $n = 2$ revealing once more that the higher mode numbers are the ones that are mostly affected by grid changes.

Figures 10.9, 10.10, and 10.11 show the real parts of the eigenfunctions for the x, r, θ-components, respectively, and a constant radial grid size of $\Delta\eta = 0.025$. All three components of the modes demonstrate high robustness for $|\eta| \geq 1.75$ as seen in the corresponding analysis of the streamwise data, but for $|\eta| < 1.75$ an increase in amplitude becomes apparent while the nodes and anti-nodes shift towards the jet center. Analogously, the modal variations with radial grid resolution are seen in Figs. 10.12, 10.13, and 10.14 for the x, r, θ-components, respectively. Lower resolutions are manifested, in the same manner as for the streamwise analysis, by increased modal amplitudes whereas the nodes and anti-nodes show no noticeable shifts along the radial coordinate.

10.8 Summary and Conclusions

The robustness of POD modes using a Lumley Projection approach was analyzed using PIV data sampled along the streamwise and crossplane of the turbulent jet. Two analyses were performed: (1) The radial extent was varied for a constant radial grid resolution and (2) The radial grid resolution was varied for a constant radial extent.

The streamwise analysis showed that the reduction of the radial extent resulted in an increase in modal energies for the most energetic modes, while higher POD modes experience an energy reduction. This indicated that the energy share was redistributed from low to high energy modes. The eigenfunctions experienced an increase in amplitudes, while the nodes and anti-nodes shifted towards the jet center.

Decreasing the radial grid resolution resulted in an increase in modal energies for $n \leq 10$, but overall affecting higher mode numbers to a higher degree than the low-energy modes. The eigenfunctions displayed an increase in amplitudes, while the radial positions of the nodes and anti-nodes remained unchanged.

For the crossplane analysis the radial extent proved to have the most significant effect on low Fourier mode numbers, $m = 0, 1, 2$. These experienced increased energy levels when the radial extent was reduced, while the energy decreased for high Fourier modes, consistent with the streamwise results. Furthermore, it was noted that the energy peak shifted from $m = 2$ to $m = 1$ for the first POD mode at $\Delta\eta = 1.0$.

Table 10.4 The percent-wise maximum and minimum energy differences for the crossplane of the first 20 Fourier modes relative to the highest radial extent, $|\eta| = 3.0$, and radial resolution, $\Delta\eta = 0.025$

Modes		Radial extent, $\|\eta\|$							Radial resolution, $\Delta\eta$							
		2.75	2.5	2.25	2.0	1.75	1.5	1.25	1.0	0.05	0.1	0.2	0.25	0.3	0.35	0.4
$n = 1$, $m \leq 19$	Max	0.5	1.3	3.3	7.5	15.3	31.7	62.6	125.0	0.0	0.2	5.0	9.3	19.4	34.7	43.3
	Min	0.4	1.0	2.6	5.6	6.9	4.3	−2.7	−16.9	−0.2	−0.7	−2.0	−2.6	−3.2	−3.0	−1.7
$n = 2$, $m \leq 19$	Max	0.5	1.3	3.2	7.7	17.2	35.1	65.4	114.3	0.0	0.3	6.7	19.5	33.5	46.1	78.6
	Min	0.3	0.8	2.0	2.2	−1.3	−7.6	−16.5	−32.0	−0.1	−0.2	−0.3	0.4	0.2	1.4	2.4

Results are presented for the first two POD modes

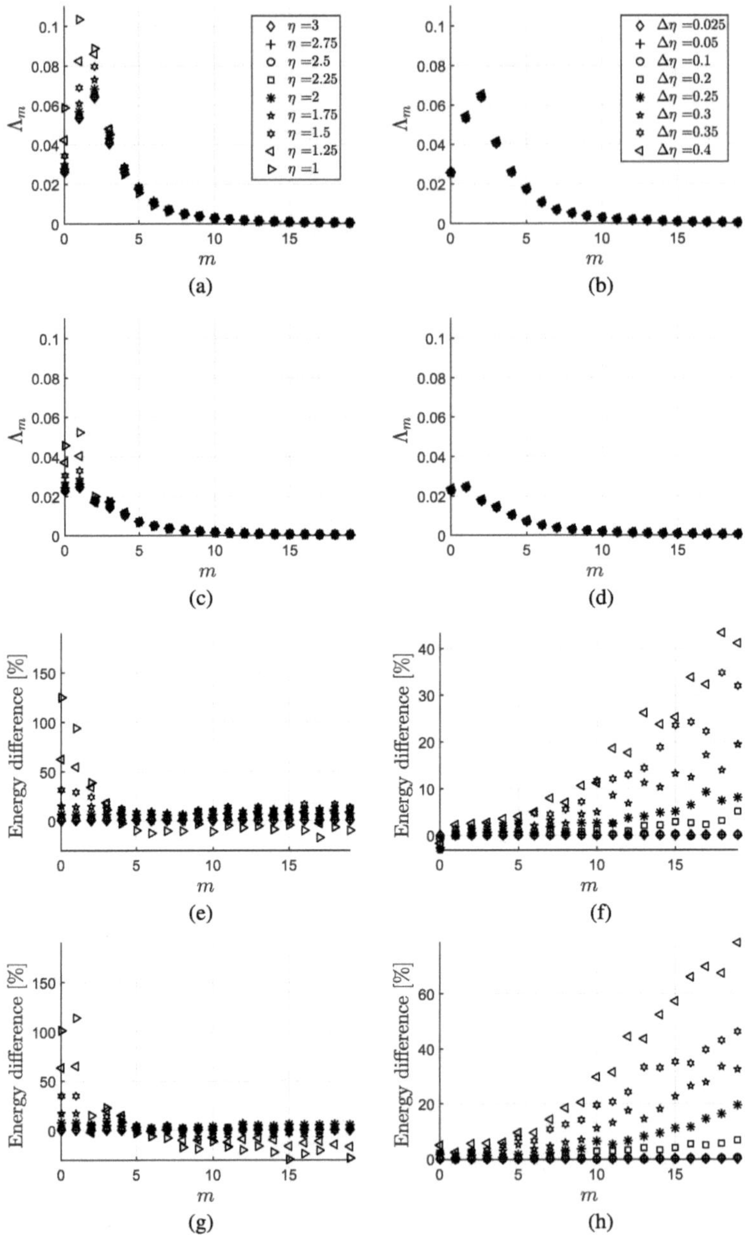

Fig. 10.8 Normalized energies of modes for various radial extents and radial resolutions for the crossplane measurements for the first two POD modes and the first 20 Fourier modes. The *left column* shows the modal variation with radial extent for $n = 1, 2$ in (**a**) and (**c**), respectively, and the energy deviations from the expected values in (**e**) and (**g**), respectively. The *right column* demonstrates the corresponding variations for various radial grid resolutions, $\Delta\eta$, for $n = 1, 2$. The legend in (**a**) is global for the *left column* and the legend in (**b**) is global for the *right* one

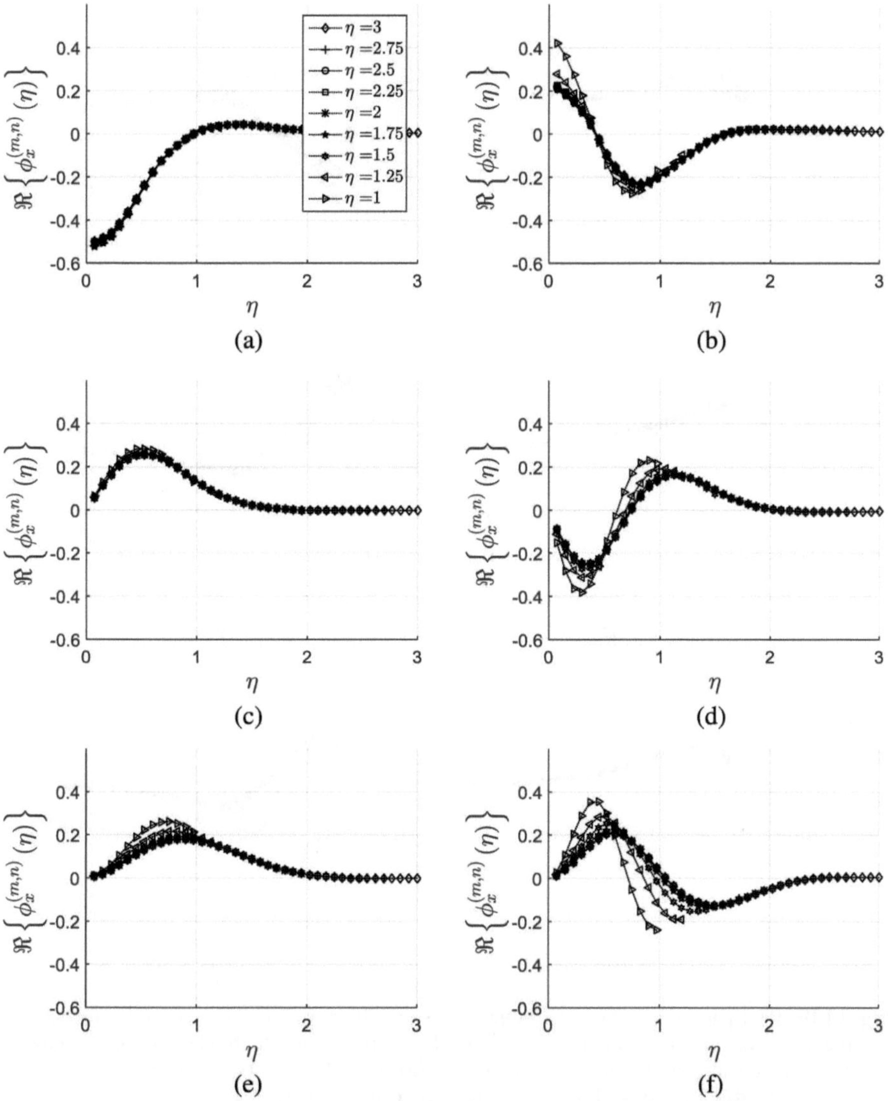

Fig. 10.9 Real parts of the first two POD modes and the first three Fourier modes for various radial extents for the crossplane measurements. The figure shows the modes for the streamwise velocity component. The legend in (**a**) is global. (**a**) $n = 1, m = 0$. (**b**) $n = 2, m = 0$. (**c**) $n = 1, m = 1$. (**d**) $n = 2, m = 1$. (**e**) $n = 1, m = 2$. (**f**) $n = 2, m = 2$

The radial resolution analysis revealed that modal energies were robust for $m \leq 10$ corresponding to an energy increase of 0.3 % for $\Delta\eta \geq 0.2$. For $\Delta\eta \geq 0.2$ mode 0 and $m \geq 5$ of $n = 1$ were mostly affected where higher mode numbers displayed the highest sensitivity to radial grid resolution.

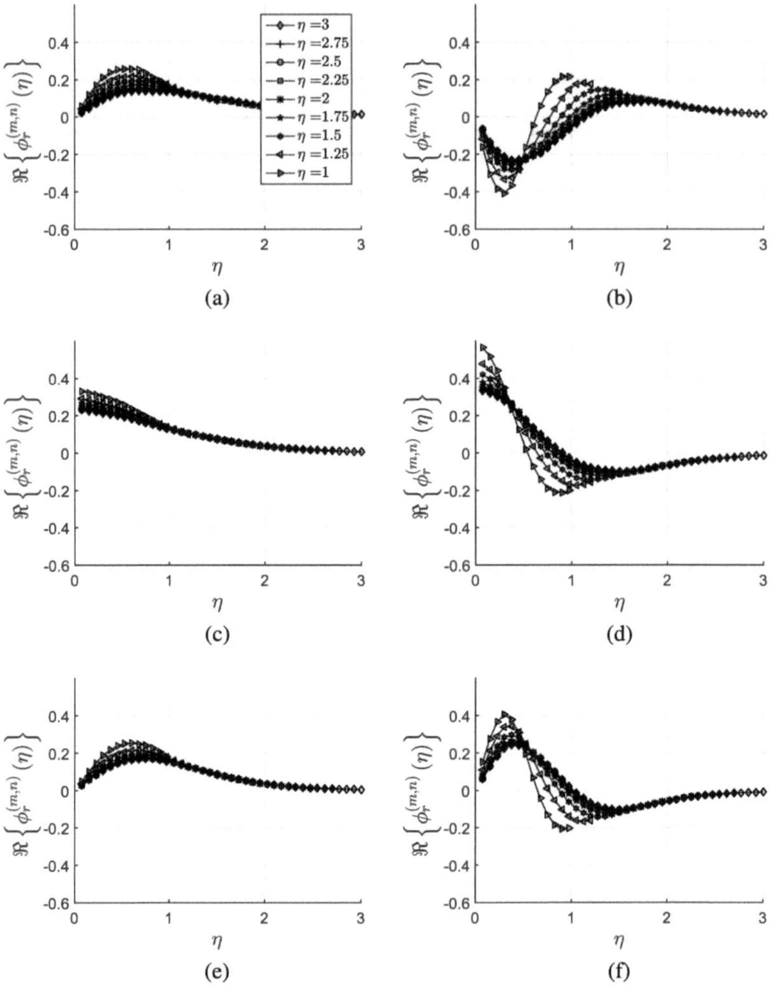

Fig. 10.10 Real parts of the first two POD modes and the first three Fourier modes for various radial extents for the crossplane measurements. The figure shows the modes for the streamwise velocity component. The legend in (**a**) is global. (**a**) $n = 1, m = 0$. (**b**) $n = 2, m = 0$. (**c**) $n = 1, m = 1$. (**d**) $n = 2, m = 1$. (**e**) $n = 1, m = 2$. (**f**) $n = 2, m = 2$

As in the streamwise analysis, the eigenfunctions demonstrated a high robustness against radial extent reductions for $|\Delta\eta| \geq 1.75$ where only minor amplitude increases were observed for $m = 0$, $m = 1$, and $m = 2$ for the first two POD modes. For $|\eta| \leq 1.5$ the effects were seen to increase the amplitudes and move the nodes and anti-nodes closer to the jet center. The radial resolution reduction resulted in an increase in amplitudes, while the radial positions of the nodes and anti-nodes remained unaffected.

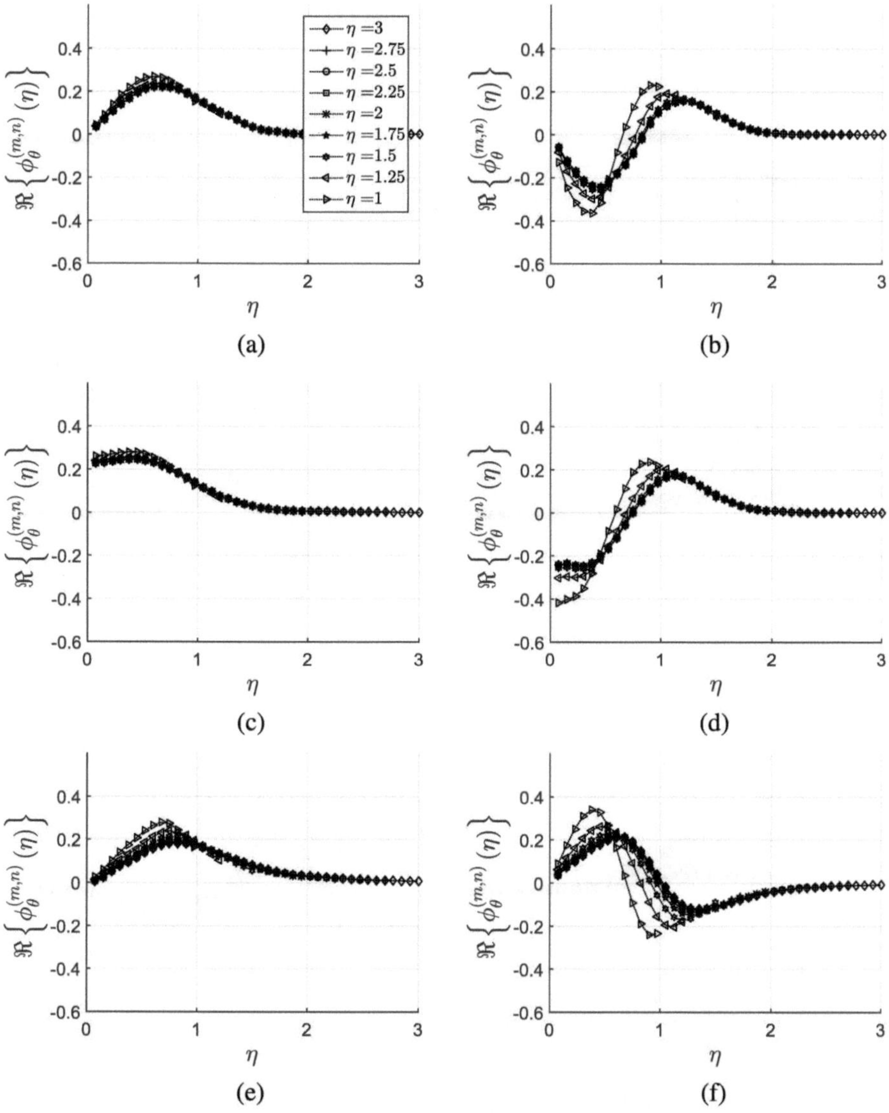

Fig. 10.11 Real parts of the first two POD modes and the first three Fourier modes for various radial extents for the crossplane measurements. The figure shows the modes for the azimuthal velocity component. The legend in **(a)** is global. **(a)** $n = 1, m = 0$. **(b)** $n = 2, m = 0$. **(c)** $n = 1, m = 1$. **(d)** $n = 2, m = 1$. **(e)** $n = 1, m = 2$. **(f)** $n = 2, m = 2$

It is the hope of the authors that the results of the current study can be used as a guideline for interpreting previous and designing future experiments in the round turbulent jet and similar types of flows.

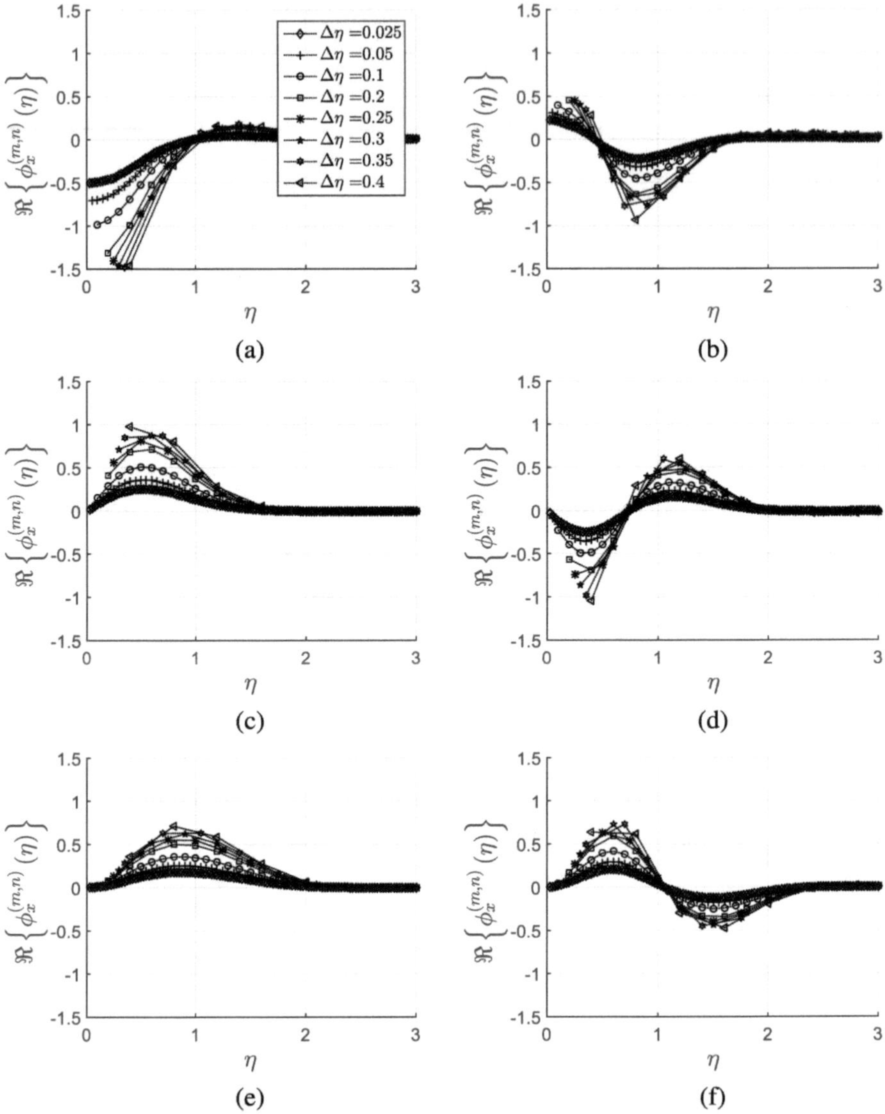

Fig. 10.12 Real parts of the first two POD modes and the first three Fourier modes for various radial resolutions for the crossplane measurements. The figure shows the modes for the streamwise velocity component. The legend in (**a**) is global. (**a**) $n = 1, m = 0$. (**b**) $n = 2, m = 0$. (**c**) $n = 1, m = 1$. (**d**) $n = 2, m = 1$. (**e**) $n = 1, m = 2$. (**f**) $n = 2, m = 2$

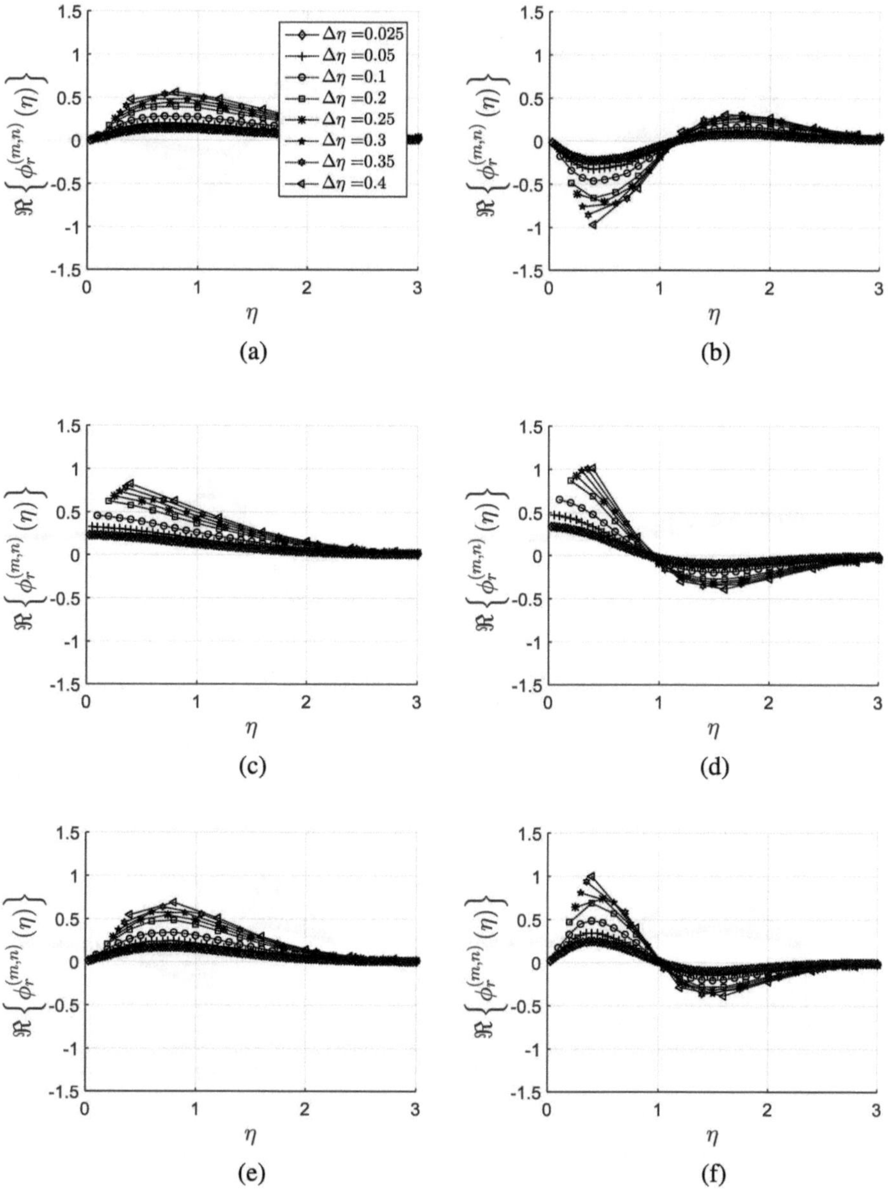

Fig. 10.13 Real parts of the first two POD modes and the first three Fourier modes for various radial resolutions for the crossplane measurements. The figure shows the modes for the radial velocity component. The legend in (**a**) is global. (**a**) $n = 1, m = 0$. (**b**) $n = 2, m = 0$. (**c**) $n = 1, m = 1$. (**d**) $n = 2, m = 1$. (**e**) $n = 1, m = 2$. (**f**) $n = 2, m = 2$

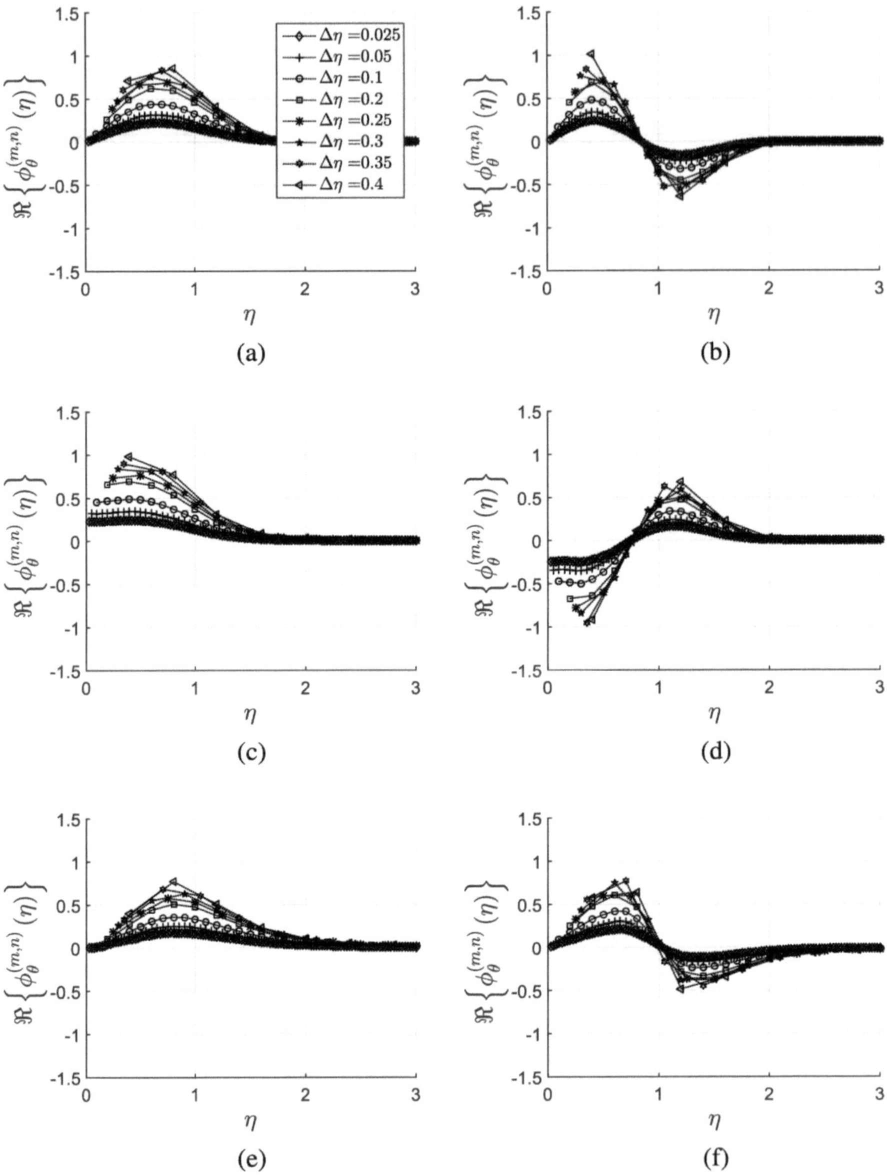

Fig. 10.14 Real parts of the first two POD modes and the first three Fourier modes for various radial resolutions for the crossplane measurements. The figure shows the modes for the azimuthal velocity component. The legend in (**a**) is global. (**a**) $n = 1, m = 0$. (**b**) $n = 2, m = 0$. (**c**) $n = 1, m = 1$. (**d**) $n = 2, m = 1$. (**e**) $n = 1, m = 2$. (**f**) $n = 2, m = 2$

References

1. N. Aubry, On the hidden beauty of the proper orthogonal decomposition. Theor. Comput. Fluid Dyn. **2**, 339–352 (1991)
2. J.H. Citriniti, W.K. George, Reconstruction of the global velocity field in the axisymmetric mixing layer utilizing the proper orthogonal decomposition. J. Fluid Mech. **418**, 137–166 (2000)
3. D. Ewing, On multi-point similarity solutions in turbulent free-shear flows. Ph.D. thesis, Lyngby, 1995
4. D. Ewing, B. Frohnapfel, W.K. George, J.M. Pedersen, J. Westerweel, Two-point similarity in the round jet. J. Fluid Mech. **577**, 309 (2007)
5. B. Frohnapfel, Multi-point similarity of the axisymmetric turbulent far jet and its implication for the POD. Master thesis, Friedrich-Alexander-Universität, Erlangen-Nüremberg, 2003
6. S. Gamard, W.K. George, D. Jung, S. Woodward, Application of a "slice" proper orthogonal decomposition to the far field of an axisymmetric turbulent jet. Phys. Fluids **14**, 2515 (2002)
7. S. Gamard, D. Jung, W.K. George, Downstream evolution of the most energetic modes in a turbulent axisymmetric jet at high Reynolds number. Part 2. The far-field region. J. Fluid Mech. **514**, 205–230 (2004)
8. W. George, Insight into the dynamics of coherent structures from a proper orthogonal decomposition. International Seminar on Wall Turbulence, Dubrovnik, 1988
9. W. George, The self-preservation of turbulent flows and its relation to initial conditions and coherent structures, in *Advances in Turbulence* (Springer, Berlin, 1989)
10. W. George, Some new ideas for similarity of turbulent shear flows. in *Proceedings of the ICHMT Symposium on Turbulence, Heat and Mass Transfer* ed. by K. Hanjalic & J.C.F. Peiera (Begell House, Lisbon, 1994), pp. 13–24
11. W.K. George, Lectures in Turbulence for the 21st Century, 2013
12. M. Glauser, S. Leib, W. George, *Coherent Structures in the Axisymmetric Turbulent Jet Mixing Layer* (Springer, Berlin, 1987)
13. S. Gordeyev, F. Thomas, Coherent structure in the turbulent planar jet. Part 1. Extraction of proper orthogonal decomposition eigenmodes and their self-similarity. J. Fluid Mech. **414**, 145–194 (2000)
14. S.V. Gordeyev, F.O. Thomas, Coherent structure in the turbulent planar jet. Part 2. Structural topology via POD eigenmode projection. J. Fluid Mech. **460**, 349–380 (2002)
15. A. Hodzic, PIV measurements on a turbulent free jet - spatial decomposition of a turbulent free jet using proper orthogonal decomposition. Master thesis, Technical University of Denmark, Kgs. Lyngby, 2014
16. P. Holmes, J. Lumley, G. Berkooz, *Turbulence, Coherent Structures, Dynamical Systems and Symmetry* (Cambridge University Press, University, 1998)
17. H.J. Hussein, S.P. Capp, W.K. George, Velocity measurements in a high-Reynolds-number, momentum-conserving, axisymmetric, turbulent jet. J. Fluid Mech. **258**(1), 31–75 (1994)
18. M.O. Iqbal, F.O. Thomas, Coherent structure in a turbulent jet via a vector implementation of the proper orthogonal decomposition. J. Fluid Mech. **571**. 281–326 (2007)
19. D. Jung, S. Gamard, W.K. George, Downstream evolution of the most energetic modes in a turbulent axisymmetric jet at high Reynolds number. Part 1. The near-field region. J. Fluid Mech. **514**, 173–204 (2004)
20. J.L. Lumley, The structure of inhomogeneous turbulent flows, in *Atmospheric Turbulence and Radio Wave Propagation* (Nauka, Moscow, 1967), pp. 166–178
21. P. Moin, R.D. Moser, Characteristic-eddy decomposition of turbulence in a channel. J. Fluid Mech. **200**, 471–509 (1989)
22. N.R. Panchapakesan, J.L. Lumley, Turbulence measurements in axisymmetric jets of air and helium. Part 1. Air jet. J. Fluid Mech. **246**, 225–247 (1993)
23. J.M. Pedersen, Analysis of planar measurements of turbulent flows. Ph.D. thesis, Technical University of Denmark, 2003

24. C.E. Tinney, M.N. Glauser, L.S. Ukeiley, Low-dimensional characteristics of a transonic jet. Part 1. Proper orthogonal decomposition. J. Fluid Mech. **612**, 107–141 (2008)
25. C.E. Tinney, L.S. Ukeiley, M.N. Glauser Low-dimensional characteristics of a transonic jet. Part 2. Estimate and far-field prediction. J. Fluid Mech. **615**, 53–92 (2008)
26. M. Wänström, Spatial decompositions of a fully-developed turbulent round jet sampled with particle image velocimetry. Ph.D. thesis, Chalmers University of Technology, 2009

Part IV
Environmental Flows and Wind Energy

Chapter 11
Large Eddy Simulation of Environmental Flows: From the Laboratory-Scale Numerical Experiments Toward Full-Scale Applications

Vincenzo Armenio

11.1 Introduction

Environmental fluid mechanics (EFM) concerns the study of the flow of air and water along with processes concerned with the quality of the environment. Among these, EFM studies the transport of chemical and biological species in water basins and in the low atmosphere.

EFM is part of the field of Fluid Mechanics and can be characterized in dependence of the scales of motion. Typically, these are of the order of few kilometres over the horizontal direction and of tens to few hundred metres along the vertical direction. Some examples are the dispersion of a pollutant jet in coastal regions or lakes and the emissions of plumes from chimneys in the low atmosphere. At environmental scales, stratification is the rule rather than the exception. In the low atmosphere, stratification is associated with vertical distribution of potential temperature and humidity; it regulates dispersion of pollutants at daily time scales associated with the diurnal cycles, from the nocturnal stable stratified condition to the diurnal convective one. In marine basins, thermal and saline stratification are always present and their variation occurs over seasonal time scales, playing a role in the transport of chemical and biological species within the basin.

In typical environmental applications, geometric features play an important role in the transport of mass, momentum, temperature and concentration. A common example is the katabatic/anabatic flow occurring in valley during the night/day, and the effect of coastline and bathymetry in mixing in coastal flows. Regarding the latter case, it is well known that man-made structures, like piers or docks, may dramatically modify the local water circulation with impact over local economy.

V. Armenio (✉)
Dipartimento di Ingegneria e Architettura, University of Trieste, Piazzale Europa 1, I-34127 Trieste, Italy
e-mail: armenio@dica.units.it

A. Pollard et al. (eds.), *Whither Turbulence and Big Data in the 21st Century?*,
DOI 10.1007/978-3-319-41217-7_11

Nowadays, numerical models are largely used for modelization of environmental flows. These models generally solve reduced forms of the Navier–Stokes (NS) equations (i.e. shallow water equations) or the non-hydrostatic Reynolds averaged NS equations using standard turbulence closures. In these models, mixing features are reproduced by turbulence models whose accuracy may be questionable in presence of combination of complex physics and geometry. Recently, in coastal hydrodynamics the need of eddy-resolving simulations for reproduction of mixing features has been highlighted in the review paper by Burchard et al. [1].

Among the available numerical methodologies to deal with turbulent flows, recently, Large Eddy Simulation (LES) has become from being promising to be reliable in a wide class of flows, especially those of interest in EFM. In LES, the larger scales of motion, energetic, anisotropic and dependent on the boundary conditions are directly resolved through a time-dependent, three-dimensional and non-hydrostatic simulation; and the smaller, more isotropic and dissipative, scales are parametrized by means of a subgrid-scale (SGS) model. Originally, LES was developed by the meteorological community for the evaluation of the characteristics of the atmospheric boundary layer under neutral or stratified conditions (see, for instance, [2]). These studies were characterized by simple geometry (usually a double periodic domain in the horizontal directions) and the use of wall-layer models to deal with real-scale values of the Reynolds number.

With the advent of the dynamic model of Germano et al. [3] and successive modifications, LES has been employed as an alternative to direct numerical simulations (DNS) for the analysis of flows characterized by complex physics or geometry. The dynamic model has shown its own ability to reproduce complex physics flows, characterized by re-laminarization and transition, non-equilibrium turbulence, rotation and stratification. The model has also proven to be able to reproduce flows characterized by massive separation, like those developing downstream of bluff bodies.

Most LES's of wall-bounded turbulent flows have been carried out by solving the flow field up to the wall. Thus, there is the need to solve the near-wall layer populated by turbulent structures responsible for the increase of the wall-shear stress with respect to the laminar case. The cost of a wall-resolving simulation has been estimated to be of the order of $Re^{2.5}$ compared to computational cost of a DNS of about $Re^{3.5}$ [4]. Due to the substantial cost of a wall-resolving LES it has been named *quasi-dns* by Spalart et al. [5].

For the reasons mentioned above, DNS's and wall-resolving LES's are not suitable to solve real-scale environmental flows with the computational capacity at present. Also, another important conceptual problem arises: real-scale environmental flows are always characterized by wall roughness. In principle, a wall-resolving LES applied to a rough wall, should be able to treat in a deterministic way the wall corrugation, but this is impossible when dealing with realistic cases. A parametrization of the rough geometry is thus required making unfeasible the use of a no-slip condition at the solid boundary.

A question may arise on the usefulness of investigations carried out with numerical tools unable to deal with real-scale values of Reynolds number, in particular in EFM where a number of processes concur to the development of the flow field. Incidentally, a noticeable answer to this question is contained in [6], where the authors have stressed the importance of DNS to gain significant insight into turbulence physics of idealized flows that cannot be easily attained in the physical laboratory. Here we follow the same philosophy, focusing on wall-resolving LES (or *quasi-DNS*) studies motivated by the need of exploiting the physics of turbulence relevant in environmental processes. Specifically, we show that wall-resolving LES is a valuable tool for studying simplified flow fields of interest for EFM at a laboratory scale. Also, we will show examples where some processes occurring in real-scale flows can be hardly reproduced in physical experiments.

However a gap still exists between idealized flows studied at laboratory scale using state-of-the-art numerical techniques and real-scale flows characterized by a number of complexities usually solved employing low-order tools. This is particularly true for coastal hydrodynamics. Most software for engineering purposes solves the Reynolds averaged Navier–Stokes (RANS) equations using low-order algorithms. When these solvers are turned onto the *LES-mode* they obviously cannot supply reasonable results since they were not designed to work under these conditions. Hence, there is the need to formulate new-generation numerical models suited for EFM and derived from state-of-the-art computational techniques for the study of fluid dynamic processes.

The research group of industrial and EFM of the University of Trieste (*IE-Fluids*), led by the author of the present paper, has been working in the direction of technological transfer of numerical methodologies suited for archetypal laboratory-scale flows to the solution of full-scale realistic problems over the last years. In the present paper, the steps taken to produce numerical models able to deal with real-scale problems taking advantage of the LES methodology are described. The models are, respectively, LES-COAST designed to evaluate turbulent mixing and hydrodynamics in coastal areas and LES-AIR for analysis of pollutant dispersion in the low atmosphere. Also, an example of application to a real site is presented.

In summary, the paper is composed of two main parts. In the first one, a brief description of the mathematical framework is given and few significant examples of laboratory-scale LES numerical experiments carried out by *IE-Fluids* are presented. In the second part, the structure of LES-COAST and LES-AIR is presented together with a brief description of the comprising parts which have led to the development of the models; an example of applications is also shown. Finally, concluding remarks are given with a viewpoint of the author on the way ahead.

11.2 The LES Mathematical Model

In LES, the separation of large scales from the small ones is carried out formally applying a low-pass filter to the Navier–Stokes equations. According to the implicit-filter methodology, the filter width is proportional to the size of the computational cells used in the simulation. As mentioned, stratification is ubiquitous of environmental flows and, usually, the density variations are small compared to the bulk density of the fluid. Also, accelerations are usually small compared to the gravitational acceleration. Under these circumstances, the Boussinesq approximation can be invoked and the equations assume the incompressible form with an additional buoyancy term accounting for gravitational effects in the fluid flow. Additionally, the transport equations for the active scalars must be considered. They are, respectively, temperature and concentration of a dissolved species (for water basins, for example, salinity or suspended sediments). The set of equations is complemented with the equation of state which gives the perturbation density as a function of temperature and concentration.

In wall-resolving LES, the no-slip condition is applied at the solid walls for the velocity field. For the scalar fields either Dirichlet or von Neuman (assigned flux) conditions can be used depending on the problem under investigation.

Here we discuss some cases where LES, carried out at a laboratory scale, has been successfully exploited to understand physical processes of interest to EFM. In all cases, the governing equations have been solved using the curvilinear-coordinate fractional step algorithm of [7]. The SGS models are dynamic with the constant averaged over the directions of homogeneity or along the trajectories of the fluid particles. The algorithm and the SGS model are described with details in [8, 9].

11.3 From the Laboratory-Scale Numerical Experiments...

Wall-resolving LES has been used by a number of researchers to exploit physical mechanisms of interest for practical applications. A discussion on relevant studies in this field is in [10]. Here we discuss two representative examples of flow fields of interest to EFM problems studied by *IE-Fluids* using laboratory-scale LES.

11.3.1 Tidal Flows

An archetypal problem of interest in environmental applications is the oscillatory (Stokes) boundary layer (SBL) which is representative of a class of flows relevant in coastal hydrodynamics. The study of SBL has implications for the comprehension of mixing properties in tidal flows. In this case, the forcing frequency is associated with the dominant tide at a certain location. In case of semi-diurnal tides (principal lunar M2, principal solar S2, larger lunar elliptic N2), the period of oscillation is $t_{tide} \sim 12\,\mathrm{h}$ and Earth rotation may play a role in the dynamics of the boundary layer; in fact, in the polar case, where the rotation effect is maximum, the Rossby number is O(1) ($Ro = \omega/f$ with ω the frequency of the tidal flow and $f = 2\Omega_E sin\theta$ the Coriolis parameter with Ω_E the Earth rotation frequency and θ the latitude). This situation is well far from that occurring in geophysical flows where $Ro << 1$ and

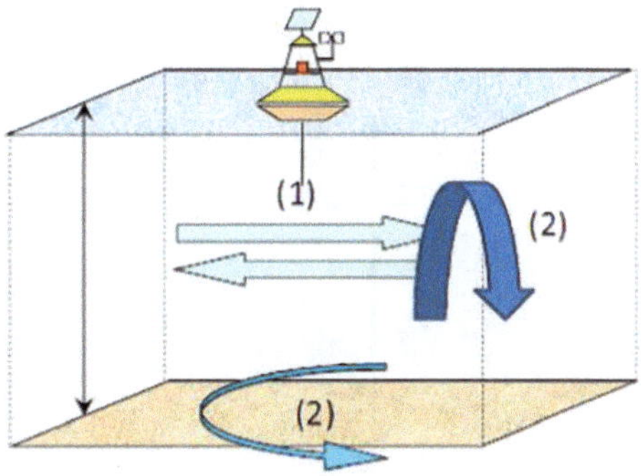

Fig. 11.1 Schematic of the tidal flow problem under the effect of Earth rotation

turbulence can be considered two-dimensional. On the other hand, since $Ro \sim O(1)$ rotation may play a role in the mixing properties of the boundary layer. Another issue of some importance is related to the horizontal, south-north component of the background rotation, always present in non-polar cases. This aspect has been investigated by few authors (see, for a discussion, [11]), since physical experiments aimed at the investigation of the effect of horizontal background vorticity cannot be performed using the standard rotating free-surface tanks.

The studies presented in [12–14] were devoted to exploit the characteristics of the SBL in the turbulent regime and the effect of Earth rotation on the fate of the oscillatory boundary layer at different latitudes (Stokes–Ekman boundary layer, SEBL). Figure 11.1 shows a schematic of the problem under investigation. Let's consider an M2 tide with period $t_{M2} = 12.42\,\text{h}$ and velocity amplitude of $U_0 = 5\,\text{cm/s}$. The free stream velocity can be represented as $U(t) = U_0 \sin \omega_{M2} t$ and the Reynolds number, based on the amplitude of motion $A_{M2} = U_0/\omega_{M2}$, is $Re_A = U_0 A_{M2}/\nu = 1.5 \times 10^7$. This gives a Reynolds number $Re_\delta = U_0\delta/\nu = 5500$ based on the Stokes BL nominal thickness $\delta = \sqrt{2\nu/\omega_{M2}}$. The Rossby number of the flow is $O(1)$ for mid-to-high latitudes. Based on the literature of the SBL, turbulence is fully developed along the whole oscillatory cycle from $Re_\delta = 3600$, so dimensional analysis suggests that the tidal flow operates under fully developed turbulent conditions. On the other hand, the dimensional analysis also suggests that Coriolis force may play a role in the dynamics of the tidal flow (Fig. 11.2).

Wall-resolving LES of this flow is quite expensive in terms of computational time, because many oscillatory periods must be reproduced to attain convergent statistics. Based on these considerations, numerical experiments have been designed in such a way to hold the Ro number of the real-scale case and reducing the value of Re_δ to a reasonable value, still remaining in a regime where equilibrium turbulence is present in most of the oscillatory cycle. The value $Re_\delta = 1790$ has been chosen, corresponding to test 8 of the experiments in [15].

In [12] the focus was on the SBL. At $Re_\delta = 1790$ the flow undergoes transition to turbulence during the early acceleration phases of the cycle (30–45°); equilibrium turbulence was observed in the range 60–120° with the boundary layer resembling a canonical, steady, turbulent BL; turbulence decay was observed from 120° up to the

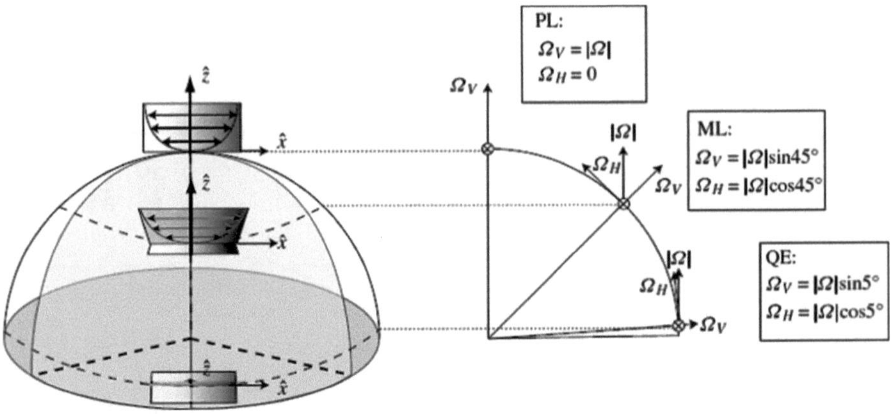

Fig. 11.2 Schematic of the tidal flow together with the background vorticity at three latitudes. Figure from [14]

new re-transition occurring during the early acceleration of the new oscillatory cycle with the flow populated of few large scale structures remnant from the previous equilibrium turbulence phases.

The effect of Earth rotation (SEBL) was considered in two successive papers: In [13] a specific flow condition was studied, where direction of the tide and latitude were representative of a water column in the Gulf of Trieste (Italy); In [14] a more general problem was studied, namely the east-west tidal flow at different latitudes. As mentioned, this is a typical study not reproducible using a rotating free-surface tank, due to the presence of the horizontal background vorticity.

The presence of the three components of the Coriolis force was shown to have two main effects: (1) the rising of a cyclic mean spanwise velocity component that, combined with the streamwise one, produces elliptic trajectories in the horizontal planes (see Fig. 11.3); (2) the breaking of the symmetry between the two semi-cycles. Both affect dramatically the time variation of the wall shear stress along an oscillatory cycle (see Fig. 11.4). Specifically, the maximum value of the wall shear stress decreases with the latitude and is minimum in the polar case. Also, unlike the other cases, in the polar case the wall-shear stress never reaches zero; the two half cycles are not symmetric in the non-polar cases, with larger values of the wall-shear stress in the second half cycle (tide flowing from east to west). Asymmetry between the two semi-cycles affects the wall shear stress by about 13–15 % in agreement u_τ/U_0 is equal to 0.033 in the polar case, to 0.044–0.045 in the mid-latitude case and to 0.045–0.048 in the quasi-equatorial case in agreement with measurements, usually between 1/30 and 1/20 (see among the others [16]).

Growth and decay of turbulent kinetic energy was observed to develop as in purely oscillatory flow (OF) with noticeable differences. Additional production terms appear in the transport equations of the Reynolds shear stresses, not present in the OF case, associated to the presence of the mean spanwise velocity and to the Coriolis force. This increases the efficiency of mixing along the whole fluid column, with respect to what it is observed in the OF case (Fig. 11.4).

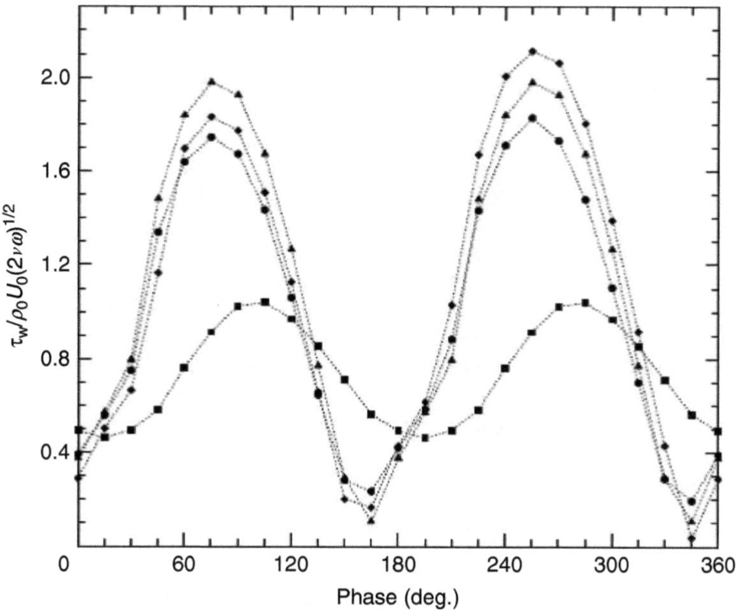

Fig. 11.3 Stokes-Ekman BL: non-dimensional wall-shear stress along the oscillatory cycle for the three rotational cases. For sake of comparison the OF case is also reported. Polar case, *squares*; Mid-latitude case, *circles*; Quasi-Equatorial case, *diamonds*; OF case, *triangles*. Figure from [14]

As a conclusion, wall-resolving LES was able to shed light on the hydrodynamic properties of a flow archetypal of the tidal motion in marine basins. The study shows that the SBL gives a good representation of a tidal current in the low latitude situations, apart the asymmetry between the two half cycles. For mid-to-high latitude cases, turbulence is strongly affected by rotation both in its intensity and in its own evolution along the cycle.

11.3.2 Dispersion Processes

Dispersion is ubiquitous of EFM. In engineering applications, dispersion problems are usually studied using Eulerian–Eulerian (EE) methodologies, in which the dispersed phase is treated through a transport equation for its concentration. This approach is not free from approximations and empiricism. In the research community, dispersion processes are also studied using a mixed Lagrangian–Eulerian technique (LE), where an aggregate of particles, treated in a Lagrangian way, are dispersed into an Eulerian turbulent field. Depending on the mass load of the dispersed phase, different levels of interaction between the two phases can be considered. The most used model treats low mass load which implies one-way

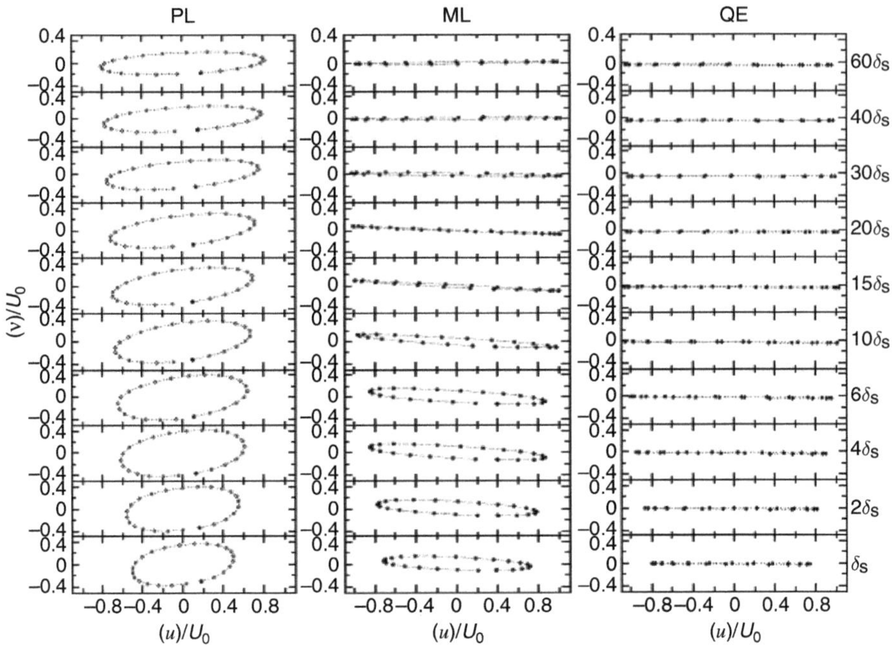

Fig. 11.4 Stokes-Ekman BL: streamwise versus spanwise velocity components at different planes for the three cases. From the *bottom to the top*, planes: δ, 2δ, 4δ, 6δ, 10δ, 15δ, 20δ, 30δ, 40δ and 60δ with δ the nominal thickness of the SBL. The *black circle* represents the value at the phase of oscillation equal to 15. The phase increases counter-clockwise. Figure from [14]

coupling methodology, namely the concentration of the particles does not modify momentum transport. From a conceptual point of view, this is equivalent to the transport of a passive scalar when using an EE approach.

The LE methodology is usually formulated considering simplified forms of the Maxey and Riley equation (see [17] for a detailed discussion). A problem arises when using LES Eulerian fields to animate the motion of Lagrangian particles, since the SGS velocities are not known. Armenio et al. [18] showed that the error can be neglected when wall-resolving LES is used since one is interested in the statistics of dispersion and diffusion of the ensemble of particles. However, in a successive paper [19], Kuerten proved that the reconstruction of the SGS velocity field, using an approximate deconvolution technique, allows to reproduce more accurately the statistics of turbulent dispersion and, in particular, the near wall accumulation due to turbophoresis.

Among the number of papers dealing with the study of dispersion of Lagrangian particles in turbulent field, here we make reference to that of Inghilesi et al. [20], inherently related to environmental processes. In [20], the rise of buoyant particles in a free surface Ekman layer was investigated. This flow is archetypal of uprising of fresh-water particles in a salty water environment under neutral and stable stratification. The flow is oversimplified with respect to a real-life case, in that periodic boundary conditions in the horizontal directions were used and topographic effects were neglected. The flow was driven by a constant wind stress at the free surface, imposed directly as a momentum flux; stable stratification was induced by a constant heat flux directly imposed at the free-surface. The effect of the SGS velocity field over particle motion was reproduced using the approximate deconvolution of [19].

Particle motion was simulated using an improved form of the simplified Maxey and Riley equation, able to reproduce buoyancy effects related to the presence of a variable density Eulerian field. Once the density field is decomposed onto a bulk density ρ_0 and a perturbation density ρ_d with $\rho_d << \rho_0$, after making non-dimensional the particle motion equation, we write

$$\frac{dv_{p,i}}{dt} = (1 - 1/\Delta\rho)\frac{1}{Fr^2}\delta_{1,2} - \frac{\text{Ri}}{\Delta\rho}\rho\delta_{1,2} + \text{DRAG} + \text{COR} \tag{11.1}$$

In Eq. (11.1) *DRAG* and *COR* represent, respectively, the drag and Coriolis forces. The term $v_{p,i}$ represents the ith-component of the particle's velocity made non-dimensional with a proper velocity scale U, t is time made non-dimensional with δ/U where δ is a length scale of the problem, and $\Delta\rho = \rho_p/\rho_0$ with ρ_p the particle density. The first two terms on the right-hand side represent, respectively, the standard non-dimensional buoyancy force associated with the density field ρ_0 and the additional contribution due to the perturbation density ρ_d associated with the stratifying agent (temperature in the present case). $\text{Fr} = U/g\delta$ is the Froude number, $\text{Ri} = g\rho_s\delta/\rho_0 U^2$ the Richardson number of the flow with ρ_s a density scale of the problem and $\rho = \rho_d/\rho_s$ is the non-dimensional perturbation density. The term proportional to Ri takes into account the buoyancy effects on the particles related to the variations of density in the flow. This term has been introduced in the particle motion equation for the first time in [20] and has been shown to affect dramatically particle dispersion when the density of the particles is comparable with the bulk density ρ_0 of the carrying fluid. This happens, for instance, when particles of fresh water are released in a salty water environment.

Two cases were studied in [20], a neutral case and a strongly stratified case, respectively. The study was carried out at $Re = \delta u_\tau/\nu = 10{,}000$ where $\delta = u_\tau/f$ is the Ekman penetration length, with f the Coriolis parameter. The grid spacing in wall units (made non-dimensional with ν/u_τ) was $\Delta x = \Delta y = 156$ in the horizontal directions and $\Delta z = 75$ in the vertical one, typical of LES where the near-wall layer is not directly solved in the simulation. This choice came from the fact that the wall-shear stress was directly imposed over the free surface as a momentum flux. The particles were continuously released over a disc of radius 0.05δ, located 0.5δ below the free surface and containing 900 particles at each releasing time.

The analysis of the Eulerian field showed that under strong stratification, the Ekman penetration length strongly decreases and a noticeable pycnocline develops. This has tremendous effects on particle dispersion as illustrated in Fig. 11.5, which shows the particle distribution at the final time of the simulation in the two cases.

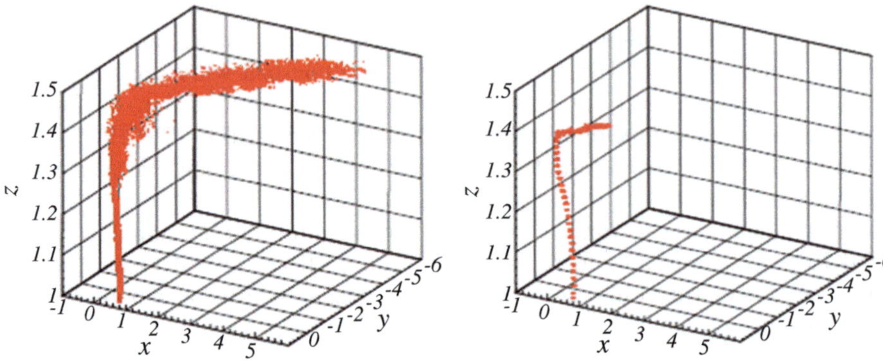

Fig. 11.5 Uprising of buoyant particles: three-dimensional view of the particle distribution at the final time of simulation: *left panel*, neutral case; *right panel*, stable stratified case. Figure from [20]

In the neutral cases, the cylindrical structure of the vertical motion is lost when the particles enter the region of turbulence mixing. The particles eventually reach the free surface and are dispersed horizontally according to the Ekman horizontal velocity field. In the stratified case, turbulence is absent in the fluid column apart a narrow layer close to the free surface; particles travel upward along very ordered trajectories originated from the disc of emission. Before reaching the free surface, particles remain entrapped below the pycnocline due to buoyancy effects and are transported horizontally along a wavy pattern in the deep layers of the current. This is a typical dispersion phenomenon occurring in water basins when a buoyant jet meets a stratified ambient. The horizontal dispersion is dramatically affected by the stratification. Since in the stratified case the particles remain confined in the deep layers of the flow where turbulence is nearly absent and the horizontal velocity is small, dilution does not occur and the cloud of particles travel at a much slower rate compared to the neutral case.

11.4 . . . Toward Full-Scale Applications

In the previous section two main examples were shown where LES was successfully employed to exploit physical processes occurring in fundamental flows of interest to EFM. However such studies are very idealized and, hence, their results are not directly applicable to real-case situations. On the other hand, as mentioned in Sect. 11.1, there is the need for computational tools for engineering analysis from one side, and on the other one, the need to study flow fields closer to those of real-scale situations. In the last decade, *IE-Fluids* has been working toward the development of state-of-the-art numerical tools to be used for environmental studies. These tools were designed to study coastal semi-closed sea basins (or lakes) and air mixing in the very low atmosphere. The ultimate goal was to take advantage of knowledge and methodologies successfully employed to study archetypal EFM problems, and to formulate novel numerical tools for the investigation of *real-life* problems. The results of these efforts are LES-COAST and LES-AIR, respectively. LES-COAST and LES-AIR share a number of modeling features. Here we show the development and a typical application of LES-COAST. Regarding specific features of LES-AIR the reader is addressed to [21, 22].

The development of the models started from a preliminary analysis on the peculiarities required to a model to be useful for studies in EFM. The model must be able to reproduce real-life geometric features, flows with realistic values of the Reynolds number, proper boundary conditions coming from field data or simulations by large scale circulation models, wall roughness, stratification and Earth rotation effects. Also, typical EFM features should be reproduced, like, among the others, scalar emissions from localized sources, the presence of low-rate chemical reactions and biological transformations.

The models were designed to use LES as the turbulence closure. The mathematical formulation is the same as that briefly discussed in Sect. 11.2. Main differences arise, respectively, in the treatment of geometric complexities, in the SGS model and in the boundary conditions. Integration of the governing equations is carried out using the same algorithm as in Sect. 11.3, namely that of [7].

In the following, we briefly describe most of the steps undertaken to arrive to the model LES-COAST and also an example of application of the model to the study of a real case.

It is well known in literature that curvilinear-coordinate, structured grids can deal with moderately complex geometry, unless complicated multi-block algorithms are generated. Coastal geometries are far from being simple and can be hardly treated using curvilinear coordinates. The alternative choice, over which most commercial codes are based, is the use of unstructured grids that may make the algorithms of integration of the NSE not fully conservative for energy. It is noteworthy that conservation properties are of primary importance when using LES methodology. A valuable alternative to the use of unstructured grids for complex geometry is the combination of structured grid solvers together with the immersed boundary method (IBM) whose excellent review is in [23]. According to this methodology, the set of governing equations is solved over the entire domain, and internal solid boundaries are set through the use of a forcing term mimicking the presence of an internal boundary. Most IBM were developed in conjunction with Cartesian grid solvers. The main advantages consist in the simplicity of implementation and the efficiency of the pressure solver designed for a simple 7-point stencil. The main drawback is that for many problems an excessive number of grid cells are lost within the solid body. Recently, Roman et al. [24] formulated an IBM in conjunction with the curvilinear coordinate algorithm of [7]. In this case, the curvilinear grid can be constructed to follow as much as possible the boundaries of the domain and the immersed boundaries can be used to model the geometric complexities. The main disadvantage consists in the fact that the Poisson equation has a 27-point stencil that makes the solution algorithm less efficient. The authors adapted existing IBM methods designed for Cartesian grids and also improved them in several aspects: the algorithm of search of the solid bodies was generalized to the case of a multibody system; a general interpolation technique (see [25]) was used to determine the variables at the grid points near the immersed boundaries.

The method was tested over several cases and the results were nearly insensitive to grid deformation from Cartesian to curvilinear. As we will show later on, the final algorithm is able to reproduce the geometric complexities typical of EFM flows.

In LES, the grid cells are usually nearly isotropic, apart in the near wall region where their aspect-ratio increases accommodating the shape of the elongated turbulent structures. The filter width is chosen implicitly to be proportional to the grid size and, in eddy-viscosity SGS models, the length-scale is chosen to be proportional to $(\Delta x \Delta y \Delta z)^{1/3}$.

In coastal hydrodynamics the area of study usually extends over few kilometres in the horizontal with depth of the order of ten-to-hundred metres. In this class of flows, dimensional considerations suggest that the horizontal velocity and length scales are much larger than the vertical ones ($U_v << U_h$, $l_v << l_h$). For instance, this is the theoretical background leading to the formulation of shallow water equations.

If we use a three-dimensional non-hydrostatic mathematical formulation, in this class of flows the grid cells can be highly anisotropic within the whole domain, even in the regions where turbulence tends to isotropy. In the LES context, this problem was addresses by different authors who basically proposed to vary the length-scale of the model according to the cell anisotropy (see [26] for a discussion).

Roman et al. [27] observed that, when using the Smagorinsky model on anisotropic grids, the SGS eddy viscosity is largely overestimated and, hence they formulated a two-eddy viscosity SGS model. Although this was a novelty in the LES framework, the idea was borrowed from the geophysical community, used to work with large scale circulation models where horizontal and vertical eddy diffusivities are employed to parametrize turbulence mixing. The authors observed that the model formulation is not Galilean invariant requiring a third eddy viscosity. However, as discussed in the paper, this is a well-known formal problem of two-eddy viscosity models, which, in practice does not affect the results of simulations. In [27] a discussion on the implementation of the model in curvilinear coordinates was also given. The two constants of the model, c_h and c_v, were calibrated considering a plane channel flow, whose turbulent statistics are well known, and running simulations varying the aspect-ratio of the cells. The calibration curves are reported in Fig. 11.6. A closure is also required for the transport equations of the scalars. In LES-COAST the Reynolds analogy is invoked for the SGS Prandtl and Schmidt number, with values ranging within the interval 0.5–1.

As mentioned in Sect. 11.1, high-Re wall-bounded flows cannot be studied by means of wall-resolving LES because of unaffordable requirements of computational resources and also for a conceptual problem related to the definition of wall-roughness when using the no-slip condition over the solid walls. Wall-layer models are hence required. Substantial literature is present on this topic, regarding different strategies (i.e. from wall-layer models to hybrid RANS-LES and Detached Eddy Simulation). These models must be able to work both over body-fitted geometry and over boundary surfaces reproduced using immersed boundaries. Although the general strategy in the two cases can be similar, the model formulation

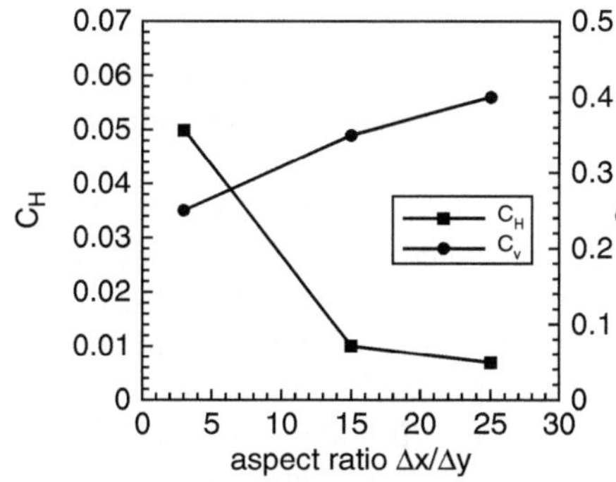

Fig. 11.6 Calibration curves of the two-eddy viscosity model versus cell aspect ratio. The friction Reynolds number of the calibration tests was $Re_\tau = 20{,}000$ and a standard wall-layer model was used. Figure from [27]

and the implementation procedure vary from the former to the latter case. In LES-COAST, a wall-layer model based on equilibrium turbulence assumption has been implemented. Two strategies were followed, depending on whether the surface is implemented through body fitted grids or immersed boundaries. In both cases the velocity at the first grid point off-the-wall is expressed as belonging to a log profile for $y_p^+ > 11$ or to a linear one in the latter case. The quantity y_p^+ is the distance from the wall of the first grid point, made non-dimensional with ν/u_τ. The non-dimensional distance from the wall can be equivalently expressed in terms of roughness height y_0 for rough walls. Once the velocity at the first grid point is known, the wall shear stress is calculated and imposed as boundary condition at the successive time step.

In case of body fitted grids, the model is applied in conjunction with a modified form of the Smagorinsky model in which the contraction of the resolved strain rate tensor is calculated in a proper way. Specifically, when the Smagorinsky model is employed in conjunction with wall-layer models which do not solve the near-wall region, the direct calculation of the leading terms of $|\overline{S}|$ is wrong. This is because the calculation of $\overline{S}_{in}$ (n is the wall-normal direction and $i = 1, 2$ the two directions parallel to the wall) requires the knowledge of the velocity at the wall, not known when the momentum flux is directly imposed as boundary condition. The problem has been overcome by calculating the terms above using analytical considerations. In case the velocity belongs to the logarithm region $\overline{S}_{in} = (u_\tau/\kappa y_p)(u_i/U_p)$, otherwise $\overline{S}_{in} = (u_\tau^2/\nu)(u_p/U_p)$.

This allows to obtain an eddy viscosity at the first grid point which adjusts consistently with the velocity profile. By simple considerations, in [21] it has been shown that in case of equispaced grids the eddy viscosity at the first grid point off the wall assumes the following RANS-like form $\nu_T = 13.7 C_s^2 y_p u_\tau$.

In case of immersed boundaries, the imposition of momentum flux directly at the cell face cannot be accommodated and an alternative strategy must be chosen. Roman et al. [28] imposed the wall stress at the immersed boundary indirectly; they set an analytical eddy viscosity ν_T from the consideration that the velocity profile is shaped according to a logarithm law, obtaining $\nu_T = C_w \kappa u_\tau d_{IB}$, where κ is the von Karman constant of the logarithm velocity profile, $u_\tau = \sqrt{\tau_w/\rho}$ is the friction velocity obtained from the knowledge of the wall shear stress, d_{IB} is the distance from the wall to the first grid point within the fluid domain and C_w is a calibration constant. The latter must take into account that the Reynolds shear stress at the location where the eddy viscosity has to be set, is a fraction of the wall shear stress and that the eddy viscosity needs to be evaluated at the cell face, where the viscous flux is evaluated. Also, in the model the velocity at the IB node must be calculated. This is done as $U_{IB} = U_{PP} - \frac{1}{\kappa}\sqrt{\tau_w/\rho} log(d_{PP}/d_{IB})$, where U_{IB} is the velocity at the first (IB) node within the fluid domain, U_{PP} is the velocity at a properly chosen projection point, close to the *IB* one, and d_{PP}/d_{IB} is the ratio between the distances of the *PP* and *IB* nodes from the solid surface. The velocity component normal to the wall is calculated through quadratic interpolation $U_{n,IB} = U_{n,PP} d_{IB}^2/d_{PP}^2$.

The model was tested on a plane channel flow at $Re_\tau = 1000$ using immersed boundaries implemented over grids deformed according to a sinusoidal shape. The tests showed that the slope of the velocity profile is completely wrong unless the analytical eddy viscosity is implemented.

At the stage of the research the wall-layer models herein discussed were not tested in presence of massive separation. Modifications of the original models aimed at reproducing accurately massive separation occurring in presence of bluff geometry were done over the latter years and satisfactory results were obtained (see [29]).

11.4.1 Free-Surface and Open Boundaries Conditions

The wind breeze produces a shear stress at the free surface of a water basin, hence generating mixing along the water column. In LES-COAST, the wind action is parametrized using a standard formulation where the stress is $\tau_w = \rho_{air} C_{10} U_{10}^2$, where ρ_{air} is the air density, U_{10} is the wind velocity 10 metres above the free surface and $C_{10} = (0.8 + 0.065 U_{10}) 10^{-3}$ is an empirical parameter calibrated over field data. Since LES needs as boundary conditions fluctuating quantities, the free-surface stress is *animated* adding a random fluctuation with variance equal to 20 % the mean value.

In standard models the stress is distributed homogeneously over the free surface, when not derived from coupled atmospheric simulations. This is reasonable for open basins, but, in coastal areas or in lakes the local topography may rule the local wind intensity and direction and, hence, local water circulation. In LES-COAST, the presence of local topography or anthropogenic structures (docks, ships, buildings ....) is considered through empirical formulations. They give an estimation of the recirculation regions developing near the obstacles, where the wind stress is practically zero. Details on the formulations adopted are in [30]. Currently, efforts are addressed at using data coming from the Weather Research and Forecasting Model (WRF) for the evaluation of the non-homogeneous wind field and, also, toward an integrated model solving the coupled air-water system.

EFM models work at local scale and, thus, they need realistic boundary conditions to be set at the edges of the computational domain. Hence, a model needs to be nested within a large circulation model (LCM) which can supply the large scale circulation at a much lower resolution. When working in the LES framework an additional problem arises, consisting in the fact that boundary conditions obtained from RANS-based models must feed an LES model which needs instantaneous fluctuating fields. This is true for both LES-AIR and LES-COAST. This is a well-known problem and a number of excellent papers have been published about this topic (see, among the others [31]). A two-step procedure to generate proper boundary conditions for local-scale LES models has been developed and presented in [30]. The procedure works as follows: First, data from an LCM (velocity components, turbulent kinetic energy, scalars, etc.) are extracted and interpolated over the boundary surfaces of the LES computational domain. Since the time step of LCM is usually much larger than that of LES models, a linear interpolation of the variables is carried out between two successive LCM time steps.

Successively a synthetic turbulence field is generated within the LES domain. Specifically, the computational domain of the LES is generated in such a way to accommodate an unphysical buffer region where the synthetic turbulent field is generated. There, the artificial field is generated by adding on the right-hand side of the momentum equations a zero-mean body force b_i', which produces a divergence-free fluctuating velocity field. The intensity of the body force is calibrated through energetic considerations, since the body force must supply production of turbulent kinetic energy, absent in the RANS flow field. Smooth transition from the buffer region is guaranteed through exponential decay of the body force terms up to the zero value in the physical part of the computational domain.

Validation tests were carried out considering a $Re_\tau = 20{,}000$ plane turbulent channel flow, animated by a RANS-like mean velocity profile at the inflow section and considering different numbers of cells in the buffer region. The results were then compared with those of an equivalent periodic simulation, which also supplied the initial velocity field in the simulations with the buffer. Figure 11.7 shows the computational domain composed of the buffer region and the physical flow region. The study showed that the basic imposition of a RANS velocity profile, even in presence of an initial turbulent field, does not allow sustaining turbulence during the simulation and after a number of time iterations the flow re-laminarizes. Conversely (see Fig. 11.8), once the fluctuating body forces are switched on, a synthetic turbulent field is triggered and a realistic level of turbulence is maintained in the physical part of the computational domain. This is well evidenced in Fig. 11.8 which shows the spanwise- and time-averaged velocity profiles and different stations off the buffer. The results of Fig. 11.8 were obtained with a buffer extended $16\Delta x$ in the streamwise direction. However equivalent results were obtained with shorter buffers.

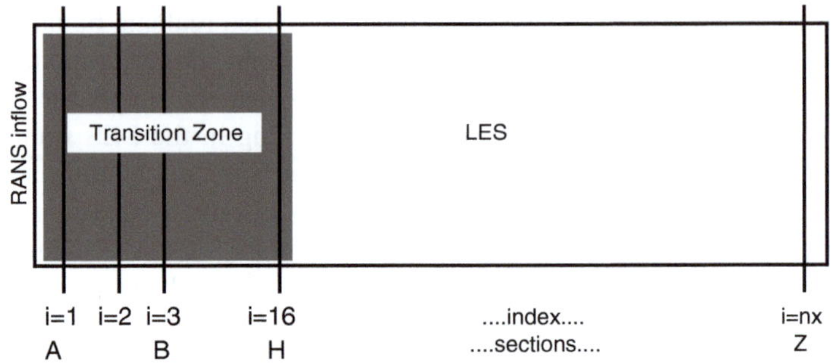

Fig. 11.7 Sketch of the computational domain for the validation of the synthetic generation of turbulence. The *left grey area* is the buffer region where the fluctuating body force field is active. Figure from [30]

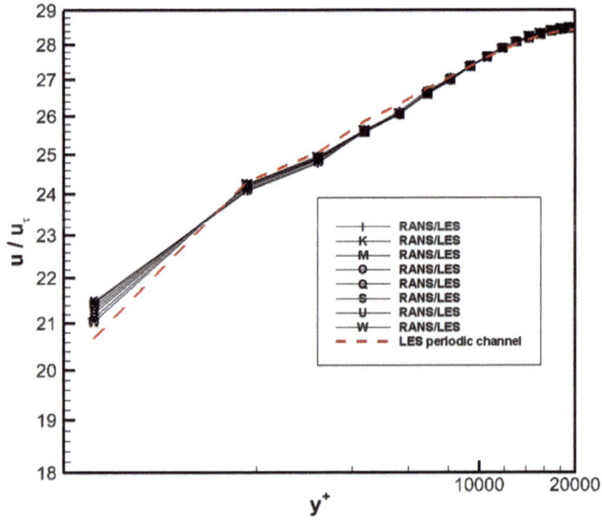

Fig. 11.8 Time- and spanwise-averaged streamwise velocity profiles at streamwise stations off the upstream buffer region. Figure from [30]

11.4.2 Application of LES-COAST to a Real Case

Over the last years, LES-COAST has been continuously improved with new modules and applied to the simulation of turbulent mixing in coastal semi-closed areas. Among the additional modules developed, dispersion of oil at sea is modeled both in a form of thin liquid film and in form of single tars [32]. The model was employed for the analysis of pollutant scenarios in case of an accident in different sites.

LES COAST has been used for the following studies: analysis of the estuary of the Tiber river, (central-west Italy); analysis of mixing and water renewal in the Gulf of Taranto (south-east Italy), in the Monfalcone bay (north-east Italy), in the Barcelona harbour (north-east Spain), in the Muggia bay (north-east Italy), in a

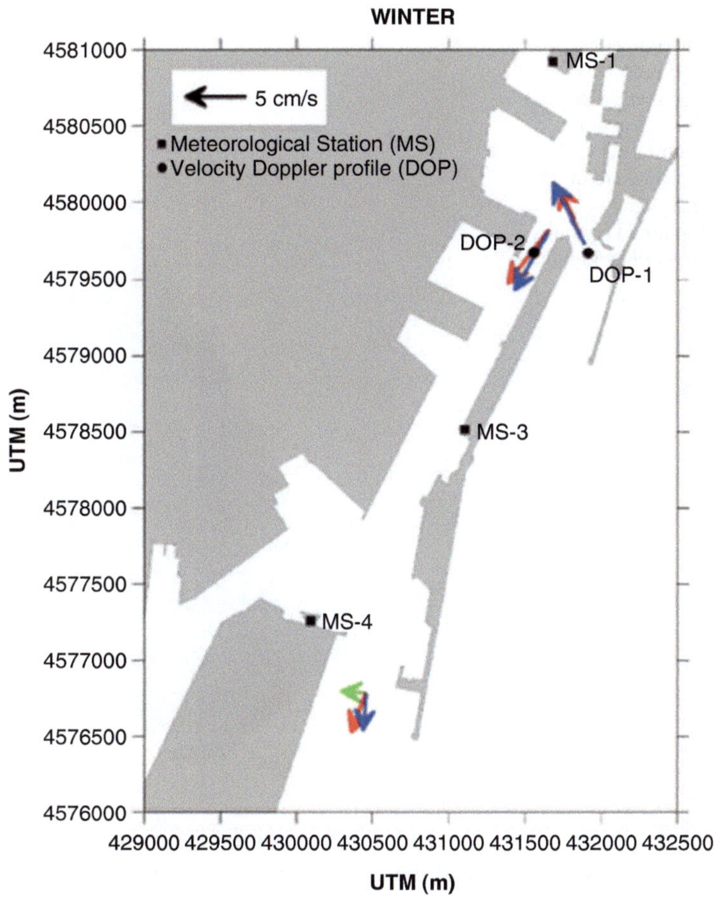

Fig. 11.9 Sketch of the Barcelona harbour with the geographical location of: the two ADCPs (DOP-1: Northern mouth and DOP-2: Central channel) used to measure the vertical profiles of the horizontal sea current; the meteorological stations (MS-1,3,4). Mean flow vectors of measured currents at 1 m (*blue*), 8 m (*red*) and 15 m (*green*) depth for the winter campaign in Barcelona harbour. Figure from [33]

coral reef (Kaneohe) bay situated north-east of Ohau, (Hawaii, USA); analysis of hydrodynamics of the Ledro lake, an alpine lake, situated in the north of Italy.

Here we show, as an example, some results of the study of mixing in the Barcelona harbour. The entire study is presented in [33] and is relative to the typical conditions occurring during the winter period. The importance of this study resides in the fact that the harbour serves different activities, from touristic to commercial and is composed of a multitude of sub-basins making the evaluation of internal currents, mixing and water renewal not trivial. The knowledge of the local hydrodynamics allows to quantify the water quality in the harbour and to set contingency plans in place in case of an environmental hazard. Figure 11.9 shows

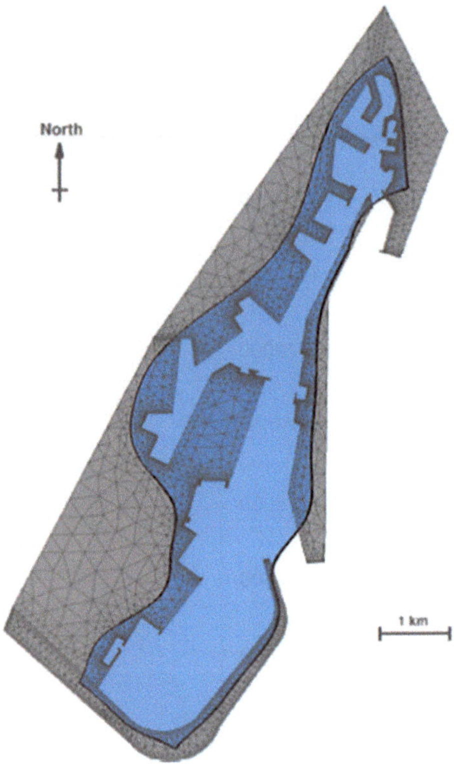

Fig. 11.10 Computational domain for the Barcelona harbour: curvilinear structured mesh (enclosed within the *thick blackline*); Immersed bodies used to model the coastline are in *grey* with *triangular* elements; The liquid region is depicted in *light blue*. Figure from [33]

the map of the harbour together with the location of the meteorological stations where the wind was recorded and the position of the instruments where the water velocity was measured. The main direction of the water current is typical of winter conditions, entering through the northern mouth of the harbour and flowing outward through the southern one. On average, the wind velocity during the winter season comes from north-northeast (25°), with a variable intensity with peaks up to 16 m/s. In winter, the water column is neutrally stratified since temperature is uniformly distributed along the vertical direction and salinity is homogeneous within the harbour.

The harbour is approximately 10 km long and 2 km wide. Its main axis is oriented 30° from the north. First, the computational domain was generated starting from the bathymetric and coastline data; a combination of curvilinear coordinates and immersed boundaries was used as shown in Fig. 11.10. Over the horizontal directions 256×1024 grid cells were employed, respectively, along the narrow and the elongated sides of the curvilinear domain, while 24 cells were used in the vertical, giving a mean cell size of $8\,\text{m} \times 10\,\text{m} \times 1\,\text{m}$. From field data, tidal flow resulted negligible and, hence, it was not considered. Also, the smallness of the area under investigation suggested neglecting the generation of surface waves. As boundary conditions, the authors used the mean inflow velocity of 0.04 m/s at

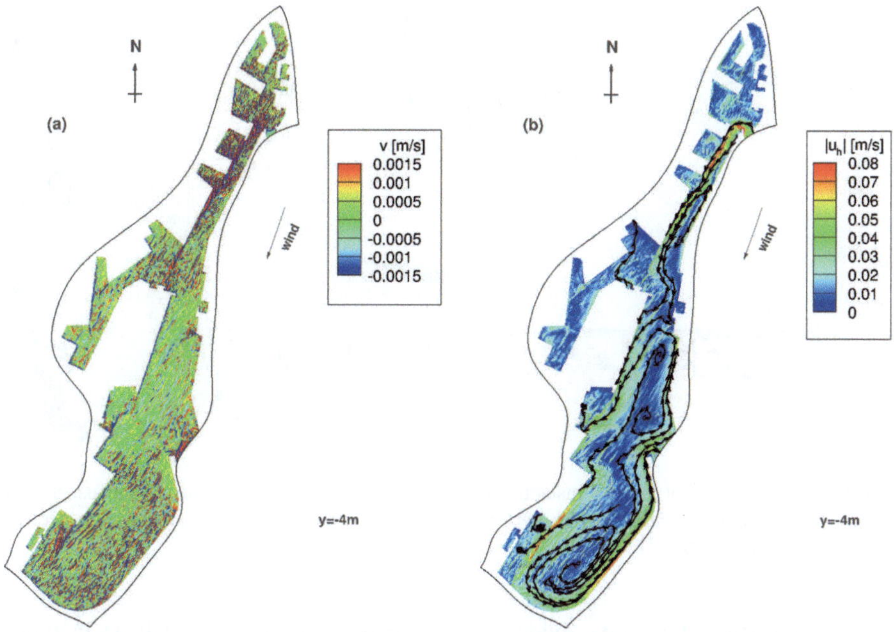

Fig. 11.11 Contour plot of: (**a**) the instantaneous vertical velocity v; (**b**) the modulus of the instantaneous horizontal velocity with instantaneous stream tracers, in a horizontal plane 4 m below the free surface. Figure from [33]

the northern mouth and a mean exit velocity from the southern mouth of 0.02 m/s satisfying conservation of mass over the entire domain. The free surface, treated as a rigid lid, was forced with a momentum flux coming from a mean wind $U_{10} = 5$ m/s, the latter being a reasonable mean value taken from the field data over the winter period. Wall-layer models were implemented over the solid surfaces of the domain. The methodology, as a whole, was discussed above. The simulation was run for about 48 h, long enough for obtaining statistical steady state.

The simulations supplied instantaneous velocity fields which were post-processed to obtain significant statistics. Here we summarize the main results. The instantaneous field 4 m below the free surface is depicted in Fig. 11.11. The contour plots of the vertical velocity (left panel) and of the horizontal one (right panel) are shown together with the instantaneous streamtracers. A first consideration is relative to the fact that the vertical velocity is one order of magnitude smaller than the horizontal one, in agreement with dimensional analysis for this class of flows. Upwelling/downwelling events are visible at the upwind/downwind sides of the harbour, together with elongated, well structured, regions of upward-downward flow in open water regions. The horizontal velocity shows a strong mean current in the central-narrow part of the harbour and a main clockwise vortex structures in the southern part of the harbour. Also, smaller vortex structures are well evident as a result of geometric complexities, whereas in the small sub-domains composing the harbour, the horizontal velocity seems very weak.

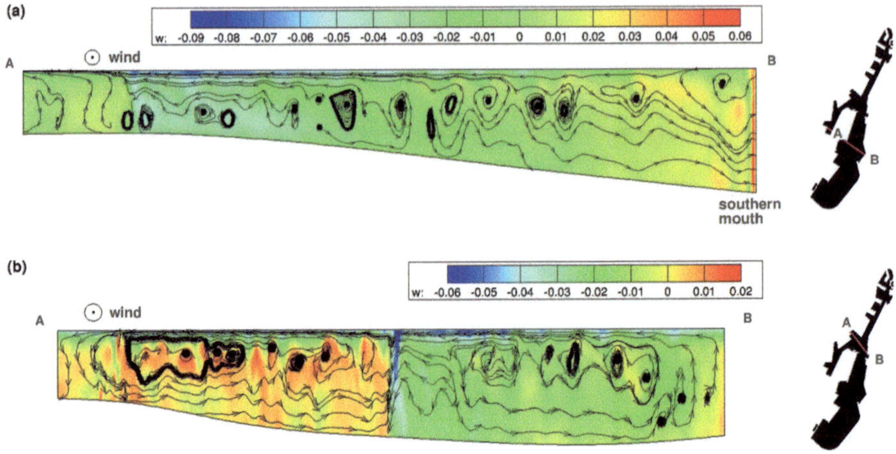

Fig. 11.12 Contour plot of the instantaneous alongshore velocity with instantaneous stream tracers (*black arrowed lines*) in two vertical cross-shore planes as indicated in the respective overviews. Figure from [33]

An interesting dynamics, evidenced in the present study as well as in other equivalent studies carried out with LES-COAST, is the presence of Langmuir-like circulations occurring beneath the free surface. Evidence is in Fig. 11.12 which shows the contour of the cross-plane velocity component and the instantaneous streamtracers, at two planes nearly orthogonal to the mean streamwise motion. They appear in form of elongated streamwise-oriented vortical structures, extending for tens of metres and spanning vertically along the whole water column. Also, their time-scale was estimated to be of the order of hours. Unlike classical Langmuir circulations which are triggered by non-linear Stokes drift, these structures develop also in absence of surface waves, not reproduced in the present simulations. In other simulations (not shown here) it was demonstrated that in presence of stable stratification, coming, for example, from the discharge of fresh water in the salty water ambient, these structures are not present, probably due to the potential energy barrier supplied by the pycnocline. These structures were observed in experiments of shallow-water wall-bounded turbulence (Nikora, 2015, Private communication), although the mechanism for their formation is still unclear.

The database generated in the simulation was used to calculate statistics according to Reynolds averaging. Among the others, the vertical and horizontal eddy viscosities were calculated. This is an important issue, since these quantities are used in standard marine hydrodynamic models. The vertical eddy viscosity was found to be more intense in the deeper water layers, peaking up to $3.5\ 10^{-3}\ \mathrm{m}^2/\mathrm{s}$ in the downwind area where significant downwelling occurs. In the internal regions, vertical eddy viscosity decreases still being significant, basically due to the vertical mixing supplied by the Langmuir-like circulations. The horizontal eddy viscosity is larger in the surface layers and peaks up to $1.5\ \mathrm{m}^2/\mathrm{s}$. It appears more uniform in space than the vertical eddy viscosity and larger values are observed where energetic horizontal boundary layers develop at the vertical boundaries. The analysis shows that these quantities, whose proper calibration is of fundamental importance for the accuracy of simulations performed with standard RANS-based marine hydrodynamic models, are substantially non-homogeneous and vary in the three directions, depending on the geometric complexities of the basin.

Finally, comparisons with available data were carried out, also considering data of simulations carried out with the Regional Ocean Model (ROMS). Field data averaged over the winter period were available at two locations within the harbour (see Fig. 11.9). The comparison between LES profiles of horizontal velocity, averaged over 33 h, and the field data is very good along the entire water column. On the other hand, a simulation carried out with ROMS on the same case exhibited horizontal velocities underestimated by one order of magnitude, posing serious concerns on the appropriateness of the use of models designed to work in open-sea large basins, when applied to the study of small scale, coastal areas.

11.5 Concluding Remarks

The present paper summarizes almost two decades of research activity in the area of modelization of problems of interest to EFM using LES. Two main aspects were discussed. First the use of wall-resolving LES for fundamental studies of interest to EFM was shown through description of significant example. Successively, efforts oriented toward the development and application of a real-scale LES model suited for the analysis of coastal hydrodynamics were described.

As shown in the present paper, over the year LES has shown to be a valuable tool to study fundamental problems and to exploit important physical aspects of processes characterizing EFM problems. This is particularly true when stratification, rotation and dispersion processes are studied. Fundamental issues, observed in field analysis, can be studied and understood using archetypal, laboratory-scale numerical experiments. In many situations, they may supply results comparable or even superior to those of correspondent physical experiments, for two main reasons: the amount of data available with LES allows calculation of statistics which can be hardly obtained in physical experiments; sometimes, numerical experiments allow studying configurations hardly reproducible in a physical laboratory (a noticeable example is the effect of background horizontal vorticity in the Stokes-Ekman boundary layer discussed in this paper). Wall-resolving LES together with DNS will get more and more importance in the next decades due to the exponential growth of computational resources and availability in managing big data; however, the dream that they will replace standard computational fluid dynamics tools in engineering will hardly come true, at least in EFM. This is also related to conceptual issues, one over the other, the presence of significant wall-roughness, which, necessarily should be treated in a deterministic way. This issue is basically related to the fact that EFM problems develop over a very wide range of spatial and temporal scales.

On the other hand, LES methodology may emerge as a valuable tool for engineering real-scale EFM problems when adapted to work for this class of problems. A significant and successful attempt to go toward application of LES to real-scale EFM problems was herein presented. The models, LES-COAST for marine applications and LES-AIR for low-atmospheric uses (the latter non discussed here), were conceived and designed to give an answer to the main needs of modeling EFM issues, like geometric complexity, wall-roughness, appropriateness of boundary conditions, difference between horizontal and vertical scales. The models, in particular LES-COAST, have been successfully applied to the study of a number of real cases depicting scenarios of mixing not reproducible with standard methodologies. Such high-resolution models may also serve to formulate new (or to calibrate existing) turbulence closures of use in standard hydrodynamic/atmospheric models. Finally, it is not difficult to imagine that, thanks to the exponential growth of computational resources in terms of CPU time and big data storage, even at the desktop or CFD-lab scale, large scale LES will become soon a standard engineering tool for the study of EFM problems.

Acknowledgements The author is indebted with the co-authors of the papers herein discussed, for their substantial contribution to the research. Over the years the research has been supported by The Italian Minister for Research, the Istituto Superiore per la Protezione e Ricerca sull'Ambiente (ISPRA), the Regional Agency for the Protection of the Environment (ARPA-FVG), the Oil terminal of Trieste (SIOT-TAL), the National Institute for Oceanography and Experimental Geophysics (OGS). The Italian Progetto Bandiera RITMARE (actions SP3-WP4-AZ3-UO07 and SP5-WP4-AZ4-UO06) is acknowledged for the financial support for the improvement of the model LES-COAST and for the composition of the present paper. The paper has been prepared under the auspices of ERCOFTAC, Special Interest Group (SIG) 5, Environmental Fluid Mechanics.

References

1. H. Burchard, P.D. Craig, J.R. Gemmrich, H. van Haren, P.P. Mathieu, H.M. Meier, W.A.M.N. Smith, H. Prandke, T.P. Rippeth, E.D. Skyllingstad, W.D. Smyth, D.J. Welsh, H.W. Wijesekera, Observational and numerical modelling methods for quantifying coastal ocean turbulence and mixing. Prog. Oceanogr. **76**, 399–442 (2008)
2. C.-H. Moeng, J.C. Wyngaard, Evaluation of turbulent transport and dissipation closures in second-order models. J. Atmos. Sci. **46**(14), 2311–2330 (1989)
3. M. Germano, U. Piomelli, P. Moin, W.H. Cabot, A dynamic subgrid-scale eddy viscosity model. Phys. Fluids A **3**(7), 1760–1765 (1991)
4. U. Piomelli, E. Balaras, Wall-layer models for large-eddy simulations. Annu. Rev. Fluid Mech. **34**, 349–374 (2002)
5. P.R. Spalart, W.-H. Jou, M. Strelets, S.R. Allmaras, Comments on the feasibility of LES for wings, and on a hybrid RANS/LES approach, in *First AFOSR International Conference on DNS/LES*, ed. by C. Liu, Z. Liu. Advances in DNS/LES, 4–8 August, Ruston, LA (Greyden Press, Columbus, OH, 1997)
6. P. Moin, K. Mahesh, Direct numerical simulation: a tool in turbulence research. Annu. Rev. Fluid Mech. **30**, 539–578 (1998)

7. J. Zang, R.L. Street, J.R. Koseff, A non-staggered grid, fractional step method for time-dependent incompressible Navier–Stokes equations in curvilinear coordinates. J. Comput. Phys. **14**, 459–486 (1994)
8. V. Armenio, U. Piomelli, A Lagrangian mixed subgrid-scale model in generalized coordinates. Flow Turbul. Combust. **65**, 51–81 (2000)
9. V. Armenio, S. Sarkar, An investigation of stably-stratified turbulent channel flow using large eddy simulation. J. Fluid Mech. **459**, 1–42 (2002)
10. S. Sarkar, V. Armenio, Direct and large eddy simulation of environmental flows, in *Handbook of Environmental Fluid Mechanics, Systems, Pollution, Modeling and Measurements*, ed. by H.J.S. Fernando, vol. 2(20) (CRC Press, Boca Raton, 2013), pp. 283–300
11. E.J. Hopfinger, P.F. Linden, The effect of background rotation on fluid motions: a report on Euromech 245. J. Fluid Mech. **211**, 417–435 (1990)
12. S. Salon, V. Armenio, A. Crise, A numerical investigation of the Stokes boundary layer in the turbulent regime. J. Fluid Mech. **570**, 253–296 (2007)
13. S. Salon, V. Armenio, A. Crise, A numerical (LES) investigation of a shallow-water, mid-latitude, tidally-driven boundary layer. Environ. Fluid Mech. **9**, 525–547 (2009)
14. S. Salon, V. Armenio, A numerical investigation of the turbulent Stokes-Ekman bottom boundary layer. J. Fluid. Mech. **684**, 316–352 (2011)
15. B.L. Jensen, B.M. Sumer, J. Fredsøe, Turbulent oscillatory boundary layers at high Reynolds numbers. J. Fluid Mech. **206**, 265–297 (1989)
16. T.B. Sanford, R.C. Lien, Turbulent properties in a homogeneous tidal bottom boundary layer. J. Geophys. Res. **104**(C1), 1245–1257 (1999)
17. V. Armenio, V. Fiorotto The importance of the forces acting on particles in turbulent flows. Phys. Fluids **13**, 2437–2040 (2001)
18. V. Armenio, U. Piomelli, V. Fiorotto, Effect of the subgrid scales on particle motion. Phys. Fluids **11**, 3030–3042 (1999)
19. J.G.M. Kuerten, Subgrid modelling in particle-laden channel flow. Phys. Fluids **18**(2) (2006). Art. No. 025108
20. R. Inghilesi, V. Stocca, F. Roman, V. Armenio, Dispersion of a vertical jet of buoyant particles in a stably stratified wind-driven Ekman layer. Int. J. Heat Fluid Flow **29**, 733–742 (2012)
21. V. Stocca, Development of a large eddy Simulation model for the study of pollutant dispersion in urban areas. Ph.D. dissertation thesis, the University of Trieste, 2010
22. A. Petronio, F. Roman, V. Armenio, F. Stel, D. Giaiotti, Large-eddy simulation model for urban areas with thermal and humid stratification effects, in *Direct and Large Eddy Simulation IX*. Ercoftac Series, vol. 20 (Springer, Berlin) 2015
23. R. Mittal, G. Iaccarino, Immersed boundary methods. Annu. Rev. Fluid Mech. **37**, 239–621 (2005)
24. F. Roman, E. Napoli, B. Milici, V. Armenio, An improved immersed boundary method for curvilinear grids. Comput. Fluids **38**, 1510–1527 (2009)
25. C. Marchioli, V. Armenio, A. Soldati, Simple and accurate scheme for fluid velocity interpolation for Eulerian–Lagrangian computation of dispersed flows in 3D curvilinear grids. Comput. Fluids **36**, 1187–1198 (2007)
26. P. Sagaut, *Large Eddy Simulations for Incompressible Flows, an Introduction* (Springer, Berlin, 1998)
27. F. Roman, G. Stipcich, V. Armenio, R. Inghilesi, S. Corsini, Large eddy simulation of mixing in coastal areas. Int. J. Heat Fluid Flow **31**, 327–341 (2010)
28. F. Roman, V. Armenio, J. Froelich, A simple wall-layer model for large eddy simulation with immersed boundary method. Phys. Fluids **21**, 101701 (2009)
29. A. Fakhari, Wall-layer modelling of massive separation in large eddy simulation of coastal flows. Ph.D. dissertation thesis, the University of Trieste, 2015

30. A. Petronio, F. Roman, C. Nasello, V. Armenio, Large eddy simulation model for wind-driven sea circulation in coastal areas. Nonlinear Process. Geophys. **20**, 1095–1112 (2013)
31. C.-H. Moeng, J. Dudhia, J. Klemp, P. Sullivan, Examining two-way grid nesting for large eddy simulation of the PBL using the WRF model. Mon. Weather Rev. **135**, 2295–2311 (2007)
32. G. Zanier, A. Petronio, F. Roman, V. Armenio, High resolution oil spill model for harbour and coastal areas. *3rd IAHR Europe Congress, Book of Proceedings*, Porto-Portugal, 2014
33. A. Galea, M. Grifoll, F. Roman, M. Mestres, V. Armenio, A. Sanchez-Arcilla, L.Z. Mangion, Numerical simulation of water mixing and renewals in the Barcelona harbour area. Environ. Fluid Mech. **12**, 1405–1425 (2012)

Chapter 12
Big Data from Big Experiments. The WindEEE Dome

Horia Hangan, Maryam Refan, Chowdhury Jubayer, Dan Parvu, and Ryan Kilpatrick

12.1 Three-Dimensional Time-Dependent Large Scale Testing Chamber: The WindEEE Dome

The Wind Engineering Energy and Environment (WindEEE) Dome at the Western University (former University of Western Ontario), Canada is a novel three-dimensional and time-dependent wind testing chamber. It is conceived as a hexagonal chamber of 25 m in diameter surrounded by a 'return circuit' of the same hexagonal shape of 40 m in diameter, see Fig. 12.1. Its aim is to create a wide variety of wind systems (e.g. tornadoes, downburst, all kind of gusts and currents, shear winds and boundary layers, etc.) at large scales and Reynolds numbers, Hangan [1].

WindEEE does that by varying the configuration, wind speed and direction of 106 fans in the dome. 100 of them are situated on the peripheral walls of the inner testing chamber distributed as follows: 8 fans situated on 5 of the 6 sides of the dome and 60 fans arranged in a matrix of 15 fans per row × 4 rows on the 6th wall. Each fan is 0.8 m in diameter and operates at approx. 25 m/s at a nominal power of 30 kW. The other 6 fans are larger (2 m in diameter with nominal power of 220 kW) and are situated above the testing chamber ceiling. These six fans produce inflow or in reverse outflow to/from a plenum that communicates with the hexagonal chamber below through a bell-mouth.

WindEEE operates in two distinct modes: (1) multi-fan wind tunnel with the 60 fan wall (14 m wide × 4 m high) pushing air inside the chamber and with the air recirculating above the chamber. In this mode any type of spatial and/or time correlations may be produced. Each of the fans on the 60 fan wall can be individually operated and can accelerate over 20 % of their nominal rpm in 1 s or can be fully accelerated from 0 to 100 % in 5 s. (2) axisymmetric mode in which 8 fans on each

---

H. Hangan (✉) • M. Refan • C. Jubayer • D. Parvu • R. Kilpatrick
WindEEE Research Institute, Western University, London, ON, Canada
e-mail: hmhangan@uwo.ca

A. Pollard et al. (eds.), *Whither Turbulence and Big Data in the 21st Century?*,
DOI 10.1007/978-3-319-41217-7_12

Fig. 12.1 Photograph of the inner chamber of the WindEEE Dome

of the 6 walls are coupled with the larger 6 fans situated above the ceiling and that communicate with the lower testing chamber through a 5 m diameter bell-mouth. In this mode WindEEE can produce, for instance, tornadoes and downbursts of up to 5 m in diameter and translating at 2 m/s over 5 m.

For tornadoes the 8×6 fans on each of the walls blow air in through a set of directional louvres creating a twist at the base. This gets coupled with a suction produced through the bell-mouth at the top producing tornado-like vortices of various flow structures. Downbursts are created by having the upper fans feeding air down into the main chamber through the bell-mouth and evacuating air through the peripheral walls therefore creating all kind of impinging jets. WindEEE can also create other type of wind systems such as gusts, currents, sheared winds and large scale boundary layers.

12.2 Large Scale Experiments

12.2.1 Multi-Fan Wind Tunnel-Like Mode

In this mode the 4 rows $\times$ 15 columns $=$ 60 fans on one of the walls, see Fig. 12.2, can be used to generate a variety of flows such as: (1) boundary layer flows; (2) sheared flows horizontally; (3) sheared flows vertically and (4) active turbulence flows.

Boundary layer flows can be generated at various scale ranging from 1/2000 to 1/1 over various rough walls. This is possible by using a combination of a contraction, passive flow conditioning elements such as spires and roughness elements, as well as active flow conditioning by manipulating the fans on the wall

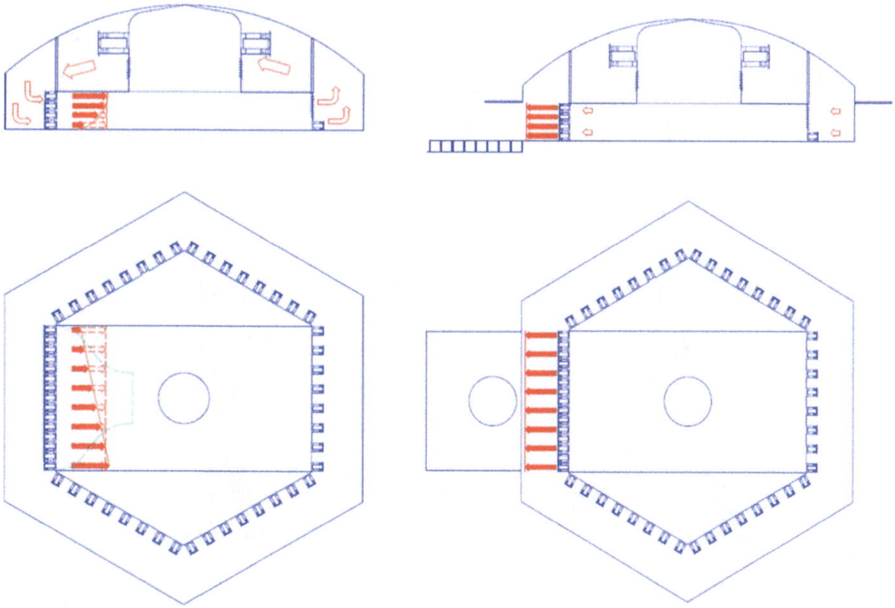

Fig. 12.2 Schematic of the straight wind tunnel and open jet modes. Representation of various vertically sheared flows, including boundary layers (*upper*) and horizontally sheared flows (*lower*). In *green* a 1/3 modular contraction (*Left*); Reversed fans, outflow open jet mode (*Right*)

in a time-dependent manner. Figure 12.3 is showing results in terms of mean and turbulence intensity profiles matched with ESDU standard profiles [2, 3] for scales 1/1 to 1/300. While the matching at the larger scales is not yet perfect, this gives a first indication of the possibility of using the WindEEE Dome as a multiscale facility which broadens the application range from typical wind engineering model scales (1/300 to 1/100) to large scale model testing such as 1/50 or larger which is needed, for example, for detailed topographic terrain testing up to prototype 1/1 testing of manufacturing products spanning a large range.

Figure 12.4 shows longitudinal component spectra for the same range of scales. Note that all spectra presented here are obtained only using the passive devices, such as spires, trips and roughness elements as well as constant vertical shear of rows of fans. For 1/1 scale as well as for the other larger scale testing we are currently manipulating the fans dynamically.

Various vertically sheared flows can be obtained. For instance, Fig. 12.5 shows inflows obtained for a complex topographic terrain for speed-up mean and turbulence intensity flow profiles matched with numerical simulation of a larger site area. The model scale is 1/1500 and covers an area of approx. 65 km^2. For each of the wind directions studied preliminary computational fluid dynamics (CFD) simulations were performed for a larger area in order to produce inflow profiles for the

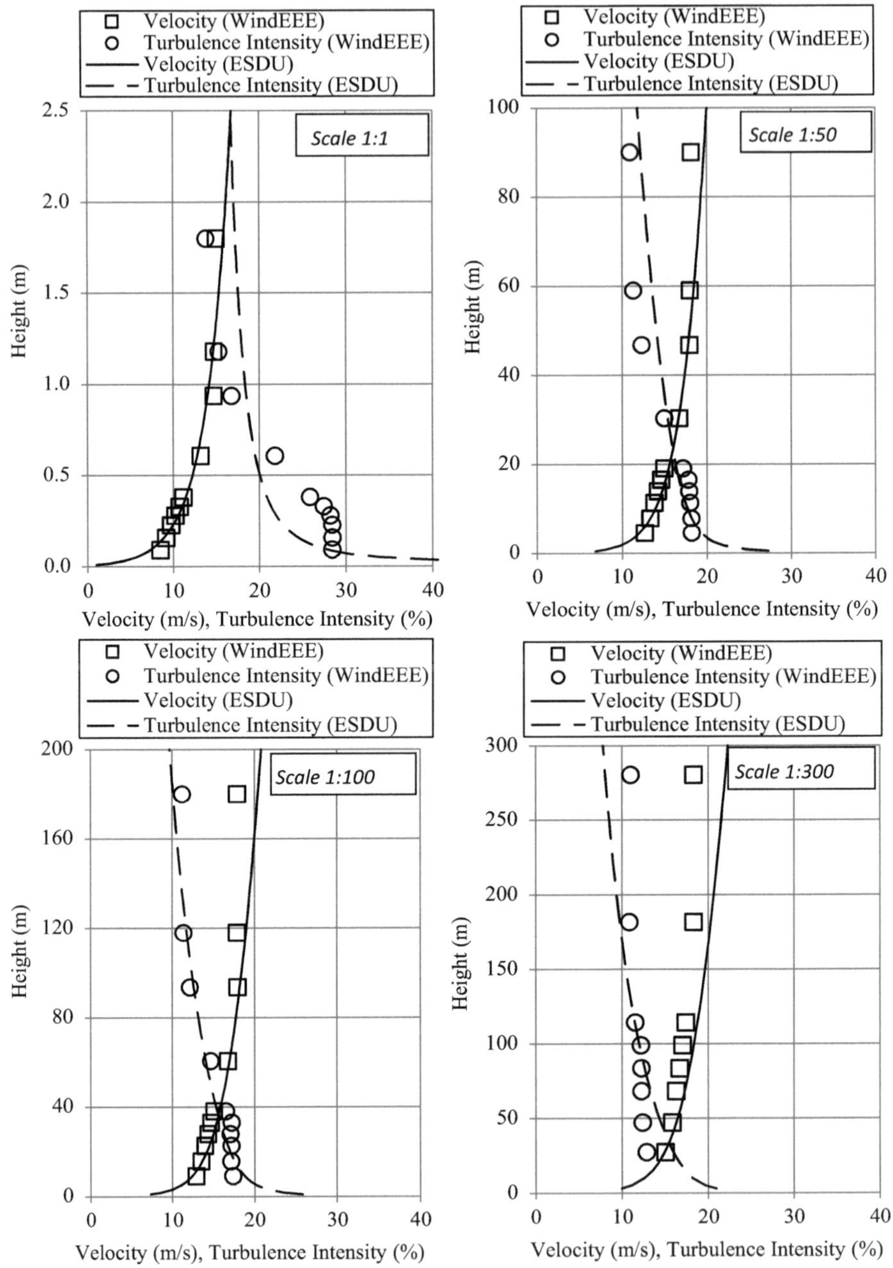

Fig. 12.3 Mean and longitudinal turbulence intensity profiles at scales between 1/1 and 1/300 measured in WindEEE (*symbols*) and compared to standard ESDU profiles

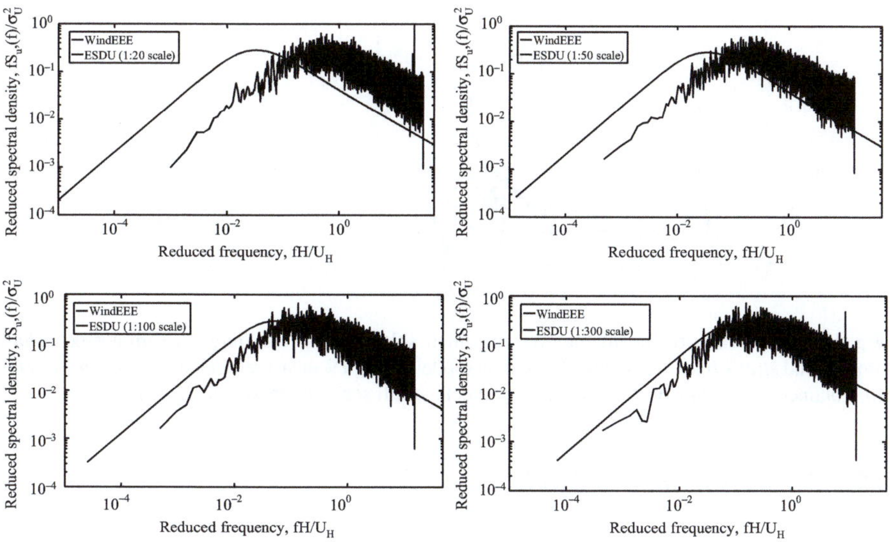

Fig. 12.4 Longitudinal non-dimensional spectra for scales between 1/20 and 1/300. Comparison between WindEEE results and ESDU standard [4] (*dotted line curve*)

Fig. 12.5 Mean and turbulence intensity profiles for speed-up flow over complex topography

area tested in WindEEE. The resulting CFD profiles at the corresponding upstream position of the WindEEE model were then matched by vertical manipulation of the 60 fan wall.

Figure 12.6 presents the preliminary capacity to produce a separation reattachment flow over a flat roof in which an array of full scale roof mounted solar panels are tested. Previous PIV measurements from the Boundary Layer Wind Tunnel Laboratory for a 1/30 scale model [5] were tentatively matched by manipulating the fans in WindEEE.

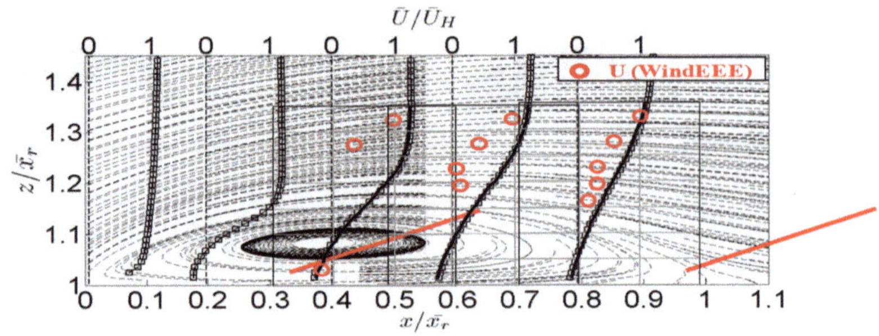

Fig. 12.6 Separation-reattachment flow in which solar panels (*red lines*) are immersed. *Black* profiles and *streak lines* are obtained based on model scale PIV measurements in BLWT, *red circles* are obtained based on Cobra probe measurements at full scale in the WindEEE Dome

12.2.2 Axisymmetric Mode

In this mode the coupling of the six large fans in the upper chamber with the inflow or outflow through the peripheral walls and fans can be used to produce a large variety of tornado-like vortices or downburst (impinging jet) flows, see Fig. 12.7. The generation of both tornado and downburst-like flows is based on replicating full scale data from various events.

12.2.2.1 Tornado Simulations

The first tornado-like vortex experiments have been performed in a 1/11 reduced scale Model WindEEE Dome (MWD) facility that has most of the WindEEE Dome geometry and functionality. The first stage was dedicated to qualitatively demonstrate that MWD can produce tornado-like vortices that can replicate the same variation of patterns with swirl ratio that previous research has identified. This has been shown by comparing flow visualizations for several Swirl ratios from Ward [6] and Church et al. [7] with smoke and Helium bubble visualizations from MWD [8]. Figure 12.8 shows flow visualizations for three swirl ratios before vortex break-down ($S = 0.12$), at touch-down ($S = 0.5$) and at the stage of multiple vortex formation ($S = 1.3$). The flow visualizations have been performed independently using smoke (left side) and Helium bubbles (right side) and they show the same features.

Flow visualizations have been repeated afterwards in the large WindEEE Dome facility and similar vortex structures have been observed for similar ranges of swirl ratios.

The surface static pressure distribution of tornado-like vortices and its variation with the swirl ratio provides useful insights into the vortex dynamics. A pressure measurement system and a floor panel with several pressure taps were used to measure the surface static pressure deficit in the WindEEE Dome. The centre

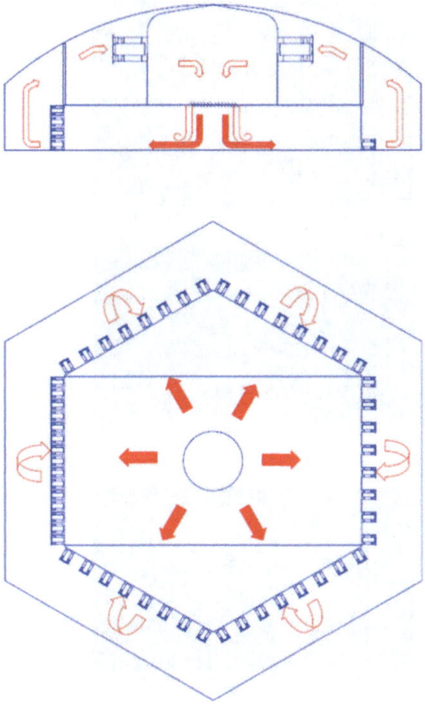

Fig. 12.7 Schematic of the axisymmetric modes of WindEEE Dome: Tornado (*left*) and Downburst (*right*)

floor panel of the simulator was equipped with 413 static pressure taps distributed on concentric circles (with a maximum diameter of 1.1 m) around the simulator centreline.

Figure 12.9 displays contours of surface pressure deficits normalized by $1/2\rho V_{ax}^2$ (where V_{ax} is the average axial velocity at the updraft). The results are presented for $S = 0.2$ and $S = 1$. It is observed that as the swirl ratio increases, the vortex core region expands and the suction in the flow intensifies.

Particle image velocimetry (PIV) has been performed to quantitatively determine the vortex flow structure for various swirls in order to compare with full scale data obtained via mobile Doppler radar measurements. Again, at a preliminary stage this has been done in the MWD [9] followed by measurements in the large WindEEE Dome where a large scale PIV technique is employed. These large PIV measurements were obtained using four cameras and cover windows of approx. 0.6×0.9 m, see Fig. 12.10. Snapshots for two swirl ratios $S = 0.2$; and $S = 1$, corresponding to vortex break-down aloft, and after touch-down phase are presented in Fig. 12.10. For low swirl ratios the vortex is meandering and the snapshot at

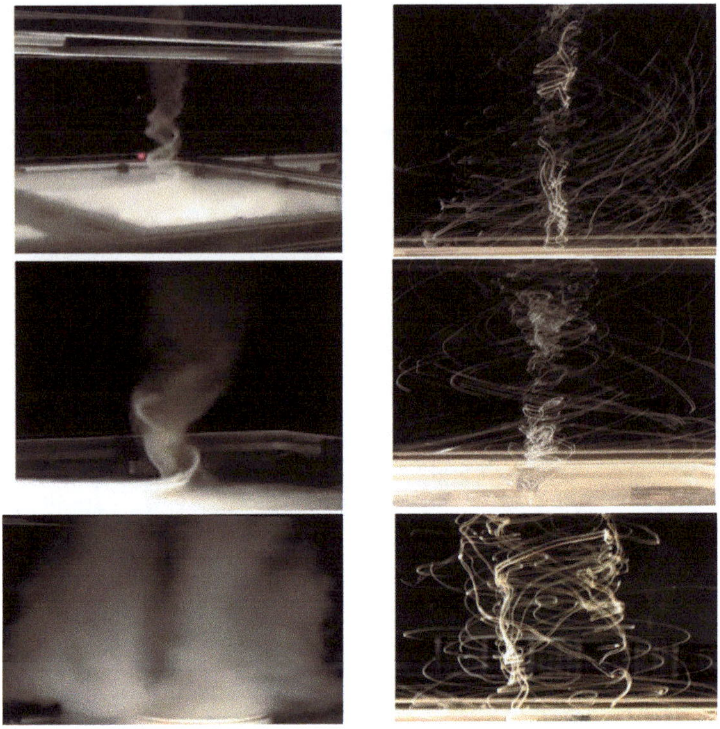

Fig. 12.8 Flow visualizations in MWD for swirl ratios (*top to bottom*) $S = 0.12$, $S = 0.5$ and $S = 1.3$ using smoke (*left column*) and Helium bubbles (*right column*)

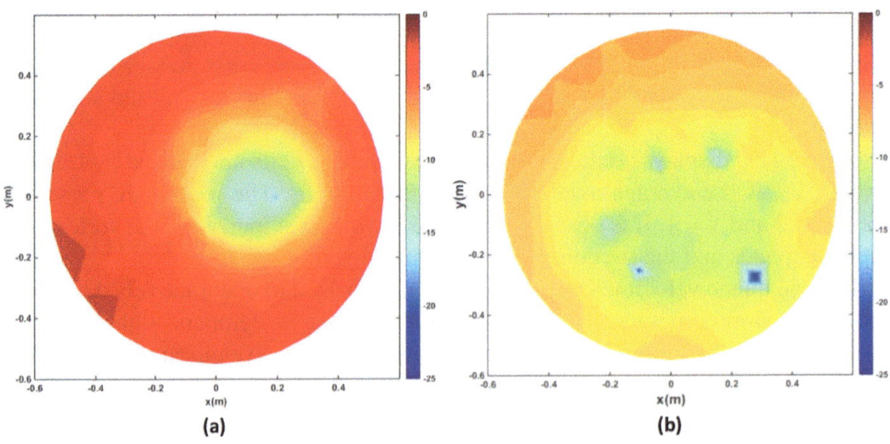

Fig. 12.9 Contours of surface pressure deficits normalized by $1/2\rho V_{ax}^2$ for (**a**) $S = 0.2$ and (**b**) $S = 1$

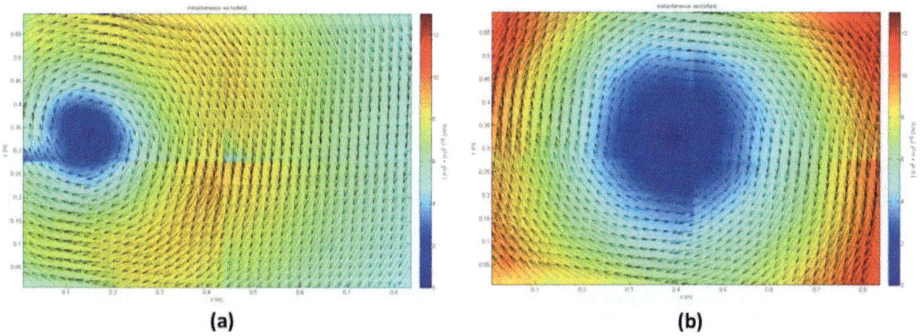

Fig. 12.10 Horizontal velocity vectors superimposed on horizontal velocity magnitude contours obtained from PIV measurements at $z = 33$ cm for (**a**) $S = 0.2$ and (**b**) $S = 1$

$S = 0.2$ captures the vortex at an off-centre position. After touch-down the vortex is less unstable and shows a clear increase in size.

In order to define a proper geometric and kinematic scaling between the real and WindEEE simulated tornadoes full scale Doppler data provided by the Center for Severe Weather Research (CSWR) in Boulder, Colorado was analysed. The ground-based velocity track display) method was used to recover velocity vectors in several quasi-horizontal slices through several tornado volumes, followed by the use of the continuity equation to determine the vertical structure in each of the volumes. Around ten tornado volumes ranging from EF0 to EF3, based on Enhanced Fujita Scale, have been analysed [9].

The large scale PIV data from WindEEE is averaged azimuthally and over time and is compared with the full scale GBDTV-retrieved data. A scaling approach is used to generate both the velocity and geometric scaling between the real and simulated tornadoes. The scaling process is detailed in Refan et al. [9] based on the initial PIV measurements in the model MWD and the same process has been used for the large WindEEE Dome data. Essentially a tornado-like vortex is characterized by two different length scales: the height and the radius at which the maximum tangential velocity occurs. These two lengths (radius and height), together with the maximum tangential velocity, are extracted from full scale data. In laboratory the swirl ratio of the physical simulation is varied and therefore both the radius and the height ratio between full scale and model scale vary. At a certain swirl ratio the two length scales intersect and therefore that swirl ratio is considered to be the match between experiment and reality, see Fig. 12.11. The matching process is shown here for several tornado volumes corresponding to various Effective Fujita Scale (EF-Scale) between EF1 and EF3. An overall 1/200 scale ratio between the full scale and the simulated tornadoes is observed.

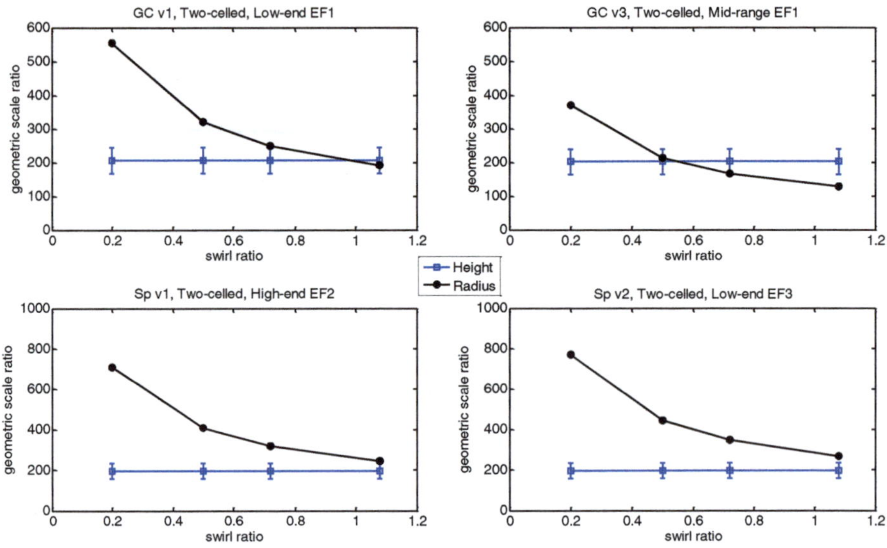

Fig. 12.11 Scale matching between various tornado volumes from Effective Fujita scale EF1 to EF3. The tornado volume event and the corresponding flow structure are marked for each plot

Fig. 12.12 Example of matching radial profiles of tangential velocity between full scale Doppler radar data and WindEEE large PIV data

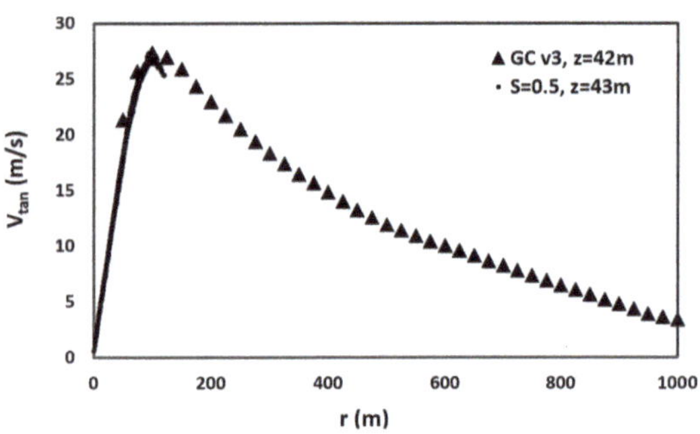

In addition the structure of the tornado vortex is checked to correspond to the a priori known structure [10] at that swirl ratio as well as corresponding to both full scale and experiment. As a result of this process a length scale and a velocity scale are determined, and the radial profiles of tangential velocity are matched, Fig. 12.12.

12.2.2.2 Downburst Simulations

As for the tornado flow, the downburst-like impinging jet flows were first qualitatively characterized using flow visualization techniques, see Fig. 12.13. The ratio between the height above the ground and the diameter (H/D) of the downburst has been varied between H/D = 0.8 and H/D = 1.2 as the flow structure is known to be

Fig. 12.13 Downburst flow visualization in WindEEE

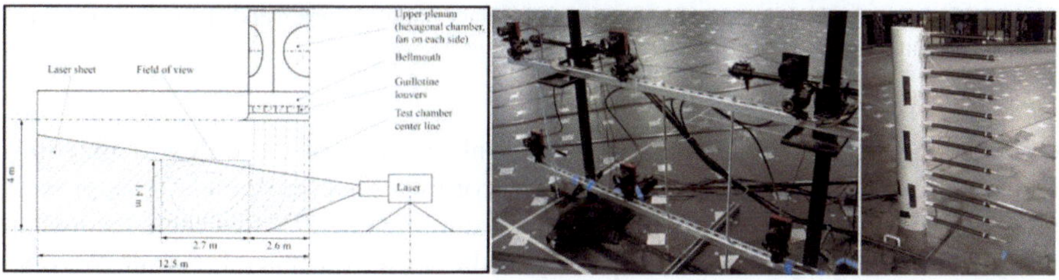

Fig. 12.14 Large scale PIV set-up. Schematic (*left*) and 6 cameras and Cobra probes (*right*)

different for H/D < 1, when the annular vortex formed by the rolling of the shear layers does not evolve entirely before impinging on the surface, as opposed to the case of H/D > 1 when the vortex is completely formed.

Large scale downbursts PIV measurements were performed in order to characterize the annular vortex dynamics resulting from the downburst-like impinging jet. These measurements cover even larger radial/vertical areas (compared to tornados) of approx. 2.7 × 1.4 m and are probably the largest scale PIV measurements taken in a laboratory set-up to date. The seeding is the same with the one used for tornado simulations. Six cameras, see Fig. 12.14, were employed to be able to characterize the main annular vortex as well as the surface layer dynamic separation reattachment. One flow realization between the release of the jet from the bell-mouth to the time that the annular vortex travels radially past the field of view of the 6 cameras takes approx. 1 s (Fig. 12.15).

12.3 Large Scale Measurement Techniques

Given the complexity both in time and space of these type of flows, as well as the large measurement volumes involved, it is envisaged that even the large scale PIV systems will not be enough to describe these type of flows. PIV measurements at these scales generate large data sets of the order of 10th of Gigabytes. For instance, for the downburst flow characterized above it takes approx. 1 s for the vortex to

Fig. 12.15 Large scale PIV snapshot capturing primary annular vortex and the surface dynamic separation reattachment

cross the approx. 3 m wide field of view. Only for one of these flow realizations and at a low sampling rate of 18 Hz a recording set corresponds to approx. 1 Gigabytes of data. In a time-resolved higher sampling rate PIV that same record can take tenths of Gigabytes. The complexity of the set-up for this large scale multiple camera PIV also needs to be taken into account. Lastly the data processing given the complexities of the PIV fields and camera windows overlapping alone makes the entire process very tedious.

Therefore a new type of particle tracking technique, the large scale particle streak velocimetry (LSPSV) method is under development at WindEEE RI aiming at measurement areas of the order of 10 sq m.

Particle tracking techniques rely on discrete particles that follow the flow. These particles also need to have good reflectivity so that their motion can be captured by the camera. In the case of the PTV and PSV methods, particle seeding density is not a major issue, Bosbach et al. [11]. A commercially available HFSB generator was used to seed the test chamber.

As an illumination method a stroboscopic lamp was chosen due to its operating simplicity and low cost. An adjustable metallic slit was placed in front of the strobe optics to control the light sheet thickness, (mean thickness of 10 cm). The resulting pulsating light sheet was used to obtain different illumination effects on a particle streak which is then used to estimate velocity. The lamp was placed close to the centre of the test chamber, as shown in Fig. 12.16.

Particle streaks are then captured, like the ones in Fig. 12.17 showing a streak field for the downburst annular vortex described above.

After a series of morphological operations the images are converted into logical 2D arrays. The velocity and position of the tracked particle are determined and pixel values are then converted into length units using camera calibration values. Vectors are then plotted, as shown in Fig. 12.17 (right). The procedure is repeated for each streak identified in the images in order to obtain a vector field. The LSPSV set-up proposed and tested herein shows the potential to be a less costly, fast, easy to use alternative to LSPIV.

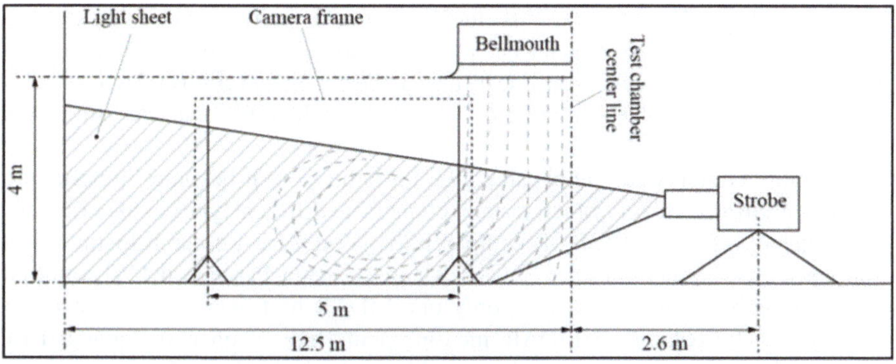

Fig. 12.16 Camera calibration relative to light sheet

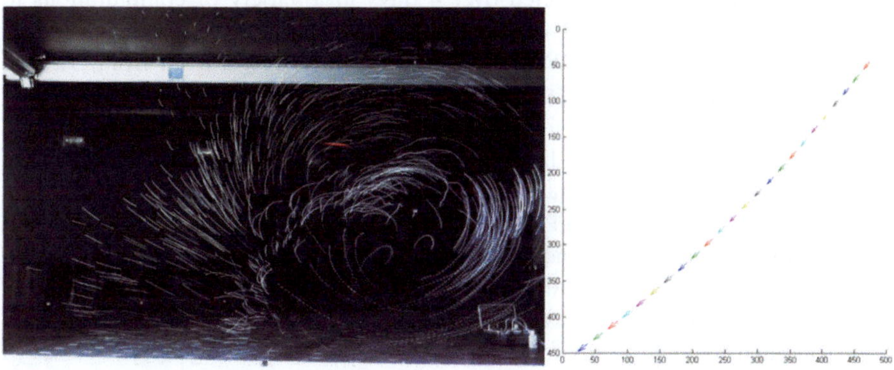

Fig. 12.17 Downburst vortex particle streaks (*left*) and a processed streak into velocity vectors (*right*)

Another typical example of large data sets are the ones resulting from surface pressure measurements. Typically in wind engineering models are instrumented with approx. 1000 pressure taps and pressures are sampled at approx. 400 Hz or higher frequency over periods of up to 6 min over 36 azimuthal angles. This amounts to more than 40 Gigabytes of data. There is a new trend in wind engineering, for instance, to go towards very large scale testing with models of buildings of the order of 1/3 full scale and other full scale testing of components such as façade components, roof components and roof mounted installations (such as solar panels). Typical model scale testing in boundary layer wind tunnels is between 1/100 and 1/600. By increasing this scale the resolution is increased by at least one order of magnitude. Simultaneous pressure, base balance and strain–stress measurements are also possible on these large scale models. Therefore we can expect combined data sets of the order of Terabytes coming out of these tests. Moreover it is structurally more and more important (and feasible) to assess the capacity of various structural systems to withstand dynamic/transitional wind loading such as the ones attributed to tornado or downburst events. This can be addressed by drawing building systems or components fragility curves that describe the probability of failure as a function of wind speed. In order to achieve this tests have to be conducted at several wind speeds therefore adding probably one more order of magnitude to the data acquisition and data processing.

The more complex the geophysical flows are the more it becomes important to combine multiple data sets which usually come from different types of experiments. For instance, a full scale LiDAR campaign that WindEEE Research Institute just completed this summer in Prince Edward Islands, Canada, Parvu et al. [12] collected over 50 Gigabytes of data running only one LiDAR unit over a couple of hours per day for 2 weeks. The LiDAR measurements have been complemented by Cobra probe measurements in order to address the turbulence characteristics of the flow. This experiment, which only collected three-component point data with four Cobra probes, was conducted over 4 days in a variety of configurations to obtain turbulence profiles and various correlations which only total to approx. 5 Gigabytes of data. Also, the WindEEE team together with a team from DTU in Denmark recently conducted large, 1/25 model scale experiments on the Bolund topographic feature in order to address the problems faced by both numerical and typical wind tunnel scale laboratory experiments to reproduce dynamic flow behaviours, in this case a dynamic separation/reattachment. Tests were therefore conducted using multi-camera PIV measurements over two model scales (1/25 and 1/100) with several inflow conditions (velocity, profile, turbulence and roughness levels) and comparison with full scale data has been provided. For one test run, i.e. 5 min of PIV sampling at time at 20 Hz about 5000 images at 12 Megabytes each was generated for each of the four cameras to a total of 240 Gigabytes. A total of approx. 30 runs were employed in this experiment bringing the total data for raw images only to approx. 6 Terabytes. Data processing added approx. 1/3 of the size of the raw data bringing the total close to 8 Terabytes. The POD analysis applied for this flow field essentially only added approx. 5 Gigabytes per run, or only 2 % of the raw data, Kilpatrick et al. [13].

The interesting breakthrough in fluid mechanics Big Data processing may come when algorithms will be developed which would be able to extract meaningful information from various data sets at various scales, for example, between full scale and laboratory and numerical data. At that point our data analysis techniques which now focus on energy (such as POD) or time-frequency analysis (such as Wavelets) or mode independency (such ICA) may evolve towards other or even multiple criteria tools. This would open a data analytics approach for fluid mechanics in which the data from multiple sources, collected with a variety of tools (differentially covering the time-space domain), at various scales and under a large variety of flow and boundary conditions can be analysed in order to obtain a holistic understanding of that type of phenomena or to gain in depth insight based on one or multiple criteria.

12.4 Conclusions

Geophysical flows in general and wind flows in particular are characterized by an extremely large spectra of scales and therefore Reynolds numbers are 3D bounded and experience time-varying initial/inflow conditions. As a result most of these flows are three-dimensional and non-stationary. On the other hand, the well-established wind or water tunnel facilities and flumes produce straight and steady fluid flow movements and are usually limited to a certain range of Reynolds numbers.

A new concept in physically modeling these flows has recently emerged. The WindEEE Dome at the University of Western Ontario in Canada is a novel large scale, three-dimensional, time-dependent and multipurpose wind testing chamber that uses a complex multi-fan system to produce:

(1) Straight sheared and time-dependent flow systems generating atmospheric boundary layer flows of scales between 1/2000 and 1/1 and any type of surface currents or complex dynamic separated flows; (2) axisymmetric flows such as tornado-like vortices and downburst flows of the order of 5 m in diameter and with translation capacity.

The first results obtained from WindEEE are analysed for a variety of these flows. The large scale of the generated flow fields also triggers the use of large scale measurement techniques such as complex, multi-camera PIV measurements. These methods generate large amounts of data and are tedious in terms of set-up, processing and data analysis.

New, particle tracking methods for large scales such as the LSPSV method proposed herein show potential for the large, three-dimensional and unsteady flows generated in WindEEE Dome.

Big Data analysis can shed new light on the way we interpret multiscale flows by use of multiple criteria in data interrogation.

References

1. H. Hangan, The Wind Engineering Energy and Environment (WindEEE) Dome at Western University, Canada. Wind Eng. JAWE **39**(4[141]), 350–351 (2014)
2. ESDU, *Strong winds in the atmospheric boundary layer. Part 1: mean hourly wind speeds* (1982)
3. ESDU, *Strong winds in the atmospheric boundary layer. Part 2: discreet gust speeds* (1983)
4. ESDU, *Characteristics of atmospheric turbulence near the ground* (1985)
5. R.N. Pratt, G.A. Kopp, Velocity measurements around low-profile, tilted, solar arrays mounted on large flat-roofs, for wall normal wind directions. J. Wind Eng. Ind. Aerodyn. **123**(Part A), 226–238 (2013)
6. N.B. Ward, The exploration of certain features of tornado dynamics using a laboratory model. J. Atmos. Sci. **29**, 1194–1204 (1972)
7. C.R. Church, J.T. Snow, G.L. Baker, E.M. Agee, Characteristics of tornado-like vortices as a function of swirl ratio: a laboratory investigation. J. Atmos. Sci. **36**, 1155–1176 (1979)

8. M. Refan, Physical simulation of tornado-like vortices. PhD Thesis, University of Western Ontario, 2014
9. M. Refan, H. Hangan, J. Wurman, Reproducing tornadoes in laboratory using proper scaling. J. Wind Eng. Ind. Aerodyn. **135**, 136–148 (2014)
10. R.R. Davies-Jones, J. Trapp, H.B. Bluestein, Tornadoes and tornadic storms. Meteorol. Monogr. **28**, 167–222 (2001)
11. J. Bosbach, M. Kühn, C. Wagner, Large scale particle image velocimetry with helium filled soap bubbles. Exp. Fluids **46**, 539–547 (2009)
12. D. Parvu, J. LoTufo, H. Hangan, Shear layer effects of cliff edge: PEIWEE experiment 2015, in *WindTech 2015 Conference*, London, ON, Canada, 2015
13. R. Kilpatrick, K. Siddiqui, H. Hangan, Analysis of turbulent coherent structures in a flow over an escarpment using proper orthogonal decomposition, in *TORQUE Conference*, 2016

Chapter 13
Visualizing Wind Farm Wakes Using SCADA Data

Shawn Martin, Carsten H. Westergaard, and Jonathan White

13.1 Introduction

The wind energy used by a turbine to produce electrical power causes a reduction in wind speed behind the turbine, also known as a wake or wind shadow [1]. The wake behind a turbine will dissipate with distance, but will affect nearby downwind turbines. For a large wind farm, these effects will accumulate, resulting in potentially significant aggregate power loss over the entire farm. In addition, wakes are turbulent, and can stress downwind turbines, possibly affecting the reliability and lifetime of turbines within the farm.

To mitigate the effects of turbine wakes in a wind farm, it is important to carefully site turbines during the design phase [1]. Research into siting wind turbines has employed semi-empirical numerical models of turbine wakes [2–5], as well as more exact models based on the Navier–Stokes equations [6]. Power losses due to wake effects have also been modeled [7, 8]. While these models provide valuable insight and guidelines, they are nevertheless incomplete and cannot address every practical concern in the siting of turbines when building a wind farm. Further, there are numerous existing wind farms, for which re-siting is not an option, but which might yield greater power production with improved understanding of on-site wake effects.

An alternative to using computational models for understanding and mitigating wake effects is to analyze data gathered from existing farms using techniques from statistics and data mining. Such approaches have yielded results in the areas of turbine failure prediction and condition monitoring, see, for example, [9–13], but have been applied to lesser degree in wake analysis [14–17].

In this paper, Supervisory Control and Data Acquisition (SCADA) data is analyzed with the goal of documenting wake effects on a functioning industrial scale

S. Martin (✉) • C.H. Westergaard • J. White
Sandia National Laboratories, Albuquerque, NM 87185, USA
e-mail: smartin@sandia.gov

A. Pollard et al. (eds.), *Whither Turbulence and Big Data in the 21st Century?*,
DOI 10.1007/978-3-319-41217-7_13

wind farm. SCADA data from this farm is summarized, corrected, transformed, and visualized in order to understand both local turbine–turbine effects and larger wind farm scale effects.

The data is inherently big data, even from only one wind farm, but certainly also if one considers that almost all wind farms daily stream this type of data to data centers around the globe for the owner and operators of wind farms to analyze performance. The primary goal of this paper is to present a novel, holistic visualization of wake effects across a wind farm. This visualization has resulted in the identification of interesting wake type effects such as channel speedup and shear, but it is only a first step in an effort to quantify power loss in existing wind farms due to wake effects. The ultimate goal is not (necessarily) to assist in the layout of new farms, but to validate and potentially improve the performance of existing farms. Once poorly performing turbines and farms are identified, efforts can be made to improve power production. Such efforts would range from pre-emptively identifying turbines likely to fail (based on their position in a farm and the wake effects acting upon them), to implementing cooperative turbine (smart-farm) strategies. Cooperative strategies work, for example, by directing upwind turbines to operate below peak efficiency so that down wind turbines have additional wind energy available, thereby optimizing aggregate power production [18].

13.2 Materials and Methods

SCADA Data Data was gathered over a 1.5 year period in 2012 and 2013 from the SCADA system at an onshore wind farm in the USA. The wind farm included 67 horizontal axis, three bladed, variable pitch turbines, along with one meteorological (met) tower. In this study, analysis was performed on a subset of the data collected from the turbines: nacelle wind speed, nacelle direction (position), rotor speed, blade pitch, and power output. The met tower collected data on temperature, air pressure, wind speed, and wind direction. The layout of the wind farm is shown in Fig. 13.1, along with a wind rose showing the prevailing wind directions.

Data was collected continuously (every 2 s) but was summarized over 10 min intervals prior to analysis. For each variable collected (e.g., wind speed), the minimum, maximum, average, and standard deviation over the 10 min interval was computed. After summarization, there were approximately 61,000 time points per turbine available for analysis.

Initial screening required an operational turbine and wind speed to lie in the range of 4–20 m/s, yielding approximately 46,000 time points per turbine, corresponding to a site average wind speed of approximately 8 m/s.

Met Tower Correction Due to sensor inaccuracies, various corrections were performed on the wind direction measurements. Following [14], the met tower was considered to have the most accurate wind direction sensor, but was adjusted for systematic bias by comparing the met tower sensor data with data gathered from two nearby turbines, as described next.

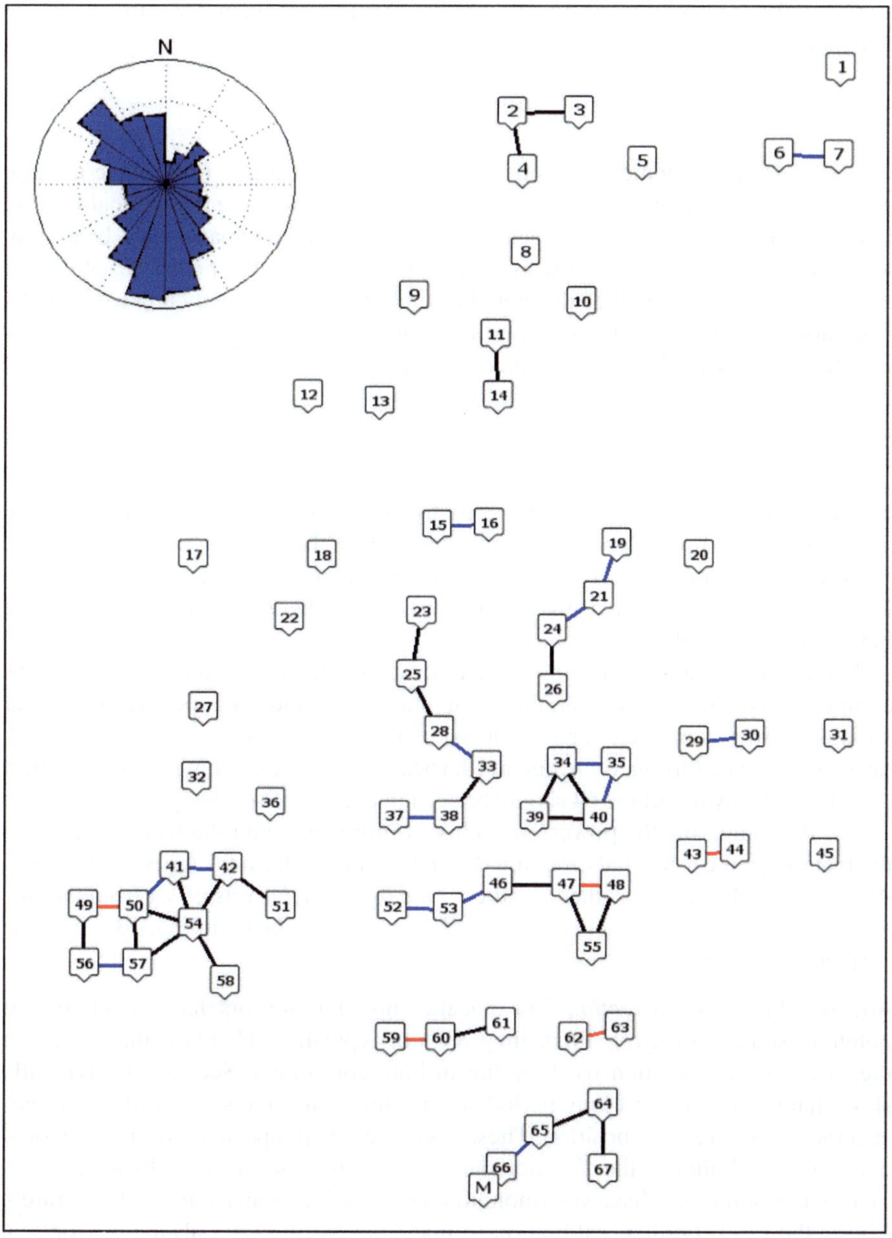

Fig. 13.1 Wind farm layout. The relative positions of the turbines are shown, with turbines numbered from 1 to 67, and the met tower marked M. The site wind rose is shown in the *upper left*. Turbines in close proximity are connected by lines: turbines within 5 rotor diameters are connected using *red lines*; turbines between 5 and 6 rotor diameters are connected using *blue lines*; and turbines between 6 and 7 rotor diameters are connected using *black lines* (turbines greater than 7 rotor diameters are less likely to experience wake conditions). Icons were taken from the Map Icons Collection (http://mapicons.nicolasmollet.com) and are licensed under Creative Commons Attribution (3.0)

Given the measured wind direction at the met tower, denoted θ_m, the goal is to find an offset θ_f such that

$$\theta_t = \theta_m + \theta_f, \tag{13.1}$$

where θ_t is the true wind direction. To find θ_f, it is necessary to estimate the true wind direction θ_t. Fortunately, θ_t can be estimated using the geographical bearing (known exactly) of two turbines near the met tower, and comparing that bearing to the power variability of the downwind turbine. In theory, the power variability of the downwind turbine will peak when the wind direction is exactly aligned with the geographical bearing, providing an estimate of θ_t.

The power variability of a turbine is computed as

$$P_v = \frac{\sigma_P}{\mu_P}, \tag{13.2}$$

where μ_P is the mean power produced by the turbine, and σ_P is the standard deviation of the power produced, both taken over the 10 min intervals. For the purpose of comparing power variability against wind direction, the power variability is averaged over time for a given wind direction, where the wind directions are binned in 1° intervals.

For the wind farm under investigation, wind direction at the met tower was compared with the power variability of nearby turbine 64. Specifically, power variability was computed for turbine 64 when turbine 64 was downwind from turbine 67. Since turbine 64 is due north (bearing 180°) of turbine 67, the greatest power variability in turbine 64 should be seen in periods of southerly winds (bearing 180°). By comparing the power variability of turbine 64 with the wind direction at the met tower, the true wind direction θ_t can be estimated, as illustrated in Fig. 13.2. According to this analysis, the measured wind direction θ_m was offset from the true wind direction θ_t by an angle $\theta_f \approx 11°$. The wind rose shown in Fig. 13.1 uses the corrected met mast data.

Nacelle Direction Correction The nacelle direction sensors have a number of potential sources of error. First, they are not typically well maintained because the direction is not often used by the turbine controllers. Second, the typically slow changes in yaw and long periods of inaction cause the sensors to experience accuracy loss in relative position. These losses can be jumps in the position or slow deterioration. Although the direction may (in some turbines) be reset by the passage of a switch, some sites have very monotonous wind direction and the turbines rarely trigger the reset. Further, calibration to magnetic North is not always performed, or is sometime re-programmed with turbine software updates. This means that there can be instant jumps in directional determination at discrete times. Finally, if no service records are available, sensors may be replaced, after which absolute direction is lost.

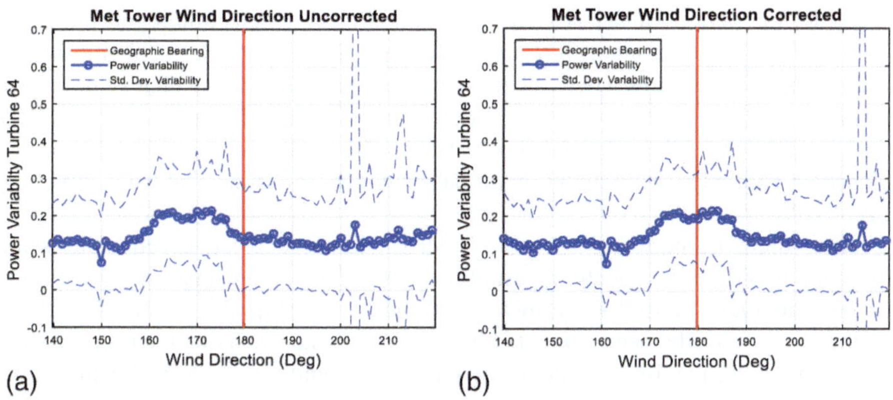

Fig. 13.2 Met tower wind direction correction. On the left (**a**), power variability of turbine 64 is plotted against the met mast measured wind direction θ_m for winds from the south. A peak in power variability is seen at approximately 169°. In reality, this peak should occur at the geographical bearing from turbine 64 to turbine 67, which is shown as a *red line*, occurring at 180°. Therefore, the wind direction offset $\theta_f = \theta_t - \theta_m \approx 11°$. On the right (**b**), power variability is plotted against the corrected wind direction, showing an alignment between the peak power variability and the geographic bearing. Throughout this figure, the power variability curves were computed across wind direction bins of 1°. The accuracy of this technique is indicated by the *dashed lines* in (**a**) and (**b**), showing the standard deviation of the power variability

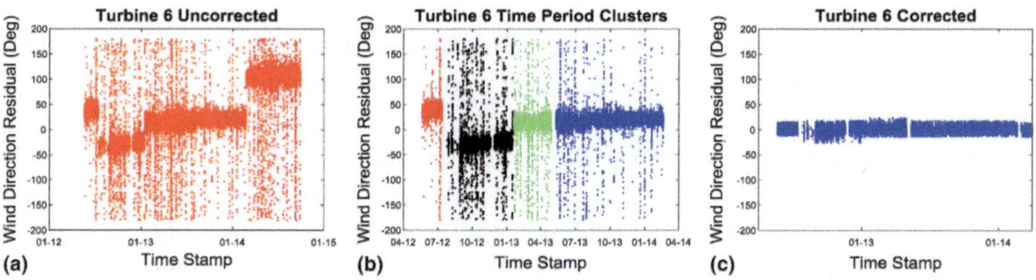

Fig. 13.3 Nacelle direction correction for turbine 6. On the left (**a**), the residual wind direction $\theta_n - \theta_m$ is shown over time for turbine 6, where θ_n is the nacelle wind direction and θ_m is met tower wind direction. In addition to the additive biases that can be observed, there are several noticeable changes in the measurements at different time points. In the middle (**b**), the time periods are clustered using a change detection algorithm so that an additive correction can be applied. The separation of the last two groups (*green* and *blue*) is due to a period of erratic measurements which can be observed in the uncorrected data as a *solid vertical line*. On the right (**c**), additive corrections are applied to the time period clusters and individual measurements significantly different from the mean are removed

The largest obstacle to overcome in correcting the nacelle wind direction is the identification of time period clusters showing significant relative change in wind direction against the corrected met tower data, now considered to be ground truth. An example of this behavior for turbine 6 is shown in Fig. 13.3a. Note that the direction measurements are present only when a turbine is operational and wind speeds exceed 4 m/s, as per the initial data screen.

Each time period cluster is identified using a change detection algorithm [19]. The change detection algorithm proceeds sequentially through each time series searching for time points where the moving average experiences a change beyond a given threshold. For the wind direction measurements, a moving average over 750 time points was used with a threshold of 20°. The results of the change detection algorithm applied to wind direction measurement data from turbine 6 are shown in Fig. 13.3b.

After the time period clusters were identified for a given nacelle, the wind direction based on the nacelle measurements was corrected using additive offsets, as was previously done for the met tower data. Using Eq. (13.1), an offset θ_f is applied to the nacelle measurements for each time cluster such that the nacelle wind direction θ_n is equal on average to the met tower wind direction θ_m for that time period cluster.

Finally, individual wind direction measurements for each nacelle were discarded if greater than one standard deviation from the mean. An example of the corrected data for a nacelle is shown in Fig. 13.3c. The nacelle correction left approximately 34,000 time points per turbine for further analysis.

Pitch Correction The final correction compared blade pitch versus wind speed to remove unusual instances of turbine operation. Normally, blade pitch should respond predictably to variations in wind speed, as dictated by the turbine controller. Unusual blade pitch response therefore indicates unusual turbine operation.

To perform this correction, instances of unusual turbine operation were removed if the mean blade pitch was more than one standard deviation from the mean blade pitch schedule, identified empirically by computing mean pitch vs. wind speed, using wind speed bins of 1 m/s. This correction was done on a per turbine basis to account for potential biases in turbine calibration, location, and wind speed measurements. The blade pitch correction for turbine 6 is shown in Fig. 13.4. Pitch correction left approximately 32,000 time points per turbine for further analysis, or approximately 222 days in operation.

Turbine–Turbine Power Curve Pair Analysis To motivate the method developed for visualizing wake effects across the entire wind farm, it is instructive to first consider the simpler case of two interacting turbines. For this exercise, consider the relatively isolated pair consisting of turbines 6 and 7, located near the northeast corner of the farm.

Turbines 6 and 7 are close enough (within 6 rotor diameters) that turbine 7 should experience wake effects given a westerly wind. However, wake effects were not observed by plotting the individual power curves for the two turbines, as shown in Fig. 13.5a. Fortunately, this observation is illusory, although the explanation is subtle.

It is important to realize that the wind speed observed by a nacelle is relative to that nacelle. In other words, a downwind nacelle will observe a wind speed slower than the true wind speed, precisely because it is in the wind shadow of the upwind turbine. For the case of two turbines, this problem can be solved by using

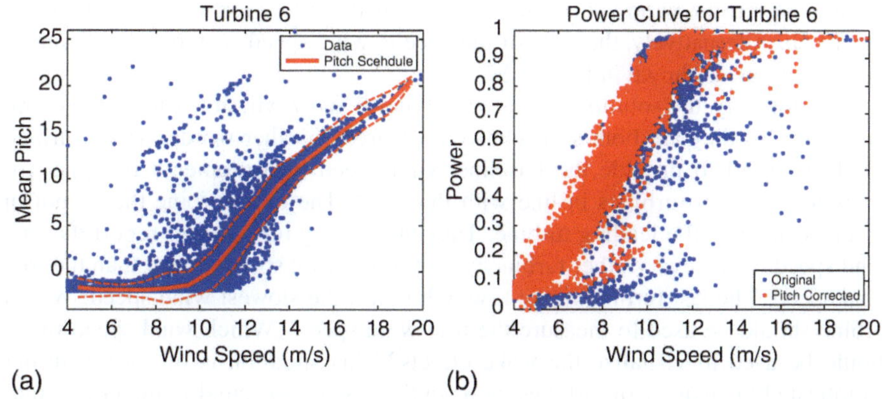

Fig. 13.4 Pitch correction. On the left (**a**), the pitch schedule is shown as the variation in mean blade pitch against wind speed. Data points more than one standard deviation from the pitch schedule (indicated by the *dashed lines*) are removed. On the right (**b**), the power curve of the corrected data is shown (on a normalized scale). Note that several abnormal operating modes were removed, including de-rated periods. These are seen as *blue points* below the power curve. The de-rated periods occur in a *blue line* of points at wind speeds from 10 to 14 m/s and power level at ~0.65

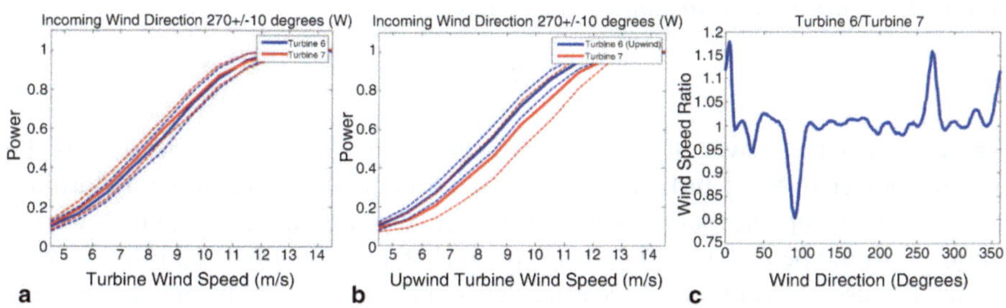

Fig. 13.5 Wake effect for turbine 7. On the left (**a**), power curves are shown for the upwind turbine 6 and the downwind turbine 7 (given westerly winds). These curves were obtained by averaging the nacelle power measurements over bins with width of 1 m/s, and the *dotted lines* denote 1 standard deviation from the mean. The curves show no wake effect because the individual nacelle wind speed measurements were used on the x-axis. In other words, both turbines are functioning normally according to the wind speeds they are observing. In the middle (**b**), power curves are shown for the same two turbines, this time using the upwind nacelle (turbine 6) sensor to measure wind speed on the x-axis. These curves show a wake effect, because turbine 7 is under-performing relative to turbine 6, given the wind speed it would observe if it were not in the wake of turbine 6. On the right (**c**), the ratio of the wind speed from turbine 6 to turbine 7 is plotted versus wind direction. The wind speeds were binned as described in the text for the normalized instant power plots. The wake effect between turbines 6 and 7 for westerly winds is observed as a peak at 270° and for easterly winds as a trough at 90° (the peak at 0° is due to the fact that turbine 7 is in the shadow of turbine 1 for northerly winds)

the upwind turbine wind speed measurement as the true wind speed, as shown in Fig. 13.5b. Alternatively, the ratio of the nacelle wind speeds can be plotted versus wind direction, as shown in Fig. 13.5c.

Unfortunately, the solutions used for turbines 6 and 7 will not scale to the entire wind farm, as they are both still relative measures. In other words, which turbine should be used to provide the baseline wind speed measurement? Imagine, for example, a row of turbines in line with the wind. The only turbine that is not in a wind shadow is the leading turbine. Thus the leading turbine will report the true wind speed, while every other turbine will report a wind speed slower than the true wind speed. The last turbine in the row will report the slowest wind speed. Which turbine should be used to measure the true wind speed? Which wind speed ratios should be used to visualize the wake effects? This situation is of course further complicated in an array of turbines, or as in the case of an actual farm, an irregular distribution of turbines.

Directional Power Performance and Variance Plots The previous discussion of nacelle wind speeds suggests a new approach for visualizing wake effects: use an average measurement over the entire farm as a baseline in order to compare individual turbines. Since wind speed and power are intimately related for a turbine, and since the power measurements were more accurate in the dataset under consideration, a normalized measure of instantaneous power was employed. For turbine i, the *normalized instant power* is defined to be

$$P_N(t) = P_i(t)/\mu_{P(t)}, \tag{13.3}$$

where $P_i(t)$ is the power of turbine i over the 10-min interval t, and $\mu_{P(t)}$ is the average power over all turbines over the same interval. For example, if turbines 6 and 7 made up the entire wind farm, then the normalized instant power of turbine 6 would be $P_N(t) = 2P_6(t)/\,(P_6(t) + P_7(t))$. The normalized instant power avoids the use of wind speed measurements and can also be averaged over time and binned against wind direction to obtain polar plots showing the performance of a given turbine against the performance of the wind farm as a whole. For the normalized instant power plots, bins centered at every integral degree (e.g., 0°, 1°, …, 359°) were used. These bins were overlapping and covered sectors 16° wide (e.g., $[-8°, 8°]$, $[-7°, 9°]$, … $[351°, 7°]$). The main reason to use overlapping bins was to smooth the resulting plots, due to a lack of data in certain wind directions. The 16° sectors correspond to anticipated wake effects for turbines separated by 7 rotor diameters. An example of normalized instant power plots, again assuming turbines 6 and 7 make up the entire wind farm, is shown in Fig. 13.6. (Note the similarities of the normalized instant power plots with the plot of relative wind direction in Fig. 13.5c.)

In Fig. 13.6, the mutual waking of the two turbines at 90° and 270° can be clearly observed. For turbine 6, there is a hint of wake deficit from turbine 1 at 30° and a stronger deficit bearing 0° for turbine 7. The difference in the deficit is smaller at turbine 6 due the larger distance to turbine 1. Turbine 7 also shows an apparent

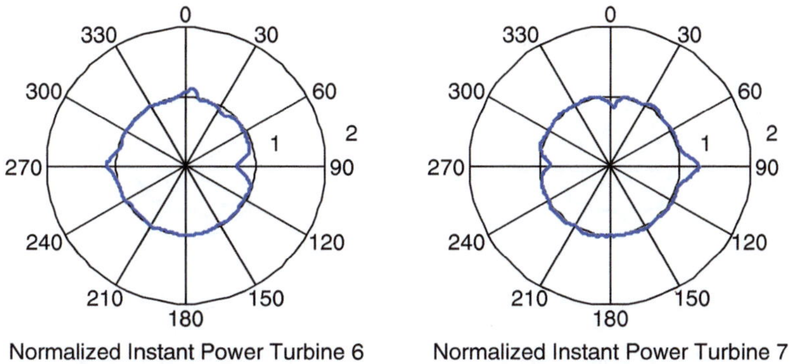

Fig. 13.6 Normalized instant power plots. These plots show normalized instant power (see Eq. 13.3) averages over time versus wind direction for turbines 6 and 7. The plots are arranged to mirror the positions of the turbines in the wind farm, so that turbine 6 is west of turbine 7. Thus, given a westerly wind, turbine 6 is upwind and turbine 7 is downwind. In this case, turbine 7 is under-performing relative to turbine 6, so that a dip is observed in the plot for turbine 7 at the angle 270° (and a corresponding bump is seen for turbine 6 at 270°). There are also bumps and dips at 90°, and the bump in turbine 6 at 0° and corresponding dip in turbine 7 occur because of the wake from turbine 1 just north of turbine 7. Both plots are obtained by averaging normalized instant power over time using wind direction bins 1° apart

increase at 90°. Note that these plots are slightly different from the plots generated for the entire wind farm (Fig. 13.8), since this example is normalized against only turbines 6 and 7 (i.e., this example assumes the wind farm consists only of turbines 6 and 7, even though the other turbines in the farm clearly affect the turbine 6 and 7 subset).

A variation on the normalized instant power computation is power variability, as defined in Eq. 13.2. Power variability averages over time can be computed and binned against wind direction (again using overlapping bins 1° apart covering ±8° sectors) as shown in Fig. 13.7.

Colored Polar Plots The normalized instant power and power variability polar plots can be used to visualize wake effects across an entire wind farm. To facilitate viewing multiple plots simultaneously, a color scale can be added to the plots. For the instant power visualization these colors highlight over-/under-performing turbines, and for the power variability visualization the colors highlight high/low power variance per turbine. The color scales are computed to be comparable across the entire farm (i.e., the same scale is used for every plot in the entire visualization). Visualizations of the wind farm are given in Fig. 13.8. Wake effects are easily seen using these images.

Note that the average power in the normalized power computation is taken over periods where most of the farm is operational, but not necessary the entire farm. Specifically, average power is computed when at least 62 turbines are operational, or no more than 5 turbines are not operating. This measure avoids ignoring periods

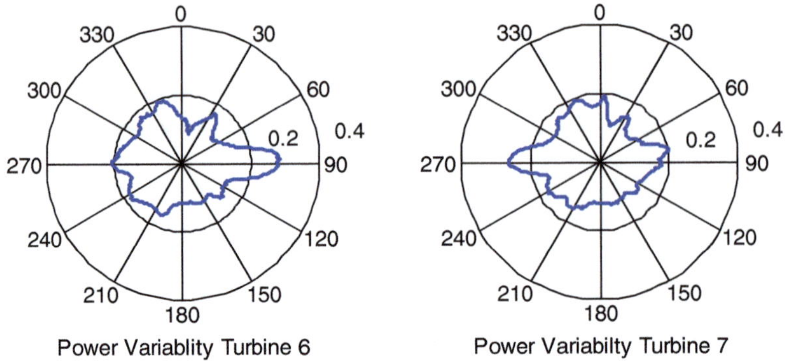

Fig. 13.7 Power variability plots. These plots show power variability (see Eq. 13.2) over time against wind direction for turbines 6 and 7. They are again arranged to mirror the positions of the turbines in the wind farm. Wake effects can be observed as increased variability, which manifest as bumps on the plots. Thus there are bumps at 90° (easterly wind) for turbine 6 and 270° (westerly wind) for turbine 7. Both plots are obtained by averaging power variability over time using wind direction bins 1° apart

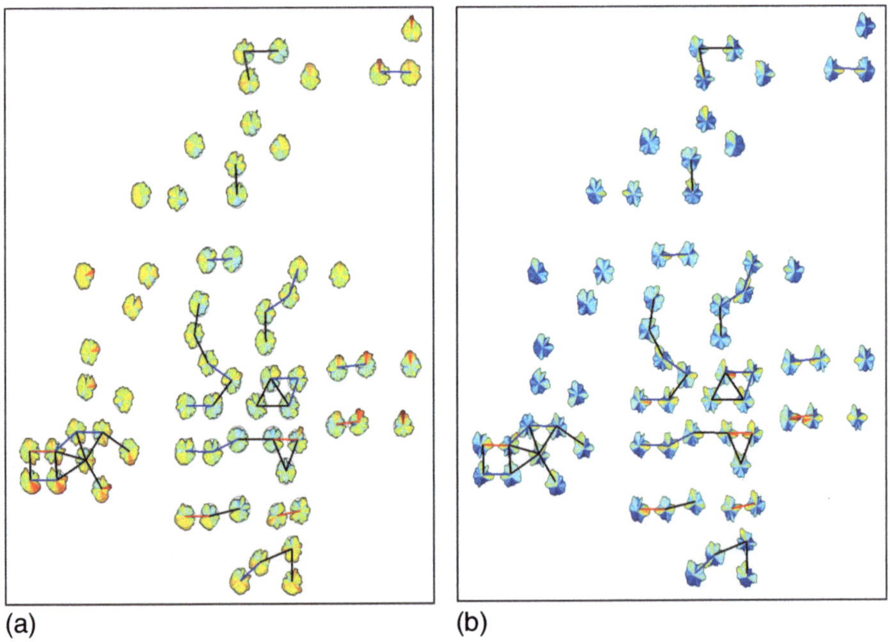

Fig. 13.8 Wake effect visualizations. On the left (**a**), visualization is shown using normalized instant power, and on the right (**b**), visualization is shown using power variability. In both cases, the rose plots are positioned in place of the number icons seen in Fig. 13.1. Further, the rose plots are colored according to the radial magnitude. For the instant power plots, over-performing turbine directions are colored *red*, while under-performing turbine directions are colored *blue*. *Grey* circles show average performance (instant normal power value of 1). For the power variability plots, high variability directions are colored *red*, and low variability directions are colored *blue*

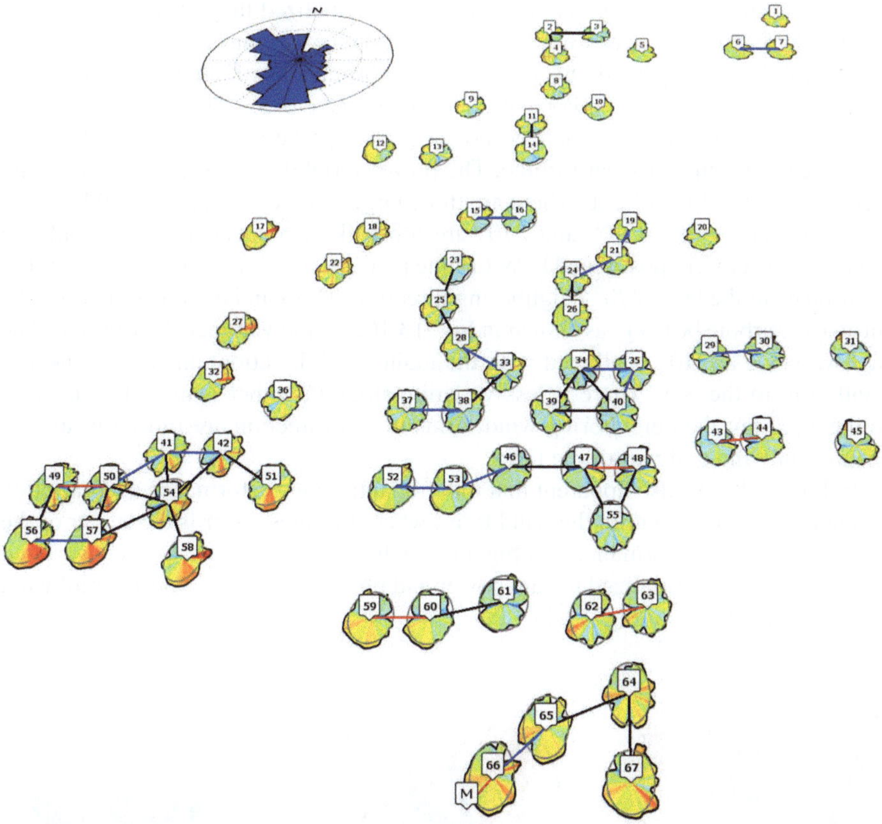

Fig. 13.9 Google Earth visualization. This visualization shows the wind farm wake effects using the normalized instant power plots, complete with the labels from Fig. 13.1. *Grey circles* show average performance (instant normal power value of 1). The wind farm terrain imagery (not shown) can also be examined for correlations between performance and local topography

when the farm is operational except for a few down turbines (or turbines with previously discarded measurements).

The visualization can also be displayed using Google Earth (http://www.google.com/earth) for improved interactivity. An example of a fully interactive visualization is shown in Fig. 13.9.

13.3 Results

In this section, a more detailed analysis of the wind farm is provided, using the normalized instant power and power variability plots, as well as other simpler curves. To start, it is interesting to understand the overall performance of the farm in relation to wind direction. Although it is not useful to take the average of the normalized instant power plots (since they are normalized they average to 1), it is useful to consider the average turbine power relative to wind direction, as shown in Fig. 13.10a. As expected, this plot is highly correlated with the wind rose for the site, albeit with an unusual spike in production from the E-SE. The power variability plots, unlike the instant normalized power plots, can be averaged, since they are computed independently per turbine. The power variability averages over the wind farm is shown in Fig. 13.10b. There are three large lobes of variability at 90°, 270°, and 340°. The lobes at 90° and 270° are most likely associated with turbulence generated by closely positioned E-W turbine pairs, as there are a significant number of turbines in the 90 – 270° neighboring direction. This can be seen comparing the turbine to turbine bearing, as shown in Fig. 13.10c, to the variance distribution. The high variance at 340° is aligned with the main wind direction, and a large nearby wind farm to the NW. There is less variation along the other main wind direction to the SSE, but the neighboring wind farms in this direction are further away and shadow only the east part of the farm.

In Fig. 13.8 it is also apparent that individual turbine performance and variation is related to position within the wind farm, where turbines towards the center of the farm have lower performance and higher variability. This is more easily visualized by averaging the normalized instant power and power variability plots over all wind directions, as shown in Fig. 13.11.

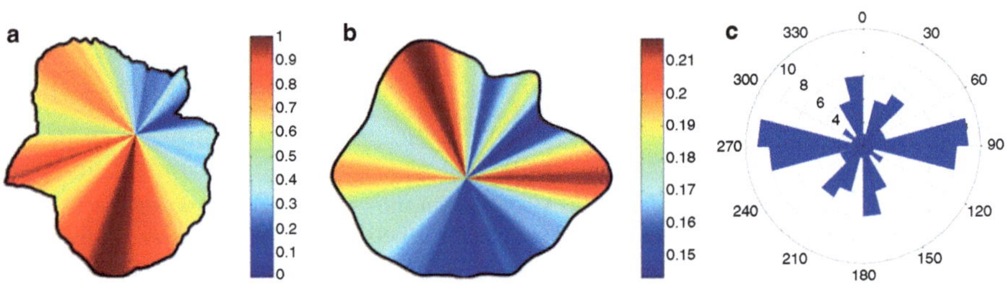

Fig. 13.10 Overall wind farm performance. On the left (**a**), the power averaged over all turbines versus wind direction is shown, maximum normalized to 1. As expected, this polar plot is highly correlated with prevailing wind directions (NW and S), although there is also an interesting spike in the E-SE direction. In the middle (**b**), the average power variability versus wind direction is shown. The power variability is stretched N-S with prevailing winds and E-W by a large number of closely positioned E-W turbine pairs. On the right (**c**), a radial histogram of counts for turbine–turbine pairs within 7 rotor diameters is shown (turbines farther than 7 rotor diameters are less likely to experience wake conditions)

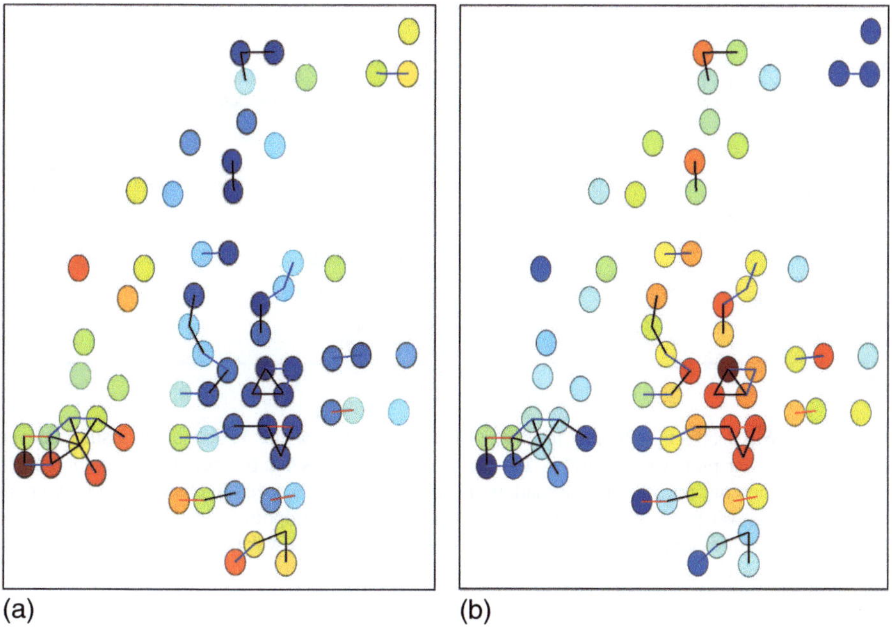

Fig. 13.11 Average turbine performance over wind farm. On the left (**a**), the wind direction averaged normalized instant power is shown for each turbine, where *red* indicates high performance, and *blue* indicates low performance. On the right (**b**), the wind direction averaged power variability is shown for each turbine, where *red* indicates high variability, and *blue* indicates low variability

Detailed Wake Analysis In addition to providing an overview of the wind farm, the normalized power and power variability plots can be used to understand wake effects on a more detailed scale. Four distinct features have been observed in the data:

- wake deficit effects,
- speedup channels from two upstream turbines,
- shear point speedup from one upstream turbine, and
- shear point speedup from multiple upstream turbines or an upstream farm.

Wakes are characterized by a power deficit in the direction of a neighboring turbine and a distinct increase in power variance. In Fig. 13.12, two distinct wake effects can be seen at turbine 11, originating from turbines 8 and 10. The first effect is the wake deficits, seen as dips in the normalized power plot of turbine 11 facing turbines 8 and 10, as well as peaks in the power variability.

The second observed effect is a speedup when a turbine is facing the midpoint of two upwind turbines. Here the power is higher than average, also seen in Fig. 13.12 when turbine 11 is facing the midpoint between turbines 8 and 10. This effect is

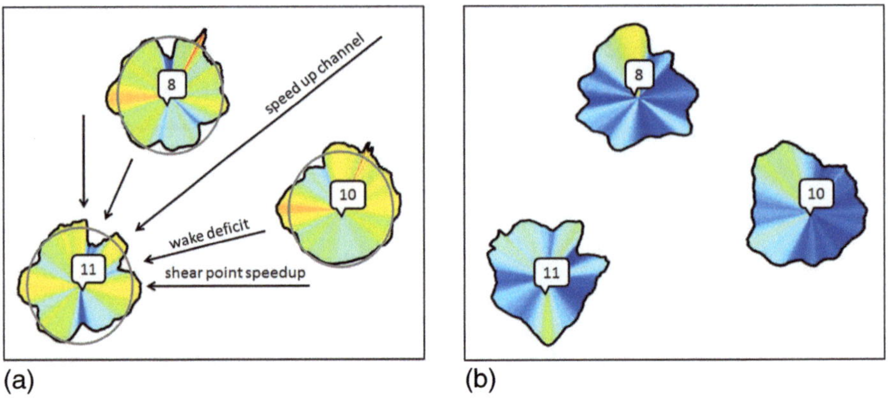

Fig. 13.12 Wake effects. On the left (**a**), three wake effects can be observed using instant normalized power plots for turbines 8, 10, and 11. Wake deficits can be seen as dips in the power production when turbine 11 is in the shadow of turbines 8 or 10; a speedup channel can be seen as a peak in the power production when turbine 11 is facing the midpoint of turbines 8 and 10; and shear point speedups can be seen when turbine 11 is tangent to the wake of turbine 10 or 8. On the right (**b**), the corresponding variability in power is shown

generally not included in the standard wind farm modeling tools, even though it is apparently significant. Somewhat surprising, the power variance in this direction is significantly lower than average.

A third effect occurs when the upstream wake is barely touching the downstream turbine, causing an increase in power. This occurs in Fig. 13.12, for example, when increased power is observed at turbine 11 just south of the downstream sector of turbine 10. Surprisingly, this is also associated with a low variance. We call this effect *shear point speedup*.

Both types of speedup effects seem to be amplified in cases of multiple upstream turbines and even a whole upstream wind farm. Finally, the three effects are also observed as products of the landscape, i.e., clusters of trees or buildings (data not shown).

Aggregate Wake Deficit The wake deficit is straightforward to detect when the wake is produced by a single upstream turbine which is relatively undisturbed. In this case, the power variance has a very clear peak in the direction of the upstream turbine, and the power produced has a corresponding minimum. Additional discussion using wake profiles to illustrate these observations is provided in the Appendix.

In a multi-turbine wake situation, however, detecting wake effects is more difficult. Nevertheless, a generic wake deficit effect can be observed across the entire wind farm. To see this effect, 854 turbine pairs were selected within 25 rotor diameters and an undistributed direct path between them (to observe potential wake effects). From these pairs, the minimum power deficit and maximum variance of the downwind turbine were recorded. These values are shown in Fig. 13.13 versus

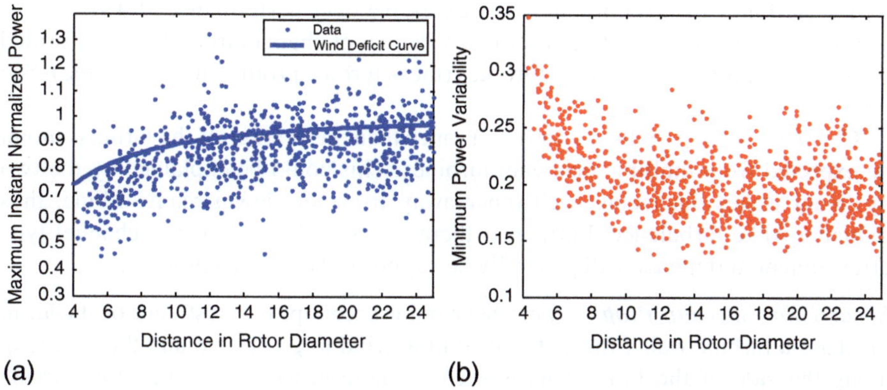

Fig. 13.13 Power deficit versus distance. Maximum instant normalized power and minimum power variance for a downstream turbine were collected for 854 turbine pairs. On the left (**a**), the maximum normalized instant power is plotted versus distance behind the upwind turbine. The semi-empirical wind velocity deficit from Eq. (13.4) is shown as a *solid curve*. On the right (**b**), the minimum power variance for the downstream turbine versus distance is shown

distance between turbine pairs. For the normalized instant power in Fig. 13.13a, the semi-empirical expression for wind velocity deficit described by Katic et al. [4] is also shown. This expression is given as

$$\frac{U_x}{U_0} = 1 - \frac{1 - \sqrt{1 - C_T}}{\left(1 + 2k\frac{x}{D}\right)^2}, \tag{13.4}$$

where U_0 is the upstream wind velocity, U_x is the wind downstream velocity at a distance x behind the turbine, C_T is the coefficient of thrust, and k is an empirical decay constant, given as 0.075 or 0.11 [4]. Reasoning that annual energy production is typically quasi-linear with wind speed averaged over time, the wind velocity deficit is compared with the normalized instant power in Fig. 13.13a to surprisingly good effect. For the wind velocity deficit, k was taken to be 0.075 and C_T was set to 0.9.

Channel Speedup Profiles It has been observed from the normalized instant power plots that there is a significant performance improvement when a turbine is facing the midpoint between two upwind turbines. The physical explanation could be that the wakes of the two upstream turbines displace airflow, which accelerates as in an ordinary channel contraction. Even minor speedup effects would boost the performance of the downstream turbine. For cases discussed in the Appendix, the boost is between 1.1 and 1.3 times average. The profiles also display a wake deficit next to the peak, as the downstream turbine faces either one of the two upstream turbines (see Figs. 13.16 and 13.17 in the Appendix).

Two turbines (17 and 58) show an excessive over performance (1.4–1.5 times average) facing what could be described as a duct or channel from the East, formed by rows of multiple neighboring turbines (these turbine profiles are not plotted). See Fig. 13.9.

Channel speedups exhibit a lower variance than the average of the farm, as might be expected from multiple upstream turbines. This could be explained by the two upstream wakes displacing each other away from the downstream turbine, thus reducing wake turbulence. Further, if there is a speedup effect in a channel type arrangement, turbulence will generally be suppressed due to contraction.

Shear Point Speedup Profiles For shear point speedup, the upstream obstacle, in this case a turbine wake, forces the incoming wind to go around and thus speedup along the side of the farm. This is seen as increased power in the downstream turbine. The shear point speedup effect is investigated for seven turbines, shown in Figs. 13.18 and 13.19 in the Appendix. Within the dataset, the increase occurs 15° to 20° from the bearing of the upstream turbine, with a magnitude of 1.1–1.22 times the average farm level. This magnitude is less than the speedup channel effect, but still significant. Again, this effect is associated with a wake presence. Effects like these have been observed in connection with buildings upstream to wind turbines, (see, e.g., Corscadden et al. [20]). Building speedup is also seen in this dataset (for example, turbine 7, bearing 80°), but is not investigated further.

Shear Speedup for Multiple Upstream Turbines Turbines 1, 6, 30, 31, 44, and 45 exhibit a narrow and a very high increased performance peak close to due north. This is due to a displacement from multiple upstream turbines. Turbines 1 and 6 face a long row of turbines in an upstream wind farm at approximately 15°. The 15° angle is comparable to what is observed for the shear point speedup from a single turbine. Several other turbines, for example, 7, 20, 29, and 43, also show the speedup, but the wake profiles are masked by a combination of speedup and waking. The wake profiles for these turbines are shown in Figs. 13.20 and 13.21 of the Appendix.

13.4 Conclusions

Based on a directional power performance analysis, four wake effects inside a 67 turbine wind farm have been identified. These effects include traditional wind shadows, channel speed ups, and shear effects. The analysis shows that a directional decomposition of turbine performance within the farm provides a holistic view of wind farm performance that can reveal more subtle features than standard power curve analysis. Further, the analysis can be made relative to subsets of the turbines in the wind farm (by normalizing against the smaller group), to highlight more subtle effects in performance. The results have been obtained even though the wind speeds have been almost completely ignored in the analysis.

Data preparation and validation was performed prior to analysis. The scrubbing of the data for sensor calibration errors and abnormal operational situations were the two most important issues [14, 22]. Furthermore, automatic processing was needed. The data analysis accurately corrected the turbine yaw position.

The wake losses identified are in reasonable agreement with classical models for wake losses. Comparing the observations to existing wake models, these only take into account the wake losses (i.e., recovery to nominal wind speed), but do not include the upside from speedup effects, due to channel or shearing points that have been observed. These new discoveries seem to be absent from the existing knowledge base and are necessary to understand in order to improve wind farm performance. Of course, these effects may be illusory and a side effect of our data analysis. Further investigation in terms of more detailed computational simulations, or suitable experiments would be necessary to provide definite evidence either way.

The suppressed variability of the speed up effect could be affecting turbine loading and reliability in a positive way, opposite the manner in which increased variability is known to have an adverse effect. Future correlation of turbine failures with directional observance could improve the understanding of wind farm operations with respect to reliability.

In the analysis, individual wake profiles and speedup effects were clearly identified. An individual turbine's average yearly operational situation is complex: even small wind direction changes can have large effects. Wakes were analyzed against nearest neighbors, but in the future, a multi-wake analysis could be extracted from the data as well, possibly using a superposition principle similar to those applied in classic wind farm models [2–4]. It may also be desirable to analyze additional data and explore different wake states based on power performance, i.e., restrict to region III rotor thrust of the power curve and analyze wind farms near nominal power, as was seen in Horns Rev data [21].

This work is a first step towards quantifying power production loss and impact on reliability of wind farm wake effects in existing wind farms. The ultimate goal of the work is not (necessarily) to assist siting new farms, but to improve the performance of existing farms. Once poorly performing turbines and farms can be accurately identified, efforts can be made to improve power production. Such efforts would range from pre-emptively identifying turbines likely to fail (based on their position in a farm and the wake effects acting upon them), to implementing cooperative turbine (smart-farm) strategies [18].

Acknowledgements This work is supported and made possible by the Department of Energy (DOE) Wind and Water Power Program. Sandia National Laboratories is a multi-program laboratory managed and operated by Sandia Corporation, a wholly owned subsidiary of Lockheed Martin Corporation, for the US Department of Energy's National Nuclear Security Administration under contract DE-AC04-94AL85000. Wind farm SCADA data was provided by a strategic industrial partner.

Appendix

Abstract

This Appendix provides additional analysis regarding the four wake effects discussed in the results section of the paper "Visualizing Wind Farm Wakes Using SCADA Data." The four effects observed are

- wake deficit effects,
- speedup channels from two upstream turbines,
- shear point speedup from one upstream turbine, and
- shear point speedup from multiple upstream turbines or an upstream farm.

For the purposes of continuity, some of the text from the paper is repeated.

Wake Deficit Profiles

To illustrate the four wake effects observed using the instant power and power variability plots, profiles for various turbines are shown. The wake deficit, for example, is straightforward to detect when the wake is produced by a single undisturbed upstream turbine. In this case, the power variance has a very clear peak in the direction of the upstream turbine, and the power produced has a corresponding minimum. Wake profiles showing wake deficits for 10 turbines pairs in terms of normalized instant power are shown in Fig. 13.14, with the corresponding power variance profiles shown in Fig. 13.15.

The detection of a wake deficit is straightforward when the wake is produced by a single undisturbed upstream turbine. In particular, the power variance signature is a very clear peak in the direction of the upstream turbines. However, in a multi-turbine wake situation, detecting wake effects is more difficult. For example, turbine 62 waked by turbine 63 shows a clear wake deficit and an increased variance, but has two distinct side lobes where the power increases over the average value outside the core of the wake. This is most likely due to wakes upstream of turbine 63 in an adjacent upstream wind farm. It is also interesting to observe that the peak variance is exactly pointing towards turbine 63, but the wake deficit is a few degrees off to the left. Turbine 10 waked by turbine 11 exhibits the opposite behavior, where the wake is symmetric, but the variance profile is asymmetric. In general, however, the variance profiles are symmetric.

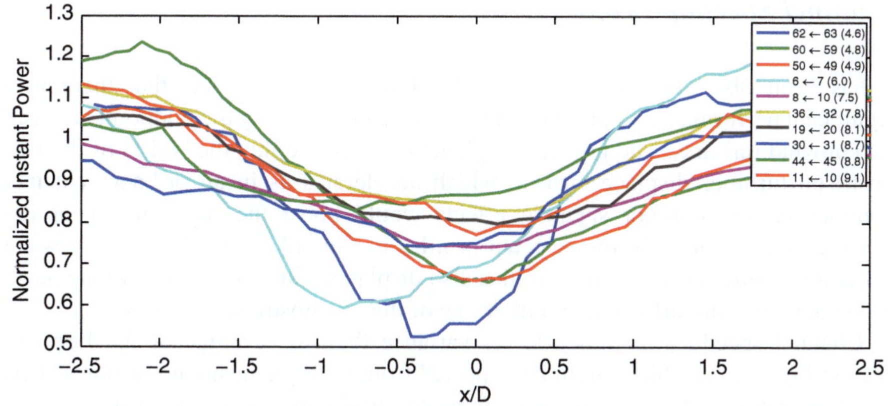

Fig. 13.14 Normalized instant power for waked turbines. The x-axis gives the wind direction degree offset normalized by the distance between the turbines (e.g., $x/D = \theta^\circ \left(\frac{\pi}{180^\circ}\right)\left(\frac{x}{D}\right)$, where θ° is the offset in degrees, x is the distance between the two turbines, and D is the rotor diameter). For example, when the offset is 0° the wind is blowing straight from the upwind to the downwind turbine. The y-axis shows the normalized instant power for the downwind turbine. The turbine pairs selected are given in the legend, where the notation $d \leftarrow u$ (r) indicates the downwind turbine (d), the upwind turbine (u), and the distance (r) in rotor diameters. See also Fig. 13.1 for turbine positions

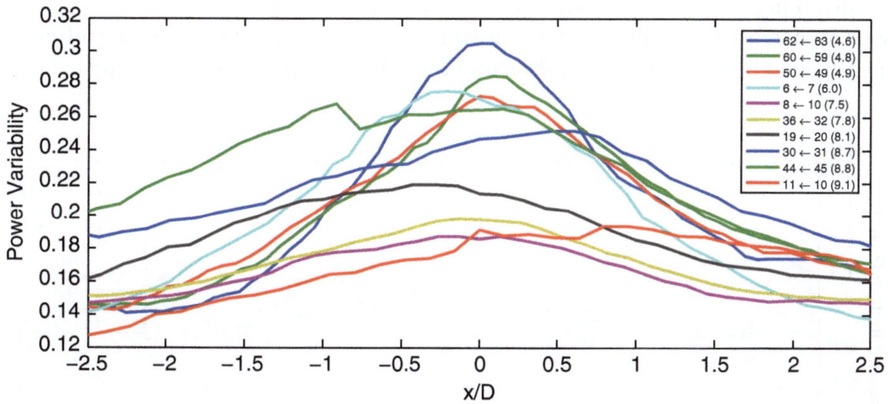

Fig. 13.15 Power variability for waked turbines. As in Fig. 13.14, the x-axis gives the normalized wind direction degree offset from the upwind turbine. The y-axis gives the power variability of the downwind turbine

Channel Speedup Profiles

It has been observed from the normalized instant power plots that there is a significant performance improvement when a turbine is facing the midpoint between two upwind turbines. The physical explanation could be that the wakes of the two upstream turbines displace airflow, which accelerates as in an ordinary channel contraction. Even minor speedup effects would boost the performance of the downstream turbine. For the selected profiles in Fig. 13.16, the boost is between 1.1 and 1.3 times average. The profiles also display a wake deficit next to the peak, as the downstream turbine faces either one of the two upstream turbines.

Channel speedups exhibit a lower variance than the average of the farm, as seen in Fig. 13.17, which might be expected from multiple upstream turbines. This could be explained by the two upstream wakes displacing each other away from the downstream turbine, thus reducing wake turbulence. Further, if there is a speedup effect in a channel type arrangement, turbulence will generally be suppressed due to contraction.

Shear Point Speedup Profiles

For shear point speedup, the upstream obstacle, in this case a turbine wake, forces the incoming wind to go around and thus speedup along the side of the farm. This is seen as increased power in the downstream turbine. The shear point speedup effect is investigated for seven turbines, shown in Figs. 13.18 and 13.19. Within the dataset,

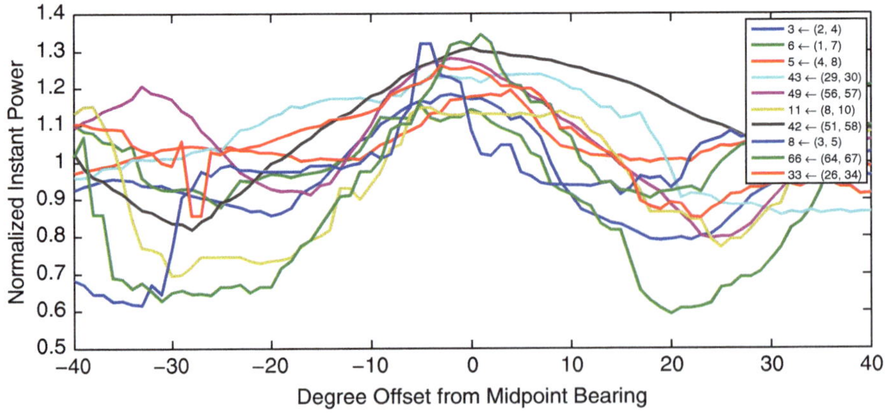

Fig. 13.16 Normalized instant power for channel speedup. The *x*-axis gives the wind direction degree difference from the downstream turbine to the midpoint between the two upstream turbines. The *y*-axis gives the normalized instant power. The channel speedup effect can be seen as the peak at 0°, flanked by wake deficits on either side of the peak

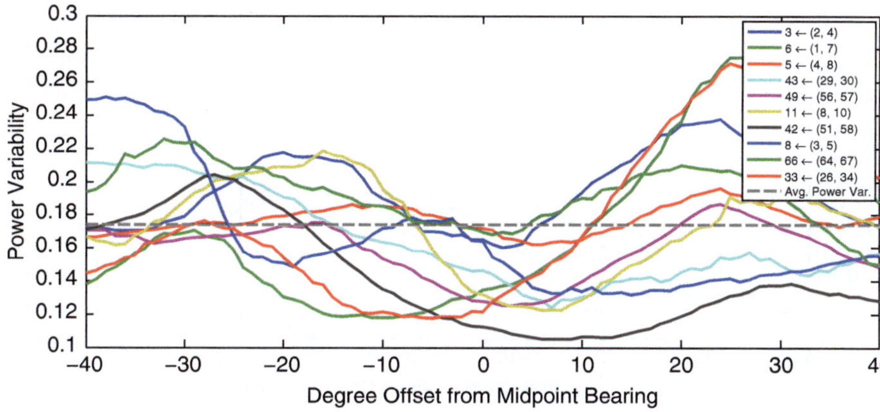

Fig. 13.17 Power variability for channel speedup. As in Fig. 13.16, the *x*-axis gives the wind offset from the midpoint between the two upstream turbines. The *y*-axis gives the power variability. The channel speedup is associated with low variability, as compared to the average variability over the farm, shown as a *gray dashed line*

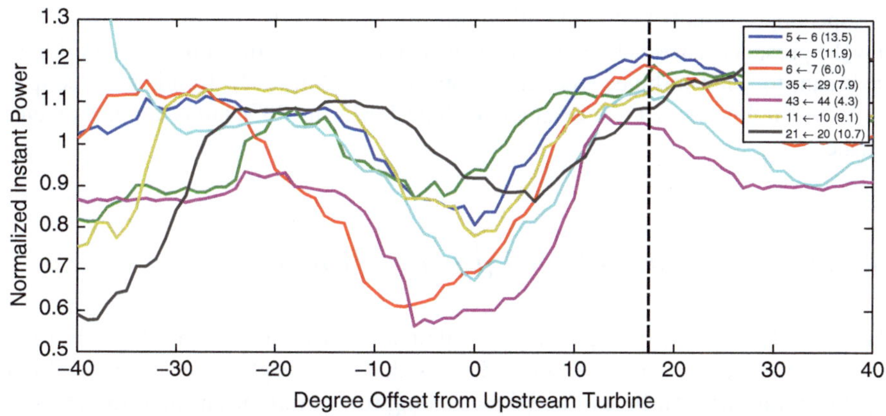

Fig. 13.18 Normalized instant power for shear speedup. The *x*-axis gives the wind direction degree offset from the bearing between and upwind and downwind turbine pair, and the *y*-axis shows the normalized instant power. Turbines were selected so that the region between 0° to 90° from the downwind turbine is undisturbed (no turbines nearby). The shear point speedup can be seen as improved power production between 15° and 20°, as indicated by the *dotted line* at 17.5°

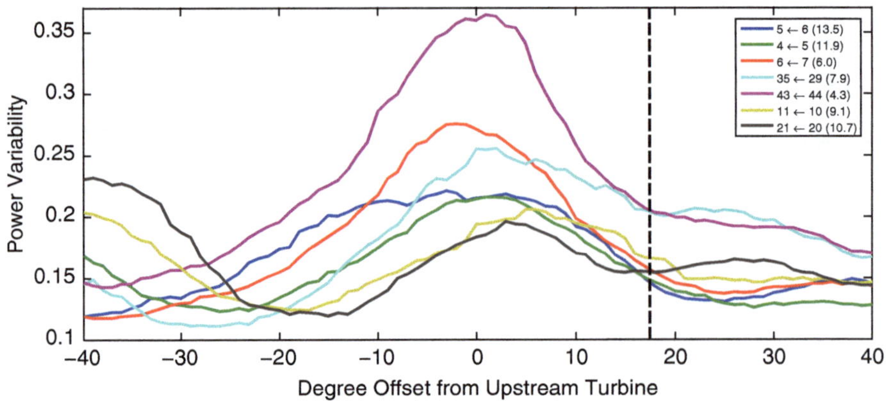

Fig. 13.19 Power variability for shear speedup. As in Fig. 13.18, the *x*-axis gives the wind direction degree offset from the upwind turbine. The *y*-axis shows power variability

the increase occurs 15° to 20° from the bearing of the upstream turbine, with a magnitude of 1.1 to 1.22 times the average farm level. This magnitude is less than the speedup channel effect, but still significant. Again, this effect is associated with a wake presence. Effects like these have been observed in connection with buildings upstream to wind turbines (see, e.g., Corscadden et al. [20]). Building speedup is also seen in this data set (for example, turbine 7, bearing 80°), but is not investigated further.

Shear Speedup for Multiple Upstream Turbines

Turbines 1, 6, 30, 31, 44, and 45 exhibit a narrow and a very high increase in performance close to due north. This is due to a displacement from multiple upstream turbines. Turbines 1 and 6 face a long row of turbines in an upstream wind farm at approximately 15°. The 15° angle is comparable to what is observed for the shear point speedup from a single turbine. Several other turbines, for example 7, 20, 29 and 43, also show the speedup, but the wake profiles are masked by a combination of speedup and waking. The wake profiles for these turbines are shown in Figs. 13.20 and 13.21.

Turbines 44 and 45 are waked by a turbine and turbine 31 is facing multiple roughness elements. These three turbines show less performance increase than turbines 1, 6, and 30. All the turbines show a high variability towards the shear point to the left and slope down to an overall low of 0.12 to the right. The increase in power variability seen in Fig. 13.21 from 25° to 40° can be attributed to landscape roughness by farm buildings and clusters of tall trees.

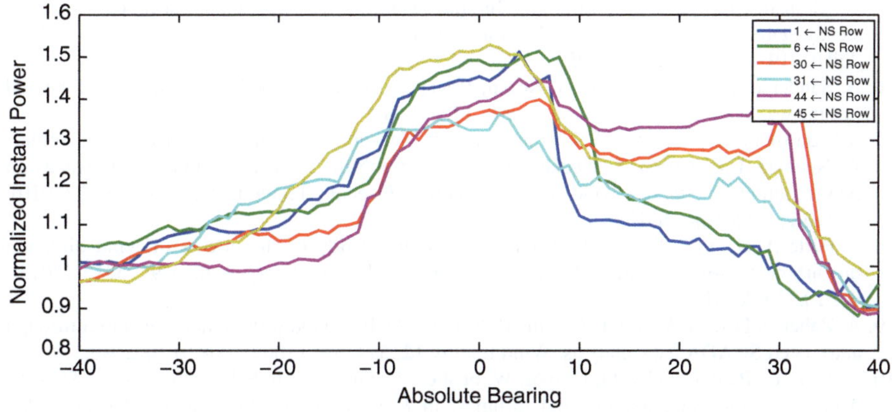

Fig. 13.20 Normalized instant power increase from multiple upstream turbines. The normalized instant power profiles are shown for six turbines offset from an upstream row of turbines. The upstream turbine rows are all arranged from North to South

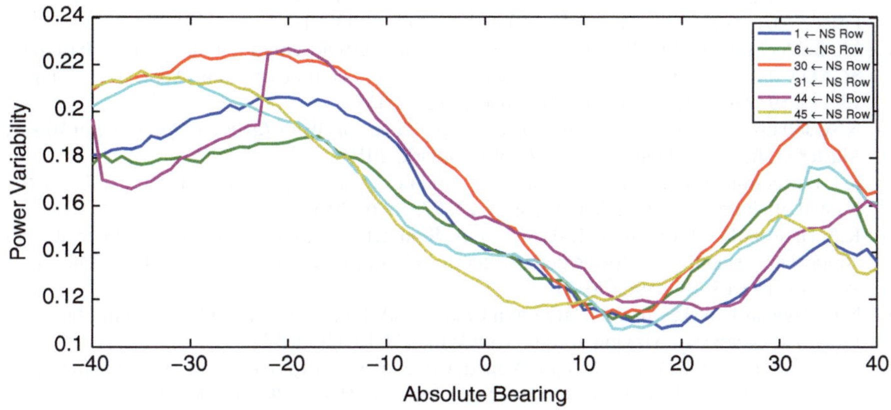

Fig. 13.21 Power variability from multiple upstream turbines. The power variability profiles are shown for six turbines offset from an upstream row of turbines. The upstream turbine rows are arranged North to South

References

1. J.F. Manwell, J.G. McGowan, A.L. Rogers, *Wind Energy Explained: Theory, Design and Application*, 2nd edn. (Wiley, 2009)
2. P.B.S. Lissaman, E.R. Bates, Energy effectiveness of arrays of wind energy conversion systems. Aerovironment Report AV FR 7050, Pasadena, CA, 1979
3. P.E.J. Vermeulen, An experimental analysis of wind turbine wakes, in *Proceedings of the Third International Symposium on Wind Energy Systems*, Lyngby, DK, 1980, pp. 431–450
4. I. Katic, J. Høstrup, N.O. Jensen, A simple model for cluster efficiency, in *Proceedings of the 1987 European Wind Energy Conference*, Rome, 1987, pp. 407–410

5. F. Gonzalez-Longatt, P. Wall, V. Terzija, Wake effect in wind farm performance: steady-state and dynamic behaviour. Renew. Energy **39**(1), 329–338 (2011). doi:10.1016/j.renene.2011.08.053
6. B. Sanderse, S.P. van der Pijl, B. Koren, Review of computational fluid dynamics for wind turbine wake aerodynamics. Wind Energy **14**, 799–819 (2011). doi:10.1002/we.458
7. R.J. Barthelmie, S.C. Pryor, S.T. Frandsen, K.S. Hansen, J.G. Schepers, K. Rados, W. Schlez, A. Neubert, L.E. Jensen, S. Neckelmann, Quantifying the impact of wind turbine wakes on power output at offshore wind farms. J. Atmos. Oceanic Technol. **27**, 1302–1317 (2010). doi:10.1175/2010jtecha1398.1
8. F. Porte-Agel, Y.-T. Wu, C.-H. Chen, A numerical study of the effects of wind direction on turbine wakes and power losses in a large wind farm. Energies **6**, 5297–5313 (2013). doi:10.3390/en6105297
9. A. Zaher, S.D.J. McArther, D.G. Infield, Y. Patel, Online wind turbine fault detection through automated SCADA data analysis. Wind Energy **12**, 574–593 (2009). doi:10.1002/we.319
10. K. Kim, G. Parthasarathy, O. Uluyol, W. Foslien, S. Shuangwen, P. Fleming, Use of SCADA data for failure detection in wind turbines, in *Proceedings of 2011 Energy Sustainability Conference and Fuel Cell Conference*, 2011; NREL/CP-5000-51653
11. A. Kusiak, W. Li, The prediction and diagnosis of wind turbine faults. Renew. Energy **36**(1), 16–23 (2011). doi:10.1016/j.renene.2010.05.014
12. S. Butler, J. Ringwood, F. O'Connor, Exploiting SCADA system data for wind turbine performance monitoring, in *2013 Conference on Control and Fault-Tolerant Systems (SysTol)*, France, 2013, pp. 389–394. doi:10.1109/systol.2013.6693951
13. M. Schlechtingen, I.F. Santos, S. Achiche, Wind turbine condition monitoring based on SCADA data using normal behavior models. Part 1: system description. Appl. Soft Comput. **13**(1), 259–270 (2013). doi:10.1016/j.asoc.2012.08.033
14. N.A. Johansen, *Verification of Simulated Fatigue Loads on Wind Turbines Operating in Wakes*, Master's Thesis, Technical University of Denmark, 2010
15. R.J. Barthelmie, L.E. Jensen, Evaluation of wind farm efficiency and wind turbine wakes at the Nysted offshore wind farm. Wind Energy **13**, 573–586 (2010)
16. K.S. Hansen, R.J. Barthelmie, L.E. Jensen, A. Sommer, The impact of turbulence intensity and atmospheric stability on power deficits due to wind turbine wakes at Horns Rev wind farm. Wind Energy **15**, 183–196 (2012)
17. N.G. Nygaard, Wakes in very large wind farms and the effect of neighboring wind farms. J. Phys. The Science of Making Torque from Wind (TORQUE 2014)
18. J.R. Marden, S.D. Ruben, L.Y. Pao, A model-free approach to wind farm control using game theoretic methods. IEEE Trans. Control Syst. Technol. **21**(4), 1207–1214 (2012)
19. C. Aggarwal, *Outlier Analysis* (Springer, New York, 2013)
20. K.W. Corscadden, W.D. Lubitzb, A. Thomson, J. McCabea, Investigation of wake effects on the energy production of small wind turbines. Wind Eng. **37**(2), 151–164 (2013). doi:10.1260/0309-524x.37.2.151
21. L.E. Jensen, Wake losses and turbulence within the horns rev off-shore wind farm, in *IEA Annex XXIII: Offshore Wind Energy Technology and Deployment, Task 23, Subtask 1, 3.1 Workshop Programme on External Conditions, Layouts, and Design of Off-Shore Wind Farms*, Risø, Denmark, 12–13 December 2005
22. P.E. Rethroe, N.A. Johansen, S.T. Frandsen, R.J. Barthelmie, K.S. Hansen, L.E. Jensen, M.A.B. Baekgaard, J.R. Kristoffersen, Systematic wind farm measurement data reinforcement tool for wake model calibration, in *European Offshore Wind Conference* (EOW 2009)

Chapter 14
Turbulent and Deterministic Stresses in the Near Wake of a Wind Turbine Array

Nicholas Michael Hamilton, Murat Tutkun, and Raúl Bayoán Cal

14.1 Introduction

The turbulent wakes of wind turbines have been discussed in some detail for individual devices (see, e.g., [14, 17]) but only recently has the interaction between wakes in large arrays been investigated. Complex flow generated by atmospheric forcing and interaction with rotating blades is still the subject of interest in making wind energy more productive and efficient to help meet the increasing global energy demands.

Large eddy simulations [3, 10–12, 15] have provided high fidelity data which were used to describe full wakes and wake interaction in infinite wind turbine arrays. The idea of the "infinite" array of wind turbines arises from observable periodicity of turbulence statistics in regularly arranged wind farms beyond the fourth row of devices [3, 4]. Periodicity in the streamwise spatial coordinate allows the convective and mean pressure gradient term to be effectively omitted from the energy balance for wind turbine wakes.

The flux of mean flow kinetic energy by turbulence, responsible for much of the resupply of momentum to the wake [2, 5, 6], is formulated with turbulent shear stresses combining fluctuations of velocity along the streamwise coordinate with either the spanwise or wall-normal directions. It is understood that the flux of

N.M. Hamilton • R.B. Cal (✉)
Department of Mechanical and Materials Engineering, Portland State University, Portland, OR 97201, USA
e-mail: rcal@pdx.edu

M. Tutkun
Department of Process and Fluid Flow Technology, IFE, 2007 Kjeller, Norway

Department of Mathematics, University of Oslo, Blindern, 0316 Oslo, Norway
e-mail: murat.tutkun@ife.no

A. Pollard et al. (eds.), *Whither Turbulence and Big Data in the 21st Century?*,
DOI 10.1007/978-3-319-41217-7_14

kinetic energy is associated with large-scale turbulence structures in the wakes [6]. The production of turbulence is also important to the energy balance in the fully developed turbine array. Although short-lived in the wake, this quantity is often representative of energy made unavailable to successive turbines in the array.

Flow effects related to the phase orientation of the rotors have been investigated for wind turbines by Hu et al. [7] and Lignarolo et al. [9] illuminating the vortex development in wakes from isolated horizontal axis wind turbines. Tip vortices produced by a two-bladed rotor were shown to persist up to 2.5 rotor diameters into the wake given a uniform inflow and turbulence intensity on the order 2 % [9]. Placing a wind turbine in the simulated atmospheric boundary layer as in [7] more accurately reproduced aerodynamic of wind turbines in the field, but only demonstrated phase-dependent structures very near the model wind turbine. Statistics over the ensemble of phase orientations of the rotor further obscured vortical structures visible in instantaneous velocity snapshots.

Significance of phase-independent standing waves in turbulent flows was formulated through the triple decomposition of the velocity field [13]. In terms of turbomachinery, phase-independent features are termed deterministic stresses, to distinguish them from the stochastic turbulence field, and have been shown to be significant to the overall energy balance, especially in terms of numerical simulation, shown by Adamczyk [1]. Further work investigating phase-related flow effects in turbomachines by Uzol [16] revealed that unsteady deterministic contribution to the stress field is of the same order of magnitude as the turbulence field. It was shown that the production of turbulence kinetic energy by deterministic stresses was the dominant source term relevant to the flow [16].

The current work undertakes the triple decomposition of velocity measurements in planes spanning a wind turbine wake in order to assess the phase-averaged and deterministic contributions to the energy balance. The flux of kinetic energy is formulated with the phase-averaged and deterministic stresses. In contrast to other turbomachinery, the rotors of a wind turbine do not impart phase variation to the flow that persist in the mean flow. Consequently, the deterministic stresses are only relevant very near the rotor blades. An extension of the triple decomposition is applied to the turbulence stress tensor showing the contribution of third-order statistics to be significant in the flow field.

14.2 Theory

The mean kinetic energy equation in a wind turbine boundary layer can be described through a slightly modified set of turbulent boundary layer equations as

$$U_j \frac{\partial \frac{1}{2} U_i^2}{\partial x_j} = -\frac{U_i}{\rho} \frac{\partial P}{\partial x_i} + \overline{u_i u_j} \frac{\partial U_i}{\partial x_j} - \frac{\partial \overline{u_i u_j} U_i}{\partial x_j} - \mathcal{F}_{x_i}. \tag{14.1}$$

Above, the convective term on the left is balanced by the pressure gradient term, the production and flux of turbulence kinetic energy, and a forcing term $\mathcal{F}_{x_i}$ representative of the thrust force added to the flow as power is extracted by the wind turbine. In many studies of wind turbine wakes [2, 5, 6, 8] the unsteady term is omitted from the consideration of the kinetic energy budget through ensemble statistics over a large set of random samples in the wake. Equation (14.1) reduces to a balance between the flux of kinetic energy, the production of turbulence, and the thrust force from the wind turbine.

The Reynolds stresses play a large role in the overall momentum balance in the wind turbine boundary layer; it contributes directly to the flux of kinetic energy and production of turbulence terms [third and fourth terms in the right-hand side of Eq. (14.1)] and can account for a relatively large portion of the total kinetic energy in the turbine canopy. In a Cartesian coordinate system, the Reynolds stress tensor is written

$$\overline{u_i u_j} = \overline{(\tilde{u}_i - U_i)(\tilde{u}_j - U_j)} \tag{14.2}$$

$$= \begin{bmatrix} \overline{uu} & \overline{uv} & \overline{uw} \\ \overline{vu} & \overline{vv} & \overline{vw} \\ \overline{wu} & \overline{wv} & \overline{ww} \end{bmatrix}$$

where the u, v, and w refer to fluctuating velocities in the streamwise, wall-normal, and spanwise directions, respectively. In the current development, a tilde implies an instantaneous quantity. Mean values are denoted with capital letters, an over bar indicates that ensemble averaging of a product is being performed. Lower case variables denote fluctuations from the mean quantities. In the following discussion and figures, the phase angle of the rotor is denoted as ϕ. The total Reynolds stress shown in Eq. (14.2) is ensemble averaged over all measurements, regardless of phase angle of the turbine rotor. Any quantity averaged over time is considered independent of phase. Angle brackets indicate that phase-dependent quantities have been averaged $\langle \tilde{a}_i \rangle_\phi = \langle \tilde{a}_i \rangle (x_i, \phi)$. Averaging phase-dependent values over the ensemble of all phase angles yields an approximation of the time-averaged quantity, $\bar{\tilde{a}}_i = \overline{\langle a_i \rangle_\phi} = A_i$. The deviation of phase-averaged quantities from their respective time-averaged values is expressed as the difference between the two quantities and is denoted with a double-prime,

$$u_i'' = U_i - \langle \tilde{u}_i \rangle_\phi. \tag{14.3}$$

These deviations may be multiplied and ensemble averaged over the period of the wind turbine rotor motion to acquire deterministic stresses [1, 13].

$$\overline{u_i'' u_j''} = \overline{\left(U_i - \langle u_i \rangle_\phi\right)\left(U_j - \langle u_j \rangle_\phi\right)}. \tag{14.4}$$

The total stress field is described by the sum of the ensemble of phase-averaged Reynolds stresses and a deterministic component that arises from deviations between phase-averaged mean velocities and the average-passage value. Similarly, the total flux of kinetic energy may be decomposed into turbulent (phase-dependent) and deterministic (phase-independent) contributions. The current literature indicates that the component of the flux tensor responsible for global kinetic energy entrainment into the array involves the streamwise–wall-normal fluctuating velocities, combined with the streamwise mean flow,

$$F_{12} = -\overline{uv}U = -\left(\overline{\langle uv\rangle}_{\phi} + \overline{u''v''}\right)\overline{\langle U\rangle}_{\phi}. \tag{14.5}$$

Locally, high-momentum flow is introduced to the wake of an individual wind turbine through streamwise–spanwise turbulence at a similar order of magnitude as above,

$$F_{13} = -\overline{uw}U = -\left(\overline{\langle uw\rangle}_{\phi} + \overline{u''w''}\right)\overline{\langle U\rangle}_{\phi}. \tag{14.6}$$

The flow is highly turbulent, combining the effects of the atmospheric boundary layer, wind turbine wake interaction, and fluid interactions with rotating solid bodies. Highly variable flow in the turbine canopy is not necessarily expected to demonstrate significant deterministic contributions to the stress field due to turbulent mixing independent of the phase orientation of the rotor. Here, the root-mean-square of the deviation of the phase-averaged stress field from the time-averaged Reynolds stresses is termed *tertiary* stresses and defined as

$$\overline{u_i'''u_j'''} = \sqrt{\overline{\left(\langle u_i u_j\rangle_{\phi} - \overline{u_i u_j}\right)^2}}. \tag{14.7}$$

Normal components of the tertiary stress tensor are the deterministic analog to the turbulence intensities along the coordinate directions. The physical significance of the off-diagonal terms is less intuitively obvious, but are taken here to be of equal importance to the flow as the deterministic stresses from Eq. (14.4).

14.3 Experiment

Measurements of a wind turbine wake in the fully developed canopy layer were made in the wind tunnel at Portland State University. The wind tunnel was furnished with a passive grid, vertical strakes, and roughness elements in the form of small-diameter chains in order to condition the turbulent inflow to more closely match the observed atmospheric boundary layer. Characteristic profiles of the approach flow to the model array are presented in Hamilton et al. [6].

Figure 14.1 shows an example turbine from the current experiment. In the figure, measurement locations accessed through $2D-3C$ stereoscopic particle image velocimetry (SPIV) are shown in relation to the model-scale wind turbine array.

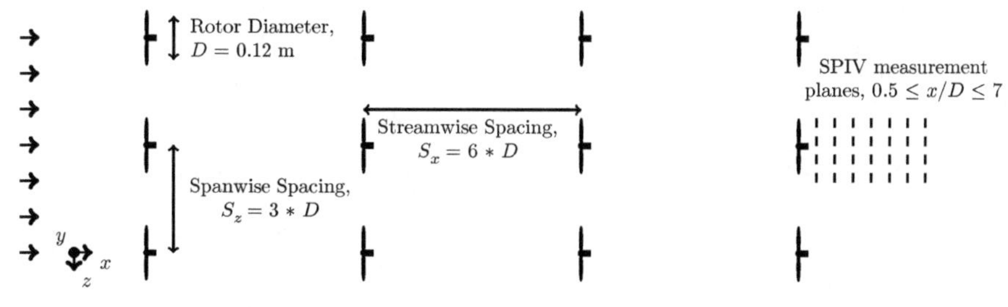

Fig. 14.1 Plan view of the model wind turbine array including spacing and location of measurement planes. Flow is from left to right. The turbine of interest is the center, fourth in the wind turbine array

Measurements were made behind a wind turbine in the fully developed region, the fourth row for arrays arranged in a Cartesian grid. Rows of turbines were spaced six rotor diameters apart (72 cm) and columns separated by 3 rotor diameters (36 cm) hub to hub.

The wind turbine models were fabricated in-house and consist of a hollow steel mast and rotor blades cut from 0.5 mm sheet steel. The blades of the turbine were given pitch and twist via die-press to ensure uniformity. Each blade was pitched approximately $\gamma_{root} = 22°$ out of the rotor plane at the root of the blade and had a 7° twist from root to tip, resulting in a pitch of $\gamma_{tip} = 15°$ at the tip of each blade. The nacelle of each turbine consisted of a DC electric motor, loaded with resistive elements to slow the rotation of the turbine blades, allowing each row of turbines to be matched to their respective peak power coefficients. Model wind turbines were mounted in steel plates (0.75 cm thick) spanning the wind tunnel. Figure 14.2 shows details of the model wind turbines.

Phase orientation of the rotor blade was located with a Monarch remote optical sensor placed outside the wind tunnel with optical access to a reflective portion of the downstream side of one blade. Detection by sensor initiated a square wave signal that in turn triggered the SPIV measurements. Four positions of the rotor were considered in the experiment, $\phi \in [0°, 30°, 60°, 90°]$, where 0° was one blade oriented vertically upward. SPIV allows access to the full Reynolds stress tensor and gradients of velocities in the measurement plane. Each SPIV measurement window is approximately 0.2 m × 0.2 m in dimension. Snapshot pairs were taken at a time delay of $\delta t = 180\,\mu$s. The Nd:Yag (532 nm, 1200 mJ, 4 ns duration) double-pulsed laser sheet varied from 1 mm at the bottom of the tunnel to approximately 2 mm at the top of the measurement window.

During the experiment, the flow was periodically reseeded with vaporized di-ethyl-hexyl-sebacat and kept at a constant level to ensure consistency of particle imaging. Cameras (4MP ImagerProX) used to collect flow snapshots were arranged below the wind tunnel and focused in the direction of the laser sheet placing them

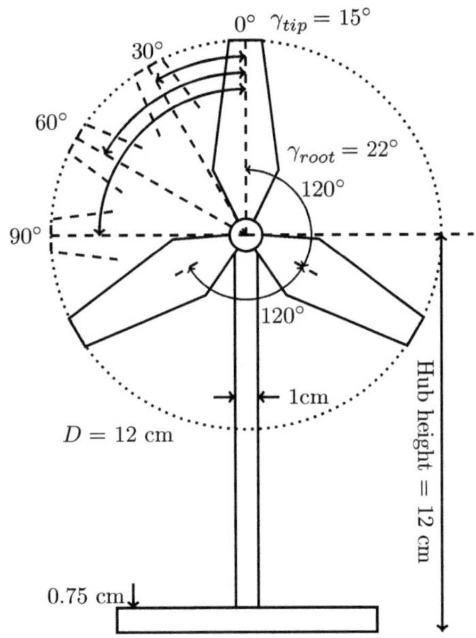

Fig. 14.2 Schematic of a fully assembled wind turbine model including the mast and a section of the mounting plate. The phase angles indicated are measured from the position of a single blade at top-dead-center

in back-scatter. Velocity vectors were calculated using a multi-pass FFT based algorithm with two passes each at 64 and 32 pixel interrogation windows. The final spatial resolution of velocity vectors was approximately 1.4 mm in both the spanwise and wall-normal directions. Measurement sets in x/D each contain 2000 SPIV snapshots to reach acceptably converged statistics with error of second-order terms approximately 3 %.

14.4 Results

Velocities and stresses have been normalized by the time-averaged inflow velocity at hub height of the model wind turbine, $U_{\text{hub}} = 4.75\,\text{m}\,\text{s}^{-1}$. The momentum deficit of the wake of the wind turbine is visible in the contours of the mean streamwise (axial) velocity U shown in Fig. 14.3. The minimum value of $U/U_{\text{hub}} = 0.1$ occurs directly following the nacelle of the model wind turbine at $x/D = 0.5$. In the conditions of this experiment, it is not uncommon to see instantaneous recirculations at $x/D = 0.5$. However, over a large set of samples any negative streamwise velocities are washed out by ensemble averaging.

In agreement with previous wind turbine array experiments, the component of the highest magnitude in the Reynolds stress tensor is the streamwise normal stress $\overline{u^2}$. Shown in Fig. 14.4a, the streamwise normal stress forms a crescent-shaped structure in time-averaged contour plots. Specific evidence of the passage of the rotor is only seen very close to the model wind turbine, at $x/D = 0.5$. Beyond that point, features

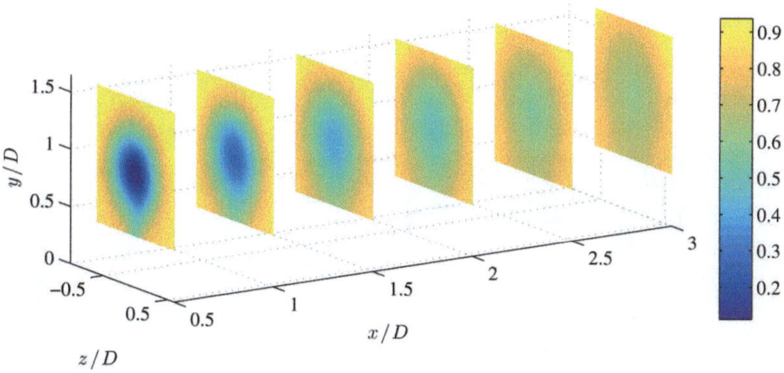

Fig. 14.3 Time-averaged streamwise velocity, U/U_{hub}. The contours above show the mean velocity averaged over all measurements and are independent of phase

evidencing the passage of the rotors are washed out by mixing and distortions of the flow from large turbulence structures. Peak values of $\overline{u^2}$ occur approximately $x/D = 1.5$ downstream from the wind turbine rotor similar to previous studies of wind turbine wakes [5, 6]. The streamwise normal stress in Fig. 14.4a is on the order of unity when normalized by the square of the hub height velocity. Wall-normal and spanwise normal stresses (not shown for brevity) are less than half the magnitude of $\overline{u^2}/U_{\text{hub}}^2$. Contributions to the overall flux of kinetic energy by $\overline{u^2}/U_{\text{hub}}^2$ are negligible due to small gradients in the streamwise direction. Time-averaged shear stresses are included in Fig. 14.4 for comparison to phase-dependent contributions below. The shear stresses are roughly antisymmetric in the vertical direction for $-\overline{uv}/U_{\text{hub}}^2$ in Fig. 14.4b and horizontally for $-\overline{uw}/U_{\text{hub}}^2$ in Fig. 14.4c.

There are two quantities contributing to the resupply of kinetic energy through the flux term in Eq. (14.1). The shear stresses $-\langle uv\rangle_\phi/U_{\text{hub}}^2$ and $-\langle uw\rangle_\phi/U_{\text{hub}}^2$ (Fig. 14.5a, b, respectively) contribute to the vertical and lateral entrainment of high-momentum flow from outside the wake. Second-order quantities reflect the growth of the wake shown in contours of the mean velocity in Fig. 14.3. The phase-averaged shear stresses show that the contributions to the stress field are prone to periodic peaks, most evident for $x/D \leq 2$.

The Reynolds shear stress composed of the streamwise and wall-normal fluctuations is discussed as its contribution to the flux and production terms of the kinetic energy budget is the most significant (see, e.g., [2, 5]). Figure 14.5a shows that $-\overline{uv}$ is positive for most locations of the near wake above hub height, with the exception being directly behind the mast of the wind turbine. The contours of $-\langle uv\rangle_{\phi=30^\circ}$ and $-\langle uv\rangle_{\phi=60^\circ}$ show regions in the near wake where the flux is of greater magnitude than $-\overline{uv}$ because the position of the rotor allows the flow to be advected into the wake. The region where $-\langle uv\rangle_{\phi=0^\circ} < 0$ is similarly more extreme than the respective region of $-\overline{uv}$. As the turbine blades rotate around the hub, the gap between individual blades is sufficient for turbulence structures to pass through undisturbed. In other regions of the flow, the blades disturb the direct advection of large structures into the wake; they are effectively cut into smaller structures by the blades.

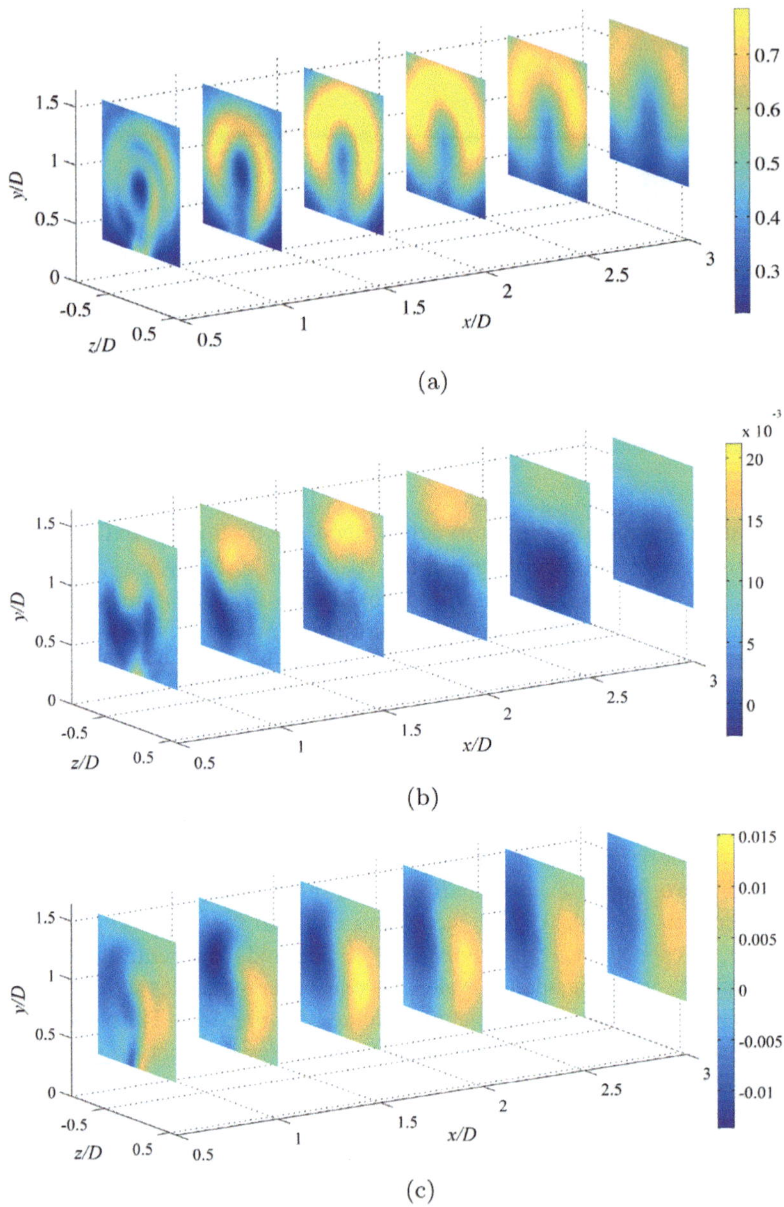

Fig. 14.4 Time-averaged Reynolds stresses, independent of phase. **(a)** $\overline{u^2}/U_{\text{hub}}^2$. **(b)** $-\overline{uv}/U_{\text{hub}}^2$. **(c)** $-\overline{uw}/U_{\text{hub}}^2$

In ideal conditions (uniform inflow of minimal turbulence intensity), tip vortices develop from the rotor and are convected downstream, illustrated by Hu et al. [7]. The current experiment places model wind turbines in an array and the inflow

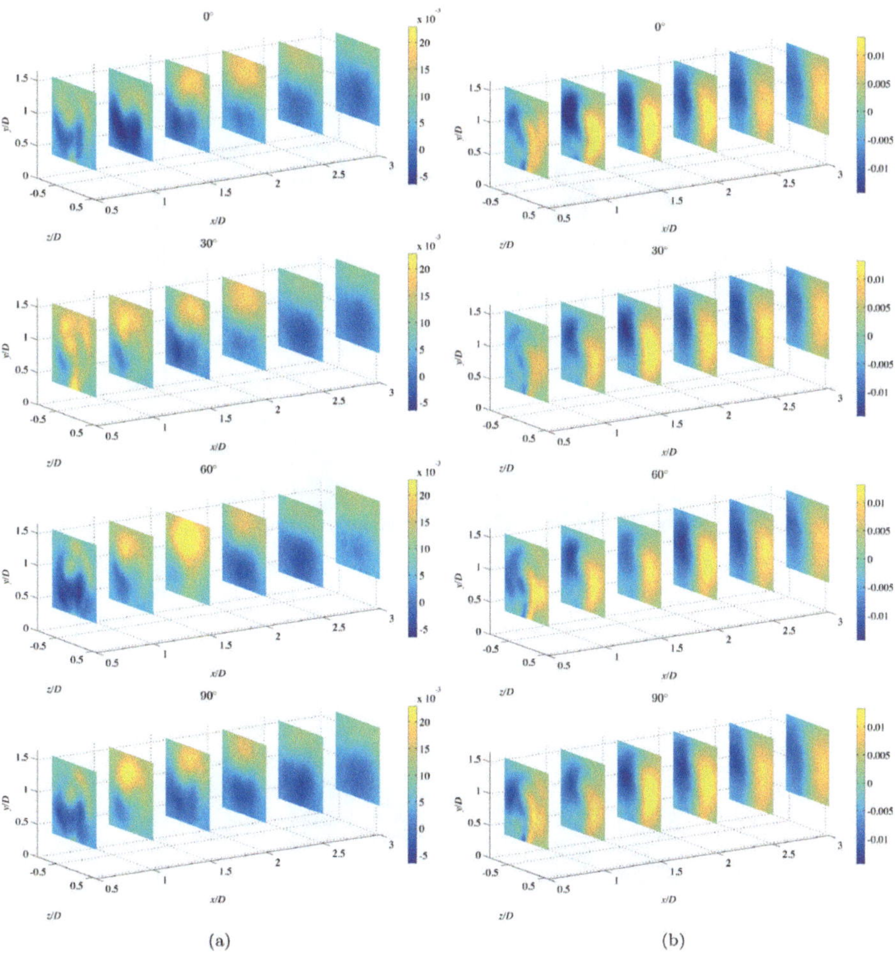

Fig. 14.5 Phase-averaged Reynolds stresses in the near wake. (a) $-\langle uv \rangle_\phi / U_{\text{hub}}^2$. (b) $-\langle uw \rangle_\phi / U_{\text{hub}}^2$

to the wind turbine investigated here is accordingly turbulent; inflow turbulence intensity to the model wind turbine is approximately 20 % (not shown). Periodicity in the flow is further complicated by the addition of the boundary layer in the wind turbine array. Accordingly, rotation of the flow is distorted and becomes difficult to visualize, especially by ensemble averaging. However, the phase-averaged vertical flux of kinetic energy, shown in Fig. 14.6, demonstrates variation between phase orientations of the rotor, indicating that there are periodic contributions to the entrainment process. Typically this quantity is discussed as bringing high-momentum flow into the wake from above the turbine canopy. The general trend of the flux of kinetic energy is into the wake from outside which may include

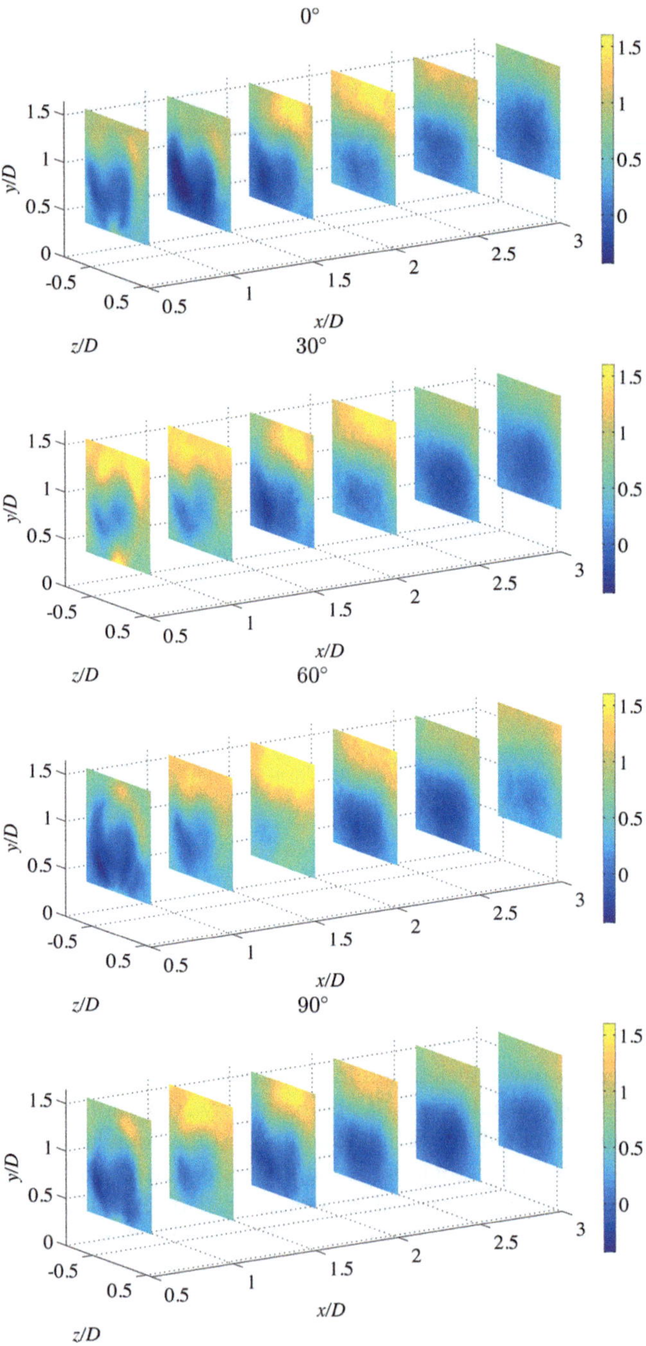

Fig. 14.6 Phase-averaged flux of kinetic energy $-\langle uvU\rangle_\phi$

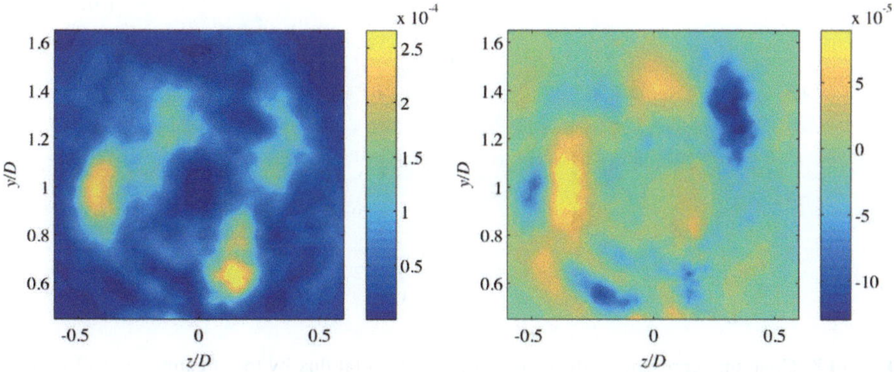

(a) (b)

Fig. 14.7 Normal and shear deterministic stresses composed of ensemble averaged deviations between phase-averaged and total mean velocities at $x/D = 0.5$. **(a)** $\overline{u''^2}/U_{\text{hub}}^2$. **(b)** $-\overline{u''v''}/U_{\text{hub}}^2$

lateral transport. Below the nacelle of the wind turbine the flux of kinetic energy is *away* from the center of the wake, most clearly for $-\langle uvU\rangle_{\phi=30°}$ at $x/D \leq 1$ and $-\langle uvU\rangle_{\phi=60°}$ at $x/D = 1.5$.

A further consideration for the flux of kinetic energy is the contribution by the deterministic stresses from Eq. (14.4), shown in Fig. 14.7. Although the turbulent stresses show significant dependence on the orientation of the rotors, mean velocities do not and the deterministic stresses are approximately two orders of magnitude smaller. The deterministic shear stress $-\overline{u''v''}$ exhibits coarse azimuthal periodicity (θ from the schematic in Fig. 14.2) around the rotor diameter. The contours in Fig. 14.7 represent the wave-like contribution added to the mean flow field. They are independent of phase orientation of the rotor blades and can be viewed as the underlying periodicity in the wake.

The flux composed with only the streamwise/wall-normal deterministic stress is shown in Fig. 14.8. The magnitude of this contribution, like the stresses above, is approximately two orders of magnitude smaller than the phase-dependent or time-averaged quantities. The magnitudes of the flux composed with the deterministic stresses indicate that the periodic contributions to the total behavior can be neglected at the first order.

That the deterministic stresses are orders of magnitude less than the phase- and time-averaged contributions is not unexpected given the nature of the turbulent flow in the wind turbine canopy. Phase-averaged contours of the streamwise velocity (not shown) are nearly identical to one another, regardless of phase of the rotor. The deterministic stresses, according to their standard definition, account for the deviation of phase-averaged velocities from the time-averaged velocities and are quite small.

However, the turbulent stress tensor demonstrates a large degree of variation with phase orientation of the rotor. Figure 14.5a, b show that the stress field varies

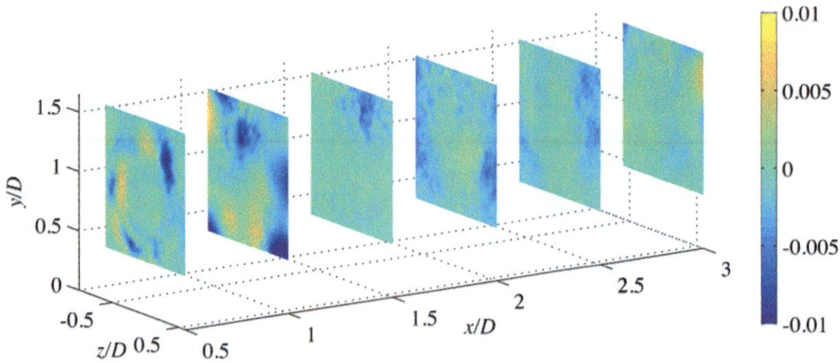

Fig. 14.8 Contours representing the contribution of the total flux by the streamwise–wall-normal deterministic stress $-\overline{u''v''}U$

considerably with phase. This variability is not accounted for in the definition of the deterministic stresses employed here. The variability of the turbulent stress tensor is assessed through the root-mean-square deviation of the phase-averaged stresses from their respective average-passage field according to Eq. (14.7), resulting in the tensor of tertiary stresses. The tertiary stresses are phase-independent and quantify the variation of the turbulence field arising from the passage of the rotor blades.

The formulation used is analogous to the definition of turbulence intensity, at least for the streamwise diagonal term $\overline{u'''^2}$ shown in Fig. 14.9a. Deterministic turbulence intensity is shown to peak at $x/D = 1$ following the same crescent shape as the turbulent stress. Very near to the model wind turbine, contours of $\overline{u'''^2}$ show similar features as the phase- and time-averaged fields.

Off-diagonal terms of the tertiary stress tensor must have a different physical interpretation; turbulence intensity is not defined for shear terms. Tertiary shear stresses are interpreted as energy added to the flow beyond what one would expect from an equivalent aperiodic flow; contributions from off-diagonal components of $\overline{u_i'''u_j'''}$ account for asymmetry and periodicity that are missed in non-rotating geometries such as porous discs often used in wind energy simulations. In the wake, shear terms of the tertiary stress tensor show magnitudes similar to the phase-averaged turbulence stresses above, which is indicative of their importance in the overall behavior of the wake. The tertiary stress composed of the streamwise–wall-normal components shows features that are quite distinct from the classically defined stresses. It is shown in the contours of $-\overline{u'''v'''}$ (Fig. 14.9b) that the deviation of phase-averaged fields from the Reynolds stress is negligible at $x/D \geq 2$. This is not the case for $-\overline{u'''w'''}$, shown in Fig. 14.9c. Tertiary stress arising from the variation in the streamwise/spanwise turbulence stress conforms more closely to the regions of extreme $-\langle uw\rangle_\phi$. The peak value of $-\overline{u'''w'''}$ occurs at $x/D = 1$ corresponding to the peak negative value of $-\langle uw\rangle_{60°}$. Tertiary stresses arise from phase-dependent variations in the turbulence field and make a significant contribution only near the turbine. Large scales of turbulence are expected to persist into the far wake but phase dependence of the Reynolds stress tensor becomes difficult to diagnose beyond $x/D = 1.5$.

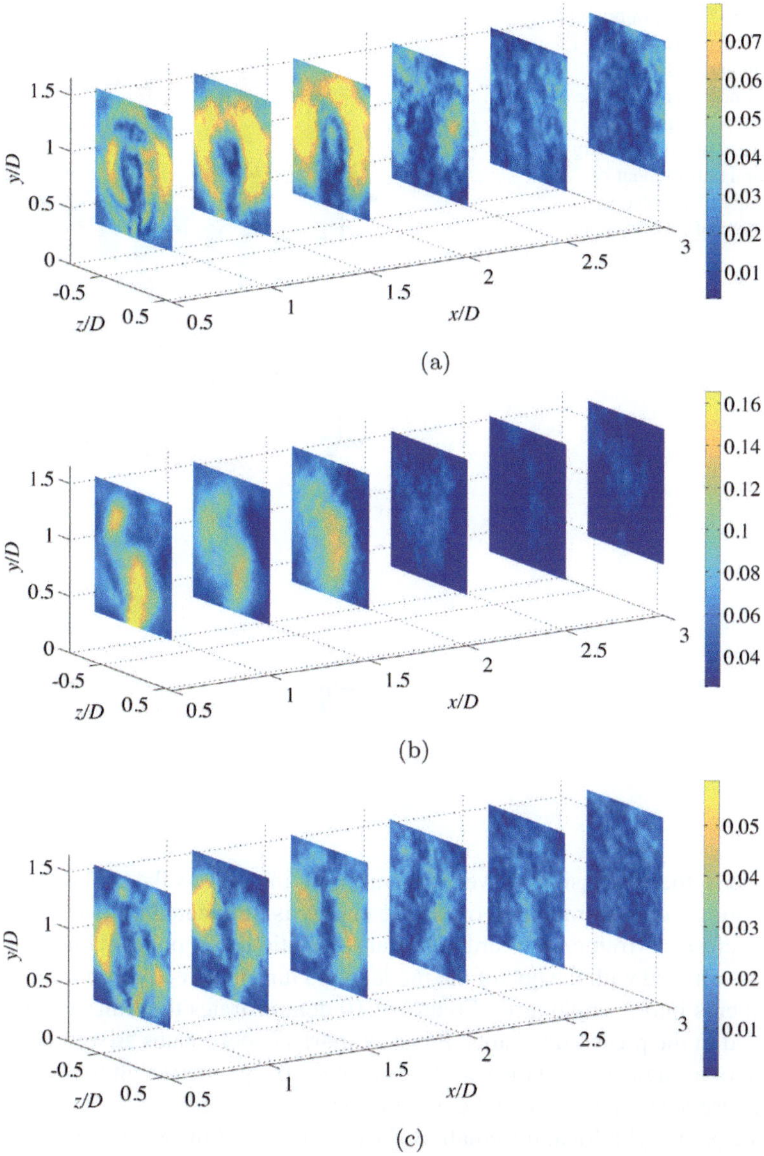

Fig. 14.9 Tertiary stresses in the near wake. Shown in the figure are the streamwise contribution, analogous to deterministic turbulence intensity, and the components that contribute to kinetic energy flux. **(a)** $\overline{u'''^2}$. **(b)** $-\overline{u'''v'''}$. **(c)** $-\overline{u'''w'''}$

While the global flux of mean flow kinetic energy into the turbine canopy by turbulence must be from above, local transport may occur laterally within the canopy. The results discussed here have been formulated in a Cartesian frame of reference,

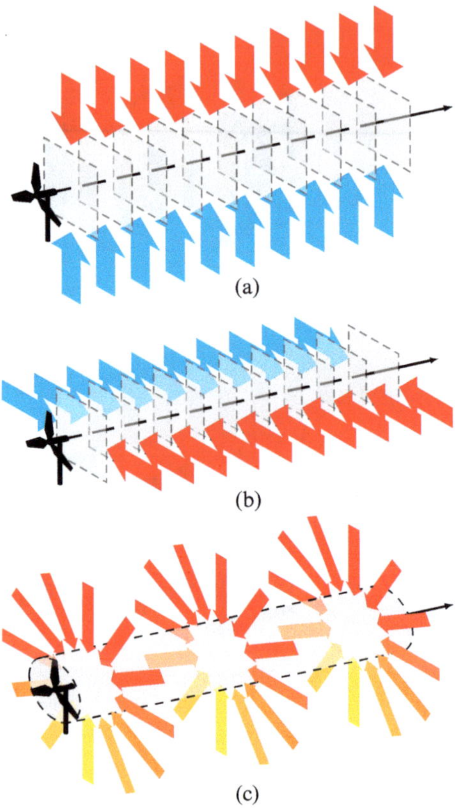

Fig. 14.10 The entrainment process described in Cartesian and polar-cylindrical coordinate systems. (**a**) Vertical entrainment of kinetic energy. (**b**) Lateral entrainment of kinetic energy. (**c**) Radial entrainment of kinetic energy

which aids the discussion of vertical transport into the wakes. The stress field contributing to the entrainment of kinetic energy is dominated by two components in this reference frame. A coordinate transformation to a polar-cylindrical frame with axis aligned with the hub of the model wind turbine effectively combines these components into a single term. Figure 14.10 demonstrates the transport processes discussed in the preceding results. Polar-cylindrical coordinates are more natural to the flow of a single wake, which is approximately axisymmetric but for the influence of the presence of the wall and the wake of wind turbine mast.

In the polar-cylindrical formulation, the component of the Reynolds stress tensor contributing to inward flux of kinetic energy is composed of fluctuations in velocity in the axial (u_x) and radial (u_r) directions. Fluctuating azimuthal velocity u_θ is contributed to redistribution of energy around the wake but does not contribute directly to wake remediation. The inward component of the flux then combines the mean axial velocity and the axial–radial turbulent stress and is denoted $F_{xr} = \overline{u_x u_r} U_x$. Figure 14.11 shows the inward flux, demonstrating that mean flow kinetic energy is transported toward the center of the wake everywhere except directly following the mast of the wind turbine. For $x/D \leq 2$, kinetic energy is moved outward, away from the center of the wake. The peak value of flux into the wake

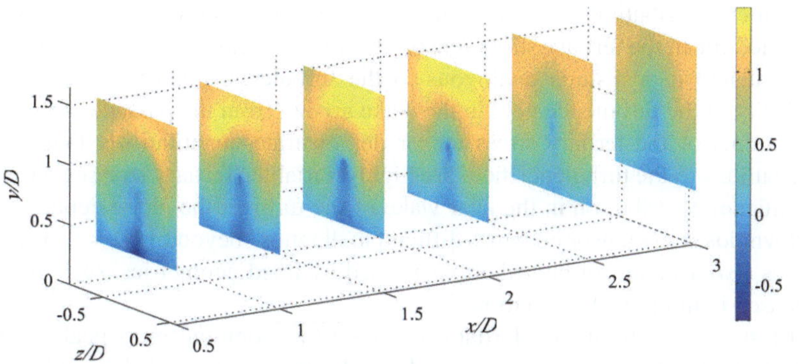

Fig. 14.11 Contours representing the radial flux of mean kinetic energy into the wake of a wind turbine, $F_{xr} = \overline{u_x u_r} U_x$

occurs trailing the top tip of the rotor at $x/D = 1.5$ and is well described by the $-\overline{uv}$ component in the Cartesian frame.

14.5 Conclusions

The data presented above result from a wind tunnel-scale experiment employing $2D - 3C$ stereo PIV measurements in a fully developed wind farm. Measurements were synchronized with the phase orientation of the rotor blades of a model wind turbine at four phase angles. Total mean values of axial velocity show the momentum-deficit area in the near wake quite clearly but little dependence on the position of the rotor blades. When the flux of kinetic energy is composed with phase-averaged flow statistics, it becomes clear that entrainment of high-momentum into the wake is subject to periodic increases according to the passage of the rotor. Following the rotor blades in certain orientations, there is flux of kinetic energy out of the wake below hub height, although evidence suggests that outward flux events are infrequent.

Deterministic stresses representing the combined deviations of phase-averaged velocities from the total mean quantities are approximately two orders of magnitude smaller than turbulent stresses. This difference in magnitude arises from the nature of the flow forcing the passage of the rotors, as opposed to other turbomachinery where the flow is forced by the blades. Although the deterministic contribution to the flux of kinetic energy is smaller in magnitude, there is some theoretical interest arising from azimuthal periodicity at the rotor diameter. That the deterministic stresses make a small contribution to the overall behavior suggests that they may merit further investigation.

An alternate decomposition quantifying deviations of phase-locked stresses from time-averaged values is of more interest in terms of the flux of kinetic energy. The third-order contribution to the entrainment of high-momentum flow into the wake is accounted for in the tertiary stress tensor, where the streamwise normal component is formulated in a fashion analogous to the turbulence intensity, and quantifies variability of the turbulence field in the wake arising from the passage of the rotors. Shear terms of the tertiary stress tensor demonstrate quantitatively that periodic contributions to the turbulence field are highly variable. Tertiary stresses only make a significant contribution in the near wake of the turbine; the phase dependence of the Reynolds stress tensor becomes difficult to diagnose beyond $x/D = 1.5$. Tertiary stresses also account for contributions missed by fixed geometries, such as porous discs used in numerical simulations.

The streamwise/spanwise tertiary stress $-\overline{u'''w'''}$ demonstrates peak values in locations highlighted by the classically defined turbulent stress. This is an indication that the phase-dependent variation of $-\overline{uw}$ yields periodic contributions to the lateral entrainment of kinetic energy into the wake. The streamwise/wall-normal tertiary stress shows peak values following the mast of the model wind turbine. While $-\overline{uv}$ contributes to the largest source of kinetic energy flux, the phase variation does not necessarily describe the periodic contributions seen for $-\overline{uw}$.

Although large-scale turbulent motions dominate in the near wake, it is the phase-dependent variations of the mean velocities that contribute to the deterministic stresses. Because mean velocities show little variation according to phase, the deterministic stresses are small when compared to turbulence. Phase-related variations in the mean velocities are difficult to detect except very near the turbine. However, variations in large-scale turbulence are easily seen throughout the near wake and contribute to the tertiary stresses. Periodicity and rotation added to the mean flow decay moving away from the turbine faster than the phase variation imparted to the turbulence field itself, ultimately leading the deterministic stresses to exhibit smaller magnitudes than the tertiary stresses. In the far wake, variations in the Reynolds stresses related to phase of the rotor also decay and turbulence tends toward a largely homogeneous state. Residual phase dependence is difficult to detect in far-wake turbulence statistics indicating that large scales of turbulence that persist do not express dependence on the passage or orientation of the rotor.

Applying a coordinate transformation to the data presented above simplifies the discussion of kinetic energy flux for an individual wake. For a single wind turbine wake, results indicate that lateral flux cannot be neglected in wake remediation. Transformation to polar-cylindrical coordinates yields a single component of the flux tensor that describes the entrainment of high-momentum flow into the center of the wake, radially inward from outside. The radial flux of kinetic energy F_{xr} shows a crescent-shaped feature similar to that of the streamwise normal stress. Antisymmetry shown by both $-\overline{uv}$ and $-\overline{uw}$ is combined in the transformation into the streamwise/radial shear stress $-\overline{u_x u_r}$.

Considering the flux of mean kinetic energy in polar-cylindrical coordinates demonstrates the tendency for inward transport, toward the core of the wake. Formulating the flux this way indicates that the wake may be represented as a sink term for energy. Analogous formulations for transport mechanisms of heat and moisture can account for the local drying of terrain seen in wind turbine arrays.

Similar to the manner in which the kinetic energy flux entrains high-momentum flow from the outside, scalar fluxes (i.e., moisture and heat) are amplified by the turbulent flow in wind turbine wakes.

Results presented here complement the growing body of the literature detailing the complex flow in wind turbine arrays. The tertiary stresses are a novel view of the variability of the turbulent wakes and would compliment other conditional statistics such as quadrant or octant analysis. Combined analysis would yield details regarding the frequency, intensity, and duration of phase-dependent turbulence events in wind turbine wakes. The streamwise–wall-normal stress is dominantly communicated through ejection events above the wind turbines canopy and through sweeps below the top tip of the rotors. These events are not clearly defined in the polar-cylindrical coordinate system demonstrated above. However, characterizing transport events radially inward and outward is of interest in detailing the entrainment of high-momentum flow toward the center of the wake.

Acknowledgements This work is in part funded by the National Science Foundation (NSF - CBET - 1034581).

References

1. J.J. Adamczyk, J. Turbomach. **122**(2), 189 (2000)
2. R.B. Cal, J. Lebrón, L. Castillo, H.S. Kang, C. Meneveau, J. Renew. Sustain. Energy **2**, 013106 (2010)
3. M. Calaf, C. Meneveau, J. Meyers, Phys. Fluids **22**, 015110 (2010)
4. L.P. Chamorro, R. Arndt, F. Sotiropoulos, Wind Energy **15**(5), 733 (2012)
5. N. Hamilton, H.-S. Kang, C. Meneveau, R.B. Cal, J. Renew. Sustain. Energy **4**(6), 063105 (2012)
6. N. Hamilton, M. Melius, R.B. Cal, Wind Energy **18**(2), 277 (2015)
7. H. Hu, Z. Yang, P. Sarkar, Exp. Fluids **52**(5), 1277 (2012)
8. J. Lebrón, L. Castillo, C. Meneveau, J. Turbul. **13**(43), N43 (2012)
9. L. Lignarolo, D. Ragni, C. Krishnaswami, Q. Chen, C.S. Ferreira, G. Van Bussel, Renew. Energy **70**, 31–46 (2014)
10. G. McBean, J. Elliott, J. Atmos. Sci. **32**(4), 753 (1975)
11. J. Meyers, C. Meneveau, Wind Energy **15**, 305 (2012)
12. J. Meyers, C. Meneveau, J. Fluid Mech. **715**, 335 (2013)
13. W. Reynolds, A. Hussain, J. Fluid Mech. **54**(02), 263 (1972)
14. H. Snel, Wind Energy **6**(3), 203 (2003)
15. N. Troldborg, G.C. Larsen, H.A. Madsen, K.S. Hansen, J.N. Sørensen, R. Mikkelsen, Wind Energy **14**(7), 859 (2011)
16. O. Uzol, Y.-C. Chow, J. Katz, C. Meneveau, Average passage flow field and deterministic stresses in the tip and hub regions of a multistage turbomachine. J. Turbomach. **125**(4), 714–725 (2003). doi:10.1115/1.1625692
17. L.J. Vermeer, J.N. Sørensen, A. Crespo, Prog. Aerosp. Sci. **39**(6), 467 (2003)

Chapter 15
Evaluation of Higher Order Moments and Isotropy in the Wake of a Wind Turbine Array

Naseem Ali, Aleksandr S. Aseyev, Matthew S. Melius, Murat Tutkun, and Raúl Bayoán Cal

15.1 Introduction

There is an increasing global demand for renewable energy sources in which wind energy plays a vital role [1]. Further improvement in the reliability and power output can be achieved by understanding the complex flow within a wind turbine array. In particular, the near- and far-wake of a turbine array would provide basis for developing optimal turbine spacing within a wind farm, thereby enhancing the overall performance. Interacting phenomena are encountered in the wake region of the turbine, notably the rotational blade effects including vortex shedding from the tip of each blade, the entraining flow from above the canopy, the effect of the surrounding turbines in the array, as well as the flow around the mast. A combination of resource constraints and turbine performance influence the spacing of wind turbines in an array [2]. In field installations, it is common to find streamwise direction turbine spacing of 6 or 7 turbine rotor diameters within a wind farm. Meyers and Meneveau [2] noted optimal downwind spacing to be as large as 15 rotor diameters. Wind farms, placed in the log region of the atmospheric boundary layer (ABL) act as surface roughness elements [3]. The power produced by the wind turbine is proportional to the cube of velocity [4]. The ABL flow above the turbine canopy is essential in recovering the momentum loss to supply further power extraction in following turbines within an array. This relationship is documented

N. Ali • A.S. Aseyev • M.S. Melius • R.B. Cal (✉)
Department of Mechanical and Materials Engineering, Portland State University, Portland, OR 97201, USA
e-mail: rcal@pdx.edu

M. Tutkun
Department of Process and Fluid Flow Technology, Institute for Energy Technology (IFE), 2007 Kjeller, Norway

Department of Mathematics, University of Oslo, Blindern, 0316 Oslo, Norway

A. Pollard et al. (eds.), *Whither Turbulence and Big Data in the 21st Century?*,
DOI 10.1007/978-3-319-41217-7_15

by Cal et al. [5], where the flux of mean kinetic energy from above the canopy is shown to be on the same order as the power extracted by the wind turbine. Velocity fluctuations due to turbulence directly relate to power output of a wind turbine. Therefore, analysis of the turbulent fluctuations and its statistics is of interest.

Central moments can be represented as $\langle u_i^p \rangle$, where u is instantaneous velocity, i is the direction of the velocity components, streamwise, wall-normal, and spanwise velocity, and $\langle \ldots \rangle$ denotes ensemble averaging. Superscript p denotes order of the moment. Mean, variance, skewness, and kurtosis are, respectively, computed for $p = 1, 2, 3$, and 4.

In understanding the third and fourth moment, a more comprehensive description of the signal distribution is achieved, thus making higher order moments relevant [6]. Skewness, A_i, quantifies the degree and direction of asymmetry of the signals and is computed as follows:

$$A_i = \frac{\langle u_i^3 \rangle}{\langle u_i^2 \rangle^{3/2}}. \tag{15.1}$$

Positive and negative skewness imply the distribution is asymmetric towards the negative and positive direction, respectively. Skewness is widely used to analyze the sweep and ejection events of the wall turbulence [7]. Furthermore, skewness is also used to identify the isotropic behavior of the flow, which is directly amalgamated to the vortex production [7]. Kurtosis, B_i, measures the degree tailedness and peakedness of the distribution and collaborates with the skewness to attain the shape of the probability density function, pdf [8]. Kurtosis is given as,

$$B_i = \frac{\langle u_i^4 \rangle}{\langle u_i^2 \rangle^2}. \tag{15.2}$$

The skewness and kurtosis of the Gaussian distributions are zero and three, respectively; generally, used as reference to quantify the distributions. Turbulent flow can be described as a combination eddies ranging from large to small scales at which energy is injected and dissipated. At sufficiently high Reynolds number, a constant flux region across the cascade exists, i.e., inertial subrange, separating the energy containing large scales and dissipative small scales, which the latter can then be assumed isotropic as detailed in Kolmogorov [9]. The inertial subrange and isotropy can be shown using energy spectra and the structure function based on power law behaviors resulting from the velocity signal [9]. Using a Fourier transformation, one dimensional wave-number spectrum, $F_{ij}(k_1)$, can be written as follows:

$$F_{11}(k_1) = C_1 \epsilon^{2/3} k_1^{-5/3}, \tag{15.3}$$

$$F_{22}(k_1) = F_{33}(k_1) = C_2 \epsilon^{2/3} k_1^{-5/3}, \tag{15.4}$$

where $F_{11}(k_1)$, $F_{22}(k_1)$, and $F_{33}(k_1)$ are, respectively, the streamwise, wall-normal, and spanwise energy spectra, k_1 is the wavenumber in the streamwise direction, C_1 and C_2 are the Kolmogorov constants, and ϵ is the turbulent energy dissipation. The $-5/3$ power law is used to identify the inertial subrange of the energy cascade. Per [9–11], local isotropy theory occurs when the ratio of $F_{22}(k_1)/F_{11}(k_1) = 4/3$. In Chamecki et al. [12], the cross-spectra, F_{21}, is identically zero under the isotropic state as a result to a marginalized effect of the shear stress. Furthermore, the cross-spectra follows a $-7/3$ power law in the inertial subrange as shown in Kaimal et al. [13]. As argued by Mestayer [14], the $-5/3$ power law should not be used to validate isotropy, but rather the ratio of the vertical and streamwise spectra.

Structure functions, S, are employed to identify the structure of the turbulent flow. Kolmogorov theory [9] highlights that the structure function of order m is only function of the separation vector, R, between two measurement points corresponding to the length scale of a structure within a flow and can be written as,

$$S(R)^m \sim R^{m/3}. \tag{15.5}$$

Based on the velocity increments, the second order structure function is defined in its most general form as,

$$S_{ij}(R) = \Big\langle \big(u_i(x+R) - u_i(x)\big)\big(u_j(x+R) - u_j(x)\big)\Big\rangle, \tag{15.6}$$

where i and j represent velocity directions and can be 1, 2, or 3 representing the longitudinal, transverse, or spanwise direction, respectively [15]. Structure functions are used to identify the inertial subrange via following an $R^{2/3}$ behavior [16]. The ratio of the second order structure function transverse and longitudinal is equivalent to $4/3$ when scales are isotropic [12]. Furthermore, the mixed structure function ($i \neq j$) approaches zero in the scales exhibiting isotropic behavior [15].

Isotropic behavior in the inertial subrange has been identified via structure functions for neutral and stratified atmospheric surface layers [10, 13], wind tunnel boundary layer [17], and plane jet [18]. Anisotropic behavior was found, using the structure functions, in a heated boundary layer [14], in a heated axisymmetric jet flow [19], in a stable stratified wind tunnel boundary layer [17], and in a neutral wind tunnel boundary layer [20]. Using a triaxial ultrasonic anemometer, Katul et al. [21] studied the velocity and temperature above a uniform dry lakebed to observe the statistical structure of the inertial subrange. In utilizing the correlation coefficient, anisotropic effects due to interaction between the large and small scales along with the thermal effect on the small scale eddy motion were determined.

In this study, the flow is considered stationary and the auto-correlation, $\rho(\tau)$, between two different times is used to describe the flow and can be computed as follows:

$$\rho(\tau) = \frac{\langle u_i(t)u_i(t+\tau)\rangle}{\left\langle u_i(t)^2\right\rangle}, \tag{15.7}$$

where t and τ are the time and time step, respectively. The energy cascade can be quantified via the three distinct scales, including the integral length scale, L_{int}, being the largest scale and determined by:

$$L_{\text{int}} = \int_0^{\infty} \rho_\tau d\tau, \tag{15.8}$$

Taylor microscale, λ, representing the scale where dissipation becomes relevant, and can be obtained through the relation:

$$\lambda^2 = \frac{\left\langle u(t)^2 \right\rangle}{\left\langle \left(\frac{\partial u(t)}{\partial t} \right)^2 \right\rangle}, \tag{15.9}$$

and finally, the Kolmogorov microscale, η, being the smallest measured scale of the flow, where the viscous effects dominate and the turbulent energy is mostly dissipated. This scale is defined as,

$$\eta = \left(\frac{\nu^3}{\langle \epsilon \rangle} \right)^{\frac{1}{4}}, \tag{15.10}$$

where ν is kinematic viscosity. Local isotropy and Taylor frozen hypothesis are employed to estimate the dissipation as follows:

$$\epsilon = 15\nu \left\langle \left(\frac{\partial u}{\partial x} \right)^2 \right\rangle. \tag{15.11}$$

Herein, statistical moments, energy spectra, and structure functions are analyzed to describe the wake of a wind turbine array at two downstream locations, namely near- and far-wake. Insight on the isotropy of the flow and detail of the inertial subrange at these locations is obtained.

15.2 Experimental Setup

The closed-loop Corrsin Wind tunnel at the Johns Hopkins University was used to conduct experiments in the wake of a wind turbine array boundary layer. The wind tunnel test section is 10 m (long) × 0.9 m (high) × 1.2 m (wide). The wind tunnel is furnished by an active grid to generate high free-stream turbulence. The active grid is constructed from rotating aluminum shaft of 19 mm diameter and distributed as seven vertical and five horizontal shafts. Each shaft is driven independently via 1/4 hp ac motor that set to change the direction of the rotation randomly. Winglets

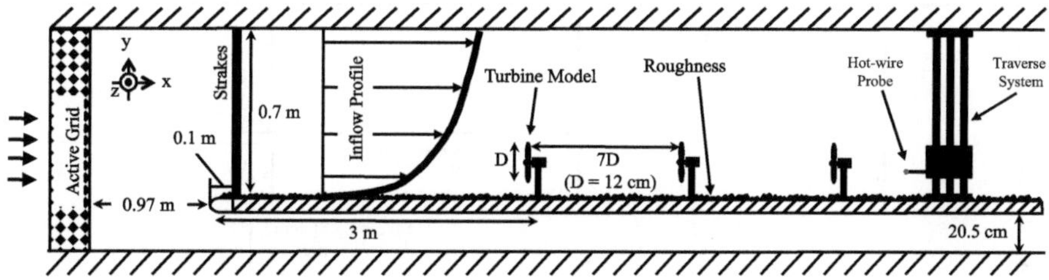

Fig. 15.1 Side view of the experiment setup including the flow conditioning elements, wind turbine array, and measurement system

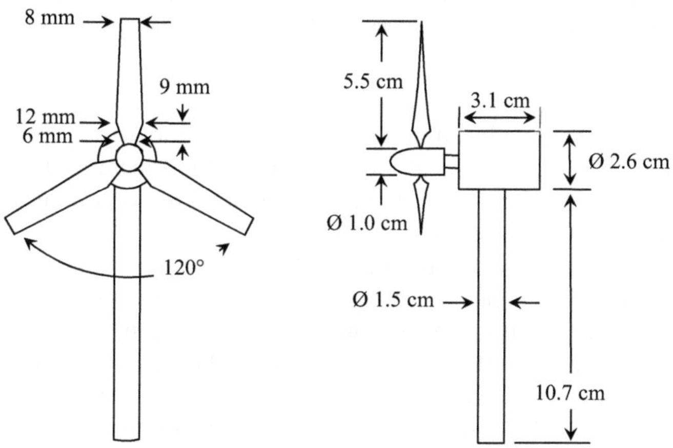

Fig. 15.2 Wind turbine model

of 3.18 mm thick aluminum plates are appended to the rods and arranged as eight horizontally and six vertically. To generate a mean velocity shear profile at the inlet, nine strakes of a thick acrylic plate are uniformly distributed on the spanwise direction of the wind tunnel. Roughness elements are used in the wind tunnel to simulate an atmospheric-like boundary layer where a 24-grit aluminum oxide sand paper coated the platform of the wind turbine array. Figure 15.1 presents the schematic of the experimental setup.

Model wind turbines are constructed from three bladed rotor of diameter, $D = 12$ cm, and the tower of 12 cm high as shown in Fig. 15.2. The rotor is manufactured from steel sheet of 0.48 mm thick and twisted 15° at the root and 10° at the tip. The dimensions of the tower are 1 cm diameter and 10.7 cm tall. The scale of the current turbine model is about 830 times from the real wind turbines. Nine of wind turbine are distributed as 3×3 array and placed at $3D$ downstream of the leading edge and $1.5D$ from the sidewalls. The array spacing was $7D$ in the streamwise and $3D$ in the spanwise directions as shown in Fig. 15.3.

An X-type hot-wire anemometry is used to collect streamwise and wall-normal velocity data for 100 s at two downstream locations $1D$ and $5D$ of the last row center

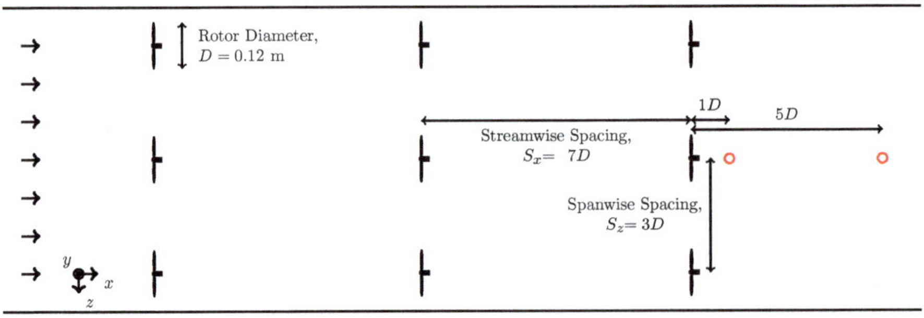

Fig. 15.3 Top view of 3 by 3 wind turbine array. The *red circles* aft of the *last row* centerline turbine represent the measurement locations. Each *circle* represents 21 vertical measurement locations

wind turbine at 21 vertical locations in order to obtain the vertical velocity profile as shown in Fig. 15.3. The vertical locations are in 1 cm increments beginning at 0.5 cm from the floor of the wind tunnel. Two platinum-coated tungsten wires of 2.5 μm in diameter are used as probe after soldering to the X-wire prong ends. The ratio between the length to diameter is about 200 and the spacing between hot-wires is 5 μm. The data are sampled at a frequency of 40 kHz and then filtered via low-pass of 10 kHz. A 12 bit simultaneous sample is used to digitize the signal and hold analog to digital converter. The hot-wire probe is calibrated via an axisymmetric jet and the thermal variation during the tests is taken into account where the velocity calibration of hot-wire is replicated for four different fixed temperatures. The hot-wires have an overheat ratio of about 1.6.

The inflow characterization including free-stream velocity is preserved constant at $9.4\,\mathrm{m\,s^{-1}}$ throughout the experiments and the consistency of velocity profile in spanwise direction is deduced via testing mean velocity and Reynolds stress as well. The results show that the maximum deviation of mean and Reynolds stress is $0.36\,\mathrm{m\,s^{-1}}$ and 1 %, respectively. More information regarding the experimental data can be found in Cal et al. [5], Hamilton et al. [22], and Melius et al. [23].

15.3 Results

Streamwise mean velocities, wall-normal mean velocities, and Reynolds stresses are plotted for all vertical locations at $1D$ and $5D$ in the wake of the center turbine at the last row of a 3×3 wind turbine array, see Fig. 15.3. This is followed by an analysis of statistical moments, energy spectra, and structure functions. Figure 15.4a shows the streamwise mean velocity profiles at $1D$ and $5D$ behind the model wind turbine. The streamwise velocity depicts a significant wake deficit at downstream position of one diameter; this deficit is recovered by five diameters. In Fig. 15.4b, wall-normal velocity profiles in the near- and far-wake regions are shown. At $1D$,

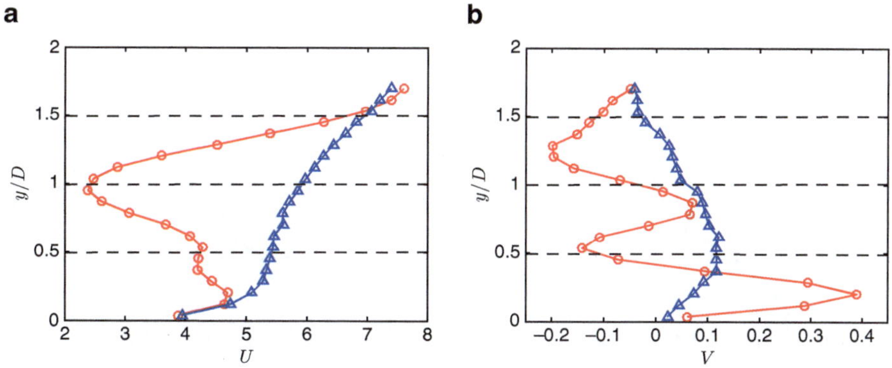

Fig. 15.4 Streamwise (**a**) and wall-normal (**b**) mean velocity plotted against normalized y/D at x/D equal to 1 and 5. *Lines* are defined as near-wake $1D$ (*red open circle*) and far-wake $5D$ (*blue open triangle*). *Dashed lines* provide a reference for the top tip ($y/D = 1.5$), hub height ($y/D = 1$), and the bottom tip ($y/D = 0.5$)

inflection points throughout the profile are observed, where minimum occur within the swept area and maximum are associated with the hub height and flow below the bottom tip. At $5D$, the inflection points no longer persist and a monotonic decrease in the wall-normal mean velocity occurs as a function of increasing height beginning at the bottom tip, although a positive sign for V persists in this region.

Figure 15.5 displays Reynolds normal stresses $\langle u^2 \rangle$ and $\langle v^2 \rangle$, as well as Reynolds shear stress $\langle uv \rangle$ for both downstream positions. $\langle u^2 \rangle$ profiles are shown in Fig. 15.5a, where peaks near the blade tips and close to the wall are observed at $x/D = 1$. At $5D$ and traversing from the wall to the upper bound of the canopy, the streamwise variance increases and peaks at top tip. Figure 15.5b displays the wall-normal Reynolds stress showing the maximum at below bottom tip at $1D$. Furthermore, this stress component remains relatively constant at $y/D > 1$. At $5D$, the wall-normal Reynolds stress increases with increasing wall-normal distance and a maxima is found in the region below top tip. Past the top tip, a decrease in the stress is evident. Interestingly, at $y/D > 1.5$, $\langle u^2 \rangle$ is approximately equal at both downstream locations. Moreover, similar values are found between hub height and bottom tip at $1D$ and $5D$. The Reynolds shear stress profiles $-\langle uv \rangle$ demonstrate a maxima at top tip due to the shear layer as shown in Fig. 15.5c in the near-wake. The stress continues to decrease with decreasing wall-normal distance through the rotor swept area until $y/D = 0.75$. Below this wall-normal position, several inflection points are found due to the effects of the bottom tip of the rotor as well as the wall. At $5D$, the shear stress is, once again, highest at top tip and decreases moving towards bottom tip, where it maintains a near zero value down to the wall. It is clear that the gradients in the far-wake profile soften in comparison to the near-wake. The shear stress is identical at both downstream locations for wall, bottom tip, mid region between the hub height and top tip as well as above the canopy.

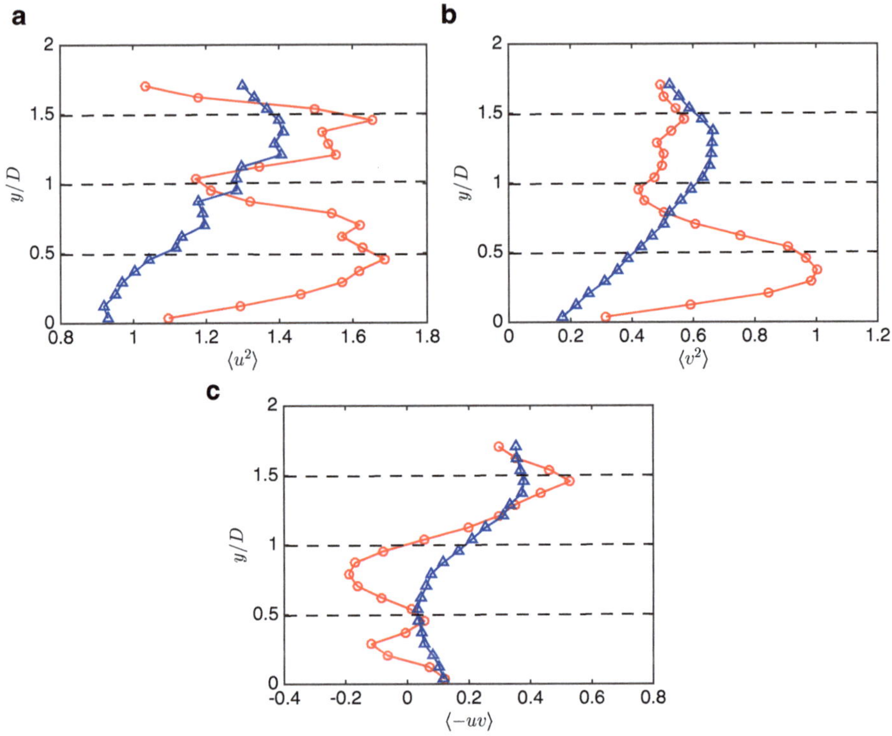

Fig. 15.5 Streamwise (**a**), wall-normal (**b**), and shear (**c**) stress plotted against normalized y/D at x/D equal to 1 and 5. *Lines* are defined as near-wake $1D$ (*red open circle*) and far-wake $5D$ (*blue open triangle*). *Dashed lines* provide a reference for the top tip ($y/D = 1.5$), hub height ($y/D = 1$), and the bottom tip ($y/D = 0.5$)

Streamwise skewness, A_1, is shown in Fig. 15.6a for the near- and far-wake. The profiles tend to contain a positive skewness slightly below the top tip at $y/D < 1.25$. Above this vertical location, the contrary is observed thus showing negative skewness and consequently preferring magnitudes below the mean. The exception is exactly at hub height for the location at $1D$ downstream where sign switching occurs due to the nacelle acting as a blunt body. In fact, several inflection points are observed throughout the profiles at $1D$, which are associated with shear layers at top and bottom tips as well as due to the large deficit generated at the hub height. Nevertheless, A_1 trends from negative to positive with decreasing wall-normal distance until hub height; thereafter, gradients of A_1 change sign with decreasing distance. At $y/D = 1$, a maxima in the A_1 is observed. At location $5D$, the jaggedness associated with the inflection points and locations of high shear is softened although the profile follows the same general trend as in the near-wake. The exception occurs at $y/D < 0.75$, where positive skewness is enhanced and points towards the remediation of the wake. The change at the hub height observed

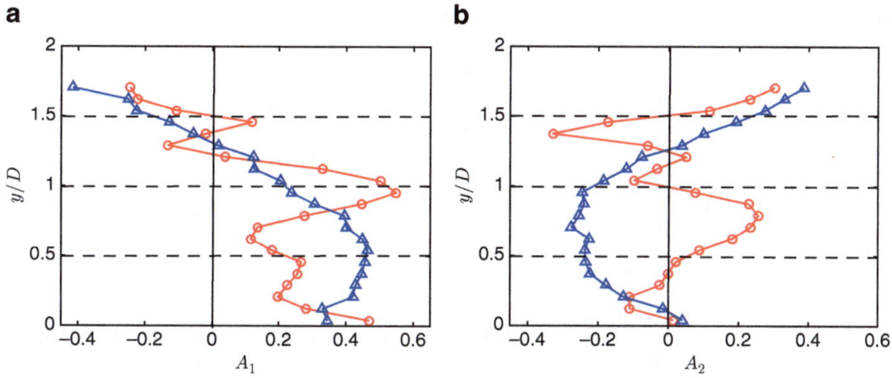

Fig. 15.6 Streamwise (**a**) and wall-normal (**b**) skewness plotted against normalized y/D at x/D equal to 1 and 5. *Lines* are defined as near-wake 1D (*red open circle*) and far-wake 5D (*blue open triangle*). *Dashed lines* provide a reference for the top tip ($y/D = 1.5$), hub height ($y/D = 1$), and the bottom tip ($y/D = 0.5$). *Vertical solid line* provides reference for zero skewness

in the near-wake is now shifted to the bottom tip in the far-wake. The crossing from positive to negative skewness is identified slightly above the hub height. This indicates a recovery of the flow as it progresses downstream.

Figure 15.6b shows the wall-normal skewness, A_2, at 1D and 5D. Wall-normal skewness is a mirror-image of streamwise skewness from above the canopy up to hub height, $y/D = 1$. In the lower part of the rotor, $A_2 > 0$ is observed and thereafter, a negative magnitudes are attained. The wall-normal component displays its sensitivity inside the canopy where extensive velocity realizations above the mean exist due to the turbine effects (top tip, bottom tip, and hub). At hub height, there is positive skewness due to the hub effect for the wall-normal component at 1D. Five rotor diameters downstream, the skewness of the wall-normal velocity also shows a profile with a monotonic change; the trend is exactly opposite to A_1. Again, the $A_2 = 0$ crossing occurs at $y/D = 1.25$.

Figure 15.7a presents streamwise kurtosis, B_1, at near- and far-wake regions. At 1D, streamwise kurtosis decreases with increasing the wall-normal distance with exception at hub height, below the top tip and above the canopy. In the near-wall region, B_1 is greater than 3, indicating a increased peakedness in the probability density function, pdf. A Gaussian normal distribution is observed at the bottom tip and hub height where the kurtosis equals to three. Kurtosis at $y/D = 0.75$ and 1.25 deviate from Gaussian distribution pointing towards a flatter pdf. Furthermore, maximum deviation from the normal distribution is observed at the top tip. At $x/D = 5$, streamwise kurtosis monotonically decreases with increasing wall-normal distance in the range of $y/D = 0 - 1.5$. Above the canopy, the kurtosis depicts a positive slope thus tending towards Gaussianity. Streamwise kurtosis tends to measure less than 3, indicating a flatter pdf.

Figure 15.7b presents wall-normal kurtosis, B_1, in both near- and far-wake regions. The wall-normal kurtosis shows an increased variation in behavior when

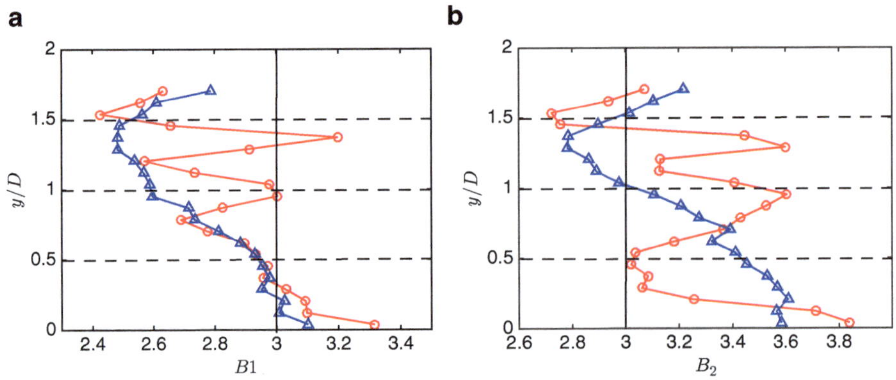

Fig. 15.7 Streamwise (**a**) and wall-normal (**b**) kurtosis plotted against normalized y/D at x/D equal to 1 and 5. *Lines* are defined as near-wake 1*D* (*red open circle*) and far-wake 5*D* (*blue open triangle*). *Dashed lines* provide a reference for the top tip ($y/D = 1.5$), hub height ($y/D = 1$), and the bottom tip ($y/D = 0.5$). *Vertical solid line* provides reference for three kurtosis

compared to the streamwise at 1*D* downstream. Effects of the top tip and hub height at 1*D* are observed to lessen the flatness, while the bottom tip force the distribution to near normal. Nevertheless, kurtosis tends to surpass three and shows a leptokurtic behavior for which an acute peak around the mean is expected. A spike (towards a less flat pdf) is noticed slightly above the wall region to a nearly similar value to the 5*D* location. Aside from the extreme locations stated previously, the behavior between wall-normal and streamwise values is similar at 1*D* and 5*D*. At 5*D*, the near-wall region begins as less flat and tends towards increased flatness crossing over the normal distribution at the hub height region (similar to the 1*D* streamwise kurtosis). Above hub height and towards the top tip, the pdf flatness increases and again reduces flatness above the canopy.

Figure 15.8a presents the pre-multiplied spectra for the streamwise velocity components at the near-wall ($y/D = 0.04$), bottom tip ($y/D = 0.5$), hub height ($y/D = 1$), and top tip ($y/D = 1.5$) at downstream locations in the near ($x/D = 1$) and far ($x/D = 5$) wake. Horizontal dashed lines are plotted for reference to identify the range of wavenumbers that fall into inertial subrange for each location. For simplicity, k is used instead of k_1 in the rest of this study. In the near-wake, the streamwise spectra representing the two locations below hub height show a relatively short inertial subrange, which is composed of wave numbers ranging approximately from $1.2 \times 10^2 < k < 10^3$. At hub height, the range of wavenumbers that the spectrum stays flat is extended to lower wavenumbers of $6 \times 10^1 < k < 8 \times 10^2$. The largest inertial subrange in the near-wake is found at the top tip, where the inertial subrange includes $3 \times 10^1 < k < 10^3$. For the streamwise spectra in the far-wake, it is shown that lowest present in the inertial subrange appears to be independent of wall-normal position, remaining at $k \approx 40$. Conversely, upper range of k that defines the inertial subrange demonstrates a slight dependence on wall-normal location.

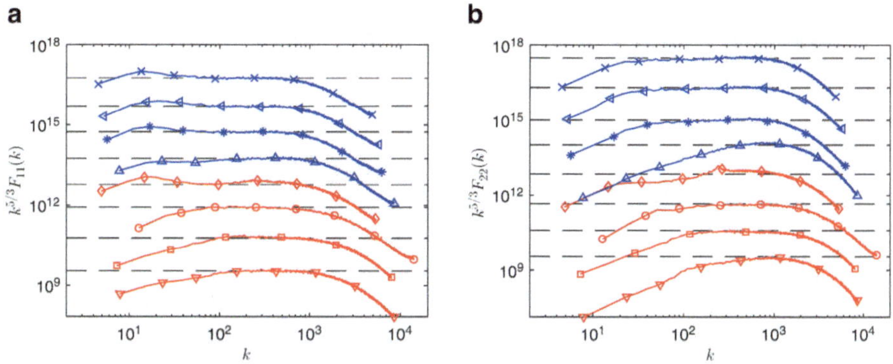

Fig. 15.8 Streamwise (**a**) and wall-normal (**b**) flattened energy spectra plotted against wavenumber. *Lines* are defined as follows: near-wall $1D$ (*red open down triangle*), bottom tip $1D$ (*red open square*), hub height $1D$ (*red open circle*), top tip $1D$ (*red open diamond*), near-wall $5D$ (*blue open triangle*), bottom tip $5D$ (*blue asterisk*), hub height $5D$ (*blue left pointing open triangle*), and top tip $5D$ (*blue times symbol*). *Dashed lines* represent the inertial subrange reference line. Plots for regions were shifted vertically for visual clarity

Figure 15.8b presents the pre-multiplied spectra for the wall-normal velocity components at the same locations as presented in Fig. 15.8a. When evaluating the spectra for the wall-normal velocity in the near-wake, the near-wall location has a limited inertial subrange of $8 \times 10^2 < k < 1.5 \times 10^3$. At the bottom tip, the inertial subrange extends into lower wave numbers, spanning $10^2 < k < 1.5 \times 10^3$. Moving away from the wall, the upper limit of the inertial subrange wavenumbers is reduced at hub height to $10^2 < k < 1.2 \times 10^3$. This trend continues at the top tip, where the range of wave numbers in the inertial subrange is reduced further to $10^2 < k < 10^3$. The lowest wavenumbers (large scales) within the inertial subrange exist in positions furthest from the wall. In the far-wake, the spectra of wall-normal velocity at the near-wall position shows approximately the same range appearing at the near-wake. At the hub height bottom tip and top tip, the range of wavenumber, k, increases significantly to include approximately the same wavenumbers in the range of $30 < k < 10^3$.

To further characterize the wake of the turbine array, the second order structure functions for the streamwise and wall-normal velocity increments are calculated. The structure functions are evaluated over a long range of scales that vary from ten times the integral length scale, L_{int}, to the Kolmogorov dissipative scale, η. In the case of second order structure function, the inertial subrange is defined by the range of incremental scales where the structure function follows a power law of $R^{2/3}$. Figure 15.9a, b shows surface plots of the streamwise and wall-normal structure functions, respectively, at the near-wall ($y/D = 0.04$), bottom tip ($y/D = 0.5$), hub height ($y/D = 1$), and top tip ($y/D = 1.5$) at downstream locations in the near ($x/D = 1$) and far ($x/D = 5$) wake. The x-axis represents the wall-normal location, the y-axis the structure function value, and the z-axis gives the spatial scale, R/D.

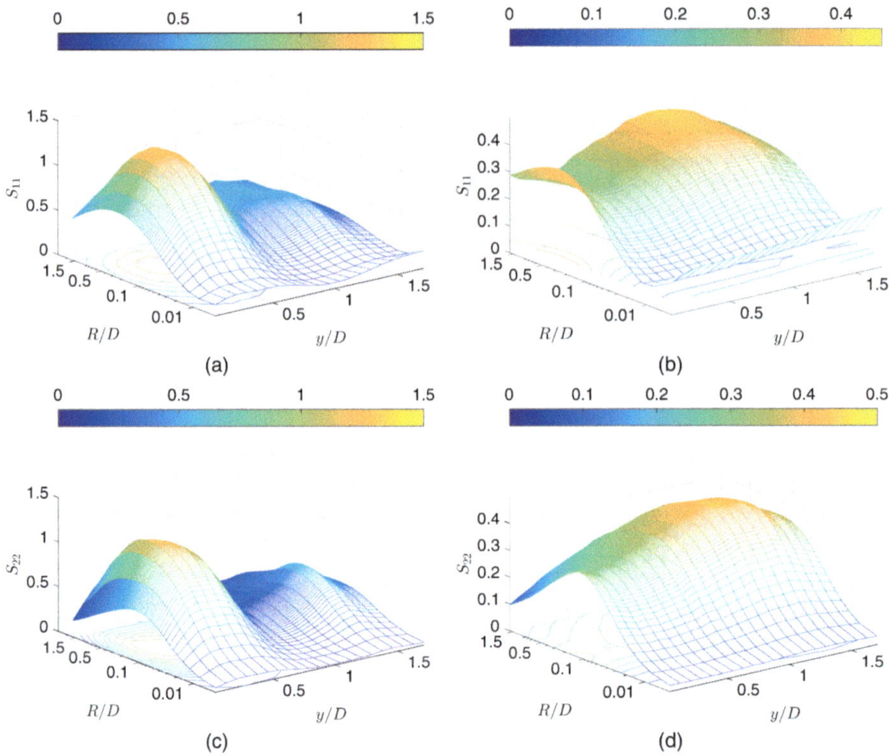

Fig. 15.9 S_{11} at $1D$ (**a**), S_{11} at $5D$ (**b**), S_{22} at $1D$ (**c**), and S_{22} at $5D$ (**d**) downstream

From Fig. 15.9a, the streamwise structure function, S_{11}, in the near-wake changes as a function of vertical location is relatively small for scales of $R/D < 10^{-1}$. However, as R/D increases to scales above $R/D = 10^{-1}$, the wall-normal dependence of S_{11} is enhanced. Below hub height, the structure function increases with increasing y/D position, reaching a maximum near the bottom tip. The S_{11} maximum then decreases as the wall-normal position approaches hub height. Above hub height, the wall-normal dependence disappears. In Fig. 15.9b, a surface plot depicts the streamwise structure function, S_{11}, at $x = 5D$ as a function of wall-normal location. For scales less than $R/D = 10^{-1}$, there is no apparent wall-normal dependence. Increasing to scales above $R/D = 10^{-1}$, the wall-normal dependence becomes more pronounced above hub height, where the maximum values increase from hub height to reach a maximum at the top tip and then decrease beyond $y/D = 1.5$. Below hub height, the structure function shows the opposite behavior, where the maximum of S_{11} decreases to a minimum at the bottom tip to then increases as the measurement location approaches the wall. Similarly, of the wall-normal structure function, S_{22}, at $x = 1D$ is shown as a function of wall-normal distance, y/D, and separation distance, R/D, in Fig. 15.9c. Wall-normal dependence lies predominantly below hub height for scales $R/D > 10^{-1}$, and the global maximum is seen near the bottom tip.

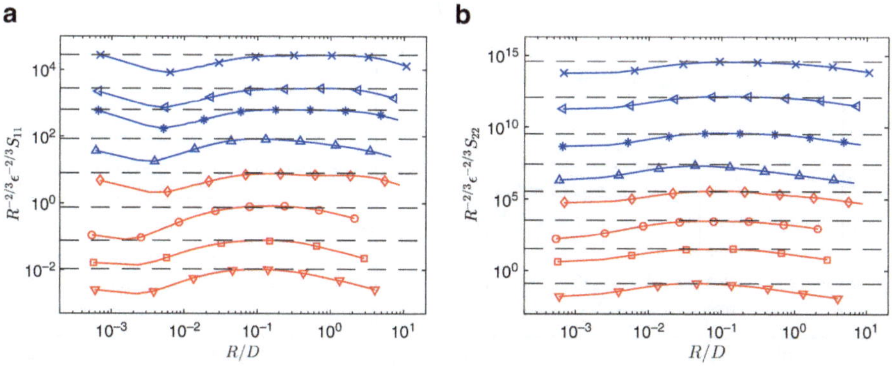

Fig. 15.10 Streamwise (**a**) and wall-normal (**b**) normalized second order structure function, where R is converted from the time step to the length. *Lines* are defined as follows: near-wall 1D (*red open down triangle*), bottom tip 1D (*red open square*), hub height 1D (*red open circle*), top tip 1D (*red open diamond*), near-wall 5D (*blue open triangle*), bottom tip 5D (*blue asterisk*), hub height 5D (*blue left pointing open triangle*), and top tip 5D (*blue times symbol*). *Dashed lines* represent the inertial subrange reference line. Plots for regions were shifted vertically for visual clarity

Above hub height, S_{22} reaches a local maximum above the top tip, but otherwise demonstrates relatively little difference between different wall-normal positions. In the far-wake at $x = 5D$, the values of S_{22} at scales of $R/D > 10^{-1}$ increase with increasing wall-normal position, reaching a maximum at the top tip and decreasing from the maximum at larger wall-normal locations, as shown in Fig. 15.9d.

Normalized structure functions, $R^{-2/3}\epsilon^{-2/3}S_{ij}$, are used to identify the inertial subrange of the energy cascade in the near- and far-wake regions. Figure 15.10a shows the normalized streamwise structure function as a function to normalized distance space, R/D. The near-wall structure function shows small range of scales that follow the 2/3 power law. The inertial subrange at the bottom tip and hub height extends wider than the subrange near-wall and shorter than the range at the top tip. At five diameters downstream, near-wall region shows the inertial subrange based on the structure function is approximately similar to the range at 1D. Thus, the hub height and bottom tip regions have the same range of the scales following the power law. A long inertial subrange is observed at the five diameter downstream of the top tip. For the normalized structure functions based on the wall-normal velocity increment, shown in Fig. 15.10b, the near-wake regions show that the near-wall locations and bottom tip have small inertial subranges compared with the top tip region. At hub height and top tip in the near-wake, the wall-normal structure functions show approximately the same range of scales that could be considered to be following the power law. In the far-wake region, an inertial subrange cannot be identified in the near-wall region. The three other locations show large inertial subranges when compared to the range of the near-wake region. Furthermore, the hub height and top tip show approximately the same inertial subrange.

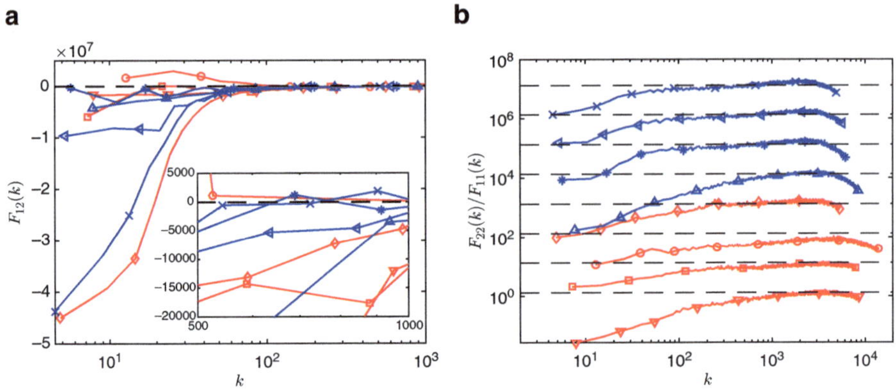

Fig. 15.11 Cross-spectra (**a**), and energy spectra ratio (**b**) plotted against wavenumber. *Lines* are defined as follows: near-wall $1D$ (*red open down triangle*), bottom tip $1D$ (*red open square*), hub height $1D$ (*red open circle*), top tip $1D$ (*red open diamond*), near-wall $5D$ (*blue open triangle*), bottom tip $5D$ (*blue asterisk*), hub height $5D$ (*blue left pointing open triangle*), and top tip $5D$ (*blue times symbol*). *Dashed lines* represent a reference for isotropic behavior. Spectra ratio plots for regions were shifted vertically for visual clarity

To test local isotropy based on the spectra analysis, the cross-spectra is computed at the near-wall, bottom tip, hub height, and top tip at $1D$ and $5D$ as shown in Fig. 15.11a. It follows that a departure from zero is an indication of anisotropic behavior. For all regions, as k increases, the spectra begin as highly anisotropic and progressively move towards isotropy. All spectra converge to null after $k \approx 5 \times 10^2$. Although all locations converge towards the isotropy line, $F_{12} = 0$, they pass approximately parallel to zero line without converging to the x axis with the exception a few points at each location, see inlay of Fig. 15.11a. This means that the flow is highly anisotropic except at few scales at each location. Also, cross-spectra, F_{12}, for all regions are negative at low wavenumber, except for the hub height at $1D$ due to the correlation between the value of streamwise and wall-normal velocity fluctuations. The spectral ratio is plotted in Fig. 15.11b to verify the local isotropic for the near-wall, bottom tip, hub height, and top tip at $1D$ and $5D$. Plots are vertically adjusted for visual clarity, where a comparable dotted line for each location is provided for reference, thus showing isotropy at $F_{22}/F_{11} = 4/3$. In the near-wall region for both downstream locations, the flow has scales behaving isotropically starting from $k \approx 1.2 \times 10^3$ and extending further in the near-wake up to $k \approx 5{\times}10^3$. In the bottom tip region, the range of isotropic scales is approximately the same range of the near-wall region. Hub height at $1D$ contains no scale indicating isotropy. For the top tip regions, the scale range for local isotropy occurs at a scale spanning $2.5 \times 10^2 < k < 8 \times 10^2$. Five diameters downstream, the bottom tip, hub height, and top tip have approximately the same range of wavenumbers, $2 \times 10^2 < k < 7 \times 10^2$, that are isotropic. Based on the relative distance and range of scales that fall on the requisite $4/3$ line, locations in the near-wake are more anisotropic at large scales than in the far-wake. The only exception happens near the

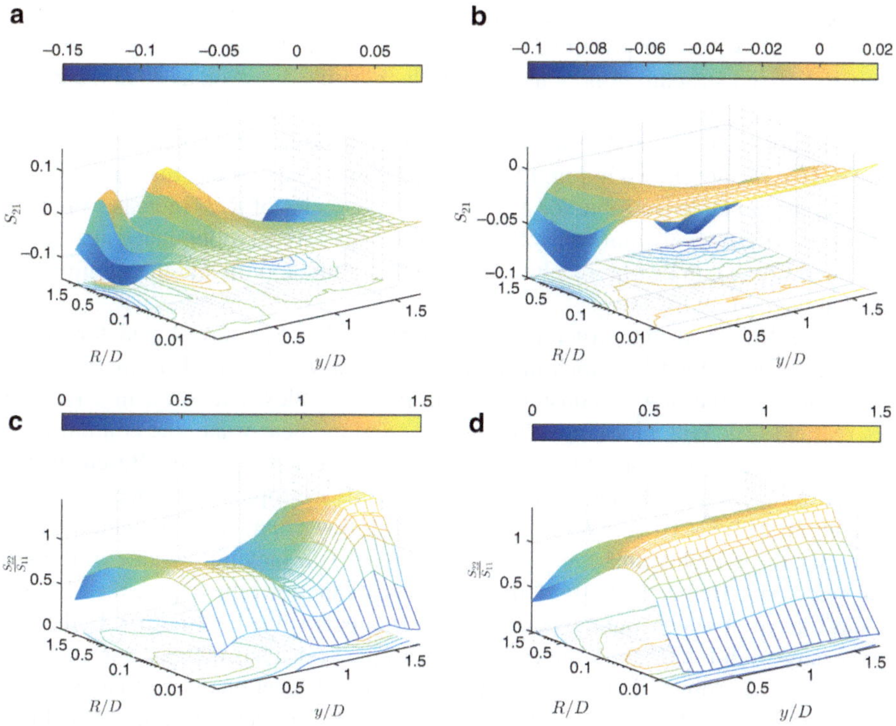

Fig. 15.12 S_{21} at $1D$ (**a**), S_{21} at $5D$ (**b**), ratio S_{22}/S_{11} at $1D$ (**c**), and ratio S_{22}/S_{11} at $5D$ (**d**) downstream

wall, where the spectral ratio appears unaffected due to the downstream location. Of equal importance to locating the isotropic scales is indeed identifying the scales at which the flow demonstrates anisotropy. The maximum anisotropy at large scales is observed near the wall at $1D$ and $5D$. The minimum anisotropy at large scale is observed at bottom tip and hub height at $1D$ and $5D$, respectively.

Figure 15.12a, b shows surface plots of the mixed structure function, S_{21}, for all 21 vertical measurement locations in the near- and far-wake. The plots include scales of R/D ranging from the largest integral scales to the smallest Kolmogorov scales found in the measurement series for $x = 1D$ and $x = 5D$ separately. Similar to Fig. 15.9, the x-axis in Fig. 15.12 represents the wall-normal location, the y-axis the structure function value, and the z-axis gives the spatial scale, R/D. For the near-wake, scales of $R/D < 3 \times 10^{-2}$, $S_{21} \approx 0$ are shown and thus, the flow can be considered isotropic for these scales at all vertical positions. The only exception is the measurement location very near to the wall in the near-wake region, which essentially becomes isotropic for scales below $R/D < 10^{-2}$. The largest range of scales which can be considered isotropic is found near to the hub height. As the value of R/D increases, S_{21} becomes highly anisotropic and demonstrates significant variation based on wall-normal location. For locations near the wall, bottom tip and top tip, the mixed structure functions become increasingly negative as R/D increases. However, for positions just below the bottom tip and directly behind the hub, the mixed structure functions grow in the positive direction as R/D increases. These trends follow the behavior of the Reynolds stresses, shown in Fig. 15.5, at these locations. The mixed structure functions for measurement locations in the far-wake, $x = 5D$, are shown in Fig. 15.12b. Similar to the near-wake, the flow is found to be approximately isotropic for $R/D < 10^{-2}$. As scales of R/D increase, the flow becomes progressively more anisotropic. Anisotropy is most pronounced near to the wall and above hub height, reaching a maximum at the top tip and the minimum is seen at the bottom tip.

In Fig. 15.12c, d, the structure function ratios, S_{22}/S_{11}, are plotted for all 21 vertical measurement locations in the near- and far-wake. Based on the structure function ratio, the flow is considered isotropic for scales that result in a ratio of $S_{22}/S_{11} = 4/3$. When looking at the results from the near-wake, the contour lines show that the flow is anisotropic for all scales below the top tip. Based on the structure function ratio, the top tip region contains isotropic turbulence for scales ranging from $1.5 \times 10^{-2} < R/D < 6 \times 10^{-1}$. It follows that, at hub height, the flow is shown to be highly anisotropic as S_{22}/S_{11} reaches a maximum value of 0.63 in the near-wake. Below the bottom tip, the structure function ratio approaches $4/3$ for a short range of scales, but never reaches the requisite value for isotropy. In the far-wake, shown in Fig. 15.12d, the extent of isotropic scales has increased throughout the wake. It is shown that at the majority of vertical locations, for scales ranging from $2 \times 10^{-2} < R/D < 2 \times 10^{-1}$ $S_{22}/S_{11} = 4/3$. In fact, all vertical locations have a small range of scales that are isotropic, the shortest range is found near to the wall. At $x = 5D$, the isotropic range increases with increasing wall-normal position up to the bottom tip, above the bottom tip the extent of isotropy remains relatively constant. Above the bottom tip, the flow is shown to be anisotropic for scales of $R/D > 2 \times 10^{-1}$.

The normalized mixed structure function is displayed in Fig. 15.13a with an inlay to clarify the scales based on region behaving isotropically. Isotropic behavior is visible when the mixed structure function, S_{21}, is equal to zero. All regions are far from isotropy for large scale, $R/D > 10^{-1}$. Thus, all regions intersect the isotropic line in one scale and then deviate away from isotropy. Figure 15.13b displays the structure function ratio of S_{22}/S_{11}, where few scales behave isotropically, thus the ratio should equal 4/3. The top tip $1D$, hub height $5D$, bottom tip $5D$, and top tip $5D$ reach approximately isotropic behavior, when $6 \times 10^{-3} < R/D < 6 \times 10^{-1}$. The remaining regions do not result in any scales behaving isotropically. The hub height at $1D$ is the furthest region from achieving isotropic behavior. The structure function ratio, S_{22}/S_{11}, demonstrates higher anisotropy in the near-wake, based on proximity to the isotropic ratio of $4/3$, at the top tip. The largest reduction in anisotropy between near- and far-wake is seen at hub height. The passage of the rotor is deemed responsible for this effect. Locations at near-wall positions capture a slight increase in anisotropy based on an increase in the departure from the isotropic ratio of $4/3$.

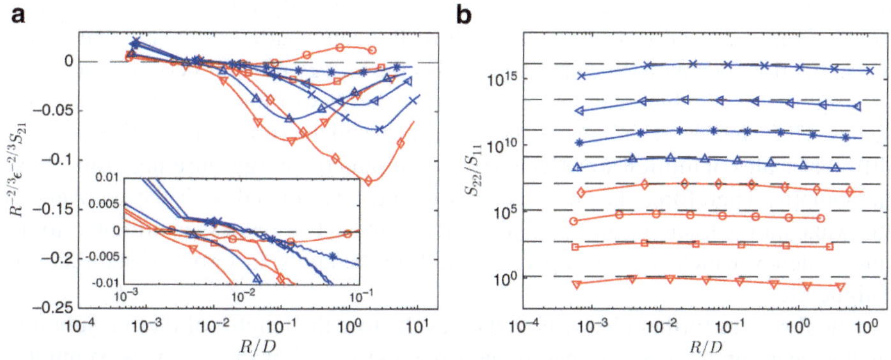

Fig. 15.13 Mixed second order structure function (**a**) and second order structure function ratio (**b**), where R is converted from the time step to the length. *Lines* are defined as follows: near-wall $1D$ (*red open down triangle*), bottom tip $1D$ (*red open square*), hub height $1D$ (*red open circle*), top tip $1D$ (*red open diamond*), near-wall $5D$ (*blue open triangle*), bottom tip $5D$ (*blue asterisk*), hub height $5D$ (*blue left pointing open triangle*), and top tip $5D$ (*blue times symbol*). Structure function ratio plots for regions were shifted vertically for visual clarity. *Dashed lines* represent a reference for isotropic behavior

15.4 Discussion and Conclusion

The use of velocity fluctuations in turbulence statistics was implemented via skewness, kurtosis, energy spectra, and structure functions. Kolmogorov principles were utilized to characterize the level of flow isotropy and the range of scales within the inertial subrange. Application of higher statistical moments described velocity fluctuation distribution based on all regions within the turbine canopy and slightly above. For these quantities, the top tip, hub height, and bottom tip showed the greatest amounts of variability. Inflection points, associated with the shear layers at the top and bottom tip, were identified using the streamwise skewness at $1D$. The variability of these quantities weaken as the flow develops downstream, shown by a negligible variation in streamwise skewness at $5D$.

Wall-normal skewness displays extensive velocity excursions above the mean at $1D$ due to effects of the top tip, bottom tip, and hub. At $5D$, these effects are not significant, where the wall-normal skewness results in a monotonic profile change. Kurtosis, in the streamwise direction, shows increasing flatness moving away from the wall. Spikes at the top tip and hub height at $1D$ lessen the flatness, while at $5D$ these effects are significantly reduced. The wall-normal kurtosis results in more variation than the streamwise direction. The top tip and hub region at $1D$ show lessened flatness while the bottom tip and near-wall region tend to a normal distribution. A steady increase in flatness moving away from the wall occurs at $5D$, where a normal distribution is observed at the hub height. Kurtosis profiles show significant variation as a function of location. This behavior is attributed to local Reynolds number in the wake and furthermore to the length scale itself, see Fig. 14

in 23. The third and fourth order moments are strongly related to the extracted power from the wind turbines as a result of the contribution of events that highly affect the flow behavior. Skewness quantifies the energy transfer between the scales since it is $u * u^2$, which can be interpreted as the flux of kinetic energy across the scales. The transferred energy between the scales and the further interaction between different scales of motion are affected by the amount of energy entrained from above the canopy. Therefore, skewness is subsequently associated with the recovery of the wake and extracted power of the wind turbines. The flatness is affiliated to the intermittency which has an impact on the flow behavior in terms of stability and fatigue life.

The flow was further characterized by evaluating the lengths of inertial subrange in the near- and far-wakes based on streamwise and wall-normal wavenumber spectra and structure functions. Streamwise spectra, for the near-wake, show two locations below hub height with relatively short inertial subranges, while the largest inertial subrange is found at the top tip as a result to Reynolds number effect. In the far-wake, the highest wavenumbers in the inertial subrange are independent of the wall-normal position, whereas the lower range increases with increasing wall-normal location. Wall-normal spectra in the near-wake indicate the largest scales in the inertial range in positions furthest from the wall. The far-wake data show short inertial subrange near the wall, though when moving away from the wall, particularly the bottom tip and hub height, the inertial subrange is significantly increased. Similarities in the streamwise and wall-normal spectra are easy to observe, in particular for top tip $1D$, bottom tip $5D$, hub height $5D$, and top tip $5D$. These are evident when comparing to the spectral ratio in testing for isotropy. A value of $4/3$ for the spectral ratio indicates energy spectra between the components must be relatively close to one another at specific scales for isotropic behavior to exist.

The second order structure functions in the streamwise direction, for the near-wake, are evaluated to compare to the spectral results. Near the wall, the inertial range is very short. Bottom tip and hub height have nearly the same length of inertial subranges, while the top tip shows the largest inertial subrange. Far-wake locations at bottom tip, hub height, and top tip show inertial subranges of nearly the same length which span over decade of scales, which are considerably longer than the inertial subrange at the near-wall. In the wall-normal component, S_{22} reaches a local maximum above the top tip, but otherwise demonstrates a relatively small dependence between wall-normal positions. A limited inertial subrange is shown in the near-wake, while moving downstream, in the far-wake, the inertial subrange has developed to span approximately a decade.

Finally, the extent of isotropy of the turbine wake was analyzed through the use of cross-spectra, spectral ratio, mixed structure functions, and structure function ratio. The turbulence scales at the top tip, bottom tip, and near-wall were found to be highly anisotropic as a result to the vortex shedding at the tip regions and sweep event near the wall. The far-wake region displays a consistently large range of isotropic scales at all vertical locations outside of the near-wall region. Anisotropy is more pronounced near the wall and above the hub height due to high shear at these locations caused. This is caused by the wall in near-wall region and by the rotor above the hub height. The anisotropy is here influenced by the turbulent kinetic energy production and the entrainment from above the canopy that increases the interaction between the scales and creates intermittent flow. These results were related to key features in wind turbine designs, providing implications to power extraction. Comparisons with downstream locations pave the way for further analysis and optimization of turbine spacing within a wind farm. All these events have the direct impact on the harvested power from the turbines in the wind farm. Its dependence on the wind farm layout requires further investigation.

Acknowledgements This work is in part funded by the National Science Foundation (NSF - CBET - 1034581). Tutkun′s work is financed by the Research Council of Norway under the FRINATEK program (project #231491).

References

1. GWEC, Global wind energy outlook 2012, November, 2012
2. J. Meyers, C. Meneveau, Optimal turbine spacing in fully developed wind-farm boundary layers. Wind Energy **15**, 305–317 (2012)
3. L.P. Chamorro, F. Porté-Agel, Turbulent flow inside and above a wind farm: a wind-tunnel study. Energies **4**, 1916–1936 (2011)
4. M.S. Melius, M. Tutkun, R.B. Cal, Solution of the Fokker-Planck equation in a wind turbine array boundary layer. Phys. D **280–281**, 14–21 (2014)
5. R.B. Cal, J. Lebron, L. Castillo, H.S. Kang, C. Meneveau, Experimental study of the horizontally averaged flow structure in a model wind-turbine array boundary layer. J. Renew. Sustain. Energy **2**, 013106, 1–25 (2010)
6. S.B. Pope, *Turbulent Flows* (Cambridge University Press, Cambridge, 2000)
7. T. Sedat, *Statistical Approach to Wall Turbulence* (Wiley, Hoboken, 2013)
8. L.T. DeCarlo, On the meaning and use of kurtosis. Psychol. Methods **2**(3), 292 (1997)
9. A.N. Kolmogorov, The local structure of turbulence in incompressible viscous fluid for very large Reynolds number. Dokl. Akad. Nauk. SSSR **30**, 299–303 (1941); A.N. Kolmogorov, Dissipation of energy in the locally isotropic turbulence. Dokl. Akad. Nauk. SSSR **32**, 19–21 (1941)
10. A.S. Monin, A.M. Yaglom, *Statistical Fluid Mechanics* (MIT Press, Cambridge, 1975), 875 pp.
11. L. Mydlarski, Z. Warhaft, On the onset of high-Reynolds-number grid-generated wind tunnel turbulence. J. Fluid Mech. **320**, 331–368 (1996)
12. M. Chamecki, N.L. Dias, The local isotropy hypothesis and the turbulent kinetic energy dissipation rate in the atmospheric surface layer. Q. J. R. Meteorol. Soc. **130**, 2733–2752 (2004)
13. J.C. Kaimal, J.C. Wyngaard, Y. Izumi, O.R. Coté, Spectral characteristics of surface-layer turbulence. Q. J. R. Meteorol. Soc. **98**, 563–589 (1972)
14. P. Mestayer, Local isotropy and anisotropy in a high-Reynolds-number turbulent boundary layer. J. Fluid Mech. **125**, 475–503 (1982)
15. S. Kurien, K.R. Sreenivasan, Anisotropic scaling contributions to high-order structure functions in high-Reynolds-number turbulence. Phys. Rev. E **62**(2), 1–7 (2000)
16. V.I. Tatarski, *Wave Propagation in a Turbulent Medium*, translated by R.A. Silverman (McGraw Hill, New York, 1961)
17. C. Van Atta, Local isotropy of the smallest scales of turbulent scalar and velocity fields. Proc. R. Soc. Lond. **434**, 139–147 (1991)

18. R.A. Antonia, B.R. Pearson, Low- order velocity structure functions in relatively high Reynolds number turbulence. Europhys. Lett. **48**(2), 163–1 (1999)
19. K.R. Sreenivasan, R.A. Antonia, D. Britz, Local isotropy and large structures in a heated turbulent jet. J. Fluid Mech. **94**, 745–775 (1979)
20. R.A. Antonia, M.R. Raupach, Spectral scaling in a high Reynolds number laboratory boundary layer. Bound. Lay. Meteorol. **65**, 289–306 (1993)
21. G.G. Katul, M.B. Parlange, J.D. Albertson, V.R. Chu, Local isotropy and anisotropy in the sheared and heated atmospheric surface layer. Bound. Lay. Meteorol. **72**, 123–148 (1995)
22. N. Hamilton, C. Meneveau, R.B. Cal, H.-S. Kang, Statistical analysis of kinetic energy entrainment in a model wind turbine array boundary layer. J. Renew. Sustain. Energy **4**, 063105 (2012)
23. M. Melius, M. Tutkun, R.B. Cal. Identification of Markov process within a wind turbine array boundary layer. J. Renew. Sustain. Energy **6**(2), 023121 (2014)

Part V
Data Manipulation

Chapter 16
Turbulent Flow Physics and Control: The Role of Big Data Analyses Tools

Andrew S. Magstadt, Pinqing Kan, Zachary P. Berger, Christopher J. Ruscher, Matthew G. Berry, Melissa A. Green, Jacques Lewalle, and Mark N. Glauser

16.1 Introduction

The turbulence community has been involved in big data analysis for several decades. The Navier–Stokes equations, which govern the dynamics of viscous fluid flow, are, in the limit of high Reynolds (*Re*) number, very high-dimensional non-linear PDEs with stochastic solutions. Analysis of such high-dimensional stochastic flow physics naturally drives the direct connection to big data. The large data sets of velocity, temperature, pressure, etc. (the haystack) need various big data analysis tools to extract low-dimensional flow physics (the needle). Over these several decades, the turbulence field has generally been at, or near, the leading edge of the latest experimental and computational tools to capture as much data as possible to understand and control this non-linear time-dependent 3D stochastic phenomena. Glauser, for example, [10], as part of his Ph.D. thesis in the mid-1980s, acquired more than 2 GB of multi-point velocity time-series data in the high-*Re* number axisymmetric jet. This data set, although by today's standards small, was one of

A.S. Magstadt (✉) • P. Kan • M.G. Berry • M.A. Green • J. Lewalle • M.N. Glauser
Department of Mechanical and Aerospace Engineering, Syracuse University, Syracuse, NY 13244, USA
e-mail: asmagsta@syr.edu; pkan@syr.edu; mgberry@syr.edu; magreen@syr.edu; jlewalle@syr.edu; mglauser@syr.edu

Z.P. Berger
Department of Mechanical and Aerospace Engineering, Syracuse University, Syracuse, NY 13244, USA

Present address: Penn State University, University Park, State College, PA 16802, USA
e-mail: zpberger@gmail.com

C.J. Ruscher
Computational Division, Spectral Energies, LLC, Dayton, OH 45431, USA
e-mail: christopher.ruscher.ctr@us.af.mil

A. Pollard et al. (eds.), *Whither Turbulence and Big Data in the 21st Century?*,
DOI 10.1007/978-3-319-41217-7_16

Table 16.1 Experimental jet campaigns at SU

Researcher	Time frame	Focus and methods	Jet details	Instrumentation	Data size
Tinney et al. [45, 46]	2001–2005	POD, LSE, noise propagation	Mach 0.85, cold, axi.	PIV in $(r-\theta)$, NF/FF-pressure	$\sim$250 GB
Hall et al. [18]	2004–2007	POD, LSE, noise propagation	Mach 0.6, hot, axi.	PIV in $(r-\theta)$, NF/FF-pressure	$\sim$250 GB
Pinier et al. [35]	2004–2007	Empirical LODS, POD	Mach 0.6, cold, axi.	DT-PIV in $(r-\theta)$, NF/FF-pressure	$\sim$500 GB
Low et al. [29]	2007–2012	Open- and closed-loop flow control	Mach 0.6–1.0, cold, axi.	TRPIV in $(r-z)$, NF/FF-pressure	$\sim$1 TB
Berger et al. [5]	2010–2014	Flow control, POD, OID	Mach 0.6, cold, axi.	TRPIV in $(r-z)$ and $(r-\theta)$, NF/FF-pressure	$\sim$1 TB
Magstadt/Berry et al. [31]	2014	POD, spatial correlations	Mach 1.1, cold, axi.	LWPIV in $(r-z)$, FF-pressure	$\sim$10 GB
Magstadt/Berry et al. [39]	2014–present	POD, spectra, flow physics	Mach 1.1–1.6, cold, complex	TR schlieren (x, y, z), LWPIV, NF/FF-pressure	>1 TB

the largest experimental turbulent databases in the world at the time and required a very large number of magnetic tapes for storage. Applying proper orthogonal decomposition (POD) on this data set to extract low-dimensional physics, required several trips to NASA Ames to solve the eigenvalue problem on the CRAY 2. Now, fast forward to this century and refer to Table 16.1, which summarizes the experiments that Glauser and his students and colleagues have been working on since the early 2000s. Note that these data sets have continued to increase in size as capability of the data acquisition systems has improved and the cost of storage has decreased—so much that currently data sets routinely reach multiple Terabytes (TB) in size. Similar trends have occurred in the computational arena as well, and other papers in this volume are focused on this aspect of the turbulence big data challenge.

The challenges of archiving, processing, and sharing these big data sets are routinely encountered. For example, for sharing TB plus data sets, in many cases it is easiest to put the data on a 5 TB removable disc drive and send the data to co-workers via next day air rather than try and transfer it through a network. In this paper, the focus is not on these particular big data challenges, but rather on the use of big data analysis tools: in particular those used for extracting key low-dimensional flow physics (the needle) from the high-dimensional big data (the haystack).

In this paper, the focus is on high-*Re* and Mach number jets, both axisymmetric and more complex. We will discuss the general efforts to apply big data analyses/modeling tools (the "right filters") to identify patterns and predictive models rather than just a posteriori trends, statistics, and distributions. These advanced tools include POD, stochastic estimation, optimal inferred decomposition, wavelet analysis, and Lagrangian coherent structure (LCS) methods which are used for understanding, modeling, and controlling such flows. Additionally, compressive sensing has also been utilized to examine high-dimensional airfoil data [2]. However, those results are not presented here and the reader is instead referred to other papers in this volume (e.g., Bai et al.) which discuss this approach in some detail.

One final connection to modern computer science tools is relevant. Machine learning is gaining some traction in the turbulence community (see Noack et al. in this volume). In the Syracuse group (Pinier et al. [34, 35]), a "machine learning-like" approach has been used, described by Perret et al. [33] and termed the lower order dynamical system (LODS) method. This assumes, for example, quadratic non-linear terms (inspired by Navier–Stokes) in a set of N differential equations (where N is low-dimensional) for the POD time-dependent coefficients, $a(t)$. Experimental data (or LES/DNS data) is obtained with sufficient time resolution so that the linear and quadratic coefficients of these equations are extracted by curve fitting or "training/learning." Such LODS can then in principle be used for understanding, prediction, and control of the turbulent flow-field. Note this is a complement to the approach of Aubry et al. [1], Glauser et al. [12], and Ukeiley et al. [47], where the linear and quadratic coefficients are obtained directly from the empirical POD eigenfunctions. It is the authors' view that using Navier–Stokes based approaches, either full (LES, DNS) or low-dimensional (such as LODS), and treating their solutions as "information" in the big data sense, when combined with machine learning methods, would be a reasonable approach moving forward in using modern big data tools to help gain insight into the difficult problem of turbulence.

16.1.1 Scope

Many researchers have generated immense data sets over the years through experimental, computational, and analytical means. Borrowing from these volumes, select examples from Syracuse University are compiled, with the focus on the broad perspective of how to handle the cumbersome amount of fluid dynamics information. While the goal of the fluid dynamicist is to get at the physics of the flow, the researcher is often confronted with difficulties in treating big data. In high-*Re* flows, this problem becomes enormous: the flow-field is turbulent, three-dimensional, unsteady, encompasses several orders of magnitude in scale, and often occupies a vast spatial domain. When experiments and simulations present the researcher with a wealth of information, knowing where to probe the flow becomes one of the greatest challenges—the so-called needle in a haystack. In an effort to reduce this struggle, we therefore communicate and discuss the effectiveness of certain tools that can be used to analyze complex field quantities.

16.2 The Axisymmetric Jet

Jet noise is a physical example that both requires and generates big data. The complex flow-field and huge spatial domain occupied by the acoustics simply *cannot* be simulated at full resolution [13]. For over 60 years, jet noise has been known to be a direct consequence of turbulence [27, 28]. Thus, producing high-*Re* flows that can be studied in a laboratory environment is critical to furthering the understanding of this deeply complex problem. The axisymmetric jet is the most fundamental of these flows, and despite years of high-quality research on this configuration, the exact mechanisms of noise generation still elude us.

To handle such difficult physics, a great deal must be known about the flow (the big data challenge). Unfortunately, one problem of turbulence is that the range of scales inherently increases with *Re* [9]. For example, in a laboratory setting, it is not uncommon for spatio-temporal fluid fluctuations to range 3–4 orders of magnitude. This poses a problem for both simulations and experiments: capturing the correct physics requires resolving the fluid mechanics in both extreme resolution and over a large domain. Such a task is intractable, however, and so one tries to minimally simulate and probe the flow while maintaining critical levels fidelity.

From 2003–2014, our anechoic chamber housed an axisymmetric, convergent nozzle to perform jet noise research. Particle imaging velocimetry (PIV) is a vital instrument for these studies, and many campaigns have been carried out, seen in Table 16.1.

Table 16.1 lists the series of jet experiments performed at Syracuse University. The third column describes the primary approaches of the researchers to handling the difficult problem. POD refers to proper orthogonal decomposition, LSE is linear stochastic estimation, LODS is for low-order dynamical systems training, and OID is observable inferred decomposition. One recognizes from the table that, as acquisition capabilities increase with time, the sizes of the data sets grow. Until recently (2014), the jet of study has been axisymmetric. Currently, a complex nozzle with multifaceted geometry is under investigation. Already, more than 1 TB exists for this configuration, and the data continue to compound.

The data generated from these experiments sum to multiple terabytes (TB). For example, the 2011 and 2013 TRPIV experiments consisted of 87 different test cases. Each run produced approximately 5 GB of raw data. An additional 10–15 GB is typically generated in post-processing these cases, which puts the total information in the 1–2 TB range. Large-window particle imaging velocimetry (LWPIV) and pressure sampling are less data intensive, but require very long run times to generate necessary statistics. Regular sizes of these data sets are approximately 1–5 GB/case.

Synchronized with near- and far-field pressure measurements, these experiments yielded large databases of various flow quantities. The largest consumers of storage space are easily the optical imaging measurements. While the additional instruments (e.g., far-field microphones, near-field pressure transducers) are critical to the research, their point-wise measurements produce far less data than field techniques such as particle imagine velocimetry (PIV) and schlieren imaging. In particular, time-resolved (TR) methods can produce a rich data set in a matter of seconds. The axisymmetric jet and TRPIV system are shown in Fig. 16.1.

Fig. 16.1 10 kHz TRPIV sampling the axisymmetric jet in Syracuse University's Skytop Turbulence Lab

16.2.1 Time-Resolved PIV

With the development of more powerful lasers, time-resolved PIV (TRPIV) systems have become more widely available to the university-level laboratories. The system selected to acquire these data sets operated a Quantronix Hawk-Duo Nd:YAG laser. In combination with a Photron FASTCAM CCD, two-component velocity measurements were taken at various planes with a sampling frequency of 10 kHz. As a majority of the research is motivated by jet noise studies, the chosen planes typically focus on the collapse of the potential core [37]. In the axisymmetric jet, a viewing window of $1.5D \times 1.5D$ was used, where D is the diameter of the jet. After performing the necessary PIV computations, a maximum vector gird resolution of 96×96 could be achieved.

The laser and cameras of Fig. 16.1 were arranged so that different planes of interest could be readily scanned. An example of one TRPIV set of test sampling different $r - z$ planes is given in Fig. 16.2. Each plane is sampled for approximately 1 second, creating near 10,000 sequential snapshots and generating over 50 GB of data. And this succession of tests is but one chapter out of a larger campaign volume. Thus, it is easy to see how quickly experiments sum to the large data sizes given in Table 16.1. At this point, one must extract physically relevant, low-dimensional information from high-dimensional databases. Thus, optimally efficient tools must be applied.

POD [30] is employed via the snapshot method [40] on the axisymmetric jet data in the $r - z$ planes as shown in Fig. 16.2. The center plane (i.e., the jet axis) is focused on here. The goal of POD is to breakdown a spatio-temporal field quantity, generally

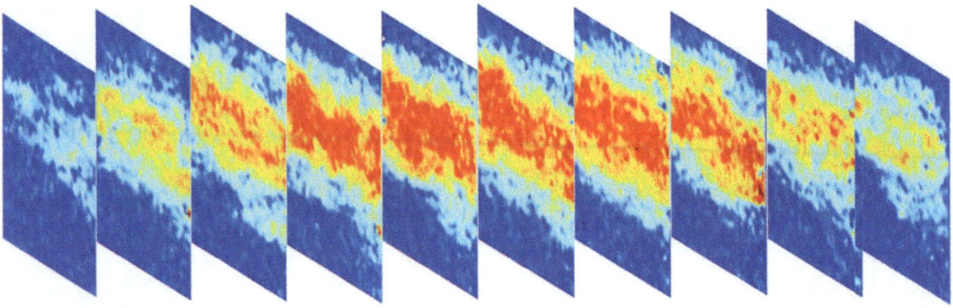

Fig. 16.2 Instantaneous streamwise velocity contours for off-center plane measurements: ordinate: r/D; abscissa: x/D

$u_i(\mathbf{x}, t)$, into time-dependent coefficients, $a_n(t)$, and spatial basis functions, $\phi_i^{(n)}(\mathbf{x})$. With 10 kHz TRPIV, the POD techniques can be used to tie certain energetic flow structures to far-field pressure signals in the time domain. After decomposing the velocity field, the time-dependent coefficients of the velocity data are correlated to the time-dependent signals from the far-field. These correlations provide insight into what structures in the flow contribute the most to noise production.

OID is a reduced-order modeling technique, developed within a Galerkin framework, to identify a cause and effect relationship for turbulent flows. The approach uses additional reduced-order models techniques (such as POD), to optimally identify acoustically weighted flow features (source) as a function of an observable subspace constructed from a discrete set of linearly related "observables" (far-field microphones). The technique provides a linear mapping between the most energetic velocity modes and the most energetic features of the far-field acoustics. One advantage of this approach is that the entire far-field is used, rather than a single microphone. Furthermore, OID focuses on modes which are highly correlated with noise rather than optimizing based on energy, as POD does.

Having isolated certain points in time that correlate highly to the far-field signal via OID, the velocity field measurements can be utilized to identify how the flow structures behave at that point in time. Figure 16.3 shows the POD reconstruction at two time instances. In this case, only the mode that correlated best with the far-field noise, Mode 6, is used for reconstruction. The instances in time with high correlations between the velocity field and the far-field pressure, Fig. 16.3a, show a contortion in the flow reconstruction. During the time stamps that show very little relation between the velocity field and the far-field pressure, Fig. 16.3b, the Mode 6 reconstruction appears very similar to a mean velocity field for the axisymmetric jet. If the TRPIV time support is used to observe the entire event, a "whiplash" like event is witnessed [3].

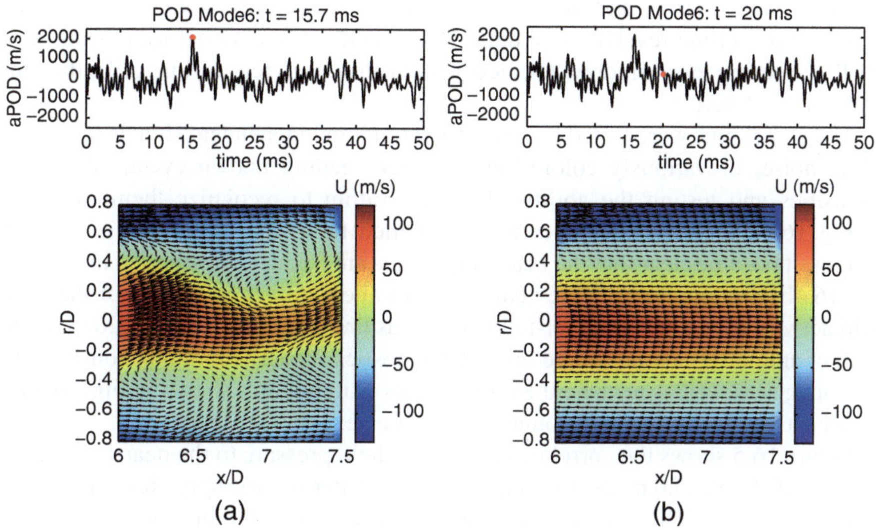

Fig. 16.3 Selective flow-field reconstruction based on loud-mode identification (i.e., using ϕ^6) of the Mach 0.6 axisymmetric jet. (**a**) Loud event. (**b**) Normal acoustic output

16.2.2 Pattern Recognition and Event Extraction

Event-extraction algorithms [24] enable pattern recognition to match individual events in noisy signals measured at distinct locations. Cross-correlations and wavelet analyses are performed on the flow-field to locate frequency bands of interest. The search-and-match is performed in the time frequency domain, implying that the data are band-pass filtered (with a frequency resolution varying from 4 to 10 per octave, over the 3 to 4 octaves that seem physically relevant to the phenomenon). The identification of coherent near-field sources is relatively easy using the far-field signature of acoustic events in the cone of propagation. The events of interest are simply marked as louder-than-background at the same frequency, so the matching consists of comparing lists of such candidate events from several microphones. Checking for consistency allows one to pin down the small differences in propagation times from the presumed source to the various microphones; and by triangulation, to estimate the location of the individual sources.

A more challenging problem is the pattern recognition involving far-field microphone signals and loud events, as above, and near-field kinematic events, which cannot be assumed to be energetic in their own right. The algorithm, called TFL (Time-Frequency-Lag) [20, 21, 26] consists of sliding the band-pass filtered signals (many of them, as above) past each other to adjust the lag, and look for maxima of their cross-correlation—not on average but for individual peaks. Relative to the length of the original signals, a multiplication factor of approximately 20 for the frequency bands of interest is used, while a factor 50 or so for physically plausible lags is used, and at least a half-dozen near-field kinematic diagnostics extracted from time-resolved PIV data. This is performed at two dozen locations in the PIV interrogation window. The needle in the haystack felt, at times, like the sand grain in the desert.

The algorithms were tested by generating random independent signals (Gaussian white noise, or variously colored noise), then seeding known events of various intensities and testing the ability of the algorithm to recognize them as shown in Fig. 16.4a [21]. As the procedure is applied to various locations in the PIV images, it became clear that the near-field events are frequently observed in clusters, Fig. 16.4b, rippling through the near field. An alternative presentation of the near-field activity for a given loud far-field event was first presented at TSFP 2013, in the form of animated movies, where several diagnostics are active simultaneously in the PIV images as well as near-field Kulite sensors (still pictures shown in Fig. 16.5a, b for an active spot and a background uncorrelated example, respectively).

Figure 16.5 shows the correlation of the far-field pressure to the near-field velocity using different diagnostic techniques (e.g., Q-criterion, vorticity, Reynolds stress, divergence, etc.). These diagnostics are calculated at every instance to preserve the temporal information, thus creating even larger databases. Their usefulness is apparent though: at two different instances in time, the stark differences in activity are observed. The left figure shows a cluster of different velocity events linked to far-field noise (matching the Fig. 16.3a), while the right demonstrates no correlation to the far-field (in agreement with Fig. 16.3b). The information from the TRPIV experiments demonstrate how intermittent are these highly correlated events and allows for the tracking of their evolution in both space and time.

A key challenge with the TRPIV measurements and their associated analyses is the reduced window size which is limited to about 1.5 diameters of the jet. As a result, a series of large-window 4 Hz PIV measurements were performed to cover a large spatial window ($\sim 6D$) as described below.

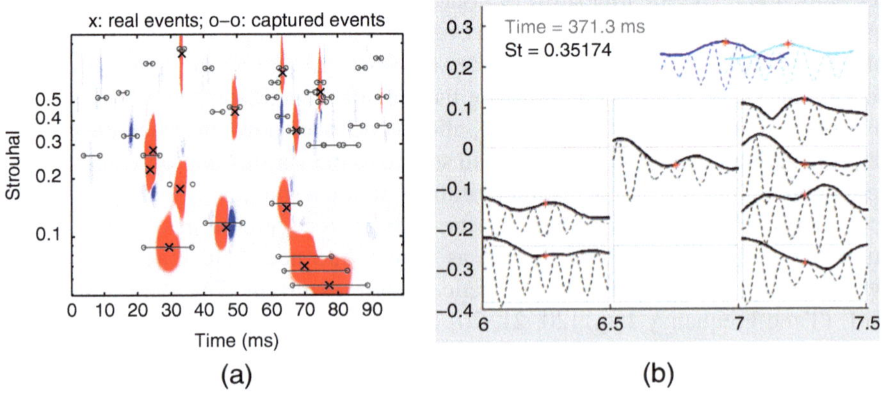

Fig. 16.4 An illustration of the event-extraction algorithm (*left*, **a**) and one instance of the event cluster extracted (*right*, **b**)

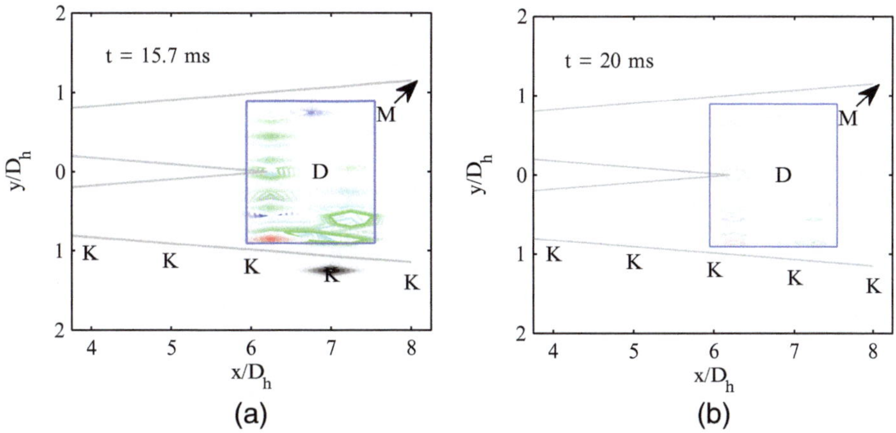

Fig. 16.5 Correlation of the far-field pressure to the near-field velocity using different diagnostic techniques [25]. (**a**) Loud event. (**b**) Normal acoustic output

16.2.3 Large-Window PIV

Referring to Table 16.1, the LWPIV cases have a relatively small data size. While the use of TRPIV or LWPIV is motivated by their own strengths and weaknesses, this instrument is, in a way, an example of selective data filtering at the experimental level. If one is interested in exclusively employing a reduced-order model (as in these experiments), sampling statistically independent realizations is beneficial. High-quality data are produced at the laboratory level, and relatively little post-processing is necessary. The trade-off to this approach is that the temporal piece of information is sacrificed and the test times are lengthy. However, as demonstrated below, this alternative approach has its advantages in ascertaining key pieces of a complicated puzzle.

The multiple-camera setup of the LWPIV captures phase-locked velocity data at three separate location and allows the merging of instantaneous vector fields. Figure 16.6 shows a snapshot of streamwise instantaneous velocities (*a*), mean velocities (*b*), and centerline velocities (*c*) for the jet operating at $M_j = 1.1$. Fluctuating mean velocities indicate the existence of shock cells typical of underexpanded jets. The jet is then further analyzed via a reduced-order model.

Because the window size of the instrument acts as a spatial filter, the LWPIV setup allows for the identification of larger structures within the jet. The large amount of independent images ($N \sim 2500$) taken during these experiments produces well-converged statistics and gives confidence that, at a minimum, the first 100 spatial modes resulting from POD are representative of fluctuating features in the flow [32]. These spatial modes (ϕ_i) can be thought of as building blocks of the flow [11], and are powerful indicators of turbulent events. The streamwise velocity

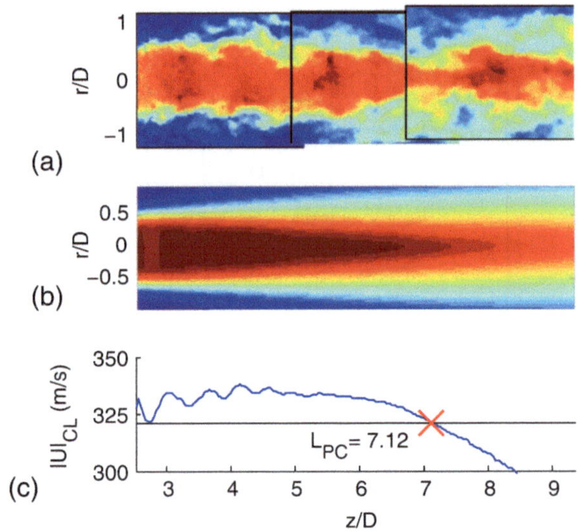

Fig. 16.6 Streamwise velocity component of the $M_j = 1.1$ jet. Separate instantaneous flow-fields (**a**) are merged and averaged (**b**), and the centerline velocity is then extracted (**c**) to illustrate shock cells and define the potential core length

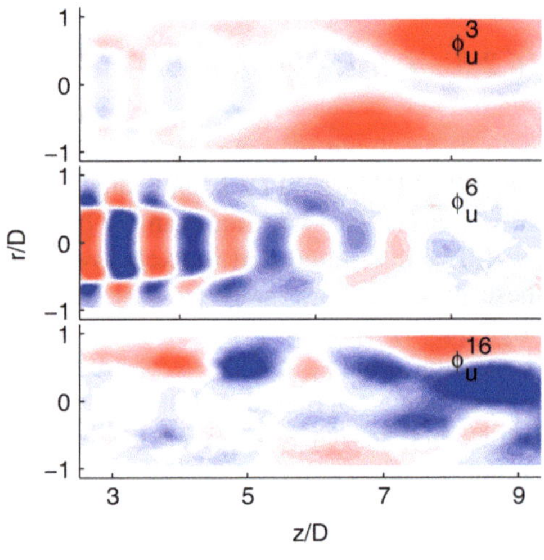

Fig. 16.7 Select eigenfunctions of the $M_j = 1.1$ data, showing modes associated with a large-scale instability wave, screech, and turbulent mixing noise

component of Modes 3, 6, and 16 is selected as examples of POD's practicality in LWPIV and plotted in Fig. 16.7. These modes have been selected through a "loud-mode" analysis, similar to the routine used in the TRPIV experiments. However, in the case of supersonic jet noise, there are additional acoustic features that can be discretely isolated (e.g., screech tones); Modes 3, 6, and 16 are related to particular characteristics of imperfectly expanded jets.

By viewing the top image in Fig. 16.7, it is clear that a large-scale wave persists across the domain of multiple cameras. This structure is thought to be related to a helical instability wave associated with imperfectly expanded jets [36, 42]. In Mode 6, regions of compression and decompression are thought to be associated with the shock cell nature. Fluctuating velocities contained within the potential core appear to be tied to axisymmetric structures in the shear layer, likely vortex rings. Spatial analyses of this mode yield wavenumbers that can be directly linked to the screech tone of this jet, thereby linking Mode 6 to the feedback mechanism [43] in supersonic jets. Finally, Mode 16 displays evidence of the structures merging around the collapse of the potential core. Rigorous analyses of the temporal coefficients correlate this (and similar) mode(s) to turbulent mixing noise; this reinforces the qualitative result of Mode 16 that increased turbulence associated with the terminating potential core contributes to acoustic production. A more in-depth discussion is available in a recent paper [32]. Note that with the TRPIV described in Sect. 16.2.1, the streamwise window of only 1.5 diameters has limited the ability to extract the rich streamwise modal structure that was extracted with the large-window PIV of over six jet diameters presented here.

As Table 16.1 demonstrates, the axisymmetric jet has been studied at multiple Mach numbers. In the search for loud modes (i.e., specifically ones from turbulence associated with the collapse of the potential core), similarities are sought across different exit velocities. Spatial correlations amongst eigenfunctions are cross-referenced with the loud modes of each case, and a few parallels are identified between modes of the Mach 1.0 and Mach 1.1 jets [31]. This suggests that the reduced-order models are extracting similar flow structures that can be tied to physics of the flow, further demonstrating the practicality of these methods. Finally, one can omit modes associated with features unique to one case (e.g., screech in the underexpanded jet) and compare it to another (e.g., the perfectly expanded jet) for additional, qualitative verification.

By removing the screech-containing modes, the remaining modes (shown to be highly spatially correlated to the sonic case) ought to reproduce a flow-field similar to the $M_j = 1.0$ jet. Figure 16.8 shows instantaneous flow-fields from the $M_j = 1.0$ case using reconstructed TPRIV data, *(a)*, and the $M_j = 1.1$ data as produced from selective reconstruction of the LWPIV *(b)*, where modes 6–9 have been omitted. The TRPIV data uses velocity information from seven different cameras and merges

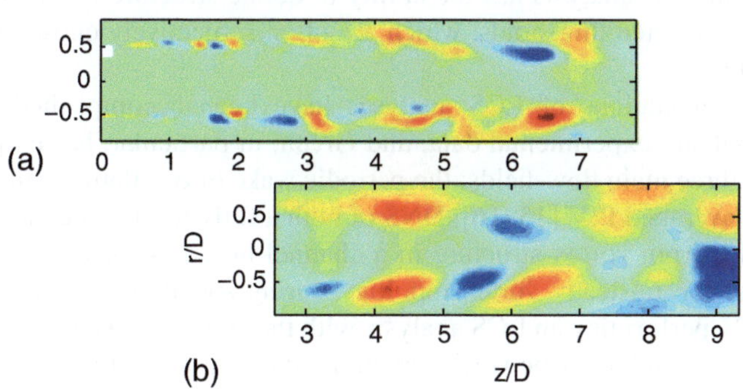

Fig. 16.8 Reconstruction of fluctuating velocities for snapshots of **(a)** $M_j = 1.0$ from TRPIV data and **(b)** $M_j = 1.1$ from LWPIV data with the shock modes omitted

the fields together through a POD modal basis, discussed in a paper by Berger et al. [4]. Identical matching would be lucky given the nature of turbulence, yet similarities between the two data sets do exist such as activity near the potential core. Most importantly, shock-related features of the underexpanded jet are almost entirely absent. Vortices clustered around the collapse of the potential core and located in the shear layer are essentially all that remain, akin to a sonic or subsonic jet.

16.2.4 Lagrangian Coherent Structures

LCS have proven to be a powerful tool for the purposes of identifying and tracking coherent structures in complex vortex-dominated fluid flows. This method has been used for both periodic and aperiodic flow-fields, and can be implemented using velocity field data from both numerical simulations and experiments. To use this method, we first calculate the finite-time Lyapunov exponent, a scalar field that measures local stretching among particle trajectories as they are advected forward in time. The local maximizing ridges of this field are the positive-time LCS (pLCS), and indicate the inherent repelling material lines of the flow-field. If this calculation is performed backward in time, we obtain the negative-time LCS (nLCS), which indicate the attracting material lines in the flow. When used together, the LCS give the relevant transport barriers in the flow, which in the case of vortex-dominated flow-fields are the vortex boundaries.

Commonly used vortex identification analyses, including vorticity or other scalar fields such as the Q or Δ criteria, are largely Eulerian: they are computed using the instantaneous velocity field and its derivatives. For this reason, the Eulerian criteria can be simpler and faster to compute, but share a few disadvantages when compared to the LCS. Computation of the FTLE field is not restricted to the spatial resolution of the velocity data and requires no velocity derivatives, and therefore LCS can provide greater detail and are often less sensitive to errors in the velocity data set. In addition, an LCS analysis has the ability to define structure boundaries without relying on preselected thresholds, which are often subjectively chosen when using Eulerian criteria.

Past implementations of LCS analyses have been accomplished using both computational and experimental data, and Green, in particular, has employed it in the study of three main flow-fields: the periodic wake of oscillating panels captured using phase-averaged PIV [15, 16], DNS of both a fully turbulent channel flow and one isolated hairpin vortex structure in a channel flow [14], and more recently, a simulation of the unsteady vortex topology during shedding in largely separated flows [19]. By performing an LCS analysis with these studies, we have been able to, not only identify and describe vortex creation, destruction, and evolution in a more quantitative manner, but also to pinpoint some of these phenomena in a way that was impossible using Eulerian criteria.

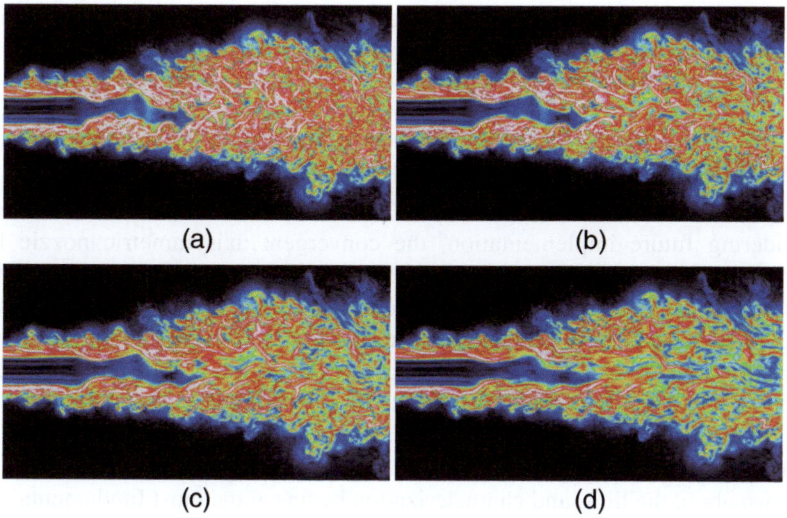

Fig. 16.9 Preliminary FTLE fields from LES of turbulent jet. $M = 0.9$, $Re = 400,000$. White ridges of the FTLE fields are the LCS. (**a**) $\Delta t^\star = \Delta t/\tau = 0.39$. (**b**) $\Delta t^\star = \Delta t/\tau = 0.77$. (**c**) $\Delta t^\star = \Delta t/\tau = 1.16$. (**d**) $\Delta t^\star = \Delta t/\tau = 1.93$

In order to implement an LCS analysis, time-resolved velocity is required such that interpolation can be used to recreate velocity fields between data sets. Preliminary FTLE analysis on the turbulent jet was done using large-eddy simulation (LES) results of a jet at a Mach number of 0.9 and Reynolds number of 400,000 [7, 8]. The LES data were saved at a frequency of 80 kHz, downsampled from a database over 10 TB in size. If a non-dimensional time scale is defined, $\tau = D/2Mc_0$, where M is the Mach number, c_0 is the speed of sound, and D is the nozzle diameter, the non-dimensional time between data sets for the LES results is $\Delta t^\star = \Delta t/\tau = 0.39$. If instead, only every other data set is used, the "sampling frequency" of the data set would be 40 kHz, and $\Delta t^\star = 0.77$. Similarly, FTLE fields calculated from data sets of $\Delta t^\star = 1.16$ and $\Delta t^\star = 1.93$ can also be generated.

The preliminary FTLE results generated using the LES data at the four non-dimensional time steps are shown in Fig. 16.9. These computations require big data, as each time step in the figure is generated from over 2 GB of output LES data. It is immediately apparent that with an increase in $\Delta t^\star$ comes a decrease in the level of LCS detail, especially near the centerline where the velocity is highest. In these regions, it is not only the level of detail that is degraded with increasing time step; the size and shape of even the larger-scale coherent structures can change. Therefore, the accurate identification of coherent structures is degraded for too large a non-dimensional time step, showing the need for finer time resolution and hence larger data sets. Furthermore, a loss of detailed structure is apparent in the high velocity region along the centerline even at a $\Delta t^\star$ as low as 0.77. Currently, work is focused on characterizing this degradation of Lagrangian calculations for temporally underresolved velocity data in a fully turbulent application.

16.3 The Complex Jet

Considering future implementation, the convergent axisymmetric nozzle is an unlikely candidate for high-performance aircraft. Engine technologies, enabling increased mission capabilities and vehicle performance, have evolved nozzles into complex geometries with intricate flow features. A nozzle representative of future engines has been designed and installed in our anechoic chamber for research into modern jet configurations. It consists of a supersonic core, airframe/aft deck integration, the addition of a tertiary stream (located just above the aft deck), and asymmetric geometry. With the large complexity associated with this jet, very little is known about the flow, and characterization begins at the most fundamental level. This new flow therefore provides a unique opportunity to implement effective means of data extraction and attempt new methods. For example, azimuthal symmetry cannot be assumed as in the axisymmetric jet. Huge volumes of data are generated to characterize this diverse flow, and this non-intuitive problem requires a robust method for identifying regions of interest in a reasonable time frame.

16.3.1 Large-Eddy Simulations

Numerically, the big data problem overwhelms even the fastest computers: they simply cannot output the data fast enough. Therefore, sections of data are preselected and output at regular intervals. Thus far, simple sections of pressure (i.e., scalar) parallel and perpendicular to the axis of the jet have been treated: even for short records, at several GB per record, internet transfer of the data cannot be taken for granted, and shipping of recording devices was found to be a cost-effective alternative.

In conjunction with the experimental work, LES have been performed on the complex jet using the JENRE code developed by the Naval Research Lab (NRL). The large data and computational requirements of CFD are well known. In this research, the experimental observations have demonstrated that the grid requires extreme fidelity to capture the finest flow details. For example, increased resolution is necessary along the inner shear layer to capture a high-frequency oscillation (discussed later). Computations have been performed on different grids [38, 39], from coarse (12 million points) and fine (40 million points), to the most current simulation, which is operating on 60 million nodes. As an example of the big data problem associated with CFD, this most recent data set is 2.5 TB (after downsampling the computational domain from a staggering 22 TB). On this scale, the traditional tools of data analysis become unrealistic.

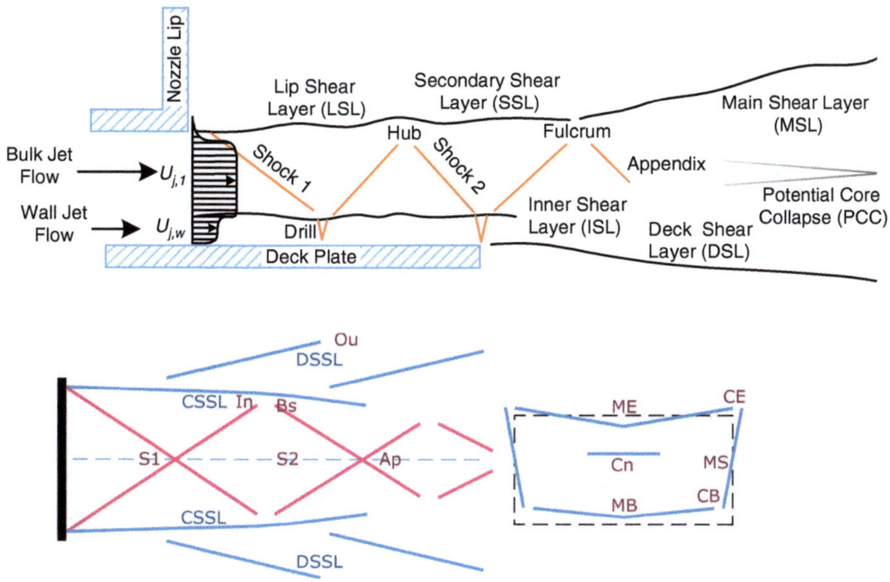

Fig. 16.10 Flow structure in the near field (*top*: major axis $(x-y)$; *bottom-left*: minor axis $(x-z)$; *bottom-right*: cross-plane $(y-z)$ plane)

Through both simulated and experimental observations, a schematic of the flow-field is established (shown in Fig. 16.10). The tertiary stream (or wall jet) is introduced internally and exhausts just above the deck plate. A series of shocks are formed, and a variety of "hot-spots" are identified. These locations are identified based on hypothesized importance; by cross-correlating quantities at each location with the surrounding flow-field, each regions is believed to reveal some information about the jet.

The simulation results of the 40 million-noded grid are probed at the various regions of interest. Figure 16.11 shows the spectra of selected points [22]. The different spectra legend entries are: CE (Corner Edge), ME (Mid Edge), Mid Base (MB), Center (Cn), Corner Base (Cb), Middle Side (MS); their relative locations are given in Fig. 16.10. Figure 16.11 compares two different downstream locations, the left at two hydraulic diameters (*YZ*2) and the right at four hydraulic diameters (*YZ*4). In the side shear layers, a weak low-frequency peak, presumably of Kelvin–Helmholtz origin, is soon overwhelmed by broad-band mixing layer turbulence. However, a 2.5 kHz (St $\sim$ 0.24) peak appears in the shocks in the plane of symmetry (Cn), and in the shear layer opposite the deck. As the probes move downstream, the flow pattern evolves: possible corner vortices exist in the high-frequency range (St $= 0.7 \sim 1.5$), while more downstream (right plot), the flow becomes more mixed and all the spectra follow the same pattern. There is no doubt that the big data is essential to unravel the complexity of this flow: in fact, the numerical simulation

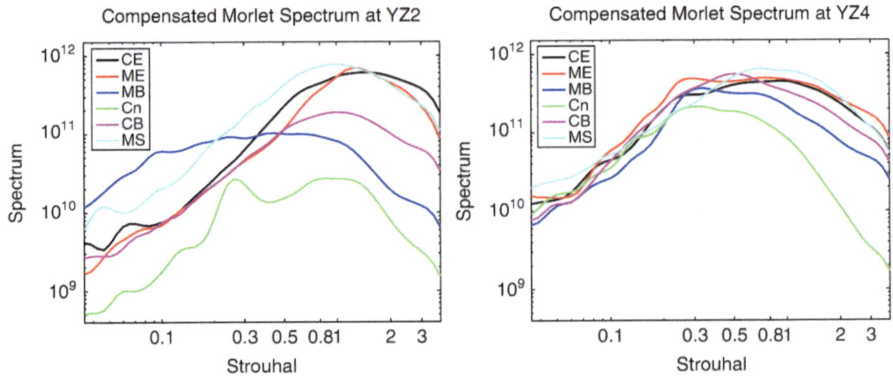

Fig. 16.11 Compensated pressure spectra in the cross-planes at two downstream locations, $x = 2D_h$ (*left*) and $x = 4D_h$ (*right*)

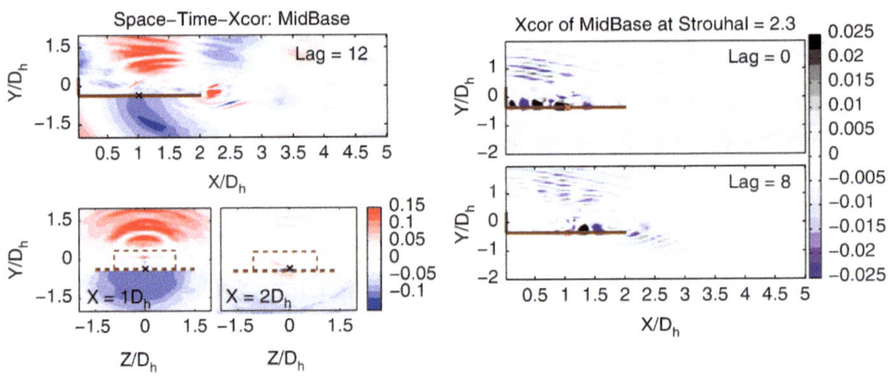

Fig. 16.12 Structures and flow interactions observed using space-time correlation (*left*) and frequency-resolved correlation (*right*)

and the experiments are viewed as complementary, and neither of them is likely to collect the necessary information in the absence of the other.

A more comprehensive view is to look at the space-time correlations (Fig. 16.12, left). A movie of this, with the time-lag monotonically increasing (presented at AIAA SciTech 2016 [22]), visualizes the propagation of various correlations that are supposedly linked to structures. At the instance shown here, the correlation level between the hot spot (marked by an "X") and pressure fields is shown in a 2D contour plot and the different orientations are shown in separate frames. At the lag time chosen, an apparent structure is observed off the edge of the deck plate. From the time evolution of the 2D fields of correlation levels, regions are observed propagating or convecting in various directions. Some of the regions can be attributed to evolving structures or radiating sound waves. For example, the structure downstream of the deck plate is seen to originate upstream and travel along the inner shear layer.

The left contours in Fig. 16.12 show the space-time correlation between the MidBase at $x/D_h = 1$ and three planes: the symmetry plane $(x - y)$ and two $y - z$ planes. The correlations at other planes have been checked as well and they are not shown since there are very limited regions of activity. Mainly three kinds of activities have been observed. The first kind of activity is very localized regions pulsing and moving downstream along the Deck Shear Layer. The second kind of activity is the concentric circular regions centered at the intersections of shocks and shear layers and at the deck edge (bottom-left). This is speculated to be part of the sound waves since they originate from fixed points (the intersections) and the radius of the wave expands as time evolves. The last group of activity is made up by two large regions that appear to move upstream, one outside the Upper Shear Layer and at $x/D_h < 2$ and the other below the deck. These regions show periodic patterns of alternating blue and red and they persist in most of the space-time correlations. The propagation of regions along the shock structures is also dimly observable. This shows up as the blue/red regions appearing/disappearing along the shocks.

To clear up the wave patterns, a continuous wavelet is applied to band-pass filter the signals before the correlation (Fig. 16.12, right). The Morlet wavelet is used to decompose the signals of the hot spots and the entire pressure fields into their frequency components. Then the signals at the same frequency level are correlated without normalizing the signals to unit variance. This is because the normalized correlations will untruthfully enhance low-frequency information. Thus for each time lag and at a specific frequency, a 2D distribution of the frequency-resolved correlations will highlight the regions that are correlated at that instant. The data produced in this process are obviously large; however, the method is powerful. In the sequence of increasing time lag, the evolution of the correlation patterns frame-by-frame is seen. This permits the identification of the frequency band in which the activity occurs. For the pulsing along the inner shear layer discussed in the left figures, the high-frequency range (St $= 2.3$) is observed to have the peak correlations, while its presence is greatly reduced at other frequencies.

Aside from the high-frequency, there is a considerable activity near 2.5 kHz (St $\sim$ 0.24). The flow patterns exhibit certain periodicity in both the raw pressure field and the space-time correlations here. This leads to phase-averaging the data, which are plotted in Fig. 16.13. The level of in-phase relation is estimated using the RMS value of the phase-averaged pressure field (right plot). There is guidance for future research in these large data sets: the dark red regions are areas to focus on for placing additional probes (both experimental and numerical) for further investigation.

It turns out that the unimpressive 2.5 kHz pulse present in the side and top shear layers and in the inner oblique shocks organizes the near-jet region as well as the far-field acoustic statistics. The phase from the data themselves is defined, taking the phase angle of the complex Morlet coefficients at the chosen frequency at a given point. This is then used as the reference for phase-averaging the entire near-field plane of symmetry. An illustrative result (Fig. 16.13, left) shows the pervasive phase-locking well beyond the edge of plate. An exploration of the best-possible reference points yields an entire patch (dark red colors in Fig. 16.13, right) as yielding large phase coherence for the entire near-jet.

This is physical information of considerable relevance to the understanding of the flow, specifically the interactions between its organizing features: the shocks, the shear layers, and the corner vortices. Additionally, these findings provide insight as to how the skeleton might be manipulated to affect specific objectives in relation to flow control for far-field acoustics or other performance statistics. For these goals to be met, the physics have to be extracted from the big data.

16.3.2 Time-Resolved Schlieren

To capture the relevant physics of this complex flow-field, the acquisition capabilities of the SU anechoic chamber were upgraded. The faster flow requires greater sampling frequencies, the asymmetric shape means that more viewing orientations must be considered, and the increased *Re* and Mach number warrant the need for improved spatial measurements (both in domain size and resolution). To meet these needs, a time-resolved schlieren instrument was installed and the data acquisition speeds of the near- and far-field pressure transducers were improved (contributing further to the big data problem). Extracting an interesting flow phenomena from the immense data set produced by these experiments exemplifies how one finds the "needle in a haystack."

Microphone data were the first to detect an interesting spike at high-frequencies, something that had not previously been observed in the original LES data nor hypothesized by researchers. Existing at 34 kHz, the tonal signature was soon found in all forms of measurements, raising questions of its origin. Only with the high sampling frequencies of the time-resolved schlieren and pressure transducers was this important flow feature discernible. However, as with the TRPIV experiments,

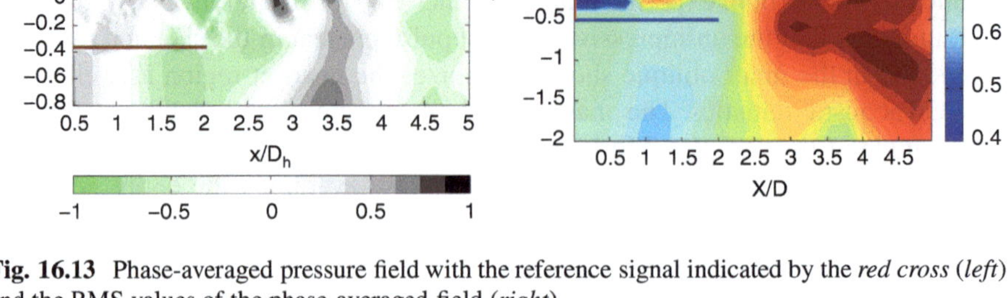

Fig. 16.13 Phase-averaged pressure field with the reference signal indicated by the *red cross* (*left*) and the RMS values of the phase-averaged field (*right*)

these fast sampling rates (up to 400 kHz) produce an enormous amount of data in a fraction of a second, and sifting through this volume is nontrivial.

The time-resolved schlieren data set produced additional challenges not previously encountered. In this case, the major bottle-neck was the number of files, rather than the size. Separate test cases contain up to 1.3×10^6 files; with this many images, indexing errors in the operating system and programs often resulted in impractically slow transfer rates, memory dumps, and sometimes crashes. Special measures must be taken to work around these issues and extract the meaningful pieces of information.

Because the jet lacks the symmetry of the axisymmetric one (here, there is only a single plane of symmetry), schlieren data are acquired at different orientations, shown in Fig. 16.14. In the sideline view, the deck plate is stainless steel and is instrumented with an array of high-speed pressure transducers (visible in the bottom-left corner), which also detected the 34 kHz signal. Rotating the jet, the deck plate was interchanged with a polycarbonate piece to view the flow immediately exiting the jet. Approximately 1.2×10^5 images were acquired in each direction for these cases.

From only two still schlieren images, the flow-field is clearly rich in detail and diverse in phenomena. A series of oblique shocks are observed leaving the jet exit in the left of the images, intersecting near $x/D_h \sim 1.1$, and traversing to the shear layers, approximately aligned with the edge of the deck plate. Reflected shocks can be seen further downstream. However, the flow appears much more turbulent once it has cleared the deck plate and the shocks are weaker. The side shear layers, found at $z/D_h \sim \pm 1$, are seen to grow in width, entraining ambient fluid, and consisting of a range of coherent structures. Acoustic waves are clear in the ambient fluid of the sideline image, with waves propagating from the nozzle exit (top) and deck plate (bottom). Fine-scale turbulent mixing noise appears at the steepest degree, approximately forming a vector from $(x = 0, y = 0.25D_h)$, angled at 70° upward. Mach wave radiation, acoustic waves generated from supersonically convecting structures, is also seen leaving the top shear layer, emanating at approximately 25° downstream [44]. Fine shocks are found below the first oblique shock, and the third stream is observed just above the deck plate as a more disorganized flow.

POD is performed on the schlieren data of both windows, and this is not a trivial task. The spatial grids of these high-res images consist of $3 \sim 4 \times 10^5$ elements. Thus, the eigenvalue problem (forming the correlation matrix) of POD is an extremely expensive computation, which cannot be performed on a desktop computer. Reducing the grid size is one solution to this, but valuable information is lost in this process. Additionally, one must downsample time-resolved data to obtain statistically independent samples (a requirement of snapshot POD). While this is always performed on the TRPIV data, there are "only" 10,000 images to deal with for each case. The larger number of files in the schlieren data previously mentioned often results in crashes because of indexing errors. It is becoming increasingly important for today's experimentalists to also be clever programmers.

Loosely speaking, the solutions used to handle these data issues essentially reduce to some type of blocking operation, a technique that has long been used

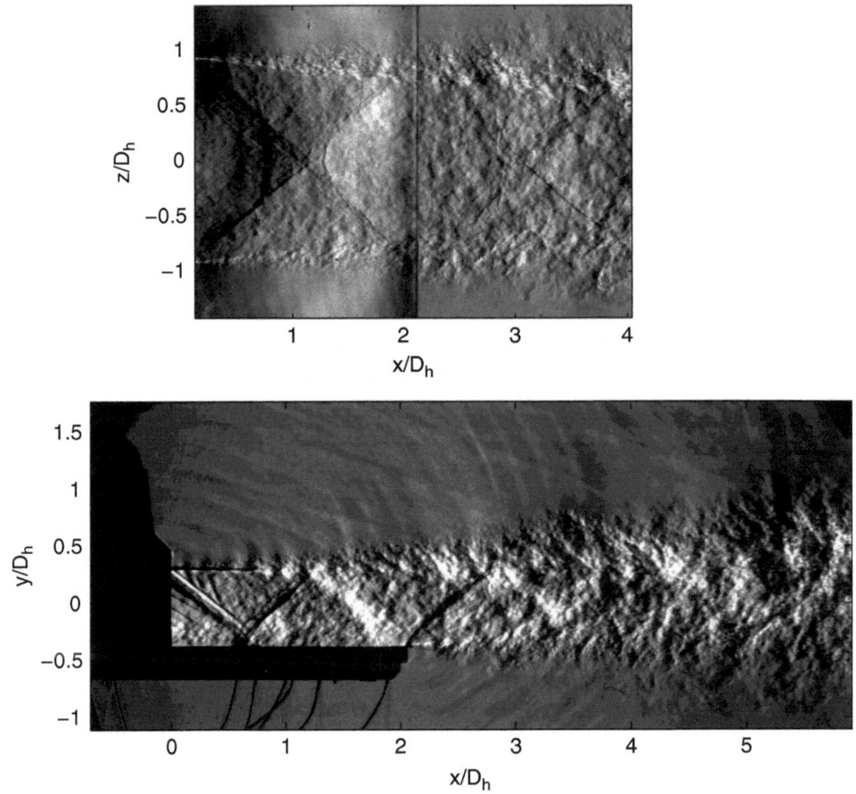

Fig. 16.14 Snapshot schlieren images showing complex shocks, turbulence, and acoustic propagation

by the computationalists. Operations are called in subsets, the desired quantity is stored, and the preallocated memory is then cleared after each iteration. There is no work-around for this—the data sets are simply too large to handle manually. For example, the POD operation must be solved in two parts: the first uses a temporally downsampled data set (to ensure statistical independence) to calculate the ϕ's, and the second then projects these computed values back onto a truncated set of the time-dependent fields to calculate the $a(t)$'s. After implementing more robust codes, the POD results are found to again bring clarity to the apparent randomness of the flow. Select spatial modes are plotted in Fig. 16.15 for the $x - z$ orientation and in Fig. 16.16 for the $y - z$ orientation.

Modes 3–8 are plotted in Fig. 16.15. Structures appear downstream of the deck plate, starting at $x/D_h > 2$. Mode 3 shows a column-type mode (similar to the first spatial Fourier mode of an axisymmetric jet [17]). Apparently, the deck plate delays this oscillatory behavior. The higher modes transition to smaller structures in the side shear layers. These are thought to be associated with corner

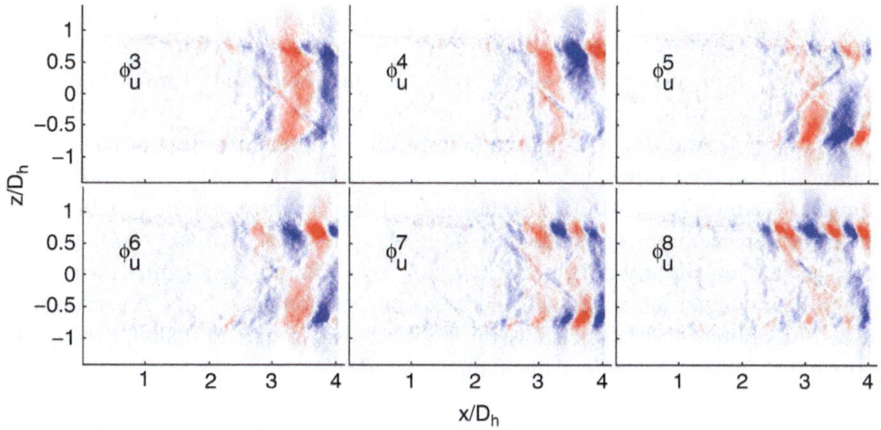

Fig. 16.15 Select spatial modes of schlieren data in the $x - z$ orientation

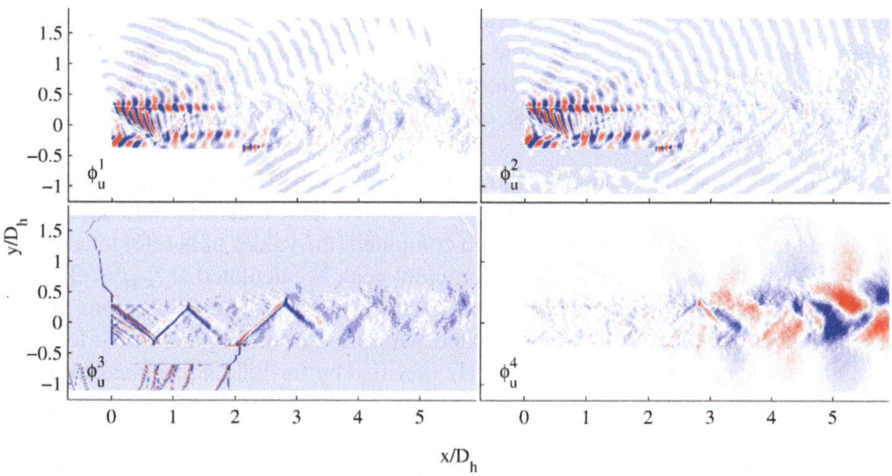

Fig. 16.16 The first spatial modes of schlieren data in the $x - y$ orientation, showing coherent structures and acoustic radiation

vortices and Kelvin–Helmholtz structures in the side shear layers. In each of the modes considered, the shocks are highlighted. Because POD extracts fluctuating quantities, the shocks must therefore be oscillating at some level. Note that the first two modes appear granular and are therefore omitted. Explanations for this are still being considered, but they may be associated with high energy density in the fluctuating shocks and/or randomly scattered light. Pre-processing to smooth optical noise may also reduce this, but again, as the computations are expensive and the data relatively fresh, this is an ongoing research effort.

The first four spatial modes of the $x - y$ planes is found in Fig. 16.16. Acoustic waves are immediately apparent in the directions previously discussed. Additionally, structures in the flow are extracted. In phase with these are a series of pulses downstream of the first oblique shock. The second mode shows the rich dynamics of a vortex train in the deck shear layer and the oscillations formed along the top shear layer which are apparently tied to the fine-scale acoustics and Mach wave radiation. Additionally, there are strong fluctuations traveling along the first oblique shock. The third stream mixing layer, first seen leaving the nozzle exit at ($x/D_h \sim 0, y/D_h \sim -0.3$), also consists of significant coherent structures that persist far downstream. The influence of the oblique shock on these vortices is also seen, as they are slightly deflected near $x/D_h \sim 0.8$ before persisting downstream. The third mode demonstrates that there is some randomness in the measurements across the oblique shocks (possibly due to the sensitive optical paths). Finally, the forth mode appears to reflect the flapping mode that was observed in the other orientation (Fig. 16.15).

With the capabilities of the schlieren setup, sampling rates up to 400 kHz were possible using reduced window sizes. This sampling rate is of the same order of magnitude as the Kolmogorov scales in this flow. The entire flow region over the deck was scrutinized at these acquisition speeds. However, only a select few have been sufficiently analyzed due to the demanding computations associated with the large data set. Thus, two windows were considered: the first is the upper nozzle lip, and the second is directly below that, capturing the flow immediately above the deck plate at the nozzle exit.

Figure 16.17 shows a snapshot (top-left) of the 400 kHz data at the nozzle lip. Time-series of 6.5×10^4 density probes are extracted at the locations indicated by the colored markers, and the spectra are then computed (the values of N refer to arbitrary indices, used for bookkeeping). The dominant peak is calculated at $f_{\max} = 34$ kHz, and the subsequent peaks are its associated harmonics. The oscillations in the density of the ambient fluid are associated with pressure waves and one infers that acoustic waves are propagating at 34 kHz (verified by far-field noise measurements [6]). This mark is also found in the shear layer (N = 4160), traveling along the first oblique shock ($N = 6054$), downstream of the oblique shock ($N = 14{,}757$), an upstream shock ($N = 4122$), and in the core of the flow ($N = 1543, N = 4100$). The high frequency of this peak makes structural vibrations and oscillations in electronic equipment lousy candidates. Additionally, resonators were sought, but a similar length scale for this frequency could not be found in the geometry. Similar results are observed in the second small window studied (bottom of Fig. 16.17). This is a snapshot of the flow above the deck plate, which captures the third stream. The oscillations along the oblique shock ($N = 8276$) appear strongest, with the probes in the wall jet shear layer ($N = 2580 - 2600$) giving the next highest peaks. Fluctuations in the upstream shock at $N = 7110$ apparently have a relatively lower amplitude, while the value increases again as that shock merges with the third stream shear layer. Because the flow is supersonic, upstream propagation of information is not permitted and transverse communication is limited by the convection speed. Thus, it is probable that an oscillation occurs where both the upstream shock and third stream shear layer are previously in contact with one another.

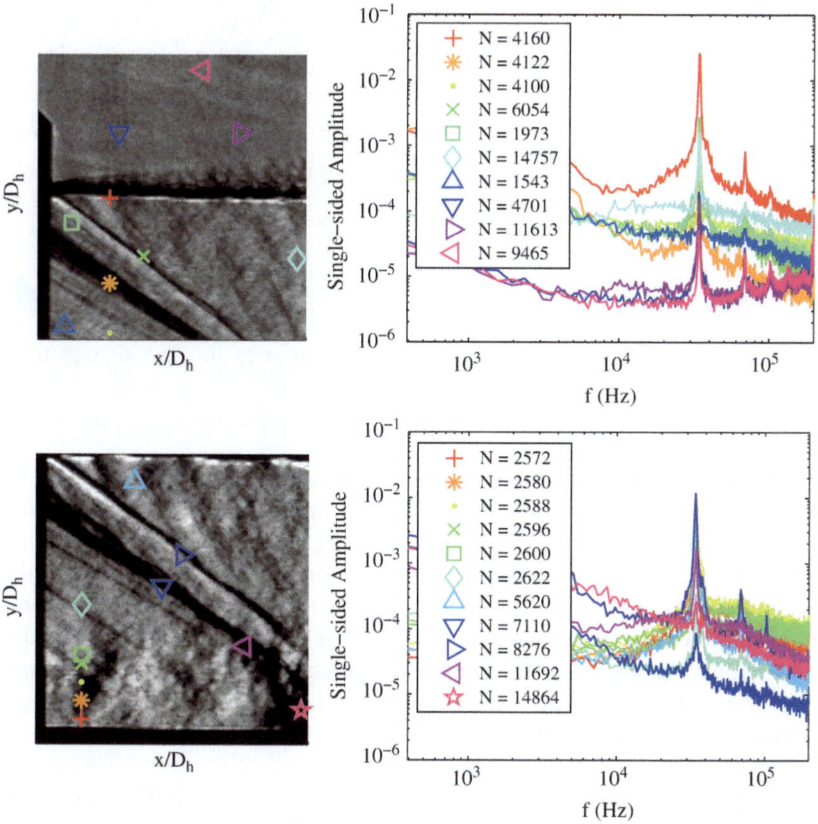

Fig. 16.17 Close-up schlieren of the upper nozzle lip (*top*) and just above the deck plate (*bottom*), taken at 400 kHz with time-series extracted at select locations and corresponding spectra computed

Fig. 16.18 Reduced window POD: upper nozzle lip

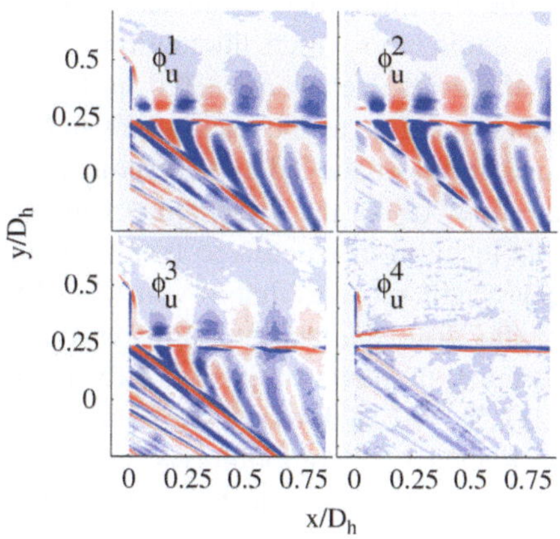

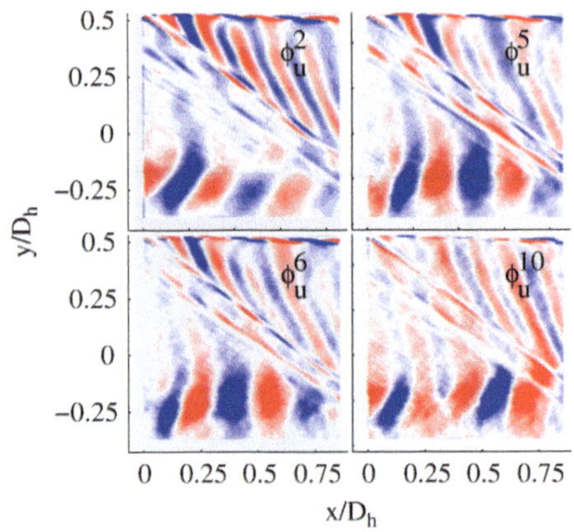

Fig. 16.19 Reduced window POD: deck shear layer

POD is again performed, and Figs. 16.18 and 16.19 show selected spatial modes of the 400 kHz data. The increased resolution allows the POD to separate the structures further. Considering Fig. 16.18, the top shear layer is divided into an inner and outer regime. Symmetric structures form on the outer side, while the inner fluctuations are clearly tied to the disturbances traveling along the oblique shock. From the first three modes, it appears that the disturbances are "shedding" off the oblique shock with the convective velocity. Spatial frequency analyses are performed by calculating wavenumbers from the POD modes and using bulk velocities to determine frequencies [32]. These calculations confirm that these waves exist at approximately 34 kHz.

Moving to the deck plate, POD of the second window, Fig. 16.19, identifies a series of coherent structures in the shear layer. Additionally, the shocks again appear to transfer disturbances. Note how the perturbations downstream of the first oblique shock are angled. This suggests that the velocity at which disturbances propagate in the oblique shock are faster than in the core flow. In a sense, the shocks act as waveguides in which fluctuations are transferred. Finally, the coherent structures in the third stream, extracted by the POD and probed via time spectra, have the same signature as the oscillations in the top shear layer. A Kelvin–Helmholtz (KH) instability is believed to be responsible for generating this 34 kHz signal throughout the flow. Using the dimension of the splitter plate and the relative bulk velocity at this location as calculated from PIV measurements, a Strouhal number of 0.26 is calculated, close to that of von Kármán vortex shedding. Given the compressibility of this flow, unequal pressures and densities of the merging streams, and reported Strouhal numbers of $St = 0.4$ for large-scale structures in supersonic shear layers [23], $St = 0.26$ does not seem unreasonable.

Recent LES computations on the 60 million-node grid have reinforced the proposed mechanism generating the 34 kHz signals, and it is illustrated in Fig. 16.20. Additionally, simulations at Ohio State University, running on 44 million nodes and producing 1.7 TB of data (downsampled from 17 TB), have resolved this mechanism in further detail [41]. A disturbance, generated from the KH instabilities, propagates

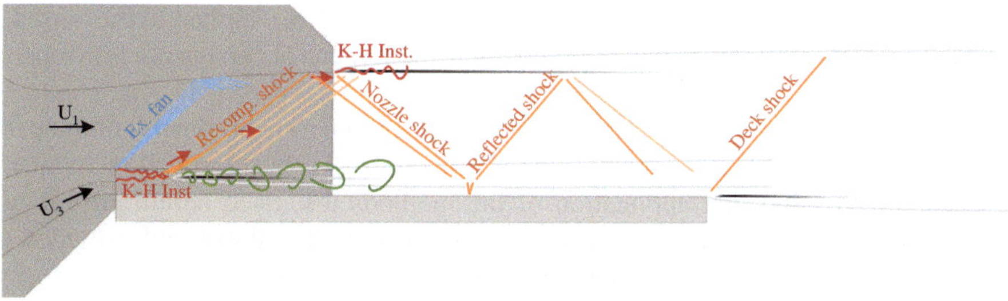

Fig. 16.20 Proposed mechanism of the 34 kHz signal

along the recompression shock to the top boundary layer in the SERN and out into the top shear layer of the jet. Along this path, it pulses the flow at this oscillation frequency (as observed throughout the high-speed schlieren data at all measured locations). Because shear layers are unstable, they are susceptible to a range of inputs. Therefore, the KH instability in the top shear layer likely amplifies the 34 kHz signal generated from the splitter plate and the high intensity observed there is explained. The vortex train found in the bottom shear layer results from the merging of the two streams of different velocities and the splitter plate (i.e., vortex shedding). This is an exciting theory, because it suggests that a single point (the splitter plate) modulates a vastly large spatial domain, *everywhere*. In the context of big data, the carefully chosen tools (e.g., experimental and computational sampling, spectral analyses, and POD) have directed researchers to one extremely important piece of information—the "needle in the haystack."

16.4 Discussion

We have shown, using high *Re* and Mach number turbulent jets, that turbulence research activities involve the collection of massive amounts of data from multi-scale computer simulations and large-scale experiments. Such data sets routinely produce multi-modal data (velocity, pressure, temperature, etc.) surpassing multiple terabytes. As an example in the experimental field, time-resolved PIV data are acquired at 10 kHz, and schlieren images are sampled at rates up to 400 kHz. Along with the fine spatial resolution and large domain of these instruments, which is necessary to capture the broad range of scales observed in high *Re* flows, immense volumes are rapidly generated. At this level, the capacity to handle the data and extract meaningful knowledge is limited with traditional tools.

In this paper, we did not focus on the big data challenges of storage and transfer, but rather on the use of analysis tools, in particular those for extracting key low-dimensional flow physics (the needle) from the high-dimensional big data (the haystack). The low-dimensional tools used include POD, stochastic estimation, optimal inferred decomposition, wavelet analysis, and LCS. Applied to the various jet databases, these methods have revealed rich phenomena occurring not previously observed.

The data requirements of complex physics and engineering problems continue to maximize the available technology, and the utilization of robust analysis techniques is essential. We find that application of the methods discussed herein affords us a great deal of insight into the physics of these complex flows and provides a reasonable path forward for further understanding and control.

Acknowledgements The authors would like to acknowledge two funding sources that made this work possible: (1) an SBIR Phase I and II project with Spectral Energies, LLC, and the AFRL turbine engine division under the direction of program manager Dr. Barry V. Kiel, and (2) an AFOSR grant, number FA9550-15-1-0435, under the guidance of program manager Dr. R. Ponnappan.

References

1. N. Aubry, P. Holmes, J.L. Lumley, E. Stone, The dynamics of coherent structures in the wall region of a turbulent boundary layer. J. Fluid Mech. **192**(1), 115–173 (1988)
2. Z. Bai, T. Wimalajeewa, Z. Berger, G. Wang, M. Glauser, P.K. Varshney, Low-dimensional approach for reconstruction of airfoil data via compressive sensing. AIAA J. **53**(4), 920–933 (2014)
3. Z.P. Berger, The effects of active flow control on high-speed jet flow physics and noise. Ph.D. thesis, Syracuse University (2014)
4. Z. Berger, M. Berry, P. Shea, M. Glauser, N. Jiang, B. Noack, S. Gogineni, E. Kaiser, A. Spohn, Analysis of high speed jet flow physics with time-resolved piv, in *52nd AIAA Aerospace Sciences Meeting* (1226) (2014)
5. Z. Berger, P. Shea, M. Berry, B. Noack, S. Gogineni, M. Glauser, Active flow control for high speed jets with large window piv. J. Flow Turbul. Control **94**(1), 97–123 (2015)
6. M.G. Berry, A.S. Magstadt, M.N. Glauser, C.J. Ruscher, S.P. Gogineni, B.V. Kiel, An acoustic investigation of a supersonic, multi-stream jet with aft deck: characterization and acoustically-optimal operating conditions, in 54th *AIAA ASM* (AIAA, San Diego, 2016), vol. 2321022
7. A.V. Cavalieri, G. Daviller, P. Comte, P. Jordan, G. Tadmor, Y. Gervais, Using large eddy simulation to explore sound-source mechanisms in jets. J. Sound Vib. **330**, 4098–4113 (2011)
8. G. Daviller, Étude numérique des effets de température dans les jets simples et coaxiaux. Ph.D. thesis, Institut P', CNRS - Université de Poitiers (2010)
9. W.K. George, Lectures in turbulence for the 21st century. University of Oslo. http://www.uio.no/studier/emner/matnat/math/MEK4300/v13/undervisningsmateriale/tb_16january2013.pdf. Accessed 15 Jan 2016
10. M.N. Glauser, Coherent structures in the axisymmetric turbulent jet mixing layer. Ph.D. dissertation, University of Buffalo (1987)
11. M. Glauser, W. George, Orthogonal decomposition of the axisymmetric jet mixing layer including azimuthal dependence, in *Advances in Turbulence*, ed. by G. Comte-Bellot, J. Mathieu (Springer, Heidelberg, 1987), pp. 357–366
12. M. Glauser, X. Zheng, C.R. Doering, The dynamics of organized structures in the axisymmetric jet mixing layer, in *Turbulence and Coherent Structures* (Springer, Dordrecht, 1991), pp. 253–265
13. M.E. Goldstein, Aeroacoustics of turbulent shear flows. Annu. Rev. Fluid Mech. **16**(1), 263–285 (1984)

14. M.A Green, C.W. Rowley, G. Haller, Detection of Lagrangian coherent structures in three-dimensional turbulence. J. Fluid Mech. **572**, 111–120 (2007)
15. M.A. Green, C.W. Rowley, A.J. Smits, Using hyperbolic Lagrangian coherent structures to investigate vortices in bioinspired fluid flows. Chaos **20**, 017510 (2010)
16. M.A. Green, C.W. Rowley, A.J. Smits, The unsteady three-dimensional wake produced by a trapezoidal pitching panel. J. Fluid Mech. **685**, 117–145 (2011)
17. A.M. Hall, An experimental investigation of low-dimensional techniques for large scale noise source characterization in a heated jet. Ph.D. dissertation, Syracuse University (2008)
18. J.W. Hall, A.M. Hall, J.T. Pinier, N.G. Mark, Cross-spectral analysis of the pressure in a mach 0.85 turbulent jet. AIAA J. **47**, 54–59 (2009)
19. Y. Huang, M.A. Green, Detection and tracking of vortex phenomena using Lagrangian coherent structures. Exp. Fluids **56**(7), 1–12 (2015). doi:10.1007/s00348-015-2001-z. http://link.springer.com.libezproxy2.syr.edu/article/10.1007/s00348-015-2001-z
20. P. Kan, J. Lewalle, G. Daviller, Comparison of near-field events and their far-field acoustic signatures in experimental and numerical high speed jets, in *International Symposium on Turbulence and Shear Flow Phenomena (TSFP-8)* (2013)
21. P. Kan, J. Lewalle, Z.P. Berger, M. Glauser, The properties and localizations of acoustic sources of high speed jet, in 53rd *AIAA ASM*, Kissimmee (2015). AIAA-2015-0737
22. P. Kan, C.J. Ruscher, J. Lewalle, M.N. Glauser, S. Gogineni, B.V. Kiel, Extracting near-field structures related to noise production in high speed jets, in 54th *AIAA ASM*, San Diego (2016). AIAA 2016-0004
23. J. Lepicovsky, K. AHUJA, W. Brown, R. Burrin, Coherent large-scale structures in high Reynolds number supersonic jets. AIAA J. **25**(11), 1419–1425 (1987)
24. J. Lewalle, K.R. Low, M.N. Glauser, Properties of individual jet noise sources identified from far-field pressure data. Int. J. Aeroacoust. **5–6**(11), 651–674 (2012)
25. J. Lewalle, P. Kan, S. Gogineni, Mach-number dependence of acoustic source properties in high speed jets – Part I: ensemble statistics of acoustically active regions, in *52nd AIAA Aerospace Sciences Meeting* (1228) (2014)
26. J. Lewalle, P. Kan, S.P. Gogineni, Mach-number dependence of acoustic source properties in high speed jets–Part I: ensemble statistics of active regions, in *52nd AIAA ASM*, National Harbor (2014). AIAA-2014-1228
27. M.J. Lighthill, On sound generated aerodynamically. I. General theory. Proc. R. Soc. Lond. Ser. A Math. Phys. Sci. **211**(1107), 564–587 (1952)
28. M.J. Lighthill, On sound generated aerodynamically. II. Turbulence as a source of sound. Proc. R. Soc. Lond. Ser. A Math. Phys. Sci. **222**(1148), 1–32 (1954)
29. K. Low, Z. Berger, S. Kostka, B. El Hadidi, S. Gogineni, M. Glauser, A low-dimensional approach to closed-loop control of a mach 0.6 jet. Exp. Fluids **54**, 1–17 (2013)
30. J.L. Lumley, The structure of inhomogeneous turbulent flows, in *Atmospheric Turbulence and Radio Wave Propagation*, ed. by A.M. Yaglom, V.I. Tatarsky (Nauka, Moscow, 1967), pp. 166–178
31. A.S. Magstadt, M. Berry, Z. Berger, P. Shea, C.J. Ruscher, S.P. Gogineni, M. Glauser, An investigation of sonic & supersonic axisymmetric jets: correlations between flow physics and far-field noise. J. Flow Turbul. Control (in review, 2016)
32. A.S. Magstadt, M.G. Berry, Z.P. Berger, P.R. Shea, M.N. Glauser, C.J. Ruscher, S. Gogineni, An investigation of sonic & supersonic axisymmetric jets: correlations between flow physics and far-field noise, in *Turbulent Shear Flow Phenomenon 9* (2015)
33. L. Perret, E. Collin, J. Delville, Polynomial identification of POD based low-order dynamical system. J. Turbul. **7**, N17 (2006)
34. J.T. Pinier, Low-dimensional techniques for active control of high-speed jet aeroacoustics. Ph.D. dissertation, Syracuse University (2007)
35. J.T. Pinier, M.N. Glauser, Dual-time piv investigation of the sound producing region of the controlled and uncontrolled high-speed jet, in *Advances in Turbulence XI: Proceeding of the 11th EUROMECH European Turbulence Conference*, vol. 117 (2007), pp. 392–394
36. A. Powell, On the mechanism of choked jet noise. Proc. Phys. Soc. Sect. B **66**(12), 1039 (1953)

37. H.S. Ribner, The generation of sound by turbulent jets, in *Advances in Applied Mechanics*, vol. 03 (Academic, New York, 1964), pp. 103–182
38. C. Ruscher, S. Gogineni, K. Viswanath, B. Kiel, M. Berry, A. Magstadt, M. Glauser, Aeroacoustic validation of a simulation for a complex nozzle, in *51st AIAA/ASME/SAE/ASEE Joint Propulsion Conference*, Orlando (2015)
39. C.J. Ruscher, A.S. Magstadt, M.G. Berry, M.N. Glauser, P.R. Shea, K. Viswanath, S.P. Gogineni, B.V. Kiel, A.J. Giese, Analysis of a supersonic 3-stream jet flow, Part I: nozzle design, experiments, and simulations. AIAA J. (in review, 2016)
40. L. Sirovich, Turbulence and the dynamics of coherent structures, part I: Coherent Structures. Q. Appl. Math. **45**, 561–571 (1987)
41. C.M. Stack, D.V. Gaitonde, L. Agostini, M.G. Berry, A.S. Magstadt, M.N. Glauser, Numerical investigation of a supersonic multistream jet with an aft-deck, in *AIAA SciTech* (2016). AIAA 2016-2058
42. C.K. Tam, On the noise of a nearly ideally expanded supersonic jet. J. Fluid Mech. **51**, 69–95 (1972)
43. C.K. Tam, Supersonic jet noise. Annu. Rev. Fluid Mech. **27**, 17–43 (1995)
44. C.K.W. Tam, Mach wave radiation from high-speed jets. AIAA J. **47**(10), 2440–2448 (2009)
45. C.E. Tinney, M.N. Glauser, L.S. Ukeiley, Low-dimensional characteristics of a transonic jet. Part 1. Proper orthogonal decomposition. J. Fluid Mech. **612**, 107–141 (2008)
46. C.E. Tinney, L.S. Ukeiley, M.N. Glauser, Low-dimensional characteristics of a transonic jet. Part 2. Estimate and far-field prediction. J. Fluid Mech. **615**, 53–92 (2008)
47. L. Ukeiley, L. Cordier, R. Manceau, J. Delville, M. Glauser, J.P. Bonnet, Examination of large-scale structures in a turbulent plane mixing layer. Part 2. Dynamical systems model. J. Fluid Mech. **441**(1), 67–108 (2001)

Chapter 17
Data-Driven Methods in Fluid Dynamics: Sparse Classification from Experimental Data

Zhe Bai, Steven L. Brunton, Bingni W. Brunton, J. Nathan Kutz, Eurika Kaiser, Andreas Spohn, and Bernd R. Noack

17.1 The Importance of Data Science in Fluid Dynamics

Fluid dynamics plays a central role in numerous scientific, industrial, and technological applications, including transportation (planes, trains, and automobiles), energy (wind, tidal, and combustion), and mixing (medicine, chocolate), to name only a few. Understanding and controlling fluid flows provides an opportunity to dramatically improve performance in these systems, resulting in lift increase, drag reduction, and mixing enhancement, all of which further important engineering goals of the modern world [1]. Rapidly developing methods in data science, largely

Z. Bai • S.L. Brunton (✉)
Department of Mechanical Engineering, University of Washington, Seattle, WA 98195, USA
e-mail: sbrunton@uw.edu

B.W. Brunton
Department of Biology, University of Washington, Seattle, WA 98195, USA

J.N. Kutz
Department of Applied Mathematics, University of Washington, Seattle, WA 98195, USA

E. Kaiser
Départment Fluides, Thermique, Combustion, Institut PPRIME, CNRS – Université de Poitiers – ENSMA, UPR 3346, F-86036 Poiters CEDEX, France

A. Spohn
Institute PPRIME, CNRS – Université de Poitiers – ENSMA, UPR 3346, Département Fluides, Thermique, Combustion, CEAT, 43 rue de l'Aérodrome, F-86036 Poitiers CEDEX, France

B.R. Noack
LIMSI-CNRS, Rue John von Neumann, Campus Universitaire d'Orsay, Bât 508, F-91403 Orsay, France

Institut für Strömungsmechanik, Technische Universit́at Braunschweig, Hermann-Blenk-Straße 37, D-38108 Braunschweig, Germany

A. Pollard et al. (eds.), *Whither Turbulence and Big Data in the 21st Century?*,
DOI 10.1007/978-3-319-41217-7_17

borne out of the computer science, statistics, and applied mathematics communities, offer a paradigm shift in our ability to measure, model, and manipulate fluid flows.

Fluid flows are often characterized by high-dimensional, multi-scale, and nonlinear phenomena that evolve on an attractor. Although the Navier–Stokes equations provide a detailed partial differential equation model, it is often difficult to use this representation for engineering design and control. An insightful quote of Richard Feynman, in his lecture on fluid mechanics, summarizes the central dilemma [2]:

> The test of science is its ability to predict. Had you never visited the earth, could you predict the thunderstorms, the volcanos, the ocean waves, the auroras, and the colorful sunset?

Instead of analyzing equations in isolation, we collect measurements of flows in relevant configurations and develop a hierarchy of models to describe critical features of the flow, rather than every subtle detail. In particular, extracting large coherent structures in fluids has provided valuable insights. The proper orthogonal decomposition (POD) [3–6], which is often formulated using the singular value decomposition (SVD) [7–9], is a form of dimensionality reduction, which takes high-dimensional data from simulations or experiments and extracts relevant low-dimensional features. In many ways, these fundamental techniques in dimensionality reduction for fluids are among the first use of data science in complex systems. Importantly, many of the most successful methods in model reduction, such as Galerkin projection onto POD modes, take a *hybrid* approach, where data is combined with our knowledge of the Navier–Stokes equations to enforce dynamics [1].

Reduced-order modeling has been especially important in obtaining computationally efficient models suitable for closed-loop feedback flow control. Many competing design constraints factor into effective control design, although one of the most important considerations is the latency in making a control decision, with larger latency imposing fundamental limitations on robust performance [10]. Thus, as flow speeds increase and flow structures become more complex, it becomes increasingly important to make fast control decisions based on efficient low-order models. A major open problem in control theory, with particular relevance for flow control, is the optimal placement of sensors and actuators for a control objective.

Powerful new techniques from data science are poised to transform the analysis of complex data from dynamical systems, such as fluids. In particular, machine learning [11, 12] provides advanced capabilities to extract features and correlations. Sparse sampling techniques, including compressed sensing [13–18], sparse regression [19–21], and sparse classification [22–24], allow for the recovery of relevant large-scale information from surprisingly few measurements.

Here, we combine machine learning and sparse sampling for efficient measurement and characterization of a fluid system. An overarching goal is to reduce the burden of data acquisition and processing. Specifically, we apply sparse classification to fluid imaging. Flow visualization, such as particle image velocimetry (PIV) [25–27], is a cornerstone in fluid mechanics, providing an understanding of flow structures and mechanisms that may be manipulated by closed-loop feedback flow control. Real-time feedback control based on PIV is becoming increasingly feasible, although it remains expensive, both in hardware cost and computational power. The methods here are designed to extract valuable data from inexpensive camera images of bubble visualizations, and they may also be used with PIV to reduce the data required for reconstruction, resulting in higher sampling rates and more inexpensive time-resolved systems. Finally, we design optimal sensor locations for categorical decisions [24], which may be eventually used for control.

17.1.1 Recent Advances in Sparsity and Machine Learning for Fluids

Advanced methods from machine learning and compressed sensing have already begun to enter fluid dynamics. Unsupervised clustering has proven effective in determining probabilistic reduced-order models based on data from the mixing layer [28], and it has also been used to determine when to switch between various POD subspaces [29]. Graph theory has recently been applied to understand the network structure underlying unsteady fluids [30]. Finally, machine learning control, based on genetic programming [31], has been applied to numerous closed-loop flow control experiments with impressive performance increases that exceed many alternative control strategies [32–34].

Sparse sensing has rapidly been embraced by fluid dynamics researchers, most likely due to the ability to sample considerably less often than suggested by the Shannon–Nyquist sampling theorem [35, 36]. Although fluids data is typically quite large, is expensive to collect and analyze, and has a large separation of spatial and temporal scales, it generally has dominant low-dimensional structures that are useful for analysis and control. Compressed sensing has been applied in a variety of contexts in fluid dynamics [37–40]. Sparsity techniques have also been applied to the computation of the dynamic mode decomposition (DMD) [41–43], including promoting sparsity for mode selection [44], spatial compressed DMD [45], and non-time-resolved sampling strategies designed for PIV [46, 47]. The DMD is rapidly developing, with data science, machine learning, and control extensions [48–53]. Sparsity methods have also been applied more broadly in dynamical systems [54–57].

17.1.2 Fluids in the Era of Big Data

Fluid dynamics is one of the original big-data communities, routinely working with rich data sets that are large, unwieldy, and require high-performance computing to analyze and visualize. Big data means many different things to different people, and it is a vital part of the growing data science movement. For some, big data implies data management: databases, data scrubbing, archiving, and reproducible analysis. For others, big data implies scaleable architectures for applying advanced data analysis on growing volumes of data. Yet another perspective sees big data as a challenge to mine and visualize hidden patterns inside unwieldy high-dimensional data.

We are in a fortunate position where there is additional high-value data to collect for increasingly complex and engineering-relevant flows. This data will continue to be generated in higher detail and greater volumes with advances in computational and experimental techniques. As a community, we should view big data as a big opportunity to coordinate our data collection and analysis efforts to solve pressing real-world challenges. Continued efforts to collect increasing volumes of data must be met with advanced data analytics to extract the most value from these data. The computational methods also inform *how* we should collect data to maximize the useful information.

17.2 Experimental Description

Experiments are conducted in a low-speed water tunnel at the Institute PPRIME, Poitiers. The closed-circuit, free surface water tunnel has a test section of 2.1 m length, 0.5 m width, and 0.34 m height. The ramp model consists of a flat plate of length $L = 100$ mm followed by a smooth ramp of height 60 mm and length 600 mm. The model is 498 mm wide and spans the width of the test section, except for 1 mm gaps between the walls and the ramp. The ramp's leadings edge divides the oncoming flow into an upper stream following the ramp contour and a stream below the model. Downstream of the ramp, a horizontal plate prolongates the separated flow to reduce the impact of temporal changes in the flow structure during forcing. The stagnation point on the leading edge is controlled by adjustable pressure losses at the outlet of the upper stream. The Reynolds number is given as $Re = UL/\nu$ with respect to the free-stream velocity U, and the kinematic viscosity ν of water. A schematic of the experimental arrangement is shown in Fig. 17.1 (left).

Beginning from the leading edge, a laminar zero pressure gradient boundary layer develops along the flat plate. Above the smooth ramp this boundary layer separates under the influence of an adverse pressure gradient which is fixed by the shape of the ramp. Downstream of the flow detachment, the newly formed separated shear layer becomes unstable and undergoes a laminar-to-turbulent transition, allowing the flow to reattach. Between the wall and the separated main flow, recirculating fluid marks the extension of the laminar separation bubble (LSB). The ramp contour follows a polynomial shape of order 7 for which Sommer [58] numerically determined the position of the LSB.

To obtain a satisfactory spatial resolution of the visualized separated flow region, the Reynolds number is fixed to $Re = 7900 \pm 100$, for which the separation bubble extends over more than 50 % of the ramp length. Locally controlled forcing is enabled by a stainless steel wire of 0.13 ± 0.01 mm in diameter and supported by an oscillating holder. The wire crosses the span of the model and is located at 90 ± 2.5 mm downstream of the leading edge. A vertical sinusoidal motion of the wire is imposed using a line servo (RS-2 modelcraft) piloted by an Arduino-Due microprocessor. The frequency is varied between 0.1 and 3 Hz to enable the actuation of the Kelvin–Helmholtz instability along the separated shear layer. In all experiments, the oscillation amplitude is set at 3 ± 1 mm and the mean vertical position of the oscillating wire is assigned to 3.5 ± 0.5 mm above the ramp model. According to the preliminary tests these settings proved to be highly efficient to excite the Kelvin–Helmholtz instability.

Flow visualizations are obtained using the hydrogen bubble technique [59]. For that purpose, a 0.050 ± 0.005 mm thick stainless steel wire deformed into a zigzag pattern is fixed in the middle of the ramp at 300 ± 5 mm downstream of the leading edge. When applying a negative potential, between 30 and 90 V, hydrogen bubbles are produced at the wire and convected downstream. A computer controlled function generator is employed to trigger the release of bubbles to obtain periodic timelines. These timelines mark the position of the separated shear layer and patches related to the rolling up of tracer particles by vortical structures during reattachment, as shown in Fig. 17.1 (right) for the baseline and controlled cases.

The images have a resolution of 2116 × 812 pixels, and they are acquired at 10 Hz. During recording of an image sequence over more than 100 seconds, uncontrollable electrochemical reactions close to the electrodes degrade the bubble production, leading to a slight increase of the diameter of the freshly released bubbles. Furthermore, during their progression in the downstream direction, the bubbles shrink. Therefore, the intensity of light reflections and contrast change in time and space during an image sequence. In the following analysis, we classify baseline and control cases using the full image data, with lighting changes, etc., and we also use an isolated data set that consists of a short sequence of images with constant lighting and bubble density. Throughout, these will be referred to as "Full Data" and "Isolated Data" with the modifiers "Baseline" or "Controlled."

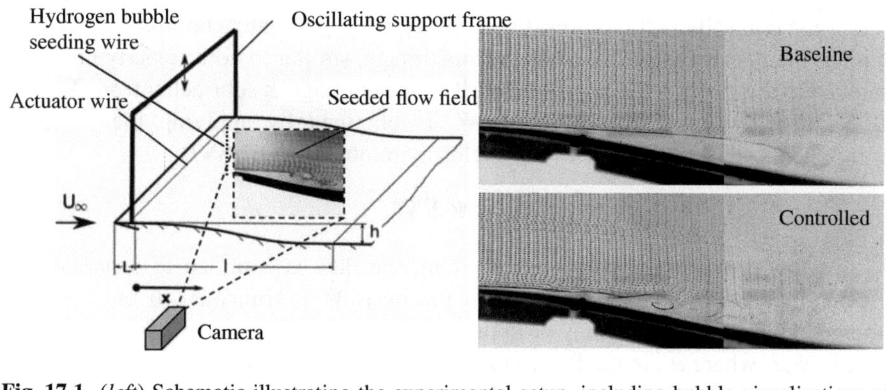

Fig. 17.1 (*left*) Schematic illustrating the experimental setup, including bubble visualizations of the separated flow past a backward facing ramp. (*right*) Bubble visualizations for flow past a ramp are shown for the baseline case (*top*) and the case with control (*bottom*)

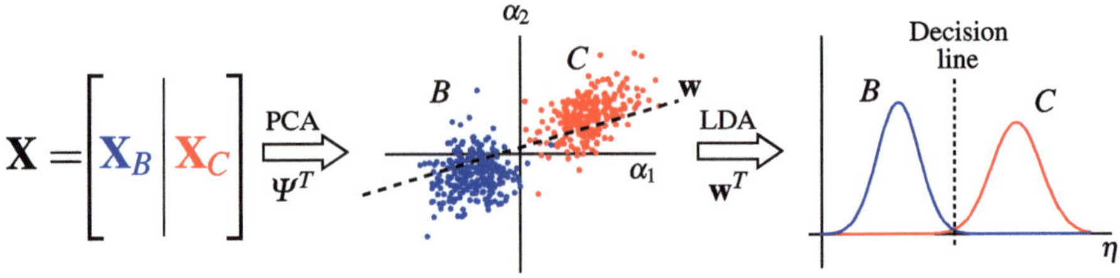

Fig. 17.2 Schematic illustrating the use of PCA (feature extraction) and LDA (supervised classifier) for the automatic classification of data into two categories B and C

17.3 Classification of Fluids Based on Image Data

Here, we demonstrate supervised learning techniques to distinguish between the baseline and controlled fluid flow fields from camera images. Supervised learning requires labeled training data, where the desired distinction (i.e., baseline vs controlled) is recorded in a vector of labels (i.e., "B" corresponds to baseline images and "C" corresponds to controlled images). In contrast, unsupervised learning, such as K-means, seeks to find natural clustering of the data in some feature space.

17.3.1 Methods: Machine Learning and Dimensionality Reduction

The methods presented here are general, and may be used to estimate other relevant flow quantities, as long as there is a labeled set of training data. Figure 17.2 shows a schematic of the supervised classification algorithm used in this work. A data matrix $\mathbf{X} = [\mathbf{X}_B | \mathbf{X}_C]$ is constructed by concatenating image vectors from the baseline ("B") and controlled ("C") cases. Each image is reshaped into a large column vector with as many rows as pixels in the original image, similar to how velocity fields are stacked in the method of snapshots [9]. The mean image is subtracted from $\mathbf{X}$.

Next, a low-rank feature space, $\boldsymbol{\Psi}$, is obtained by applying the principal components analysis (PCA), which is closely related to POD/SVD:

$$\mathbf{X} = \boldsymbol{\Psi}\boldsymbol{\Sigma}\mathbf{V}^*. \tag{17.1}$$

In this low-dimensional coordinate system, the data is assumed to separate into clusters according to the labels. Often the basis $\boldsymbol{\Psi}$ is truncated to only contain energetic modes. A state $\mathbf{x}$ may be approximated in this truncated coordinate system as $\mathbf{x} \approx \boldsymbol{\Psi}\boldsymbol{\alpha}$, where $\boldsymbol{\alpha}$ are the PCA/POD coordinates of $\mathbf{x}$ in $\boldsymbol{\Psi}$.

Finally, it is possible to identify the direction $\mathbf{w}$ in feature space that optimally separates the data clusters using the linear discriminant analysis (LDA) [11, 12]. Once the discriminant vector $\mathbf{w}$ is determined, it is possible to project images into a decision space by taking the inner product of the image PCA coordinates $\boldsymbol{\alpha}$ with $\mathbf{w}$.

$$\eta = \mathbf{w}^T \boldsymbol{\alpha} = \mathbf{w}^T \boldsymbol{\Psi}^T \mathbf{x}. \tag{17.2}$$

The value of η determines whether the image $\mathbf{x}$ is classified as category "B" or "C".

The performance of a classifier is determined using cross-validation, whereby the data is randomly partitioned into a training set (80 %) and a test set (20 %). The classifier is built using only training data and it is then used to predict labels in the test set; the percentage of correctly identified test labels determines the accuracy of the classifier. Many rounds of cross-validation are performed on different 80 %/20 % random shuffling of the data.

There are many alternatives to the choices above. First, if the data does not cluster in a PCA feature space, then *feature engineering* will be critical to determine the transformations that isolate features to distinguish the data. Next, there is a host of advanced supervised learning techniques including quadratic discriminant analysis, support vector machines, and decision trees, to name a few [11, 12]. However, we prefer to use PCA/LDA because of the ease of implementation and their usefulness with optimization algorithms in later sections. And most importantly, the data is well-separated with LDA in a PCA feature space.

PCA is often computed using an SVD, which is a spatial-temporal decomposition of data $\mathbf{X}$ into a hierarchy of spatial coherent structures, given by the columns of $\boldsymbol{\Psi}$, and temporal coherent patterns, given by the columns of $\mathbf{V}$. The importance of each mode is quantified by the entries of the diagonal matrix $\boldsymbol{\Sigma}$. For high-dimensional data the SVD may be computed using the method of snapshots [9]:

$$\mathbf{X}^*\mathbf{X} = \mathbf{V}\boldsymbol{\Sigma}^2\mathbf{V}^* \quad \Longrightarrow \quad \mathbf{X}^*\mathbf{X}\mathbf{V} = \mathbf{V}\boldsymbol{\Sigma}^2. \tag{17.3}$$

Thus $\boldsymbol{\Sigma}$ and $\mathbf{V}$ may be obtained by an eigendecomposition of the symmetric matrix $\mathbf{X}^*\mathbf{X}$. Afterwards, the modes $\boldsymbol{\Psi}$ may be constructed as: $\boldsymbol{\Psi} = \mathbf{X}\mathbf{V}\boldsymbol{\Sigma}^{-1}$. Note that $\boldsymbol{\Psi}$ and $\mathbf{V}$ are both unitary matrices.

17.3.2 Classification Results on High Resolution Image Data

Figure 17.3 shows the results of PCA on the high-resolution full image sequence data. The modal variance decays somewhat slowly, and the modes and coefficients are shown below. Mode 2 corresponds to a lighting change observed in the full image sequence, which can also be seen in the spikes in the temporal coefficients in both the baseline and controlled data. When performing PCA on the isolated image sequence, there is no longer a mode corresponding to a change in lighting, and the modal energy decays more rapidly.

Figure 17.4 shows the baseline and controlled data projected into the first three PCA coordinates, for both the full image sequence data and the isolated image sequence data. In both cases, the baseline and control sequences are well separated,

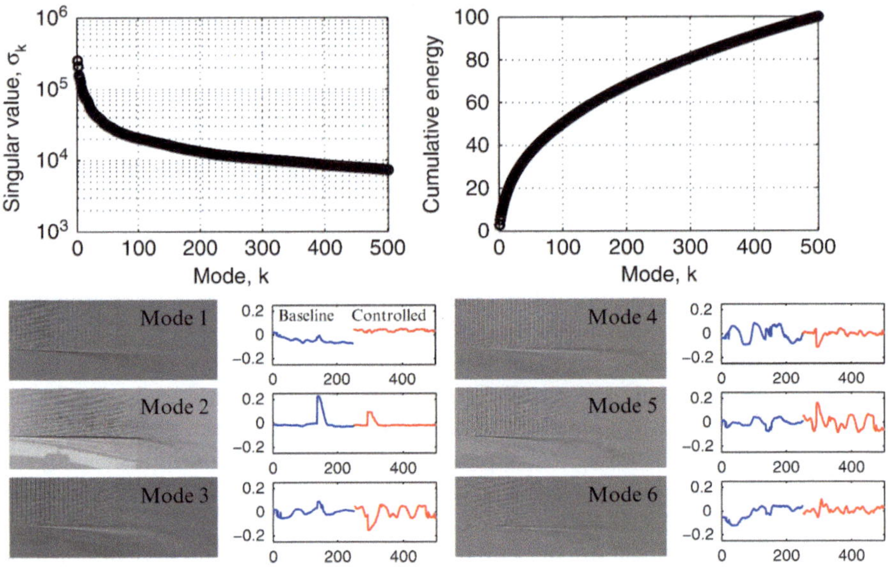

Fig. 17.3 PCA results on full image sequence data. The singular values (*top*) indicate the energy of each mode. The PCA modes (*left*) and coefficients (*right*) show dominant spatial/temporal features

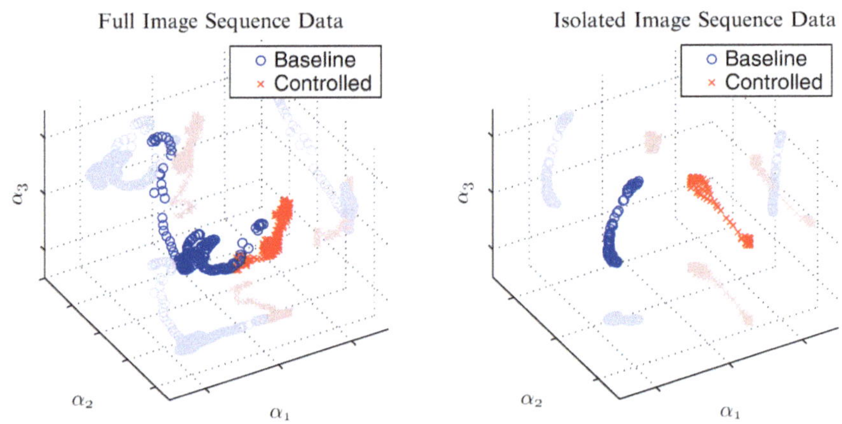

Fig. 17.4 Data plotted in the first three PCA coordinates $\boldsymbol{\alpha} = (\alpha_1, \alpha_2, \alpha_3)^T$. The full data (*left*) is reasonably well-separated. The isolated data (*right*) is very well separated

although the separation is better for the isolated images, which have more uniform conditions. Figure 17.5 shows the separating plane determined by LDA. Table 17.1 quantifies the performance of LDA classification in a PCA space with 5 modes and with 10 modes. With 10 PCA modes, the LDA classifier is perfect in both the isolated and full image sequences. Using only 5 modes, the full image sequence has around 4 % error.

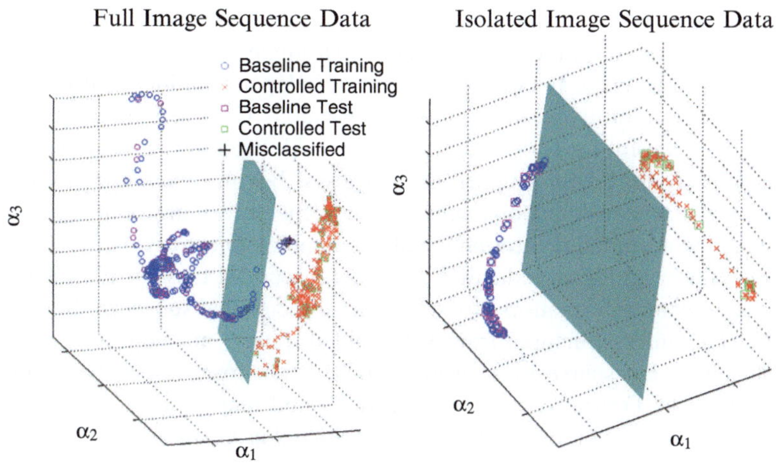

Fig. 17.5 The LDA separating plane is shown for one instance of cross-validation. Although all controlled data are correctly classified, any *purple squares* to the *right* of the plane are misclassified, and are also labeled with *black crosses*

Table 17.1 Performance of LDA classification in a PCA feature space with 5 and 10 modes on the full image sequence data and the isolated image sequence data.

		Full image sequence (%)	Isolated image sequence (%)
Error	5 Modes	3.82 ± 1.79	0.00
	10 Modes	0.00	0.00

17.4 Sparse Classification on Compressed/Subsampled Data

After demonstrating in the previous section that flows may be classified accurately using full-resolution images, here we show that similar classification may be achieved using heavily subsampled or compressed image data. This is important to reduce the data acquisition and processing required for high-level decisions. Reducing processing is important for mobile applications, where on-board computations are power constrained, and for control, where the fastest decision is desirable.

17.4.1 Methods: Sparsity and Low-Rank Structures

In this section, we assume that we take subsampled or compressed measurements $\mathbf{Y}$, which are related to the full-resolution data $\mathbf{X}$ by:

$$\mathbf{Y} = \mathbf{C}\mathbf{X}. \tag{17.4}$$

The matrix $\mathbf{C} \in \mathbb{R}^{p \times n}$ is a measurement matrix. It may consist of p random rows of the identity matrix, which would correspond to p single-pixel measurements at those locations. Alternatively, $\mathbf{C}$ may be a matrix of independent, identically distributed Gaussian or Bernoulli random variables. Random Gaussian measurements are generically powerful for signal reconstruction [15], but single-pixel measurements are particularly useful for engineering purposes. Beyond their use in classifying images, we may consider point sensor placement on a wing or in the ocean or atmosphere to accumulate information about complex time-varying flows.

Even with a significant reduction in the data, accurate classification is possible, since the relevant information exists in a low-dimensional subspace. Interestingly, the ability to infer structures from subsampled data is not new [60]. Nearly all natural images are sparse in a discrete Fourier transform (DFT) basis, meaning that most of the Fourier coefficients are small and may be neglected; this is the foundation of image compression. Fluid velocity fields are also sparse in the Fourier domain [38].

If the data $\mathbf{X}$ is sparse in a basis $\boldsymbol{\Psi}$ (either DFT or PCA), then we may write:

$$\mathbf{Y} = \mathbf{C}\mathbf{X} = \mathbf{C}\boldsymbol{\Psi}\mathbf{S}, \tag{17.5}$$

where the columns of $\mathbf{S}$ are sparse vectors (i.e., mostly zero), and the basis $\boldsymbol{\Psi}$ is a unitary matrix. Compressed sensing is based on the observation that under certain conditions on the measurement matrix $\mathbf{C}$, the projection $\mathbf{C}\boldsymbol{\Psi}$ will act as a near isometry on sparse vectors [14–16]. This means that inner products of the columns of $\mathbf{Y}$ will be similar to the inner products of corresponding columns of $\mathbf{S}$. Further, since $\boldsymbol{\Psi}$ is unitary in the case of a DFT or PCA basis, these inner products of columns of $\mathbf{Y}$ will also resemble inner products of columns of $\mathbf{X}$. Thus, using the method of snapshots, we recover the dominant correlations in the data $\mathbf{X}$ from the SVD of $\mathbf{Y}$:

$$\mathbf{Y}^*\mathbf{Y} \approx \mathbf{X}^*\mathbf{X} = \mathbf{S}^*\mathbf{S}. \tag{17.6}$$

17.4.2 Classification Results on Subsampled Data

Figure 17.6 shows the PCA projection of the baseline and controlled data for random single-pixel subsampling of the data. In the top row, $p = 1718$ random pixels are used, which account for 0.1 % of the total pixels in the image. Decreasing the number of random pixels causes the clusters to merge, making classification more difficult.

Figure 17.7 shows the cross-validated classification error versus the number of random sensors chosen. In both the top and the bottom plots, LDA classification is applied in a PCA feature space with 10 modes, and 1000 instances were used for cross-validation. For the isolated image sequence data, the median error is 0 % for as few as 34 random sensors, and for the full image sequence data, the median error

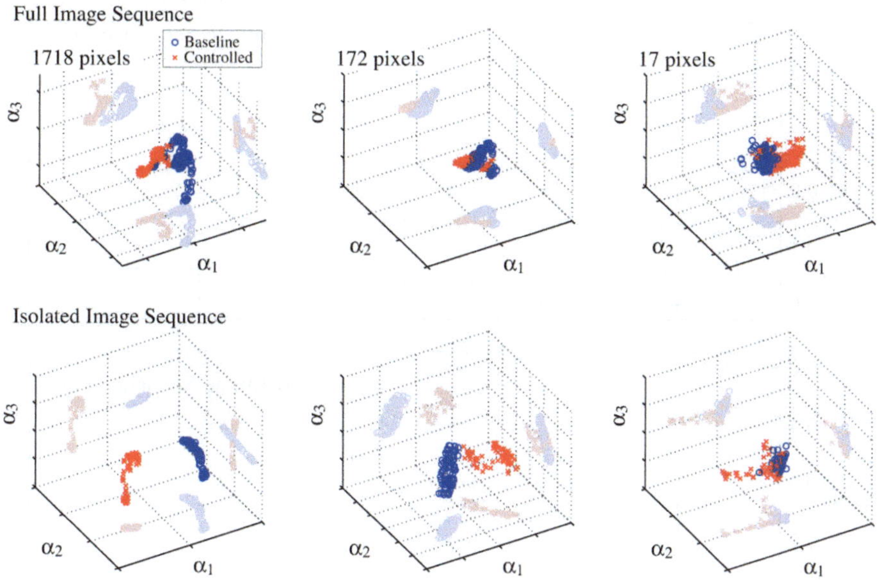

Fig. 17.6 Subsampled data plotted in the first three PCA coordinates for the full image sequence (*top*) and isolated image sequence (*bottom*). The number of random single-pixel sensors range from 1718 (*left*), to 172 (*middle*), to 17 (*right*). With more compression, the clusters begin to merge

is 0 % for 344 random sensors. As might be expected, it is easier to classify baseline and control images in the isolated image sequence, because it is more uniform and coherent. However, depending on the 80 %/20 % partition used for cross-validation, the classification error may be nearly 50 %.

The ability to perform accurate classification with $p \sim \mathcal{O}(10)$–$\mathcal{O}(100)$ randomly selected single-pixel sensors has significant implications in the data-driven processing and control of fluid systems from optical measurements. First, less spatial data must be collected, reducing data transfer and making improved temporal sampling rates possible. Second, all computations are done in a low-dimensional subspace, making it possible to make control decisions with low latency.

17.5 Optimal Sensor Placement and Enhanced Sparsity

In the previous section, we demonstrated that machine learning may be applied to heavily subsampled data, although performance was degraded at large compression ratios. Here, we demonstrate an algorithm that optimizes sensor locations for categorical decisions, resulting in accurate classification with an order of magnitude less sensors than achieved with random placement [24].

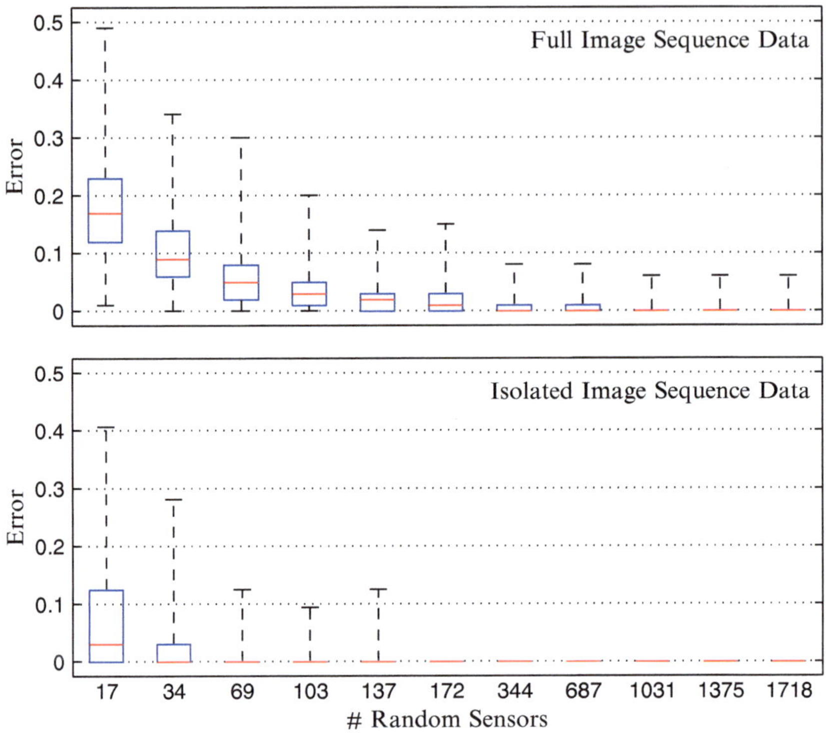

Fig. 17.7 Error vs. number of random single-pixel sensors on full image sequence (*top*) and isolated image sequence (*bottom*) for 10 PCA modes. 1000 instances are used for cross-validation. The *red line* is the median, and the *dashed lines* and *blue box* boundaries denote quartiles of the distribution

17.5.1 Methods: Optimal Sensor Placement

One of the cornerstone advances in compressed sensing is that it is now possible to solve for the sparsest solution vector to an underdetermined system of equations

$$\mathbf{A}\mathbf{x} = \mathbf{b}, \tag{17.7}$$

using convex optimization. Previously, solving for the sparsest vector $\mathbf{x}$ would involve a combinatorial brute-force search to find the $\mathbf{x}$ with smallest ℓ_0 norm, where $\|\mathbf{x}\|_0$ is equal to the number of nonzero elements in $\mathbf{x}$. However, it is now known that we may approximate the sparsest solution with *high probability* by minimizing the ℓ_1 norm, $\|\mathbf{x}\|_1 = \sum_{k=1}^{N} |x_k|$, which is a convex minimization. Therefore, it is now possible to solve increasingly large systems in a way that scales favorably with Moore's law of exponentially increasing computer power. There are a number of technical restrictions on the sizes of $\mathbf{x}$ and $\mathbf{b}$ as well as the spectral properties of the matrix $\mathbf{A}$ [14–16].

Recently, the ℓ_1 convex-minimization architecture has been leveraged to solve for optimal sensor placement for categorical decision making [24]. This optimization seeks to find a small number of pixels that are able to capture as much information as possible about the position of an image in the decision space. Specifically, we seek to find the sparsest vector $\mathbf{s} \in \mathbb{R}^n$ that satisfies the following relationship:

$$\mathbf{s} = \underset{\mathbf{s}'}{\operatorname{argmin}} \|\mathbf{s}'\|_1 \text{ such that } \boldsymbol{\Psi}^T \mathbf{s} = \mathbf{w}. \tag{17.8}$$

The vector $\mathbf{s}$ is the size of a full image, but it contains mostly zeros. Since $\mathbf{w}$ is in an r-dimensional feature space, Eq. (17.8) may be solved with a vector $\mathbf{s}$ with at most r nonzero components. Thus, it is possible to sample the image data at these r critical pixel locations, and perform classification in an r-dimensional subspace. This is called the sparse sensor placement optimization for classification (SSPOC) algorithm. We will demonstrate that accurate classification may be achieved using an order of magnitude fewer sensors, as compared with randomly placed sensors.

17.5.2 Classification on Optimized Sensors

Figure 17.8 shows the PCA clustering of data using six optimal sensor locations (top) and six randomly chosen pixels (bottom). The cluster separation with optimal sensors is striking, when compared with the clusters from random sensors. The cross-validated classification performance is shown in Fig. 17.9. The optimal six sensor locations provide a significant improvement over random.

Figure 17.10 shows the ensemble of sensor locations determined by the SSPOC algorithm over 100 instances of cross-validation. A number of interesting features are found in this data, including sampling of the boundary layer profile and the shear layer. The boundary layer sampling is more pronounced in the isolated image sequence data. In the baseline case, the shear layer remains steady and is nearly horizontal, as opposed to the controlled case, where the Kelvin–Helmholtz instability causes vortex roll-up to occur much sooner (see Fig. 17.11).

In the image sequence of the controlled case, the disturbance propagation can be observed close to the ramp wall before the flow actually separates. This may explain why so few sensors are along the separation line in the isolated image sequence.

17.6 Conclusions and Discussion

In this analysis, we demonstrate that methods from machine learning and sparse sampling may be applied to classify fluid flows from inexpensive camera images. In particular, we use LDA clustering techniques in a POD/PCA reduced subspace to classify images of a transitional separation bubble with and without forcing.

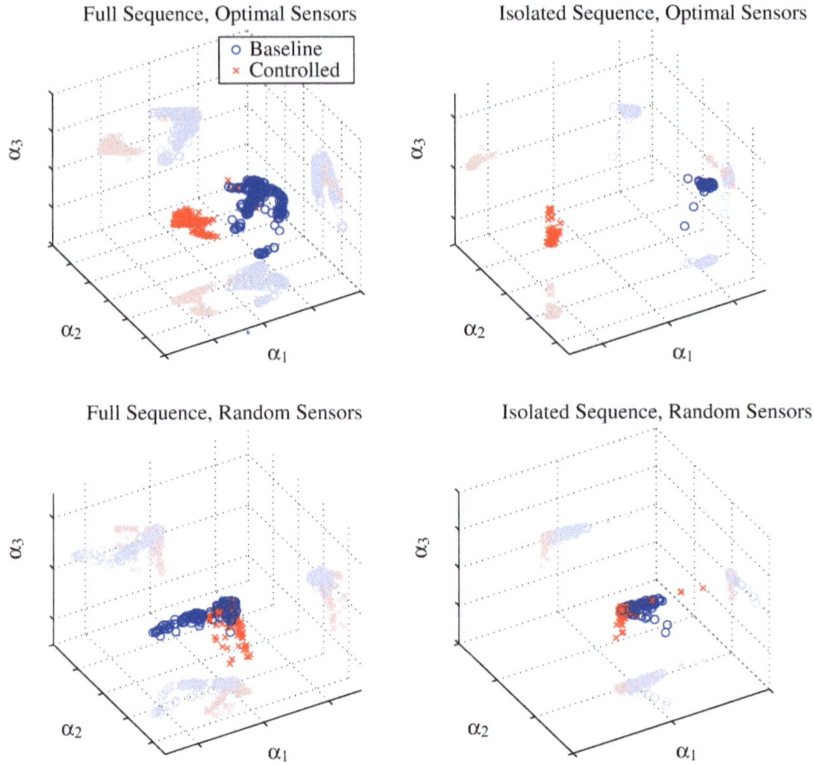

Fig. 17.8 PCA clustering of data using optimal sensors (*top*) and using random sensors (*bottom*)

Sparsity techniques are used to demonstrate that similar classification performance can be obtained with many fewer pixel measurements. Finally, a sparse sensor optimization algorithm is used to determine the fewest pixel sensors required for classification. We find that a small handful of sensors (between 5 and 10) result in a median cross-validated classification performance of $\geq 97\,\%$.

There are numerous avenues to extend this work in fluid dynamics. First, it would be natural to apply these methods to multi-way classification in flows with more distinct states. It may also be possible to estimate the phase of a periodic or quasi-periodic flow for use in a closed-loop feedback control strategy. The sparse estimation of bifurcation regimes may also be useful for parameterized reduced-order modeling techniques [28, 29]. It is also imperative that we do not apply machine learning naively to fluids data without respecting the constraints and dynamics imposed by the flow physics. Modern tools from data science tell us how to measure and on what space to perform computations, but they do not guarantee that the basic physics will be enforced, such as conservation of energy. A hybrid approach, combining data methods with physics, will likely be essential in future robust nonlinear flow control strategies.

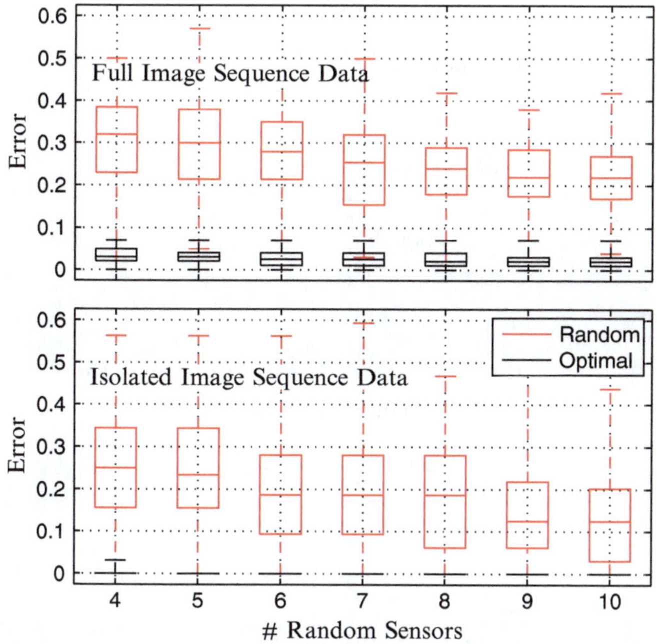

Fig. 17.9 Comparison of cross-validated error using optimal sensor locations (*black*) and random sensors (*red*) on the full image sequence (*top*) and the isolate image sequence (*bottom*). Here, the LDA classification is done directly in the pixel space

The proposed sparse classification algorithm has been demonstrated on a relatively low-dimensional flow with large-scale coherent structures. This geometry is certainly relevant for many practical flows, such as are found in small aircraft and bio-locomotion. However, these methods also scale to more complex high-dimensional data. For example, similar algorithms have been applied to sparse face recognition using images, and these examples require significantly more POD modes for accurate signal reconstruction. In the context of higher-dimensional data, a hierarchical categorization may be favorable, where a series of descriptors are identified for a given flow. Modifying the cost function for sparse sensor selection to include multiple categorization tasks is an interesting and important direction of future work.

Although the basic algorithms scale to larger data sets with a greater variety of features, there are a number of challenges that may occur specifically for high-dimensional fluids. First, the method described above is able to classify flows based on large spatially non-localized features, relying on the incoherence of point measurements with spatial Fourier modes. However, as flow velocities increase, a number of flow structures become smaller and faster; if these spatially compact features are important to classification, they will be challenging to locate and identify using point measurements. Richer random-projection measurements may

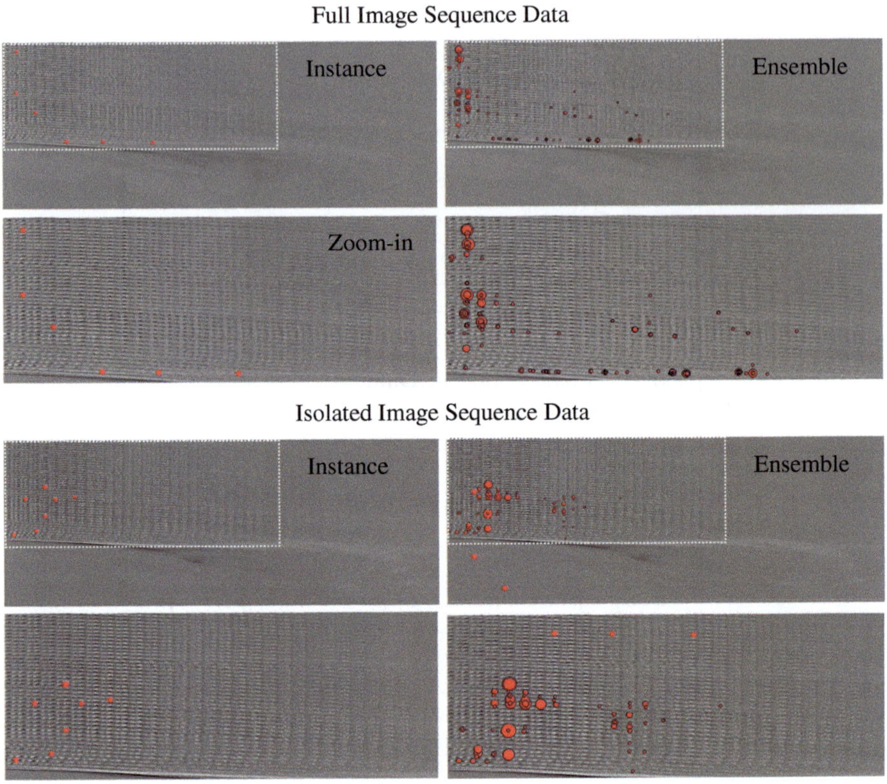

Fig. 17.10 Optimal sensor locations (*red*) for full image sequence data (*top*) and isolated image sequence data (*bottom*). A single cross-validation instance is shown on the *left*, and the ensemble of sensor locations are shown on the *right*. In each case, the *second row* provides a zoom-in near the ramp. The size of the *circle* denotes how frequently this location was chosen in the ensemble

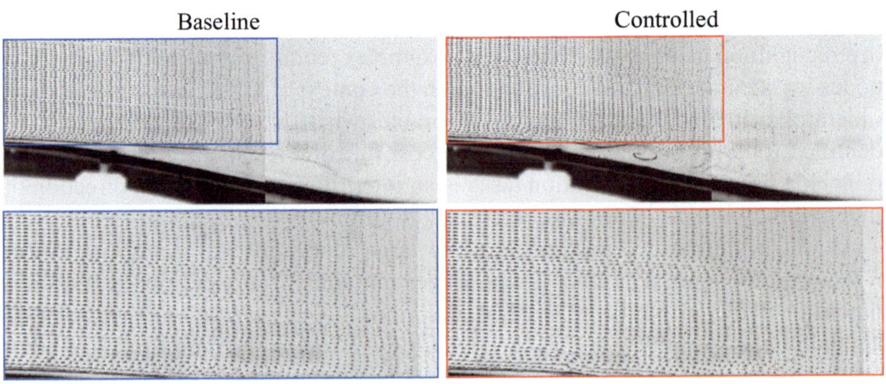

Fig. 17.11 Bubble visualizations for flow past a ramp with zoom-in around inlet

need to be employed in these cases. In addition, coherent structure identification methods such as POD or DMD assume a separation of variables, and they don't capture traveling vorticity and local features well. This limits the applicability of these methods for convective turbulence, where structures do not repeatably occupy the same *pixels*. One potential solution to this problem is to perform classification on the magnitude of the spatial Fourier transform, which removes the phase information.

The simultaneous explosion of data, the miniaturization of sensing and actuation hardware, and the renaissance of techniques in applied mathematics make this an exciting time for data-driven control in fluid dynamics. Data streams are becoming increasingly large and inexpensive, both from experimental measurement and from CFD. In the era of Big Data, it is tempting to continually gather more data and apply the same analyses to larger data matrices. With increasingly large data sets, innovative methods to distill meaningful features from data will become more important. Furthermore, bio-inspired engineering and control design will likely favor low-dimensional computations that evolve on subspaces or manifolds that capture relevant information for control and decision tasks using subsampled data. The prospect of big data in fluid dynamics is promising, and we must continue to embrace *smart* data analysis techniques to complement our *big* data.

Acknowledgements We would like to thank Mark Glauser for valuable suggestions that have improved this work, especially encouraging us to elaborate on the connection to big data. We gratefully acknowledge discussions with Josh Proctor about sparsity methods in machine learning. SLB and ZB acknowledge generous support from the Department of Energy (DOE DE-EE0006785). SLB also acknowledges support from the Air Force Office of Scientific Research (FA9550-14-1-0398) and from the University of Washington Department of Mechanical Engineering. SLB and BWB acknowledge sponsorship by the UW eScience Institute as Data Science Fellows. EK, AS, and BRN acknowledge additional support by the ANR SepaCoDe (ANR-11-BS09-018) and ANR TUCOROM (ANR-10-CEXC-0015).

References

1. S.L. Brunton, B.R. Noack, Closed-loop turbulence control: progress and challenges. Appl. Mech. Rev. **67**, 050801–1–050801–48 (2015)
2. R.P. Feynman, R.B. Leighton, M. Sands, *The Feynman Lectures on Physics*, vol. 2 (Addison Wesley, Reading, 2013)
3. W.K. George, Insight into the dynamics of coherent structures from a proper orthogonal decomposition, *In Symposium on Near Wall Turbulence in Dubrovnik*, (Dubrovnik, 1988)
4. M.N. Glauser, S.J. Leib, W.K. George, *Coherent Structures in the Axisymmetric Turbulent Jet Mixing Layer*. (Springer, Berlin, 1987)
5. G. Berkooz, P. Holmes, J.L. Lumley, The proper orthogonal decomposition in the analysis of turbulent flows. Annu. Rev. Fluid Mech. **23**, 539–575 (1993)
6. P.J. Holmes, J.L. Lumley, G. Berkooz, C.W. Rowley, *Turbulence, Coherent Structures, Dynamical Systems and Symmetry*, 2nd edn. Cambridge Monographs in Mechanics (Cambridge University Press, Cambridge, 2012)
7. G. Golub, W. Kahan, Calculating the singular values and pseudo-inverse of a matrix. J. Soc. Ind. Appl. Math. Ser. B Numer. Anal. **2**(2), 205–224 (1965)

8. G.H. Golub, C. Reinsch, Singular value decomposition and least squares solutions. Numer. Math. **14**, 403–420 (1970)
9. L. Sirovich, Turbulence and the dynamics of coherent structures, Parts I–III. Q. Appl. Math. **XLV**(3), 561–590 (1987)
10. S. Skogestad, I. Postlethwaite, *Multivariable Feedback Control: Analysis and Design*, 2 edn. (Wiley, Hoboken, 2005)
11. C.M. Bishop et al., *Pattern Recognition and Machine Learning*, vol. 1 (Springer, New York, 2006)
12. J.N. Kutz, *Data-Driven Modeling & Scientific Computation: Methods for Complex Systems & Big Data* (Oxford University Press, Oxford, 2013)
13. D.L. Donoho, Compressed sensing. IEEE Trans. Inf. Theory **52**(4), 1289–1306 (2006)
14. E.J. Candès, J. Romberg, T. Tao, Robust uncertainty principles: exact signal reconstruction from highly incomplete frequency information. IEEE Trans. Inf. Theory **52**(2), 489–509 (2006)
15. E.J. Candès, T. Tao, Near optimal signal recovery from random projections: universal encoding strategies? IEEE Trans. Inf. Theory **52**(12), 5406–5425 (2006)
16. E.J. Candès, Compressive sensing, in *Proceedings of the International Congress of Mathematics*, 2006
17. R.G. Baraniuk, Compressive sensing. IEEE Signal Process. Mag. **24**(4), 118–120 (2007)
18. J.A. Tropp, A.C. Gilbert, Signal recovery from random measurements via orthogonal matching pursuit. IEEE Trans. Inf. Theory **53**(12), 4655–4666, (2007)
19. T. Hastie, R. Tibshirani, J. Friedman, T. Hastie, J. Friedman, R. Tibshirani, *The Elements of Statistical Learning*, vol. 2 (Springer, Berlin, 2009)
20. G. James, D. Witten, T. Hastie, R. Tibshirani, *An Introduction to Statistical Learning* (Springer, Berlin, 2013)
21. R. Tibshirani, Regression shrinkage and selection via the lasso. J. R. Stat. Soc. Ser. B (Methodological) **58**, 267–288 (1996)
22. J. Wright, A. Yang, A. Ganesh, S. Sastry, Y. Ma, Robust face recognition via sparse representation. IEEE Trans. Pattern Anal. Mach. Intell. (PAMI) **31**(2), 210–227 (2009)
23. L. Clemmensen, T. Hastie, D. Witten, B. Ersbøll, Sparse discriminant analysis. Technometrics **53**(4), 406–413 (2011)
24. B.W. Brunton, S.L. Brunton, J.L. Proctor, J.N. Kutz, Optimal sensor placement and enhanced sparsity for classification. arXiv preprint arXiv:1310.4217, 2013
25. D.P. Hart, High-speed PIV analysis using compressed image correlation. J. Fluids Eng. **120**, 463–470 (1998)
26. S. Petra, C. Schn orr, TomoPIV meets compressed sensing. Pure Math. Appl. **20**(1–2), 49–76 (2009)
27. C.E. Willert, M. Gharib, Digital particle image velocimetry. Exp. Fluids **10**(4), 181–193 (1991)
28. E. Kaiser, B.R. Noack, L. Cordier, A. Spohn, M. Segond, M. Abel, G. Daviller, J. Östh, S. Krajnovic, R.K. Niven, Cluster-based reduced-order modelling of a mixing layer. J. Fluid Mech. **754**, 365–414 (2014)
29. D. Amsallem, M.J. Zahr, C. Farhat, Nonlinear model order reduction based on local reduced-order bases. Int. J. Numer. Methods Eng. **92**(10), 891–916 (2012)
30. A.G. Nair, K. Taira, Network-theoretic approach to sparsified discrete vortex dynamics. J. Fluid Mech. **768**, 549–571 (2015)
31. J.R Koza, *Genetic Programming: On the Programming of Computers by Means of Natural Selection*, vol. 1 (MIT Press, Cambridge, 1992)
32. N. Gautier, J.-L. Aider, T. Duriez, B.R. Noack, M. Segond, M. Abel, Closed-loop separation control using machine learning. J. Fluid Mech. **770**, 442–457 (2015)
33. T. Duriez, V. Parezanovic, J.-C. Laurentie, C. Fourment, J. Delville, J.-P. Bonnet, L. Cordier, B.R. Noack, M. Segond, M. Abel, N. Gautier, J.-L. Aider, C. Raibaudo, C. Cuvier, M. Stanislas, S.L. Brunton, Closed-loop control of experimental shear flows using machine learning, in AIAA Paper 2014–2219, *7th Flow Control Conference*, 2014

34. V. Parezanovic, J.-C. Laurentie, T. Duriez, C. Fourment, J. Delville, J.-P. Bonnet, L. Cordier, B. R. Noack, M. Segond, M. Abel, T. Shaqarin, S.L. Brunton, Mixing layer manipulation experiment – from periodic forcing to machine learning closed-loop control. J. Flow Turbul. Combust. **94**(1), 155–173 (2015)
35. H. Nyquist, Certain topics in telegraph transmission theory. Trans. AIEE **47**, 617–644 (1928)
36. C.E. Shannon, A mathematical theory of communication. Bell Syst. Techn. J. **27**(3), 379–423 (1948)
37. I. Bright, G. Lin, J.N. Kutz, Compressive sensing and machine learning strategies for characterizing the flow around a cylinder with limited pressure measurements. Phys. Fluids **25**, 127102–1–127102–15 (2013)
38. Z. Bai, T. Wimalajeewa, Z. Berger, G. Wang, M. Glauser, P.K Varshney, Low-dimensional approach for reconstruction of airfoil data via compressive sensing. AIAA J. **53**(4), 920–933 (2014)
39. J.-L. Bourguignon, J.A. Tropp, A.S. Sharma, B.J. McKeon, Compact representation of wall-bounded turbulence using compressive sampling. Phys. Fluids (1994–present) **26**(1), 015109 (2014)
40. I. Bright, G. Lin, J.N. Kutz, Classification of spatio-temporal data via asynchronous sparse sampling: application to flow around a cylinder. arXiv:1506.00661, 2015
41. C.W. Rowley, I. Mezić, S. Bagheri, P. Schlatter, D.S. Henningson, Spectral analysis of nonlinear flows. J. Fluid Mech. **641**, 115–127 (2009)
42. P.J. Schmid, Dynamic mode decomposition of numerical and experimental data. J. Fluid Mech. **656**, 5–28 (2010)
43. J.H. Tu, C.W. Rowley, D.M. Luchtenburg, S.L. Brunton, J.N. Kutz, On dynamic mode decomposition: theory and applications. J. Comput. Dyn. **1**(2), 391–421 (2014)
44. M.R. Jovanović, P.J. Schmid, J.W. Nichols, Low-rank and sparse dynamic mode decomposition. Center for Turbulence Research, 2012
45. S.L. Brunton, J.L. Proctor, J.N. Kutz, Compressive sampling and dynamic mode decomposition. arXiv preprint arXiv:1312.5186, 2014
46. J.H. Tu, C.W. Rowley, J.N. Kutz, J.K. Shang, Spectral analysis of fluid flows using sub-Nyquist-rate PIV data. Exp. Fluids **55**(9), 1–13 (2014)
47. F. Gueniat, L. Mathelin, L. Pastur, A dynamic mode decomposition approach for large and arbitrarily sampled systems. Phys. Fluids **27**(2), 025113 (2015)
48. J. Gosek, J.N. Kutz, Dynamic mode decomposition for real-time background/foreground separation in video. (2013, submitted for publication). https://arxiv.org/abs/1404.7592
49. M.O. Williams, C.W. Rowley, I.G. Kevrekidis, A kernel approach to data-driven Koopman spectral analysis. arXiv preprint arXiv:1411.2260, 2014
50. M.O. Williams, I.G. Kevrekidis, C.W. Rowley, A data-driven approximation of the Koopman operator: extending dynamic mode decomposition. arXiv:1408.4408, 2014
51. M.S. Hemati, M.O. Williams, C.W. Rowley, Dynamic mode decomposition for large and streaming datasets. Phys. Fluids **26**(11), 111701 (2014)
52. M.O. Williams, C.W. Rowley, I. Mezić, I.G. Kevrekidis, Data fusion via intrinsic dynamic variables: an application of data-driven Koopman spectral analysis. Europhys. Lett. **109**(4), 40007 (2015)
53. J.L. Proctor, S.L. Brunton, J.N. Kutz, Dynamic mode decomposition with control: Using state and input snapshots to discover dynamics. SIAM J. Appl. Dyn. Syst. **15**(1), 142–161 (2016)
54. H. Schaeffer, R. Caflisch, C.D. Hauck, S. Osher, Sparse dynamics for partial differential equations. Proc. Natl. Acad. Sci. USA **110**(17), 6634–6639 (2013)
55. A. Mackey, H. Schaeffer, S. Osher, On the compressive spectral method. Multiscale Model. Simul. **12**(4), 1800–1827 (2014)
56. S.L. Brunton, J.H. Tu, I. Bright, J.N. Kutz, Compressive sensing and low-rank libraries for classification of bifurcation regimes in nonlinear dynamical systems. SIAM J. Appl. Dyn. Syst. **13**(4), 1716–1732, (2014)

57. J.L. Proctor, S.L. Brunton, B.W. Brunton, J.N. Kutz, Exploiting sparsity and equation-free architectures in complex systems (invited review). Eur. Phys. J. Spec. Top. **223**(13), 2665–2684 (2014)
58. F. Sommer, Mehrfachlösungen bei laminaren Strömungen mit Druckinduzierter Ablösung: eine Kuspen-Katastrophe. VDI Fortschrittsbericht, Reihe 7, Nr. 206, VDI Verlag Düsseldorf (Dissertation Bochum), pp. 429–443, 1992
59. F.A. Schraub, S.J. Kline, J. Henry, P.W. Runstadler, A. Littell, Use of hydrogen bubbles for quantitative determination of time-dependent velocity fields in low-speed water flows. J. Fluids Eng. **87**(2), 429–444 (1965)
60. M.N. Glauser, W.K. George, Application of multipoint measurements for flow characterization. Exp. Thermal Fluid Sci. **5**(5), 617–632 (1992)

Chapter 18
Conversion of Measured Turbulence Spectra from Temporal to Spatial Domain

Preben Buchhave and Clara M. Velte

18.1 Introduction

The spatial structure of a turbulent velocity field is of great theoretical interest as its kinematics describe the distribution of spatial scales and its dynamics describe their evolution from large energy carrying scales to smaller scales and finally to dissipation.

However, the overwhelming number of turbulence measurements results in time records from stationary probes, either hot-wire probes (hot-wire anemometers, HWA) or laser beam probes (laser Doppler anemometers, LDA). The spatial structure of the turbulent velocity field is then inferred by "Taylor's hypothesis," as first presented in [1], assuming a "frozen" velocity field carried past the probe with the local mean velocity. However, Taylor's hypothesis breaks down at higher turbulence intensities and can then only be applied with additional corrections, see, for example, [2–4].

The spatial structure of the flow field can also be probed by simultaneous measurements at two or more points in space, e.g., by two LDAs or two HWAs. However, these measurements are still subject to distortions due to the varying local velocity. Only by quite advanced techniques such as moving probes has the fluctuating convection velocity issue previously been beaten, but only for

P. Buchhave (✉)
Intarsia Optics, Sønderskovvej 3, 3460 Birkerød, Denmark
e-mail: buchhavepreben@gmail.com

C.M. Velte
Department of Mechanical Engineering, Technical University of Denmark, Nils Koppels Allé, Bldg. 403, 2800 Kgs. Lyngby, Denmark
e-mail: clara@alumni.chalmers.se

A. Pollard et al. (eds.), *Whither Turbulence and Big Data in the 21st Century?*,
DOI 10.1007/978-3-319-41217-7_18

computing spatial derivatives of fluctuating velocities [5]. This strategy is, however, not practical for measuring the frequency contents of the flow since the limited record lengths attainable are typically too short to acquire useable spectra.

Whole flow field measurements can supply the instantaneous spatial velocity field and are great for the measurement of large scales. Until now, however, these methods lack the resolution needed for the most interesting small scales near the dissipation range. Furthermore, flow field methods, e.g., particle image velocimetry (PIV) require local homogeneity of the velocity field for meaningful statistical results, while time records merely require stationarity. The power spectrum derived from a temporal record is not wrong, of course. It is just one way of viewing the distribution of turbulence power over frequency as measured by a stationary probe. Power spectra from temporal records show the power of the fluctuations of the velocity of the fluid as it passes the probe. It is different from the energy spectrum of the spatial structures found from a velocity measured as a spatial record of the velocity field at one instance in time, which is basis for fundamental turbulence theory (see, e.g., [6]).

The measurement we propose here is best described as a measurement of the fluid velocity at a chosen point in the flow field, the measurement point (MP) or more realistically in a small measurement volume (MV) surrounding the MP. However, we convert the temporal record of the velocity through the MV to a spatial record of the velocity along the instantaneous streakline through the MP. We do this by measuring simultaneously both the desired velocity component and the velocity vector magnitude. We illustrate the method in Fig. 18.1.

The desired velocity component is recorded at the MP as a function of time, either by a time record with constant sampling intervals, as, e.g., in the case of a digitized HWA record, or by a randomly sampled time record as, e.g., in the case of a burst-mode LDA. At the same time, we measure or infer the magnitude of the instantaneous velocity vector. This quantity is used to convert the time record to a spatial record consisting of elements of the instantaneous streakline through the MP as explained below. Since a spectrum must be computed as an average over

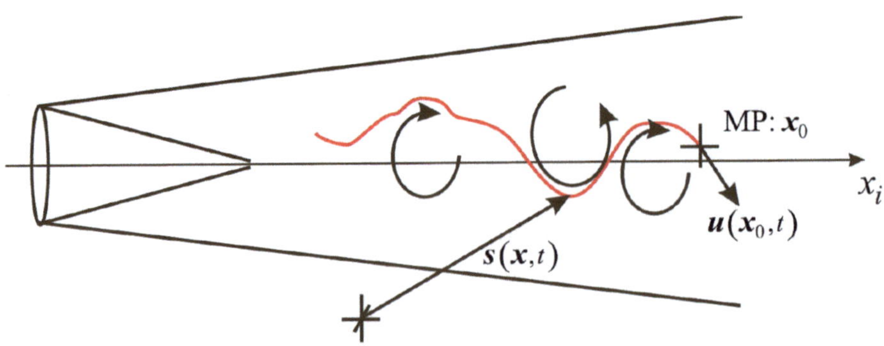

Fig. 18.1 A streakline passing through the measurement point (x_0, y_0) in a jet flow at time t. *Red*: Streakline, $s(\mathbf{x}, t)$. MP: position of measurement point, $\mathbf{x}_0$

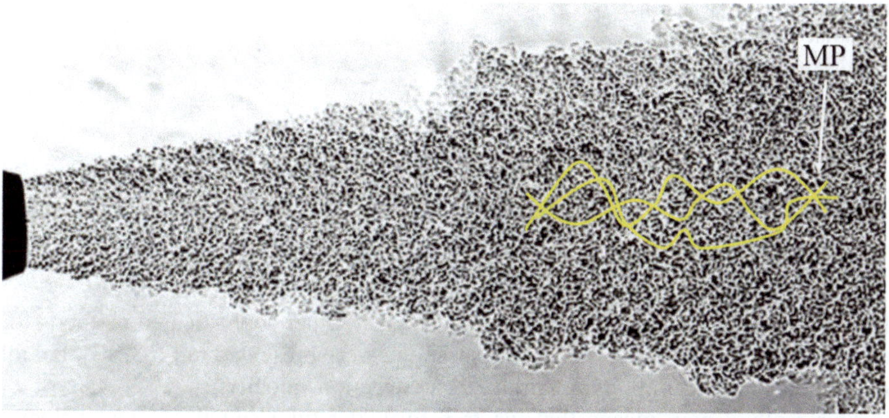

Fig. 18.2 Schlieren image of the spatial structure of a free jet in air. *Yellow*: streaklines passing through the measurement point. From M. van Dyke: An album of fluid motion

many records (for example, by block averaging), we can view the measurement as illustrated in Fig. 18.2. This figure shows the spatial structures in a jet flow as recorded by an instantaneous Schlieren image. Block averaging would consist in an average of spectra from a number of consecutive time records corresponding to different streaklines passing through the MP.

The problem with a temporal record of the velocity measured as the fluid streams past the MP is that it does not accurately describe the spatial structure of the turbulence that is of interest to turbulence theory and to turbulence modeling. The small spatial structures are convected past the MP by the instantaneous large velocity. This convection of the small scales causes small spatial structures to appear at different temporal frequencies in the power spectrum depending on the instantaneous convection velocity. In other words, strong aliasing may distort the computed spectrum.

In the past, multiple attempts were made to pass from the temporal (or frequency) spectrum to a spatial (or wavenumber) spectrum, the simplest and best known one based on the so-called Taylor's hypothesis (TH) [1] according to which the small scales are carried past the probe with the local mean velocity assumed to be so great that the small scales do not have time to change while being swept past the MP. This is also called the "frozen turbulence model". However, this method works only for relatively low intensity turbulence (turbulence intensity defined as the rms fluctuating velocity relative to the mean velocity). In cases of high intensity turbulence, as, e.g., at the outer limits of the free turbulent jet or in atmospheric boundary layers, the TH is useless and even misleading as the instantaneous velocity fluctuates greatly. However, TH may work satisfactorily in some atmospheric flow measurements where the time/length scales of interest are sufficiently small [7].

TH has been examined in numerous studies, experimentally such as [8] and [9] and in a free round jet in [5, 10, 11]. Other studies assume theoretical models

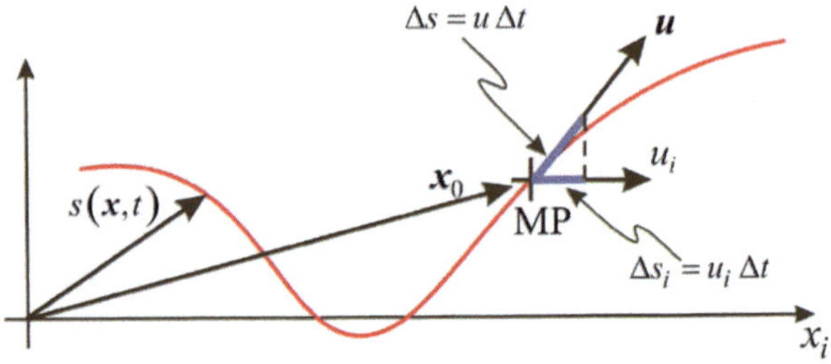

Fig. 18.3 The instantaneous streakline element through the MV

for the large-scale velocity structure such as [12]. Other sophisticated methods for transforming the time spectrum to a spatial one were presented in [2, 3, 13], but all relying on some assumption regarding the convection velocity.

Our method now means that we transform the time signal of, e.g., velocity component u_i recorded at the MP with a sampling interval Δt and measured together with the magnitude of the three-dimensional velocity $u = |\mathbf{u}|$ to a function of the length of the streakline element $\Delta s = u\Delta t$ of a fluid particle passing through the MP. In other words, we change the measurement from one using temporal sampling intervals to one using spatial sampling intervals, but still recorded at the MP. The method is illustrated in Fig. 18.3.

$\Delta s = u\Delta t$ is the (scalar) streakline element in the direction of the instantaneous velocity $\mathbf{u}$. $\Delta s_i = u_i\Delta t$ is the projection of this element onto the i-direction. Thus, Δs requires an independent measurement of the velocity magnitude u, whereas Δs_i can be computed based on the measured velocity component u_i.

A similar approach was attempted by Pinton and Labbé [14] in a von Karman swirling flow in which there exists no steady well-defined mean velocity for applying TH. A "local" TH was therefore implemented, reminiscent of the streakline approach presented herein. However, the lengths of the spatial/temporal elements in this transformation were chosen to be of the order of the revolution time of the disc, or correspondingly the integral timescale of the flow. This is quite contrary to the approach in the current work, where we apply the instantaneous velocity magnitude and where we desire a high enough sample rate to resolve the smallest scales of interest.

18.2 Spatial Sampling

The method we present here requires in addition to the measurement of the desired velocity component $u_i(t)$, the measurement of the magnitude of the velocity vector, $u(t) = |\mathbf{u}(t)|$ at the MP. In the case of a HWA, this could be realized by a spherical probe mounted next to the HWA. In case of the burst-mode LDA, the particle transit time or the so-called residence time, Δt_r, which is now measured in most commercial instruments, can be used to infer the velocity magnitude. $u(t)$ can of course also be derived from a three-component LDA measurement. The residence time is used

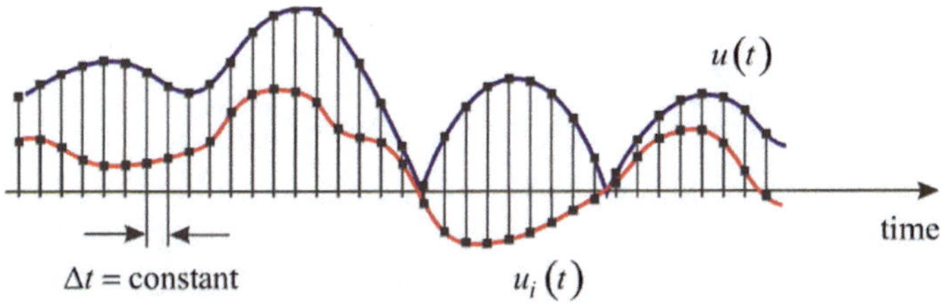

Fig. 18.4 Sampling the velocity time signal with constant temporal sample rate. The *black curve* displays the velocity magnitude while the *red curve* displays the measured single velocity component

for "residence time weighting" (RTW) of time records thus compensating for the velocity–sample rate correlation or velocity bias and allowing the computation of bias-free time averages [15, 16]. However, the residence time in our temporal-to-spatial scheme is now also used for transforming the temporal spectrum to a spatial one.

18.2.1 Example: Regular Sampling (HWA)

Assume the measurement of a temporal record with the constant sampling interval Δt, with velocity component samples $u_{i,n}(t)$ and velocity magnitude $u_n(t)$, as illustrated in Fig. 18.4 (this would apply to a hot-wire anemometer, HWA, able to measure both a velocity component and the velocity magnitude):

We choose to use in the following the so-called direct method or "periodogram" to compute the spectrum, $S_i(f)$, of the velocity component u_i, but the results apply to other computational methods as well:

$$S_i(f) = \frac{1}{T}\hat{u}_i(f)\hat{u}_i(f)^* \tag{18.1}$$

where T is the length of the time record and $\hat{u}_i(f)$ is the Fourier transform of $u_i(t)$. In case of the HWA the samples are equidistant, so we can use the fast Fourier transform (FFT):

$$\hat{u}_i(f) = \frac{T}{N}\sum_{n=1}^{N} e^{-i2\pi f t_n} u_{i,n}, \qquad \text{where} \qquad t_n = \sum_{n'=1}^{n} \Delta t = n\Delta t. \tag{18.2}$$

Knowing u_n, we can compute the corresponding length of the instantaneous streakline element through the MP in the time Δt: $\Delta s_n = u_n \Delta t$. We can then construct a spatial sampling record, illustrated in Fig. 18.5, from the spatial sampling intervals, Δs_n:

$$s_n = \sum_{n'=1}^{n} \Delta s_{n'} = \sum_{n'=1}^{n} u_{n'}\Delta t \tag{18.3}$$

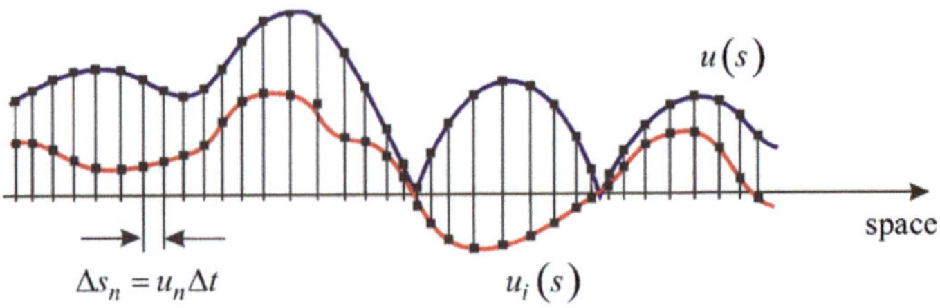

Fig. 18.5 Spatial sampling when sampled with constant temporal sample rate. The *black curve* displays the velocity magnitude while the *red curve* displays the measured single velocity component

Using this record, we can compute the spatial energy spectrum (we shall apply the term energy spectrum for the spectrum in the spatial domain) for the i'th velocity component, $S_i(k)$, by the direct method:

$$S_i(k) = \frac{1}{L}\hat{u}_i(k)\hat{u}_i(k)^*, \tag{18.4}$$

where L is the length of the spatial record and $\hat{u}_i(k)$ is the Fourier transform of $u_i(s)$. Note that this is a one-dimensional spectrum as a function of the wave number k, which is a scalar.

Since the samples are no longer equidistant, we cannot use the FFT, but must use the discrete Fourier transform (DFT):

$$\hat{u}_i(k) = \frac{N}{L}\sum_{n=1}^{N} e^{-i2\pi k s_n} u_{i,n} \qquad \text{where} \qquad s_n = \sum_{n'=1}^{n} u_{n'}\,\Delta t. \tag{18.5}$$

The small-scale spatial structures are now displayed correctly and the spatial spectrum $S_i(k)$ is now computed without the scrambling caused by the convection velocity.

18.2.2 The Method Applied to LDA

One might be a bit skeptical when it comes to applying the method to the burst-mode LDA. The LDA presents some unique signal processing challenges. Firstly, the LDA data arrive randomly with a sampling rate that contains a sampling rate—velocity correlation. Secondly, the random sampling results in random temporal sampling intervals corresponding to the interval between arrival times for the particles passing the laser beam.

Furthermore, the average data rate is often relatively low compared to the highest frequencies in the velocity signal. We then cannot convert a small constant temporal sampling interval, Δt, to a variable spatial one, Δs, as we did in the HWA case. The only information we have about the temporal sampling interval is the time between samples, and if this time is of the order of the temporal scales of the velocity, we cannot accurately compute the path length along a streamline simply by using the measured velocity magnitude and the time between samples. On the other hand, the measured residence time allows us to estimate the magnitude of the velocity vector in addition to the measured velocity component that we get from the Doppler modulation.

The temporal power spectrum in the case of the burst-mode LDA is again given by (using the direct method):

$$S_i(f) = \frac{1}{T}\hat{u}_i(f)\hat{u}_i(f)^*, \tag{18.6}$$

where T is the length of the time record and $\hat{u}_i(f)$ is the Fourier transform of $u_i(t)$.

The samples are not equidistant, so we must use the DFT (using residence time weighted formulas):

$$\hat{u}_i(f) = \frac{N}{T}\sum_{n=1}^{N} e^{-i2\pi f t_{a,n}}\Delta t_{r,n}u_{i,n} / \sum_{n=1}^{N}\Delta t_{r,n} \tag{18.7}$$

where $\Delta t_{r,n}$ is the measured residence time and $t_{a,n}$ is the measured arrival time.

We now use the known diameter of the measurement volume, d_{MV}, and the measured residence time to get a measure of the velocity magnitude: $u_n = d_{MV}/\Delta t_{r,n}$. We can compute the corresponding length of the instantaneous streakline passing through MP in the time $\Delta t_n = t_{a,n} - t_{a,n-1}$: $\Delta s_n = u_n \Delta t_n$. We then construct a spatial sampling record from the temporal random sampling intervals, Δt_n:

$$s_n = \sum_{n'=1}^{n}\Delta s_{n'} = \sum_{n'=1}^{n} u_{n'}\Delta t_{n'} = \sum_{n'=1}^{n}(d_{MV}/\Delta t_{r,n'})\left(t_{a,n'} - t_{a,n'-1}\right). \tag{18.8}$$

Using this record, we can compute the spatial energy spectrum for the i'th velocity component, $S_i(k)$, by the direct method:

$$S_i(k) = \frac{1}{L}\hat{u}_i(k)\hat{u}_i(k)^*, \tag{18.9}$$

where L is the length of the spatial record and $\hat{u}_i(k)$ is the Fourier transform of $u_i(s)$. Note that this is again a one-dimensional spectrum as a function of the wave number k, which is a scalar.

Since the samples are not equidistant, we cannot use the FFT, but must use the DFT:

$$\hat{u}_i(k) = \frac{N}{L}\sum_{n=1}^{N} e^{-i2\pi k s_n}\Delta t_{r,n}u_{i,n} / \sum_{n=1}^{N}\Delta t_{r,n} \tag{18.10}$$

where L is the record length. The small-scale spatial structures are now again displayed correctly and the spatial spectrum $S_i(k)$ is now computed without distortion due to the convection velocity.

18.3 Tests with a Computer Generated Large Eddy

To illustrate the method, we generate a spatial structure consisting of a large vortex or eddy and superimpose on this convection velocity a small, harmonically oscillating spatial structure that is carried along with the velocity of the large eddy. The center of the large eddy travels with a constant velocity, $u_c = 2\,\mathrm{m\,s^{-1}}$, along the flow direction. The geometry is shown in Fig. 18.6.

18.3.1 HWA Signal

We compute the power spectrum of a 1 s long time record $u_i(t)$, modulated with a spatial period of 0.005 m corresponding to a spatial frequency of $200\,\mathrm{m^{-1}}$ and sampled with a sample rate of 1000 sps. This spectrum is shown in Fig. 18.7a. The original spatial modulation of $200\,\mathrm{m^{-1}}$ appears scrambled in the temporal spectrum due to the changing convection velocity. Assuming measurement of the velocity magnitude at the MP and using the conversion described above, we obtain a spatial spectrum $S_i(k)$ as shown in Fig. 18.7b. The original spatial harmonic of $200\,\mathrm{m^{-1}}$ has been restored.

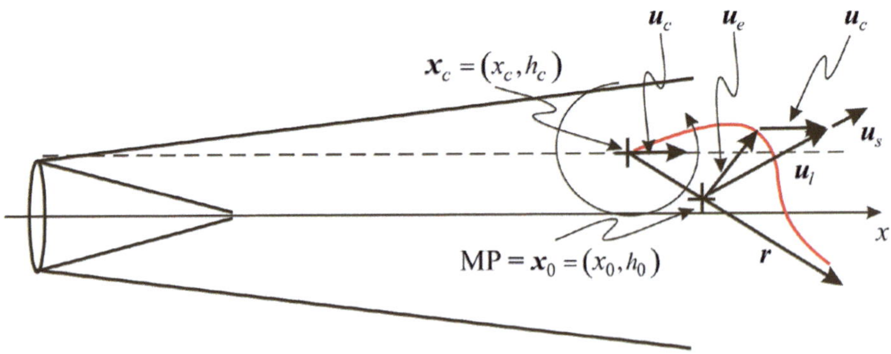

Fig. 18.6 The figure shows the large eddy with velocity $\mathbf{u}_l$ (*red curve*) plus the small harmonic oscillator $\mathbf{u}_s$. The small harmonic oscillation is described by a spatial velocity function: $\mathbf{u}_s(s) = u_{s,0}\cos(2\pi s/\lambda)$, where λ is the spatial period along s of the small structure

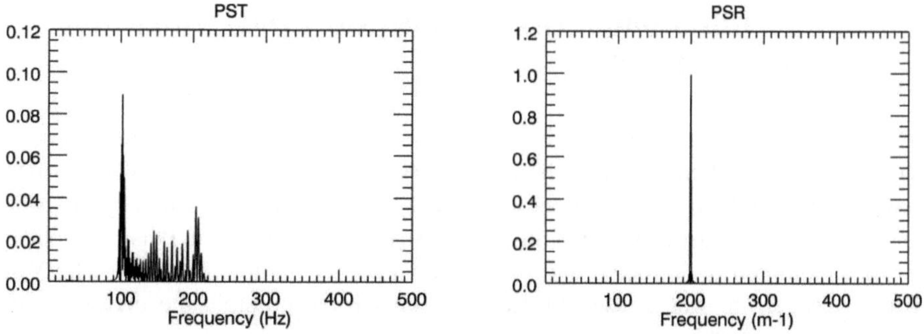

Fig. 18.7 Spectra of the large eddy and the small spatial modulation measured with HWA. PST: Temporal spectrum. The spatial 200 m^{-1} modulation appears as aliased frequencies in the temporal spectrum. PSR: Using the conversion method the spatial spectrum is restored. The slight broadening of the 200 m^{-1} peak is due to the low frequency content of the large eddy

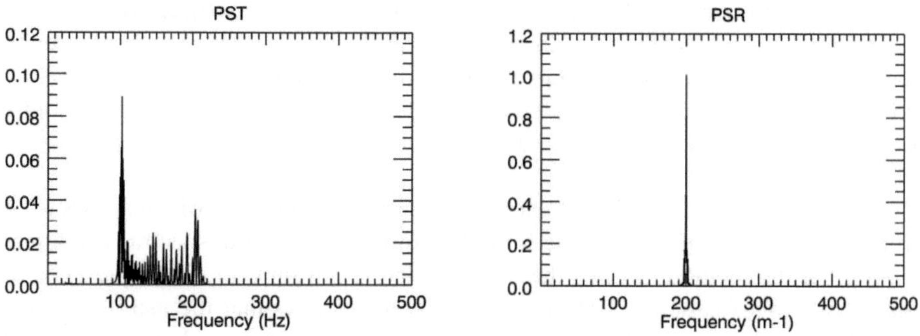

Fig. 18.8 Spectra of the large eddy and the small spatial modulation measured with LDA. PST: Residence time weighted temporal power spectrum. The spatial 200 m^{-1} modulation is aliased in the temporal spectrum. PSR: Using the conversion method the spatial energy spectrum is restored. The slight broadening of the 200 m^{-1} peak is due to the low frequency content of the large eddy

18.3.2 LDA Signal

We assume a uniform particle distribution in space. The LDA probe will see a random temporal sample rate correlated with the velocity magnitude. However, the overall result of the temporal-to-spatial conversion is similar to the HWA result, see Fig. 18.8.

18.4 Tests with Computer Generated LDA Data with a Von Karman Spectrum

The computer generated data has been described in [17]. Briefly, the random spatial samples were grabbed from a high rate primary velocity record with a Von Karman power spectrum by a Poisson process. Since we are sampling the velocity in space and assuming the particles to be uniformly distributed in space, the Poisson process is not modulated by velocity, and there is no velocity–sample rate bias. The Von Karman spectrum was chosen to model as closely as possible the measured jet spectrum described below:

$$S_{vK}(f) = \frac{1}{62.5} \cdot \frac{1}{\left[1 + (f/45)^2\right]^{5/6}} \cdot \exp\left[-(f/2500)^{4/3}\right]. \tag{18.11}$$

The integral time scale of this process can be found from $l = S(0)/2\text{var}(u)$.

We now assume that the Von Karman velocity fluctuations are convected through the LDA MV by 3-D Gaussian large-scale eddies. The turbulence intensity for the complete fluctuating convection velocity signal was 54 %. The resulting randomly sampled time record results in an aliased time spectrum (dark blue) as shown in Fig. 18.9 LHS.

The mean velocity was $1\ \text{m}\,\text{s}^{-1}$ and the record length was 1 s, which allows us to use the same scale on the abscissa for both temporal and spatial spectra. The figure shows that the method restores the Von Karman spectrum even in the presence of large-scale low frequency 3-D fluctuations. It also shows that the use of Taylor's hypothesis in this case with a 50 % turbulence intensity does not adequately convert the time record to a spatial one.

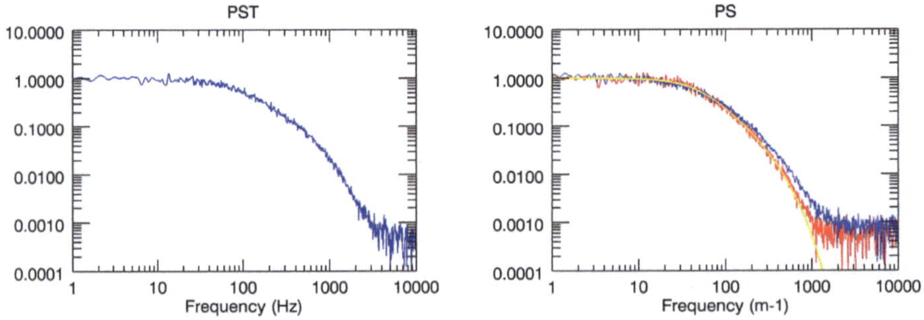

Fig. 18.9 Spectra of the Von Karman turbulence convected past the LDA MV by a large low frequency Gaussian fluctuation. Left: *blue*: The temporal spectrum. Right: *yellow*: The original Von Karman model spectrum. *Red*: The spatial spectrum restored. *Blue*: The spatial spectrum restored by the conventional Taylor's hypothesis

18.5 Measurements

We now try our method on LDA measurements in a free, round, turbulent jet in air. The measured velocity is the streamwise component. This experiment is described in detail in [16]. The jet data for these measurements were: Jet exit diameter: 10 mm, measurement location: 30 diameters downstream, 0, 13, 26, 39, 52 mm off axis, center mean velocity: 6.2 m s^{-1}, center line turbulence intensity: 7.5 %, Re = 20,000, integral time scale 0.0048 s. The active measurement volume was 75 × 700 μm and the average data rate was approx. 6400 s^{-1}. The scales are estimated as: Kolmogorov scale = 53 μm, Taylor scale = 2.2 mm at the center line. 4.000.000 data points were used.

Figure 18.10 shows the measured temporal spectra. The power spectra are shifted towards higher frequencies as the mean velocity increases. The total power also increases as the local mean velocity increases, but these spectra are normalized to the same values at low frequency in order to be able to compare the shape of the spectra. The spectra show high similarity for the high frequency fluctuations in agreement with the qualitative impression from the Schlieren photograph in Fig. 18.1. The dip at high frequency is due to dead time [18, 19], which has not been removed by deconvolution, as is otherwise possible, see [17].

After application of our conversion method, we find the spatial spectra shown in the left graph of Fig. 18.11. These should be compared to the true spatial (PIV) spectra in Fig. 18.11 right. The red color shows power spectra based on the spatial record obtained by our method. The blue curve shows power spectra based on Taylor's method. The spectra based on the method using the measured magnitude of the velocity collapse to nearly overlapping curves whereas Taylor's method fails for the 26, 39, and 52 mm off-axis positions. The turbulence intensity at these positions are 23 %, 65 %, and 420 %, respectively.

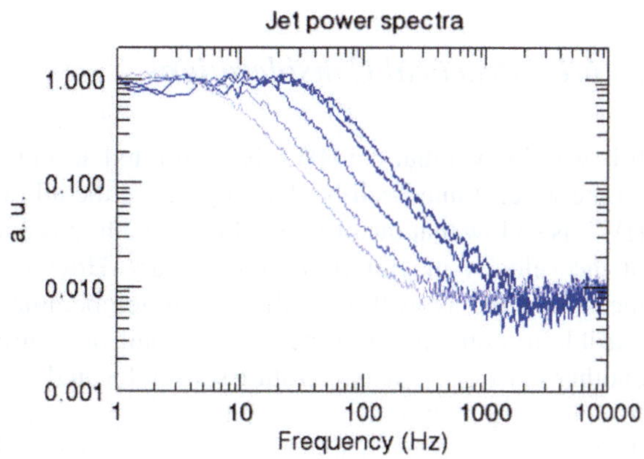

Fig. 18.10 Power spectra of jet measurements at different off-axis positions. From *heavy blue* to *light blue*: off-axis position 0, 13, 26, 39, 52 mm

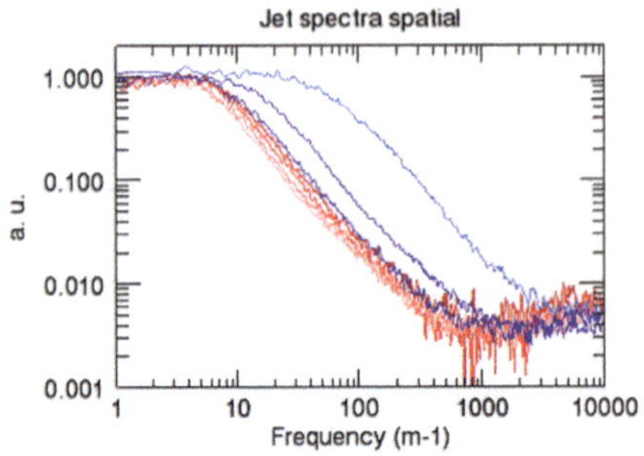

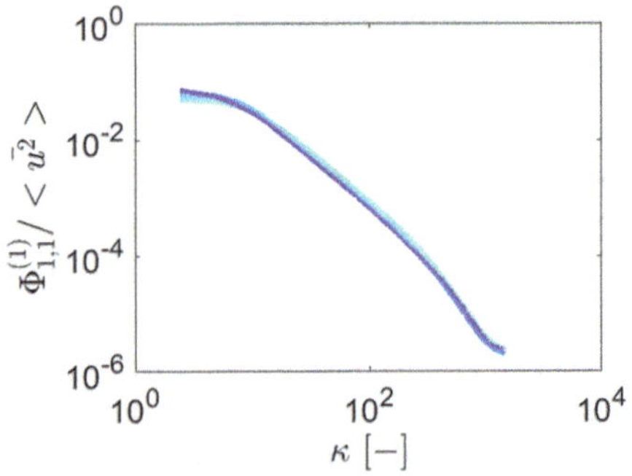

Fig. 18.11 *Left*: Restored spatial spectra. *Red*: Our method. Off-axis position *heavy red* to *light red*: 0, 13, 26, 39, 52 mm. *Blue*: Taylor's hypothesis. Off-axis position *heavy blue* to *light blue*: 26, 39, 52 mm. *Right*: Spatial spectra from homogeneous directions in the jet using PIV [20]. From *light blue* to *heavy blue*: 0, 13, 26, 39, 52 mm. When normalizing the spectra by their respective energy, they collapse onto the same curve

18.6 Discussion

Apart from the practical questions related to this method (the possibility for measuring the magnitude of the 3-D velocity vector with a HWA and with the LDA), a number of more principal questions deserve a further investigation:

- How do we interpret the first-order statistics (e.g., mean values) computed from the data with and without residence time correction?
- There are a number of different ways of measuring and calculating the turbulent power spectrum. How do they relate to the classical definitions of turbulent power spectra and to the classical kinetic energy spectrum?

18.6.1 Practical Considerations

It is well known that the HWA has a limited acceptance angle for the flow incident on the wire. It may well be that in practice the allowed turbulence intensity for the HWA is so low that the ordinary Taylor method is adequate and that a measurement of the velocity magnitude is unnecessary. However, the method applies to other measurements as well, e.g., fluctuating temperature or concentration. It might be feasible to construct a probe with a small omnidirectional (spherical) sensor and another sensor, e.g., a temperature sensor, located close by. It would then be possible to convert a temporal temperature or concentration measurement to the spatial domain and remove the aliasing effect connected with a power spectrum in the time domain.

As we have demonstrated above, it is feasible for an LDA to measure both a velocity component (by the ordinary Doppler effect) and the velocity magnitude (by means of the measured residence time). We can then convert a time record of a flow with unlimited turbulence intensity (even a flow with no mean convection velocity) to a spatial record and compute a spatial energy spectrum. The practical concerns related to this method arise in connection with especially two effects: Firstly, the random sampling causes the usual additional noise known from randomly sampled records. However, an additional concern is related to the measured random time between samples, which must be used to estimate the spatial sampling intervals from the temporal intervals. It is obvious that if the time between samples is much longer than the typical time scales of the velocity, information about the spatial record between measurements is lost as only a zeroth order interpolation replaces the true streakline element. The method thus requires a relatively high average sample rate. This, however, is already a criterion for obtaining high quality power spectra c.f. [15, 16]. Secondly, the measured residence time fluctuates greatly due to the different particle trajectories through the MV. However, the same concern applies to other LDA measurements that use the residence time, and experience shows that this noise does not add appreciably to the noise already present from the random sampling. Both noise effects are reduced by the block averaging that is necessary anyway to obtain low noise spectra.

18.6.2 Interpretation of First-Order Statistics

We consider here the mean values computed from the data as other statistical quantities must be interpreted the same way. The straightforward way of measuring the mean velocity is to simply use an arithmetic mean:

$$\text{Temporal mean value, HWA}: \quad \overline{u}_i = \frac{\sum_{n=1}^{N} u_{i,n}}{\sum_{n=1}^{N} 1} = \frac{1}{N}\sum_{n=1}^{N} u_{i,n}. \tag{18.12}$$

This gives the correct temporal mean value for a regularly sampled HWA signal (however, the spatial structures are distorted due to the convection effect). When we apply our conversion method to the regularly sampled time signal, the resulting spatial record shows what we may call an "inverted velocity bias." This is because the measured velocities will now be close together on the spatial record where the velocity magnitude is low and further apart where the velocity magnitude is high. However, even if the velocities in this spatial record are unevenly spaced, they are the same values as in the time record, and the velocity mean value is still the arithmetic mean:

$$\text{Spatial velocity mean}: \quad \langle u_i \rangle = \frac{\sum_{n=1}^{N} u_{i,n}}{\sum_{n=1}^{N} 1} = \frac{1}{N}\sum_{n=1}^{N} u_{i,n}. \tag{18.13}$$

The small spatial structures, however, are now spaced correctly on the spatial record.

The mean value computed from a randomly sampled LDA signal leads to an overweight of high velocity samples ("velocity bias"). As shown theoretically in [16] and by computer generated data in [17], weighting by the measured residence time gives correct time mean values for any sample rate:

$$\text{Temporal velocity mean value}: \qquad \overline{u}_i = \frac{\sum_{n=1}^{N} \Delta t_{r,n} u_{i,n}}{\sum_{n=1}^{N} \Delta t_{r,n}}. \tag{18.14}$$

The question is now: When this temporal record is converted to a spatial record consisting of a sum of spatial intervals of the instantaneous streakline through the MP, but the measured velocities remain the same, how do we then compute the velocity mean value? Since the measured velocities are now spaced by the original spatial distribution, which we assume uniform in space, it is tempting to consider the simple arithmetic average as a measure of the velocity mean value. However, the measurement is still based on the biased sampling in time, and therefore the mean value must still be corrected with the measured residence times. The arithmetic mean of the measured velocities must instead be interpreted as a measure of the volume flow through the MV. This is easily seen by reference to Fig. 18.3 above. The streakline element Δs multiplied by the MV cross section is the volume of fluid carried through the MV in the time element Δt. Thus the sum

$$\overline{\rho} = \frac{\sum_{n=1}^{N} u_{i,n} \Delta t}{\sum_{n=1}^{N} \Delta t} = \frac{\sum_{n=1}^{N} \Delta s_{i,n}}{T}. \tag{18.15}$$

is a measure of the average volume flow density through the MV. To get the correct velocity mean we must again weigh the data in the spatial record by the residence time, Δt_r:

$$\text{Spatial velocity mean value}: \qquad \langle u_i \rangle = \frac{\sum_{n=1}^{N} \Delta t_{r,n} u_{i,n}}{\sum_{n=1}^{N} \Delta t_{r,n}}. \tag{18.16}$$

We thus see a principal difference between the regularly sampled time signal from, e.g., a HWA and the random data of a burst-mode LDA. We can say that the HWA samples uniformly in time whereas the LDA samples uniformly in space. LDA data measured in the time domain must be corrected with residence time, also when the record is converted back to a spatial record.

18.6.3 Interpretation of the Measured Spectra

Again, we must distinguish between the regularly sampled time signal from a HWA and the spatially sampled LDA. Taking first the HWA, it is generally accepted that the small spatial velocity structures are swept past the stationary probe by the large convective velocity components. The distinction between small-scale structures and large-scale convective structures is somewhat artificial, however. In reality, the "sweeping" is done by the total instantaneous velocity vector. In any case, the result is that the temporal power spectrum is aliased; the small structures are smeared out over a large temporal frequency domain, especially when the turbulence intensity is greater than about 40 % as we have seen in the examples above. To analyze the computed spectra further, we must distinguish between the various velocities and restored spatial records. The desired spectrum is for the measured velocity component u_i (or possibly the temperature fluctuations or concentration fluctuations). As we have seen above, we can construct two spatial records: One based on the measurement of the magnitude of the three-dimensional velocity vector, $u = |\mathbf{u}|$, and one based on the measured velocity component, u_i. The energy spectra computed for the velocity component, u_i, based on these two spatial records, s and s_i are both one-dimensional. The first is given by

$$\begin{aligned} 1\text{a}: \quad & S_i(k_i) = \frac{1}{L}\hat{u}_i(k_i)\hat{u}_i(k_i)^* \quad \text{with} \quad \hat{u}_i(k_i) = \frac{N}{L}\sum_{n=1}^{N} e^{-i2\pi k_i s_{i,n}} u_{i,n} \\ & \text{and} \quad s_{i,n} = \sum_{n'=1}^{n} u_{i,n'}\Delta t. \end{aligned} \tag{18.17}$$

where k_i is a scalar wave number in the direction of the measured velocity component u_i. The other is expressed by

$$\begin{aligned} 2\text{a}: \quad & S_i(k) = \frac{1}{L}\hat{u}_i(k)\hat{u}_i(k)^* \quad \text{with} \quad \hat{u}_i(k) = \frac{N}{L}\sum_{n=1}^{N} e^{-i2\pi k s_n} u_{i,n} \\ & \text{and} \quad s_n = \sum_{n'=1}^{n} u_{n'}\Delta t. \end{aligned} \tag{18.18}$$

where k is a scalar wavenumber in the direction of the instantaneous streakline. Before we discuss these spectra any further, we shall consider the LDA case. The corresponding LDA energy spectra are given by (residence weighted formulas, see [15]):

$$\begin{aligned} 1\text{b}: \quad & S_i(k_i) = \frac{1}{L}\hat{u}_i(k_i)\hat{u}_i(k_i)^* \quad \text{with} \quad \hat{u}_i(k_i) = \frac{N}{L}\sum_{n=1}^{N} e^{-i2\pi k_i s_{i,n}} \Delta t_{r,n} u_{i,n} \Big/ \sum_{n=1}^{N} \Delta t_{r,n} \\ & \text{and} \quad s_{i,n} = \sum_{n'=1}^{n} u_{i,n'}\Delta t_{n'}. \end{aligned} \tag{18.19}$$

and

$$2b: \quad S_i(k) = \frac{1}{L}\hat{u}_i(k)\hat{u}_i(k)^* \quad \text{with} \quad \hat{u}_i(k) = \frac{N}{L}\sum_{n=1}^{N} e^{-i2\pi k s_n}\Delta t_{r,n}u_{i,n} / \sum_{n=1}^{N}\Delta t_{r,n}$$

$$\text{and} \quad s_n = \sum_{n'=1}^{n} u_{n'}\Delta t_{n'}. \tag{18.20}$$

There is a possibility for one more energy spectrum based on the velocity magnitude or rather on the 3-D velocity vector in the direction of the instantaneous streakline through the MP:

$$3: \quad S(k) = \frac{1}{L}\hat{u}(k)\hat{u}(k)^* \quad \text{with} \quad \hat{u}(k) = \frac{N}{L}\sum_{n=1}^{N} e^{-i2\pi k s_n}\Delta t_{r,n}u_n / \sum_{n=1}^{N}\Delta t_{r,n}$$

$$\text{and} \quad s_n = \sum_{n'=1}^{n} u_{n'}\Delta t_{n'}. \tag{18.21}$$

Note: This is again a one-dimensional spectrum; the measured 3-D velocity vector is always in the direction of the streakline.

We now suggest the following interpretation of the high wavenumber part of these spectra, assuming that the small spatial structures are isotropic. We can think of the small scales as an isotropic superposition of plane waves. Expressions 1 and 2 above are spatial energy spectra of the measured velocity component, u_i. The "a" and "b" indices refer to the HWA and LDA cases, respectively. Expressions 1a and 1b are converted by the use of the magnitude of the measured velocity component, whereas 2a and 2b use the measured magnitude of the instantaneous 3-D velocity vector. Expression 3 is the energy spectrum of the measured 3-D velocity vector magnitude using this quantity also for the conversion process.

Spectrum 1 is then the energy spectrum of the measured velocity component in the direction of the i-axis on a spatial record, which is composed of the projection of the streakline element onto the i-direction. We consider this equivalent to the usual one-dimensional energy spectrum of the i-component of the velocity along a line through the MP in the i-direction, assuming local homogeneity for the flow along the measurement line. As the plane wave decomposition of the small-scale structure is projected onto the i-direction, the projection of the plane waves onto the i-axis will appear at an increased wavelength, and the one-dimensional spectrum will be direction-aliased. Spectrum 2 is again the energy spectrum of the measured i-component of the velocity, but in this case referred to the spatial record composed of steak line elements in the direction of the 3-D flow velocity. Here we measure the transport along the streakline of the i-component of the small-scale velocity structures through the MP. This one-dimensional spectrum is also aliased. Finally, we associate the third spectrum with the energy spectrum of the velocity magnitude along a spatial record, which is always in the same direction as the velocity vector.

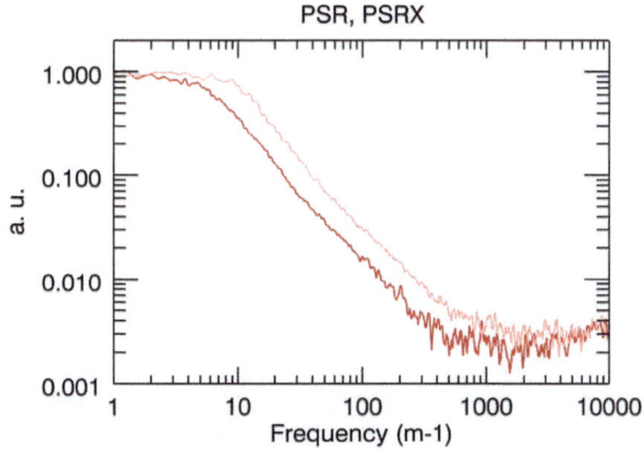

Fig. 18.12 LDA measurements of power spectrum, direct method, uniform random sampling in space, 100 record block average. *Red*: Reconstructed spatial record s. *Light red*: Reconstructed spatial record s_x

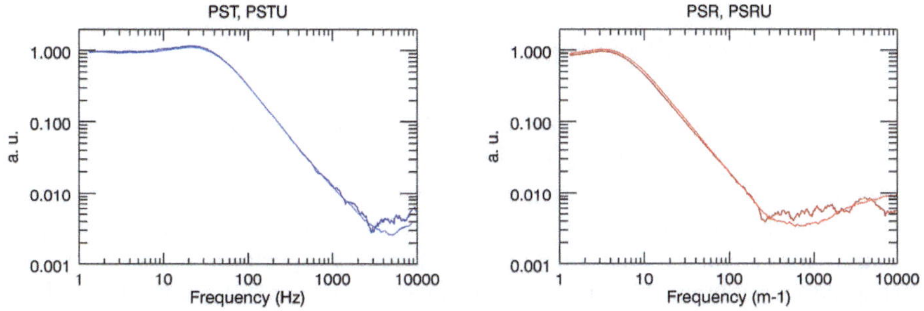

Fig. 18.13 Comparison of spectra 2b and 3. Jet measurements, $y = 0$ mm. *LHS*: Temporal spectra. *RHS*: Spatial spectra. *Light color*: Case 2b, *Dark color*: Case 3

This must mean that the final wave front of the spatial structures is always normal to the direction of the streakline element, and that the streakline element and the velocity vector are always co-parallel. There can be no velocity component normal to the streakline. Spectrum 3 must therefore express the total kinetic energy. This spectrum is also one-dimensional and will therefore again be aliased. Note, however, that we obtain the total kinetic energy by a single measurement; not as in the classical case as a sum of energy spectra along three orthogonal coordinate axes.

We have computed two spatial spectra 2a and 2b from the measured jet LDA data. Figure 18.12 shows the result for an off-axis position of 39 mm, indicating that the spectrum based on a projected k-vector component is shifted towards higher frequencies as compared to the one based on the 3-D streakline.

Figures 18.13 and 18.14 display a comparison between the 2b and 3 spectra, the difference being that the spectra are based on a single velocity component and the full velocity vector, respectively. The left column shows the temporal spectra and the right column the spatial ones. At the jet centerline, these two spectra are expected to display the same behavior due to the (relatively) low turbulence intensities, which is seen to be true for both the temporal and spatial spectra. The dip seen in the spectra arise due to dead time effects. Interestingly, this concurrence still persists as far out as 39 mm (1.5 jet half-widths) from the jet centerline.

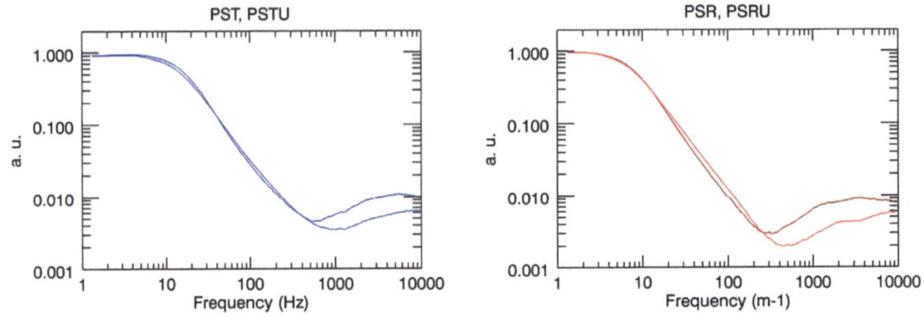

Fig. 18.14 Comparison of spectra 2b and 3. Jet measurements, $y = 39\,\text{mm}$. *LHS*: Temporal spectra. *RHS*: Spatial spectra. *Light color*: Case 2b. *Dark color*: Case 3

18.7 Linking to Big Data

We propose the following interpretations of the computed spectra in Sect. 18.6.3: The spatial isotropic small-scale spectra obtained by the use of the measured instantaneous velocity magnitude are interpreted as the classical one-dimensional spectrum of the measured velocity component (spectrum 2b) whereas we interpret spectrum 3 as the one-dimensional total kinetic energy spectrum. Interestingly, this last spectrum is obtained by just a single measurement and not as the sum of three measurements along orthogonal axes.

The results of our investigations are relevant to the "Big Data" theme of this conference by helping to improve the accuracy of experimental databases, in particular concerning the important question as to how energy is cascaded from the energy creating eddies to the high frequency dissipative scales. Direct solution of the fundamental equations has proven infertile up to now, and the future response to fluid mechanics problems may be a combination of physical guide lines and a huge database of experiments, all under the auspices of a "Google-like" artificial intelligence. Obviously, such solutions require large databases of accurate experimental results.

18.8 Conclusion

A novel method for converting temporal turbulent flow measurement records into spatial ones, bypassing the adverse fluctuating convection velocity effect, is proposed. The method is based on the principle of converting time records into spatial piecewise streakline records using the instantaneous measured velocity magnitude. By measuring along a streakline in time from a fixed point in space, the seldom fulfilled requirement of statistical homogeneity can be replaced by the considerably more relaxed requirement of statistical stationarity. The method requires that the flow scales of interest are well resolved in space/time, in particular the smallest scales that are expected to deviate most from the true spatial representation. Unless this temporal-spatial conversion is performed correctly, the structures of the original signal are scrambled in its transformed counterpart. This was exemplified using computer simulations of a large eddy with a superimposed high spatial frequency modulation, where the original signal could be restored using the proposed method. Computer generated data from a von Karman spectrum model of turbulence were used to demonstrate the kind of adverse energy scrambling effects that are to be expected in a real turbulent flow measurement and how well the proposed method is able to alleviate them. The method was finally demonstrated on measured data from an axisymmetric turbulent jet measured at different distances from the centerline. This is a direct test of the method since the turbulence intensity increases dramatically with radial distance from the jet center, a requirement well known to be challenging for the classical Taylor's hypothesis not least due to the fluctuating convection velocity effect. This is also evident in the present comparison between this and our proposed method, where the latter displays exactly the same behavior as the true spatial spectra measured with PIV along homogeneous flow directions.

We propose interpretations of the computed spectra. The spatial isotropic small-scale spectra obtained by the use of the measured instantaneous velocity magnitude are interpreted as the classical one-dimensional spectrum of the measured velocity component (Spectrum case 2b) and whereas we interpret Spectrum case 3 as the one-dimensional total kinetic energy spectrum. Interestingly, this last spectrum is obtained by just a single measurement and not as the sum of three measurements along orthogonal axes.

References

1. G. Taylor, The spectrum of turbulence. Proc. R. Soc. Lond. (1938)
2. J.L. Lumley, Interpretation of time spectra measured in high intensity shear flows. Phys. Fluids **8**, 1056 (1965)
3. F.H. Champagne, The fine-scale structure of the turbulent velocity field. J. Fluid Mech. **86**, 67–108 (1978)
4. J.C. Wyngaard, S.F. Clifford, Taylor's hypothesis and high-frequency turbulence spectra. J. Atmos. Sci. **34** (1977)
5. W.K. George, H.H. Hussein, S.H. Woodward, An evaluation of the effect of a fluctuation convection velocity on the validity of Taylor's hypothesis, in *Proceedings of the 10th Australasian Fluid Mechanics Conference*, Melbourne, vol. 11.5 (1989)
6. G.K. Batchelor, *The Theory of Homogeneous Turbulence* (Cambridge University Press, Cambridge, 1953)
7. D. Schlipf, D. Trabucchi, O. Bischoff, M. Hofsäß, J. Mann, T. Mikkelsen, A. Rettenmeier, J.J. Trujillo, M. Kühn, Testing of frozen turbulence hypothesis for wind turbine applications with a scanning LIDAR system, in *15th International Symposium for the Advancement of Boundary Layer Remote Sensing*, Paris, June 28–30, 2010
8. A.K.M. Uddin, A.E. Perry, I. Marusic, On the validity of Taylor's hypothesis in wall turbulence. J. Mech. Eng. Res. Dev., 19–20 (1997)

9. H. Sadeghi, A. Pollard, Axial velocity spectra scaling in a round, free jet. Turb. Heat Mass Transf. **7** (2012)
10. J. Mi, R.A. Antonia, Corrections to Taylor's hypothesis in a turbulent circular jet. Phys. Fluids **6**, 1548 (1994)
11. K.B.M.Q. Zaman, A.K.M.F. Hussain, Taylor hypothesis and large-scale coherent structures. J. Fluid Mech. **112**, 379–396 (1981)
12. M. Wilczek, H. Xu, Y. Narita, A note on Taylor's hypothesis under large-scale flow variation. Nonlin. Proc. Geophys. **21**, 645–649 (2014)
13. J.C. Del Alamo, J. Jimenez, Estimation of turbulent convection velocities and corrections to Taylor's approximation. J. Fluid Mech. **640**, 5–26 (2009)
14. J.-F. Pinton, R. Labbé, Correction to the Taylor hypothesis in swirling flows. J. Phys. II France **4**, 1461–1468 (1994)
15. P. Buchhave, W.K. George, J. Lumley, The measurement of turbulence with the laser-Doppler anemometer. Ann. Rev. Fluid Mech. **11**, 443–504 (1979)
16. C.M. Velte, W.K. George, P. Buchhave, Estimation of burst-mode LDA power spectra. Exp. Fluids **55**, 1674 (2014)
17. P. Buchhave, C.M. Velte, Reduction of noise and bias in randomly sampled power spectra. Exp. Fluids **56**, 79 (2015)
18. C.M. Velte, P. Buchhave, W.K. George, Dead time effects in laser Doppler anemometry measurements. Exp. Fluids **55**, 1836 (2014)
19. P. Buchhave, C.M. Velte, W.K. George, The effect of dead time on randomly sampled power spectral estimates. Exp. Fluids **55**, 1680 (2014)
20. Hodžić, PIV measurements on a Turbulent Free Jet, MSc dissertation, Technical University of Denmark, Department of Mechanical Engineering, 2014

Part VI
General Aspects of Turbulence

Chapter 19
Effects of Unsteady Coanda Blowing on the Wake and Drag of a Simplified Blunt Vehicle

D. Barros, J. Borée, B.R. Noack, A. Spohn, and T. Ruiz

19.1 Introduction

An ongoing challenge for manufacturers worldwide concerns the reduction of road vehicles' drag due to increasing need for fuel saving and diminution of greenhouse gas emissions. The aerodynamic drag of such vehicles is mainly the result of significant pressure differences between their front and rear surfaces. In particular, square back geometries like cars with vertical rear hatch panels, buses, or trucks have been extensively studied in the past [1] and are still subject of intense research to elucidate the main flow features [2, 3]. The domain of recirculating flow formed behind these geometries is surrounded by free shear layers, which act as a noise amplifier developing through the Kelvin–Helmholtz instabilities, while the absolute wake instability results in self-sustained oscillations [4].

D. Barros (✉)
Institut Pprime UPR-3346, CNRS – Université de Poitiers – ENSMA, Futuroscope, Chasseneuil, 86360, France

PSA Peugeot Citroën, Centre Technique de Vélizy, 78943 Vélizy-Villacoublay Cedex, France
e-mail: diogo.barros@ensma.fr

J. Borée • A. Spohn
Institut Pprime UPR-3346, CNRS – Université de Poitiers – ENSMA, Futuroscope, Chasseneuil 86360, France

B.R. Noack
LIMSI – CNRS, UPR 3251, Campus Universitare dÓrsay, Rue John Von Neumann, Bat 508, F-91405 Orsay Cedex, France

Institut für Strömungsmechanik, Technische Universität Braunschweig, Hermann-Blenck-Straße 37, 38108 Braunschweig, Germany

T. Ruiz
PSA Peugeot Citroën, Centre Technique de Vélizy, 78943 Vélizy-Villacoublay Cedex, France

A. Pollard et al. (eds.), *Whither Turbulence and Big Data in the 21st Century?*,
DOI 10.1007/978-3-319-41217-7_19

In recent years, flow control turned out to be an efficient way to modify bluff body wakes with the aim to decrease drag [5]. In fully 3D flows past rectangular bluff bodies, for example, base pressure recovery can be obtained by the use of boat-tailing surfaces installed on the trailing edges [3] or by producing flow deflection through steady jets associated or not with a Coanda effect [6–8]. The combination of fluidic actuation with boat-tailing has been recently tested to improve flow attachment along the extended surface and better deviating the separated flow [9, 10]. For practical applications, the tested control techniques show limitations related to power cost and geometrical constraints. Here we explore the possibility to reduce both constraints by coupling unsteady pulsed jets with a Coanda effect to deviate more efficiently the flow. Recent studies applied fluidic actuation to virtually manipulate the flow consequently increasing the blunt body base pressure [11, 12]. Depending on the forcing frequency and amplitude of periodic jets, which are released in the downstream direction along the four trailing edges, the time-averaged wake can be virtually shaped with reduced cross-section leading to a decrease of the pressure drag. In the present work, we combine the high frequency jet actuation discussed in [11, 12] with an unsteady Coanda effect to further reduce the pressure drag.

19.2 Experimental Setup

The experiments are performed in a subsonic wind tunnel with $2.6 \times 2.4\,\mathrm{m}^2$ cross-sectional area. Figure 19.1 shows the details of the test section configuration. The blunt body model with height $H = 0.297\,\mathrm{m}$, width $W = 0.350\,\mathrm{m}$, and length $L = 0.893\,\mathrm{m}$ is the same as the numerically simulated in [13] and similar to the square back Ahmed body [1]. The leading edges of the model are rounded with radius $R = 0.085\,\mathrm{m}$. The model is mounted on a false floor to control the upstream boundary layer thickness. Four profiled supports fixed the geometric ground clearance G to $0.05\,\mathrm{m}$. Most of the results are presented for an upstream velocity U_o of $15\,\mathrm{m\,s^{-1}}$ resulting in a Reynolds number of $Re_H = 3 \times 10^5$, but other upstream conditions are also tested and discussed.

In order to evaluate the pressure drag changes obtained by control, 17 pressure taps are installed on its rear surface. The pressure measurements are normalized as $C_p = \frac{p-p_o}{q_o}$, where p_o and q_o are the upstream static and dynamic pressure. Particle image velocimetry (PIV) is performed in the wake of the model to capture the essential modifications of the forced wake. Sequences of 1000–1500 velocity fields are used to compute first and second order statistics of the streamwise and transverse (respectively x and y directions) velocity components with a resulting spatial resolution of 1 % of the model's height. In addition, a PIV setup was used to perform a zoomed view close to the upper trailing edge of the model. The resulting spatial resolution of this mounting is 0.3 % of the model's height. Further details on the entire experimental apparatus can be found in [14].

The periodic wake forcing is performed by pulsed jets along the four trailing edges of the model, as presented in Fig. 19.1b. The jet velocity V_j and frequency F_i are driven periodically by the use of solenoid valves located inside the model, as

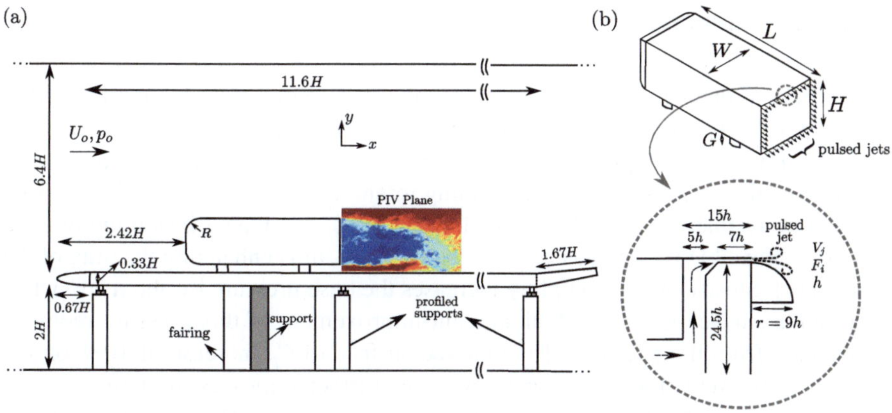

Fig. 19.1 Experimental setup. (**a**) Wind tunnel, flat plate, and model dimensions. The PIV field of view is displayed in the symmetry plane of the configuration. (**b**) Model geometry and description of the pulsed jet system with the Coanda surface

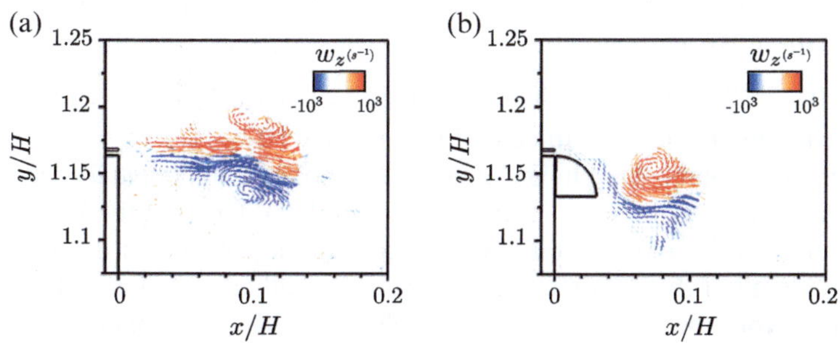

Fig. 19.2 Vorticity field ω_z of the pulsed jets during the stroke phase: (**a**) without Coanda surface and (**b**) with Coanda surface. Both cases show the evolution of the actuator jets in quiescent air at $\mathrm{St}_H = 0.4$

explained in [15]. The width of the exit cross-section of the jet slit is $h \sim 1$ mm. A Coanda surface composed of a quarter of a disk with radius $r = 9$ mm can be fixed flush to the exit slit along all trailing edges. Measurements performed with and without the presence of this small appendix will be analyzed in what follows. We define the momentum coefficient of the jet as $C_\mu = \frac{s_j \overline{V_j^2}}{S U_o^2}$, where s_j and S are, respectively, the jet slit cross-sectional area and the frontal surface of the model. For spectral analysis, the actuation frequencies are considered in dimensionless form by defining the Strouhal number $\mathrm{St}_H = \frac{F_i H}{U_o}$. An example of the vorticity field issued during the stroke phase of actuation at $\mathrm{St}_H = 0.4$ is illustrated in Fig. 19.2 in the case of quiescent surrounding air.

19.3 Results and Discussions

The effects of the actuation frequency St_H on the base pressure are summarized in Fig. 19.3. Clearly the actuation frequency has a strong impact on the base pressure changes. First, low-frequency forcing at $\mathrm{St}_H \sim 0.4$ considerably decreases the pressure of the rear surface, corresponding to an increase of the model's pressure drag by more than 10 %. This Strouhal number St_H is situated inside the interval [0.1, 1] responsible for decreasing the base pressure, thus enhancing the drag of the model [16]. Steady blowing slightly increases the base pressure by about 2 % when no Coanda surface is added. A significant improvement of this pressure recovery is obtained from the addition of the curved surface (a C_p increase of 10 % on the rear surface). Actuation at higher frequencies further improves the base pressure recovery even without the Coanda surface at $\mathrm{St}_H \sim 12$. The frequency of this excitation corresponds to approximately 60 times the discrete oscillatory wake mode (see the next paragraph) and about 11 times the most amplified estimated frequency of the upper shear layer [17]. Hence, actuation at this time-scale is decoupled from the natural flow instabilities and is denominated high frequency (HF) actuation. By applying HF Coanda blowing, the base pressure increases over 30 % at $Re_H = 3 \times 10^5$ when the effective jet velocity is $\sqrt{\overline{V_j^2}} = 11.6\,\mathrm{m\,s^{-1}}$ with an associated momentum coefficient of $C_\mu = 7.5 \times 10^{-3}$.

PIV measurements of the mean streamwise velocity component $\overline{u}$ shed light into the spatial evolution of the wake. In Fig. 19.4 (a, left), far from the bluff body, the contours $\overline{u} = 0.25$ indicate that HF Coanda forcing decreases the wake width, while the length of the recirculation zone remains rather unchanged. This can also be deduced from the streamlines of the velocity field in this plane colored by $\overline{u}$: actuation leads to a thinner wake. Close to the model, Fig. 19.4b shows for the forced case a rapid convergence of the maximum mean vorticity inside the shear layers towards the center of the wake, which points to important modifications of the shear development leading to a thinner wake by mean flow vectoring. At the same time, the thickness of the shear layers becomes larger while the peak value of vorticity decreases more than 15 % between $x/H = 0.08$ and $x/H = 0.16$.

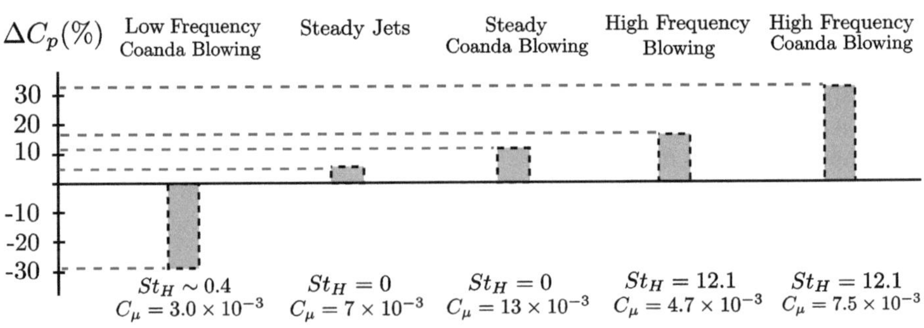

Fig. 19.3 Evolution of base pressure for different forcing parameters at $Re_H = 3 \times 10^5$. Low-frequency forcing considerably decreases the base pressure of the model, while high frequency actuation increases the rear pressure over 30 % using the Coanda surface

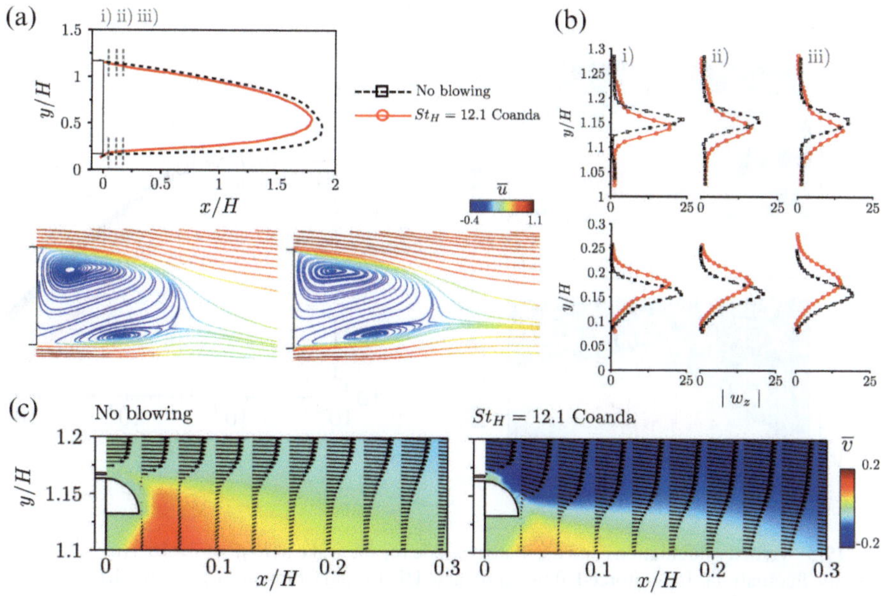

Fig. 19.4 (**a**) Streamwise velocity iso-contour $\overline{u} = 0.25$ for the unforced flow and for HF Coanda blowing ($St_H = 12.1$) together with the streamlines of the symmetry plane velocity field. (**b**) Profiles of time-averaged vorticity at three streamwise x locations close to the edge. (**c**) Upper shear layer zoom and transverse velocity field $\overline{v}$ for the unforced flow (*left*) and HF Coanda blowing (*right*). The present Reynolds number is $Re_H = 3 \times 10^5$

The changes in flow curvature are particularly visible in a zoomed view near the upper trailing edge of the model, as shown in Fig. 19.4c. We compare the unforced reference flow (i.e., no blowing) to HF Coanda blowing. The picture shows the mean transverse velocity $\overline{v}$ contours and velocity vectors near the flow separation at the upper shear layer. The deflection of the actuated flow is clearly visible. The flow attachment along the Coanda surface implies a significant flow deflection modifying the curvature of the mean velocity field consequently altering the pressure gradients across the separating shear layer. This flow deviation acts as a *pneumatic boat-tailing* leading to higher base pressure when HF Coanda blowing is applied. Additionally, we observe a noteworthy reduction of the upward velocity close to the rear surface of the model, indicating that the intensity of the recirculating flow has been reduced.

The velocity fluctuations and spectral content of both wakes give additional information. Figure 19.5a presents contour maps of the transverse velocity fluctuations $\overline{v'^2}$ in the wake. We note an overall damping of these fluctuations for $x/H > 0.2$. At $x/H = 0.2$, $\overline{v'^2}_{max}$ reduced by 16 % and at $x/H = 0.5$ the reduction is even increased to 26 % in both shear layers, suggesting a modified development of shear flow instabilities. The distribution of $\overline{v'^2}$ remains, however, quite similar along the wake for both natural and forced cases. Therefore the dynamics of the

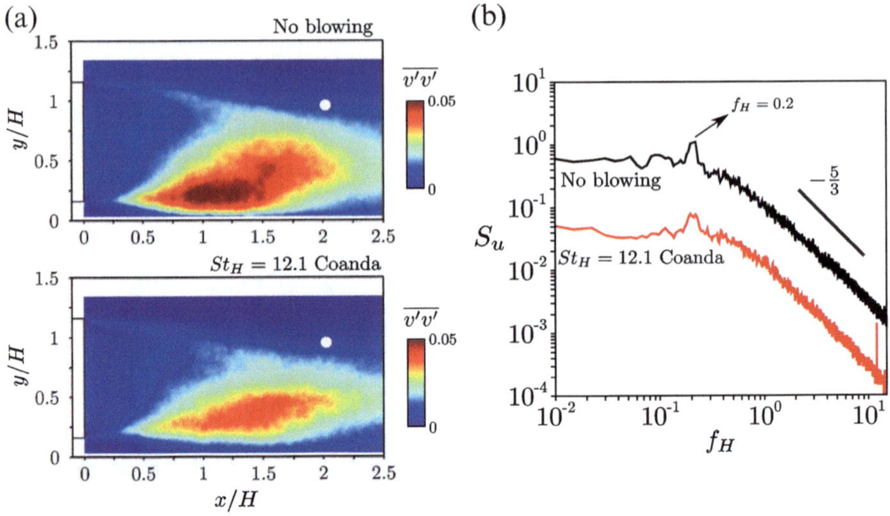

Fig. 19.5 Velocity fluctuations and oscillatory mode for the wake at $Re_H = 3\times10^5$. (**a**) Transverse velocity fluctuations for unforced flow (*top*) and HF Coanda blowing (*bottom*). (**b**) Effects of actuation on the global vortex shedding frequency

near wake flow is not strongly affected spatially by forcing. This is also confirmed by the power spectral density of a velocity signal taken by a single hot-wire at $(x/H = 2.1, y/H = 0.9)$. In Fig. 19.5b, the reference flow exhibits a discrete peak at the normalized frequency $f_H = \frac{fH}{U_o} \sim 0.2$, which is the vortex shedding mode commonly observed for this square back geometry [2, 16]. The high frequency forced wake exhibits the same mode, indicating similar global oscillatory shedding for both flows at the same frequency.

The unsteady, high frequency shear layer forcing affects the wake in mainly two ways. Without the Coanda effect (see [11, 15]), it enables a narrowing of the recirculating flow leading to aerodynamic wake shaping. In the present results, actuation with a small radius of curvature along the trailing edges increases the base pressure by further flow deviation. Both cases show significant modifications of the turbulent properties of the wake. The injection of small-scale vorticity along the shear layers has a stabilizing effect on the velocity fluctuations, similarly to the dissipative small-scale actuation discussed in [18]. Understanding the way this high frequency forcing affects the shear layer development is crucial to our comprehension of the bluff body drag changes. Time-resolved analysis from velocity fields acquired in the vicinity of the pulsed jets would be fundamental to characterize how the vorticity is convected along the shear flow. Besides, parametric tests to unravel the coupling between the unsteady forcing and the Coanda effect

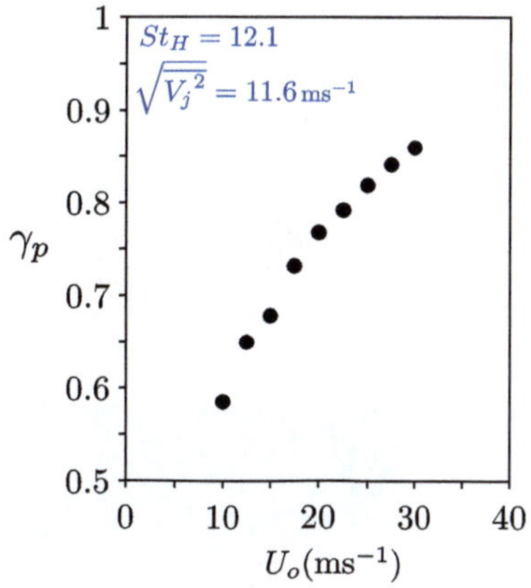

Fig. 19.6 Effect of upstream flow velocity U_o on the base pressure recovery $\gamma_p = 1 - \Delta C_p(\%)/100$ when HF Coanda blowing is applied at $\mathrm{St}_H = 12.1$ with a fixed jet amplitude $\sqrt{\overline{V_j^2}} = 11.6\,\mathrm{m\,s^{-1}}$

should be conducted to gain further information on the pressure drag mechanisms involved here.

An example is the influence of the Reynolds number on the flow deviation and the pressure drag changes promoted by unsteady Coanda actuation. To investigate it, we vary the upstream flow velocity U_o and measure the base pressure recovery parameter $\gamma_p = 1 - \Delta C_p(\%)/100$, as shown in Fig. 19.6. There is a noteworthy augmentation of γ_p with increasing U_o. For example, by varying U_o from 15 to $30\,\mathrm{m\,s^{-1}}$, γ_p is modified from 0.67 to approximately 0.85, halving the base pressure recovery $\Delta C_p(\%)$.

The decrease of the actuation performance with U_o might be associated with a smaller flow deviation by the Coanda effect, among other effects which remain to be clarified. Figure 19.7 shows the contour maps of the time-averaged cross-stream velocity $\overline{v}$ with the velocity streamlines close to the upper edge in the cases of the unforced flow ($U_o = 15\,\mathrm{m\,s^{-1}}$) and for the HF Coanda blowing with variable U_o. A notable deviation of the flow when $U_o = 10\,\mathrm{m\,s^{-1}}$ is observed, leading to more than 40 % base pressure recovery. By the analysis of $\overline{v}$ magnitudes, we note a smooth reduction of the amount of vectored flow for increasing U_o: the result is a smaller deviation of the shear layers directly reducing the fluidic boat-tailing effect and the base pressure recovery.

19.4 Concluding Remarks

In the present work, we combined periodic shear layer forcing and flow deflection by the Coanda effect to increase the base pressure of a square back bluff body. Base pressure increases of more than 30 % can be achieved by this combination. The unsteady Coanda effect produces flow deflection near the base of the blunt body as well as a decrease of the velocity fluctuations in the wake. The deflections are responsible for the formation of convex curved mean streamlines with strong

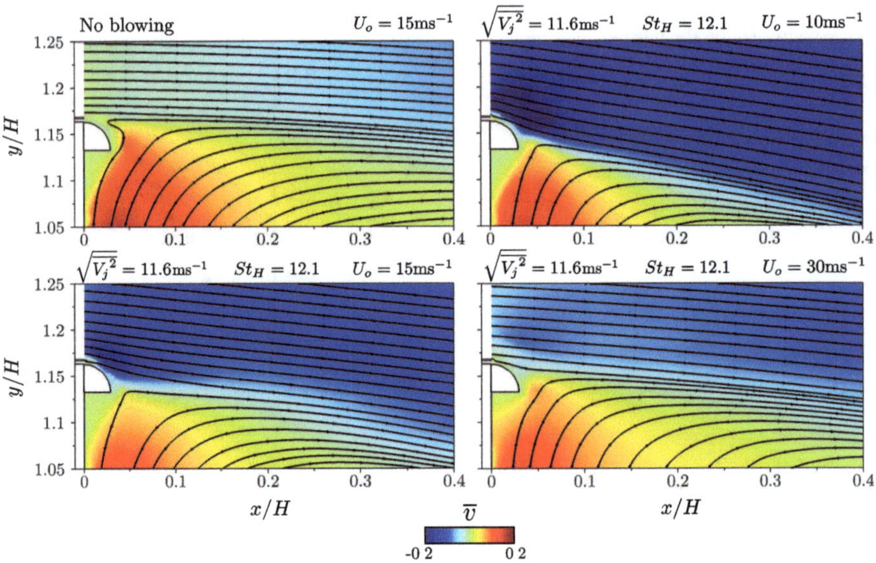

Fig. 19.7 Impact of upstream flow velocity U_o on the flow deviation close to the upper edge of the model when HF Coanda blowing is applied at $\text{St}_H = 12.1$ for a fixed jet amplitude $\sqrt{\overline{V_j^2}} = 11.6\,\text{m}\,\text{s}^{-1}$

transverse pressure gradients, which finally lead to a significant increase of base pressure. Hence, the HF Coanda actuation works like aerodynamic form shaping but needs minimal geometrical modifications to achieve substantial pressure recovery. It is expected to accumulate two main advantages. First, it enables a pneumatic boat-tailing effect due to an unsteady separation delay and to the interaction of pulsed jets with the separated flow. Besides, the high frequency forcing may act as a dissipative actuation of the turbulent shear layers in a similar way as proposed by Vukasinovic et al. [18]. These effects are highly dependent on the upstream Reynolds number: an increase of the flow velocity leads to a smaller flow deviation from the unsteady Coanda actuation at a fixed jet amplitude. The dynamics leading to the aforementioned mean flow modifications are presently under investigation. The low and high frequency dynamics of forced wakes suggests the use of time-resolved PIV to analyze more deeply the physical mechanisms of such actuation. A sufficiently long time exposition is needed and will significantly increase the amount of data for converged averaging in the wake. This will enable to educe the pulsed jet dynamics and the associated introduction of vorticity into the flow. Obtaining a complete picture of the physics of these recirculating flows forced at high frequency is important to optimize control strategies [19].

Acknowledgements The thesis of D.B is supported financially by PSA—Peugeot Citroën and ANRT in the context of the OpenLab Fluidics between PSA - Peugeot Citroën and Institut Pprime (fluidics@poitiers). The authors acknowledge the funding of the Chair of Excellence - Closed-loop control of turbulent shear flows using reduced-order models (TUCOROM)- supported by the French Agence Nationale de la Recherche (ANR). We warmly thank the support during the experiments by J.M. Breux.

References

1. S.R. Ahmed, G. Ramn, G. Faltin, Some salient features of the time averaged ground vehicle wake. SAE Technical Report. No. 840300, Society of Automotive Engineers Inc., Warrendale, PA (1984)
2. M. Grandemange, M. Gohlke, O. Cadot, Turbulent wake past a three-dimensional blunt body. Part 1. Global modes and bi-stability. J. Fluid Mech. **722**, 51–84 (2013)
3. H. Choi, J. Lee, H. Park, Aerodynamics of heavy vehicles. Ann. Rev. Fluid Mech. **46**, 441–468 (2014)
4. P. Huerre, P.A. Monkewitz, Local and global instabilities in spatially developing flows. Ann. Rev. Fluid Mech. **22**(1), 473–537 (1990)
5. H. Choi, W.P. Jeon, J. Kim, Control of flow over a bluff body. Ann. Rev. Fluid Mech. **40**, 113–139 (2008)
6. R.J. Englar, Advanced aerodynamic devices to improve the performance, economics, handling and safety of heavy vehicles. SAE Technical Report. No. 2001-01-2072, Society of Automotive Engineers (2001)
7. R.P Littlewood, M.A. Passmore, Aerodynamic drag reduction of a simplified squareback vehicle using steady blowing. Exp. Fluids **53**(2), 519–529 (2012)
8. J. Pfeiffer, R. King, Multivariable closed-loop flow control of drag and yaw moment for a 3d bluff body, in *Proceedings of the 6th AIAA Flow Control Conference* (2012)
9. S. Chaligné, T. Castelain, M. Michard, D. Juvé, Active control of the flow behind a two-dimensional bluff body in ground proximity. C. R. Méc. **341**(3), 289–297 (2013)
10. H.J. Schmidt, R. Woszidlo, C.N. Nayeri, C.O. Paschereit, Drag reduction on a rectangular bluff body with base flaps and fluidic oscillators. Exp. Fluids **56**(7), 1–16 (2015)
11. A.R. Oxlade, J.F. Morrison, A. Qubain, G. Rigas, High-frequency forcing of a turbulent axisymmetric wake. J. Fluid Mech. **770**, 305–318 (2015)
12. D. Barros, J. Borée, B.R. Noack, A. Spohn, T. Ruiz, Bluff body drag manipulation using pulsed jets and Coanda effect . J. Fluid Mech. (2016). In print
13. J. Östh, B.R. Noack, S. Krajnović, D. Barros, J. Borée, On the need for a nonlinear subscale turbulence term in POD models as exemplified for a high-Reynolds-number flow over an Ahmed body. J. Fluid Mech. **747**, 518–544 (2014)
14. D. Barros, Wake and drag manipulation of a bluff body using fluidic forcing. Ph.D. thesis, École Nationale Supérieure de Mécanique et d'Aérotechnique (ENSMA), 2015
15. D. Barros, T. Ruiz, J. Borée, B.R. Noack, Control of a three-dimensional blunt body wake using low and high frequency pulsed jets. Int. J. Flow Control **6**(1), 61–74 (2014)
16. D. Barros, J. Borée, B.R. Noack, A. Spohn, Resonances in the forced turbulent wake past a 3D blunt body. Phys Fluids (1994-present) 28(6), 065104 (2016)
17. C.M. Ho, P. Huerre, Perturbed free shear layers. Ann. Rev. Fluid Mech. **16**, 365–422 (1984)
18. B. Vukasinovic, Z. Rusak, A. Glezer, Dissipative small-scale actuation of a turbulent shear layer. J. Fluid Mech. **656**, 51–81 (2010)
19. S.L. Brunton, B.R. Noack, Closed-loop turbulence control: progress and challenges. Appl. Mech. Rev. **67**(5), 050801 (2015)

Chapter 20
Challenges for Large Eddy Simulation of Engineering Flows

C. Fureby

20.1 Introduction

Large eddy simulation (LES) is a mathematical method for turbulent fluid flow used in computational fluid dynamics. It was first proposed in 1963 by J. Smagorinsky to study atmospheric boundary layer flow [1], and many of the salient features of LES were first explored by Deardorff [2]. During the last 3 decades the use of LES has increased considerably, primarily due to the rapid increase in computational capacity, and is currently used in a variety of applications such as aerodynamics [3–5], hydrodynamics [6, 7], combustion and turbomachinery [8–12], acoustics [13, 14], geophysics [15], weather and climate modeling [16, 17], and astrophysics [18, 19]. LES is based on the incompressible, compressible, or reactive Navier–Stokes equations (NSE), depending on subject area, and use low-pass filtering [20] to eliminate the small-scale flow physics not resolved on the computational grid. This means that the large-scale flow physics is resolved, whereas the effects of the unresolved small-scale flow physics must be handled by subgrid models. It is commonly agreed [21] that at least 85 % of the total kinetic energy should be resolved in order for the small-scale flow physics to be universal enough to be amenable to modeling based on our present understanding and parameterization of Kolmogorov turbulence [22–24] and its effects on chemical reactions and interfaces [25–27]. Properly used, LES allows for better fidelity than alternative approaches such as Reynolds Averaged Navier–Stokes (RANS) methods [28, 29], at a lower computational cost compared to direct numerical simulation (DNS) [30, 31]. For practical engineering systems the use of LES requires large grids, and for wall-bounded flows either wall-resolved LES or wall-modeled LES can be used.

C. Fureby (✉)
Defense Security Systems Technology, Swedish Defense Research Agency – FOI, SE 147 25 Tumba, Stockholm, Sweden
e-mail: fureby@foi.se

A. Pollard et al. (eds.), *Whither Turbulence and Big Data in the 21st Century?*,
DOI 10.1007/978-3-319-41217-7_20

In wall-resolved LES the grid is successively refined as the wall is approached in order to resolve all dynamically relevant flow scales, whereas in wall-modeled LES a wall model is used to embody the effects of the near-wall flow scales eliminated by the low-pass filtering [32, 33].

Presently there exist several different LES methodologies such as detached eddy simulation (DES) [34], and its sequels delayed DES (DDES) [35], improved delayed DES (IDDES) [35], implicit LES (ILES) [36], and two scale LES (TSLES) [37]. For DES, DDES, and IDDES a carefully selected RANS model is used as a built-in wall model, which together with a blending function transitions into an LES model outside of the near-wall region. This approach leads to large savings in grid size and computational time, whilst delivering rational predictions. ILES, on the other hand, is based on the observation that the leading order truncation error from certain finite volume discretization algorithms is of the same mathematical form as that of most subgrid models [38]. Using a selected flux reconstruction function, efficient and accurate ILES models result. TSLES is founded on a combination of a conventional LES and a simplified (one-dimensional) DNS in the wall normal direction, resulting in an efficient model for high Reynolds (Re) number wall-bounded flows.

This paper attempts to summarize the current state of the art in LES of practical engineering systems, and to highlight some of the challenges that need to be addressed before a truly predictive model framework is available. As LES is employed in a wide range of applications an effort is made here to provide a compact but still comprehensive description of the LES methodology and an overview of how LES is used to study practical engineering systems. Both theoretical aspects of LES, such as subgrid modeling and numerical methods, and examples of LES computations, ranging from building block flows to real engineering systems, will be presented and discussed. However, due to the broad scope of the paper, most technical details will not be provided explicitly but only in the references cited. Towards the end of the paper an attempt will be made to summarize the remaining modeling and operational challenges.

20.2 Large Eddy Simulation Models and Numerical Methods

The incompressible, compressible, and reactive NSE are the balance equations of mass, momentum, and energy supplemented with appropriate constitutive equations and thermal- and caloric equations-of-state [39], describing convection, diffusion, and chemical reactions (when appropriate). LES is based on a separation of scales, which is achieved via spatial low-pass filtering. Physical processes occurring on scales larger than the filter width, Δ, are resolved, whereas physical processes occurring on scales smaller than Δ are represented by subgrid models. Depending on how the low-pass filtering, variable decomposition, and associated subgrid modeling are performed, different "flavors" of LES result.

20.2.1 Conventional or Explicit LES Models

In conventional or explicit LES the dependent variables are decomposed into resolved and subgrid components, respectively, with the resolved components being obtained from low-pass filtering of the dependent variables [20]. When the low-pass filtering is applied to the incompressible, compressible, and reactive NSE this operation results in an almost identical set of equations but with additional stress and flux terms resulting from the variable decomposition and the associated low-pass filtering [8, 9, 20, 21]. More specifically, for the incompressible NSE's the filtering of the momentum equation gives rise to the subgrid stress tensor $\mathbf{B} = \overline{\mathbf{v} \otimes \mathbf{v}} - \overline{\mathbf{v}} \otimes \overline{\mathbf{v}}$, with $\mathbf{v}$ being the velocity and the overbars denoting the filtering operation, G∗. This subgrid stress tensor thus contains information about the physics occurring on scales smaller than the filter width, Δ, which in most practical situations corresponds to the characteristic grid size, h, and must therefore be modeled by subgrid models. For the compressible and reactive NSE's the situation is more complicated due to these equations being more complicated, and due to the fact that the fluid properties usually depend on the density, ρ, and/or the temperature, T. In this case we often use Favre-filtering, e.g., [28], to simplify the variable decomposition, and as a result of that the filtering of the momentum equation gives rise to the subgrid stress tensor $\mathbf{B} = \overline{\rho}\left(\widetilde{\mathbf{v} \otimes \mathbf{v}} - \tilde{\mathbf{v}} \otimes \tilde{\mathbf{v}}\right) + \tilde{\mathbf{S}} - \overline{\mathbf{S}}$, in which the tildes represent the Favre-filtering and $\mathbf{S}$ the viscous stress tensor. Similarly, filtering of the (total) energy equation gives rise to the subgrid flux vector $\mathbf{b}_E = \overline{\rho}\left(\widetilde{\mathbf{v}\mathrm{E}} - \tilde{\mathbf{v}}\tilde{\mathrm{E}}\right) + \left(\tilde{\mathbf{S}} - \overline{p}\mathbf{I}\right)\tilde{\mathbf{v}} - \overline{(\mathbf{S} - p\mathbf{I})\,\mathbf{v}} + \tilde{\mathbf{h}} - \overline{\mathbf{h}}$, in which E is the total energy, p the pressure, and $\mathbf{h}$ the heat flux vector. For reacting flows we also need to consider the species concentrations, being determined by separate transport equations the filtering of which gives rise to additional subgrid flux vectors of the form $\mathbf{b}_\mathrm{i} = \overline{\rho}\left(\widetilde{\mathbf{v}Y_i} - \tilde{\mathbf{v}}\tilde{Y}_i\right) + \tilde{\mathbf{j}}_i - \overline{\mathbf{j}}_i$, where j_i are the mass diffusion vectors. An additional complication for reactive flows is the reactions themselves, being determined by a reaction mechanism, [40], of varying complexity from a 1-step global reaction mechanism, with only a few species, to detailed reaction mechanisms of thousands of reactions and hundreds of species. The sum of the (formation and destruction) rates of these species appears as a source term, $\overline{\dot{w}_i} = P_{ij}\overline{\dot{w}_j}$, in the species transport equations, in which $\dot{w}_j$ are the reaction rates and P_{ij} the stoichiometric matrix, representing the reaction mechanism. This term is very complex as the reaction rates are non-linear functions of the species concentrations involved in the reactions and the temperature. Modeling the filtered reaction rates is the key challenge in LES of reacting flows in general and combustion in particular.

For LES based on finite volume formulations, as nowadays often is the case, the low-pass filtering is often performed implicitly, on the grid, using the finite volume discretization operator as the filtering operation. This has not only the advantage of simplifying the LES formulation considerably, but also the disadvantage of combining physics and numerics so that they cannot be entirely separated. Whether or not this is important is a very controversial issue.

As indicated by Pope [21], the grid resolution plays an important role in LES as the filtering is performed to remove scales smaller than $\Delta \approx \mathrm{h}$. Following Kolmogorov [22], dimensional analysis suggests that the smallest scales in turbulent flow are the Kolmogorov scales $\ell_{\mathrm{K}} \approx (\nu^3/\varepsilon)^{1/4}$, in which ε is the dissipation [41], and by further assuming equilibrium between dissipation and production, $P \propto v_{\mathrm{I}}^3/\ell_{\mathrm{I}}$, where v_{I} and ℓ_{I} are the integral scales, it follows that $\ell_{\mathrm{I}}/\ell_{\mathrm{K}} \approx Re_{\mathrm{I}}^{3/4}$, where Re_{I} is Reynolds number based on the integral scales. This suggests that DNS requires approximately $N_{\mathrm{DNS}} \approx (\ell_{\mathrm{I}}/\ell_{\mathrm{K}})^3 \approx Re_{\mathrm{I}}^{9/4}$ grid cells to be well resolved. For high Re number flows the grid resolution therefore quickly becomes prohibitive. A similar estimate can be made for LES by assuming that LES resolves the Taylor scales [42], so that $\Delta\,(\approx h) < \ell_{\mathrm{T}}$. The Taylor scales typically reside between the integral and Kolmogorov scales, but towards the end of the inertial sub-range, and can be approximated from the assumption that $\varepsilon = v_{\mathrm{I}}^3/\ell_{\mathrm{I}} \approx 15\nu v_{\mathrm{I}}^2/\ell_{\mathrm{T}}^2$ in homogeneous isotropic turbulence. This finally results in that $\ell_{\mathrm{I}}/\ell_{\mathrm{T}} \approx Re_{\mathrm{I}}^{1/2}$, which then indicates that a well-resolved LES computation need on the order of $N_{\mathrm{LES}} \approx (\ell_{\mathrm{I}}/\Delta)^3 \approx Re_{\mathrm{I}}^{3/2}$ grid cells. Compared to that of a DNS computation, this is much more manageable, even for very high Re_{I}, although it will require substantial computational resources, particularly since LES is founded on simulating the evolution of the flow over time, whilst gathering statistics during the simulation.

As an illustration of the filtering process, Fig. 20.1a shows a slice through a DNS velocity field of homogeneous isotropic turbulence obtained on a 1024^3 grid, whereas Fig. 20.1b shows the corresponding filtered velocity field on a 64^3 grid. Figure 20.1c shows results from an LES computation on the same 64^3 grid using the localized dynamic k-equation model [43]. The most apparent difference is the lack of small-scale details both in the filtered field and in the

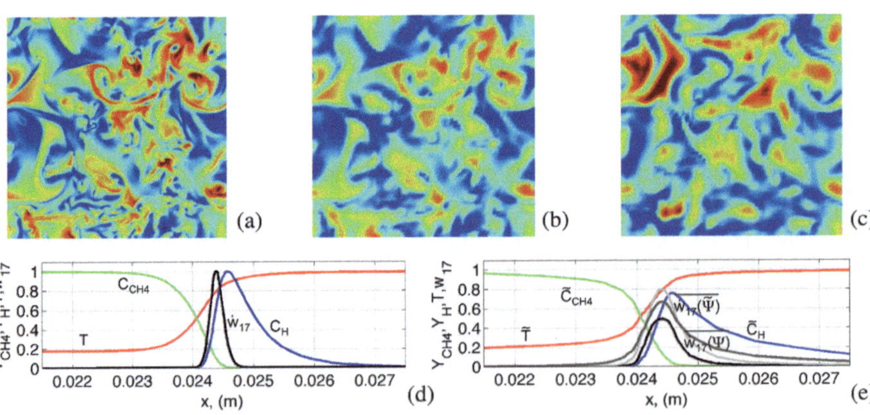

Fig. 20.1 Illustration of key aspects in LES. Representation of flow fields: (**a**) a slice through a DNS flow field, (**b**) the same DNS flow field filtered by a box-filter with $\Delta = 16\mathrm{h}_{\mathrm{DNS}}$, (**c**) LES predictions of the same flow field. Representation of source terms: (**d**) unfiltered flame profiles and reaction rate profiles and (**e**) filtered flame profiles and reaction rate profiles

LES predictions. The larger flow structures are, however, very similar for all three cases, suggesting that a carefully performed LES, on a comparatively coarse grid, have potential to reproduce the large-scale energy containing flow structures. Note also that when refining the LES grid, the LES results will monotonically approach the result of the DNS computation. Figure 20.1d, e similarly illustrates the challenge of accurately modeling the highly non-linear reaction rates for the elementary reaction $CH_4 + H = CH_3 + H_2$, in the Smooke and Giovangigli CH_4-air reaction mechanism [44]. More specifically, Fig. 20.1d shows the unfiltered temperature, T, H concentration, C_H, CH_4 concentration, C_{CH4}, and reaction rate, $\dot{w}_{17} = A_{17}\exp(-T_{a,17}/T)\, C_{CH4}C_H$. In the rate-expression $\dot{w}_{17}$, A_{17} and $T_{a,17}$ are the pre-exponential factor and activation temperatures, respectively. Similarly, Fig. 20.1e shows the corresponding filtered variables, $\tilde{T}$, $\tilde{C}_{CH4}$ and $\tilde{C}_H$ together with the filtered reaction rate $\overline{\dot{w}_{17}(T, C_{CH4}, C_H)}$. By comparing profiles it is clear that the comparatively small difference in raw and filtered dependent variables give rise to a significant difference between the raw and filtered reaction rates. In Fig. 20.1e we have also included a few efforts to represent the filtered reaction rates using only resolved data, such as $\dot{w}_{17}\left(\tilde{T}, \tilde{Y}_{CH4}, \tilde{Y}_H\right)$ and $\overline{\dot{w}_{17}\left(\tilde{T}, \tilde{Y}_{CH4}, \tilde{Y}_H\right)}$. As evident, neither of these approximations comes close to approximating the filtered reaction rate, suggesting that highly specified models are required to represent some of these non-linear source and sink terms occurring particularly in reactive and combusting flows.

20.2.1.1 Subgrid Flow Models

The subgrid stress and flux terms, $\mathbf{B}$ (for incompressible, compressible, and reactive flows), $\mathbf{b}_E$ (for compressible and reactive flows), and $\mathbf{b}_i$ (for reactive flows), need to be modeled in order to model the effects of the small-scale, unresolved, flow physics on the resolved flow physics, and to close the LES equations. Based on the intrinsic nature of these subgrid stress and flux terms it is clear that they share many features, and primarily that the compressible (and reactive) versions of these terms can be interpreted as extensions of the incompressible subgrid stress tensor, resulting solely from filtering the convective term in the momentum equation. It is thus natural to consider the non-linear convective contribution to the subgrid stress and flux terms as the most significant, or even dominating, part as it represents the influence of turbulent transport at various scales. This is true for incompressible flows, but may not be as straightforward for compressible flows with shocks and for reactive flows with heat release.

The convective part of the subgrid stress and flux terms can be expanded by inserting the decomposition of the unfiltered variables into resolved and subgrid components, resulting in the replacement of this term by three terms having clear physical interpretations. For the incompressible LES equations, $\mathbf{B} = \left(\overline{\bar{\mathbf{v}} \otimes \bar{\mathbf{v}}} - \bar{\bar{\mathbf{v}}} \otimes \bar{\bar{\mathbf{v}}}\right) + \left(\overline{\bar{\mathbf{v}} \otimes \mathbf{v}'} - \bar{\bar{\mathbf{v}}} \otimes \overline{\mathbf{v}'} + \overline{\mathbf{v}' \otimes \bar{\mathbf{v}}} - \overline{\mathbf{v}'} \otimes \bar{\bar{\mathbf{v}}}\right) + \left(\overline{\mathbf{v}' \otimes \mathbf{v}'} - \overline{\mathbf{v}'} \otimes \overline{\mathbf{v}'}\right) = \mathbf{L} + \mathbf{C} + \mathbf{R}$, in which $\mathbf{L}$ is the Leonard stress tensor, responsible for interactions between the smallest resolved scales, $\mathbf{C}$ the cross stress tensor, responsible for interactions between the smallest resolved scales and the subgrid scales, and $\mathbf{R}$ the Reynolds stress tensor, responsible for the interactions between the subgrid scales. The subgrid stress tensor and flux vectors in the compressible and reactive LES equations can be handled in the same way, thus also resulting in Leonard, cross, and Reynolds stress tensors and flux vectors such that $\mathbf{B} = \mathbf{L} + \mathbf{C} + \mathbf{R} + \tilde{\mathbf{S}} - \overline{\mathbf{S}}$, $\mathbf{b}_E = \mathbf{l}_E + \mathbf{c}_E + \mathbf{r}_E + \tilde{\mathbf{S}}\tilde{\mathbf{v}} - \overline{\mathbf{S}\mathbf{v}} + \overline{p\mathbf{v}} - \bar{p}\tilde{\mathbf{v}} + \tilde{\mathbf{h}} - \overline{\mathbf{h}}$, and $\mathbf{b}_i = \mathbf{l}_i + \mathbf{c}_i + \mathbf{r}_i + \tilde{\mathbf{j}}_i - \overline{\mathbf{j}}_i$, respectively. The Leonard terms do not need any closure modeling, and can be obtained directly by applying a second level of filtering, whereas both the cross and Reynolds terms require modeling as these terms involve also interaction with or between the unresolved scales. In contrast to the incompressible subgrid stress tensor, the compressible and reactive subgrid stress tensor contain a term $\tilde{\mathbf{S}} - \overline{\mathbf{S}}$ arising from the replacement of T by $\tilde{T}$ in the dynamic viscosity, μ, particularly important close to shocks and reacting flows. Similarly, does the total energy subgrid flux vector $\mathbf{b}_E$ and the species subgrid flux vector $\mathbf{b}_i$ contain terms $\tilde{\mathbf{S}}\tilde{\mathbf{v}} - \overline{\mathbf{S}\mathbf{v}}$, $\overline{p\mathbf{v}} - \bar{p}\tilde{\mathbf{v}}$, $\tilde{\mathbf{h}} - \overline{\mathbf{h}}$, and $\tilde{\mathbf{j}}_i - \overline{\mathbf{j}}_i$, respectively, arising from the non-linear dependencies on the dependent variables, primarily T, of the constitutive equations. In many studies these contributions are not always mentioned although in some cases of paramount influence. Transport equations for the subgrid stress tensor, $\mathbf{B}$, and flux vectors, $\mathbf{b}_E$ and $\mathbf{b}_i$, can also be formulated following the work of Deardorff [2], and Fureby et al. [45]. For the incompressible LES equations, $\partial_t \mathbf{B} + \nabla \cdot (\mathbf{B} \otimes \tilde{\mathbf{v}}) = -\left(\mathbf{B}\nabla\tilde{\mathbf{v}} + \nabla\tilde{\mathbf{v}}^T\mathbf{B}\right) - \mathbf{J} + \boldsymbol{\Phi} - \mathbf{E}$, in which the first term on the right-hand side denotes production by resolved strain, $\mathbf{J}$ diffusion, $\boldsymbol{\Phi}$ correlation between pressure and strain, and $\mathbf{E}$ destruction by viscous action. The trace this equation becomes the exact equation of the subgrid kinetic energy, $k = \frac{1}{2} tr\mathbf{B}$.

The subgrid stress tensor and flux vectors are modeled using subgrid models, and according to Sagaut [20], these models can be divided into *functional* and *structural* models depending on if they are indented to mimic the kinetic energy cascade from large to small scales, or if they are intended to mimic the structure of the subgrid flow physics. Most subgrid models are derived for incompressible LES, and then simply extended to compressible and reactive LES, and in the subsequent overview of subgrid models, this approach is adopted.

Functional Models

Functional models are the most widely used class of subgrid models and are [20] formulated as $\mathbf{B} \approx -2\nu_k \overline{\mathbf{D}}_D$ in which $\overline{\mathbf{D}}_D$ is the deviatoric part of the rate-of-strain tensor, $\overline{\mathbf{D}}$, and ν_k the (kinematic) subgrid viscosity. Li et al. [46] recently showed, utilizing matrix exponentials, that this class of models constitutes the first order approximation to the transport equation for $\mathbf{B}$, and how these models can be improved by including also non-linear terms. This class of models includes the Smagorinsky model (SMG) [1, 47] $\nu_{kk} = c_D \Delta^2 \|\overline{\mathbf{D}}\|$ and $k = c_I \Delta^2 \|\overline{\mathbf{D}}\|^2$, in which the model coefficients c_I and c_D can be prescribed [1, 20] ($c_I \approx 0.07$ and $c_D \approx 0.12$), or dynamically computed [20, 47]. Another functional subgrid model is the wall-adapting local eddy (WALE) viscosity model [48] $\nu_k = c_w^2\Delta^2\Omega$, in which $\Omega = (\mathbf{G}_\mathrm{D}^2)^{3/2}/\left[(\mathbf{D}^2)^{5/2} + (\mathbf{G}_\mathrm{D}^2)^{5/4}\right]$, and $\mathbf{G} = \frac{1}{2}\left(\nabla\overline{\mathbf{v}}^2 + (\nabla\overline{\mathbf{v}}^2)^T\right)$, and in which the model coefficient, c_W, is calibrated to 0.50 by comparison with experimental data [49]. Yet another functional model is the one-equation eddy viscosity model (OEEVM), [43, 50], with $\nu_\mathrm{k} = c_\mathrm{k}\Delta k^{1/2}$. In this model, the subgrid kinetic energy, k, results from solving a modeled transport equation of the form $\partial_t(k) + \nabla\cdot(k\overline{\mathbf{v}}) = -\mathbf{B}\cdot\overline{\mathbf{D}} + \nabla\cdot(\nu_k\nabla k) - \varepsilon$, in which $\varepsilon = c_\varepsilon k^{3/2}/\Delta$ is the subgrid dissipation. The model coefficients, c_k and c_ε, can be assumed constant as originally proposed in [50] ($c_\mathrm{k} \approx 0.07$ and $c_\mathrm{I} \approx 1.05$), or can be evaluated dynamically (using scale similarity arguments) as in the LDKM of Kim and Menon [43].

Structural Models

Structural models incorporate a wide range of subgrid flow models, see [20] for a comprehensive survey, aiming at describing the structure of the subgrid stress tensor and flux vectors instead of the effects of these terms on the resolved flow. The most straightforward structural model is the scale similarity (SS) model of Bardina et al. [51], in which the subgrid stress tensor is modeled as $\mathbf{B} \approx \overline{\overline{\mathbf{v}}\otimes\overline{\mathbf{v}}} - \overline{\overline{\mathbf{v}}}\otimes\overline{\overline{\mathbf{v}}}$. The disadvantage of this model, although showing high correlations with experimental data [52], is that it is not dissipative enough to represent all the effects of the small-scale turbulence. Instead, it was observed that by combing the scale similarity model with an eddy viscosity model, [53], a more accurate and robust subgrid flow model was obtained. This so-called mixed model (MM) can be formulated as $\mathbf{B} \approx \overline{\overline{\mathbf{v}}\otimes\overline{\mathbf{v}}} - \overline{\overline{\mathbf{v}}}\otimes\overline{\overline{\mathbf{v}}} - 2\nu_k\overline{\mathbf{D}}_D$ in which ν_k is the kinematic subgrid viscosity of any of the aforementioned subgrid viscosity models. An interesting extension to this model is the approximate deconvolution model (ADM) [54, 55], in which G is inverted using iterative methods based on polynomial approximations of the filter kernel [55] to obtain an approximation of the unfiltered fields $\mathbf{v}^* = G^{-1} * \overline{\mathbf{v}}$, which then are inserted back into the subgrid stress tensor such that $\mathbf{B} \approx \overline{\mathbf{v}^*\otimes\mathbf{v}^*} - \overline{\mathbf{v}}^*\otimes\overline{\mathbf{v}}^*$. Using the N$^\mathrm{th}$ order van-Cittert deconvolution algorithm [54, 55], $\mathbf{v}^* \approx G_N^{-1} * \overline{\mathbf{v}} = \Sigma_{n=0}^N (1-G)^n * \overline{\mathbf{v}}$. However, as for the SS model this model of $\mathbf{B}$ is not sufficiently dissipative to represent the effects of the small-scale turbulence, and therefore the ADM model must be combined with an explicit dissipation term or a subgrid viscosity ν_k.

Subgrid Models for Compressible and Reactive Flows

The compressible and reactive LES equations are, as mentioned, more complex than the incompressible LES equations, involving a subgrid stress tensor $\mathbf{B} = \overline{\rho}\left(\widetilde{\mathbf{v}\otimes\mathbf{v}} - \tilde{\mathbf{v}}\otimes\tilde{\mathbf{v}}\right) + \tilde{\mathbf{S}} - \overline{\mathbf{S}}$, and two subgrid flux vectors $\mathbf{b}_E = \overline{\rho}\left(\widetilde{\mathbf{v}E} - \tilde{\mathbf{v}}\tilde{E}\right) + \tilde{\mathbf{S}}\tilde{\mathbf{v}} -$

$\overline{\mathbf{Sv}}+\overline{p\mathbf{v}}-\bar{p}\tilde{\mathbf{v}}+\tilde{\mathbf{h}}-\overline{\mathbf{h}}$ and $\mathbf{b}_i = \overline{\rho}\left(\widetilde{\mathbf{v}Y_i} - \tilde{\mathbf{v}}\tilde{Y}_i\right)+\tilde{\mathbf{j}}_i-\overline{\mathbf{j}}_i$. Besides the parts resulting from the filtering of the convective terms in the momentum, energy and species equations contributions also arise from the non-linear dependencies on the dependent variables of the constitutive equations. This should be reflected in the modeling of the subgrid stress and flux terms but closure models for these terms are mainly based on simple extensions of the subgrid closure models adopted for incompressible LES. The most common subgrid models for compressible or reactive flows are based on extended subgrid viscosity models in which $\mathbf{B} \approx -2\mu_k\tilde{\mathbf{D}}_D$, where now μ_k is the subgrid viscosity and $\mathbf{b}_E \approx (\mu_k/Pr_t)\nabla\tilde{E}$ and $\mathbf{b}_i \approx (\mu_k/Sc_{t,i})\nabla\tilde{C}_i$, in which $\tilde{\mathbf{D}}_D$ is the deviatoric part of $\tilde{\mathbf{D}}$, and Sc_{it} and Pr_t are the turbulent Schmidt and Prandtl numbers, respectively. These models depend on μ_k, which typically is prescribed using simple compressible extensions of the models described in section "Functional Models". Also mixed models have been successively used to compute compressible and reactive flows [8, 56], in which $\mathbf{B} = \overline{\rho}\left(\widetilde{\tilde{\mathbf{v}} \otimes \tilde{\mathbf{v}}} - \tilde{\tilde{\mathbf{v}}} \otimes \tilde{\tilde{\mathbf{v}}}\right)-2\mu_k\tilde{\mathbf{D}}_D$, $\mathbf{b}_E = \overline{\rho}\left(\widetilde{\tilde{\mathbf{v}}\tilde{E}} - \tilde{\tilde{\mathbf{v}}}\tilde{\tilde{E}}\right) - \frac{\mu_k}{Pr_t}\nabla\tilde{E}$, and $\mathbf{b}_i = \overline{\rho}\left(\widetilde{\tilde{\mathbf{v}}\tilde{Y}_i} - \tilde{\tilde{\mathbf{v}}}\tilde{\tilde{Y}}_i\right) - \frac{\mu_k}{Sc_t}\nabla\tilde{C}_i$. These models are, however, not particularly well suited for regions involving strong shocks, and have no built-in mechanism to model the effects of the non-linear constitutive equations in compressible and reactive flows. Research focusing on developing subgrid closure models with these attributes is very welcome.

Only a few subgrid models are designed for compressibility effects, in which $\nabla \cdot \tilde{\mathbf{v}}$ becomes large, and one such model is the model of Cook and Cabot [57]. They developed a subgrid viscosity model for shock–turbulence interactions. This model is based on the shock-capturing method of Neumann and Richtmyer [58], and the SMG model and takes the form $\mathbf{B} \approx -\beta_k(\nabla \cdot \tilde{\mathbf{v}})\mathbf{I} - 2\mu_k\tilde{\mathbf{D}}_D$, in which β_k is the subgrid bulk viscosity. More specifically, they propose to use $\beta_k = C_\beta^r\overline{\rho}\Delta^{(r+2)}\left|\nabla^r\tilde{\mathbf{D}}\right|$ and $\mu_k = C_\mu^r\overline{\rho}\Delta^{(r+2)}\left|\nabla^r\tilde{\mathbf{D}}\right|$, for $r = 2, 4, 6$. The use of the β_k term is the key to capturing shocks without destroying vorticity as β_k can be made large (to smooth shocks) without harming small-scale turbulence in regions where $\nabla \cdot \tilde{\mathbf{v}} \approx 0$. By setting $r>0$ the subgrid viscosity keys directly on the ringing, rather than indirectly on gradients, thus eliminating the need for ad hoc limiters and dynamic procedures to turn off μ_k in special cases and uniform shear.

20.2.1.2 LES Combustion Models

As discussed in Sect. 20.2.1 reacting flows are considerably more complicated than incompressible and compressible flows due to the chemical reactions. For example, combustion of hydrogen (H_2) in air involves 9 species and 19 reactions following O'Conaire et al. [59], whereas combustion of methane (CH_4) in air involves 53 species and 325 elementary reactions using the GRI3.0 mechanism [60]. Combustion of jet fuels, such as kerosene, is notably more complicated due to that kerosene consists of many large hydrocarbon species with an average composition of $C_{12}H_{23}$, for which Ranzi et al. [61] have developed a detailed reaction mechanism with 451 species and 17,848 reactions. Detailed reaction mechanisms are too large to be useful in simulating turbulent combustions since besides the equations of mass, momentum, and energy, equations needs also to be solved for each of the species. Instead, global [62] or skeletal reaction mechanisms [63] are usually used in LES of combustion, e.g., [64–66]. Yet another alternative is to use flamelet models in which the flame is assumed infinitely thin and the chemistry is decoupled from the flow simulation [67]. The use of skeletal reaction mechanisms, involving 10–30 species and 20–100 reaction steps, offers considerable advantage in accuracy of laminar flame data [64], and increased fidelity in modeling the turbulence chemistry interactions. An additional complexity of reacting flows is the increased range of scales, often extending beyond the (turbulent) Kolmogorov scales. This results in stiff equations, associated with small spatial scales reminiscent of thin, possibly discontinuous, fronts of intermediate, and radical species [68].

The topology of turbulent flames depends on the Re number together with the Damköhler number, $Da = \tau_I/\tau_c$, and the Karlowitz number, $Ka = \tau_K/\tau_c$, in which τ_I represents the integral time scale, τ_K the Kolmogorov time scale, and τ_c the chemical time scale [67, 69]. Figure 20.2 shows two examples of turbulent premixed CH_4-air flames in homogeneous isotropic turbulence at Re = 967 and 126, Da = 0.22 and 3.60, and Ka = 101 and 2.5, respectively, together with their representations in the Williams diagram [69]. For $Da_t > 1$ reactions occur faster than reactants are mixed by the turbulence and *reaction sheets* form, whereas for $Da_t < 1$ the turbulent scales mix reactants, which leads to *distributed reaction zones*. If both scales are of similar size, $Da_t \approx 1$, turbulence–chemistry interactions dominate the flame process. For non-premixed combustion τ_c represents a measure for the speed at which the chemistry progresses, since the flame propagation speed is limited by mixing. For premixed combustion, however, τ_c relates directly to the flame structure which helps to distinguish between further sub-regimes such as wrinkled flamelets, corrugated flamelets, thin reaction zones, broken reaction zones, and well-stirred

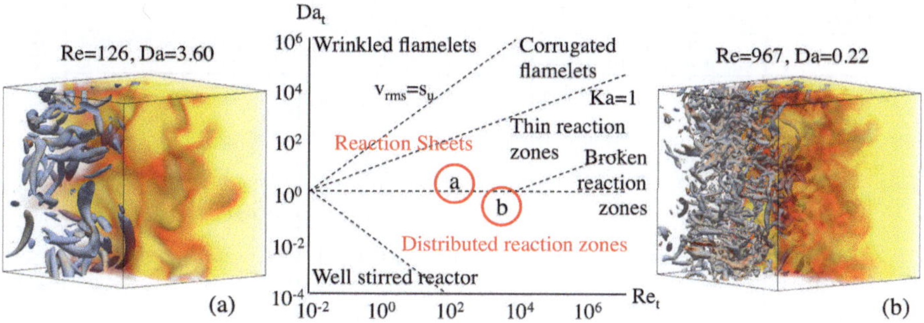

Fig. 20.2 Planar CH_4-air flames propagating in homogeneous isotropic turbulence at different Re, Da, and Ka numbers, and the Williams diagram

reactors. For Ka < 1 the smallest Kolmogorov scales are larger than the chemical scales, that limit the interaction between turbulence and chemistry. The turbulent scales can merely alter the shape of the flame front. Depending on the velocity of the turbulent eddies, v_{rms}, compared to the laminar flame speed, s_{u}, wrinkled or corrugated flamelets exist. For small v_{rms} the flame front can only be wrinkled, whereas for large v_{rms} a strong interaction with the flame front occurs, which leads to convoluted/disturbed flame fronts typical of high intensity turbulent flames studied by Poludnenko and Oran [70].

Besides being able to represent the combustion chemistry properly the ability to resolve the thin wrinkled reaction fronts is the other major challenge in LES of turbulent combustion. In the LES equations, this is implicitly manifested through the filtered reaction rate, $\overline{\dot{w}_j}$, the sensitivity of which is described in Fig. 20.1d, e. Since this is a challenging issue for LES considerable amounts of research have been devoted to develop models for turbulent combustion, e.g., [9, 67, 71]. An overview of different LES combustion models is presented in Fig. 20.3, parameterized by the type of combustion (*premixed*, *non-premixed*, and *partially premixed*) and how the chemistry is modeled (*infinitely fast*, *flamelet*, or *finite rate chemistry*). This division is somewhat arbitrary, but provides some guidance to which features particular LES combustion models are designed to manage. In Fig. 20.3 the most common LES combustion models are included.

The class of infinitely fast chemistry LES models are all based on the assumption that $\tau_{\mathrm{c}} \ll \tau_{\mathrm{k}}$ and τ_{I}, which sometimes is referred to as "mixed is burnt." Models of this class were the first to be used in LES of turbulent combustion, borrowed from turbulent combustion RANS, but lack the ability of accounting for the effects of combustion chemistry. The most well known of these models are the Bray Moss Libby [72] and eddy break-up [73] models for premixed turbulent combustion, used, e.g., by Porumbel and Menon [74], and the conserved scalar equilibrium model

	Premixed	*Non-premixed*	*Partially premixed*
Inifinitely fast chemistry	Bray Moss Libby model, [74]		
	Eddy Break Up model, [75]	Conserved Scalar Equilibrium model, [77]	
Flamelet	c or G equation flamelet, [81-83]	z-equation flamelet, [78-79]	G/c/-z-equation flamelet, [84-85]
Finite rate chemistry	Thickened Flame (TF) model, [86]		
	Presumed Filtered Density Function (FDF) models, [87]		
	Partially Stirred Reactor (PaSR) Models, [26] (1st gen. MSM)		
	Conditional Moment Closure (CMC) model, [88]		
	Transported FDF models, [89-90] (Lagrangian & Eulerian stochastic fields)		
	Linear Eddy Model (LEM), [91] (2nd gen. MSM)		

COMPUTATIONAL COST

Fig. 20.3 Overview of different LES combustion models. For detailed information about a specific model we refer to the original reference for further information

[75] used, e.g., by Cook and Riley [76]. Here, the species equations are replaced by equations for a mixture fraction, $\tilde{z}$, in non-premixed combustion, a reaction progress variable, $\tilde{c}$, in premixed combustion, and both in partially premixed combustion. Closure models are then postulated for the reacting source terms in the $\tilde{c}$-equation [72–74], whereas the temperature, $\tilde{T}$, is obtained from models of the burnt and unburnt states.

Flamelet models can be considered an extension of infinitely fast chemistry models assuming that the flame is thin compared to the length scales of the flow, and the flame behaves like an interface between fuel and oxidizer (in non-premixed combustion), [76, 77], or between reactants and products (in premixed combustion) [78–81]. Due to the scale separation it is convenient to decouple the flow simulation from that of the chemistry, which can be separately modeled using zero- and one-dimensional laminar flames, resulting in the flamelet library, which in turn can be combined with $\tilde{c}$ or a kinematic $\tilde{G}$-field as proposed in [79], $\tilde{z}$, or both $\tilde{c}$ (or $\tilde{G}$) and $\tilde{z}$, as used in [82, 83], to estimate the species concentrations, $\tilde{C}_i$, and the temperature, $\tilde{T}$. Different degrees of fidelity can be employed when combining the flamelet library with $\tilde{c}$ (or $\tilde{G}$), $\tilde{z}$ or both. For non-premixed combustion it is important to include also the scalar dissipation rate, χ, as a parameter to model the degree of departure from the equilibrium state, whereas for premixed combustion the flame wrinkling, Ξ, is a key quantity to include the increase in flame surface area, Σ, caused by the turbulence. These quantities have, however, to be modeled since they represent information pertinent to the subgrid scales as elaborated on in [76–83].

Finite rate chemistry models are the widest class of LES combustion models and are based on solving the full set of reactive LES equation for a given set of species related by a global, skeletal, or detailed reaction mechanism. The main advantage of this model is that the same model can be used in all combustion regimes, whereas the main drawback is the closure of the filtered reaction rates, $\overline{\dot{w}_j}$. Many different approaches have been developed and tested for the closure of $\overline{\dot{w}_j}$, the non-linearity of which is very complicated as illustrated in Fig. 20.1d, e. The thickened flame model [84] is a relatively simple and popular model based on the assumption that a flame can be artificially thickened by decreasing the reaction rate, $\overline{\dot{w}_j} \sim \dot{w}_j/F$, and increasing the diffusivity, $D_i \sim D_iF$, in order to keep s_u constant. Turbulence effects are included through an efficiency factor, E_F, such that $D_i \sim FE_FD_i$. Presumed filtered density functions (P-FDF), [85], are based on the fact that $\overline{\dot{w}_j} = \int_{\Psi} \wp\left(\Psi; \tilde{\Psi}, \Psi'\right) \dot{w}_j\left(\Psi\right) d\Psi$, where Ψ denotes the dependent variables and $\wp$ the probability density function to be defined. Typically, a Dirac distribution is chosen for the density, a multi-variate β-distribution for the species, and a clipped Gaussian distribution for the temperature [87]. The PDFs are parameterized by the LES fields, $\tilde{\Psi}$, and their variances, Ψ', which can be estimated in different ways. The partially stirred reactor (PaSR) model [26] is a first generation multi-scale model in which a simplified set of equations are solved in each LES cell to estimate the state of the reacting fine structures, which is then used to estimate $\overline{\dot{w}_j}$. This model has seen widespread use in gas turbines and ram- and scramjets.

The conditional moment closure (CMC) model [86] is rather involved method, using z as an additional independent variable, typically requiring a distinct CMC grid for the solution of the CMC equations necessary to close the reaction rate terms in the LES equations. Transported PDF models, e.g., [65, 87], are also quite complex models in which transport of the one-point, one-time Eulerian PDF for chemical compositions is modeled either by the evolution of a large number of notional particles as in the Lagrangian stochastic fields methods, [87], or by solving for N stochastic species concentration representations as in the Eulerian stochastic fields methods [65]. These methods are both very appealing since the reaction rates occur in unclosed form. Finally, we have the linear eddy model [88], which is a second generation multi-scale method in which a grid-within-the-grid approach is used for solving the species equations down to the Kolmogorov scales. Models for subgrid stirring are used to account for the interaction of the smallest scales with the flame. This method has been demonstrated to be both physically realistic and accurate.

20.2.2 Implicit LES Models

In implicit LES (ILES) the unfiltered incompressible, compressible, or reactive flow equations are solved using non-oscillatory finite volume methods (FVM) [36, 89]. ILES was found to be accurate and robust, in particular when based on monotonicity preserving convection schemes, such as the flux corrected transport [90] algorithm. Using modified equations analysis (MEA) Grinstein and Fureby [91] were able to derive expressions for the implicit (or built-in) subgrid models in a finite volume framework using a hybrid flux formulation. More specifically, can the implicit subgrid stress tensor be expressed as $\mathbf{B} = \mathbf{C}(\nabla\mathbf{v})^T + (\nabla\mathbf{v})\,\mathbf{C}^T + \chi^2\,(\nabla\mathbf{v})\,\mathbf{d}\otimes(\nabla\mathbf{v})\,\mathbf{d}$ for incompressible flow and $\mathbf{B} = \rho\left[\mathbf{C}(\nabla\mathbf{v})^T + (\nabla\mathbf{v})\,\mathbf{C}^T + \chi^2\,(\nabla\mathbf{v})\,\mathbf{d}\otimes(\nabla\mathbf{v})\,\mathbf{d}\right]$ for compressible and reacting flows, in which $\mathbf{d}$ denotes the vector connecting neighboring cell centers, $\mathbf{C} = \chi\,(\mathbf{v}\otimes\mathbf{d})$ and χ a non-linear function of the numerical flux-limiter, Γ, used to switch between the underlying high-order and low-order convective flux reconstruction schemes. Similarly, the implicit subgrid terms in the energy and species equations can be expressed as $\mathbf{b}_E = \rho\left[\mathbf{C}\nabla E + \chi^2\,(\nabla E\cdot\mathbf{d})\,(\nabla\mathbf{v})\,\mathbf{d}\right]$ and $\mathbf{b}_i = \rho\left[\mathbf{C}\nabla C_i + \chi^2\,(\nabla C_i\cdot\mathbf{d})\,(\nabla\mathbf{v})\,\mathbf{d}\right]$, respectively. These implicit subgrid models are of the same form as the explicit subgrid models, but with the subgrid viscosity being a fourth-rank tensor instead of a scalar. This allows these models to better handle simultaneous flow and grid anisotropies but makes them dependent of the selection of flux limiters, with monotonicity preserving flux limiters performing the best [92]. ILES is not extendable to chemical reacting flows since the reaction rates are unaffected by the handling of the convective terms and should thus be avoided unless combined with any of the aforementioned LES combustion models.

20.2.3 Near-Wall LES Models

The flow just adjacent to a solid wall is characterized by successively smaller and smaller flow structures which also tend to be highly anisotropic, Marusic et al. [93] and the references therein. Incompressible flow is typically characterized by the viscous, buffer, log-law and wake regions, and the streamwise elongated and Ω-shaped interacting vortex structures occupying this near-wall region. For LES of high Re-number flows, the grid cells will be large in comparison to the size of the near-wall flow structures, and usually the anisotropy of the flow structures are not reflected in the grid topology which is typically being graded towards the wall to resolve as much as possible of the velocity profile. Almost irrespectively of which LES model is adopted a near-wall subgrid model or a modification of the subgrid model to accommodate the specifics of the near-wall flow physics is needed unless the grid is sufficiently fine to resolve the scales of the flow in the boundary layer [93, 94]. Piomelli [95] provides a comprehensive review of different LES wall models but in addition to the models covered therein Fureby [96] and Duprat et al. [97] presented alternative methods that are more amendable to seamless implementation in an LES code together any type of subgrid flow model. Another way of dealing with the flow physics near the wall and the conflicts in grid resolution and anisotropic flow physics is to use a DES, DDES, or IDDES model, [34, 35], in which a RANS model is gradually blended in as the wall is approached. These methods have not only some appealing features but also some esoteric issues associated with blending two inherently different methods explicitly manifested in the definition of the filtering and the selection of RANS-LES interface characteristics. Figure 20.4 compares predictions from RANS, DES, and LES for the flow around a contoured ramp [98, 99], revealing the LES in general predicts a more accurate mean flow, with a smaller recirculation region, and that LES is capable of predicting the developing turbulent boundary layer upstream of the contoured ramp. Structured grids between 2.5 and 156 million cells were used in [99].

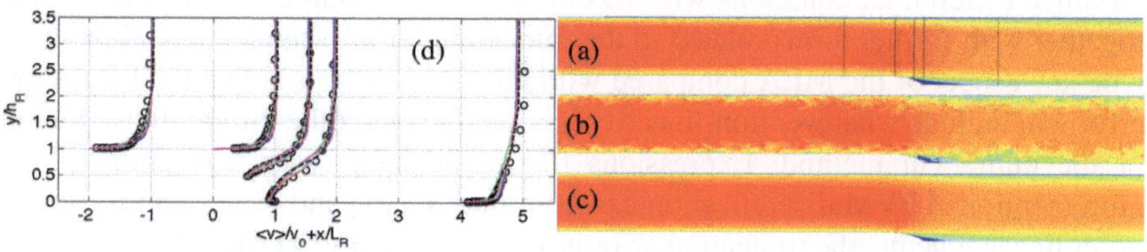

Legend: (O) experiments, (—) RANS-SST-kω, (—) LES-WALE, (—) DES-SA, (—) LES-OEEVM, (—) LES-LDKM and (—) LES-ILES.

Fig. 20.4 Wall-bounded flows: flow over a contoured ramp [98, 99] on a 20 million cell grid. (**a**) RANS, (**b**) instantaneous LES, (**c**) time-averaged LES, and (**d**) profiles at the cross sections in (**a**)

20.2.4 Numerical Methods for Complex Geometry LES and ILES

Unstructured FVM is most appropriate for LES of engineering flows since they allow greater flexibility in generating and adapting the grids. Central to FVM's is that the values of the variables, $\aleph$, are represented by control volume averages, $\aleph_P = \frac{1}{\delta V_P}\int_{\Omega_P} \aleph dV$, representing implicit filtering of the incompressible, compressible, or reacting NSE. The discretized FV equations can be derived from Reynolds transport theorem so that $\partial_t(\aleph_P) + \frac{1}{\delta V_P}\Sigma_f\left[\mathbf{F}_f^C - \mathbf{F}_f^D\right] = \mathbf{f}_P$, in which $\mathbf{F}_f^C$ and $\mathbf{F}_f^D$ are the convective and diffusive flux functions, respectively, whereas $\mathbf{f}_P$ is the source terms. These equations need to be integrated in time and rules must be set for how to reconstruct the fluxes $\mathbf{F}_f^C$ and $\mathbf{F}_f^D$ from $\aleph_P$. For high-speed flows, explicit time integration, e.g., [100] is preferred, whereas for low-speed flows, requiring a particular treatment of the pressure–velocity system [101], semi-implicit multi-step methods [102] are generally more appropriate. For the convective fluxes, $\mathbf{F}_f^C$, linear or cubic flux reconstruction schemes (of second and fourth order accuracy, respectively) are recommended, but may not always work in complex engineering applications. The best remedy is then to use flux-limiter based [91] reconstruction schemes of the form $\mathbf{F}_f^C = \psi\mathbf{F}_f^L + (1-\psi)\,\mathbf{F}_f^U$ in which the convective flux function is computed as a blend of a non-dissipative linear/cubic flux function, $\mathbf{F}_f^L$, and a more dissipative upwind biased flux function, $\mathbf{F}_f^U$, with ψ being a blending function that depends on the local flow [91, 103]. For the diffusive fluxes, $\mathbf{F}_f^D$ linear interpolation algorithms are often applied. For LES, the time-step should be limited as to allow flow structures to be transported across the computational grid, resulting in that the maximum Courant number should be less than $\sim$0.5.

One way to estimate the accuracy of the numerical method is to estimate the leading order truncation error using the modified equations approach (MAE) [104]. Briefly stated, given the differential equations of interest and the numerical method to be used, the MAE provides the differential equations solved numerically. These modified differential equations will thus include the original differential equations together with further terms related to the truncation error of the numerical method. One key aspect of the MEA compared with other approaches is that the truncation error will include whatever non-linearity is related to either the differential equation or the numerical method. Expressions for the leading order truncation of the aforementioned FVM discretization can be found in [91] and [103], and for a 2nd order linear scheme, the truncation error is $\mathbf{T}_v = \frac{1}{8}\left[\mathbf{v}\otimes\left(\left(\nabla^2\mathbf{v}\right)(\mathbf{d}\otimes\mathbf{d})\right)\right]_{sym} + \ldots$, where $\mathbf{d}$ is the grid spacing, revealing the nature of the scheme and how it may interact with the explicit subgrid model.

20.3 Examples of Applied LES Predictions and Validation

Next, we will present a few selected high-fidelity studies of real engineering problems performed with LES. These cases are selected to demonstrate the capability and issues of LES when used to study high Re number, geometrically and physically, complex flow problems. For a more extensive survey of engineering LES applications see Fureby [105].

Vehicle aerodynamics is one of the areas currently believed to benefit most from progressing from the current industry standard of RANS to LES. The main reasons for that are the sensitivity of the RANS models to the turbulence model and the absence of unsteady flow characteristics in the RANS results. Even for idealized conditions the unsteady turbulent flow occurring behind wing mirrors, around the wheels, in the engine room, and underneath and behind the vehicle is imperative to understand and predict from a resistance and aero-acoustic point of view. As an example of recent LES of vehicle aerodynamics, Fig. 20.5 shows results from an LES study of a full scale Mazda Atenza, including not only the outer shape but also the engine room and the underbody, Tsubokura et al. [106]. The grid contains about 38 million tetrahedral cells with prismatic boundary layer cells and $y^+ \approx 100$, thus requiring wall-modeled LES. Here, the SMG model is used together with a traditional wall-damping model. The flow over the hood, windscreen, and roof is rather steady whereas the wake formed after the surface flow has detached from the roof trailing edge is inherently unsteady. The flow behind the vehicle, in the engine compartment and underneath the vehicle, is also unsteady, with the underbody flow strongly influencing the wake behind the vehicle and thus also the pressure drag. Comparison between experimental flow visualizations, RANS (not shown)

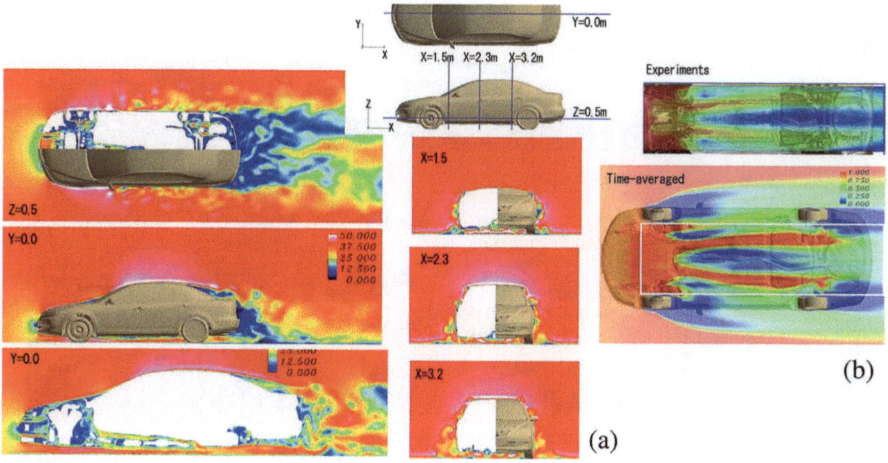

Fig. 20.5 Vehicle aerodynamics: (**a**) instantaneous representations of the velocity magnitude around the vehicle and (**b**) comparison between experimental data and time-averaged LES predictions on a plane 100 mm above ground. Figures obtained from Tsubokura et al. [106]

and LES prediction, reveals that both the instantaneous and time-averaged flows are well captured by the LES. RANS captures the time-averaged flow reasonably well, but cannot capture the unsteadiness and therefore not the pressure and velocity fluctuations crucial for noise predictions. The LES is about one order of magnitude more expensive than the RANS, but provides more information about the turbulent flow and the noise.

Submarine hydrodynamics is another area that benefits from progressing from the contemporary industry standard of RANS to LES. Naval ships, and submarines in particular, need not only to be designed to have low pressure and viscous drag to minimize fuel consumption and increase range, but also to minimize the signatures, i.e., the footprints that the hull leaves in the water volume. The signatures consist of many elements including hull and appendage wakes, propeller slipstream, turbulence, and flow induced noise, many of which inaccessible to RANS due to their unsteady nature. During design and modifications, RANS is useful to assist in finding a hull-form with low drag, and to compute maneuvering characteristics, but regarding the unsteady flow features LES provides the only viable alternative. Figure 20.6a shows an example from Fureby et al. [107], in which a generic submarine hull with a five bladed propeller, having a displacement of about 3500 t, is examined using LES. An unstructured grid of about 500 million cells with $y^+ \approx 20$ is employed. Approximately 100 million cells are used for the propeller, in order to resolve the blade tip-vortices and the unsteady flow shed off the propeller blade trailing edges. In spite of the "streamlined" hull shape, the flow is very complicated with unsteady wakes developing behind the fin, stern appendages, and around the casing. Particular flow features include the horseshoe vortices originating at the junctions between the fin and the casing, and between

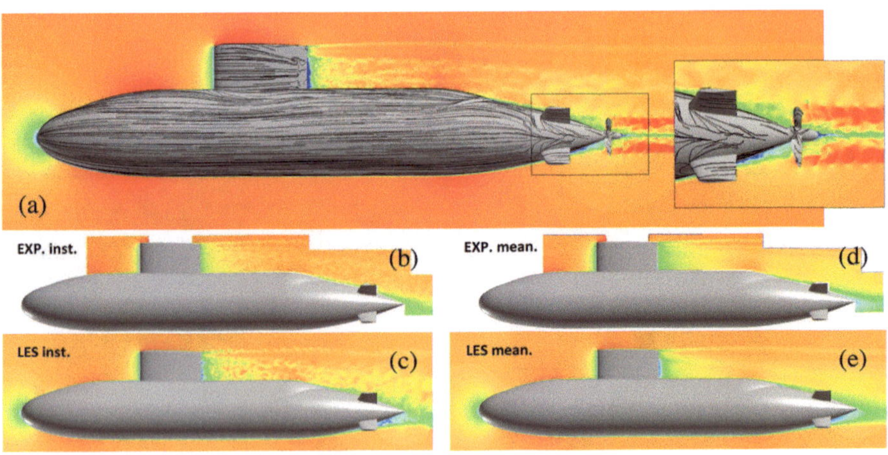

Fig. 20.6 Submarine hydrodynamics: (**a**) instantaneous velocity from LES of a fully appended submarine in full scale, (**b**) experimental instantaneous velocity, (**c**) corresponding time-averages in model scale, (**d**) LES instantaneous velocity, and (**e**) corresponding time-averages in model scale

the hull and the stern appendages. The inflow to the propeller is particularly interesting, being composed of several unsteady wakes and vortex systems, resulting in unsteady loading of the propeller and the generation of noise, and occasionally even tonals. The propeller slipstream is also unsteady and subject to instabilities. In previous investigations [108, 109] combined experimental and computational studies in model scale (1:52), using both RANS and LES, were used to characterize the flow at straight ahead and yaw conditions, and to reciprocally validate both the experimental and computational (RANS and LES) capabilities. Figure 20.6b–e shows selected comparisons, revealing that LES can capture both the instantaneous and time-averaged flow, and thus also the rms fluctuations, being significant for signature predictions.

Gas turbine combustion is an area currently developing rapidly due the widespread use of combustion LES, e.g., [9, 110] and the references therein. It is recognized that combustion RANS is not capable of providing information about unsteady combustion events, such as thermoacoustic oscillations, but it is also appreciated that the turbulence chemistry interactions, pertinent to turbulent combustion, cannot be modeled using RANS-based models. A complicating factor is that most real aero-propulsion and power-generation gas turbines use annular combustors with multiple burners or fuel-spray nozzles to facilitate a compact design that can easily be combined with a turbine and a compressor into an axisymmetric gas turbine engine. From a computational perspective this is appealing since then it is sufficient to simulate a sector of the combustor. This is, however, not true, since the fuel manifold is usually not symmetric and the pressure field in the combustor may exhibit combined longitudinal, azimuthal, and radial pressure waves, the simulation of which requires the whole combustor to be computed [111, 112]. Such simulations will be large as the combustor geometry is rather complicated, including multiple burners or fuel-spray nozzles and a flame tube with often more than 100 film-cooling, dilution, and mixing holes. Another complicating factor is that most aero-propulsion gas turbines burn kerosene, which is a mix of hydrocarbons with an average molecular formula of $C_{12}H_{23}$, challenging to model. In Fig. 20.7 some results from a recent study of a generic aero-propulsion gas turbine combustor of Zettervall et al. [113] are included. An unstructured grid of about 75 million cells is used to represent the full annular geometry including film-cooling, dilution and mixing holes, and the fuel-spray nozzles. The kerosene–air combustion chemistry is (for the first time) modeled using a 57-step skeletal reaction mechanism, the performance of which is favorably compared with detailed and global [61, 113] reaction mechanism in Fig. 20.7a. Figure 20.7b, c shows the pressure fluctuations on the inner side of the flametube, revealing an azimuthal pressure wave moving around the combustor. The individual flames in Fig. 20.7d have a very interesting structure with C_2H_4 (yellow) being located outside of $C_{12}H_{23}$ (green), outside of which HCO (red) is located. Figure 20.7e, f shows the distribution of temperature, velocity, CO and CO_2, revealing a complex flow pattern typical of the Rich-burn, Quick-mix, Lean-burn (RQL) combustor concept in which the flames burn in a rich region, after which air is injected to dilute the reaction products so that final combustion can take place in a lean region to minimize

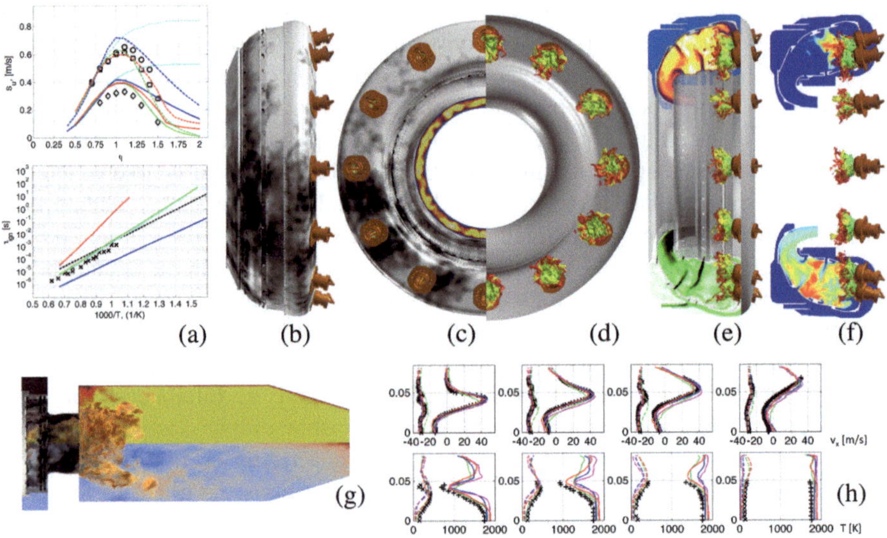

Fig. 20.7 Gas turbine combustion: (**a**) flame speed (*top*) and ignition delay time (*bottom*) for selected kerosene air reaction mechanisms, (**b**) and (**c**) pressure fluctuations on the inner side of the flame tube, (**d**) end view of the flames colored by $C_{12}H_{23}$ (*green*), C_2H_4 (*yellow*), and HCO (*red*), (**e**) temperature and velocity magnitude, (**f**) CO and CO_2 concentrations, (**g**) velocity, temperature, and flame, and (**h**) comparison with experimental data from the validation study, [64]

emissions. No experimental data to support these simulations are available, and instead we rely on simplified laboratory combustor configurations such as the one used in [64, 87] to validate the combustion model. Figure 20.7g, h presents selected results from [64] to demonstrate the accuracy of the combustion LES model.

Supersonic ramjet combustion is probably the area that has benefitted the most from being able to perform combustion LES. Supersonic ramjet combustion is a very interesting propulsion concept but since the combustor is aerodynamically controlled everything hinges on how the fuel is injected, mixes with the supersonic high temperature air flowing through the combustor, and self-ignites and subsequently burns, all in less than 0.5 ms. These processes and their interactions are critical but extremely challenging to study. Experiments and ground testing [114–116] provide critical information but simulations using RANS [117] and LES [118–121] give invaluable information about how these processes function and interact. Figure 20.8 presents selected results from a new study of the HyShot II combustor by Nordin-Bates et al. [121]. Figure 20.8a is a composite figure of the combustor flow showing the pressure distribution on the body wall (top) and an iso-surface of the second invariant of the velocity gradient colored by (from left to right) velocity, temperature, OH, and heat release (middle) and a detail of how the reflected shocks interact with one of the central the H_2-rich jet plumes. As the H_2 is injected some H_2 is peeled of and becomes ingested in the horseshoe-vortex system and thus transported downstream between the injectors. The H_2 is gradually being mixed

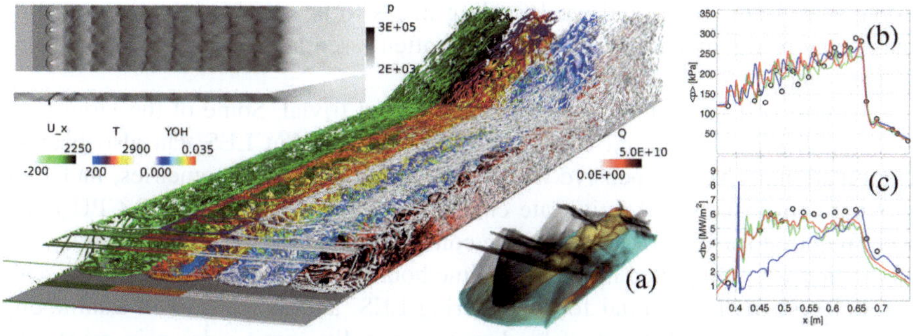

Fig. 20.8 Supersonic combustion ramjets: (**a**) composite figure showing the distribution of body wall-pressure and an iso-surface of the second invariant of the velocity gradient tensor colored, from *left* to *right*, by the axial velocity (*white-green-black*), temperature, (*blue-red-yellow-white*), OH concentration, Y_{OH}, (*white-blue-green-yellow-red*), and heat release, Q, (*white-red-black*)

with the supersonic hot air, but only when the normal and oblique reflected shocks hits the plume and horseshoe-vortex structures consisting of partially mixed H_2 and air combustion occurs in pockets that are convected downstream whilst igniting other sufficiently well mixed regions. Figure 20.8b, c shows comparisons of pressure and heat flux on the body and cowl walls, respectively, between RANS, LES, and experimental data, from which it can be concluded that both RANS and LES can be used to accurately simulate supersonic combustion, but LES is preferable since it captures more of the detailed physics of mixing and combustion than RANS, preferably when used with skeletal reaction mechanisms.

20.4 Computational Cost and Resource Estimates for LES

Since LES is based on resolving the most energetic turbulent flow structures, ideally of the order of the Taylor scales, the grids used in LES are usually comparatively large, and as LES is an unsteady computational approach, in which the flow evolves in time, LES is expensive in comparison to RANS, but still affordable in comparison to DNS. The progress of LES is therefore naturally tied to the development of modern parallel computing based on message passing interface libraries and domain-decomposed models in which the case is divided into smaller parts that are solved by the local processor. With appropriate programming, very high scalability may be reached, for example, using 640,000 cores in LES of the flow past a ship hull using 32 billion cells, [122], or using 4096 cores in LES of fully annular gas turbine combustion using 100 million cells, [123], or using 15,600 cores in DNS of combustion using 6.9 billion cells [31]. It is essential to note that parallel scalability needs to be achieved independent of the computer architecture since there are many systems available, with hardware and software being upgraded every year. Both strong and weak scaling needs to be considered; strong scaling may be more significant for LES since it is often essential to complete a run during a given time.

The cost of LES in terms of CPU hours is not trivial. Some of the large-scale DNS studies have required enormous resources [31] but LES can also become computationally very expensive when considering complex geometries, including moving components and finite rate chemistry. For example, 112,000 CPU hours were required for LES of a gas turbine combustion but due to good scalability the simulation was completed in 160 real-time hours [123]. Availability of large-scale parallel systems is essential for engineering LES, and as the cost of commodity PC clusters, using off-the-shelf hardware, is rapidly coming down in price, this may provide an alternative at least to intermediate size systems, using up to ~10,000 cores. Systems of this size will in any case require dedicated infrastructure, including power, cooling, storage, as well as personnel. The frontiers in this field are rapidly evolving, and new hardware and software solutions are rapidly emerging, providing new possibilities and challenges for LES.

As examples of engineering LES we consider submarine hydrodynamics and gas turbine combustion using 500 and 100 million cells, respectively. Submarine hydrodynamics is described by five variables (p, $\mathbf{v}$, and k) and 11 statistical variables ($\langle p \rangle$, p_{rms}, $\langle \mathbf{v} \rangle$, and $\mathbf{R}$) resulting in a total memory requirement of 5.6 GB per time-step. A typical time-step for such a simulation is 0.2 ms and hence 125,000 time-steps are needed per hull flow past. For a simulation about 20 hull flow pasts are required, thus resulting in $2.5 \cdot 10^6$ time-steps. A typical LES is often run with running averages and purged time-steps, typically requiring 10–20 time-steps to be saved, resulting in memory requirement of about 60 GB, including also the grid. However, when animations are to be prepared this memory requirement rapidly increases to TB levels. Combustion using a skeletal reaction mechanism with 20 species is thus described by 25 variables and 50 statistical variables, and is often run for about $2 \cdot 10^6$ time-steps, requiring 5.3 GB per time-step. Such a combustion LES will hence also require on the order of 60 GB. Carefully designed operational procedures are thus crucial for successful LES of engineering flows.

20.5 Outstanding Challenges and Concluding Remarks

Predictive modeling of turbulent flows is becoming increasingly important for the development of land vehicles, aircrafts, ships, medical equipment, turbomachinery, power plants, process devices, etc. The increase in computational power in the past decade has made many of these flow configurations numerically accessible. Nevertheless, the interaction of turbulence with physical processes, such as chemical kinetics and phase change processes, is a great challenge. Significant advances in modeling turbulent flow are now possible with the development of LES and similar methods. The philosophy of LES is to explicitly solve for the large (energetic) scales of the flow, directly affected by boundary conditions, whilst modeling the small (less energetic) scales of the flow. For historical and practical reasons, the development of LES models is based on ideas from RANS models, resulting in improved predictions, due mainly to that in LES the large turbulence scales are resolved. However, to meet the future demands in more accurate, fast, reliable, and detailed predictions of engineering applications some improvements to the LES subgrid modeling will be necessary. In this author's opinion, the most critical areas that need to be addressed are:

- enhanced subgrid flow models for high Re number (geometrically complex) flows valid in both homogeneous isotropic turbulence and in wall-bounded turbulence;
- improved subgrid models for compressible flows, including shock–shock, shock boundary layer, and shock–interface interactions;
- upgraded models for the low-pass filtered reaction rates in chemically reactive flows;
- improved models for spray combustion and other phase change processes in reactive flows.

In addition the different communities using and developing LES should start to engage in cross-disciplinary activities such as systematic validation of models and methods.

Acknowledgement The presented work was supported by the Swedish Armed Forces and the Swedish Defense Material Agency and Drs B. Andersson, D. Clarke, E. Fedina, F. Grinstein, D. Jones, M. Liefvendahl, S. Karl, S. Menon, K. Nordin.Bates, J. Tegnér, D. Norrison, K. Petterson, V. Wheatley, V. Sabelnikov, and N. Zettervall are acknowledged for providing valuable support.

References

1. J. Smagorinsky, General circulation experiments with the primitive equations. Mon. Weather Rev. **91**, 99–164 (1963)
2. J. Deardorff, A numerical study of three-dimensional turbulent channel flow at large Reynolds numbers. J. Fluid Mech. **41**, 453–480 (1970)
3. M. Islam, F. Decker, E. de Villiers, A. Jackson, J. Gines, T. Grahs, A. Gitt-Gehrke, J. Comas I Font, Application of detached-eddy simulation for automotive aerodynamics development, SAE 2009-01-0333 (2009)
4. D.R. McDaniel, R.M. Cummings, K. Bergeron, S.A. Morton, J.P. Dean, Comparisons of computational fluid dynamics solutions of static and maneuvering fighter aircraft with flight test data. J. Aerosp. Eng. **223**, 323–339 (2008)
5. H. Hemida, N. Gil, C. Baker, Large-eddy simulation of train slipstream. J. Fluids Eng. **132**, 051103 (2010)
6. C. Fureby, Large eddy simulation of ship hydrodynamics, in *27th Symposium on Naval Hydrodynamics*. Invited topical Review, 2008, pp. 389–408
7. S. Bhushan, P. Carrica, J. Yang, F. Stern, Scalability studies and large grid computations for surface combatant using CFDShip-Iowa. Int. J. High Perform. Comput. Appl. **25**, 466–487 (2011)

8. C. Fureby, LES modeling of combustion for propulsion applications. Phil. Trans. R. Soc. A **367**, 2957–2969 (2009)
9. S. Menon, C. Fureby, Computational combustion, in *Encyclopedia of Aerospace Engineering*, ed. by R. Blockley, W. Shyy (Wiley, 2010)
10. S. Apte, V. Yang, Unsteady flow evolution and combustion dynamics of homogeneous solid propellant in a rocket motor. Combust. Flame **131**, 110–131 (2002)
11. C. Rutland, Large eddy simulations for internal combustion engines – a review. Int. J. Engine Res. **12**, 421–452 (2001)
12. S.J. Eastwood, P.G. Tucker, H. Xia, C. Klostermeier, Developing large eddy simulation for turbomachinery applications. Phil. Trans. R. Soc. A **367**, 2999–3013 (2009)
13. W.Z. Shen, J.N. Sörensen, Aeroacoustic modelling using large eddy simulation. J. Phys. Conf. Ser. **75**, 012085 (2007)
14. C.A. Wagner, T. Hüttl, P. Sagaut, *Large-Eddy Simulation for Acoustics* (Cambridge University Press, Cambridge, 2007)
15. P.K. Smolarkiewicz, L.G. Margolin, Studies in geophysics, in *Implicit Large Eddy Simulation: Computing Turbulent Fluid Dynamics*, ed. by F.F. Grinstein, L. Margolin, B. Rider (Cambridge University Press, Cambridge, 2007), pp. 413–438
16. M.J.P. Cullen, A.R. Brown, Large eddy simulation of the atmosphere on various scales. Phil. Trans. R. Soc. A **367**, 2947–2956 (2009)
17. R. Rotunno, Y. Chen, W. Wang, C. Davis, J. Dudhia, G.J. Holland, Large-eddy simulation of an idealized tropical cyclone. Am. Meteorol. Soc. **90**, 1783–1788 (2009)
18. V.N. Gamezo, A.M. Khokhlov, E. Oran, A.Y. Chtchelkanova, R.O. Rosenberg, Thermonuclear supernovae: simulations of the deflagration stage and their implications. Science **299**, 77–81 (2003)
19. D.H. Porter, P.R. Woodward, Using PPM to model turbulent stellar convection, in *Implicit Large Eddy Simulation: Computing Turbulent Fluid Dynamics*, ed. by F.F. Grinstein, L. Margolin, B. Rider (Cambridge University Press, Cambridge, 2007), pp. 439–469
20. P. Sagaut, *Large Eddy Simulation for Incompressible Flows* (Springer, Berlin/Heidelberg, 2001)
21. S.B. Pope, Ten questions concerning the large-eddy simulation of turbulent flows. New J. Phys. **6**, 35–59 (2004)
22. A.N. Kolmogorov, A refinement of previous hypotheses concerning the local structure of turbulence in a viscous incompressible fluid at high Reynolds number. J. Fluid Mech. **13**, 82–85 (1962)
23. A.S. Monin, A.M. Yaglom, *Statistical Fluid Mechanics: Mechanics of Turbulence*, vol. 2 (MIT Press, Cambridge, 1971)
24. A. Tsinober, *An Informal Conceptual Introduction to Turbulence*. Fluid Mechanics and Its Applications Series, vol. 92 (Springer, Dordrecht, 2009)
25. J. Chomiak, Basic considerations in the turbulent flame propagation in premixed gases. Prog. Energy Combust. Sci. **5**, 207–221 (1979)
26. V. Sabelnikov, C. Fureby, LES combustion modeling for high Re flames using a multi-phase analogy. Combust. Flame **160**, 83–96 (2013)
27. P. Lombardi, L.P. De Angelis, S. Banerjee, Direct numerical simulation of near-interface turbulence in coupled gas-liquid flow. Phys. Fluids **8**, 1643–1666 (1994)
28. D.C. Wilcox, *Turbulence Modelling for CFD*, 2nd edn. (DCW Industries, 2000)
29. T. Poinsot, D. Veynante, *Theoretical and Numerical Combustion* (R. T. Edwards, Philadelphia, 2001)
30. S. Hoyas, J. Jimenez, Reynolds number effects on the Reynolds stress budgets in turbulent channels. Phys. Fluids **20**, 101511 (2008)
31. E.R. Hawkes, O. Chatakonda, H. Kolla, A.R. Kerstein, J.H. Chen, A petascale direct numerical simulation study of the modelling of flame wrinkling for large-eddy simulations in intense turbulence. Combust. Flame **159**, 2690–2703 (2012)
32. U. Piomelli, Wall-layer models for large-eddy simulations. Annu. Rev. Fluid Mech. **34**, 349–374 (2008)

33. N.V. Nikitin, F. Nicoud, B. Wasistho, K.D. Squires, P.R. Spalart, An approach to wall modeling in large eddy simulation. Phys. Fluids **12**, 1629–1632 (2000)
34. P.R. Spalart, W.H. Jou, M. Strelets, S.R. Allmaras, Comments on the feasibility of LES for wings, and on a hybrid RANS/LES approach, in *Advances in DNS/LES, 1st AFOSR International Conference on DNS/LES* (Greyden Press, 1997)
35. M.L. Shur, P.R. Spalart, M.K. Strelets, A.K. Travin, A hybrid RANS-LES approach with delayed-DES and wall-modelled LES capabilities. Int. J. Heat Fluid Flow **29**, 1638–1649 (2008)
36. J.P. Boris, On large eddy simulations using subgrid turbulence models, in *Wither Turbulence? Turbulence at the Crossroads*, ed. by J.L. Lumly. Lecture Notes in Physics, vol. 357 (Springer, Berlin/Heidelberg, 1990), pp. 344–353
37. A.G. Gungor, S. Menon, A new two-scale model for large eddy simulation of wall-bounded flows. Prog. Aerosp. Sci. **46**, 28–45 (2010)
38. C. Fureby, F.F. Grinstein, Large eddy simulation of high Reynolds-number free and wall bounded flows. J. Comput. Phys. **181**, 68–97 (2002)
39. E.S. Oran, J.P. Boris, *Numerical Simulation of Reactive Flow* (Cambridge University Press, Cambridge, 2001)
40. C.K. Westbrook, F. Dryer, Chemical kinetic modeling of hydrocarbon combustion. Prog. Energy. Combust. Sci. **10**, 1–57 (1984)
41. U. Frisch, *Turbulence* (Cambridge University Press, Cambridge, 1995)
42. H. Tennekes, J.L. Lumley, *A First Course in Turbulence* (MIT Press, Cambridge, 1972)
43. W.-W. Kim, S. Menon, A new incompressible solver for large-eddy simulations. Int. J. Numer. Methods Fluids **31**, 983–1017 (1999)
44. M.D. Smooke, V. Giovangigli, Formulation of the premixed and nonpremixed test problems, in *Lecture Notes in Physics: Reduced Kinetic Mechanisms and Asymptotic Approximations for Methane-Air Flames*, vol. 384, ed. by M.D. Smooke (Springer-Verlag, New York, 1991)
45. C. Fureby, G. Tabor, H.G. Weller, A.D. Gosman, Differential subgrid stress models in large eddy simulations. Phys. Fluids **9**, 3578–3582 (1997)
46. Y. Li, L. Chevillard, G. Eyink, C. Meneveau, Matrix exponential-based closures for the turbulent subgrid-scale stress tensor. Phys. Rev. E **79**, 016305 (2009)
47. M. Germano, U. Piomelli, P. Moin, W.H. Cabot, A dynamic sub grid scale eddy viscosity model. Phys. Fluids A **3**, 1760–1765 (1991)
48. F. Nicoud, F. Ducros, Subgrid-scale stress modeling based on the square of the velocity gradient tensor. Flow Turbul. Combust. **62**, 183–200 (1999)
49. G. Comte-Bellot, S. Corrsin, Simple Eulerian time correlation of full- and narrow-band velocity signals in grid generated, 'isotropic' turbulence. J. Fluid Mech. **48**, 273–337 (1971)
50. A. Yoshizawa, K. Horiuti, A statistically-derived subgrid scale kinetic energy model for large eddy simulation of turbulent flows. J. Phys. Soc. Jpn. **54**, 2834–2839 (1985)
51. J. Bardina, J.H. Ferziger, W.C. Reynolds, Improved subgrid scale models for large eddy simulations, AIAA 1980-1357 (1980)
52. S. Liu, C. Meneveau, J. Katz, On the properties of similarity subgrid scale models as deduced from measurements in a turbulent jet. J. Fluid Mech. **275**, 83–119 (1994)
53. S. Stolz, N.A. Adams, An approximate deconvolution procedure for large eddy simulation. Phys. Fluids **11**, 1699–1702 (1999)
54. N.A. Adams, S. Stolz, Deconvolution methods for subgrid-scale approximations in LES, in *Modern Simulation Strategies for Turbulent Flow*, ed. by B. Geurts (Edwards Publisher, 2001), pp. 21–41
55. W.J. Layton, R. Lewandowski, Residual stress of approximate deconvolution models of turbulence. J. Turbul. **7**, N46 (2009)
56. G. Erlerbacher, M.Y. Hussaini, C.G. Speziale, T.A. Zang, Toward the large eddy simulation of compressible turbulent flows. J. Fluid Mech. **238**, 155–185 (1992)
57. A. Cook, W.H. Cabot, Hyperviscosity for shock-turbulence interactions. J. Comput. Phys. **203**, 379–385 (2005)

58. J. von Neumann, R.D. Richtmyer, A method for the numerical calculations of hydrodynamical shocks. J. Appl. Phys. **21**, 232–237 (1950)
59. M. O'Conaire, H.J. Curran, J.M. Simmie, W.J. Pitz, C.K. Westbrook, A comprehensive modeling study of hydrogen oxidation. In. J. Chem. Kinet. **36**, 603–622 (2004)
60. F. Frenklach, H. Wang, C.L. Yu, M. Goldenberg, C.T. Bowman, R.K. Hanson, D.F. Davidson, E.J. Chang, G.P. Smith, D.M. Golden, W.C. Gardiner, V. Lissianski, http://www.me.berkeley.edu/gri_mech
61. E. Ranzi, A. Frassoldati, A. Stagni, M. Pelucchi, A. Cuoci, T. Faravelli, Reduced kinetic schemes of complex reaction systems: fossil and biomass-derived transportation fuels. In. J. Chem. Kinet. **46**, 512–542 (2014)
62. C.K. Westbrook, F.L. Dryer, Simplified reaction mechanisms for the oxidation of hydrocarbon fuels in flames. Combust. Sci. Technol. **27**, 31–43 (1981)
63. A.S. Tomlin, T. Turanyi, M.J. Pilling, in *Mathematical Tools for the Construction, Investigation and Reduction of Combustion Mechanisms*, vol. 35 (Elsevier, Amsterdam, 1997), pp. 293–437. Chapter 4
64. G. Bulat, E. Fedina, C. Fureby, W. Meier, U. Stopper, Reacting flow in an industrial gas turbine combustor: LES and experimental analysis. Proc. Combust. Inst. **35**, 3175–3183 (2015)
65. J. Kim, S.B. Pope, Effects of combined dimension reduction and tabulation on the simulations of a turbulent premixed flame using a large-eddy simulation/probability density function method. Combust. Theor. Model. **18**, 388–413 (2014)
66. W.P. Jones, A.J. Marquis, F. Wang, Large eddy simulation of a premixed propane turbulent bluff body flame using the Eulerian Stochastic field method. Fuel **140**, 514–525 (2015)
67. H. Pitsch, Large eddy simulation of turbulent combustion. Annu. Rev. Fluid Mech. **38**, 453–482 (2006)
68. Z.S. Li, B. Li, Z.W. Sun, X.S. Bai, M. Aldén, Turbulence and combustion interaction: high resolution local flame front structure visualization using simultaneous single-shot PLIF imaging of CH, OH, and CH_2O in a piloted premixed jet flame. Combust. Flame **157**, 1087–1096 (2010)
69. F.A. Williams, *Combustion Theory* (Benjamin/Cummings, Menlo Park, 1985)
70. A. Poludnenko, E. Oran, The interaction of high-speed turbulence with flames: global properties and internal flame structure. Combust. Flame **157**, 995–1011 (2010)
71. T. Echekki, E. Mastorakos, *Turbulent Combustion Modeling: Advances, New Trends and Perspectives* (Springer Science & Business Media, Dordrecht, 2010)
72. K. Bray, P. Libby, J. Moss, Flamelet crossing frequencies and mean reaction rates in premixed turbulent combustion. Combust. Sci. Technol. **41**, 143–172 (1984)
73. B.F. Magnussen, B.H. Hjertager, On mathematical models for turbulent combustion with special emphasis on soot formation and combustion. Proc. Combust. Inst. **16**, 719–729 (1977)
74. I. Porumbel, S. Menon, Large eddy simulation of bluff body stabilized premixed flame, AIAA 2006-0152 (2006)
75. R. Bilger, Turbulent flows with non-premixed reactants, in *Turbulent Reacting Flows.* Topics in Applied Physics (Springer, Berlin/Heidelberg, 1986), pp. 65–113
76. A.W. Cook, J.J. Riley, A subgrid model for equilibrium chemistry in turbulent flows. Phys. Fluids **6**, 2868–2871 (1994)
77. N. Branley, W.P. Jones, Large eddy simulation of a turbulent non-premixed flame. Combust. Flame **127**, 1914–1934 (2001)
78. N. Peters, *Turbulent Combustion* (Cambridge University Press, Cambridge, 2000)
79. C. Fureby, Large eddy simulation of combustion instabilities in a jet-engine afterburner model. Combust. Sci. Technol. **161**, 213–243 (2000)
80. E.R. Hawkes, R.S. Cant, Implications of a flame surface density approach to large eddy simulation of premixed turbulent combustion. Combust. Flame **126**, 1617–1629 (2001)
81. R. Knikker, D. Veynante, Experimental study of the filtered progress variable approach for LES of premixed combustion, in *Advances in LES of Complex Flows*, ed. by R. Friedrich, W. Rodi (Kluwer Academic Publishers, 2002), pp. 353–366

82. C. Duwig, C. Fureby, Large eddy simulation of unsteady lean stratified premixed combustion. Combust. Flame **151**, 85–103 (2007)
83. C.D. Pierce, P. Moin, Progress-variable approach for large eddy simulation of non-premixed turbulent combustion. J. Fluid Mech. **504**, 73–97 (2004)
84. O. Colin, F. Ducros, D. Veynante, T. Poinsot, A thickened flame model for large eddy simulation of turbulent premixed combustion. Phys. Fluids **12**, 1843–1863 (2000)
85. P. Gerlinger, Investigation of an assumed PDF approach for finite rate chemistry. Combust. Sci. Technol. **175**, 841–872 (2003)
86. S. Navarro-Martinez, A. Kronenburg, F. Di-Mare, Conditional moment closure for large eddy simulations. Flow Turbul. Combust. **75**, 245–274 (2005)
87. G. Bulat, W.P. Jones, A.J. Marquis, Large eddy simulation of an industrial gas turbine combustion chamber using the sub-grid PDF method. Proc. Combust. Inst. **34**, 3155–3157 (2013)
88. S. Menon, A.R. Kerstein, The linear-eddy model, in *Turbulent Combustion Modeling*, ed. by T. Echekki, E. Mastorakos. Fluid Mechanics and Its Applications Series, vol. 95, 2010, pp. 221–247
89. J.P. Boris, F.F. Grinstein, E.S. Oran, R.J. Kolbe, New insights into large eddy simulation. Fluid Dyn. Res. **10**, 199–228 (1992)
90. J.P. Boris, D.L. Book, K. Hain, Flux-corrected transport. II Generalizations of the method. J. Comput. Phys. **18**, 248–283 (1975)
91. F.F. Grinstein, C. Fureby, On flux-limiting-based implicit large eddy simulation. ASME J. Fluids Eng. **129**, 1483–1492 (2007)
92. D. Drikakis, C. Fureby, F.F. Grinstein, D. Youngs, Simulation of transition and turbulence decay in the Taylor-Green vortex. J. Turbul. **8**, N20 (2007)
93. I. Marusic, B.J. McKeon, P.A. Monkewitz, H.M. Nagib, A.J. Smits, K.R. Sreenivasan, Wall-bounded turbulent flows at high Reynolds numbers: recent advances and key issues. Phys. Fluids **22**, 065103 (2010)
94. D.R. Chapman, Computational aerodynamics development and outlook. AIAA J. **17**, 1293–1313 (1979)
95. U. Piomelli, Wall-layer models for large eddy simulations. Prog. Aerosp. Sci. **44**, 437–446 (2008)
96. C. Fureby, On LES and DES of Wall Bounded Flows, Ercoftac Bulletin No 72, Marsh Issue (2007)
97. C. Duprat, G. Balarac, O. Métais, P.M. Congedo, O. Brugiére, A wall-layer model for large-eddy simulations of turbulent flows with/out pressure gradient. Phys. Fluids **23**, 0151901 (2011)
98. S. Song, J.K. Eaton, Reynolds number effects on a turbulent boundary layer with separation, reattachment, and recovery. Exp. Fluids **36**, 246–258 (2004)
99. C. Fureby, S. Zhu, D. Jones, Large eddy simulation of the flow over a contoured ramp, Turbulence Shear Flow Phenomena 9 (Melbourne, 2015)
100. S. Gottlieb, C.-W. Shu, Total variation diminishing Runge-Kutta schemes. Math. Comput. **67**, 73–85 (1998)
101. R.I. Issa, Solution of the implicitly discretized fluid flow equations by operator splitting. J. Comput. Phys. **62**, 40–65 (1986)
102. C. Hirsch, *Numerical Computation of Internal and External Flows* (Wiley, New York, 1999)
103. D. Drikakis, C. Fureby, F.F. Grinstein, M. Liefendahl, ILES with limiting algorithms, in *Implicit Large Eddy Simulation: Computing Turbulent Fluid Dynamics*, ed. by F.F. Grinstein, L. Margolin, B. Rider (Cambridge University Press, Cambridge, 2007), pp. 94–129
104. C.W. Hirt, Heuristic stability theory for finite difference equations. J. Comput. Phys. **2**, 339–355 (1968)
105. C. Fureby, Large eddy simulation: a useful tool for engineering fluid dynamics, in *Proceedings of the 18th Australasian Fluid Dynamics Conference*, Launceston, 2012, pp. 980–1002
106. M. Tsubokura, T. Kobayashi, T. Nakashima, T. Nouzawa, T. Nakamura, H. Zhang, K. Onishi, N. Oshima, Computational visualization of unsteady flow around vehicles using high performance computing. Comput. Fluids **38**, 981–990 (2009)

107. C. Fureby, B. Anderson, D. Clarke, L. Erm, M. Giacobello, S. Henbest, M. Giacobello, D. Jones, M. Nguyen, M. Johansson, M. Jones, C. Kumar, S.-K. Lee, P. Manovski, D. Norrison, K. Petterson, G. Seil, B. Woodyatt, S. Zhu, Unsteady flow about a generic submarine – a modelling capability. MAST (2015)
108. B. Anderson , M. Chapuis, L. Erm, C. Fureby, M. Giacobello, S. Henbest, D. Jones, M. Jones, C. Kumar, M. Liefvendahl, P. Manovski, D. Norrison, H. Quick, A. Snowden, A. Valiyf, R. Widjaja, B. Woodyatt, Experimental and computational investigation of a generic conventional submarine hull form, in *29th Symposium on Naval Hydrodynamics*, Gothenburg, Sweden, 2012
109. C. Fureby, B. Anderson, D. Clarke, L. Erm, M. Giacobello, S. Henbest, D. Jones, M. Nguyen, M. Johansson, M. Jones, C. Kumar, S.-K. Lee, P. Manovski, D. Norrison, K. Petterson, G. Seil, B. Woodyatt, S. Zhu, Experimental and numerical study of a generic conventional submarine at 10-degrees yaw, in *30th Symposium on Naval Hydrodynamics*, Hobart, Australia, 2014
110. L.Y.M. Gicquel, G. Staffelbach, T. Poinsot, Large eddy simulations of gaseous flames in gas turbine combustion chambers. Prog. Energy Combust. Sci. **38**, 782–817 (2012)
111. G. Staffelbach, L. Gicquel, G. Boudier, T. Poinsot, Large eddy simulation of self-excited azimuthal modes in annular combustors. Proc. Combust. Inst. **32**, 2909–2916 (2009)
112. E. Fedina, C. Fureby, S.A. Borzov, V.N. Gusev, T.V. Stepanova, Combustion LES of CESAR multi-burner annular combustor, AIAA 2011-0785 (2011)
113. N. Zettervall, E. Fedina, K. Nordin-Bates, E. Heimdal Nilsson, C. Fureby, Combustion LES of a multi-burner annular aero-engine combustor using a skeletal reaction mechanism for Jet-A air mixtures, AI-AA-2015-4020 (2015)
114. A.D. Gardner, K. Hannemann, J. Streelant, A. Paull, Ground testing of the HyShot supersonic combustion flight experiment in HEG and comparison with flight data, AIAA 2004-3345 (2004)
115. M.K. Smart, N.E. Hass, A. Paull, Flight data analysis of the HyShot II flight experiment. AIAA J. **44**, 2366–2375 (2006)
116. J.M. Schramm, S. Karl, K Hannemann, J. Streelant, Ground testing of the HyShot II Scramjet configuration in HEG, AIAA 2008-2547 (2008)
117. S. Karl, K. Hannemann, J. Streelant, A. Mack, CFD analysis of the HyShot supersonic combustion flight experiment configuration, AIAA 2006-8041 (2006)
118. A. Ingenito, D. Cecere, E. Giacomazzi, Large eddy simulation of turbulent hydrogen-fuelled supersonic combustion in an air cross-flow. Shock Waves **23**, 481–494 (2013)
119. M. Chapuis, E. Fedina, C. Fureby, K. Hannemann, S. Karl, J.M. Schramm, A computational study of the HyShot II combustor performance. Proc. Combust. Inst. **34**, 2101–2109 (2012)
120. J. Larsson, Large eddy simulations of the HyShot II scramjet combustor using a supersonic flamelet model, AIAA 2012-4261 (2012)
121. K. Nordin-Bates, N. Zettervall, C. Fureby, Understanding scramjet combustion using LES of the HyShot II combustor, AIAA Hypersonics 2015, AIAA 2015-3615 (2015)
122. T. Nishikawa, Y. Yamade, M. Sakuma, C. Kato, Fully resolved large eddy simulation as an alternative to towing tank tests – 32 billion cells computation on K computer, in *16th Numerical Towing Tank Symposium*, Mülheim, 2–4 September 2013
123. M. Boileau, G. Staffelbach, B. Cuenot, T. Poinsot, C. Berat, LES of an ignition sequence in a gas turbine engine. Combust. Flame **154**, 2 (2008)

Chapter 21
Coarse Grained Simulation and Turbulent Material Mixing

F.F. Grinstein, A.J. Wachtor, and J.R. Ristorcelli

21.1 Introduction

Turbulent mixing modeling and predictability rely on availability of *big data* generated in computational and laboratory experiments. Research challenges involve developing effective strategies for data generation, characterization, and analysis—as well as data curation, storage, mining, sharing, transfer, and visualization. Laboratory studies typically demonstrate the end outcome of complex nonlinear three-dimensional physical processes with many unexplained details and mechanisms. Flow experiments based on numerical simulations carried out with precise control of initial and boundary conditions are ideally suited to provide insights into the underlying dynamics of laboratory observations. Fundamental issues to be addressed involve characterizing and modeling the specific flow conditions at the subgrid scale (SGS) level—within a computational cell or instrumentation resolution, and at the supergrid (SPG) scale—at initialization and boundaries of computational *and* laboratory experiments.

Direct numerical simulation (DNS) resolving all relevant physical space/time scales is prohibitively expensive in the foreseeable future for most practical flows and regimes of interest at moderate-to-high Re. On the other end of the simulation spectrum are the Reynolds-Averaged Navier–Stokes (RANS) approaches—which focus on statistical moments for an ensemble of realizations and model the turbulent effects. In *coarse grained* simulation (CGS) large energy containing structures are

Invited paper at the Symposium on *Whither Turbulence and Big Data In the 21st Century*, honoring W.K. George on occasion of his 70th birthday, Cargese, Corsica, France, April 20–24, 2015.

F.F. Grinstein (✉) • A.J. Wachtor • J.R. Ristorcelli
Los Alamos National Laboratory, Los Alamos, NM 87545, USA
e-mail: fgrinstein@lanl.gov

A. Pollard et al. (eds.), *Whither Turbulence and Big Data in the 21st Century*,
DOI 10.1007/978-3-319-41217-7_21

resolved, smaller structures are spatially filtered out, and unresolved (presumably more-universal) dynamics and consequences are described with SGS models.

In the absence of an established universal theory of turbulence, the construction of SGS models is pragmatic and primarily based on rational use of empirical information. In the CGS paradigm small-scales are presumed enslaved to the dynamics of the largest, or put in other words, the spectral cascade rate of energy (the rate limiting step) is determined by the initial and boundary condition constrained large-scale dynamics. CGS focuses on explicit SGS models in classical LES [1], relies on SGS modeling and filtering provided by *physics capturing* numerical algorithms in implicit LES (ILES) [2], or is generally based on using mixed explicit/implicit SGS modeling. The CGS strategy of separating resolved and unresolved physics constitutes the viable intermediate approach between DNS and RANS to address practical geometries and multiphysics.

Because SPG choices *select* flow solutions, emulating particular flow realizations demands precise characterization of initial and asymptotic conditions, as well as conditions at solid and other relevant boundaries [3]. Long term effects of SPG specifics have been extensively reported in laboratory experiments [4–8]; their practical impact is clearly stated by George in [7]: *Unlike the theoretician, the experimentalist already knows the solution, for it is the flow he has realized. His objective is to find which equations and which boundary and initial conditions his solution corresponds to, and then to compare them and his results to those dealt with by the theoretician.* In particular, the sensitivity of turbulent flows to initial conditions (IC) is now well established. Far field (late time) characteristics of turbulent flows *remember* their near-field (initial) features [6, 9–11], and the mechanism by which transition from IC to asymptotic flow occurs involves unsteady large-scale coherent structure dynamics; transition can be captured by CGS but not by single-point closure turbulence modeling [12, 13] typical in RANS. *The long standing view that an IC independent universal turbulence state is eventually achieved in the far field (or at late time)*—e.g., [14], *has been replaced by the recognition that different self-similar end states are possible depending on IC* [9].

Cascade pathways driven by vortex instabilities, self-induced deformations, stretching, and reconnections have been demonstrated [15, 16], and the dynamics of transition to turbulence has been examined in terms of well-defined prototypes such as the Taylor–Green vortex [17, 18]. These processes are fundamental building blocks to be captured by CGS. Inherent to any CGS approach is the fact that the smallest characteristic resolved turbulence scale is determined by the resolution cutoff wavelength prescribed by an explicit or implicit spatial-filtering process, and practical convergence metrics must be formulated in terms of a suitable effective Re [19, 20].

Small-scale enslavement ideas relevant to *under resolved* simulation of high-Re scalar mixing driven by *under resolved* velocity field in isotropic forced turbulence [19] are revisited in Sect. 21.2. Turbulence metrics are used to show that a well-designed ILES can accurately capture the mixing transition and asymptotic self-similar behaviors as function of effective Re determined by grid resolution; we find that *realizability constraints are required to accurately address scalar mixing SGS modeling*.

In many applications of interest SGS and SPG issues are intertwined and cannot be dealt with independently. This is typical when studying shock driven turbulence generated via the Richtmyer–Meshkov instability (RMI) [22], where ILES has been proven the effective CGS *strategy* [23–29] because of its unique combination of shock and turbulence emulation capabilities. In Sect. 21.3, we examine the effects of initial material interfacial morphology and spectral content on transitional and late-time turbulent mixing in ILES of fundamental shock-tube experiments. We find that *usefully complete IC characterization, modeling, and parameterization is crucially needed to ensure repeatability of computational and laboratory experiments for late time predictability assessments*.

21.2 Enslaved Isotropic Turbulent Scalar Mixing

Mixing of a passive scalar by a fluctuating flow field is a classical problem in turbulence and relevant in many industrial flow applications. Overholt and Pope [30] conducted DNS of mixing of a passive scalar in the presence of a mean scalar gradient, by forced, spatially periodic, isotropic turbulence. In this flow, a statistically steady-state scalar variance is achieved by balancing scalar variance production and dissipation. This problem was first investigated in the LES context by Pullin [21], who proposed a model for the flux of a passive scalar by the SGS motions; the LES predicted that the normalized scalar variance asymptotically approaches a constant value as a function of Re_λ for Schmidt number $Sc \approx 1$, in agreement with laboratory observations [31]. The velocity-to-scalar dissipation time-scale ratio was also reported to be asymptotically constant in [21], but comparisons with the DNS [30] available at the time were inconclusive as to whether such a result captured physical behavior. Analysis of results of subsequent theory [32] and DNS [33] showed that the time-scale ratio should exhibit continued growth with increasing *Re*—consistent with the earlier DNS [30].

Extensive evaluations of ILES in fundamental forced and decaying isotropic homogeneous turbulence have been reported based on both, Euler and Navier–Stokes equations [34–36]. At moderately high *Re*—when convective time scales are much smaller than those associated with molecular diffusion—the primary concern is capturing the convectively driven material interpenetration mixing processes (entrainment and stirring due to velocity gradient fluctuations). Predicted asymptotic behaviors of the scalar variance and squared velocity-to-scalar Taylor micro-scales ratio are specifically used here to benchmark the performance of ILES against the previously reported theoretical, laboratory, and DNS studies.

ILES uses a 4th-order FV, multi-dimensional, flux-corrected transport (FCT) algorithm [19] based on the Euler equations,

$$\rho_{,t} + \left(\rho u_j\right)_{,j} = 0 \tag{21.1}$$

$$\left(\rho u_i\right)_{,t} + \left(\rho u_i u_j\right)_{,j} + p_{,i} = f_i \tag{21.2}$$

$$E_{,t} + \left[(E+p)\, u_j\right]_{,j} = 0 \tag{21.3}$$

$$\left(\rho\theta\right)_{,t} + \left(\rho\theta u_j\right) j = f_\theta \tag{21.4}$$

where ρ is the mass density, u_i the velocity, p the pressure, f_i the momentum forcing, $E{=}\rho u_j u_j/2{-}p/\left(\gamma{-}1\right)$ is the total energy density, f_θ the scalar forcing, $()_{,i}{\equiv}\partial\,()\,/\partial x_i$, and summation over repeated *roman* indices is implied. The passive scalar θ is forced by imposing a uniform mean scalar gradient, α_i, prescribed along the *i*-direction ($\alpha_i = \alpha_1\delta_{1i}$ chosen here), such that $f_\theta = -\rho\alpha_i u_i$.

Effects of grid resolution on the flow and scalar mixing are investigated at turbulent Mach numbers 0.13 and 0.27 based on *rms*-velocity-fluctuation u'. The simulation model effectively models convection-driven flow with high *Re* and $Sc \approx 1$. An isotropic turbulence simulation strategy [37] was implemented by which low wave number forcing can be enforced separately for solenoidal and dilatational components on the momentum equations. The configuration studied assumes triply periodic boundary conditions on a cubical domain with unit box length and uniformly spaced 32^3, 64^3, 128^3, and 256^3 grids. The simulated flow is characterized by volume–time-averaged quantities: u', velocity and scalar Taylor micro-scales λ and λ_θ,

$$\lambda_\theta = \frac{1}{3}\sum_{\beta=1}^{3}\sqrt{\langle\theta^2\rangle/\left\langle\left(\theta_{,\beta}\right)^2\right\rangle},\quad \lambda = \frac{1}{3}\sum_{\beta=1}^{3}\sqrt{\left\langle u_\beta^2\right\rangle/\left\langle\left(u_{\beta,\beta}\right)^2\right\rangle}.$$

On dimensional grounds, an effective viscosity $\nu_{\rm eff}$ can be estimated as ratio of dissipation and mean computed squared strain rate magnitude [34],

$$\nu_{\rm eff} = \varepsilon_s/\left[2\left\langle s_{ij}s_{ij}\right\rangle_T\right], \tag{21.5}$$

where ε_s denotes the dissipation imposed by the forcing scheme, $<>_{\rm T}$ volume–time average, and $s_{ij} = \left(u_{i,j} + u_{j,i}\right)/2$ is the strain tensor. For a given grid resolution, effective $Re_{\rm eff} = u'\lambda/\nu_{\rm eff}$ is thus directly evaluated from the developed simulation data.

Figure 21.1 exemplify the developed vorticity field for the 128^3 and 256^3 resolutions, depicting flow dominated by elongated structures characteristic of high-*Re* isotropic turbulence; Fig. 21.2 shows typical scalar visualizations for

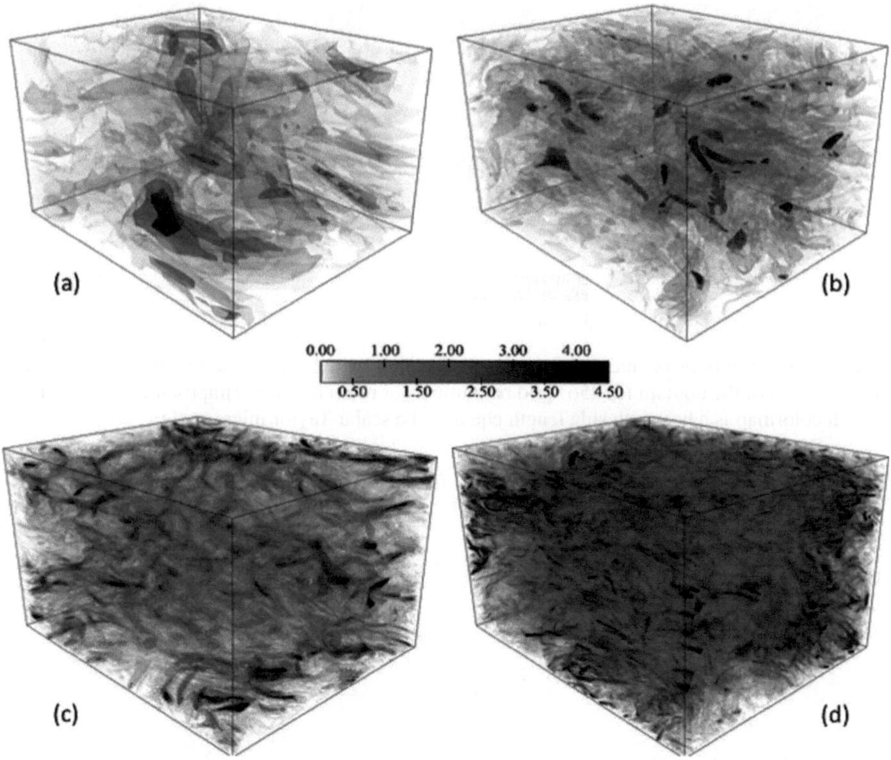

Fig. 21.1 Isosurfaces of vorticity magnitude scaled by vorticity rms vs. resolution; (**a**) 32^3, (**b**) 64^3, (**c**) 128^3, (**d**) 256^3

the 256^3 resolution case. Figure 21.3 demonstrates scalar and velocity spectra as function of grid resolution. The scalar spectra exhibit longer inertial ranges and more pronounced spectral bumps than their velocity counterparts at corresponding resolutions—consistent with [33]; the latter results directly reflect on the $\lambda^2/\lambda_\theta^2$ results discussed further below.

Probability distribution function (PDF) analysis is reported for DNS data (e.g., [38, 39]) and extensively used as validation tool for ILES [16, 19, 34, 35, 40, 41]. It is exemplified in Fig. 21.4a in terms of the PDFs of the vorticity magnitude. Of considerable interest is noting that the vorticity PDFs associated with the DNS data exhibit Re independence near $\omega \sim 2.5\omega'$ and approach a high *Re* limit above the mixing transition threshold $Re_\lambda \approx 100$ [42] (Fig. 21.4b)—which are captured with ILES. PDF trends with increasing ILES resolution are the same as for the increasing

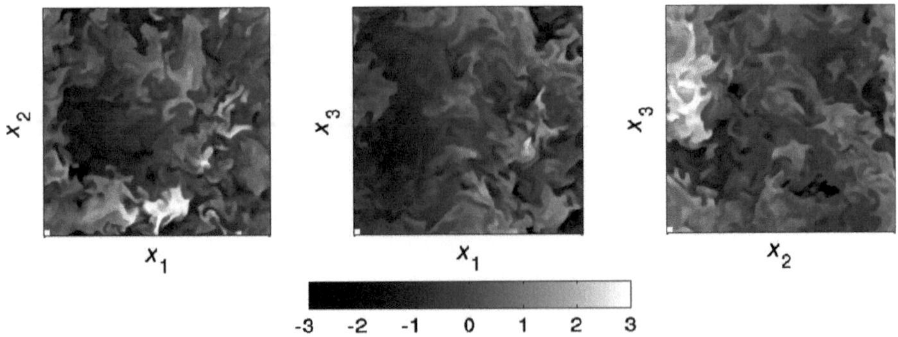

Fig. 21.2 Distributions of fluctuating scalar field scaled by the rms of the fluctuating scalar field in mid-planes of the domain for 256^3 grid resolution; for reference, superimposed on the lower left of each colormap is a box with side length equal to the scalar Taylor micro-scale

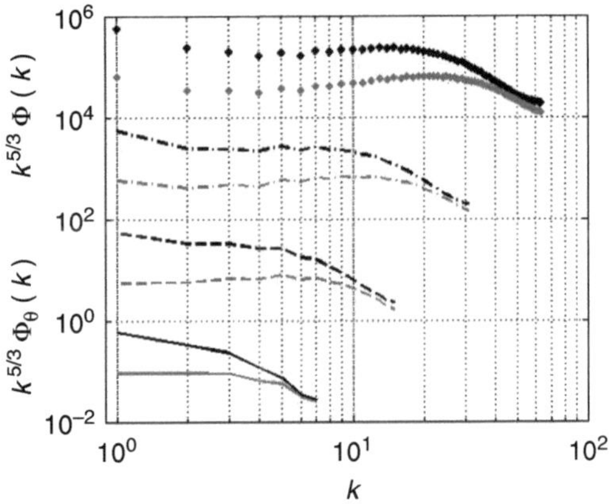

Fig. 21.3 Compensated velocity (*black*) and scalar variance spectra (*gray*). *Solid line*: 32^3; *Dashed line*: 64^3; *Dash-dot line*: 128^3; *Diamond*: 256^3; $Ma = 0.27$ case

Re of the DNS data (Fig. 21.4a) suggesting an Re_{eff} for ILES determined by grid resolution. More detailed analysis of the velocity and scalar simulation data can be found separately [19, 20].

The non-dimensionalized scalar variance, $\langle\theta^2\rangle_T/(\alpha_1 L_\varepsilon)^2$, where $L_\varepsilon = (u')^3/\varepsilon_s$, is compared with predictions of state-of-the-art mixing LES using a *stretched vortex* SGS stress model [21]. Figure 21.5 shows that the ILES scalar variance results asymptotically reach nearly constant behavior above the cited mixing transition threshold for $Re_\lambda = Re_{\text{mix}} \sim 100$ [42], in close agreement with laboratory experiments [31] and previous LES [21] using an explicit scalar SGS model. Good or bad SGS physics can be implicitly built into the simulation model depending on the choice of numerics and its particular implementation. In particular, the unsuccessful results denoted as "LES without scalar SGS model" in Fig. 21.5 reflect that the implicit SGS scalar modeling provided by the particular (dispersive) numerics used in [21] is inadequate by itself to predict the expected asymptotic behavior. Lower scalar variance predictions with present ILES are mainly attributed to differences

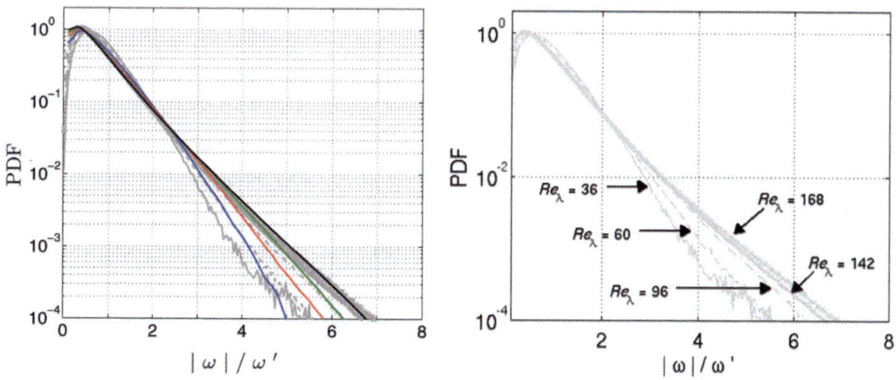

Fig. 21.4 PDF vorticity magnitude analysis shows trends with varying Re predicted with DNS as function of grid resolution. ILES—*Blue*: 32^3; *Red*: 64^3; *Green*: 128^3; *Black*: 256^3; superimposed to DNS by Jimenez et al. [38] (*Gray*)—*Solid*: $Re_\lambda = 36$; *Dashed*: $Re_\lambda = 60$; *Dash-dot*: $Re_\lambda = 96$; *Diamond*: $Re_\lambda = 142$; X: $Re_\lambda = 168$; $\omega' = \omega_{rms}$ and results are for the ILES $Ma = 0.27$ case. (**a**) ILES results superimposed to those from DNS; (**b**) DNS results only shown

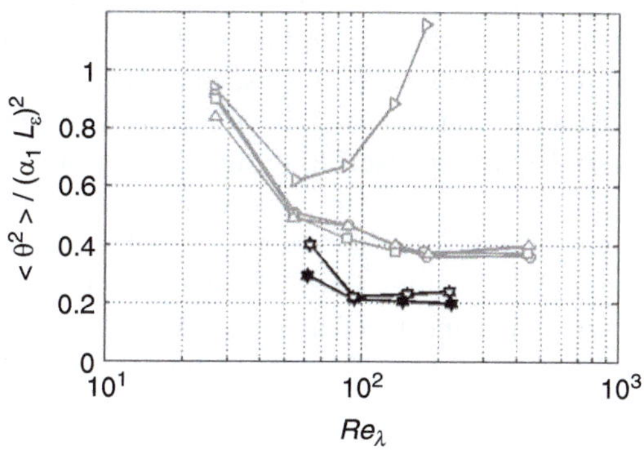

Fig. 21.5 Non-dimensional scalar variance as a function of Re_λ. *Gray open square/triangle/circle*: LES [21]; *Gray open right-triangle*: LES without scalar SGS model [21]; *Black full star*: ILES $Ma = 0.27$; *Black open star*: ILES $Ma = 0.13$, vs. Re_λ

in the forcing schemes, and somewhat less to compressibility effects—since the solenoidal velocity component is responsible for the generation of a passive scalar flux [43] and subsequent scalar stirring. Otherwise, scatter of previous and current results around presumably constant asymptotic scalar variance behaviors depict similar overall fluctuations.

The ratio $\lambda^2/\lambda_\theta^2$—directly proportional to the velocity-to-scalar dissipation time-scale ratio, r, through $\lambda^2/\lambda_\theta^2 = Sc\ (5/3)\ r$—is plotted in Fig. 21.6 for the current ILES, the ($Sc = 0.7$) LES of [21], DNS [30, 33], for which $Sc = 0.7$ and $Sc = 1.0$, respectively. Following the theory in [32], using small scale isotropy assumptions, and rearranging the stationary forced equations for the scalar variance and scalar dissipation, we derived an expression for $\lambda^2/\lambda_\theta^2$ [19] which exhibits its leading linear dependence with Re_λ for $Sc = 1.0$ and high *Re*—also plotted in Fig. 21.6. This result is noteworthy on several accounts. For one it shows that the squared ratio of Taylor micro-scales $\lambda^2/\lambda_\theta^2$ is a function of *Re* in developed turbulence. That the eddy

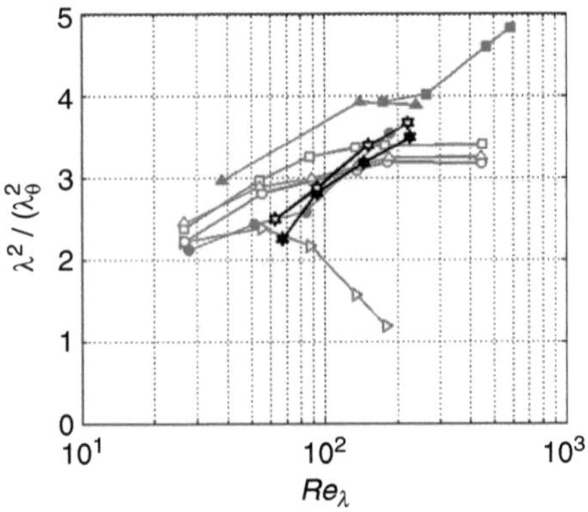

Fig. 21.6 Ratio $(\lambda/\lambda_\theta)^2$ as a function of Re_λ. *Gray full circle*: DNS [30]; *Gray full square*: DNS [33]; *Gray open square/triangle/circle*: LES [21] with scalar SGS model; *Gray open right-triangle*: LES [21] without scalar SGS model; *full star*: ILES $Ma = 0.27$; *open star*: ILES $Ma = 0.13$; *full black-line*: theory [19, 32]

turnover rate of turbulent kinetic energy and scalar variance are proportional to each other and not a function of *Re* has been a fundamental assumption underlying many models in turbulence phenomenology—e.g., [21].

Early DNS [30] shows continued growth of $\lambda^2/\lambda_\theta^2$ with increasing Re_λ, while the LES results using an explicit scalar SGS model [21] are asymptotically constant. Moreover, when the explicit scalar SGS model in [21] is turned off, the ratio rapidly decreases with Re_λ—again reflecting that the (dispersive) numerics used in [21] is inadequate by itself to ensure correct physical asymptotic behavior. In contrast, ILES (also without an explicit scalar SGS term) exhibits neither decreasing nor asymptotically constant behavior, but shows continued growth over the simulated range of effective Re_λ very consistent with early [30] and recent DNS [33]. Asymptotic linear dependence on Re_λ of $\lambda^2/\lambda_\theta^2$ is thus predicted based on theory [19, 32], DNS [30, 33], and ILES [19].

21.2.1 Realizability Constraints for SGS Modeling

ILES has been linked to the finite scale Navier–Stokes (NS, FSNS) framework for physical observables [44, 45]—involving averaged quantities over characteristic length and temporal scales (L and T). FSNS reframes the computational questions of turbulence. In the continuum, the usual turbulence questions relate to the velocity at a point, a concept limited computationally by FVs, mesh spacing Δx, and timestep Δt, and there are associated convergence issues as well as issues of consistent definition of FV averaged velocities. The question FSNS proposes to answer is different: what is the mean velocity in an FV (volume L^3) over finite time interval T? Motivation for FSNS relates to directly addressing the fundamental nature of physical observables: measurement devices and computations always involve finite space/time scales. The demonstrated connections between ILES and the solution of FSNS equations [44, 45] provide a rationale for ILES [46]: *ILES works because it solves the equations that most accurately represent the dynamics of FVs of fluid, i.e., governing the behavior of measurable physical quantities on the computational cells.*

Recent ensemble moment equation analysis of FSNS [47] provided an objective understanding of ILES of high Re NS turbulence. The analysis shows that passive scalar mixing by FSNS turbulence is mathematically consistent with the mixing fluid physics in NS turbulence; specifically, the NS stirring term in the scalar gradient variance equation appears in a form consistent with the material conservation of the original NS equations. A notable result in [47] was showing that popular eddy viscosity SGS modeling does not exhibit this fundamental conserved invariant property present in NS and FSNS.

On a similar note (in this section), we find that the asymptotic growth with Re_λ of the ratio of eddy turnover rates of turbulent kinetic energy and scalar variance (predicted by theory and DNS) can be captured with a (numerically) well-designed ILES—whereas LES using state-of-the-art explicit SGS modeling practices predicts constancy. We interpret these recent results [19, 47] as pointing at required realizability constraints for SGS modeling which appear naturally built into ILES and FSNS—but not in the mainstream classical SGS modeling strategies.

21.3 Predictability Challenges in Shock Driven Turbulence

Shock driven turbulence generated via the RMI is of fundamental interest in inertial confinement fusion, understanding the collapse of the outer cores of supernovas, explosives, and scramjets—where multiscale resolution issues become compounded with the inherent sensitivity of turbulent flows to IC. Beyond the RMI literature surveyed in [22], investigation of IC effects on RMI has been subject of many experimental [25, 48–50], computational [23, 24, 26–28, 50–54, 56], and theoretical studies [26, 55]. A critical feature of the RMI impulsive driving is that the turbulence decays as dissipation removes kinetic energy from the system. RMI add the complexity of shock waves and other compressible flow effects to the basic physics associated with mixing. Because RMI are shock driven, resolution requirements make DNS unpractical even on the largest supercomputers. State-of-the-art simulations use hybrid methods [56]—switching between shock capturing schemes and conventional LES depending on local flow conditions, or ILES—combining shock and turbulence emulation capabilities [23–29, 52, 53].

Our particular ILES strategy used the LANL RAGE code [57]. RAGE solves the multi-material compressible conservation equations for mass density, momenta, total energy, and partial mass densities, using a 2nd-order Godunov scheme, adaptive mesh refinement, a variety of numerical options for gradient terms—limiters. Material interface treatments were not used, and the van Leer limiter option was chosen. The simulation model is nominally inviscid, effectively models high-Re, Sc$\sim$1, miscible convection-driven flow. Simulations of planar [24, 27] and gas-curtain [28] shock-tube experiments were examined, involving high (SF_6)

and low-density (air) gases, with sharp and diffusive initial material interfaces, respectively, and Atwood number $A = 0.67$.

The accurate and reliable simulation of material interfaces is an essential aspect in simulating the material mixing fluid dynamics. Interfaces between fluids can be miscible or immiscible, with the character changing during the evolution of a system. This causes numerical approximations to the flow physics to be extremely challenging because the basic features of the approximations need to be inherently dynamic, i.e., adapt to the evolution of the materials. For example, in many applications of interest (e.g., inertial confinement fusion capsules) an interface may begin as sharp and immiscible, but evolves into a state where it mixes at an atomic level with neighboring material due to the effects of temperature and diffusion, or as it becomes a plasma. *Detailed material interface characteristics are IC for the flow instabilities* (e.g., for RMI), and among the most crucial issue associated with these details is the sensitivity of late-time consequences to IC specifics: small variations in the initial state of the interface can result in quite significant changes to even the integral character of a mixing layer at late times [24, 58, 59].

A typical shock-tube configuration used in the investigations involves significantly different density gases—e.g., low (air) and high (SF6) in Fig. 21.7, presumed IC at the material interface initially separating the gases, periodicity in the transverse (*y*,*z*) directions, and allowing eventual shock reflection off an end-wall to emulate reshock. The shock propagates in the (*x*) direction through the contact discontinuity, and the detailed evolution of the shocked mixing layer is investigated.

The surface displacement of the material interface in the RMI experiments [48] has been modeled using well-defined modes often combined with random perturbation components. In the early simulation work [23] the shocked planar RMI phase was examined in detail using local S+L deformations, combining a short (preferred, S) *egg crate* mode—chosen to represent the result of pushing the membrane through the wire mesh, and superimposed distortion of the wire mesh on a longer characteristic shock-tube scale L—chosen to be the (transverse) periodic dimension of the computational domain. In subsequent papers, the reshocked mixing layer was also simulated, and the IC modeling strategies included S+ϕ deformation [24, 56], where ϕ is a spatial broadband based distribution intended to break the presumed characteristic interface symmetries; simulations were performed

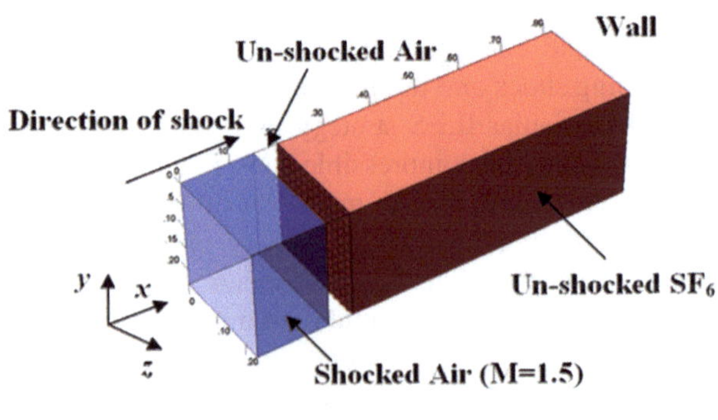

Fig. 21.7 Typical planar shock-tube computational domain; lengths in m

for a variety of random perturbations superimposed to the S mode [24], or having *only* a spatial randomly generated distribution ϕ as IC [27, 60]—the case revisited here.

We deal with inherently unsteady transitional and decaying turbulent flow. Suitable metrics for analysis of turbulence for this kind of flows are not established, and state-of-the-art analysis largely relies on using (unsteady) versions of diagnostics designed for developed homogeneous isotropic (steady) regimes. Mixing analysis can be carried out in terms of suitable variables and ensemble averaging,

$$\langle f \rangle (x) = \tfrac{1}{A}\int\!\!\int f(x,y,z)dydz, \quad A = \int dydz,$$
$$Y_{SF_6} = \rho_{SF_6}/\rho, \quad \theta = 4Y_{SF_6}\left[1 - Y_{SF_6}\right],$$
$$\psi(x) = \langle Y_{SF_6} \rangle, \quad M(x) = 4\psi(x)\left[1 - \psi(x)\right], \quad \delta = \int dxM(x),$$

where < > denotes ensemble (transverse plane) averaging, ρ_{SF_6} is the SF_6 mass density, Y_{SF6} is the SF_6 mass fraction, θ and $M(x)$ are local and averaged mixedness, and δ provides an integrated measure of mix thickness.

The initial interfacial morphology can be statistically characterized by the initial *rms* slope, $\eta_o = \kappa_o\delta_o \sim \langle \nabla\chi\nabla\chi \rangle^{1/2}$, where $\chi(y,z)$ is the local deviation of the initial material interface around the mean interface location, $\kappa_o = 2\pi/\lambda_o$, λ_o is a representative wavelength of the multiscale perturbation in the initial interface, and $\delta_o = \delta\,(t = 0)$ denotes the initial interface thickness (Fig. 21.8). Thus, high value of η_o denotes highly corrugated interfaces with high *rms* slope. The quantity κ_o is used in the study of homogeneous stochastic processes, where it is called the mean *zero-crossing frequency* [61]. In practice, the initial material interface value of κ_o is computed by checking for sign changes of the mass density fluctuation over lines within transverse planes and averaging their occurrences (Fig. 21.9); for $t > 0$, $\kappa = \kappa(t)$ is similarly evaluated within the mixing region. Mathematically, $\kappa(t)$ is related to the Taylor micro-scale [27].

A variety of IC perturbations and grid resolutions (using up to two levels of adaptive mesh refinement) were considered. The various cases can be organized into two distinct categories having significantly different (low/high) initial *rms* slope (η_0) with prescribed spectral content but different interfacial thickness (Table 21.1).

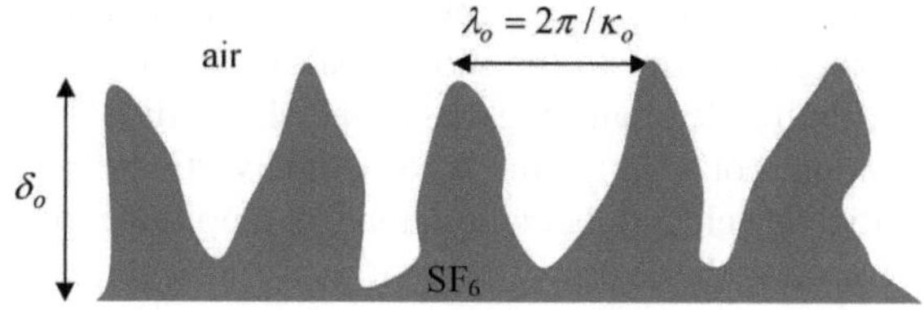

Fig. 21.8 Schematic of the initial interface

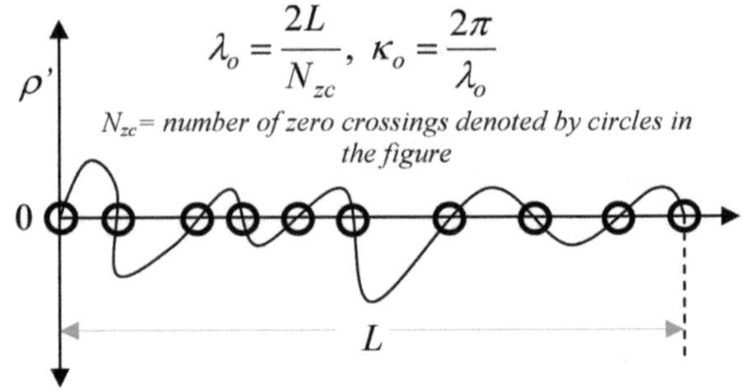

Fig. 21.9 Zero-crossings of ρ'; L is the transverse dimension of the computational domain

Table 21.1 Planar shock-tube simulations

$(L_{\min}, L_{\max})$	$L\left(\frac{1}{24}, \frac{1}{6}\right)$	$L\left(\frac{1}{12}, \frac{1}{4}\right)$	$L\left(\frac{1}{6}, \frac{1}{3}\right)$	$L\left(\frac{1}{4}, \frac{1}{2}\right)$	$L\left(\frac{1}{24}, \frac{1}{6}\right)$	$L\left(\frac{1}{12}, \frac{1}{4}\right)$	$L\left(\frac{1}{6}, \frac{1}{3}\right)$	$L\left(\frac{1}{4}, \frac{1}{2}\right)$
δ_o(cm)	0.5 (low η_o)				5 (high η_o)			
k_o (cm^{-1})	π	$\pi/2$	$\pi/4$	$\pi/6$	π	$\pi/2$	$\pi/4$	$\pi/6$
$\eta_o = k_o\delta_o$	$\pi/2$	$\pi/4$	$\pi/8$	$\pi/12$	$10\pi/2$	$10\pi/4$	$10\pi/8$	$10\pi/12$
Mode #	402	90	18	7	402	90	18	7

21.3.1 Different Linear and Nonlinear RMI Scalings

Despite the similarity in problem geometry, linear and nonlinear RMI share no dynamical or statistical features. We have recognized the nonlinear RMI as occurring when a high-η_o interface is first-shocked [27] and in reshock situations [60]. Both configurations involve high initial *rms*-slope η_o, both exhibit the same scalings and collapse of data [27, 60], and are also very different from the low-η_o (linear) RMI. A simple way to demonstrate the very large physical differences between linear and nonlinear RMI is with the first-shocked problem (Fig. 21.10, from [27]) and its trends with η_o. Instantaneous visualizations of Y_{SF_6} at the selected time, $t = 3000$ μs shown in Fig. 21.10, suggest material (interpenetration) mixing increasing with η_o.

Following Richtmyer's classical single-mode analysis [62], passage of a shock through the material interface has the effect of having baroclinic vorticity deposited as function of η_o ($\sim \kappa_o \dot{a} = \Delta U A^+ \kappa_o \eta_{oo}$); here, the temporal growth rate of the instability is expressed as $d\delta_{MZ}/dt = \dot{a} = \Delta U A^+ \kappa_o a_o$, with, a_o being the initial amplitude of the interface perturbation, ΔU the change in velocity of the mixing layer due to the shock, and A^+ is the post-shock Atwood number. For *low* η_o, the modes mainly grow in the shock direction in a *ballistic* (non-interacting) fashion. For *high* η_o, more baroclinic vorticity is generated; with the characteristic initial wavelength being less than its amplitude, perturbation *crests* and *troughs* are closer together, vortex interactions are stronger, and new modes are created through nonlinear processes.

Figure 21.10 shows growth of the mixing layer thickness as function of time for the two different initial interfacial perturbations shown on the top left. For the flatter interfaces (low *rms*-slope, $\eta_o < 1$), as one increases the initial η_o, one increases

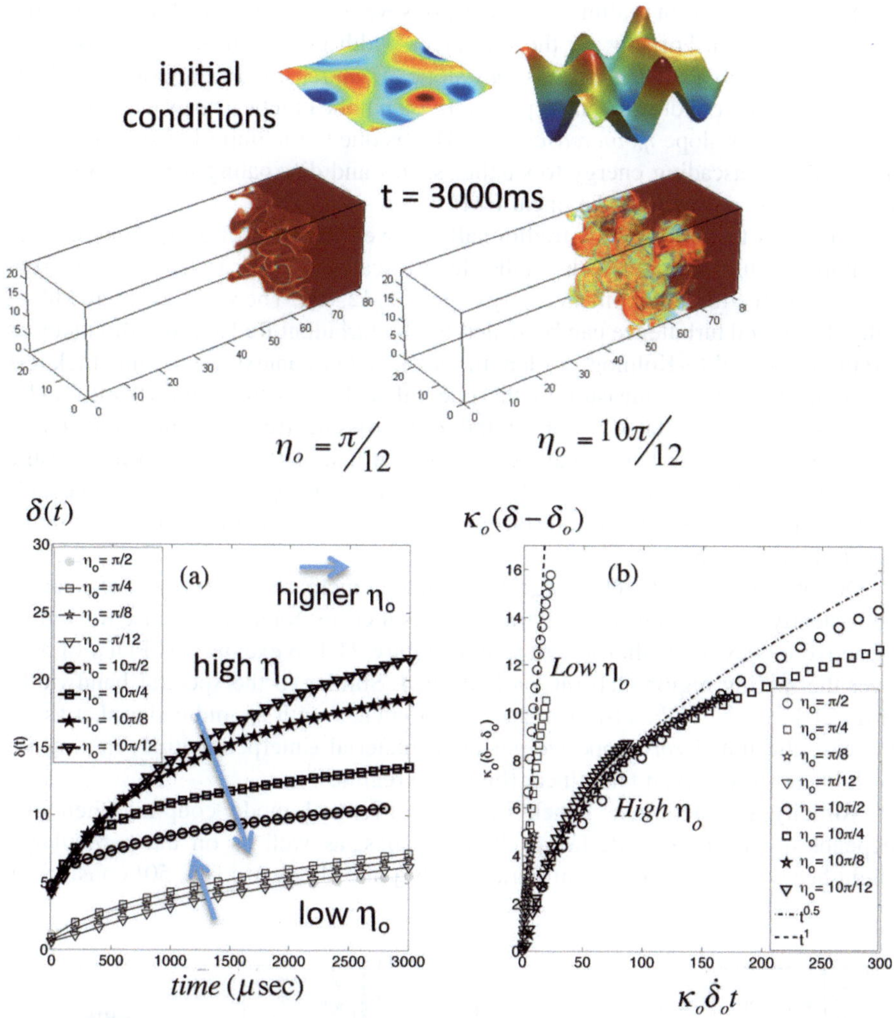

Fig. 21.10 The bipolar RM behavior [27, 60]; IC are indicated on very *top* and instantaneous mix visualizations 3000 s after first-shock on the *middle*; the evolving mix widths vs. time plotted on the *bottom* in dimensional (**a**) and non-dimensional form (**b**), depict distinctly different instability behaviors depending on initial *rms* interface slope η_o (linear RMI for low η_o, nonlinear RMI for high η_o)

the growth rate of the mixing layer width as indicated by the ascending arrow on the lower set of lines. The shocked nearly flat interface problem and scalings correspond to the linear RMI (Richtmyer's analysis above). In contrast, for nonlinear RMI associated with highly corrugated initial interface (high $\eta_o > 1$), opposite trends are seen as initial η_o increases. The radically different data collapse of mix widths for the two IC classes is apparent in Fig. 21.10: low η_o shows linear scaling while high

η_o shows a square-root scaling. This is one of several very major differences in fluid physics observed. For low η_o, the mix-layer width grows linearly and enstrophy and kinetic energy all *increase* as *rms*-slope η_o *increases*. In the nonlinear RMI regime mix-layer width and enstrophy *increase* but the kinetic energy and mix width *decrease* as *rms*-slope η_o *increases* [27]. This is due to the importance of nonlinear mechanisms cascading energy to smaller scales and dissipating it more rapidly—which do not contribute to the linear RMI.

Transition to turbulence is traditionally viewed in terms of a rapid increase in the population of motions with smaller length scales, which can lead to an inertial subrange in the turbulent kinetic energy spectra [42, 64]. The spectral bandwidth of fully developed turbulence can be scaled by the turbulent Re [63], usually taken as a ratio of integral-to-Kolmogorov length scales. In our context, we use the thickness of the layer, $\delta(t)$ as a measure of the integral scale, and the mass-density Taylor micro-scale $\lambda(t)$ —related to the spatial zero crossing frequency through $\lambda(t) = 2\pi/\kappa(t)$—as proxy for the small scales. We can use $\eta(t) = \kappa(t)\delta(t)$ as measure suggestive of the *spectral bandwidth* (Fig. 21.11a), with $\eta(t)$ playing two roles: (1) as η_o it describes the initial interface and (2) $\eta(t)$ is proportional to the evolving spectral bandwidth.

Our observations [27, 60] suggest that sudden increase in $\eta(t)$ (*and* $\kappa(t)$) can be consistently used to indicate flow transition as it corresponds to increase in number of active modes of smaller length scales. Figure 21.11b exemplifies PDFs of Y_{SF_6} over the mixing-region slab (at $t = 2500$ μs). Similar to the spectral bandwidths (Fig. 21.11a), the PDFs also increase monotonically with η_o, indicating that as we increase the initial *rms* slope we get more material (interpenetration) mixing for both linear (low- η_o) and nonlinear (high- η_o) regimes.

Vorticity production at shock time and eventual mode coupling thereafter depends on the initial interfacial characteristics, as well as on the particular A (light/heavy or heavy/light configuration) [65] and shock *Ma* [25, 50] considered.

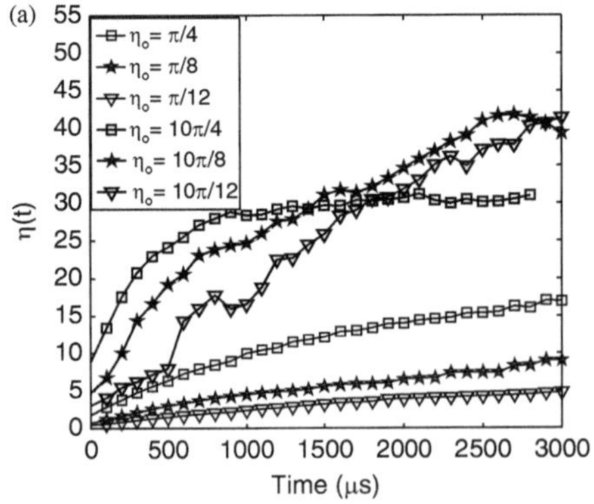

Fig. 21.11 (**a**) Spectral bandwidth $\eta(t)$; (**b**) PDFs of Y_{SF_6} at $t = 2500$ μs

However, the initial *rms* slope of the material interface η_o is a relevant parameter determining whether the flow is in the linear ballistic regime, or in nonlinear mode coupling regime.

21.3.2 Late-Time Predictability and Sensitivity to IC

Important practical consequences of our results are addressed separately [60]: (1) reshock effects on mixing and transition can be emulated at first-shock if η_o is high enough; (2) simple turbulence models cannot handle both classes of IC, i.e., state-of-the-art moment closure models cannot predict RMI flow with sequential shocks.

Turbulent flows are now recognized to have a sensitive dependence on IC [6, 9, 12]. The shock-accelerated material interface with high initial η_o is (by far) the most challenging example of acute sensitivity to IC known. This is due to the large amplification a shock has on the various parameters of the initial interface and the fact that the reshock problem actually comprises a group of instabilities—beyond just the linear RMI. Our work [27, 60] consistently exhibits extreme sensitivity to material interfacial conditions at reshock time raising potential late-time predictability issues. The impossibility of very-long-range weather forecasting [66], appropriately comes to mind here. Because of chaotic variability associated with unavoidable small perturbations (uncertainties) of presumed SGS and SPG conditions, it may become impossible—even within a mathematically well-posed dissipative flow simulation framework—to provide realistic late-time solutions good enough to address the specific questions of interest (e.g., the very-long-range weather forecasting).

Figure 21.12 (from [60]) shows the results of shocking a planar interface twice: first-shocked at $t = 0$, at which time η_o is in the linear RMI regime, and reshocked at $t \sim 3700$ ms, at which time the material interface has high η_o. We plot as functions of time, mix-layer width, bulk Re, and turbulent eddy-viscosity—used to scale nonlinear-to-linear interactions of the flow. After reshock, mixed layer response and growth trends are hugely different. Note that all the flows have similar values of layer thickness and eddy viscosity at reshock time and that there is a rapid divergence of metrics trajectories upon reshock, indicating that a simple IC characterization in terms of mix width and kinetic energy—as done in the state-of-the-art engineering turbulence models, is insufficient.

The predictability problem manifests itself as a rapid divergence of the trajectories and depends on subtleties of the balance of various possible instability mechanisms (beyond linear RMI) during enstrophy deposition and generation in high-η_o interfaces in ways that are unknown. Such sensitivity to IC at reshock is also observed in other configurations [28, 67] and raises questions regarding repeatability of numerical and laboratory experiments [28]. This leads to very important questions regarding nonlinear RMI predictability when using an incomplete set of IC metrics, and leads of course, to the question of what a usefully complete set of metrics might be to assure predictability.

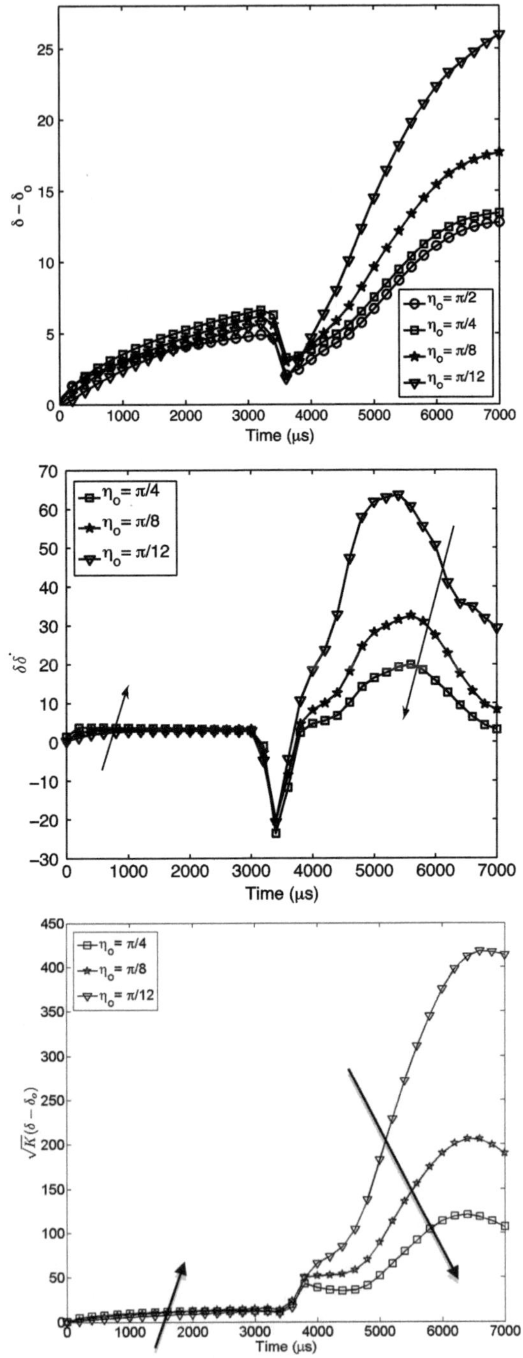

Fig. 21.12 Mix width thickness (*top*), bulk Re (*middle*) and turbulent eddy viscosity (*bottom*) for shocked/reshocked planar shock-tube ILES (reshocked at $t \sim 3700$ s) as function of initial *rms* slope η_o (see [60]). *Arrows* indicate the direction of increasing η_o, and K denotes turbulent kinetic energy

21.4 Summary

Throughout this presentation, we have emphasized the inherently intrusive nature of *coarse grained observations in computational and laboratory experiments*, intimately linked to their SGS and SPG specifics. Turbulent mixing modeling and predictability rely on availability of big data from computational and laboratory experiments. Fundamental issues of big data *generation, reduction, and organization* involve *characterizing and modeling* the specific flow conditions at the SGS and SPG scales. SGS modeling issues have motivated intense research in the last four decades. More recently, late-time (or far field) sensitivity of observations to (SPG) initial and other boundary conditions has also been recognized as a crucial aspect to be addressed.

Non-dissipative SGS issues are presumably less important for high enough Re (large-scale separation), but need to be carefully addressed when relatively short scale separations are involved. Largely uncharacterized small-scale turbulence processes remain in the turbulent variable-density context, including effects of SGS perturbations to initial material interfaces; effects of interacting shocks and sharp material interfaces with turbulence; baroclinic production of vorticity and other effects of small scale density variations; and exothermicity effects of chemical (or thermonuclear) reactions resulting from molecular scale material mixing.

Robust simulations for dissipative turbulent phenomena exhibiting enslavement of small-scale dynamics are in principle achievable with suitable SGS modeling, enough scale separation, and well-resolved IC. However, late-time predictability assessments for high-Re phenomena cannot be robust when inherent resolution sensitivities are present—*while nature controls the flow physics independently*. This is a particularly serious concern when coarse grained observations are constrained by *characterization and modeling* of (intertwined) SGS and SPG specifics.

If the IC information contained in filtered-out smaller and SGS scales can significantly alter the evolution of the larger scales of motion (and practical integral measures), then the utility of coarse grained observations for their prediction is not rationally or scientifically justifiable [68]. The *big data* purpose is qualified by the very specific research questions of interest to be addressed and by the predictability metrics in the associated verification, validation, and uncertainty quantification process. In practice, *only certain statistical predictions and big data types will be useful in this context.* Ensemble averaging over a suitably complete set of realizations covering the relevant IC variability is a data reduction strategy of choice.

Acknowledgements LANL is operated by the LANS, LLC for the US DOE NNSA under Contract No. DE-AC52-06NA25396.

References

1. P. Sagaut, *Large Eddy Simulation for Incompressible Flows*, 3rd edn. (Springer, New York, 2006)
2. F.F. Grinstein, L.G. Margolin, W.J. Rider (eds.), *Implicit Large Eddy Simulation: Computing Turbulent Flow Dynamics* (Cambridge University Press, New York, 2007), 2nd printing 2010
3. F.F. Grinstein (ed.), Boundary conditions for large eddy simulation, Special Section. AIAA J. **42**, 437–492 (2004)
4. A.K.M.F. Hussain, M.F. Zedan, Effects of the initial condition on the axisymmetric free shear layer: effect of the initial fluctuation level. Phys. Fluids **21**, 1475–1481 (1978)
5. E. Gutmark, C.M. Ho, Preferred modes and the spreading rates of jets. Phys. Fluids **26**, 2932 (1983)
6. I. Wygnanski, F. Champagne, B. Marasli, On the large-scale structures in two-dimensional, small-deficit, turbulent wakes. J. Fluid Mech. **168**, 31–71 (1986)
7. W.K. George, Governing equations, experiments, and the experimentalist. Exp. Therm. Fluid Sci. **3**, 557–566 (1990)
8. G. Li, E.J. Gutmark, Experimental study of boundary condition effects on non-reacting and reacting flow in a multi-swirl gas turbine combustor. AIAA J. **44**, 444 (2006)
9. W.K. George, The self-preservation of turbulent flows and its relation to initial conditions and coherent structures, in *Advances in Turbulence*, ed. by W.K. George, R.E. Arndt (Hemisphere, New York, 1989)
10. M.D. Slessor, C.L. Bond, P.E. Dimotakis, Turbulent shear-layer mixing at high Reynolds numbers: effects of inflow conditions. J. Fluid Mech. **376**, 115–138 (1998)
11. P. Ramaprabhu, G. Dimonte, M.J. Andrews, A numerical study of the influence of initial perturbations on the turbulent Rayleigh–Taylor instability. J. Fluid Mech. **536**, 285–319 (2005)
12. W.K. George, L. Davidson, Role of initial conditions in establishing asymptotic flow behavior. AIAA J. **42**, 438–446 (2004)
13. W.K. George, M. Tutkun, Mind the gap: a guideline for large eddy simulation. Phil. Trans. R. Soc. A **367**(1899), 2839–2847 (2009)
14. A.A. Townsend, *Structure of turbulent shear flow* (Cambridge University Press, New York, 1976)
15. F.F. Grinstein, Self-induced vortex ring dynamics in subsonic rectangular jets. Phys. Fluids **7**, 2519–2521 (1995)
16. F.F. Grinstein, Vortex dynamics and entrainment in regular free jets. J. Fluid Mech. **437**, 69–101 (2001)
17. U. Frisch, *Turbulence* (Cambridge University Press, New York, 1995)
18. D. Drikakis, C. Fureby, F.F. Grinstein, D. Youngs, Simulation of transition and turbulence decay in the Taylor-Green Vortex. J. Turbul. **8**(020), 1–12 (2007)
19. A.J. Wachtor, F.F. Grinstein, C.R. DeVore, J.R. Ristorcelli, L.G. Margolin, Implicit large-eddy simulation of passive scalar mixing in statistically stationary isotropic turbulence. Phys. Fluids **25**(2), 1–19 (2013)
20. Y. Zhou, F.F. Grinstein, A.J. Wachtor, B.M. Haines, Estimating the effective Reynolds number in ILES. Phys. Rev. E **89**, 013303 (2014)
21. D.I. Pullin, A vortex-based model for the subgrid flux of a passive scalar. Phys. Fluids **12**, 2311 (2000)
22. M. Brouillette, The Richtmyer-Meshkov instability. Annu. Rev. Fluid Mech. **34**, 445–468 (2002)
23. R.H. Cohen, W.P. Dannevik, A.M. Dimits, D.E. Eliason, A.A. Mirin, Y. Zhou, D.H. Porter, P.R. Woodward, Three-dimensional simulation of a Richtmyer-Meshkov instability with a two-scale initial perturbation. Phys. Fluids **14**, 3692–3709 (2002)
24. F.F. Grinstein, A.A. Gowardhan, A.J. Wachtor, Simulations of Richtmyer-Meshkov instabilities in planar shock-tube experiments. Phys. Fluids **23**, 034106 (2011)

25. E. Leinov, G. Malamud, Y. Elbaz, A. Levin, G. Ben-dor, D. Shvarts, O. Sadot, Experimental and numerical investigation of the Richtmyer-Meshkov Instability under reshock conditions. J. Fluid Mech. **626**, 449–475 (2009)
26. B. Thornber, D. Drikakis, R.J.R. Williams, D.L. Youngs, The influence of initial conditions on turbulent mixing due to Richtmyer-Meshkov instability. J. Fluid Mech. **654**, 99–139 (2010)
27. A.A. Gowardhan, J.R. Ristorcelli, F.F. Grinstein, The bipolar behavior of the Richtmyer-Meshkov instability. Phys. Fluids **23**(Letters), 071701 (2011)
28. A.A. Gowardhan, F.F. Grinstein, Numerical simulation of Richtmyer-Meshkov instabilities in shocked gas curtains. J. Turbul. **12**(43), 1–24 (2011)
29. B.M. Haines, F.F. Grinstein, J.R. Fincke, Three-dimensional simulation strategy to determine the effects of turbulent mixing on inertial-confinement-fusion capsule performance. Phys. Rev. E. **89**(5), 053302 (2014)
30. M.R. Overholt, S.B. Pope, Direct numerical simulation of a passive scalar with imposed mean gradient in isotropic turbulence. Phys. Fluids **8**(11), 3128–3148 (1996)
31. D.R. Dowling, P.E. Dimotakis, Similarity of concentration field of gas phase turbulent jets. J. Fluid Mech. **218**, 109–141 (1990)
32. J.R. Ristorcelli, Passive scalar mixing: analytic study of time scale ratio, variance, and mix rate. Phys. Fluids **18**(7), 1–17 (2006)
33. T. Gotoh, T. Watanabe, Y. Suzuki, Universality and anisotropy in passive scalar fluctuations in turbulence with uniform mean gradient. J. Turbul. **12**(48), 1–27 (2011)
34. C. Fureby, F.F. Grinstein, Monotonically integrated large eddy simulation of free shear flows. AIAA J. **37**, 544–556 (1999)
35. C. Fureby, F.F. Grinstein, Large eddy simulation of high Reynolds-number free and wall bounded flows. J. Comput. Phys. **181**, 68–97 (2002)
36. J.A. Domaradzki, Z. Xiao, P. Smolarkiewicz, Effective eddy viscosities in implicit large eddy simulations of turbulent flows. Phys. Fluids **15**, 3890–3893 (2003)
37. M.R. Petersen, D. Livescu, Forcing for statistically stationary compressible isotropic turbulence. Phys. Fluids **22**(11), 1–11 (2010)
38. J. Jimenez, A.A. Wray, P.G. Saffman, R.S. Rogallo, The structure of intense vorticity in isotropic turbulence. J. Fluid Mech. **255**(65–90), 832 (1993)
39. Y. Li, C. Meneveau, Origin of non-gaussian statistics in hydrodynamic turbulence. Phys. Rev. Lett. **95**, 1–4 (2005)
40. D.H. Porter, P.R. Woodward, A. Pouquet, Inertial range structures in decaying turbulent flows. Phys. Fluids **10**, 237–245 (1998)
41. Y. Li, E. Perlman, M. Wan, Y. Yang, R. Burns, C. Meneveau, R. Burns, S. Chen, A. Szalay, G. Eyink, A public turbulence database cluster and applications to study Lagrangian evolution of velocity increments in turbulence. J. Turbul. **9**, N31 (2008)
42. P.E. Dimotakis, The mixing transition in turbulent flows. J. Fluid Mech. **409**, 69–98 (2000)
43. G.A. Blaisdell, N.N. Mansour, W.C. Reynolds, Compressibility effects on the passive scalar flux within homogeneous turbulence. Phys. Fluids **6**, 3498–3500 (1994)
44. L.G. Margolin, W.J. Rider, A rationale for implicit turbulence modeling. Int. J. Numer. Methods Fluids **39**, 821–841 (2002)
45. L.G. Margolin, Finite-scale equations for compressible fluid flow. Phil. Trans. R. Soc. A **367**, 2861–2871 (2009)
46. F.F. Grinstein, L.G. Margolin, W.J. Rider, A rationale for implicit LES, Chapter 2, in *Implicit Large Eddy Simulation: Computing Turbulent Fluid Dynamics*, ed. by F.F. Grinstein, L.G. Margolin, W.J. Rider (Cambridge University Press, New York, 2010), 2nd printing
47. J.R. Ristorcelli, Material conservation of passive scalar mixing in finite scale Navier-Stokes fluid turbulence, LA-UR-15-20837, Chapter 4, in *Coarse Grained Simulation and Turbulent Mixing*, ed. by F.F. Grinstein (Cambridge University Press, 2016)
48. M. Vetter, B. Surtevant, Experiments on the Richtmyer-Meshkov instability of an air/SF6 interface. Shock Waves **4**, 247–252 (1995)
49. J.W. Jacobs, J.M. Sheeley, Experimental study of incompressible Richtmyer-Meshkov instability. Phys. Fluids **8**, 405–415 (1996)

50. G.C. Orlicz, B.J. Balakumar, C.D. Tomkins, K.P. Prestridge, A Mach number study of the Richtmyer-Meshkov instability in a varicose, heavy-gas curtain. Phys. Fluids **21**, 064102 (2009)
51. J.A. Greenough, E. Burke, The effect of initial conditions on late time asymptotics and mixing for multimode Richtmyer-Meshkov instability, in *9th International Workshop on the Physics of Compressible Turbulent Mixing*, Cambridge, UK, 19–23 July 2004
52. O. Schilling, M. Latini, High-order WENO simulations of three-dimensional reshocked Richtmyer-Meshkov instability to late times: dynamics, dependence on initial conditions, and comparisons to experimental data. Acta Math. Sci. **30B**(2), 595–620 (2010)
53. D.L. Youngs, Rayleigh–Taylor and Richtmyer-Meshkov Mixing, Chapter 13, in *Implicit Large Eddy Simulation: Computing Turbulent Fluid Dynamics*, ed. by F.F. Grinstein, L.G. Margolin, W.J. Rider (Cambridge University Press, New York, 2010), 2nd printing, pp. 392–409
54. N.J. Nelson, F.F. Grinstein, Effects of initial condition spectral content on shock driven turbulent mixing. Phys. Rev. E **92**, 013014 (2015)
55. K.O. Mikaelian, Extended model for Richtmyer-Meshkov mix. Phys. D **240**, 935–942 (2010)
56. D.J. Hill, C. Pantano, D.I. Pullin, Large-Eddy simulation and multiscale modelling of a Richtmyer-Meshkov instability with reshock. J. Fluid Mech. **557**, 29–61 (2006)
57. M. Gittings, R. Weaver, M. Clover, T. Betlach, N. Byrne, R. Coker, E. Dendy, R. Hueckstaedt, K. New, W.R. Oakes, D. Ranta, R. Stefan, The RAGE radiation-hydrodynamic code. Comput. Sci. Discovery **1**, 015005 (2008)
58. G. Dimonte, Nonlinear evolution of the Rayleigh-Taylor and Richtmyer-Meshkov instabilities. Phys. Plasmas **6**(5), 2009–2015 (1999)
59. D. Drikakis, F.F. Grinstein, D. Youngs, On the computation of instabilities and symmetry-breaking in fluid mechanics. Prog. Aerosp. Sci. **41**(8), 609–641 (2005)
60. J.R. Ristorcelli, A.A. Gowardhan, F.F. Grinstein, Two classes of Richtmyer-Meshkov instabilities; a detailed statistical look. Phys. Fluids **25**, 044106 (2013)
61. K.R. Sreenivasan, A. Prabhu, R. Narasimha, Zero-crossings in turbulent signals. J. Fluid Mech. **137**(251–272) (1983)
62. R.D. Richtmyer, Taylor instability in shock acceleration of compressible fluids. Commun. Pure Appl. Math. **13**, 297–319 (1960)
63. H. Tennekes, J.L. Lumley, *A First Course in Turbulence* (MIT Press, Cambridge, 1972)
64. J. Holford, S. Dalziel, D. Youngs, Rayleigh-Taylor instability at a tilted interface in laboratory experiments and numerical simulations. Laser Part. Beams **21**, 419–423 (2003)
65. M. Lombardini, D.J. Hill, D.I. Pullin, D.I. Meiron, Atwood ratio dependence of Richtmyer-Meshkov flows under reshock conditions using large-eddy simulations. J. Fluid Mech. **670**, 439–480 (2011)
66. E.N. Lorenz, Deterministic nonperiodic flow. J. Atmos. Sci. **20**, 130–141 (1963)
67. S. Balasubramanian, G.C. Orlicz, K.P. Prestridge, B.J. Balakumar, Experimental study of initial condition dependence on Richtmyer-Meshkov instability in the presence of reshock. Phys. Fluids **24**, 034103 (2012)
68. F.F. Grinstein, On integrating large eddy simulation and laboratory turbulent flow experiments, by invitation. Phil. Trans. R. Soc. A **367**, 2931–2945 (2009)

Chapter 22
Non-classical/Exponential Decay Regimes in Multiscale Generated Isotropic Turbulence

M. Meldi and P. Sagaut

22.1 Introduction

Among the different aspects dealing with homogeneous isotropic turbulence (HIT) studies, the sensitivity to initial conditions is a timely subject of discussion. In particular, the quantification of production effects on the resulting decay regime has proven elusive. In fact, the presence of epistemic uncertainties in grid turbulence experiments and the limits in actual computational resources available to perform direct numerical simulation (DNS) prevent the isolation of the production mechanisms. Because of its relevance for a number of environmental/industrial applications, this subject has received particular attention in the last two decades. The most simple approach to consider production effects on turbulence decay is to assume that they influence the initial state only. The physical production mechanisms are thus accounted in the initial conditions. In the case of HIT free decay, this implies that the resulting regime can be exactly described by the Lin equation, which represents the counterpart of the Karman–Howarth equation in the spectral space:

$$\frac{\partial E(k,t)}{\partial t} + 2\nu k^2 E(k,t) = T(k,t) \qquad (22.1)$$

M. Meldi (✉)
Department of Fluid Flow, Heat Transfer and Combustion, Institut PPRIME, CNRS - Université de Poitiers - ENSMA, UPR 3346, SP2MI - Téléport, 211 Bd. Marie et Pierre Curie, B.P. 30179, F86962 Futuroscope Chasseneuil Cedex, France
e-mail: marcello.meldi@ensma.fr

P. Sagaut
Aix Marseille Université, CNRS, Centrale Marseille, M2P2 UMR 7340, 13451 Marseille, France

A. Pollard et al. (eds.), *Whither Turbulence and Big Data in the 21st Century?*,
DOI 10.1007/978-3-319-41217-7_22

where k is the wave number, t is the time, ν the kinematic viscosity of the flow, $E(K, t)$ the energy spectrum and $T(k, t)$ is the non-linear energy transfer. One of the most interesting aspects of the resulting flow configuration is that partial/complete self-similarity of the energy spectrum can be observed [6, 18, 30]. A number of theoretical analyses [3, 4, 17] indicate as well that the resulting decay regime can be described by a power law evolution of the main flow statistical quantities, such as the turbulent kinetic energy $\mathcal{K}$, the energy dissipation rate ε, the integral length scale L and the Taylor micro-scale λ. For each physical quantity, the power law exponent driving the decay is determined by the parameter σ, which describes the behaviour of the spectrum at the large scales, i.e. $E(k, t) \propto k^{\sigma}$, $k < L^{-1}$. The value of the parameter σ is tied to the conservation of physical invariants, such as the linear momentum. Two classical cases have been proposed in the literature by Saffman [27] ($\sigma = 2$) and by Batchelor [1] ($\sigma = 4$).

A number of papers reported in the open literature support the idea that production effects can be efficiently reproduced by an initial condition state, once a precise estimation of the virtual origin in time/space is obtained. This implies that the production mechanisms do not leave a permanent signature on the turbulent flow. In particular, detailed grid turbulence analysis by Krogstad and Davidson [11, 12] indicates that a power law decay regime for the turbulent kinetic energy is observed independently of the shape of the grid used. More theoretical and experimental analysis by the same authors [5, 10] suggest that the expected slope of the energy spectrum at the large scales must be $\sigma = 2$, so that grid turbulence should be of the Saffman type.

On the other hand, studies over the last two decades indicate that time-lasting anomalous behaviours of HIT statistical quantities are observed in grid turbulence when using exotic grid elements. A particular interest has been devoted to the study of grid turbulence produced by fractal elements. Numerous studies by the Imperial College group [8, 13, 15, 29] report an exponential/fast power law decay which does not comply with the classical theoretical model in the open literature. Still, the possible emergence of exponential regimes in HIT decay has been recently demonstrated in theoretical works by in HIT decay has been recently demonstrated in theoretical works by George and Wang [7] and by Oberlack and Zieleniewicz [24]. In the last years, numerous research works on the subject have been published [9, 23, 32]. Comparing these observations, it is not possible to deduce if the emergence of these anomalous regimes is tied to a strong signature of the production mechanisms or to other physical reasons, such as anisotropy or post-processing issues. The latter are particularly important when the data sampling is sparse/incomplete and the prediction of the behaviour of the physical quantities relies on reconstruction techniques.

The objective of the present contribution is to derive more information about the signature effect of production mechanisms by the use of a new version of the eddy-damped quasi-normal Markovian (EDQNM) model recently proposed by Meldi et al. [19]. The classical version of the EDQNM model allows for a detailed representation of turbulence statistical moments with spectral accuracy in the case of free HIT decay. The improved model allows for the implementation of time-lasting production effects, which are related to the features of the grid geometry in the physical space. In particular, the conditions leading to the emergence of an exponential regime for $\mathcal{K}$ will be thoroughly investigated. The paper is divided as follows. In Sect. 22.2 the model is briefly described and the set-up of the computations is detailed. In Sect. 22.3 the EDQNM results are introduced and commented. Finally, the concluding remarks are drawn in Sect. 22.4.

22.2 EDQNM Model and Numerical Set-up

The EDQNM model [14, 25, 28] in the new version proposed by Meldi et al. [19] is here briefly introduced. The model is a turbulence closure based on the discretisation of a modified Lin equation, which accounts for production effects $F(k,t)$:

$$\frac{\partial E(k,t)}{\partial t} + 2\nu k^2 E(k,t) = T(k,t) + F(k,t) \tag{22.2}$$

The fourth order cumulants in the non-linear energy transfer evolution equation, which account for the effects of the deviation from the actual velocity pdf from the Gaussian pdf, are represented as a linear eddy damping term. This closure proves to be more precise than the standard quasi-normal (QN) approximation, where the fourth order cumulants are neglected, and it prevents the emergence of non-physical solutions. The model has shown characteristics of reliability, robustness and efficiency when used for the analysis of homogeneous turbulence [2, 22, 31].

The term $F(k,t)$ accounts for forcing effects affecting the turbulent flow evolution. In the case of classical free HIT decay, $F(k,t) = 0 \; \forall \, k, t$. If $F(k,t) = F(k)$, the statistically steady forced HIT case is obtained. In the case of time evolution of F, the resulting turbulent regime will be driven by the interactions between the physical dissipation and the forcing term. In a recent work, Meldi et al. [19] proposed a form for $F(k,t)$ term which mimics the statistical effects of fractal/multiscale forcing:

$$F(k,t) = \begin{cases} 2f\,\gamma_R\,(k\,L_b)^\beta\,a_k\,k\,\sqrt{2\,\pi\,E(k,t)}\,, & t < \tau \\ 2f\,\gamma_R\,(k\,L_b)^\beta\,a_k\,k\,\sqrt{2\,\pi\,E(k,t)}\,(1 + \alpha\,kL_\gamma\,(t-\tau))^{-2}\,, & t \geqslant \tau \end{cases} \tag{22.3}$$

The forcing term $F(k,t)$ behaves like a statistically steady forcing term $F(k) = 2f\,\gamma_R\,(k\,L_b)^\beta\,a_k\,k\,\sqrt{2\,\pi\,E(k,t)}$ for $t < \tau$, where τ is a characteristic time set by the user. Meldi et al. [19] imposed $\tau = 3t_0$, where $t_0 = \mathcal{K}(0)/\varepsilon(0)$ is the initial turnover time. Meldi et al. [19] showed that, in EDQNM computations, a statistically steady state of forced HIT was reached for $t \approx 0.3\tau$. For $t > \tau$, the forcing term evolves in time following a law $(1 + \alpha\,kL_\gamma\,(t-\tau))^{-2}$. The parameters β and α in Eq. (22.3) govern the statistically steady term and the time evolving function, respectively. The EDQNM closure has been here used to produce a database, choosing the value of the parameters as $\beta = 0.1, 0.3, 0.5, 0.7$ and $\alpha = 10^i$, $i = 2, 1, \ldots, -4$.

All the simulations have been initialised imposing the conditions $\mathcal{K}(0) = 1$ and $Re_\lambda(0) = 10^3$. For every simulation of the database, the initial energy spectrum imposed is a functional form inspired by the proposals by Pope [26] and Meyers and Meneveau [20]:

$$E(k) = C_k\, \varepsilon^{2/3}\, k^{-5/3} f_L(kL) f_\eta(k\eta) \tag{22.4}$$

with:

$$f_L(kL) = \left(\frac{kL}{[(kL)^{3/2} + c_L)]^{2/3}} \right)^{5/3+\sigma} , \quad f_\eta(k\eta) = \exp(-\xi([(k\eta)^4 + c_\eta^4]^{1/4} - c_\eta)) \tag{22.5}$$

where $c_\eta = 0.4$, $\xi = 5.3$ and c_L has been chosen to obtain $L(0) = 1$.

The results are sampled in the range $t \in [T_1, T_2]$, where $T_1 = 10t_0$ and $T_2 \rightarrow Re_\lambda(T_2) = 100$. The choice of T_2 is made so that the analysis is restricted to high to moderate Reynolds numbers. The computational mesh encompasses 11 decades in the spectral space. For every simulation, the elements of the computational mesh are distributed following a geometrical progression so that $k_{i+1}/k_i = 10^{1/20}$, being k_i a generic mesh element. The largest resolved mode is chosen so that $k_N = 10/\eta(0)$, where η is the Kolmogorov scale. The smallest mesh element has been chosen so that $k_1 = 10^{-6}/L(0)$. Moreover, it has been verified for every simulation that $k_1 \times L(T_2) < 10^{-2}$. The fulfilment of this condition implies that saturation effects due to lack of resolution at the large scales are excluded [21]. The EDQNM computations have been sampled in time by a geometrical progression of ratio $2^{0.1}$. Thus, the numerical data have been sampled more than 30 times par time decade. Moreover, the forcing effect is not applied to the whole spectral domain investigated, but it is restricted to one decade around the initial value of the integral length scale $L(0)$. A complete summary of the set-up parameters of the EDQNM simulations is reported in Table 22.1.

The time evolution of the energy spectrum $E(k, t)$ for simulation 3 of the database ($\alpha = 10^0$, $\beta = 0.5$) is shown in Fig. 22.1. For $t < \tau$, the energy spectrum converges to the steady fractal forcing shape (black continuous line) from the initial functional form prescribed (white square markers). For $t \geq \tau$, the forcing term evolves following a time law which is governed by the parameter α. The dissipation and the production effects are no longer in equilibrium, and the energy spectrum exhibits a clear time evolution. A first qualitative look at the spectrum allows to appreciate an inertial region with a slope of $-5/3$ for every t. On the other hand, the forced region around the integral length scale L exhibits a larger plateau if compared with the classical functional form prescribed for $t = 0$.

Table 22.1 Main parameters defining the database of the EDQNM computations used in the present analysis

Computation	σ	β	α	$Re_\lambda(0)$	$Re_\lambda(T_1) - Re_\lambda(T_2)$	Forced range $[k_{\min}; k_{\max}]$
1	2	0.5	10^2	10^3	1464 − 100	$[10^{-1/2}\,k_L(0);\ 10^{1/2}\,k_L(0)]$
2	2	0.5	10^1	10^3	1472 − 100	$[10^{-1/2}\,k_L(0);\ 10^{1/2}\,k_L(0)]$
3	2	0.5	10^0	10^3	1569 − 100	$[10^{-1/2}\,k_L(0);\ 10^{1/2}\,k_L(0)]$
4	2	0.5	10^{-1}	10^3	2179 − 100	$[10^{-1/2}\,k_L(0);\ 10^{1/2}\,k_L(0)]$
5	2	0.5	10^{-2}	10^3	2866 − 100	$[10^{-1/2}\,k_L(0);\ 10^{1/2}\,k_L(0)]$
6	2	0.5	10^{-3}	10^3	3074 − 100	$[10^{-1/2}\,k_L(0);\ 10^{1/2}\,k_L(0)]$
7	2	0.5	10^{-4}	10^3	3103 − 100	$[10^{-1/2}\,k_L(0);\ 10^{1/2}\,k_L(0)]$
8	2	0.5	10^{-5}	10^3	3105 − 100	$[10^{-1/2}\,k_L(0);\ 10^{1/2}\,k_L(0)]$
9	2	0.1	10^0	10^3	1569 − 100	$[10^{-1/2}\,k_L(0);\ 10^{1/2}\,k_L(0)]$
10	2	0.3	10^0	10^3	1569 − 100	$[10^{-1/2}\,k_L(0);\ 10^{1/2}\,k_L(0)]$
11	2	0.7	10^0	10^3	1569 − 100	$[10^{-1/2}\,k_L(0);\ 10^{1/2}\,k_L(0)]$

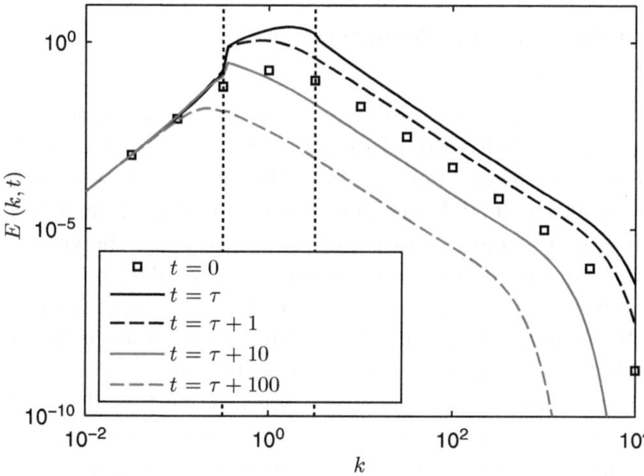

Fig. 22.1 Time evolution of the energy spectrum $E(k, t)$ computed by the EDQNM model. The results have been sampled for simulation 3 of the database

22.3 Results

The analysis of the results of the EDQNM database is now addressed. The sensitivity of the evolution of HIT statistical quantities to the parameters β and α will be investigated.

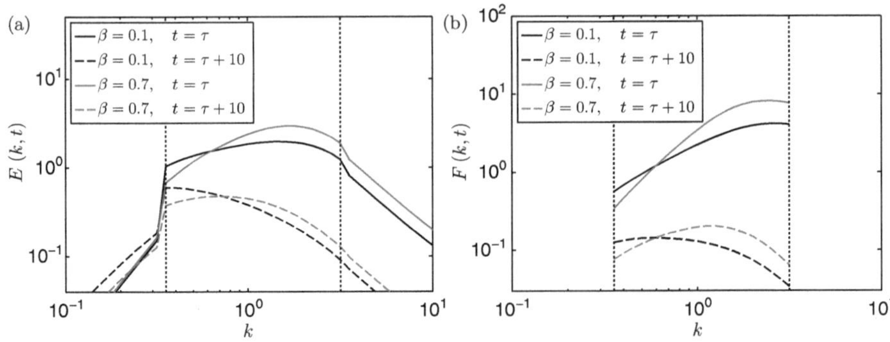

Fig. 22.2 (**a**) Energy spectrum $E(k, t)$ and (**b**) production term $F(k, t)$ computed by the EDQNM model. The results have been sampled for simulation 9 of the database ($\beta = 0.1$) and for (**b**) simulation 11 of the database ($\beta = 0.7$)

22.3.1 Sensitivity to the Parameter β

The parameter β drives the shape of the forcing term $F(k, t)$. In particular, we expect to observe a power law behaviour of $F(k, t)$ in steady fractal forced HIT, as shown in DNS simulations by Mazzi and Vassilicos [16]. The observation of the EDQNM results is consistent with the DNS results, as shown in Fig. 22.2a, b. Here, the shape of $E(k, t)$ and $F(k, t)$ in the forced region is, respectively, shown for simulations 8 and 10 of the database. For higher β values, the distribution of energy tends to move towards the smallest forced scales, as well as the slope of $F(k, t)$ increases. In particular, the analysis of the data shows that the behaviour of the forcing can be approximated by the law $F(k, t) \propto k^{1.33}$ for $\beta = 0.1$ and $F(k, t) \propto k^{2.2}$ for $\beta = 0.7$. This approximation drops in accuracy just towards the boundaries of the forced range, because of continuity reasons. As the forcing evolves in time, one can observe a decrease of intensity, as expected. At the same time, though, it appears that the forcing peak tends to move progressively towards larger scales. While the parameter β is responsible for the distribution of F in the forced range, it marginally influences its integral over the spectral space. Thus, the total intensity of the lasting production mechanisms is almost independent of the parameter β. This is the reason why β has a weak influence over the evolution of HIT statistical quantities, as already shown by Meldi et al. [19].

22.3.2 Sensitivity to the Parameter α

The sensitivity to the parameter α is now investigated. As previously stated in Sect. 22.2, this parameter governs the time evolution of the forcing term $F(k, t)$. The first quantity analysed is the turbulent kinetic energy $\mathcal{K}$, which is reported in

Fig. 22.3a. It is possible to observe that, for values of the parameter $\alpha \geq 1$, the turbulent kinetic energy decreases following a convex curve for a short transient after τ, while the classical theoretical behaviour is observed for $t \gg \tau$. This result is justified by the observation of Fig. 22.2a. In fact, for $t \approx \tau$, the energy spectrum exhibits very energetic modes in the range $[10^{-1/2}\,k_L,\ 10^{1/2}\,k_L]$, when compared to the wave numbers which are outside the forced range. For $t > \tau$, the forcing term decreases very fast in time, so that the resulting regime is actually a free decay starting from a steady fractal turbulence spectrum. The local interaction is here strong because of the shape close to $L(\tau)$, so that a higher amount of energy is transferred towards the small scales. This is the reason why a fast initial decay for $\mathcal{K}$ is observed. Moreover, this effect does not last in time, as a classical behaviour is achieved after a sufficiently long evolution time. The results for $\alpha \gg 1$ are thus in agreement with the experimental findings by Krogstad and Davidson [11, 12] and in this case the production effects can be well represented as by initial conditions. On the other hand, $\mathcal{K}$ evolution in time can be described by a concave curve for $\alpha \ll 1$. More precisely, an exponential decay is observed for $t \leq 200$, as shown in Fig. 22.4. This exponential decay progressively turns into a power law evolution for $t \gg \tau$ but, surprisingly, the power law exponent is higher in magnitude than the expected classical value. This result is produced by a local equilibrium between the energy dissipation rate ε and the integral of the forcing term $F(k,t)$, so that HIT decay is governed by the latter. The ratio of the magnitude of the two dynamic effects is shown in Fig. 22.5a. For very low values of α, the ratio $\int_{-\infty}^{+\infty} F(k,t)dk/\varepsilon(t)$ is almost constant, which can be physically interpreted as a strong signature of the production effect over HIT evolution. On the other hand, for $\alpha \gg 1$ the ratio is almost zero, which is consistent with the previous free decay observation. The exponential phase is observed for a progressively longer time for lower α values, with an asymptotic permanent exponential regime for $\alpha \to 0$. In fact, the data in Fig. 22.4 indicate that the exponential regime lasts for a time $t - \tau \approx 10, 100, 1000$ for $\alpha = 10^{-3}, 10^{-4}, 10^{-5}$, respectively.

We now address the results in Fig. 22.3b, where the ratio L/λ is reported. For $\alpha \ll 1$ and $\alpha \gg 1$ the results are qualitatively in agreement with the observations for $\mathcal{K}$. In particular, the EDQNM results $L/\lambda = \text{const.}$ and an exponential decay for $\mathcal{K}$, which are observed for $\alpha \ll 1$, are in agreement with the theoretical works by George and Wang [7] and by Oberlack and Zieleniewicz [24]. On the other hand, a non-classical regime is observed for $\alpha = 10^0$. For this specific value, the evolution of $F(k,t)$ is not sufficiently strong to completely drive the decay, but at the same time is not negligible with respect to the physical dissipation (see Fig. 22.5a). While the evolution of $\mathcal{K}$ does not seem to be significantly affected by this phenomenon, the behaviour of the characteristics length scales appears to be more sensitive to the parameter α. This difference can be clearly observed in the analysis of the well-known turbulence coefficient $C_\varepsilon = L\varepsilon/\mathcal{K}^{3/2}$. For $\alpha \ll 1$ and $\alpha \gg 1$, i.e. when HIT decay is governed by a dominant evolution effect, C_ε is an invariant of the decay regime. On the other hand, a time evolution is observed for $\alpha \approx 1$. This evolution seems to be consistent with the experimental observation of $C_\varepsilon \propto Re_\lambda^{-1}$ in fractal grid turbulence by Valente and Vassilicos [33, 34].

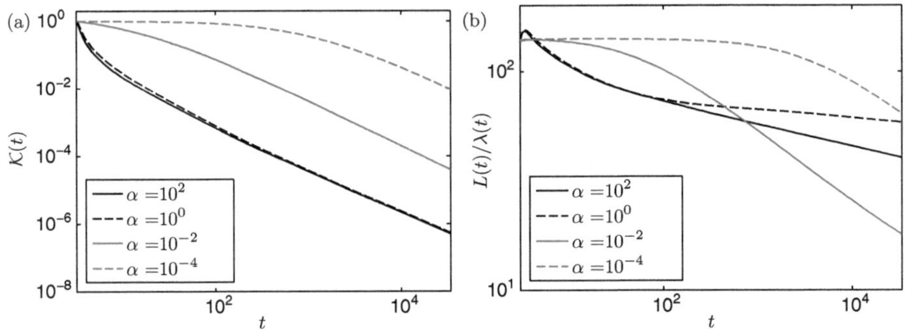

Fig. 22.3 Time evolution of HIT statistical quantities, computed by the EDQNM model. The sensitivity to the parameter α (simulations 1, 3, 5, 7 of the database) is highlighted. (**a**) The turbulent kinetic energy $\mathcal{K}$ and (**b**) the ratio between the integral length scale L and the Taylor micro-scale λ is reported, respectively

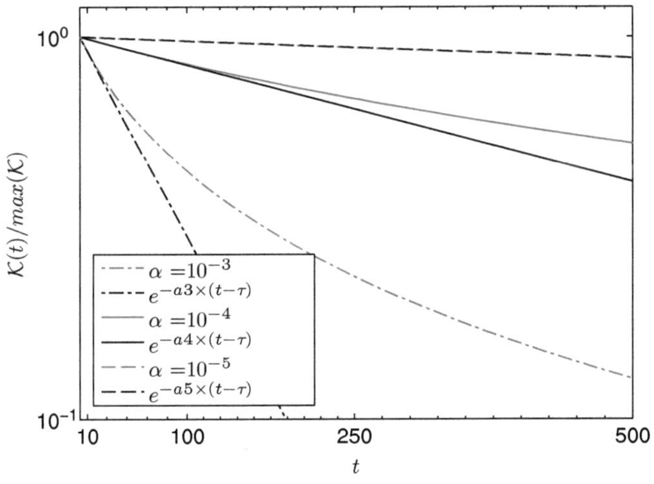

Fig. 22.4 Time evolution of the turbulent kinetic energy $\mathcal{K}$ for $\alpha \ll 1$ (decay regime governed by the time evolution of the forcing term $F(k, t)$). Here, data from simulation 6 ($\alpha = 10^{-3}$), 7 ($\alpha = 10^{-4}$) and 8 ($\alpha = 10^{-5}$) of the database are compared with exponential decay curves

22.4 Conclusions

New insights about the sensitivity of the HIT decay regime to lasting production effects have been investigated in the present work. The analysis has been performed by the observation of an EDQNM simulation database, which has been produced using the recent proposal by Meldi et al. [19]. This version allows to consider the statistical effects of multiscale/fractal grid production mechanisms in the evolution of HIT statistical quantities.

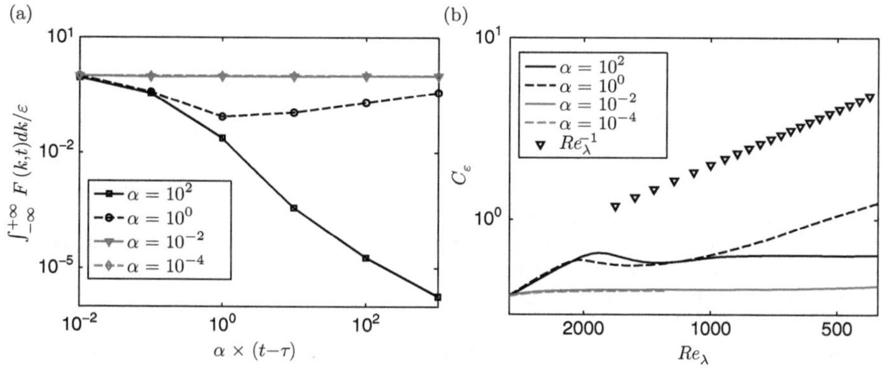

Fig. 22.5 (**a**) Time evolution of the ratio between the integral of the forcing term $F(k,t)$ and the energy dissipation rate ε. (**b**) Turbulence parameter $C_\varepsilon = L\varepsilon/\mathcal{K}^{3/2}$. The data have been sampled from simulation 1, 3, 5, 7 of the EDQNM database

The sensitivity to the two model parameters β and α has been investigated. It has been observed that the shape of the energy spectrum in the forced range is sensitive to the parameter β, which governs the slope of the forcing term $F(k,t)$ in the spectral space. While this parameter locally affects the energy distribution in the forced range, it does have a weak global effect on the decay regime as the main HIT statistical quantity do not exhibit a sensitivity to β.

The sensitivity to the parameter α has been as well investigated. This parameter governs the time evolution of the forcing term $F(k,t)$. The observation of the sensitivity of the physical quantities to this parameter allowed for the identification of three main classes. For $\alpha \gg 1$, the decay of the forcing $F(k,t)$ is faster than the physical dissipation rate. A classical power law regime is observed after a fast decay transient, which is caused by the strong local interactions due to the peculiar energy distribution close to L of the steady fractal energy spectrum. Thus, turbulence decay forgets this initial effect after a suitable transient. For $\alpha \ll 1$ an initial exponential regime is observed, where the theoretical predictions by George and Wang [7] and by Oberlack and Zieleniewicz [24] are satisfied. This regime lasts longer for smaller α values. After this transient, a power law decay is established. However, the magnitude of the power law exponent is significantly higher than the classical value derived in the case of free HIT decay and it is determined by the time evolution law of $F(k,t)$. In the last case, which is for $\alpha \approx 1$, there is not a dominant effect (production or physical dissipation) driving HIT time evolution. One of the main results is that the physical quantities show a different sensitivity to the parameter α. Thus, a number of anomalous results are observed, such as a time evolution of the turbulence coefficient C_ε.

While EDQNM analyses are relatively inexpensive when compared to DNS, they are relevant for the issue of big data. In fact, turbulence closures allow for a systematic statistical analysis of a wide spectrum of physical configurations. Once this information is gathered, turbulent regimes showing peculiar characteristics can be identified and studied by the use of more powerful tools.

References

1. G.K. Batchelor, *The Theory of Homogeneous Turbulence* (Cambridge University Press, Cambridge, England, 1953)
2. C. Cambon, N.N. Mansour, F.S. Godeferd, Energy transfer in rotating turbulence. J. Fluid Mech. **337**, 303–332 (1997)
3. T.T. Clark, C. Zemach, Symmetries and the approach to statistical equilibrium in isotropic turbulence. Phys. Fluids **10**(11), 2846–2858 (1998)
4. G. Comte-Bellot, S. Corrsin, The use of a contraction to improve the isotropy of grid-generated turbulence. J. Fluid Mech. **25**, 657–682 (1966)
5. P.A. Davidson, The minimum energy decay rate in quasi-isotropic grid turbulence. Phys. Fluids **23**(8), 085108 (2011)
6. W.K. George, The decay of homogeneous isotropic turbulence. Phys. Fluids A **4**(7), 1492–1509 (1992)
7. W.K. George, H. Wang, The exponential decay of homogeneous turbulence. Phys. Fluids **21**(2), 025108 (2009)
8. R. Gomes-Fernandes, B. Ganapathisubramani, J.C. Vassilicos, Particle image velocimetry study of fractal-generated turbulence. J. Fluid Mech. **711**, 306–336 (2012)
9. R.J. Hearst, P. Lavoie, Decay of turbulence generated by a square-fractal-element grid. J. Fluid Mech. **741**, 567–584 (2014)
10. P.Å. Krogstad, P.A. Davidson, Is grid turbulence Saffman turbulence? J. Fluid Mech. **642**, 373–394 (2010)
11. P.Å. Krogstad, P.A. Davidson, Freely-decaying, homogeneous turbulence generated by multi-scale grids. J. Fluid Mech. **680**, 417–434 (2011)
12. P.Å. Krogstad, P.A. Davidson, Near-field investigation of turbulence produced by multi-scale grids. Phys. Fluids **24**(3), 035103 (2012)
13. S. Laizet, J.C. Vassilicos, DNS of fractal-generated turbulence. Flow Turbul. Combust. **87**, 673–705 (2011)
14. M. Lesieur, *Turbulence in Fluids*, 4th edn. (Springer, New York, 2008)
15. N. Mazellier, J.C. Vassilicos, Turbulence without Richardson–Kolmogorov cascade. Phys. Fluids **22**(7), 075101 (2010)
16. B. Mazzi, J.C. Vassilicos, Fractal-generated turbulence. J. Fluid Mech. **502**, 65–87 (2004)
17. M. Meldi, P. Sagaut, On non-self-similar regimes in homogeneous isotropic turbulence decay. J. Fluid Mech. **711**, 364–393 (2012)
18. M. Meldi, P. Sagaut, Further insights into self-similarity and self-preservation in freely decaying isotropic turbulence. J. Turbul. **14**, 24–53 (2013)
19. M. Meldi, H. Lejemble, P. Sagaut, On the emergence of non-classical decay regimes in multiscale/fractal generated isotropic turbulence. J. Fluid Mech. **756**, 816–843 (2014)
20. J. Meyers, C. Meneveau, A functional form for the energy spectrum parameterizing bottleneck and intermittency effects. Phys. Fluids **20**(6), 065109 (2008)
21. V. Mons, J.-C. Chassaing, T. Gomez, P. Sagaut, Is isotropic turbulence decay governed by asymptotic behavior of large scales? An eddy-damped quasi-normal Markovian-based data assimilation study. Phys. Fluids **26**, 115105 (2014)
22. V. Mons, M. Meldi, P. Sagaut, Numerical investigation on the partial return to isotropy of freely decaying homogeneous axisymmetric turbulence. Phys. Fluids **26**, 025110 (2014)
23. K. Nagata, Y. Sakai, T. Inaba, H. Suzuki, O. Terashima, H. Suzuki, Turbulence structure and turbulence kinetic energy transport in multiscale/fractal-generated turbulence. Phys. Fluids **25**, 065102 (2013)
24. M. Oberlack, A. Zieleniewicz, Statistical symmetries and its impact on new decay modes and integral invariants of decaying turbulence. J. Turbul. **14**(2), 4–22 (2013)
25. S.A. Orszag, Analytical theories of turbulence. J. Fluid Mech. **41**, 363–386 (1970)
26. S.B. Pope, *Turbulent Flows* (Cambridge University Press, Cambridge, 2000)
27. P.J. Saffman, The large-scale structure of homogeneous turbulence. J. Fluid Mech. **27**, 581–593 (1967)

28. P. Sagaut, C. Cambon, *Homogenous Turbulence Dynamics* (Cambridge University Press, Cambridge, 2008)
29. R.E. Seoud, J.C. Vassilicos, Dissipation and decay of fractal-generated turbulence. Phys. Fluids **19**, 105108 (2007)
30. C.G. Speziale, P.S. Bernard, The energy decay in self-preserving isotropic turbulence revisited. J. Fluid Mech. **241**, 645–667 (1992)
31. J. Tchoufag, P. Sagaut, C. Cambon, Spectral approach to finite Reynolds number effects on Kolmogorov's 4/5 law in isotropic turbulence. Phys. Fluids **24**(1), 015107 (2012)
32. A. Thormann, C. Meneveau, Decay of homogeneous, nearly isotropic turbulence behind active fractal grids. Phys. Fluids **26**, 025112 (2014)
33. P.C. Valente, J.C. Vassilicos, The decay of turbulence generated by a class of multiscale grids. J. Fluid Mech. **687**, 300–340 (2011)
34. P.C. Valente, J.C. Vassilicos, Universal dissipation scaling for nonequilibrium turbulence. Phys. Rev. Lett. **108**, 214503 (2012)

Chapter 23
A Minimal Flow Unit for Turbulence, Combustion, and Astrophysics

P. Orlandi, S. Pirozzoli, M. Bernardini, and G.F. Carnevale

23.1 Introduction

At the end of volume 2 of his Lectures in Physics, Feynman [1], speaking about an equation for life, Feynman asserted that "we very possibly already have the equation ..." He was referring to the Schödinger equation. His point was that although that equation appears very simple, it nonetheless has the ability to describe incredibly complex phenomenon if only we have the resources to analyze it sufficiently. Similarly, the simplicity of the Navier-Stokes equations (NS) governing the motion of viscous flows hides a wide spectrum of complex phenomena. Feynman claimed that there are no terms missing from the NS equations and that the only difficulty, at that time, was the lack of mathematical power to analyze them except at very low Reynolds number, in which case, the flow would be laminar. High Reynolds number flow, on the other hand, is typically turbulent. Such flow is very complex, so much so that the very definition of turbulence lacks universality. Many specialists in this area offer their own version of a definition of turbulence. Tsinober [2] (first edition, p. 237) listed several of these definitions. Behind each one of these definitions is the certainty that the Navier–Stokes (NS) equations completely describe turbulent flow, and that the boundary conditions are very important. Over the years since Feynman made his comments, great effort has been devoted to finding the most efficient way to numerically simulate the NS equations. This together with the incredible improvement in computing hardware has allowed us to achieve the ability to obtain

P. Orlandi (✉) • S. Pirozzoli • M. Bernardini
Dipartimento di Ingegneria Meccanica e Aerospaziale, Sapienza Università di Roma, Via Eudossiana 16, 00184 Roma, Italia
e-mail: paolo.orlandi@uniroma1.it

G.F. Carnevale
Scripps Institution of Oceanography, University of California San Diego, La Jolla, CA 92093, USA

A. Pollard et al. (eds.), *Whither Turbulence and Big Data in the 21st Century?*,
DOI 10.1007/978-3-319-41217-7_23

solutions at Reynolds numbers close to those available in laboratory turbulent flows. Numerical simulations allow us to perfectly control the boundary conditions of a flow and to evaluate any quantity of interest. Many scientists and engineers have been attracted by the advantages of numerical simulation, and their efforts have resulted in major improvements in our understanding of turbulence.

In great part, the early focus of turbulence research was on the analysis of statistically stationary turbulence, and the most studied flows were isotropic turbulence in a periodic domain and wall-bounded flows. With periodic boundary conditions, the effects of solid boundaries are eliminated, and, in order to achieve a statistical steady state, various kinds of forcing were applied at the large scales. The aim of these simulations was to understand the energy cascade mechanism and to try to reproduce the results from Kolmogorov's theory of turbulence. In these studies, the transient evolution of the flow from initial conditions to the statistically stationary turbulent state was typically not studied. To motivate approaching the study of turbulence with a rather different strategy, we note that Moffatt et al. [3] asserted that stretched vortices are the "sinews of turbulence." It is interesting in this regard to investigate the formation of a turbulent flow starting from an initial condition with well-defined vortical structures. Such an initial vortex distribution that evolves toward fully developed turbulent flow is defined here as the minimal flow unit (MFU). This assumption is analogous to that by Jiménez and Moin [4] which, to understand wall-bounded flows, introduced the concept of the Minimal Channel, that is, a channel with the smallest lateral and streamwise dimensions capable maintaining a turbulent flow. That configuration, designed to require minimal computational resources, was widely used to manipulate the boundary conditions and, thus, to understand the physics of drag control. Similarly, the MFU proposed here could be of major utility in understanding the complex physics of 3D turbulent flows, where energy is transferred from large to small scales. The passage from initial well-defined vortical structures to a turbulent state seen in the evolution of an MFU cannot be obtained from DNS of forced isotropic turbulence.

The MFU is defined by the initial vorticity distribution. There are several attractive options for this distribution. The Taylor–Green vortex has been often used in inviscid [5] and viscous simulations [6]. This initial condition has kinetic energy concentrated at one wave number. Therefore, the evolution to a turbulent state requires significant energy transfer from large scale to small before there is any significant energy in the small scale spectra. Other options were mentioned by Orlandi and Carnevale [7] and Orlandi et al. [8] and are not repeated here.

Sir Horace Lamb famously said "I am an old man now and when I die and go to heaven there are two matters on which I hope for enlightenment. One is quantum electrodynamics, and the other is the turbulent motion of fluids. And about the former I am rather optimistic." [9]. As Moin and Kim [10] noted, Lamb could not foresee the development of supercomputers. Lamb could then not imagine that initial conditions derived from his dipole [11, p. 738] could be useful in investigating several fundamental aspects of turbulence and in giving qualitative answers to the behavior of the Euler and the NS equations. Lamb dipoles are 2D form-preserving solutions of the Euler equations. To create our MFU, we consider the extension of the 2D Lamb dipole in the direction perpendicular to the plane of the dipole. The resulting structure is a dipole comprising two parallel vortex tubes. We can define the axis of the dipole as the center of the circular dipole also extended in the direction perpendicular to the plane of the dipole. The MFU consists of two such 3D Lamb dipoles arranged so that their axes are perpendicular to each other, and directed so that they propagate toward each other [7]. To introduce specific coordinates, we take the axes of the two dipoles to lie along the x_2 and x_3 coordinate axes, respectively. They are oriented to approach each other along the x_1 axis. Initial transients are minimized in this configuration. Since each dipole individually self-advects without any change in form, there is no transient evolution in the form of the vortex as there would have been if we had created the dipoles from say Gaussian vortex tubes. Such transients have caused problems in early efforts to show a tendency toward a singularity and attempts were made to avoid their effects by modifying or filtering [12] the initial vorticity field. Furthermore, the velocity field of a dipole, which has zero circulation, falls off rapidly with distance compared to the velocity field of a single vortex with circulation. Thus, two Lamb dipoles can be placed initially very close together, shortening the time before significant interaction takes place and avoiding early unwanted transient growth of vorticity outside the dipoles themselves before the actual collision occurs. Additionally, Lamb dipoles have an initial energy spectrum with k^{-6} power law at high wavenumbers, decreasing the time to populate small scales in the spectrum and facilitating comparison with the subsequent evolution of the spectrum toward a power-law form.

With these initial conditions, the NS equations can be solved in a three-dimensional box with periodic boundary conditions, or on a reduced domain that takes advantage of the two reflection symmetries at lower computational cost. Details of the simulations and of the numerical methods were given in Orlandi et al. [8]. The only controlling parameter is the Reynolds number, here defined as $\nu = Ua/\nu$, with U the propagation velocity and a the radius of the dipoles. When $\nu = 0$ the Euler equations are solved, and in that case it is interesting to investigate whether a finite-time singularity (FTS) occurs. Indeed Orlandi et al. [8] found that several requirements necessary to demonstrate the occurrence of an FTS were easily found with this MFU. More interesting was the observation by Orlandi and Carnevale [7] that the vorticity component along the intermediate strain-rate eigen-direction is proportional to the corresponding eigenvalue of the strain-rate tensor ($\widetilde{S_2}$). This analysis suggests that the transport equation for $\widetilde{\omega}_2$ can be approximated by a model equation having an FTS.

With this dipole MFU, we have investigated how the spectra and time evolution of relevant statistics vary with the Reynolds number. An important theoretical question is whether, given any finite viscosity ν, no matter how small, the viscous terms can prevent the formation of small regions with infinite velocity gradients from developing. The numerical answer is rather difficult; however, simulations of the dipole MFU at different Reynolds numbers with the appropriate resolution may help to understand the behavior of some fundamental properties of turbulent flows. In particular, since the MFU accounts for the evolution of a single "atom" of turbulence, it can be used to understand the link between the energy spectrum and the geometry of the vortices.

If there is a trend toward a singularity with decreasing viscosity, then, based on experience with the Euler equation, we expect that near singularities that form should be localized in small regions embedded within sheet-like structures in the vorticity field. As the sheet rolls-up, viscosity acts producing a sea of rod-like structures, which affect the energy-containing scales and modify process of energy cascade. The process continues with the formation of an inertial the range with the Kolmogorov's power law, and an exponential range at the highest wavenumbers. In this period of the evolution, total energy and enstrophy decay in time with a power-law behavior.

Numerical experiments can be designed to understand the difference between genuine and "passive turbulence" [13], i.e., the difference between the behavior of active fields such as velocity and vorticity and passively advected scalar and vector fields in a turbulent flow. Tsinober and Galanti [14] simulated forced isotropic turbulence with both passive scalar and passive vector fields. They found similarities between the evolution of the vorticity field and the passive vector field. On the other hand, the evolution of the vorticity field statistics proved very different from that of the scalar gradient. The differences between the statistics ω_i and $G_i = \partial\theta/\partial x_i$ (where θ is a passive scalar field) are related to the different alignment of the vorticity field and the passive-gradient field with respect to the principal axes of the strain-rate tensor S_{ij}. Below, the analysis of Tsinober and Galanti [14] in isotropic forced turbulence will be applied to the evolution of the dipole MFU to shed light on the differences in the evolution of the vorticity ω_i, the passive vector B_i, and the components G_i of the passive scalar gradient. The analysis is facilitated by looking at the behavior of the components along the principal axes of the strain-rate tensor. Let $\widetilde{\omega}_i$, $\widetilde{B}_i$, and $\widetilde{G}_i$ be the components of vorticity, passive vector, and scalar gradient vector in the direction of the eigenvectors of S_{ij} that have eigenvalues $\widetilde{S}_1 > \widetilde{S}_2 > \widetilde{S}_3$. In a series of simulations of the evolution of passive scalars for the dipole MFU, we used two passive scalars defined by the fields $\theta^{(2)}$ and $\theta^{(3)}$ that at $t = 0$ satisfy $\theta^{(2)} = \omega_2/\omega_{\max}$ and $\theta^{(3)} = \omega_3/\omega_{\max}$. The evolution of these two separate scalars allowed us to see how they mix subject to different values of the viscosity. The scalar gradients G_i were evaluated using $\theta = \theta^{(2)} + \theta^{(3)}$. For the passive vector at $t = 0$, we took $B_i = \omega_i/\omega_{\max}$. It was, therefore, particularly interesting to follow the growth of B_1 and ω_1 which started from zero at $t = 0$. During the vorticity intensification phase $|\widetilde{\omega}_2|/|\omega|$ and $|\widetilde{B}_2|/|B|$ are larger than $|\widetilde{G}_2|/|G|$, and $|\widetilde{G}_1|$ and $|\widetilde{G}_3|$ are both larger than $|\widetilde{G}_2|$. The difference of the alignments of these fields with the principal axes of the strain-rate tensor is the reason for the different power laws for E_v and E_θ, the energy and passive scalar spectra, respectively. These differences are evident in the evolution equations too since the stretching terms are similar for ω_i and B_i but of opposite sign for G_i.

Related problems of great practical relevance include turbulent combustion for passive scalars and magneto-hydrodynamics (MHD) of passive vectors.

23.2 Physical and Numerical Method

23.2.1 Governing Equations

The momentum and continuity equations for incompressible flows considered here are

$$\frac{Du_i}{Dt} = -\frac{\partial p}{\partial x_i} + I\left(\frac{\partial B_i B_l}{\partial x_l} - \frac{\partial B_l^2/2}{\partial x_i}\right) + \frac{I_V}{Re}\frac{\partial^2 u_i}{\partial x_l^2}; \qquad \frac{\partial u_i}{\partial x_i} = 0, \tag{23.1}$$

where the coefficient I is zero for non-conducting conditions and unity for conducting flows, and I_V is zero for inviscid flows and unity for viscous flows. The term with the coefficient I is the Lorentz force term. The time evolution of the components of the magnetic field B_i, or equivalently a divergenceless passive vector field, is given by

$$\frac{DB_i}{Dt} = \frac{\partial B_l u_i}{\partial x_l} + \frac{I_V}{Re\mathrm{Pr}_M}\frac{\partial^2 B_l}{\partial x_l^2}; \qquad \frac{\partial B_i}{\partial x_i} = 0, \tag{23.2}$$

The equation for passive scalars is

$$\frac{D\theta^{(i)}}{Dt} = \frac{I_V}{Re\mathrm{Pr}_M}\frac{\partial^2 \theta^{(i)}}{\partial x_i^2}, \tag{23.3}$$

in which case the superscript i accounts for scalars with different initial distributions as discussed above. Pr_M is either the magnetic or the scalar Prandtl number, here set to unity. Physical quantities are reported here in units such that the initial speed U and the radius of the dipoles are both unity. The centers of the dipoles are initially separated by a distance $d = \pi$. All the simulations have been performed exploiting the reflection symmetry in the x_2 and x_3 directions to reduce the computational load. The length of the domain in the periodic direction x_1 is taken to be $L = 2\pi$.

The governing equations are solved by a numerical method consisting of a second-order finite-difference scheme with staggered velocities. This method, which conserves the total kinetic energy in the inviscid flow limit, is described in [15], assuming the necessary modifications to solve the B_i equations in [16]. All the simulations have been performed with a uniform discretization in all directions, with effective 2048^3 resolution considering the reflection symmetries. As demonstrated by Orlandi et al. [8] this grid is not sufficiently fine to evaluate the trends of the Euler equations toward the FTS and to demonstrate that the Navier–Stokes equations do not have an FTS. However, this resolution is satisfactory to show that an incompressible passive vector behaves as the vorticity and that large differences for the energy and enstrophy time evolution are observed in the viscous and inviscid cases.

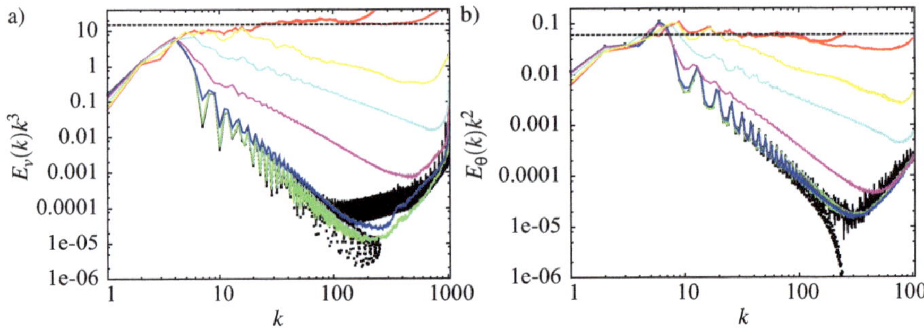

Fig. 23.1 Compensated velocity (**a**) and scalar (**b**) spectra from a 2048^3 simulation at $t = 0, 0.5, 1, 1.5, 2, 2.5, 2.75$, and from a 512^3 simulation at $t = 0, 2.75$

23.3 Results

23.3.1 Inviscid Simulations

Inviscid simulations have been performed to get insight into the possible occurrence of an FTS for the Euler equations, and to see how passive scalars behave in such conditions. The results obtained from a 2048^3 resolution simulation were described in [8] and in [16]. Here the evolution of the compensated E_v and E_θ spectra are reported in Fig. 23.1 to show how wide the inertial range is at a time close to that of the hypothesized FTS. Theoretical considerations suggest that close to an FTS, E_v should develop an infinite k^{-3} range. Orlandi and Carnevale [7] and Orlandi [17] observed a scale similarity in the corresponding flow structures that developed. As mentioned above, the vortical structures are expected to be different from the scalar structures. Indeed, the spectra shown in Fig. 23.1b for E_θ exhibit a long k^{-2} range, corresponding to the formation of sheet-like structures.

In Fig. 23.1, at $t = 2.75$, when the simulation was stopped, there is a small pile-up of energy near the maximum resolved wavenumber. If the simulation were continued, the continued transfer of energy to the small scales would produce an equipartition type k^2 range at the small scale end of the spectrum. This kind of simulation was studied by Chichowlas et al. [18]. They showed that between the large energy-containing scales and the small scale equipartition range, a $k^{-5/3}$ inertial range developed. In our view, the formation of the $k^{-5/3}$ spectrum is better studied in very well-resolved DNS, as discussed in the next section.

23.3.2 Viscous Simulations

In our previous paper [16] the evolution of $E_\theta(k)$ and $E_v(k)$ were shown, and it was observed that for viscous simulations, at the extrapolated FTS time $t_s = 2.75$, they differ from those in Fig. 23.1 because of the appearance of an exponential decay

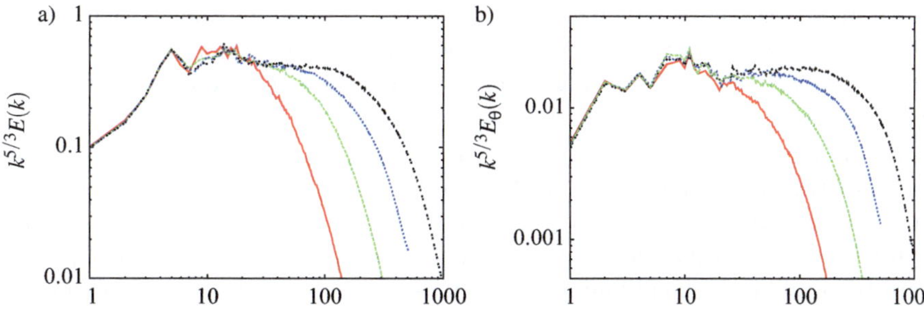

Fig. 23.2 Compensated velocity (**a**) and scalar (**b**) spectra at $t = 4.0$, *red* $\nu = 1000$, *green* $\nu = 2500$, *blue* $\nu = 5000$, and *black* $\nu = 10{,}000$

range. Only when this range is well resolved it is possible to assert that a true direct numerical simulation (DNS) is achieved. Depending on the Reynolds number and the resolution, a satisfactorily wide power law (k^{-n}) can be observed. Despite the different values of n at $t = t_s$ the $E_\theta(k)$ and $E_v(k)$ spectra afterwards evolve toward wide power laws with a $k^{-5/3}$ at high ν. Later on, due to the decrease of R_λ (the Reynolds number based on the Taylor micro-scale), the power laws fit a $k^{-3/2}$ spectrum. The experiments of [19] have indicated that $k^{-5/3}$ spectrum is approached only when R_λ approaches 10^4. As R_λ falls below this value, the inertial range narrows and the power law becomes shallower. For R_λ, close to those found here (two or three hundred), the power law is close to $-3/2$, in accord with DNS results at comparable R_λ. In Fig. 23.2 the compensated spectra at $t = 4.0$ show that both have a quite long inertial range at $\nu = 10{,}000$ and that this was achieved earlier for $E_\theta(k)$ than for $E_v(k)$. In [16] the spectra in Kolmogorov units were also compared with those of DNS of forced isotropic turbulence [20]. In Kolmogorov units, the nondimensional wavenumber k^* is obtained by multiplying the wave number k by $\eta = (\nu^3/\epsilon)^{1/4}$ with $\epsilon = 2\nu \int Ek^2 dk$, and $E^*(k^*)$ is obtained by multiplying $E(k)$ by $(\nu^5\epsilon)^{-1/4}$. Donzis et al. [20] stated that the value of the Kolmogorov constant C_K is weakly dependent on the Reynolds number, and its value is approximately 1.58, which matches reasonably well the values obtained in the present MFU calculations. Regarding the scalar spectra, Sreenivasan [21] suggested that the appropriate value of the Obukhov–Corrsin constant for one-dimensional spectra is $C_\theta \approx 0.4$, which, assuming flow isotropy yields $C_O = (5/3)C_\theta \approx 2/3$. In the present MFU simulations this constant shows fluctuations in time, with $0.75 \leq C_O \leq 1$, in fair agreement with the value quoted by Sreenivasan [21].

In forced turbulence, a significant "bottleneck" occurs between the inertial and the exponential ranges, highlighted by the formation of a bump in the spectrum between these two ranges. Donzis et al. [20] analyzed this effect in DNS of forced turbulence, and showed that dependence on the Reynolds number is weak. However, those authors did not make any attempt to relate the bottleneck to the shape of the flow structures contributing to it. In the present DNS, we have observed that the amplitude of the "bump" is small at low Reynolds number, and at $\nu = 10{,}000$

the amplitude is comparable with that of forced DNS only in a small time interval around $t = 6$, that is well after $t_s = 2.75$, the extrapolated singularity time found for the same dipole MFU in inviscid flow [16]. It has been in addition observed that the bump begins at $k^* = 0.035$, as was shown in Fig. 9d in [16]. This value corresponds to the value of η/λ at $\nu = 10{,}000$. The MFU evolution, starting from well-defined vortical structures, allows us to understand which vortical structures contribute to the bottleneck effect. Orlandi [17] observed, through contour plots of ω calculated by filtering the velocity field, that in the spectral range where the bottleneck occurs there are a relatively small number of ribbon-like structures immersed in a field of a large number of rod-like structures. On the other hand, Fig. 23.2b shows that the bottleneck is well developed in the passive scalar spectra at $t = 4$. When plotted in Kolmogorov scales, in the bottleneck region at $t = 6$ the passive scalar spectra all collapse nicely onto one curve for values of R_λ ranging from 96 to 293 (see Fig. 10d [16]).

Differences in the evolution of the energy and passive scalar spectra at high wave numbers should depend on the different evolution of the enstrophy (Ω) and the scalar gradient variance (χ). The evolution equations for these are

$$\frac{\partial \Omega}{\partial t} = \langle \omega_i \omega_j S_{ij} \rangle - \frac{1}{\nu} \left\langle \frac{\partial \omega_i}{\partial x_j} \frac{\partial \omega_i}{\partial x_j} \right\rangle, \tag{23.4}$$

$$\frac{\partial \chi}{\partial t} = -\langle G_i G_j S_{ij} \rangle - \frac{1}{\nu \mathrm{Pr}} \left\langle \frac{\partial G_i}{\partial x_j} \frac{\partial G_i}{\partial x_j} \right\rangle. \tag{23.5}$$

Differences between these two equations were previously discussed by Tsinober and Galanti [14], who noted that the production terms have opposite sign. The analysis of the production terms if carried out in the principal axes of strain-rate tensor could give insight into the physics of the relevant flow structures. Let $\widetilde{S}_\ell$ be the eigenvalues of the tensor. The vorticity and the scalar gradient components along the principal axes are, respectively, $\widetilde{\omega}_\ell$ and $\widetilde{G}_\ell$. Similarly, the projected production terms are $\widetilde{P}_{\Omega_\ell} = \langle \widetilde{S}_\ell \widetilde{\omega}_\ell^2 \rangle$ and $\widetilde{P}_{\chi_\ell} = -\langle \widetilde{S}_\ell \widetilde{G}_\ell^2 \rangle$ (no summation). In Fig. 23.3a, instead of Ω and χ, we have plotted the kinetic energy dissipation rate ($\epsilon = 2\nu\Omega$) and the scalar dissipation rate ($\epsilon_\theta = \alpha\chi$). Starting from different values, these quantities grow, reaching a maximum value independent of ν, and after some oscillation in time they decay algebraically, as in decaying isotropic turbulence. Here we find that the decay rate is dominated by a power law t^n with exponent $n \approx -2.6$, close to that measured in the decay of grid turbulence. This behavior is in agreement with one of the basic hypotheses in turbulence that the mean decay rate of kinetic energy is constant for $\nu \to 0$. In the time interval around the peak of ϵ, a $k^{-5/3}$ power-law spectral range is observed. Similar behavior is observed for ϵ_θ (Fig. 23.3b) with its maximum value independent of α. Once an exponential range is formed, ϵ_θ starts to decay algebraically with a smaller exponent ($n \approx -2.0$) than that for the decay of ϵ.

The effects of the strain field on the scalar gradient as compared to the effects on the vorticity field can be appreciated through the time evolution of the components

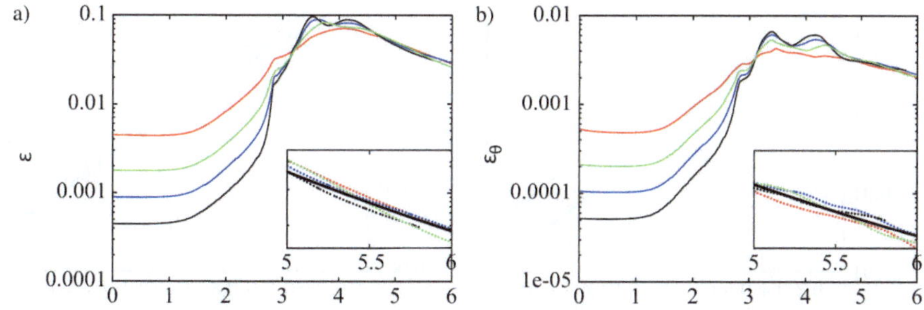

Fig. 23.3 Evolution of (**a**) $\epsilon = \nu\Omega$, (**b**) $\epsilon_\theta = \alpha\chi$ (*red* $\nu = 1000$, *green* $\nu = 2500$, *blue* $\nu = 5000$, and *black* $\nu = 10{,}000$). In the *inset* of (**a**) the *black solid line* is $\approx t^{-2.6}$, in the *inset* of (**b**) the *black solid line* is $\approx t^{-2.0}$

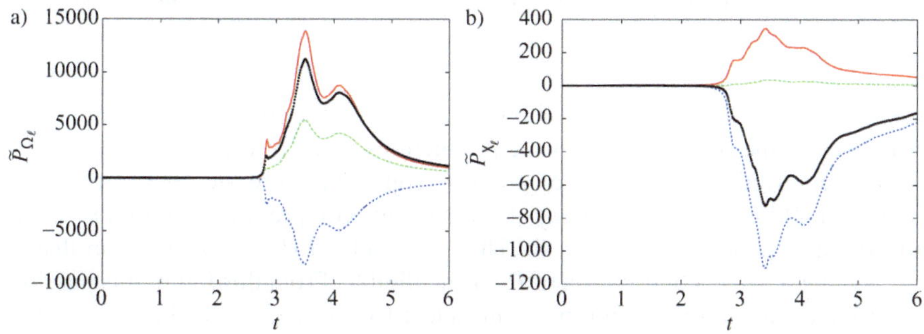

Fig. 23.4 Evolution of components of (**a**) enstrophy production, (**b**) scalar variance production in the principal axes of the strain-rate tensor at $\nu = 10{,}000$ (*red* $\ell = 1$, *green* $\ell = 2$, *blue* $\ell = 3$, *solid black* total)

of $\widetilde{P}_{\Omega_\ell}$ and $\widetilde{P}_{\chi_\ell}$, shown in Fig. 23.4a and b. Although early growth of the enstrophy production cannot be appreciated here because of the scale of the graph, it is evident that there is rapid growth starting near $t_s \approx 2.75$, the extrapolated FTS for inviscid flow. The growth continues with somewhat reduced rate until the maximum is achieved at $t \approx 3.5$, when the exponential range is fully formed, and when production becomes smaller than dissipation. During the evolution, the magnitude of the production component $\widetilde{P}_{\Omega_1}$ due to the extensional strain rate $\widetilde{S}_1$ is greater than that of the negative contribution due to the compressed strain rate $\widetilde{S}_3$. However, as was observed in isotropic turbulence [22], the $\widetilde{P}_{\Omega_2}$ term due to the vorticity component aligned with the intermediate $\widetilde{S}_2$ contributes to the total $\widetilde{P}_{\Omega_\ell}$ by the same amount as $\widetilde{P}_{\Omega_1} + \widetilde{P}_{\Omega_3}$. This term is contributing to the formation of the so-called worms, which were studied in detail by [25] in forced isotropic turbulence. The comparison between viscous and inviscid visualizations shows that viscosity destroys the self-similar sheet-like structures produced by inviscid vortex interactions, and instead produces unstable ribbon-like structures that roll up into rod-like structures.

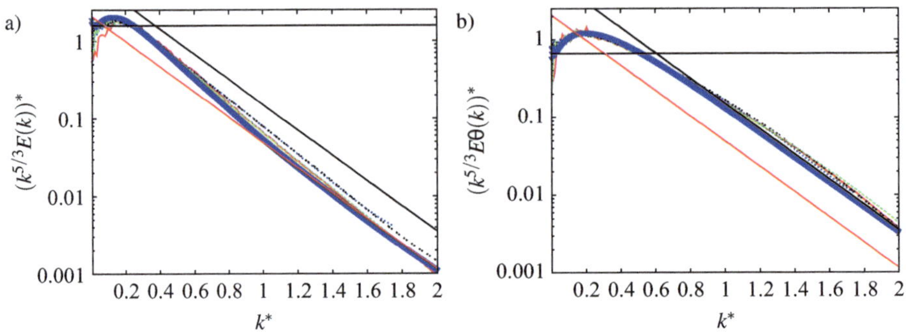

Fig. 23.5 Compensated velocity (**a**) and scalar (**b**) spectra Kolmogorov scaled at $t = 4.0$, *red* $\nu = 1000$, *green* $\nu = 2500$, *blue* $\nu = 5000$, and *black* $\nu = 10{,}000$. The *thick blue curves* are the results of [20]. The *straight red line* is $0.05\exp(-3.5(k^* - 1.))$; the *straight black line* $1.5\exp(-3.5(k^* - 1.))$

Figure 23.4b, for passive scalar production, shows a completely different scenario. Due to the opposite sign of the productions terms in Eqs. (23.3) and (23.5), the total contribution should be of opposite sign. Therefore the intensification of ϵ_θ or χ should be due to the effect of the negative compressive strain rate $\widetilde{S}_3$, and indeed this is depicted in Fig. 23.4b. The magnitude of $\widetilde{P}_{\chi_3}$ is greater than that of $\widetilde{P}_{\chi_1}$, and the positive contribution of $\widetilde{P}_{\chi_2}$ is negligible. From this figure it is possible to speculate that the strain field forms thin sheet-like structures which, convected by the velocity field, roll-up but are not intensified along the $\widetilde{S}_2$ direction and, therefore, are not active, unlike the "worms" generated in the vorticity field. Visualizations of these differences were reported by Orlandi et al. [16]. Although it is rather difficult to visualize the very small scales in the exponential range, a rough idea can be obtained by looking at the exponential range in Kolmogorov scales, using a semi-log representation. A further indication that the present MFU simulations agree well with the simulations of forced isotropic turbulence is given in Fig. 23.5a, b, showing that the behavior of $E_v^*(k^*)$ and $E_\theta^*(k^*)$ is similar to that obtained by Yeung et al. [23] at $R_\lambda = 700$. The two completely different simulations show that the scalar and the velocity structures differ in the range of wavenumber between the bottleneck region (here at $k^* \approx 0.1 - 0.2$) and $k^* = 1$. The scalar spectrum adjusts immediately to a true exponential decay. On the other hand, the velocity spectrum, which is larger than the passive scalar spectrum at $k^* < 1$, requires a faster decay to form a true exponential range.

23.3.3 *Passive Vector*

The main issue to investigate here is whether passive vectors follow the same evolution as the vorticity field. Since they obey the same evolution equations, the only difference resides in the initial conditions. In particular, a passive vector field can have non-zero divergence at the initial time, whereas vorticity is by definition divergence-free. The general observation that passive vectors need not behave as the vorticity field leads to the conclusion that real turbulence (i.e., with a feedback of the vorticity to the velocity field) is less general than "passive" turbulence, in which there is a one-way effect of the velocity field on the passive vector field. This was discussed by Moffatt [13] who stated that "if generality brings to linearity as a bonus, no further motivation is needed for the detailed study of [the equation for **B**]." Here, **B** is the passive vector, and Moffatt substituted his equation number for [the equation for **B**]. Here we are speaking of Eq. (23.2).

The mutual interaction between velocity and magnetic fields driven by the Lorentz force plays a fundamental role in astrophysical flows. Consider solar activity, which is driven by strong flow and current sheets. Kitiashvili et al. [24] investigated this phenomenon with fully three-dimensional simulations. They observed the formation of strong vortex tubes in the inter-granular lanes. These vortex tubes interact strongly with nearby vortices, and this interaction may lead to their disruption. However, the resolution in those simulations involving many interacting vortices did not allow the investigation of the complex physics of the interaction between vorticity and magnetic fields in detail. Here again, the use of the dipole MFU can be beneficial.

The addition of the evolution equation for **B** increased the computational burden significantly in our numerical simulations. With our current resources this limited us to simulations with a resolution of 1024^3 with which we could fully resolve the viscous flow only up to $\nu = 3000$. Here we are interested in seeing how the individual components ω_i and B_i behave. Given the reflection symmetry of the problem, it will be sufficient to examine the $i = 1$ and $i = 3$ components. Simulations of **B** as a passive vector, i.e., simulations with $I = 0$ in Eq. (23.1), allow us to see how close the vorticity field and passive vector field remain if initially the same. Theoretically, they should be identical. Figure 23.6 shows the evolution of the maximum absolute values of these components for the dipole MFU. In addition to the simulations with **B** as a passive vector, we have also performed simulations of the full MFD equations, that is Eqs. (23.1) and (23.2) with $I = 1$. In Fig. 23.6, we present plots of the maximum values of the components of $\boldsymbol{\omega}$ and **B** versus time for inviscid non-conducting ($I = 0$) flow, viscous non-conducting flow, and viscous conducting ($I = 1$) MHD flow. Comparison of the results of the non-conducting and conducting flows can help us understand the effect on the evolution of the magnetic field acting or not on the velocity field through the Lorentz force.

The dipoles in the MFU are placed at distance $d = \pi$ apart, their radius is $a = 1$ and their speed is $U = 1$. Thus it takes until $t = (\pi - 2)/2 \approx 1/2$ before the dipoles just touch at their outer boundaries. During this period and somewhat after only the outer regions of the dipoles interact. Figure 23.6a and b shows that, during this phase, from $t = 0$ to $t \approx 1.5$, the evolution of all the components is practically the same independent of whether the flow is conducting or non-conducting, with $\nu = \infty$ or $\nu = 3000$. In this phase, the early enstrophy evolution is characterized by exponential growth. This growth is due to vortex stretching and tilting. We can

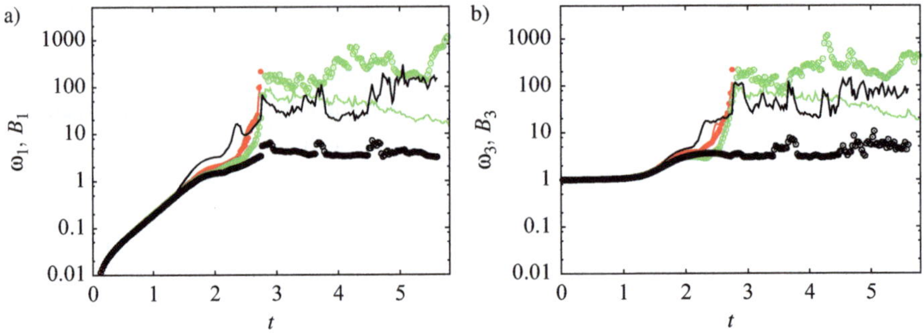

Fig. 23.6 Evolution of the maximum absolute values of $\omega_i/\omega_{\max}|_0$ and $B_i/B_{\max}|_0$ components with (**a**) $i = 1$, (**b**) $i = 3$. *Solid lines* are used for ω_i and *symbols* for B_i (*red* inviscid passive vector, *green* viscous passive vector at $\nu = 3000$, and *black* full MHD ($I = 1$ in Eq. (23.1)) at $\nu = 3000$)

see the effect on the evolution of the components along the x_1 direction in Fig. 23.6a. These components are zero initially, but, after a short transient phase, they enter an exponential phase that lasts from about $t = 0.5$ to about $t = 1.5$.

Following the exponential growth stage, in the inviscid case, all components of $\boldsymbol{\omega}$ and $\mathbf{B}$ increase much more rapidly behaving as $(t_s - t)^{-1}$. During this singular like phase of the growth all of the components have nearly the same values. This singular-like behavior is better displayed by plotting the data as in [7]. In the viscous non-conducting case, the growth of the components of $\boldsymbol{\omega}$ and $\mathbf{B}$ is reduced with respect to the inviscid case. In the conducting (MHD) case, after $t \approx 1.5$, the ω_i become larger than the $\mathbf{B}_i$, reflecting the effect of the Lorentz force acting on the vorticity field.

In both of the non-conducting simulations, inviscid and viscous, differences develop between the ω_i and B_i components. A possible reason for this is that small differences between $\boldsymbol{\omega}$ and $\mathbf{B}$ introduced initially because the B_i were calculated on the staggered grid from interpolated values of the ω_i. These differences would grow rapidly once the flow becomes turbulent. For the MHD flow, for which $\boldsymbol{\omega}$ and $\mathbf{B}$ are *not* expected to remain equal because of the presence of the Lorentz force in Eq. (23.1), Fig. 23.6 shows that after $t \approx 1.5$ the values of ω_i are higher than those of the B_i.

For both conducting and non-conducting viscous flow, there is a tendency toward equilibration after $t = t_s \approx 2.75$ as seen in both panels of Fig. 23.6. The levels that the values of the components reach in this period depend on the value of the viscosity, higher values associated with lower viscosity. The oscillations or fluctuations observed in the later parts of the graphs as the viscous flows are tending to equilibrate are possibly due to the marginal resolution of the small scales. The values of the maxima and their fluctuations in time can be affected by the insufficient resolution. These fluctuations are reduced or disappear by plotting the evolution of quantities averaged over the whole domain instead of the maxima over the domain. Consider the evolution of $q = \langle u_i^2 \rangle$ (twice the total kinetic energy, $\Omega = \langle \omega_i^2 \rangle$

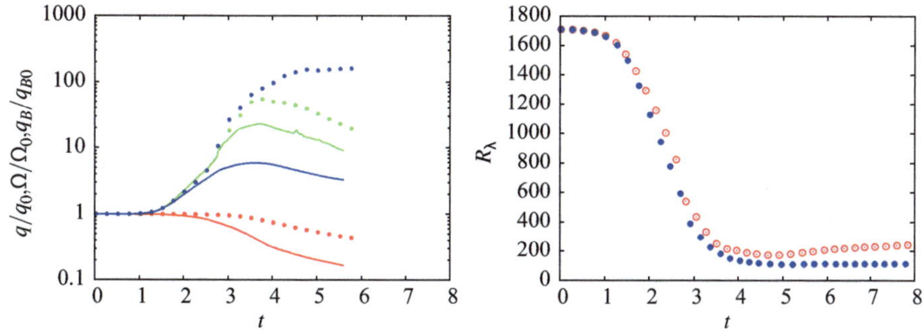

Fig. 23.7 (**a**) Evolution of q (*red*), Ω (*green*), and q_B (*blue*): *lines* conducting; *symbols* non-conducting. (**b**) R_λ: *open symbols* non-conducting; *closed symbols* conducting

(enstrophy), and $q_B = \langle B_i^2 \rangle$ (magnetic energy). These are plotted in Fig. 23.7a. In Fig. 23.7a, we compare the non-conducting case results for q_B (thin blue line) and q (thin green line). Recall that initially $B_i = \omega_i/\omega_{\max}$. We see that they are very similar, and conclude that they indicate a strong similarity between the statistics of the evolution of the vorticity and passive vector fields.

In Fig. 23.8a, we show the energy spectra for the inviscid non-conducting case (open symbols), in computational units, at times $t = 0$, 2.75, and 8. By $t = t_s \approx 2.75$ the inviscid spectrum exhibits a long k^{-3} power-law inertial range. The small energy pile-up near the maximum k is due to the finite resolution constraining the inviscid flow from cascading energy to even higher k. The low energy levels in the pile-up suggest that the largest and intermediate scales are not affected by the lack of additional resolution, and therefore, the physics driving the formation of power-law spectrum is reproduced well by the simulation. Comparison between the inviscid spectra (open symbols) and those at $\nu = 3000$ (solid curves) in Fig. 23.8a shows that large ($k < 10$) and intermediate ($10 < k < 40$) scales are not affected very much by viscosity at $t = 2.75$ On the other hand, an approximately exponential tail in the spectrum appears at small scales in the $\nu = 3000$ case. The formation of small structures in the spectral range of exponential fall-off is the feature which in the successive evolution leads to flow with the characteristics of fully developed turbulence. From the spectra colored in cyan the values of $R_\lambda = \nu \int E dk \sqrt{20/(6 \int E k^2 dk)}$, $R_\lambda = 210$, can be evaluated. At these low values of R_λ, a genuine $k^{-5/3}$ spectral range cannot be achieved. Instead, the spectrum at low and intermediate wavenumbers is much shallower than k^{-3} (dashed black straight line), although not quite as shallow as $k^{-5/3}$ (solid black straight line).

In the conducting case (Fig. 23.8b), at $t = 2.75$, the inviscid and $\nu = 3000$ energy spectra are similar for $k < 20$, while for $k > 20$, the amplitude of the viscous spectrum is significantly lower than the inviscid. The energy drop in the fully turbulent range is high, and comparison between Fig. 23.8a and b shows that the largest scales are those losing more energy. To get an overall impression of the

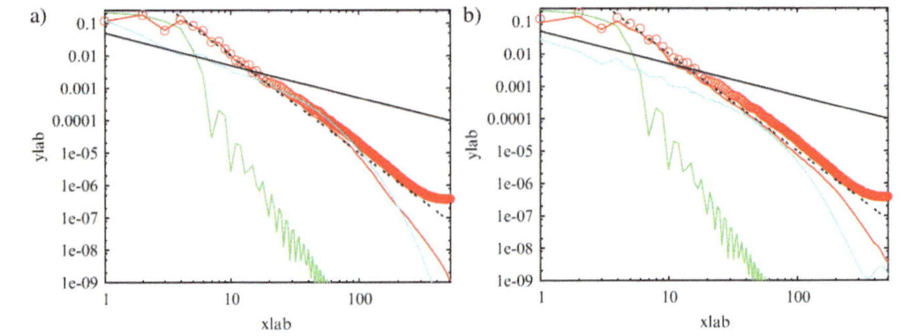

Fig. 23.8 Three-dimensional energy spectra (**a**) non-conducting, (**b**) conducting: at $t = 0$ (*green*), $t = 2.75$ (*red*), $t = 8$ (*cyan*): *open symbols* inviscid at $t = 2.75$; *black solid* $k^{-5/3}$; *dashed* k^{-3}

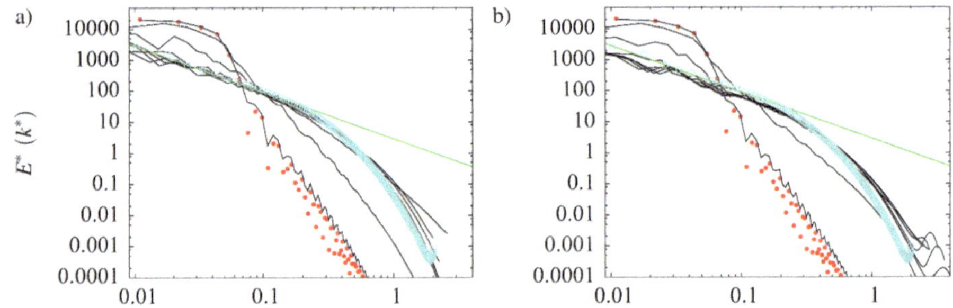

Fig. 23.9 Three-dimensional energy spectra in Kolmogorov units for the $\nu = 3000$ simulation for (**a**) non-conducting, (**b**) conducting flow: at $t = 0.5$ (*red symbols*) and *black line* from $t = 1$ to $t = 8$ with $\Delta t = 2$: *green line* $k^{-3/2}$, (*cyan*) *open symbols* forced isotropic DNS [25] at $R_\lambda = 126$

differences in the behavior of the two flows, conducting and non-conducting, it is worth looking at the time evolution of R_λ. The two dipoles are not modified as they approach each other, even up to the time ($t = 3.7$ in Fig. 23.5a) when the maximum Ω is reached. Only in the turbulent stage are differences in the vortical structures between conducting and non-conducting conditions found, leading to an R_λ (see Fig. 23.7b) in the conducting case half as high as in the non-conducting case.

Figure 23.9 shows that in Kolmogorov units, the spectra for both conducting and non-conducting flows have an exponential decay range that fits well that obtained in DNS of forced isotropic turbulence [25]. A further interesting result is that at low wavenumbers the spectra fit a $k^{-3/2}$ power law, as was also found by Lee et al. [26] in the non-conducting case. Some effect of the initial conditions is retained during the whole evolution at small wavenumbers (Fig. 23.9a). For conducting flows, well-defined spectra at intermediate and small wavenumbers are rapidly achieved (Fig. 23.9b). The small bump at high wavenumber that appears in Fig. 23.9b could be due to folding and stretching at small scales caused by the magnetic field. However, it may be also due to the insufficient resolution. Hence, more refined simulations are needed to draw solid conclusions. The results

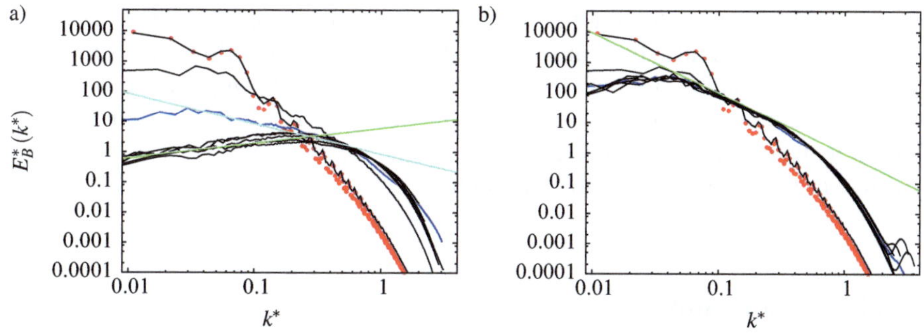

Fig. 23.10 Three-dimensional magnetic energy spectra in Kolmogorov units (**a**) non-conducting, (**b**) conducting: $t = 0.5$ (*red symbols*) from $t = 1$ to $t = 8$ in increments of $\Delta t = 2$ *black lines*. The *green straight lines* in both panels are k^{-2}. The *blue curves* are the spectra at the time when the FTS occurs for inviscid flows

in Fig. 23.9 demonstrate that a turbulent state with spectra having inertial and exponential ranges typical of isotropic turbulence is achieved. The differences in the spectra at the large scales persist for a long time as they slowly lose energy through triadic interactions.

For non-conducting flows, some controversy exists as to whether B_i and ω_i, having the same transport equation, should behave in a similar way. Moffatt [13] points out that if the initial conditions are identical for $\boldsymbol{\omega}$ and $\mathbf{B}$, then according to the evolution equations for the vorticity and magnetic fields, both would have to evolve equivalently. This argument, however, does not tell us what would happen if there were small errors in the initial conditions or small roundoff or other additional noise during the simulations that would introduce small differences between $\boldsymbol{\omega}$ and $\mathbf{B}$. In a turbulent flow, these errors would grow exponentially and very different behavior could be observed. The evolution of the maxima of the components of $\boldsymbol{\omega}$ and $\mathbf{B}$ shown in Fig. 23.6 suggests that the vorticity and passive vectors do not evolve similarly. Since we have small differences between $\boldsymbol{\omega}$ and $\mathbf{B}$ in our initial conditions, it is not surprising that a sensitive measure like the maximum values of these fields would behave differently in turbulent flow even though their evolution equations are the same. On the other hand, the evolution of the globally averaged quantities, the kinetic and magnetic energies do evolve very similarly as shown in Fig. 23.7a. Additionally, the spectral evolution of the two fields is remarkably similar. In Fig. 23.9a, we see that in the $\nu = 3000$ case, at large scales, the kinetic energy spectrum tends to $k^{-3/2}$, which is probably the "correct" power law at low R_λ as discussed above. Thus the enstrophy spectra at large scales tends to a $k^{1/2}$ spectrum. In Fig. 23.10a, we see that the magnetic energy spectrum also tends to a $k^{1/2}$ spectrum. This is strong evidence in support of the equivalence between $\mathbf{B}$ and $\boldsymbol{\omega}$, at least in a statistical sense.

For the conducting case, the magnetic energy spectrum differs significantly from the enstrophy spectrum, as expected, because of the effect of the Lorentz force.

Figure 23.10b for conducting flows shows a k^{-2} range forms at large scales. This same power law has also been observed in the solar magnetic field [27]. To conclude, the present simulations for both insulating and conducting flows generate in the fully turbulent regime a velocity field with a characteristic $k^{-3/2}$ spectral range, and a magnetic fields with a k^{-2} spectral range; hence, the associated typical vortical and magnetic structures should also be different.

23.4 Conclusions

Simulations of fully turbulent flows are usually performed by introducing random forces at low wavenumber in spectral space, so that in few large eddy turnover times a statistical steady state can be achieved. In such simulations, spectra with an inertial and an exponential range are obtained [25]. These numerical experiments cannot explain why an inertial range forms, and the large range of scales does not allow us to easily understand the separate effects of large and small structures. A minimal flow unit (MFU) with well-defined initial vortical structures is an ideal choice to help us to understand the transition from a dynamics dominated by a few vortices to a fully turbulent state. In the MFU, the structural metamorphosis from one shape to an other can be followed in time and space. We have found that the MFU can be of great help in understanding the differences between structures, spectra, and statistics obtained by inviscid and viscous simulations. The addition of a passive scalar to the MFU allowed us also to draw additional conclusions which could be of interest for combustion-related problems [16].

A natural extension of the previous work is the consideration of passive vectors, to understand the differences in the evolution with respect to those of the vorticity vector. This is a fundamental problem discussed by Tsinober [2], and which requires DNS to reach solid conclusions. In this respect, it has been found that small differences between vorticity and passive vectors at $t = 0$ may produce substantial differences in the time evolution of the maxima, but the kinetic and magnetic energy spectra evolve nearly identically. This is a typical result of all fully turbulent flows, in which small initial disturbances lead to vastly different trajectories in phase space.

From a more practical point of view, the study of active vectors is relevant to the behavior of the magnetic field in conducting fluids where it affects the velocity field through the Lorentz force. This is the basic physics acting in astrophysical phenomena, as, for instance, in the solar corona, which is of extreme importance in everyday life. The MFU can be useful to understand the reasons for the formation of current sheets. Even in this case, it may be of interest to verify whether the inviscid MHD equations lead to an FTS. This issue cannot be sorted out in the present study, and more refined simulations are required. At limited resolution, viscous conducting MFU simulations have shown that in a short time kinetic energy spectra with an inertial and an exponential range similar to that for non-conducting conditions are generated (see Fig. 23.9). On the other hand, the magnetic field in the conducting flow at large scales shows a k^{-2} power-law spectra typical of sheet-like structures and at small scales a well-defined exponential range.

The formation of large-scale helical structures suggests that an interesting continuation of this study should be focused on the analysis of hydrodynamic helicity and magnetic helicity. It is expected that if these increase, the suppression of the energy transfer from large to small scales can be better understood.

The present work fits well within the framework of the ongoing intensive research efforts aimed at the use of big data analysis for improved understanding and prediction of flow turbulence. In fact, the availability of a simple and controllable database as the one developed here clearly facilitates the verification and validation of statistical inferences regarding, for instance, the relationship between coherent structures and wavenumber spectra, as compared to classical simulations based on forced isotropic turbulence. The presence of more "complex" physics here than in ordinary hydrodynamic simulations further makes the present database also an invaluable resource for analyzing future dedicated turbulence models for combustion and astrophysics.

Acknowledgements We acknowledge that the results reported in this paper have been achieved using the PRACE Research Infrastructure resource FERMI based at CINECA, Casalecchio di Reno, Italy, within Project "Euler and Navier–Stokes singularities." GFC acknowledges support from National Science Foundation grant OCE 11-29059.

References

1. R. Feynman, *Lectures on Physics*, vol. 2 (Addison-Wesley, Reading, MA, 1963)
2. A. Tsinober, *An Informal Introduction to Turbulence* (Kluwer Academic Publishers, Dordrecht, 2001)
3. H.K. Moffatt, S. Kida, K. Ohkitani, Stretched vortices: the sinews of turbulence; large-Reynolds-number asymptotics. J. Fluid Mech. **259**, 241–264 (1994)
4. J. Jiménez, P. Moin, The minimal flow unit in near wall turbulence. J. Fluid Mech. **225**, 213–241 (1991)
5. C. Chichowlas, M.E. Brachet, Evolution of complex singularities in Kida-Pelz and Taylor-Green inviscid flows. Fluid Dyn. Res. **36**, 239–248 (2005)
6. M.E. Brachet, Direct simulation of three-dimensional turbulence in the Taylor-Green vortex. Fluid Dyn. Res. **8**, 1–8 (1991)
7. P. Orlandi, G.F. Carnevale, Nonlinear amplification of vorticity in inviscid interaction of orthogonal lamb dipoles. Phys. Fluids **19**(5), 057106 (2007)
8. P. Orlandi, S. Pirozzoli, G.F. Carnevale, Vortex events in Euler and Navier-Stokes simulations with smooth initial conditions. J. Fluid Mech. **690**, 288–320 (2012)
9. S. Goldstein, Fluid mechanics in the first half of this century. Ann. Rev. Fluid Mech. **1**, 1–29 (1969)
10. P. Moin, J. Kim, Tackling turbulence with supercomputers. Sci. Am. **276**, 62–68 (1997)
11. H. Lamb, *Hydrodynamics* (Cambridge University Press, Cambridge, 1932)
12. R.M. Kerr, Evidence for a singularity of the three-dimensional, incompressible Euler equations. Phys. Fluids A **5**, 1725–1746 (1993)
13. H.K. Moffatt, Some developments in the theory of turbulence. J. Fluid Mech. **106**, 27–47 (1981)

14. A. Tsinober, B. Galanti, Exploratory numerical experiments on the differences between genuine and passive turbulence. Phys. Fluids **15**, 3514–3531 (2003)
15. P. Orlandi, *Fluid Flow Phenomena: A Numerical Toolkit* (Kluwer, Dordrecht, 2000)
16. P. Orlandi, S. Pirozzoli, M. Bernardini, G.F. Carnevale, A minimal flow unit for the study of turbulence with passive scalars. J. Turbul. **15**, 731–751 (2014)
17. P. Orlandi, Energy spectra power laws and structures. J. Fluid Mech. **623**, 353–374 (2009)
18. C. Chichowlas, F. Debbasch, M. Brachet, Evolution of complex singularities and Kolmogorov scaling in truncated three-dimensional Euler flows, in *IUTAM Symposium on Elementary Vortices and Coherent Structures: Significance in Turbulence Dynamics*, ed. by S. Kida (2006), pp. 319–328
19. L. Mydlarski, Z. Warhaft, On the onset of high-Reynolds-number grid-generated wind tunnel turbulence. J. Fluid Mech. **320**, 331–368 (1996)
20. D.A. Donzis, K.R. Sreenivasan, P.K. Yeung, The Batchelor spectrum for mixing of passive scalars in isotropic turbulence. Flow Turbul. Combust. **85**, 549–566 (2010)
21. K.R. Sreenivasan, The passive scalar spectrum and the Obukhov-Corrsin constant. Phys. Fluids **8**, 189–196 (1996)
22. W.T. Ashurst, A.R. Kerstein, R.M. Kerr, C.H. Gibson, Alignment of vorticity and scalar gradient with strain rate in simulated Navier-Stokes turbulence. Phys. Fluids **30**, 23–43 (1987)
23. P.K. Yeung, D.A. Donzis, K.R. Sreenivasan, High-Reynolds-number simulation of turbulent mixing. Phys. Fluids **17**, 081703 (2005)
24. I.N. Kitiashvili, A.G. Kosovichev, N.N. Mansour, A.A. Wray, Dynamics of magnetized vortex tubes in the solar chromosphere. arXiv:1201.5442v1 [astro-ph.SR] (2012)
25. J. Jiménez, A.A. Wray, P.G. Saffman, R.S. Rogallo, The structure of intense vorticity in isotropic turbulence. J. Fluid Mech. **255**, 65–90 (1993)
26. E. Lee, M.E. Brachet, A. Pouquet, P.D. Mininni, D. Rosenberg, Lack of universality in decaying magnetohydrodynamic turbulence. Phys. Rev. E **81**, 016318 (2010)
27. A.A. Schekochihin, S.C. Cowley, T.A. Yousef, MHD turbulence: nonlocal, anisotropic, nonuniversal? in *IUTAM Symposium on Computational Physics and New Perspectives in Turbulence* (2008), pp. 347–354

Chapter 24
Linear Stability Analysis of Compressible Channel Flow over Porous Walls

Iman Rahbari and Carlo Scalo

24.1 Introduction

In the present work we investigate the effects of porous walls modeled with a simple Darcy-like boundary condition derived as a particular case of linear acoustic impedance boundary condition (IBC) with zero reactance. The latter can be directly written in the time domain as

$$p' = R\, v_n' \tag{24.1}$$

where R is the impedance resistance, and p' and v_n' are the fluctuating pressure and wall-normal component of the velocity at the boundary (positive if directed away from the fluid side), normalized with the base density and speed of sound (unitary dimensionless base impedance). The present investigation is limited to a local linear stability analysis (LSA) and is inspired by the hydro-acoustic instability observed by Scalo et al. [20], who performed numerical simulations of compressible channel flow over a complex three-parameter broadband wall-impedance [21]. In the latter, the effective wall permeability is a broadband function of frequency tuned to allow maximum transpiration only for the large-scale turbulent energy containing eddies. Despite the lack of the frequency-dependency in (24.1), the results shown in the present investigation provide a useful theoretical framework for the understanding of hydro-acoustic instabilities and a guide for future investigations in the manipulation of compressible boundary layer turbulence.

The effects of porous walls on shear flows have been a topic of formidable research effort, especially in the low-Mach-number limit. Lekoudis [15] performed LSA of an incompressible boundary layer over two types of permeable boundaries:

I. Rahbari (✉) • C. Scalo
School of Mechanical Engineering, Purdue University, West Lafayette, IN 47907-2045, USA
e-mail: irahbari@purdue.edu

A. Pollard et al. (eds.), *Whither Turbulence and Big Data in the 21st Century?*,
DOI 10.1007/978-3-319-41217-7_24

one modeling a perforated surface over a large chamber (with stabilizing effects); the second one, modeling pores over independent cavities (with negligible effects). Jiménez et al. [12] performed incompressible channel flow simulations over active and passive porous walls observing the creation of large spanwise-coherent Kelvin–Helmholtz in the outer layer responsible for frictional drag increase; surprisingly, the near-wall turbulence production cycle was not found to be significantly altered [11]. More recently, Tilton and Cortelezzi [22] investigated channel flow coated with finite-thickness homogeneous porous slabs. Two unstable modes were observed, one symmetric and one anti-symmetric, originating from the left branch of eigenvalue spectrum. Similar results have been found in the present investigation, which also supports the findings of Scalo et al. [20] who observed a complete reorganization of the near-wall turbulence, following the application of the tuned wall-impedance.

One of the earliest works in LSA of compressible shear flow has been carried out by Malik [17], who investigated high-speed flat plate boundary layers up to freestream Mach numbers of 10. Duck et al. [10] performed a similar investigation in the case of viscous compressible Couette flow, revealing fundamental differences with respect to external flow cases. Hu and Zhong [8, 9] investigated supersonic viscous Couette flow at finite Reynolds numbers finding two unstable acoustic inviscid modes when a region of locally supersonic flow, relative to the phase speed of the instability wave, is present. Their structure was deemed consistent with two unstable modes shown by Mack [16], which are sustained by the acoustic interaction between the walls and the sonic line. Malik and co-workers [18] have also used modal and non-modal stability to study the effect of viscosity stratification on the stability of compressible Couette flow. Energy transfer from the mean flow to perturbations occurs at two locations: near the top wall and in the bulk of flow domain, associated with two different unstable modes. More recent efforts are focused on the transitional hypersonic boundary layer stability [3, 5, 23–25], where porous walls are used as means to delay transition via acoustic energy absorption (allowed by a frequency-selective permeability).

The present work aims to inform future investigations of wall-bounded compressible boundary layers over porous walls. Comparison against previous LSA studies of channel and Couette flow is first discussed. Impermeable-wall laminar and turbulent channel flow data used as a base flow in the present study are then presented. Finally, the structure of the unstable modes triggered by the porous walls is discussed along with their predicted effect on the structure of near-wall turbulence.

24.2 Problem Formulation

The fully nonlinear governing equations are first introduced, followed by their linearized counterpart, used for the present LSA. All quantities reported hereafter, including results from previous works are normalized with the channel's half-width (or total height in the case of Couette flow), the bulk density (constant for channel flow), and the speed of sound, temperature, and dynamic viscosity at an isothermal wall.

24.2.1 Governing Equations

The dimensionless governing equations for conservation of mass, momentum, and total energy are, respectively,

$$\frac{\partial}{\partial t}(\rho) + \frac{\partial}{\partial x_j}(\rho u_j) = 0 \tag{24.2}$$

$$\frac{\partial}{\partial t}(\rho u_i) + \frac{\partial}{\partial x_j}(\rho u_i u_j) = -\frac{\partial}{\partial x_i}p + \frac{\partial}{\partial x_i}(\tau_{ij}) + f_1\delta_{1i} \tag{24.3}$$

$$\frac{\partial}{\partial t}(\rho E) + \frac{\partial}{\partial x_j}\left[u_j(\rho E + p)\right] = \frac{\partial}{\partial x_j}(u_i\tau_{ij} + q_j) + f_1 u_1 \tag{24.4}$$

where x_1, x_2, and x_3 (alternatively x, y, and z) are, respectively, the streamwise, wall-normal, and spanwise coordinates, u_i the velocity components in those directions. Thermodynamic pressure, density, and temperature are related by the equation of state $p = \gamma^{-1}\rho T$, where γ is the ratio of specific heats. E is the total energy per unit mass and Re_a the Reynolds number based on the reference length, (either the channel's half-width or wall-to-wall distance for Couette flow), and speed of sound at the isothermal wall temperature, which is related to the bulk Mach and Reynolds numbers via $Re_b = M_b Re_a$. The viscous stress tensor and conductive heat fluxes are

$$\tau_{ij} = \frac{2\mu}{Re_a}\left[S_{ij} - \frac{1}{3}\frac{\partial u_m}{\partial x_m}\delta_{ij}\right], \qquad q_j = \frac{1}{\gamma - 1}\frac{1}{Re_a}\frac{\mu}{\mathrm{Pr}}\frac{\partial T}{\partial x_j} \tag{24.5}$$

where S_{ij} is the strain-rate tensor, μ the dynamic viscosity, respectively, given by $S_{ij} = \frac{1}{2}\left(\frac{\partial u_j}{\partial x_i} + \frac{\partial u_i}{\partial x_j}\right)$, $\mu = T^n$ where $n = 0.75$ is the viscosity exponent and Pr is the Prandtl number.

The complete set of governing equations (24.2)–(24.4) are solved for compressible turbulent channel flow with impermeable, isothermal walls in discretized form on a Cartesian domain with the structured code *Hybrid*, originally developed by Johan Larsson for numerical investigation of the fundamental canonical shock–turbulence interaction problem [13, 14]. The code has been scaled up to 1.97 million cores in recent remarkable computational efforts by Bermejo-Moreno et al. [2]. A solution-adaptive strategy is employed to blend high-order central polynomial and WENO schemes in presence of strong flow gradient [1, 19]. A fourth order discretization both in space and time has been adopted in the present work. Results from the high-fidelity simulations are used to generate the base flow (see Sect. 24.4.1) for the present LSA.

24.2.2 Linearized Equations

Decomposing a generic instantaneous quantity, $a(x, y, z)$ into a base state, $\mathcal{A}(y)$, and a two-dimensional fluctuation, $a'(x, y, t)$, assuming ideal gas and retaining only the first order fluctuations yields the following equations for conservation of mass, streamwise and wall-normal momentum, and energy [17]:

$$\frac{\gamma}{\mathcal{T}}\frac{\partial p'}{\partial t}-\frac{1}{\mathcal{T}^2}\frac{\partial T'}{\partial t}+\frac{1}{\mathcal{T}}\frac{\partial u'}{\partial x}+\mathcal{U}\left(\frac{\gamma}{\mathcal{T}}\frac{\partial p'}{\partial x}-\frac{1}{\mathcal{T}^2}\frac{\partial T'}{\partial x}\right)+\frac{1}{\mathcal{T}}\frac{\partial v'}{\partial y}-\frac{1}{\mathcal{T}^2}\frac{\partial \mathcal{T}}{\partial y}v'=0 \quad (24.6)$$

$$\begin{aligned}\left(\frac{\partial u'}{\partial t}+\mathcal{U}\frac{\partial u'}{\partial x}+v'\frac{\partial \mathcal{U}}{\partial y}\right)\frac{1}{\mathcal{T}} &= -\frac{\partial p'}{\partial x}+\frac{\mathcal{M}}{Re_a}\left[l_2\frac{\partial^2 u'}{\partial x^2}+l_1\left(\frac{\partial^2 v'}{\partial x\partial y}\right)+\frac{\partial^2 u'}{\partial y^2}\right.\\ &+\frac{1}{\mathcal{M}}\frac{d\mathcal{M}}{d\mathcal{T}}\frac{d\mathcal{T}}{dy}\left(\frac{\partial u'}{\partial y}+\frac{\partial v'}{\partial x}\right)+\frac{1}{\mathcal{M}}\frac{\partial \mathcal{M}}{\partial \mathcal{T}}\left(\frac{\partial^2\mathcal{U}}{\partial y^2}T'+\frac{\partial \mathcal{U}}{\partial y}\frac{\partial T'}{\partial y}\right)\\ &\left.+\frac{1}{\mathcal{M}}\frac{\partial^2\mathcal{M}}{\partial \mathcal{T}^2}\frac{\partial \mathcal{T}}{\partial y}\frac{\partial \mathcal{U}}{\partial y}T'\right]\end{aligned} \quad (24.7)$$

$$\begin{aligned}\left(\frac{\partial v'}{\partial t}+\mathcal{U}\frac{\partial v'}{\partial x}\right)\frac{1}{\mathcal{T}} &= -\frac{\partial p'}{\partial y}+\frac{\mathcal{M}}{Re_a}\left[\frac{\partial^2 v'}{\partial x^2}+l_1\left(\frac{\partial^2 u'}{\partial x\partial y}\right)+l_2\frac{\partial^2 v'}{\partial y^2}+\cdots\right.\\ &\left.+\frac{1}{\mathcal{M}}\frac{\partial \mathcal{M}}{\partial \mathcal{T}}\left(\frac{\partial T'}{\partial x}\frac{\partial \mathcal{U}}{\partial y}\right)+\frac{1}{\mathcal{M}}\frac{\partial \mathcal{M}}{\partial \mathcal{T}}\frac{\partial \mathcal{T}}{\partial y}\left\{l_0\left(\frac{\partial u'}{\partial x}\right)+l_2\frac{\partial v'}{\partial y}\right\}\right]\end{aligned} \quad (24.8)$$

$$\begin{aligned}&\left(\frac{\partial T'}{\partial t}+\mathcal{U}\frac{\partial T'}{\partial x}+v'\frac{\partial \mathcal{T}}{\partial y}\right)\frac{1}{\mathcal{T}}=(\gamma-1)\left[\frac{\partial p'}{\partial t}+\mathcal{U}\frac{\partial p'}{\partial x}\right]+\cdots\\ &+\frac{\mathcal{M}}{Re_a\mathrm{Pr}}\left[\frac{\partial^2 T'}{\partial x^2}+\frac{\partial^2 T'}{\partial y^2}+\frac{2}{\mathcal{K}}\frac{\partial \mathcal{K}}{\partial \mathcal{T}}\frac{\partial \mathcal{T}}{\partial y}\frac{\partial T'}{\partial y}+\left(\frac{1}{\mathcal{K}}\frac{\partial \mathcal{K}}{\partial \mathcal{T}}\frac{\partial^2\mathcal{T}}{\partial y^2}+\frac{1}{\mathcal{K}}\frac{\partial^2\mathcal{K}}{\partial \mathcal{T}^2}\left(\frac{\partial \mathcal{T}}{\partial y}\right)^2\right)T'\right]\\ &+\cdots+\frac{(\gamma-1)\mathcal{M}}{Re_a}\left[+2\frac{\partial \mathcal{U}}{\partial y}\left(\frac{\partial u'}{\partial y}+\frac{\partial v'}{\partial x}\right)+\frac{1}{\mathcal{M}}\frac{\partial \mathcal{M}}{\partial \mathcal{T}}\left(\frac{\partial \mathcal{U}}{\partial y}\right)^2\right]\end{aligned} \quad (24.9)$$

where $l_j = j - 2/3$, $\mathcal{M} = \mathcal{T}^n$, $\mathcal{K} = C_p\,\mathcal{M}/\mathrm{Pr}$, where $\mathcal{M}$, $\mathcal{K}$, $\mathcal{T}$, and $\mathcal{U}$ are the base dynamic viscosity, conductivity, temperature, and streamwise velocity. In the derivation of Eqs. (24.6)–(24.9) it was assumed that $\mathcal{P} = 1/\gamma$. The base density is derived from the base pressure and temperature via the equation of state with gas constant equal to $1/\gamma$, consistently with the normalization adopted. The effect of eddy viscosity on the fluctuations is neglected.

Assuming harmonic two-dimensional fluctuations

$$a'(x, y, t) = \hat{a}(y)e^{i(\alpha x-\omega t)} = \hat{a}(y)\,e^{i\alpha(x-c\,t)} \quad (24.10)$$

where α is the wavenumber in the streamwise direction, and $\omega = \alpha\,c$ and $c = c_r + i\,c_i$ are the complex frequency and complex wave (or phase) speed. Applying the spectral collocation method based on Chebyshev polynomials along the wall-normal direction yields the generalized eigenvalue problem

$$\mathbf{A}\,\boldsymbol{\Psi} = \omega \mathbf{B}\,\boldsymbol{\Psi} \tag{24.11}$$

with $\boldsymbol{\Psi} = \left\{\hat{\mathbf{u}}; \hat{\mathbf{v}}; \hat{\mathbf{p}}; \hat{\mathbf{T}}\right\}$ where the generic column vector $\hat{\mathbf{a}}$ is the collection of the discretized complex amplitudes of the generic fluctuating quantity, a'. Four conditions at each boundary are needed to solve (24.11): no-slip conditions for the streamwise velocity component, $\hat{u} = 0$; isothermal conditions for temperature fluctuations, $\hat{T} = 0$; impedance boundary conditions $\hat{p} = \mp R\,\hat{v}$ for $y \pm 1$ (upper and lower wall, respectively); Non-homogeneous Neumann conditions for wall-normal derivative of pressure derived by applying (24.8) at the boundary.

Including the aforementioned boundary conditions on the left-hand side of (24.11) results in a singular **B** matrix. Therefore, eigenvalues obtained by directly solving the general eigenvalue problem in this form yield eigenfunctions contaminated by numerical noise, despite eigenvalues still being accurate. This problem can be solved by recasting (24.11) into

$$\omega^{-1}\boldsymbol{\Psi} = \mathbf{A}^{-1}\,\mathbf{B}\,\boldsymbol{\Psi} \tag{24.12}$$

and solving the eigenvalue problem directly for the matrix $\mathbf{C} = \mathbf{A}^{-1}\mathbf{B}$ [6]. While this technique considerably improves the quality of the eigenfunctions, results for high wavenumbers ($\alpha > 2$) still exhibit numerical issues such as top-down anti-symmetry, which has limited the current investigation to $\alpha \leq 2$.

The base velocity and temperature are chosen as $\mathcal{U} = \overline{u}$ and $\mathcal{T} = \overline{T}$ where $\overline{(\,)}$ indicates Reynolds-averaged quantities, which are obtained from companion high-fidelity Navier–Stokes simulations (see Sect. 24.4.1).

In the following, validation against LSA studies available in literature is shown (Sect. 24.3) and new results from the present LSA analysis are discussed (Sect. 24.4).

24.3 Comparison Against Previous LSA

Validation is first carried out against LSA results for compressible Couette flow by Hu and Zhong [8, 9], where isothermal conditions, $T_\infty = 1$, and the tangential velocity, U_∞, are imposed at the top wall, and no-slip and adiabatic conditions at the bottom wall. Reynolds number and Mach number based on the topwall velocity are Re_∞ and M_∞. All quantities are normalized with speed of sound based on top wall temperature, the bulk density and the total wall-to-wall distance. To enable the direct comparison with Hu and Zhong [8, 9], the dynamic viscosity has been changed to:

$$\mathcal{M} = \mathcal{T}^{1.5}\frac{1 + C}{\mathcal{T} + C}, \quad C = 0.5 \tag{24.13}$$

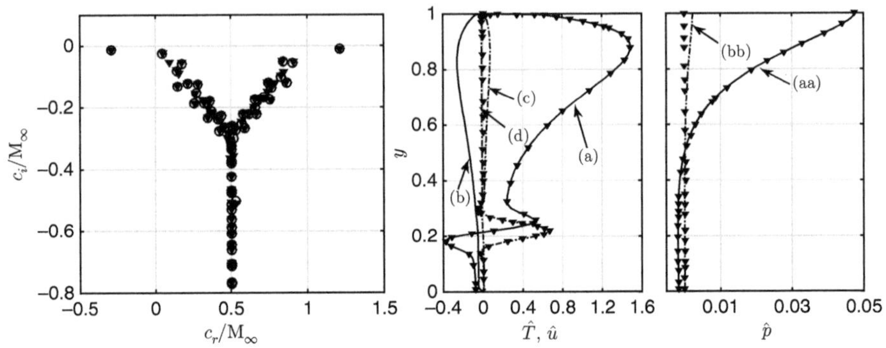

Fig. 24.1 Comparison of complex phase velocity spectrum for compressible Couette flow at $M_\infty = 2$, $Re_\infty = 2 \times 10^5$, $\alpha = 1$ using N=100 grid points (*open circle*) with results from Hu and Zhong [9] (*filled downward triangle*). Real part (*solid line*) and imaginary part (*dash dotted line*) of velocity, $\hat{u}$, temperature, $\hat{T}$, and pressure, $\hat{p}$, eigenfunctions of the most unstable mode at $M_\infty = 5$, $Re_\infty = 5 \times 10^5$, $\alpha = 3$ with *(a)*: $Re\left\{\hat{T}\right\}$, *(b)*: $Re\{\hat{u}\}$, *(c)*: $Im\left\{\hat{T}\right\}$, *(d)*: $Im\{\hat{u}\}$, *(aa)*: $Re\{\hat{p}\}$, *(bb)*: $Im\{\hat{p}\}$

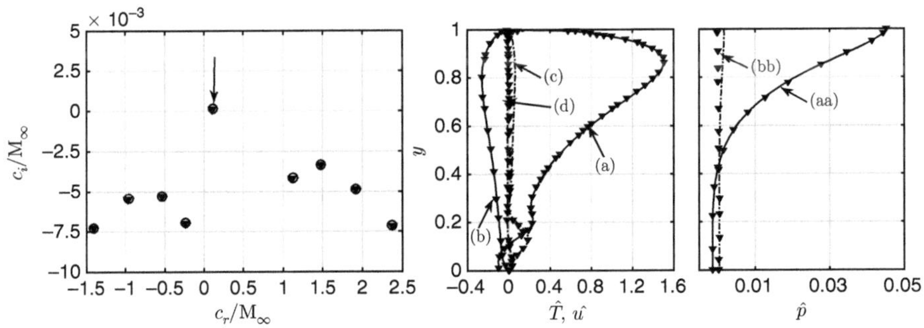

Fig. 24.2 Comparison of complex phase velocity spectrum for laminar compressible Couette flow at $M_\infty = 5$, $Re_\infty = 1 \times 10^5$, and $\alpha = 2.5$ using $N = 100$ grid points (*filled downward triangle*) against Hu and Zhong [8]. Real part (*solid line*) and imaginary part (*dash dotted line*) of velocity, $\hat{u}$, temperature, $\hat{T}$, and pressure, $\hat{p}$, eigenfunctions of the most unstable mode (indicated with *downward arrow*) with *(a)*: $Re\left\{\hat{T}\right\}$, *(b)*: $Re\{\hat{u}\}$, *(c)*: $Im\left\{\hat{T}\right\}$, *(d)*: $Im\{\hat{u}\}$, *(aa)*: $Re\{\hat{p}\}$, *(bb)*: $Im\{\hat{p}\}$

and Pr = 0.72 (only for the sake of this comparison). Excellent agreement is found as shown by the eigenvalue spectra and eigenfunctions (Figs. 24.1 and 24.2). A grid convergence study of the eigenvalues (Table 24.1) also includes comparisons with more recent LSA work on the same flow by Weder [26].

Validation has also been carried out against Friedrich and Bertolotti [7], who performed a stability analysis of (impermeable isothermal wall) compressible turbulent channel flow at $Re_c = 4880$ based on the centerline velocity, $M_b = 3$, $\alpha = 1$ using parabolized stability equation (PSE) with a multi-domain spectral dis-

Table 24.1 Comparison of complex wave speed, c, of the most unstable mode for compressible Couette flow with previously published results for various Re_{∞}, M_{∞}, and α

Hu and Zhong [8]		Weder [26]		Current study	
c_r/M_{∞}	c_i/M_{∞}	c_r/M_{∞}	c_i/M_{∞}	c_r/M_{∞}	c_i/M_{∞}
$Re_{\infty} = 5 \times 10^6, M_{\infty} = 5, \alpha = 2.1$, Mode I					
+0.972869314676	−0.003456356315	+0.972869178324	−0.003456358661	+0.972869280518	−0.003456323886
+0.972869272448	−0.003456466520	+0.972869198693	−0.003456318849	+0.972869272445	−0.003456466743
+0.972869272450	−0.003456466522	+0.972869201921	−0.003456324875	+0.972869272355	−0.003456466505
$Re_{\infty} = 5 \times 10^6, M_{\infty} = 5, \alpha = 2.1$, Mode 0					
+0.040730741952	+0.000876050503	+0.040726205831	+0.000884153855	+0.040730596292	+0.000874754264
+0.040722854287	+0.000885530891	+0.040722853306	+0.000885531375	+0.040722854373	+0.000885530566
+0.040722853034	+0.000885531421	+0.040722853219	+0.000885531398	+0.040722853032	+0.000885531421
$Re_{\infty} = 2 \times 10^5, M_{\infty} = 2, \alpha = 0.1$, Mode I					
+1.213965119859	−0.011585118523	+1.213965119851	−0.011585118549	+1.213965119851	−0.011585118547
+1.213965119817	−0.011585118448	+1.213965119852	−0.011585118548	+1.213965119852	−0.011585118547
+1.213965119854	−0.011585118558	+1.213965119851	−0.011585118548	+1.213965119852	−0.011585118547
$Re_{\infty} = 2 \times 10^5, M_{\infty} = 2, \alpha = 0.1$, Mode 0					
−0.291572925106	−0.013821128462	−0.291572925110	−0.013821128465	−0.291572925109	−0.013821128464
−0.291572925140	−0.013821128536	−0.291572925116	−0.013821128473	−0.291572925109	−0.013821128464
−0.291572925108	−0.013821128457	−0.291572925112	−0.013821128467	−0.291572925109	−0.013821128464

Grid convergence study shown for $N = 100$, $N = 200$, and $N = 300$, respectively, top, middle, and bottom rows for each case. Deviating digits from Hu and Zhong [8] are underlined. Values of c have been normalized with the top-wall Mach number, M_{∞}, to allow direct comparison

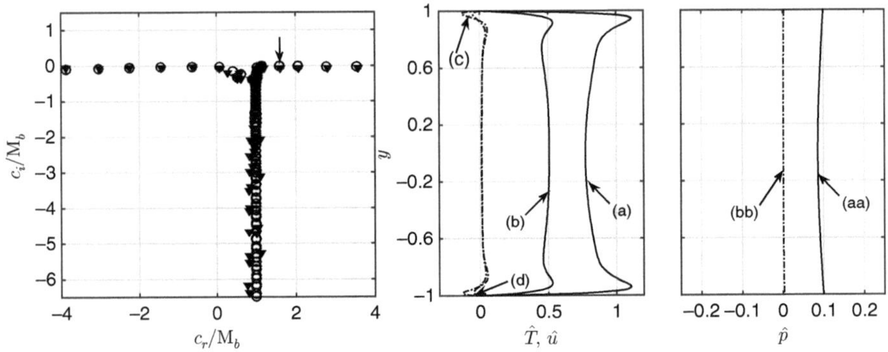

Fig. 24.3 Complex phase velocity spectrum for compressible channel flow at $M_b = 0.5$, $Re_c = 4880$, and $\alpha = 1$ using $N = 165$ grid points (*open circle*) compared with data from Friedrich and Bertolotti [7] (*filled downward triangle*). Real part (*solid line*) and imaginary part (*dash dotted line*) of eigenfunctions of the most unstable mode indicated with a *downward arrow*. *(a)*: *Re* $\left\{\hat{T}\right\}$, *(b)*: *Re* $\{\hat{u}\}$, *(c)*: *Im* $\left\{\hat{T}\right\}$, *(d)*: *Im* $\{\hat{u}\}$, *(aa)*: *Re* $\{\hat{p}\}$, *(bb)*: *Im* $\{\hat{p}\}$

cretization method. The base flow was taken from the Reynolds-averaged velocity and temperature profiles in Coleman et al. [4]. The eigenvalue spectrum is compared with Fig. 24.3 where eigenfunctions of the most unstable mode (not originally plotted in their paper) are also included. Despite the fundamental difference between our local LSA approach and their PSE, the agreement is still acceptable. Friedrich and Bertolotti [7] did not explicitly mention what law for viscosity was used. The results in Fig. 24.3 have been reproduced assuming $\mu = \mathcal{T}^n$, with $n = 0.7$, consistent with the simulations by Coleman et al. [4].

24.4 Compressible Channel Flow with Porous Walls

In the following results from impermeable-wall compressible turbulent channel flow high-fidelity simulations are first discussed (Sect. 24.4.1). The mean velocity and temperature profiles extracted from the latter are used as base flow in the present LSA with porous walls (Sect. 24.4.2).

24.4.1 Impermeable-Wall Turbulent Channel Flow

The base velocity, density, and temperature profiles for the present LSA investigations have been taken from laminar and turbulent numerical simulations of fully compressible impermeable isothermal-wall channel flow.

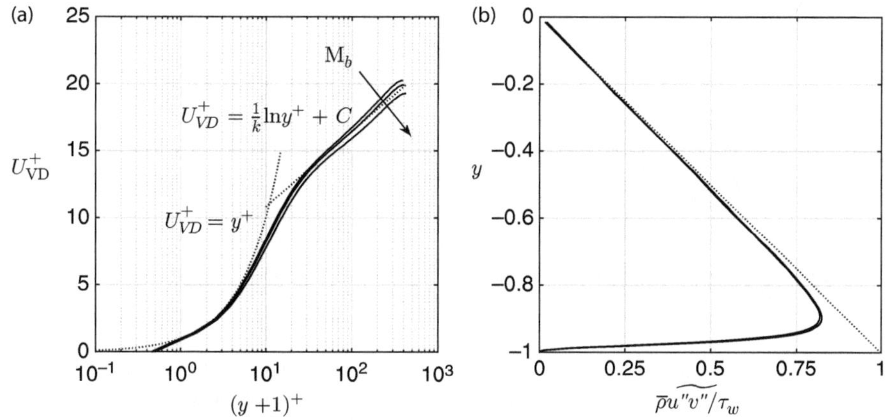

Fig. 24.4 Profiles of turbulent mean velocity following Van Driest transform with reference law of the wall and log-law $k = 0.41$ and $C = 5.2$ (*dotted line*) (**a**) and Reynolds stresses normalized by the average wall-shear stress, τ_w (**b**)

The computational domain considered for the turbulent simulations in this study is $L_x \times L_y \times L_z = 8 \times 2 \times 4$ which is discretized with a number of control volumes $N_x \times N_y \times N_z = 256 \times 128 \times 192$ resulting in a quasi-DNS resolution of $\Delta x^+ \sim 12.27 - 13.22$, $\Delta z^+ \sim 8.17 - 8.81$, $\Delta y^+{}_{\min} = 0.45 - 0.47$, over the range of bulk Mach numbers investigated. The superscript $^+$ indicates classic wall-units, $\delta_\nu = u_\tau \rho_w / \mu_w$, where $u_\tau = \sqrt{\tau_w/\rho_w}$ is the friction velocity and ρ_w and μ_w are the density and dynamic viscosity evaluated at the wall and the wall-shear stress is $\tau_w = \mu_w \, \partial\overline{u}/\partial y|_w$ where $\overline{(\,)}$ indicates Reynolds averaging.

Mean velocity profiles in Fig. 24.4a are plotted using Van Driest transform following:

$$U^+_{\mathrm{VD}} = \int_0^{U^+} \left(\frac{\rho}{\rho_w}\right)^{1/2} dU^+ = \frac{1}{\kappa} \ln y^+ + C \tag{24.14}$$

In this set of simulations, friction Reynolds number, $Re_\tau = u_\tau \delta/\nu$, varies from 385.6 to 417.4. Collapse of both mean velocity and Reynolds stress profiles is obtained, showing that the state of near-wall turbulence has been successfully kept constant while varying the bulk Mach number.

The laminar base flow has been generated with the same code with $N_y = 192$ control volumes in the wall-normal direction, for values of Re_b and M_b matching the turbulent simulations, and is shown in Fig. 24.5.

24.4.2 Linear Stability Analysis with Porous Walls

In the following we explore the effects of porous walls on the linear stability of the two-dimensional perturbation (24.10) using the impermeable channel flow results generated in Sect. 24.4.1 as the base flow. Since turbulent eddy viscosity is neglected

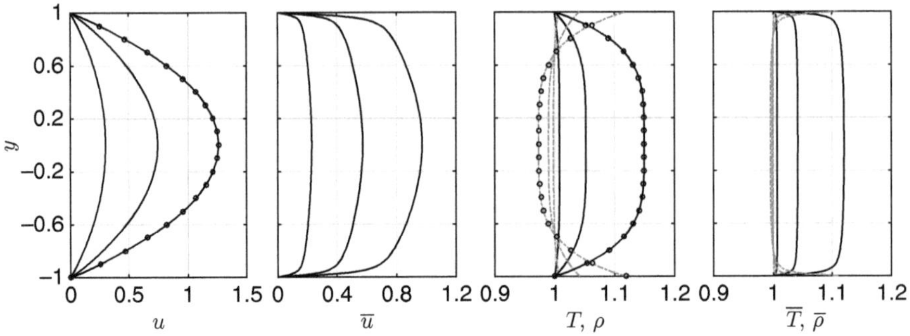

Fig. 24.5 Profiles of laminar velocity, $u(y)$, temperature, $T(y)$, and density $\rho(y)$ and Reynolds-averaged turbulent velocity, $\overline{u}(y)$, temperature, $\overline{T}(y)$, and density $\overline{\rho}(y)$ at $Re_b = 6900$ and $M_b = 0.2$, 0.5, and 0.85. *Symbols* show laminar results calculated at $Re_b = 6900$ and $M_b = 0.85$ obtained separately by discretizing the governing equations retaining only the streamwise velocity component and wall-normal gradients

in (24.7)–(24.9), the nature of the instability between laminar and turbulent cases only differs due to the base flow. In particular, the profile of the wall-normal gradient of the mean streamwise velocity plays a fundamental role in shaping the unstable modes triggered by the wall permeability and determining the location of the energy production regions. In the following, the trajectory of the eigenvalues of the dominant unstable modes is analyzed, together with their induced excess Reynolds shear stresses (RSSs).

24.4.2.1 Unstable Modes Eigenvalues Trajectory

A parametric study has been performed by varying the impedance resistance, R, from $R \to \infty$ (zero permeability) to $R = 0.01$ (high permeability). Two modes, Mode 0 and Mode I, are made unstable ($c_i > 0$) for a sufficiently low value of resistance $R < R_{cr}$ (Fig. 24.6) for both laminar and turbulent base flows. In all cases, the phase speed of the unstable modes increases monotonically with increasing the permeability, exceeding the bulk velocity.

In the laminar case, for $\alpha = 1$, the two modes originate from different regions of the eigenvalue spectrum. Mode I at $R \to \infty$ exhibits a very slow decay rate, requiring low values of permeability (high values of resistance, $R_{cr} = 1$) to become unstable. For near-impermeable conditions, Mode 0 is instead initially located close to the junction of Y-shape spectrum (phase velocity approximately equal to the bulk velocity) and becomes unstable at much lower values of resistance, $R_{cr} \sim 0.05$ (higher values of permeability). However, the differences in their instability dynamics and spatial structures (discussed later) are not due to their different location in the eigenvalue spectrum. As α increases (not shown), the two starting points, and the corresponding trajectories, become closer such that for $\alpha = 2$ they start from two adjacent nodes at the middle of the left branch of

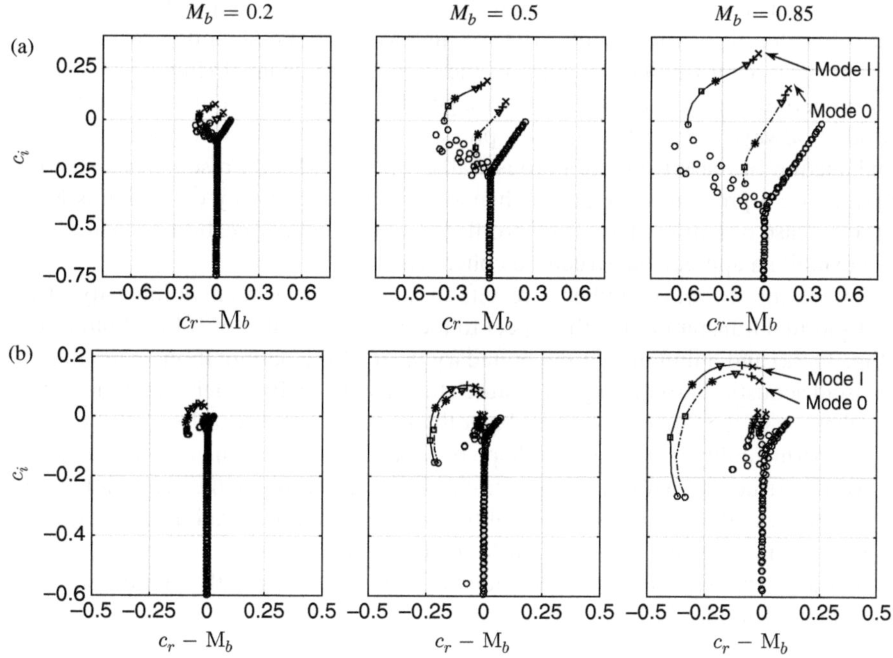

Fig. 24.6 Trajectory of Mode 0 and Mode I for α=1 in the complex phase velocity spectrum of compressible channel flow with impedance boundary conditions (24.1) using laminar (**a**) and turbulent (**b**) base flow at $Re_b = 6900$, $M_b = 0.85$ traced by varying the impedance resistance in the range $R = 10 - 0.01$ for $N = 200$ grid points. In increasing order of permeability, $R \to \infty$ (zero permeability): (*open circle*), $R = 1$: (*open square*), $R = 0.5$: (*asterisk*), $R = 0.1$: (*open downward triangle*), $R = 0.05$: (*plus*), $R = 0.01$: (*times symbol*)

Y-spectrum. Decreasing α yields the opposite effect, with trajectories originating from opposite ends of the Y spectrum for $\alpha = 0.1$ (not shown).

The impermeable-wall spectrum for $\alpha = 1$ does not exhibit a clear Y-shape for the turbulent case. The trajectories of the modes follow very similar paths in the range of $\alpha = 0.1 - 2$ (not shown) tested and share very similar values of critical impedance resistance (or wall permeability), suggesting their coexistence. This behavior is due to the *flatness* of the mean velocity profile, which is even more accentuated in the case of fully developed turbulence over tuned wall-impedance, as observed by Scalo et al. [20].

The growth rates of the unstable modes, in case of laminar base flow, increase monotonically for decreasing R in the range investigated. However, when using turbulent base flow, the growth rate saturates and then decreases for the same range of R. For the turbulent case only, two new unstable modes (with a similar pairing behavior as Mode 0 and Mode I) originate from the central part of the spectrum. They are not present in the laminar base flow case, travel at roughly the bulk velocity, ($c_r \sim M_b$) and become unstable, for very low values of resistance $R_{cr} \sim 0.05$. A detailed study of them is deferred to future work.

The transition from subcritical to supercritical permeability does not qualitatively alter the structure of Mode 0 and Mode I; both modes retain a very pronounced acoustic-like structure in the core of the channel (Fig. 24.7), even in the impermeable-wall limit. For the range of α and Mach numbers investigated, Mode 0 always manifests as a bulk pressure mode, causing symmetric expulsion and suction of mass from the porous walls, whereas Mode I resembles a standing wave, with a pressure node at the channel centerline. The only exception is Mode 0 in the case of laminar base flow profiles at subcritical permeabilities, which does not exhibit an appreciable structural coherence.

Increasing the wavenumber α results in an increase of relative intensity of the wall-normal transpiration with respect to the centerline value of $|\hat{v}|$ for both modes (Fig. 24.7a) and a thinning of the boundary layer of the perturbation, extending the inviscid acoustic-like core region. Future studies will verify if such effects are due to a reduction of the Stokes boundary layer thickness, $\delta_s = \sqrt{2\nu/\omega_r} = \sqrt{2\nu/(\alpha\, c_r)}$.

As permeability increases, the shape of Mode I deviates from a typical standing-wave-like structure, with two new pressure nodes appearing in the near-wall region. An increase in the relative intensity of the wall-normal transpiration with respect to the centerline vertical velocity $\hat{v}_c$ in Mode 0 (Fig. 24.7b) is also observed. Varying bulk Mach number, in the range considered, does not significantly alter the structure of both modes (Fig. 24.7c).

24.4.2.2 Perturbation-Induced Reynolds Shear Stress

The modifications to the RSS distribution that would result from the application of porous walls can be qualitatively predicted by analyzing the normalized, perturbation-induced excess shear stresses (Fig. 24.8)

$$\widetilde{R}_{12} = \frac{Re\{\hat{u}\hat{v}^*\}}{\int_{-1}^{0} Re\{\hat{u}\hat{v}^*\}\, dy}. \tag{24.15}$$

In the case of turbulent base flow, both modes generate an anti-symmetric RSS distribution concentrated in base flow's viscous sublayer (Fig. 24.4b) with negligible effects in the turbulent core ($-0.9 < y < 0.9$).

When employing a laminar base flow, normalized RSS for Mode 0 do not change considerably as permeability varies. Moreover, the corresponding eigenmodes are structurally incoherent for subcritical permeabilities, as also pointed out in Sect. 24.4.2.1. Results for Mode I show that its RSS peak approaches the wall for increasing permeability (decreasing R value). With a turbulent base flow, the RSS shape for both modes exhibits very similar features. Decreasing the R value brings the RSS peak closer to the wall with normalized values evanescent in the channel core (in contrast with the laminar base flow results). It should be noted that increasing the permeability (lowering value of R) results in a higher growth rate of the disturbance, and will therefore yield a more intense augmentation of the total turbulent RSS (Fig. 24.8).

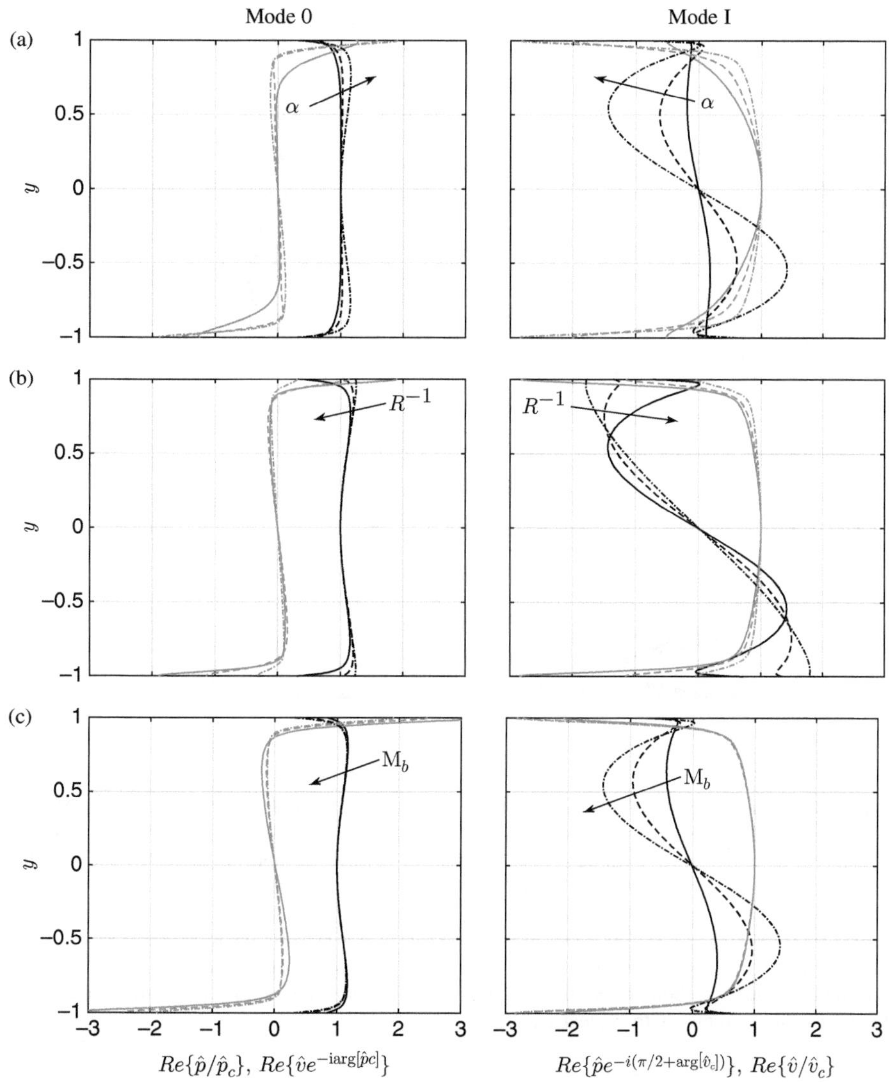

Fig. 24.7 Comparison between Mode 0 (*left*) and Mode I (*right*) showing pressure, $\hat{p}$, (*black*) and wall-normal velocity, $\hat{v}$ (*grey*) eigenfunctions with base flow taken from turbulent mean profiles at $Re_b = 6900$. Resistance $R = 0.01$ and $M_b = 0.85$, for wavenumbers $\alpha = 0.1$ (*solid line*), $\alpha = 0.5$ (*dashed line*), and $\alpha = 1.0$ (*dash dotted line*) (**a**); Wavenumber $\alpha = 1$ and $M_b = 0.85$, for resistances $R = 1.0$ (*dash dotted line*), $R = 0.1$ (*dashed line*), and $R = 0.01$ (*solid line*) (**b**). Wavenumber $\alpha = 1$ and $R = 0.01$, for bulk Mach numbers $M_b = 0.2$ (*solid line*), $M_b = 0.5$ (*dashed line*), and $M_b = 0.85$ (*dash dotted line*) (**c**). In all Mode 0 plots, all pressure eigenfunctions have been normalized with their value at the centerline, $\hat{p}_c$; the resulting phase shift has been applied to the eigenfunctions of wall-normal velocity. In all Mode I plots, the vertical velocity eigenfunctions have been normalized with their value at the centerline, $\hat{v}_c$, the pressure eigenfunctions' phase has been shifted accordingly and then further changed by 90° to highlight the standing-wave-like structure in the channel core

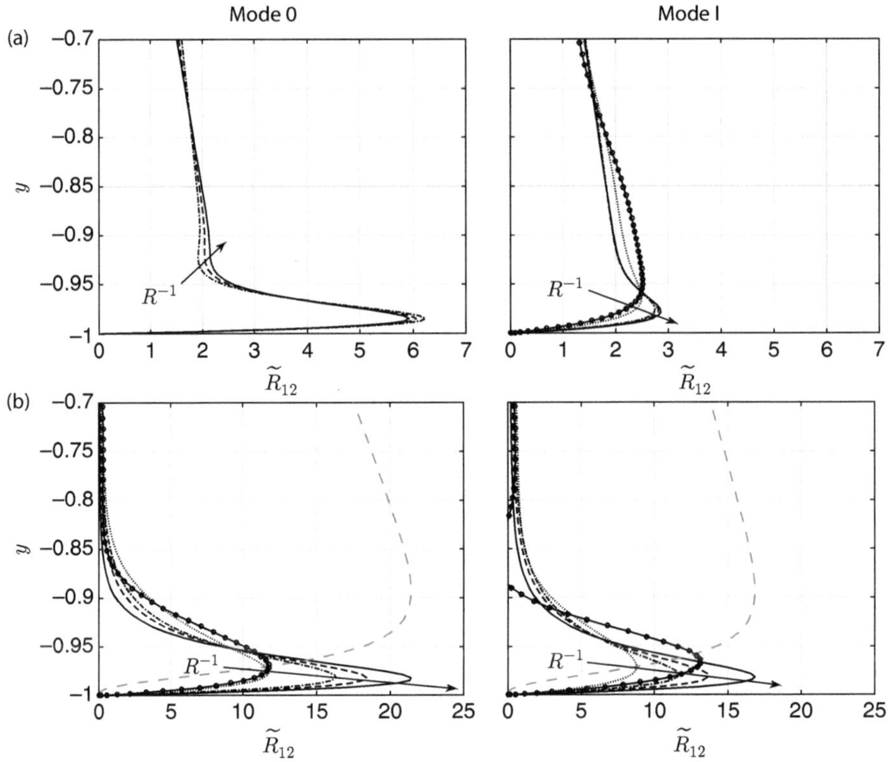

Fig. 24.8 Near-wall profiles of normalized perturbation-induced RSS, $\widetilde{R}_{12}$ (24.15) for Mode 0 (*left*) and Mode I (*right*) predicted by LSA for N =200 grid points, for different values of resistance, R, and $\alpha = 1$ with base flow taken from laminar (**a**) and turbulent (**b**) channel flow calculations at $Re_b = 6900$, $M_b = 0.85$. *Arrows* indicate increasing values of permeability (decreasing value of resistance) with $R = 1$ (*open circle*), $R = 0.5$ (*dotted line*), $R = 0.1$ (*dash dotted line*), $R = 0.05$ (*dashed line*), $R = 0.01$ (*solid line*). Results for $R = 1$ and $R = 0.5$ are omitted for Mode 0 with a laminar base profile, since corresponding eigenmodes do not exhibit a coherent structure. Turbulent RSS from the impermeable-wall calculations (Fig. 24.4) are reported here (in arbitrary units) with *dashed grey lines*

The predicted shape of the perturbation-induced RSS profiles is consistent with the structure of the Kelvin–Helmholtz instability observed by Scalo et al. [20], sustained by the wall-normal acoustic resonance generated by the assigned wall-impedance.

24.5 Discussion and Future Work

In the present paper, LSA is employed to predict the effects of wall permeability on a laminar and turbulent compressible channel flow up to transonic bulk Mach numbers, for a fixed bulk Reynolds number of $Re_b = 6900$. Two-dimensional streamwise-periodic disturbances are assumed and porous walls are modeled by linear acoustic IBC with zero reactance. This results in a purely real impedance, defined uniquely by its resistance, R, inversely proportional to the wall permeability. The resistance has been varied in the range, $R = 1 - 0.01$.

Increasing the permeability, or equivalently the inverse of the resistance R^{-1}, destabilizes the flow by triggering two modes; one characterized by bulk (quasi-uniform) pressure oscillations in the core, with gradients of fluctuating quantities concentrated in the near-wall viscous region (Mode 0); the other resembling an acoustic standing wave resonating in the channel's core, with similar near-wall behavior (Mode I). Mode 0 is characterized by symmetric u', T', and $p'-$ eigenmodes and anti-symmetric v'. The latter is associated with an anti-symmetric, periodic, expulsion, and suction of mass from the boundaries driven by the bulk pressure fluctuations in the core. On the other hand, Mode I exhibits anti-symmetric u', T', and p' profiles with symmetric v' distribution. For both modes, increasing the wall permeability causes the perturbation-induced RSS peak to shift closer to the porous walls, for both laminar and turbulent base flows.

For very high values of permeability (low values of resistance $R < 0.01$) and only in the case of a turbulent base flow, a saturation of the growth rate is observed. Values of $R < 0.1$ may, however, be unrealistic (see measurements of the acoustic response of perforated panels [21]) and are also expected to yield nonlinear effects that are not captured by a linear IBC formulation.

Although the current analysis is limited to purely real acoustic impedances (i.e., classic Darcy-like formulation for porous walls), the results obtained qualitatively confirm the structure of the hydro-acoustic instability observed by Scalo et al. [20] generated by complex impedance boundary conditions. In their case, large spanwise-coherent Kelvin–Helmholtz rollers are found to be confined near the wall where Reynolds shear stresses are significantly augmented by the instability, while the outer layer remained unaffected. This result is in qualitative agreement with the perturbation-induced Reynolds shear stress distribution predicted by the current analysis and in contrast with the findings by Jiménez et al. [12], valid for the incompressible limit, where Kelvin–Helmholtz rollers were observed in the outer layer and the near-wall turbulence was unaltered. A more accurate comparison between our LSA and the results by Scalo et al. [20] will be obtained in future studies, which will incorporate a broadband reactance (the impedance's imaginary part) in the eigenvalue problem.

Eigenvalue tracking shows that the modes made unstable by permeability preserve the characteristic structure they had when stable. Only in the laminar case, Mode I is at the verge of instability for zero permeability. As the wavenumber α is increased, trajectory of the unstable modes becomes close to each other, and near-wall gradients of fluctuating quantities intensify.

Future work will focus on direct numerical simulation of channel flow with general IBC and new companion LSA efforts aimed at finding the optimal impedance distribution for effecting flow control, with improved numerics and physical modeling of the boundary conditions. Investigations will also be carried out considering boundary layers over flat plates, to rule out confinement effects.

24.6 From Small Data to Big Data (and Vice Versa)

This work has been centered around the use and generation of *small* data, in the form of a classic LSA, in order to acquire a *general* understanding of the response that it is expected from compressible channel flow turbulence under the effects of porous walls. This problem can only be truly explored *in detail* with large-scale, high-fidelity simulations, that is, *big* data. Simplified low-order models, however, prove to be essential in guiding the design and interpretation of companion large-scale numerical simulations, since they provide us with a rough but robust understanding of the physical problem under investigation at a considerably reduced cost. Moreover, low-order models allow for a true parametric study, spanning, for example, several orders of magnitude of variation of a key parameter in the problem, such as, in this case, the wall permeability. High-fidelity simulations, on the other hand, can only populate a few key regions of the parameter space with limited modeling assumptions, and therefore used to calibrate and/or verify the results from the low-order modeling efforts.

When possible, a synergistic interaction between small data and big data should therefore be sought. A "blind big data"—or a "blind small data"—only approach may suffer from severe limitations. The development, analysis, and adoption of the two approaches should go hand in hand in order to cross-verify results, especially when experimental data is not available. However, with the advances in computational power, the definition of small and big data is changing rapidly. For example, turbulent simulations performed on a computational grid of $N^3 = 1024^3$ points, can nowadays (2016) be considered *average*-size data. In spite of the gradual shift in the demarcation line between small and big data, we believe that the importance of low-order models (used here as an example of small data) will never (and should never) cease to retain a prominent role in the design, analysis, and support of results from large-scale numerical investigations.

Acknowledgements The authors would like to thank Prof. Johan Larsson for very kindly providing us with code, *Hybrid*, that was used to perform the high-fidelity turbulent simulation. The authors have also significantly benefited from a generous computational allocation on Purdue's eighth on-campus supercomputer, *Rice*.

References

1. N.A. Adams, K. Shariff, A high-resolution hybrid compact-EDO scheme for shock-turbulence interaction problems. J. Comput. Phys. **127**(1), 27 (1996)
2. I. Bermejo-Moreno, J. Bodart, J. Larsson, B.M. Barney, Solving the compressible Navier-Stokes equations on up to 1.97 million cores and 4.1 trillion grid points, in *IEEE International Conference on High Performance Computing* (2013)
3. G.A. Bres, M. Inkman, T. Colonius, A.V. Fedorov, Second-mode attenuation and cancellation by porous coatings in a high-speed boundary layer. J. Fluid Mech. **726**, 312 (2013)

4. G.N. Coleman, J. Kim, R.D. Moser, A numerical study of turbulent supersonic isothermal-wall channel flow. J. Fluid Mech. **305**, 159–183 (1995)
5. N. De Tullio, N.D. Sandham, Direct numerical simulation of breakdown to turbulence in a Mach 6 boundary layer over a porous surface. Phys. Fluids **22**, 094105 (2010)
6. V. Esfahanian, Computation and stability analysis of laminar flow over a blunt cone in hypersonic flow. Ph.D. Thesis, The Ohio State University, 1991
7. R. Friedrich, F.P. Bertolotti, Compressibility effects due to turbulent fluctuations. Appl. Sci. Res. **57**, 165–194 (1997)
8. S. Hu, X. Zhong, Linear instability of compressible plane couette flows, in *AIAA Aerospace Sciences Meeting and Exhibit, Reno, NV*, pp. 1–16 (1997)
9. S. Hu, X. Zhong, Linear stability of viscous supersonic plane couette flow. Phys. Fluids **10**(3), 709–730 (1998)
10. M.Y. Hussaini, P.W. Duck, G. Erlebacher, On the linear stability of compressible plane couette flow. J. Fluid Mech. **258**, 131–165 (1994)
11. J. Jiménez, A. Pinelli, The autonomous cycle of near-wall turbulence. J. Fluid Mech. **389**, 335–359 (1999)
12. J. Jiménez, M. Uhlmann, A. Pinelli, G. Kawahara, Turbulent shear flow over active and passive porous surfaces. J. Fluid Mech. **442**, 89–117 (2001)
13. J. Larsson, S.K. Lele, Direct numerical simulation of canonical shock/turbulence interaction. Phys. Fluids **21**(12), 126101 (2009)
14. J. Larsson, I. Bermejo-Moreno, S.K. Lele, Reynolds- and mach-number effects in canonical shock–turbulence interaction. J. Fluid Mech. **717**(2), 293–321 (2013)
15. S.G. Lekoudis, Stability of boundary layer over permeable surfaces, in *AIAA Aerospace Sciences Meeting and Exhibit, Huntsville, AL*, pp. 1-8 (1978)
16. L.M. Mack, On the inviscid acoustic-mode instability of supersonic shear flows. Theor. Comput. Fluid Dyn. **2**(2), 97–123 (1990)
17. M.L. Malik, Numerical methods for hypersonic boundary layer stability. J. Comput. Phys. **86**, 376–413 (1990)
18. M. Malik, J. Dey, M. Alam, Linear stability, transient energy growth, and the role of viscosity stratification in compressible plane couette flow. Phys. Rev. E **77**, 1–15 (2012)
19. S. Pirozzoli, Conservative hybrid compact-WENO schemes for shock-turbulence interaction. J. Comput. Phys. **178**, 81–117 (2002)
20. C. Scalo, J. Bodart, S.K. Lele, Compressible turbulent channel flow with impedance boundary conditions. Phys. Fluids **27**, 035107 (2015)
21. C.K.W. Tam, L. Auriault, Time-domain impedance boundary conditions for computational aeroacoustics. AIAA J. **34**(5), 917–923 (1996)
22. N. Tilton, L. Cortelezzi, Linear stability analysis of pressure-driven flows in channels with porous walls. J. Fluid Mech. **604**, 411–445 (2008)
23. A. Wagner, Passive hypersonic boundary layer transition control using ultrasonically absorptive carbon-carbon ceramic with random microstructure. Ph.D. thesis, Katholieke Universiteit, Leuven, 2014
24. A. Wagner, K. Hannemann, M. Kuhn, Experimental investigation of hypersonic boundary-layer stabilization on a cone by means of ultrasonically absorptive carbon-carbon material. AIAA Paper 2012–5865 (2012)
25. V. Wartemann, H. Lüdeke, N.D. Sandham, Numerical investigation of hypersonic boundary-layer stabilization by porous surfaces. AIAA J. **50**, 1281 (2012)
26. M Weder, Linear stability and acoustics of a subsonic plane jet flow. M.Sc. thesis, 2012

Chapter 25
Dissipation and Topological Features Conditioned by Velocity Level-Crossings in Wall Turbulence

Sedat Tardu

25.1 Introduction

Palm statistics are the statistics of a given quantity under the condition that another stochastic process crosses a fixed level. A relatively easy flow quantity to understand connected to the level-crossing statistics in wall turbulence is the local instantaneous production $P = -uv\frac{\partial \overline{U}}{\partial y}$. Here, u and v are, respectively, the fluctuating streamwise and wall normal velocity and $\frac{\partial \overline{U}}{\partial y}$ is the mean shear. The conditional level-crossing statistics are determined when, for example, u crosses a fixed amplitude $\ell_u \sigma_u$, where $\sigma_u = \sqrt{\overline{uu}}$ is the root mean square of streamwise velocity fluctuations. Figure 25.1 adapted from Tardu and Bauer [14] shows the distribution of the production $\overline{P}_{\ell_u}$ conditioned by the u crossings and scaled by the local mean $\overline{P}$ versus the threshold ℓ_u at the edge of the viscous sublayer, at a distance scaled by the viscosity and the shear (friction) velocity of $y^+ = 5$. The profiles are compared with the simplest reference model assuming joint normality between u and v in which case $\frac{\overline{P}_{\ell_{u_i}G}}{\overline{P}} = \ell_{u_i}^2$ although it is known that u and v deviate strongly from normality next to the wall. Note first that the $\frac{\overline{P}_{\ell_u}}{\overline{P}}$ profiles are significantly asymmetric with respect to the threshold ℓ_u and are obviously far being Gaussian. Strong contributions of the ejections or of the quadrant Q_{II} events ($v > 0$, $u < 0$) make $\frac{\overline{P}_{\ell_u}}{\overline{P}}$ over Gaussian at the $\ell_u < 0$ side, pointing at the preponderance of the interaction of intense u<0 with $v > 0$. The striking feature of the results presented in Fig. 25.1 is that, at the $\ell_u > 0$ crossings, the mean production is curiously *negative* within the range $0 < \ell_u < 1.5$ and stays weak at larger thresholds compared to the $\ell_u < 0$ side. The only categories

S. Tardu (✉)
Laboratoire des Ecoulements Géophysiques et Industriels (LEGI), Université Grenoble Alpes, B.P. 53 X, 38041 Grenoble, Cédex 9, France
e-mail: Sedat.Tardu@legi.grenoble-inp.fr

A. Pollard et al. (eds.), *Whither Turbulence and Big Data in the 21st Century*,
DOI 10.1007/978-3-319-41217-7_25

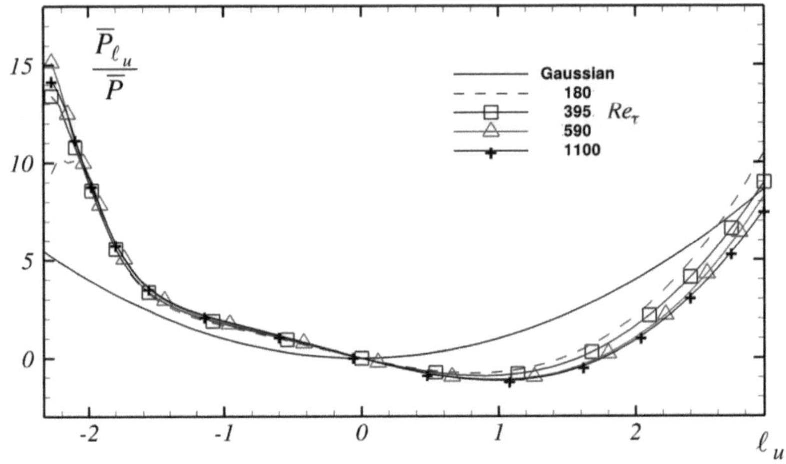

Fig. 25.1 Production conditioned by up and down crossings of the streamwise velocity u at $\ell_u\sigma_u$ levels, at the edge of the viscous sublayer at the distance$y^+ = 5$ from the wall and at four different Reynolds numbers

of events that can induce *noise contamination* and render $\overline{P}_{\ell_u}$ negative when $u > 0$ are the Q_I interactions ($u > 0, v > 0$). The Q_I events induce incoherence at the edge of the viscous sublayer and render blurred the sweeps marked by the u-crossings. Enhancing this incoherence and weakening the effect of $\ell_u < 0$ events can reduce significantly the production process in its early stage and lead to appreciable drag reduction. Note that such information can hardly be obtained from the classical conditional statistics such as the quadrant analysis.

It has been argued for a while that the zero-crossings of fluctuating velocity signals should largely contribute to the dissipation in wall bounded flows [9]. These arguments are based on hot wire measurements at the outer edge of the inner layer and assume local isotropy. We will revisit this topic by summarizing some salient features of the dissipation level-crossing statistics next to the wall. The second point that will be analyzed here is the topology of the wall bounded flows. Despite many studies on this subject, a clear image of the dominant topology in different sublayers of the near wall turbulent field is still lacking. The statistical means of the velocity gradient tensor invariants conditioned by the velocity level-crossings are analyzed in an attempt to clarify this point.

Direct numerical simulations (DNS) of four fully developed turbulent channel flows performed at Karman numbers $Re_\tau = \frac{h\bar{u}_\tau}{\nu} = h^+$=180, 395, 590, and 1100 in particularly large computational domains similar to Hoyas and Jimenez [6] will be used. Here, h is the channel half width, $\bar{u}_\tau$ is the shear velocity, ν is the viscosity, and $()^+$ stands for quantities scaled by the inner variables $\bar{u}_\tau$ and ν. The details on the numerical code can be found in Bauer et al. [2] and Tardu and Bauer [14].

25.2 Level-Crossing Statistics

Palm averages are conditional statistics. Consider two stochastic stationary signals: $u(t)$ and $q(t)$. The aim is to determine the statistics of $q(t_i)$ conditioned by the events $u\left(t_i\right) = \ell_u\sigma_u$. One is at first tempted to determine the conditional mean $\overline{q}_{\ell ub}$ by computing simply

$$\overline{q}_{\ell_u b} = \frac{\sum_{i=1}^{N_{\ell u}} q\left(t_i\right) \left|u = \ell_u\sigma_u\right.}{N_{\ell u}} = E\left\{q \left|u = \ell_u\sigma_u\right.\right\} \tag{25.1}$$

i.e., by collecting all the $q(t_i)$ when the signal u crosses the level $\ell_u\sigma_u$, and performing the statistics over this specific data set. We recall that$E$ stands for the expected value and | refers to the conditional event. This procedure is correct if and only if the number of samples N_{ℓ_u} is statistically independent of the quantity q. However,N_{ℓ_u} in an interval $[0, T]$ depends on the absolute derivative of $|u'| = |du/dt|$ and is given by:

$$N_{\ell_u} = Tf_{\ell_u} = Tp\left(u = \ell_u\sigma_u\right) E\left\{\left|u'\right| \left|u = \ell_u\sigma_u\right.\right\} \tag{25.2}$$

(see [14] for the details). Consequently, if $q(t)$ and $|u'|$ are correlated, then N_{ℓ_u} depends on the quantity that is of interest and Eq. (25.1) leads to biased moments. Thus, the average $\overline{q}_{\ell_u}$ at level-crossing points is the normalized mean of q *weighted by* the absolute velocity derivative, i.e.,

$$\overline{q}_{\ell_u} = \frac{E\left\{\left|u'\right| q \left|u = \ell_u\sigma_u\right.\right\}}{E\left\{\left|u'\right| \left|u = \ell_u\sigma_u\right.\right\}} = \frac{\sum_{i=1}^{N_{\ell u}} \left|u'\right|\left(t_i\right) q\left(t_i\right) \left|u = \ell_u\sigma_u\right.}{\sum_{i=1}^{N_{\ell u}} \left|u'\right|\left(t_i\right) \left|u = \ell_u\sigma_u\right.} \tag{25.3}$$

One has

$$\overline{q}_{\ell_u} = \overline{q}_{\ell_u b} = E\left\{q \left|u = \ell_u\sigma_u\right.\right\} \tag{25.4}$$

if and only if $q$and u' are statistically independent, i.e.:

$$E\left\{q \left|u'\right| \;\middle|\; u\right\} = E\left\{q \;\middle|\; u\right\} E\left\{\left|u'\right| \;\middle|\; u\right\} \tag{25.5}$$

That is obviously not the general case. To our knowledge, this point has not been clearly stated in previous investigations dealing with level-crossing problems in turbulence. The average quantity given by (25.1) will be called the biased mean hereafter. Interesting readers are referred to Tardu and Bauer [14] for a short review on the Palm statistics and to the references they give.

The contribution at level-crossings to the expected mean $\overline{q} = E(q)$ is conventionally defined

$$c_q = \frac{\sum_{i=1}^{N_{\ell u}} q_i \,|u = \ell_u \sigma_u}{\sum_{j=1}^{N} q_j} \tag{25.6}$$

which is the ratio of the conditional sum at level-crossings, to the total unconditional $\sum q$.

Hereafter we will use mixed notations for convenience. The ith velocity component will be denoted by u_i, standing for the streamwise fluctuating velocity u for $i = 1$, the wall normal component v ($i = 2$), and the spanwise velocity w for $i = 3$. In the same way, the short hand notation for the spatial coordinates is x_i, where $i = 1, 2, 3$ correspond to the streamwise x, the wall normal y, and the spanwise z directions, respectively. The turbulent intensity of u_i is denoted by $\sigma_{u_i} = \sqrt{\overline{u_i u_i}}$.

One of the aims of the present paper is to investigate the level-crossing characteristics in homogeneous planes $y =$ constant, at a fixed time t, in the streamwise x and spanwise z directions through DNS in a turbulent channel flow. There are several ways to sample the data in this case. For example, one can sample the *ith* velocity component $u_i(x, y, z; t)$ in the homogeneous x direction along constant z lines. In this case u_i is still a one-dimensional signal, and Eq. (25.3) applies directly by replacing the derivative u' by the partial derivative of u_i along x, i.e., $u' = \partial u_i / \partial x$. One can also fix the sampling direction along the spanwise direction z, and obtain the statistics of a quantity q at $u_i(x, y, z; t)$ level-crossings by taking $u' = \partial u_i / \partial z$ in this case.

These concepts have to be completed by contour crossings in the $x-z$ plane. The expected mean of a given quantity q at level-crossing contours is a straightforward generalization of Eq. (25.3) and reads for:

$$\overline{(q)}_{\ell_{u_i}} = \frac{E\left\{q \times \left|\vec{\nabla}\, u_i\right| \,|u_i = \ell_{u_i} \sigma_{u_i}\right\}}{E\left\{\left|\vec{\nabla} u_i\right| \,|u_i = \ell_{u_i} \sigma_{u_i}\right\}} \tag{25.7}$$

where the norm of the local gradient $\left|\vec{\nabla}\, u_i\right|$ is given by

$$\left|\vec{\nabla}\, u_i\right| = \sqrt{\left(\frac{\partial u_i}{\partial x}\right)^2 + \left(\frac{\partial u_i}{\partial z}\right)^2} \tag{25.8}$$

Unless clearly indicated, all the statistics presented here are based on the contour crossings.

25.3 Results

25.3.1 Dissipation

To begin with, Fig. 25.2 shows the profiles of the dissipation $\overline{\varepsilon}^{+} = \nu \overline{\partial u_i/\partial x_j (\partial u_i/\partial x_j + \partial u_j/\partial x_i)}$ scaled by the inner variables as a function of the distance y^+ to the wall at the four Reynolds numbers investigated here. The dissipation at $y^+ < 20$ is strongly Re_τ dependent in particular in the viscous sublayer, essentially because of the indirect effect of the large-scale outer eddies [6]. Our data confirms the results of Antonia et al. [1] at larger Reynolds numbers that one form of the axisymmetric approximations provided by George and Hussein [4] is suitable for ε in the whole layer, and that the dissipation approaches the local isotropy with $\overline{\varepsilon}_{iso} = 15\nu\overline{(\partial u/\partial x)^2}$ only at $y^+ \geq 100$.

The dissipation is a quantity that provides a good example of why care has to be taken to determine the conditional Palm statistics. Consider homogeneous isotropic turbulence and suppose to simplify that u and $\partial u/\partial x$ are normal (but in reality $\partial u/\partial x$ is not Gaussian). According to Eq. (25.3) the unbiased correct mean of the dissipation at $u = \ell_u \sigma_u$ crossings is $(\overline{\varepsilon}_{iso})_{\ell_u} = 15\nu E\left(|\partial u/\partial x|^3\right)/E\left(|\partial u/\partial x|\right)$, which leads to $(\overline{\varepsilon}_{iso})_{\ell_u}/\overline{\varepsilon}_{iso} = 2$ and is independent of ℓ_u. The biased mean (25.4), however, is $(\overline{\varepsilon}_{iso})_{b\ell_u}/\overline{\varepsilon}_{iso} = 1$ and underestimates the correct conditional average by a factor two.

It has been argued for a while that the zero-crossings of the streamwise fluctuating velocity should largely contribute to the dissipation in, at least, isotropic turbulence [9] and in wall bounded flows as well. The zero-crossing frequency is also taken into account in some models to connect the integral scale to the

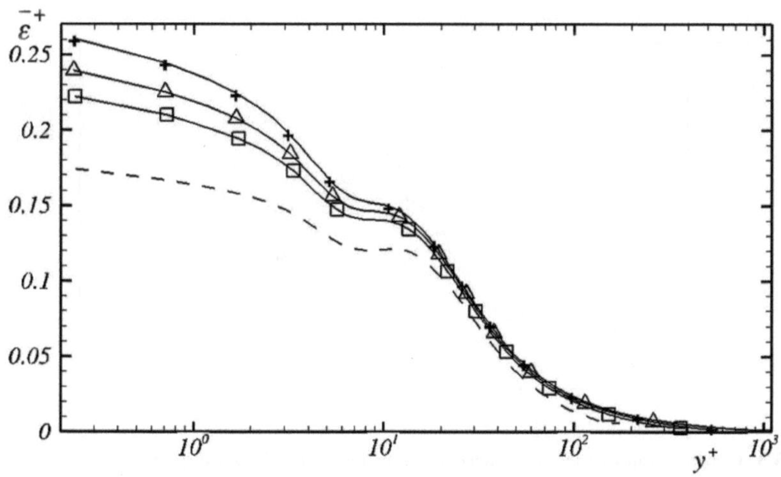

Fig. 25.2 Mean dissipation versus the wall normal distance in wall units. See Fig. 25.1 for legend

Taylor scale [10]. That the zero-crossing frequency of u is related to $\overline{(\partial u/\partial x)^2}$, does obviously not imply that the dissipation is locally important during the zero-crossings, and this point has not been entirely elucidated, even though it was argumented that the average distance between consecutive zero-crossings, which is related to the Taylor scale $\lambda_{ux}^{+} = \sqrt{\sigma_u^{+2}/\overline{(\partial u^{+}/\partial x^{+})^2}}$, provides a length scale for the dissipation. The wall flow, on the other hand, can be considered locally isotropic only far away from the wall, below the inner layer $y^{+} \geq 100$. Kailasnath and Sreenivasan [9] claimed out that the average dissipation rate is maximum when $u = 0$ near the edge of the inner layer at $y^{+} = 110$. The clear formulation of the dissipation rate conditioned by the u-level-crossings is not given by these authors. The statistics they report are biased in terms of the Palm characteristics provided in Sect. 25.2. It turns out that these arguments are not valid in the viscous sublayer where the dissipation $\overline{\varepsilon} = \nu\, \overline{\partial u_i/\partial x_j\ (\partial u_i/\partial x_j + \partial u_j/\partial x_i)}$ reaches its maximum and deviates strongly from the local isotropy and wherein $\overline{\varepsilon}^{+} \propto \overline{\frac{\partial u}{\partial y}^{+2}}$ [1]. Figure 25.3 shows the contribution c_{ε_u} of the fluctuating streamwise velocity level-crossings at $y^{+} = 1$ at four different Reynolds numbers. It is first seen that the contribution hardly exceeds 10 % and secondly that two peaks clearly emerge from Fig. 25.3 at, respectively, $\ell_u = -1$ and $\ell_u = 2$ which are far from zero.

Curiously, the velocity component that contributes much to the dissipation next to the wall is not the streamwise but the wall normal one. It is seen in Fig. 25.4 that the contribution c_{ε_v} of v crossings is about 30 % near $\ell_v = 0$. The contribution of the spanwise velocity c_{ε_w} is similar but twice smaller (not shown). In addition, the contribution to the dissipation of the level-crossings of v and w in the streamwise direction is entirely irrelevant compared to z. The spanwise direction z is indeed much more active than x. Tardu and Bauer [14] have shown that the level-crossing frequency in space follows reasonably well a Gaussian distribution in the whole

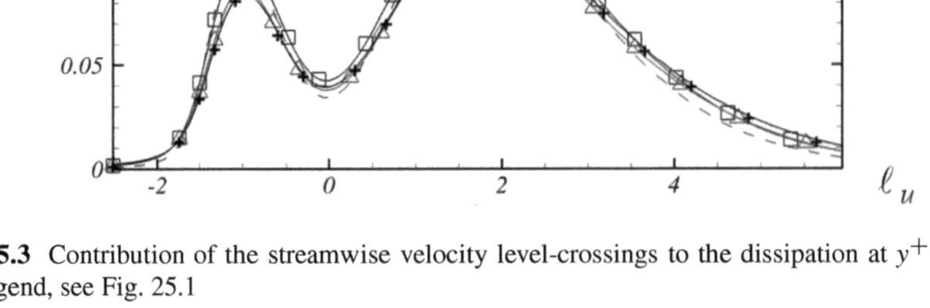

Fig. 25.3 Contribution of the streamwise velocity level-crossings to the dissipation at $y^{+} = 1$. For legend, see Fig. 25.1

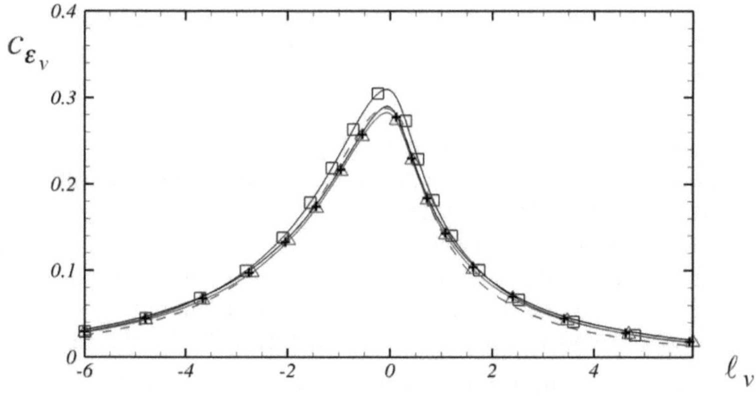

Fig. 25.4 Contribution of the wall normal velocity level-crossings to the dissipation at $y^+ = 1$. For legend, see Fig. 25.1

layer. The frequency of the level-crossings $u_i = \ell_{u_i}\sigma_{u_i}$of a Gaussian signal can be written in wall units as a function of the corresponding Taylor scale as

$$f^+_{\ell_{uiG}} = \frac{1}{\pi\lambda^+_{u_i}} e^{-\frac{\ell^2_{u_i}}{2}} = \frac{\sqrt{\overline{(\partial u_i/\partial x_j)^2}}}{\pi\sigma_{u_i}} e^{-\frac{\ell^2_{u_i}}{2}} \tag{25.9}$$

where $\lambda^+_{u_i}$ is the Taylor scale, as, for example, of u along the streamwise direction $\lambda^+_{ux} = \sigma^+_u/\sqrt{\overline{(\partial u^+/\partial x^+)^2}}$. That is, the level-crossing (activity) frequency is inversely proportional to $\lambda^+_{u_i}$. Careful analysis of the ensemble of $\lambda^+_{u_i}$ distributions conducted by Tardu and Bauer [14] revealed that the smallest Taylor scales in the viscous sublayer are, respectively, $\lambda^+_{vz} = \sigma^+_v/\sqrt{\overline{(\partial v^+/\partial z^+)^2}}$ and λ^+_{wz}. Both scales are remarkably independent of the Reynolds number in the viscous sublayer and they are approximately 3η near the wall (η is the Kolmogorov scale). Thus, an appropriate intermediate length scale ℓ_ε of the dissipation next to the wall is plausibly a function of $\ell^+_\varepsilon\left(\lambda^+_{vz}, \lambda^+_{wz}\right)$. The process in the viscous sublayer is not really important, since after all, the dissipation in this region is mainly governed by the diffusion of kinetic energy, which is directly related to the enstrophy at the wall. The important mechanism takes place in an intermediate region that extends from the edge of the viscous sublayer up to $y^+ = 100$ where local isotropy is approximately reached. The interesting result we obtained in this region is the remarkable independence of the mean isotropy conditioned by the *spanwise* velocity level-crossings with respect to the wall distance. More clearly, the conditional mean scaled by the local isotropy $\overline{\varepsilon}_{\ell_w}/\overline{\varepsilon}$ is a weak function of y^+ and $\overline{\varepsilon}_{\ell_w}/\overline{\varepsilon} = f\left(\ell_w, Re_\tau\right)$ within $5 \leq y^+ \leq 100$.

Detailed results on these characteristics will be published elsewhere. Let us indicate, however, that above the edge of the inner layer $y^+ > 100$ where the isotropic dissipation ε_{iso} constitutes a good approximation to ε, the conditional distributions $\overline{\varepsilon}_{\ell_{u_i}}/\overline{\varepsilon}$ coincide reasonably well for fixed amplitudes of the three components of the velocity. Another interesting result is obtained in the equilibrium region wherein the mean production is equal to the mean dissipation. The conditional statistics shows that there is no equilibrium at fixed amplitudes $\ell_{u_i}\sigma_{u_i}$, i.e., $\overline{\varepsilon}_{\ell_{u_i}} \neq \overline{P}_{\ell_{u_i}}$. This is expected since obviously the production is zero at the stagnation $\ell_{u_i} = 0$ points while $\overline{\varepsilon}_{\ell_{u_i}=0} \neq 0$. The contribution to the dissipation is largest roughly at $\ell_{u_i} = 0$ when$y^+ > 100$, but it hardly exceeds 20 %, which means that the role played by the scales smaller than the average zero-crossing intervals is far from being negligible. Since, the conditional production distributions agree well with a simple joint-normal model according to Tardu and Bauer [14] in this region, one has $\frac{\overline{P}_{\ell_{u_i}}}{\overline{P}} = \ell_{u_i}^2$. Besides, the mean conditional dissipation behaves like $\frac{\overline{\varepsilon}_{\ell_{u_i}}}{\overline{\varepsilon}} = a + b\ell_{u_i}^2$, with $a \approx (\overline{\varepsilon}_{iso})_{\ell_u}/\overline{\varepsilon}_{iso} = 2$ at the zero-crossings.

25.3.2 Topology

The local topology of wall turbulence is so far analyzed in relation to an observer moving with the local velocity [11]. All the flow points are then critical by definition and the local topology is related to the invariants of the velocity gradient tensor, namely $Q = \frac{1}{2}\left(\Omega_{ij}\Omega_{ij} - S_{ij}S_{ij}\right)$ and $R = -\frac{1}{3}\left(3\Omega_{ij}\Omega_{jk}S_{ki} + S_{ij}S_{jk}S_{ki}\right)$. Here $\Omega_{ij} = \left(A_{ij} - A_{ji}\right)/2$ is the rate of rotation tensor, $S_{ij} = \left(A_{ij} + A_{ji}\right)/2$ is the symmetric rate of strain tensor, with $A_{ij} = \partial U_i/\partial x_j$ and $U_i = \overline{U}_i + u_i$. The first term $\Omega_{ij}\Omega_{ji}$ entering in the expression of the invariant Q is proportional to the enstrophy $\omega_i\omega_i$ and $S_{ij}S_{ij}$ is related to the dissipation rate of the kinetic energy. The topological features are function of positions in the $Q-R$ plane. The nature of the critical points depends on the discriminant $\Delta = (27/4)R^2 + Q^3$ (Fig. 25.5). Most of the near wall coherent structures detection schemes are based on the $Q > 0$ criteria or its extensions ([3, 15, 16]; see [13] for a review), except the λ_2 technique of Jeong and Hussain [8] that nonetheless gives similar results at least qualitatively and coincides reasonably well, even with Lagrangian techniques [5]. There are two main points to be stressed here. First, in finite dimensional chaotic systems the critical points are well defined, and although turbulence can be interpreted as a dynamical system of infinite dimensions, it is unclear why all the points have to be defined as critical. The hidden hypothesis in relation to coherent structures detection strategies is to suppose that the compact structures have a constant or nearly constant advection velocity. Second, the topology associated with fixed amplitudes of the Eulerian velocity field in different sublayers of wall turbulence has not been yet addressed.

One of the simplest ways to analyze the local flow pattern that would be seen by an observer travelling with the fluid particles is to determine the mean invariants conditioned by given amplitudes of the velocity components u_i. The idea behind this methodology is that if one observes a common behavior independently of the fixed amplitudes of all three components of the velocity field, one can then decide

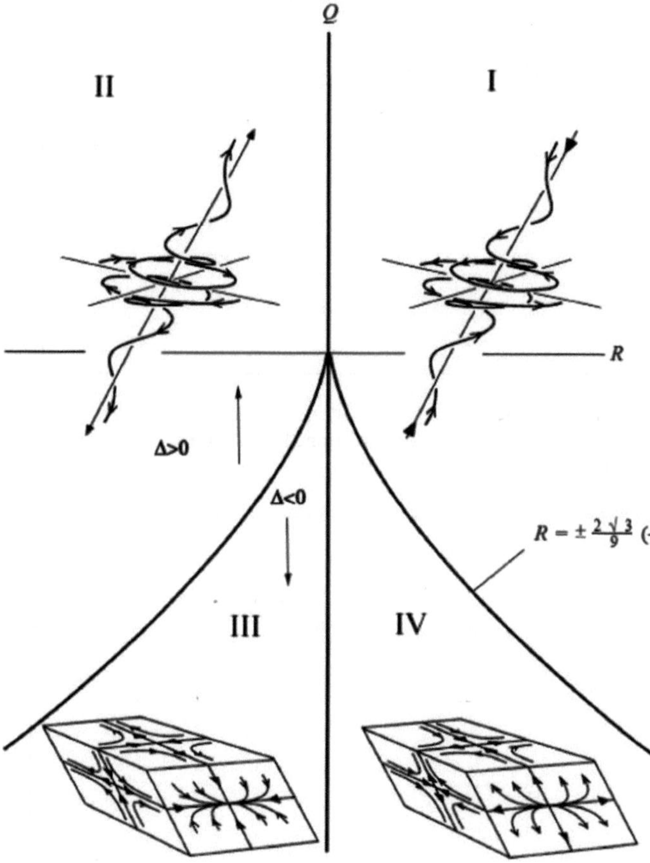

Fig. 25.5 Local topology in the $Q - R$ plane according to Chong et al. (1990). I: Unstable focus compressing; II: Stable focus stretching; III: Stable node/saddle/saddle; IV: Unstable node/saddle/saddle

for a dominant topology from a statistical point of view. All the features shown in Fig. 25.5 coexist of course randomly in time and space but the statistical means of the invariants conditioned by $u_i = \ell_{u_i}\sigma_{u_i}$ provide the prevailing topological characteristics.

Figure 25.6 shows the mean discriminant $\overline{\Delta}_\ell$ scaled in inner variables in the high buffer layer at $u_i = \ell\sigma_{u_i}$ level-crossings in the high buffer layer $10 \leq y^+ \leq 30$ when the invariant $R > 0$ (see Fig. 25.5). It is seen that $\overline{\Delta}_\ell > 0$ for all the amplitudes $u_i = \ell\sigma_{u_i}$. Similar results are obtained at $R < 0$, but $\overline{\Delta}_\ell$ are then nearly an order of magnitude larger. Thus, the statistically dominant topology in the high buffer layer is either stable/focus/stretching, or unstable/focus/compressing, but the stretching prevails in agreement with Schoppa and Hussein [12]. Careful examination of the data revealed that the either stable or unstable focuses are predominant everywhere,

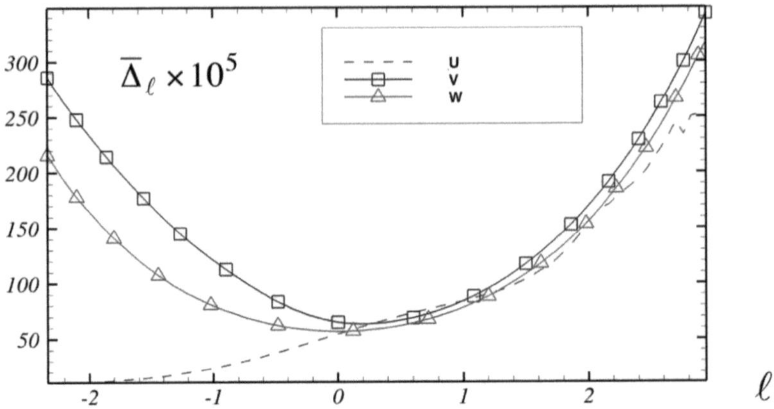

Fig. 25.6 Discriminant $\overline{\Delta}_\ell$ in wall variables conditioned by $u_i = \ell\sigma_{u_i}$ crossings at $R > 0$, $Re_\tau = 1100$ within the high buffer layer $10 \leq y^+ \leq 30$

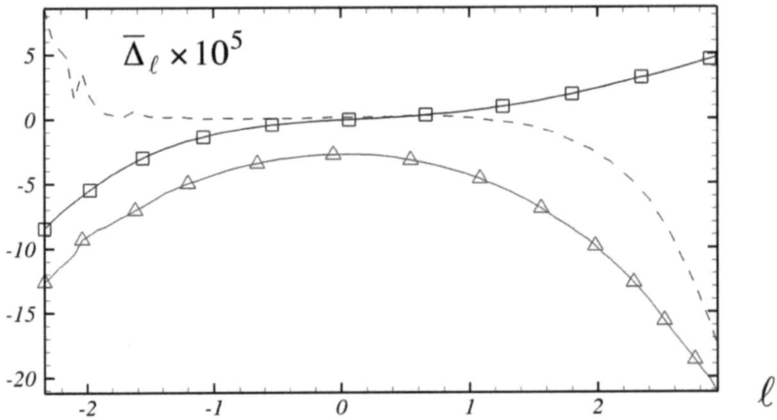

Fig. 25.7 Discriminant $\overline{\Delta}_\ell$ in wall variables conditioned by $u_i = \ell\sigma_{u_i}$ crossings at $R > 0$, $Re_\tau = 1100$ within the viscous sublayer $0 \leq y^+ \leq 5$. For legend, see Fig. 25.6

except in the viscous sublayer. Figure 25.7 shows the distribution of $\overline{\Delta}_\ell$ at the $R > 0$ side within the viscous sublayer $0 \leq y^+ \leq 5$. It is seen that $\overline{\Delta}_\ell \leq 0$ in a wide range of the threshold ℓ, pointing at the dominance of unstable node-saddle-saddle topology. Expressed in a different way, the local dissipation of kinetic energy (straining) is much larger than the local enstrophy (Fig. 25.5). This behavior is expected since the dissipation reaches its maximum in the viscous sublayer, mainly because the local dissipation of kinetic energy (straining) is much larger than the local enstrophy.

The conditional mean of the invariant $\overline{Q}_\ell$ expressed in inner variables is negative at the level-crossings of both u, v, and w, in the entire layer at the unstable $R > 0$ half-plane and $\overline{Q}_\ell > 0$ at the opposite stable $R < 0$ side (not shown). Thus the unstable focus topology is connected to $\Omega_{ij}\Omega_{ji} < S_{ij}S_{ij}$, i.e., the enstrophy locally dominates the dissipation, and vice versa. In other words, the fluctuating velocity components are statistically associated with local straining and enstrophy when the local topology is, respectively, unstable and stable. Detailed results will be published elsewhere.

25.4 Conclusion

The conditional mean dissipation at the zero-crossings of the three components of the velocity field is roughly twice the local $\overline{\varepsilon}$ above the edge of the inner layer $y^+ = 100$, where the local isotropy assumption $\overline{\varepsilon} = \overline{\varepsilon}_{iso}$ is approximately valid. The Gaussianity is never reached in this region, in the sense that the level-crossing statistics still depend on the thresholds ℓ_{u_i}, and this is essentially due to the non-normality of the velocity derivatives. The spanwise level-crossings of the wall normal velocity dominate the dissipation in terms of contribution in the viscous sublayer. The smallest Taylor scales at $y^+ \leq 5$ are those of v and w, namely λ^+_{vz} and λ^+_{wz} in the spanwise direction along which the level-crossing activity is largest. An interesting result concerns the mean dissipation conditioned by the level-crossings of the spanwise velocity which varies like $\frac{\overline{\varepsilon}_{\ell_w}}{\overline{\varepsilon}} = a + b\ell_w^2$ is remarkably independent of the y position from $y^+ = 100$ to the edge of the viscous sublayer. These results are currently under analysis to establish a persistent model for the dissipation in the inner layer. The Hunt's model is also considered for this purpose [7].

The analysis of the conditional velocity gradient tensor invariants statistics allowed determining the global topological features of the turbulent wall layer. The dominant topology above the viscous sublayer is the focus topology wherein the stable and unstable focuses are, respectively, associated with the enstrophy and straining. In the viscous sublayer, the unstable nodes straining marking the strong dissipation of the kinetic energy in this layer cohabit with the stable focuses. Only the salient features of these characteristics could be summarized here, and the details will be published elsewhere.

The DNS performed in large computational domains at four Reynolds numbers in the low to mild range $180 \leq Re_\tau \leq 1100$ generated data of about 5 TB. Even the post-processing requires massively parallelized codes in this case. Big data complexity due to their volume, variety, and variability is inherent in sharing the data within the turbulence community. The rapid data transfer is a problem in itself and building smaller samples and typical subroutines to facilitate the analysis are necessary. For instance, the data providers may build up generic codes to help to analyze simple flow statistics directly on line, but more complicated problems require closer collaborations. Effort can also be made to put the data in some commonly defined standard formats. Managing the data requests means both in terms of time and computational resources and that is a problem right at the start.

Acknowledgments This work was granted access to the HPC resources of IDRIS under the allocation 100226 of Grand Equipement National de Calcul Intensif (GENCI). Technical aid of Dr. F. Bauer is greatly appreciated.

References

1. R.A. Antonia, J. Kim, L.W. Browne, Some characteristics of small-scale turbulence in a turbulent duct flow. J. Fluid Mech. **233**, 369–388 (1991)
2. F. Bauer, S. Tardu, O. Doche, Efficiency of high order DRP schemes in direct numerical simulations of incompressible turbulent flows. Comput. Fluids **107**, 123–140 (2015)
3. P. Chakraborty, S. Balachandar, R.J. Adrian, On the relationships between local vortex identification schemes. J. Fluid Mech. **535**, 189–214 (2005)
4. W.K. George, J.H. Hussein, Locally axisymmetric turbulence. J. Fluid Mech. **233**, 1–23 (1991)
5. M.A. Green, C.W. Rowley, G. Haller, Detection of Lagrangian coherent structures in three-dimensional turbulence. J. Fluid Mech. **572**, 111–120 (2007)
6. S. Hoyas, J. Jimenez, Reynolds number effects on the Reynolds stress budgets in turbulent channels. Phys. Fluids **20**, 101511 (2008)
7. J.C.R. Hunt, J.F. Morrison, Eddy structure in turbulent boundary layers. Eur. J. Mech. B. Fluids **19**, 673–694 (2000)
8. J. Jeong, F. Hussain, On the identification of a vortex. J. Fluid Mech. **285**, 69–94 (1995)
9. P. Kailasnath, K.R. Sreenivasan, Zero crossings of velocity fluctuations in turbulent boundary layers. Phys. Fluids **5**(11), 2879–2885 (1993)
10. N. Mazelier, J.C. Vassilicos, The turbulence dissipation constant is not universal because of its universal dependence on large-scale flow topology. Phys. Fluids **20**, 015101 (2008)
11. A.E. Perry, M.S. Chong, A description of eddying motions and flow patterns using critical-point concepts. Annu. Rev. Fluid Mech. **19**, 125–155 (1987)
12. W. Schoppa, F. Hussein, Coherent structure generation in near-wall turbulence. J. Fluid Mech. **453**, 57–108 (2002)
13. S. Tardu, *Transport and Coherent Structures in Wall Turbulence* (Wiley, ISTE, 2014), 459 p. ISBN-10: 1848213956
14. S. Tardu, F. Bauer, Level crossing statistics and production in low Reynolds number wall turbulence. J. Turbul. **16**(9), 847–871 (2015)
15. J. Zhong, T. Huang, R.J. Adrian, Extracting 3D vortices in turbulent fluid flow. IEEE Trans. Pattern Anal. Mach. Intell. **20**, 193–199 (1996)
16. J. Zhou, R.J. Adrian, S. Balachandar, Autogeneration of near wall vortical structures in channel flow. Phys. Fluids **8**, 288–290 (1996)

Part VII
Four Perspectives on Big Data and the Turbulence Community

Chapter 26
Cyberinfrastructure to Empower Scientific Research

Thomas Hacker and Shirley Dyke

26.1 Introduction

Scientific research in the past century was based on theory, experiment, and computational simulation. This three front approach was built upon the dual approach of theory and experiment that was developed from the eighteenth and nineteenth centuries. During the twentieth century, computational simulation provided the ability to simulate physical systems at increasingly larger temporal scales and smaller spatial scales. This capability allowed researchers to investigate a wide range of simulated conditions to understand the behavior of physical systems. Over the past decade, the ability to collect and process vast amounts of numerical and visual data from sensors and experiments has led to the emergence of "big data" as a new fourth paradigm [1] powering science. Today, the power of the Internet coupled with the dynamic open source movement and commercial software has created a global ecosystem of computing and communication capabilities that were only dreamed of in the twentieth century. These new capabilities are constantly evolving over time, and consequently require continual effort by the community of providers and users to learn to use these emerging technologies.

Cyberinfrastructure is playing an increasingly important role in empowering collaborative activities as well as inspiring new types of research within widely distributed communities of researchers. Two examples of this are the Large Hadron Project Compact Muon Solenoid (CMS) project [2] and the recently completed

T. Hacker (✉)
Department of Computer & Information Technology, Purdue University, West Lafayette, IN, USA
e-mail: tjhacker@purdue.edu

S. Dyke
School of Mechanical Engineering, Lyles School of Civil Engineering, Purdue University, West Lafayette, IN, USA
e-mail: sdyke@purdue.edu

A. Pollard et al. (eds.), *Whither Turbulence and Big Data in the 21st Century*,
DOI 10.1007/978-3-319-41217-7_26

George E. Brown Network for Earthquake Engineering Simulation (NEES) project [3]. Cyberinfrastructure is an essential element of both of these projects, and helps to share the essential nature of the research of these projects.

Considering the experience of the scientific computing and cyberinfrastructure community in successfully developing, deploying, and using cyberinfrastructure for projects such as NEES and CMS, there are challenges faced by the community of cyberinfrastructure researchers and developers, and by scientists who rely on cyberinfrastructure, specifically:

(a) how can we create and sustain a stable technology framework that can provide the scaffolding for integrating and providing relevant and reliable services for researchers?

And with this framework in place:

(b) how can we continue to add new services, upgrade existing ones, and retire outdated services in an efficient and effective manner?; and
(c) how can we coordinate and leverage efforts to design and develop cyberinfrastructure within different domain science communities?

The next section of this paper will discuss these challenges in detail.

26.2 A Framework for Cyberinfrastructure

A framework, loosely defined, is a persistent and resilient structure that provides structure, precedence, and order for software elements as well as the relationships between those elements. For example, a computer desktop is an example of a framework that facilitates interaction between a user and software, data, and services provided by the operating system and applications installed on the computer. Although the software, data, and operating system change over time, the desktop framework paradigm remains fairly stable. Certainly, the framework evolves over time, but does not undergo radical shifts in usability or functionality that would pose a significant impediment to users. Similar to this, a stable cyberinfrastructure framework is needed that can be easily learned and remains fairly stable over time to facilitate the discovery and use of cyberinfrastructure services through this framework.

There are several characteristics needed for this framework: usability, stability, configurability, expandability, expressive power, security, and mobility.

In terms of usability, such a framework should provide simple functions governed by organizing principles and rules that can be easily learned and applied throughout the entire system. Blair-Early [4] describes this as the need for "consistency." If a cyberinfrastructure system is comprised of a conglomeration of elements without a consistent set of usability guidelines, users would need to learn a different (and possibly constantly changing) set of rules for every component and element. This poor consistency would inhibit use and put a considerable burden on users. Another crucial factor is stability, which is the reliability of a component in terms of availability as well as functionality. If a component is frequently unavailable due to software or hardware failures or misconfiguration, then users will not be able to rely on that element for their work. More importantly, if the component provides inconsistent, incomplete, or incorrect results, users will not use the service provided by the component.

Configurability, expandability, and the expressive power fostered (or potentially hampered) by the framework are essential for exploiting the full potential of the software and services. Configurability is the ability for users to customize the environment or features of the services to meet their individual need. Expressive power is a property of a system that allows users to articulate a wide range and variety of combinations of tasks, data flows, and analysis using simple expressions or constructs. The combination of configurability, expandability, and expressive power are essential elements of a framework. If users cannot easily and succinctly express an adequately broad range of abstract operations built upon combinations of simpler operations, it would be very difficult to effectively use the system. Another characteristic, security, is a mandatory element today for any cyberinfrastructure. Building security capabilities into the fundamental framework is necessary to ensure data provenance, integrity, and privacy. Finally, mobility, loosely defined, represents the ability of the framework to function as a widely distributed system that can integrate technologies such as cloud computing, distributed data, and computing resources, and facilitate access to these services for a global user community.

There are many examples of cyberinfrastructure systems over the past decade that provide a framework with many of these characteristics.

The National Data Service (NDS) [5] has created an initial implementation of a cyberinfrastructure framework (the NDS Labs) based on OpenStack https://www.openstack.org/ and Docker https://www.docker.com/ (both open source) that enables users to create and distribute containers containing applications developed by the community. One of the goals of the NDS is to provide a framework focused on data collection, publishing, and reuse. The envisioned and prototype framework of the NDS is focused on a general framework for infrastructure and services rather than a specific domain specific solution. In terms of stability and usability, NDS is still in a prototype stage, and relies on the services provided by Docker and OpenStack. In terms of configurability, expandability, and expressive power, through the use of Docker, the NDS has a significant potential to facilitate a wide range of applications from many science domains. What is less clear is the degree of container level security and mobility, since these needs will most likely will be based on the characteristics of the specific application that uses NDS.

Globus [6] (formerly known as Globus Online) is a data oriented service that is focused on data movement, and data publication and discovery. In terms of framework characteristics, Globus provides good usability, stability, mobility, and security. On the other hand, since it is focused primarily on data movement and sharing, it is not highly configurable (without directly modifying the source code, or using the Globus Toolkit) beyond the uses envisioned by the designers, but could be used as an integral data movement and publishing component in a larger system built up from other cyberinfrastructure components.

The Ocean Observatories Initiative (OOI) [7] operates a network of sensors in the ocean to collect sensor data on water conditions, with the goal of distributing these data through a cyberinfrastructure to researchers. The current cyberinfrastructure consists of a website through which users can select and download preliminary datasets containing measurements collected from sensors. The OOI today is still under development [8]. Since the CI today is focused on a dataset browser, it has low configurability, expandability, or expressive power from a user's perspective. Stability, security, and mobility are difficult to assess, since the only functionality available to users today from the OOI website is data download.

A related parallel effort to OOI in Canada is the Ocean Networks Canada (ONC) project [19]. ONC provides access to data collected from a variety of instruments located on and under the ocean on the Canadian Arctic, Pacific, and Atlantic coasts. The ONC cyberinfrastructure provides public access to sensor data, with additional capabilities provided to authorized users. In terms of usability, accessing the CI is straightforward, and data can be plotted interactively on the ONC website, or downloaded in a variety of formats. The CI seems to be stable. The configurability, expandability, and expressive power are difficult to assess using only public access. However, the CI (at http://www.oceannetworks.ca) notes the availability of a facility named *CodeRunner* that allows registered users to upload and run their own processing scripts on the ONC CI.

RedCloud [9] provides a Eucalyptus based Infrastructure-as-a-Service (IaaS) cloud computing environment which allows users to create their own small scale computing infrastructure to which they can attach virtual disks holding data. RedCloud is an example of a cyberinfrastructure that does not seek to provide a customized interface and experience tailored to a specific scientific discipline. It is instead a suite of general infrastructure elements that can be remotely used by researchers to build their own customized environment. In terms of usability, users must learn to work with the Eucalyptus cloud computing management interface, which can be complex. The freedom given to users within the cloud computing environment provides a significant level of configurability, expandability, and expressive power. Security and mobility, however, beyond the basic IaaS provided by the RedCloud infrastructure, depends on the implementation of these features by users developing their own applications on RedCloud.

The Compact Muon Solenoid (CMS) project cyberinfrastructure is unique in that was developed primarily by the physics research community as a part of the overall project to develop and operate the CMS detector at the Large Hadron Collider at CERN [10]. The cyberinfrastructure is centered on the quick collection, filtering, and transfer of data collected from the CMS combined with the analysis of these data with a global distributed network of high performance computing resources. Computing is distributed into three tiers, the top being at CERN, the second level at a country level, and the third (called Tier-2) is distributed globally. In terms of usability and stability, the cyberinfrastructure is comprised of an integration of custom developed and existing software (such as Hadoop). The software infrastructure takes some effort to learn. However, the configurability, expandability, and expressive power are significant, since many users will access CMS computing capabilities by developing or using applications on Linux, and by submitting these applications to a high performance computing system to access computational resources. In terms of security, CMS uses a distributed authentication certificate infrastructure that provides good authorization and authentication security. In the area of mobility, CMS computing was designed from conception to be built on a widely distributed computing infrastructure, and thus today computational resources are pooled across the CMS project to allow users to use resources beyond their campus.

A final example is the HUBzero framework used by the NEES project for the NEEShub [3]. The NEEShub is designed to serve the needs of civil engineering researchers involved in earthquake engineering research. The NEES cyberinfrastructure includes a web portal interface (the NEEShub), a data repository that contains experimental data, simulation results, and reports from earthquake engineering research (the NEES Project Warehouse), and a suite of software tools that run within the NEEShub environment and user desktop systems. In terms of stability, the NEES cyberinfrastructure over the past 6 years was highly reliable, and served a large global community of users. In the area of security, the NEES cyberinfrastructure allows users to create project spaces, and to control access to these spaces to members of their team. In terms of mobility, the NEES cyberinfrastructure operates as a central service and repository, thus it doesn't exploit the availability of distributed cloud resources distant from the central service. In the areas of configurability, expandability, and expressive power, the NEEShub provides facilities to allow users to create and their own applications and tools that can run on Linux or Windows, and then provide these tools to the user community through the NEEShub, which can been run on the NEEShub as Software-as-a-Service.

These projects provide instances of efforts that implement in part or in whole the framework needed for a cyberinfrastructure that is responsive to the needs of the domain science community it seeks to serve.

26.3 A Lifecycle for Cyberinfrastructure Services

The second major challenge faced by the cyberinfrastructure community is the provision of reliable and stable services while simultaneously continuing to develop, prototype, and roll out newer more innovative services. Certainly, as described in the previous section, a stable and reliable framework to deliver and develop these services. The aspects of this problem are: the motivation for creating new services, and retiring old services; determining membership of the group that decides which new services to add; and the need for a deliberate process to identify and deploy new services for the user community.

The first aspect of this challenge is to identify and understand the underlying motivations for creating new services, and to retire aged services. The most important motivation is to meet the current requirements of users (identified by members of the community, with the overall goal of improving their productivity).

To be useful and effective, new services must meet current needs by providing new capabilities that help users to become more productive or more efficient. The secondary motivation is to reduce operational costs, but not at the expense of providing a highly reliable suite of services.

Another aspect is the membership of the group that steers cyberinfrastructure development and deployment. This sociological question is a critical problem that is frequently overlooked. Computer scientists and technologists can often make the mistaken assumption that they fully understand the needs of the user community, and fail to meaningfully involve representatives of the user community in every essential aspect of the effort. The inclusion of domain experts in the group of computing specialists can provide a "reality check" at the conceptual and prototyping stage, as well as providing critical feedback during the development of the service for likely adoption and widespread use. These interactions help the group to steer efforts away from tasks that might be inherently interesting to the technologists, but have little if any actual impact on the final cyberinfrastructure service. This group must find a balance between following and leading the scientific user community to be served. Hence, the inclusion of respected user community leaders from the science domain is essential. On users' side, another problem arises when users seek to be passive consumers or customers of the infrastructure and its products rather than *active contributors* in shaping and influencing the provided services as well as providing domain specific content such as data, documents, software tools, and education materials.

Another important issue is the need to evolve and update services over time as new technologies become available and user and operational needs change. This concern can be managed by one of two paths: responding on demand when a crisis occurs, such as a change in software license or the immediate need to meet an unforeseen need; or alternatively, to create and follow a process to proactively seek out new technologies for assessment, and potentially, for adoption. There are many potential problems with the former, such as not having adequate time to perform a comprehensive assessment of potential solutions. Instead, a deliberate process is necessary to identify new services, assess them, put them into production, and ultimately retire them once they are no longer useful.

This technology assessment and decision process needs to address several fundamental questions. First, to identify the specific needs for which the services are intended, and also to identify the communities who will use these services. There is always a tension between meeting internal needs for project administration and infrastructure versus focusing on the external needs of the user community. Often, internal needs, support services, or new technologies that are interesting to cyberinfrastructure providers can inadvertently take a higher priority than the external needs of the user community. It is imperative that the external needs of the user community take priority over internal needs as much as it practicable to do so. The articulation of external needs should also be collected and prioritized by representative members of the user community, rather than by project personnel. This process provides a degree of separation between the project and user community to ensure that the cyberinfrastructure is truly driven by the needs of a given user community. Second, the budgetary reality of any project is that resources (personnel time, funds, and equipment) are limited, and consequently the problem is to maximize the delivery of highly used and impactful services to the user community within the constraint of available resources. The operational reality of this constraint is that among the many options and choices available for new services and technologies to meet user needs, a deliberate choice about what cannot be done must be taken along with decisions about what must be done. Finally, under constrained resources, the assessment process also needs to identify services that should be retired due to underutilization, excessive maintenance costs, or obsolescence.

The NEES network cyberinfrastructure provided an example of how this process was implemented. One of the primary principles guiding the CI development and deployment was that all external facing services were designed to meet the requirements of the civil engineering community. These requirements were collected and prioritized by a committee of civil engineering researchers and educators, and were updated on a frequent basis as older requirements were met, and new requirements were identified. The requirements were delivered to the cyberinfrastructure team, who formally responded through a product contract that described the new services and infrastructure to be developed that were matched with the highest priority requirements identified by civil engineers. The product contract was then reviewed and approved by the requirements assessment committee as well as project leadership prior to the start of the development work described in the product contract. NEES followed this process at least 6 times over a 6-year project period, and consistently delivered the services described in the product contract on schedule. We found this process useful for making deliberate choices about the services we provided, as well as services we could not provide given the finite resources available for cyberinfrastructure development and operation.

As a result of following this process, we observed a growing number of users over the 6 years of the project. As of August 2015 the NEEShub logged 156,869 total users over the period from September 2014 through August 2015. As of June 2015, there were 157 fully curated projects consisting of 1742 experiments stored in the NEEShub, with over 2.1M project files stored in the project data repository. Fully assessing the success of a cyberinfrastructure depends ultimately on its long term impact on the community, which takes time to be fully realized. Based on NEEShub usage information, it is clear that the global user community found it beneficial to access the services and resources available through the NEEShub.

26.4 Coordinating Cyberinfrastructure Design and Development Efforts Across Domain Science Communities

There have been many projects over the past decade seeking to develop and operate cyberinfrastructure for individual domain science communities. One challenge facing the cyberinfrastructure community is that opportunities to fully leverage and share development efforts and code among these projects have not been fully explored or exploited.

There are many underlying challenges that impede efforts to coordinate cyberinfrastructure design and development efforts. First, it is often the case that differing science domains have unique approaches, perspectives, and practices that guide the ways in which the members of these communities organize and share information and data. These disciplinary differences make it difficult to directly share software or interfaces because the underlying ways in which researchers work can be fundamentally different. For example, searching a genome database for a DNA sequence match is fundamentally different than searching for archived experimental data that describes the seismic response of a specific type of building structure. Both problems are superficially similar in that both use computing infrastructure, networks, and require a search function. In detail, however, a genome search uses a specific search algorithm (e.g., Smith–Waterman) that might be entirely different from a keyword search that may be used for discovering similar types of experiments performed on buildings. Moreover, the user interfaces, which include pop-down guides, database selector, search status bars, and the display of results, are very likely to be completely different.

The challenge, then, is to identify those elements that can be used across various systems to provide a robust and reliable base that can be extended and customized for the specific needs of each domain science community. There are a few examples of projects over the past few years that have sought to create a common platform that is not specifically tied to a science domain. One example of the HUBzero project at Purdue University [21], which created a simulation framework and web based access to computational resources. Many cyberinfrastructure projects, such as nanoHUB (http://nanohub.org) and the NEEShub [3], were built in all or in part using the common HUBzero framework. Another example is Agave [22], which exports services through a "foundational API" that can be used to build a suite of user services and mediate access to infrastructure. A final example is HPCCloud [20], which is a system that can be used for scientific applications and creates a cloud based high performance computing job management and execution environment that can create on-demand HPC clusters using cloud resources.

26.5 A Vision for the Next Generation of Cyberinfrastructure

Clearly, we have made good progress in NEES in developing a process to provide a useful cyberinfrastructure for a broad user community based on the expressed needs of the users.

To look forward to anticipate future needs, it is useful to consider the question: what will the next generation of cyberinfrastructure look like? Specifically, what kinds of new services and technologies are likely to be needed and used in the next decade by the user community? Considering these trends, and where the community has been recently in regard to cyberinfrastructure, it is helpful to consider several emerging areas, trends, and early stage technologies that will create opportunities and challenges for the community. These trends include cloud computing, changes in physical laboratories connected to the Internet, the Internet of Things (IoT), big data, virtualization, and communication technologies.

Cloud computing, in a general sense, seeks to provide computing and storage as a true utility service (similar to power, gas, or water) in which a user may simply tap into an existing infrastructure rather than creating and operating their own infrastructure. The original vision for this was from the Multics system in the 1960s [17]. In a more specific sense, the types of reliable and ubiquitous services now available include distributed file sharing and storage (i.e., Google Drive, Dropbox), computing (Amazon AWS, Google Cloud Platform [11]), and Windows Azure), and services (Microsoft Online, Virtual Office). Also available to the community is a rich set of infrastructure tools and services (such as Docker, VMware, and CloudStack) that can be used to create a domain focused cyberinfrastructure if a research group wishes to operate their own cloud computing services.

Similar to the development of cyberinfrastructure in the past, the next stage of cyberinfrastructure will likely build on these capabilities to provide a "science utility" that can be paid for as it is used as a replacement for making significant capital investments in infrastructure on a local level. One example of this is Science Exchange [12], which provides users a service to search for laboratories who can conduct experiments for a fee. The concept of "science as a service" [13] is a potential direction that could be taken. One example of a cyberinfrastructure that seeks to provide the computing framework to support this service is the Agave Platform [14]. The question is whether a model of the integration of physical laboratories with cyberinfrastructure can be extended to a wide range of scientific disciplines with varying practices and needs. One example of this integration was the NSF NEES project. The NEES project featured a collection of distributed civil engineering laboratories across the USA at which civil engineering researchers conducted physical experiments. The results of these experiments were uploaded and disseminated through the NEEShub Project Warehouse.

Changes are taking place in the ways in which equipment and physical laboratories are shared. Large-scale projects, such as the Large Hadron Collider CMS at CERN, NEES, the Argonne National Laboratory Advanced Photon Source [23], and the Large Synoptic Survey Telescope (LSST) [24], seek to provide a shared facility for the benefit of the research community. These lab locations can be in difficult to access areas (such as the top of a mountain in Chile for the LSST, or in the ocean for the OOI) to provide access to unique conditions that are not amenable to easy access. A knitting together of common facilities such as these would facilitate the increased sharing of resources. The emerging area of the IoT is another trend closely tied with infrastructure. The ability to use a wide variety and types of sensors actively within an active experiment or passively within a structure (for tasks such as structural health monitoring) that can stream data across a network opens many new possibilities for conducting remote experiments and collecting data. The availability of internet-connected sensors is also opening new opportunities for discovery. One example of this is the CAM2 project at Purdue University [15, 18, 25]. CAM2 is a project focused on developing a platform that allows users to create collections of cameras from over 50,000 available Internet-connected cameras. Using CAM2, users can trigger the collection of data from these cameras, and deploy analysis software on cloud computing resources to analyze the data from these cameras as images from the cameras are collected.

Some of the challenges that will need to be addressed by the research community that will likely arise from these trends are: how can physical laboratories be repositioned to allow remote researchers to effectively use laboratory resources through cyberinfrastructure?; what will be the classes of needs and uses from the domain science communities, and how will these map to a hierarchy of resources for a "science utility"?; and who will fund and operate hybrid facilities, and how will users gain access to these facilities and pay for use?

Another major trend is in the area of data management, curation, annotation, and dissemination. The collection and analysis of data from experiments and simulation is the first step, once papers are published based on these data, the raw and processed data can now be curated, preserved, and potentially disseminated for the benefit of the community, and potentially reanalyzed at a later date. There is also an emerging need to allow producers or users of data to attach annotation to these data to facilitate quick comprehension of the information represented by the data as well as potential future use of the data by others Data repositories, such as those described at the Registry of Research Data Repositories [16], provide a wealth of data for their respective domain science communities. Some of the current challenges with these data repositories are in securing adequate and stable funding sources to keep these repositories in operation. Along with this concern, another question is how institutional libraries will be involved in the operation of domain specific repositories. This question is an area in which the sociology of how overlapping communities (in which individuals are members of more than one community, for example, a faculty member on a university campus as well as a member of a large distributed science projects) will also play an important role in addressing these challenges.

From our work with NEES, it is clear that there is no "one size fits all" solution to these data problems. The common elements, such as author names, dates, and file sizes, are quite simple. However, above this basic level, there are a significant number of complexities in areas such as data schemas, naming conventions, and measurement units that are very specific to the scientific domain. The complexity involved in providing a data management system that can address this will likely lead to a large collection of unique individualized centers that each support a relatively narrow scientific domain.

Another area of challenges involve institutional polices. One good example of this challenge is the issue of floating software licenses for commercial software. The existing common framework for institutional licenses for software often constrain use to the members of the institution, or even physically to institutional resources.

What is needed is the ability for a group of institutions or a distributed community to be able to use a floating license tied to the membership of individual to one of more of these groups, rather than to a physical institution only. Another policy consideration is whether institutions would allow outside researchers to remotely use lab facilities at an institution. A question to consider is whether similar laboratories at a different institution could or should compete for users (and their usage fees) based on prices and services.

The final challenge is in the area of sociology. Human beings are an integral part of any cyberinfrastructure system, and more efforts are needed to understand how communities can fully benefit from interacting through cyberinfrastructure. Some of the questions in this area are: what are common needs across many different communities, and what are the different specialized needs for each community?; and how should domain scientists be involved in the specification, development, and deployment of cyberinfrastructure? Positive examples of work in this area include the NEES and LHC CMS projects.

Considering all of these challenges, the question is: what will be the fifth paradigm? We discussed the four paradigms (theory, experiment, simulation, and data). What will be the next paradigm? Each paradigm historically was built upon and synthesized the capabilities and models of existing paradigms. The question here is what will the synthesis of theory, experiment, simulation, and data produce? The new environment for research and discovery, powered by Internet and low cost ubiquitous communication technologies are helping to eliminate distance and time zones as barriers to collaboration and research. Some possible new elements of a new fifth paradigm might include the increased emergence of virtualized laboratories which break down the need to physically be an institution to conduct or participate in research. The growth in the capabilities of cyberinfrastructure and the IOT is now beginning to make this possible. Physical laboratories might be shared within a research community, and new types of labs could be created at remote or difficult to access locations. A virtual laboratory that could be "plugged in" to physical laboratories to utilize and build upon physical resources (sensors, equipment, and computational) that could be shared, copied, and carried about by researchers. This concept includes the ability to preserve physical and simulation experiments that can be preserved, automatically analyzed, and "played back" in the future as a record of the research as well as a resource for future research. Another potential aspect is the creation of high fidelity computational instruments that could quickly and thoroughly explore both the physical universe (through the collection and preservation of experimental data) and a synthetic model of the physical universe created through computational simulation. The high fidelity representation of this virtual universe as a hybrid model of the real world coupled with simulations based on theory has the potential to provide a space that can be quickly and efficiently explored with algorithms and analytics developed by researchers. This vision has the potential to allow researchers to quickly search for interactions and conditions that were previously unknown.

26.6 Conclusion

Cyberinfrastructure, which includes high performance computing, networks, data, and communication services, is fundamentally changing the ways in which scientific research is conducted. Projects such as NEES and LHC CMS are pioneering examples of an emerging hybrid model of the combination of physical laboratories with cyberinfrastructure.

References

1. T. Hey, S. Tansley, K.M. Tolle (eds.), *The Fourth Paradigm: Data-Intensive Scientific Discovery*, vol. 1 (Microsoft Research, Redmond, 2009)
2. The Compact Muon Solenoid experiment at the CERN Large Hadron Collider. http://cms.web.cern.ch
3. T.J. Hacker, R. Eigenmann, S. Bagchi, A. Irfanoglu, S. Pujol, A. Catlin, E. Rathje, The NEEShub cyberinfrastructure for earthquake engineering. Comput. Sci. Eng. **13**(4), 67–78 (2011)
4. A. Blair-Early, M. Zender, User interface design principles for interaction design. Des. Issues **24**(3), 85–107 (2008)
5. The National Data Service project. http://www.nationaldataservice.org/
6. The Globus project. http://www.globus.org
7. The Ocean Observatories Initiative (OOI) project. http://oceanobservatories.org/
8. M. Kelly, Message from the Director, April 2015. http://oceanobservatories.org/message-from-the-director-april-2015/
9. Cornell University Red Cloud project. https://www.cac.cornell.edu/services/cloudservices.aspx
10. D. Bonacorsi, CMS Collaboration, The CMS computing model. Nucl. Phys. B Proc. Suppl. **172**, 53–56 (2007)
11. Google Cloud Platform. https://cloud.google.com/
12. The Science Exchange. https://www.scienceexchange.com
13. R. DiResta, Science as a service: what happens when you apply software-as-a-service principles to science? January 20, 2013, O'Reilly Radar. Available online at http://radar.oreilly.com/2013/01/science-as-a-service.html
14. R. Dooley, J. Stubbs, Dynamically provisioning portable gateway infrastructure using docker and agave, in *Proceedings of the 2014 Annual Conference on Extreme Science and Engineering Discovery Environment*, ACM, July 2014, p. 55
15. A.S. Kaseb, E. Berry, Y. Koh, A. Mohan, W. Chen, H. Li, Y. Lu, E.J. Delp, A system for large-scale analysis of distributed cameras, in *2014 IEEE Global Conference on Signal and Information Processing (GlobalSIP)*, IEEE, December 2014, pp. 340–344
16. Registry of Research Data Repositories. http://www.re3data.org
17. F.J. Corbató, V.A. Vyssotsky, Introduction and overview of the multics system, in *Proceedings of the November 30–December 1, 1965, Fall Joint Computer Conference, Part I*, ACM, November 1965, pp. 185–196
18. W. Chen, A. Mohan, Y. Lu, T. Hacker, W. Ooi, E. Delp, Analysis of large-scale distributed cameras using cloud. IEEE Cloud Comput. **2**(5), (2015) ISSN 2325-6095, IEEE
19. C.R. Barnes, M.M.R. Best, F.R. Johnson, B. Pirenne, NEPTUNE Canada: installation and initial operation of the world's first regional cabled ocean observatory, in *Seafloor Observatories* (Springer, Berlin, Heidelberg, 2015), pp. 415–438

20. P. O'Leary, M. Christon, S. Jourdain, C. Harris, M. Berndt, A. Bauer, HPCCloud: a cloud/web-based simulation environment, in *Proceedings of the IEEE 7th International Conference on Cloud Computing Technology and Science*, Vancouver, Canada, 30 November–3 December 2015
21. M. McLennan, R. Kennell, HUBzero: a platform for dissemination and collaboration in computational science and engineering. Comput. Sci. Eng. **12**(2), 48–53 (2010)
22. R. Dooley, M. Vaughn, D. Stanzione, S. Terry, E. Skidmore, Software-as-a-service: the iPlant foundation API, in *5th IEEE Workshop on Many-Task Computing on Grids and Supercomputers (MTAGS)*, November 2012
23. Argonne National Laboratory Advanced Photon Source Facility. Described at http://www1.aps.anl.gov
24. Large Synoptic Survey Telescope. Described at http://www.lsst.org
25. W. Chen, Y.H. Lu, T.J. Hacker, Adaptive cloud resource allocation for analysing many video streams, in *Proceedings of the IEEE 7th International Conference on Cloud Computing Technology and Science*, Vancouver, Canada, 30 November–3 December 2015

Chapter 27
Turbulence in the Era of Big Data: Recent Experiences with Sharing Large Datasets

Charles Meneveau and Ivan Marusic

27.1 Introduction

The notion of "Big Data" has in recent years become ubiquitous in many fields, ranging from the natural to social and political sciences. Big Data means different things to different people. In the area of turbulence, its meaning is clear: the ability of both numerical simulations and experiments to generate a huge amount of data has been outpacing our ability to efficiently analyze it. A recognized visionary of the big data revolution, Jim Gray writes in his "Fourth Paradigm of Science" [1], that data analysis in general leads to unforeseen insights and generates hypotheses that guide modeling, new theory, and further experimentation. Turbulence research, especially when based on large numerical simulations, has been challenged to fully realize the true power of data-driven discovery. The largest simulations of the fundamental laws governing fluid flows use tens of millions of hours of CPU time [2–6], yet most of the analysis must be performed in memory of compute nodes while the simulation is running because the simulation state is too large to be transferred over networks or stored on traditional file systems. Even transferring and analysis of a few representative snapshots at select time steps is challenging. It is difficult, if not impossible, to confront model predictions and observations with the exponentially increasing amounts of such simulation data. The push has been to run ever larger direct numerical simulations (DNS) of turbulent flows

C. Meneveau (✉)
Department of Mechanical Engineering, and Institute for Data Intensive Engineering and Science (IDIES), The Johns Hopkins University, 3400 N. Charles Street, Baltimore, MD 21218, USA
e-mail: meneveau@jhu.edu

I. Marusic
Department of Mechanical Engineering, University of Melbourne, Melbourne, VIC 3010, Australia
e-mail: imarusic@unimelb.edu.au

A. Pollard et al. (eds.), *Whither Turbulence and Big Data in the 21st Century?*,
DOI 10.1007/978-3-319-41217-7_27

at ever increasing Reynolds numbers. Based on Moore's law continued progress in computing technologies means that within 10–20 years DNS of wall-bounded turbulence will reach Reynolds numbers obtained in the largest laboratory facilities. However, at $Re_\tau = 10^4$ (such as in a large wind tunnel), the DNS would generate 23 Terabytes of data at each time step, while for $Re_\tau = 10^5$ (such as in the Princeton superpipe) the DNS file size grows to nominally 23 Petabytes per time step, and the full simulation would require an estimated 10^7 time steps [7]. While it is not feasible to keep data from all time steps (for $Re_\tau = 10^5$ that would be 0.2 Yottabytes (10^{24})!), it is clear that although these simulations may be realizable in the coming decades, the processing and data-handling tools required to deal with these large datasets will need to keep pace. Deciding what and when it is stored requires a high level appreciation of the physics, and invariably this process is an iterative and evolving one. Continuing future efforts will likely benefit greatly from collaborations with computer scientists in devising efficient and tailored data-handling tools [8].

Big Data is also generated from physical simulations of turbulence that involve laboratory and field study experiments, where very high Reynolds numbers are accessible [7], and rapid developments in camera, laser, and computer processing technologies have enabled major advances in planar and volumetric particle image velocimetry (PIV) techniques [9]. However, in this case as in DNS, since the dynamical range of length and time scales of the turbulent motions increases with Reynolds number, acquiring fully time-resolved and spatially resolved information requires extremely large datasets [10, 11]. Modern time-resolved tomographic PIV experiments of turbulence routinely generate multiple Terabytes of data per minute.

Meaningful progress in the field requires not only that we run sophisticated and massive DNS and large-scale experiments, but also that the results of these simulations persist for further discovery. Once the relevant and large datasets are stored it is increasingly important to create usable scientific contributions from numerical and physical simulations accessible to a broader pool of users. Data arising from the largest simulations must be released publicly, shared, reanalyzed, and archived over extended periods of time. But simply hosting the simulation output files for download, as is done by most projects today, is not good enough once the data volumes exceed a few Terabytes. Once the data volume is too large, one has to move much of the analysis to where the data reside rather than the traditional approach, which moved the data to our local computers.

In the following section we describe a unique BigData Open Numerical Laboratory, the Johns Hopkins Turbulence Databases (JHTDB), which presents persistent storage and public access to a select set of DNS data. The system preserves the significant computation effort of simulation and enables further experiments leveraging the data to accelerate discovery.

27.2 An Open Numerical Laboratory for Turbulence: JHTDB

27.2.1 System Description

The system is part of a set of big data prototypes for open numerical laboratories including open laboratories in other areas of science such as astronomy [12, 13], computational cosmology [14, 15], and neuroscience [16]. JHTDB (see http://turbulence.pha.jhu.edu) is an open numerical laboratory that addresses the increasingly untenable situation of the large HPC-generated data being inaccessible to the vast majority of researchers in turbulence. Its primary goal is to expose large-scale turbulence data to the research community while at the same time providing easy-to-use client interfaces for retrieving and interacting with the data.

Fundamental to JHTDB's approach is the ability of the user community to interact easily and flexibly with these massive amounts of data. A Database Web Service has thus been built that handles requests over the web for velocities, pressure, various space-derivatives of velocity and pressure, interpolation functions, particle tracking for trajectories, etc. We index the data using a space-filling fractal curve (Z-index) and apply the "move the program to the data" philosophy [12], a fundamental tenet in the design of large-scale scientific databases. A first dataset was placed in the open laboratory in 2008 [17], and consisted of a DNS of forced isotropic turbulence on 1024^3 grid points. A total of 1024 time steps were captured, encompassing the time evolution during a single large-eddy turnover time.

In Fig. 27.1, a schematic of the JHTDB system is shown, with remote users indicated at the top of the figure. They are separated from the databases by the Web server which processes incoming requests and returns the requested data. The first component of JHTDB is the Web server (front-end) that provides the layer with which remote clients can interact. The client interfaces remove the need for the end user to know details of the data storage, such as the distribution of data across database nodes and how they are indexed. The database cluster (backend), composed of a networked database system running Microsoft SQL Server, is the second component. The cluster contains the datasets and provides a scalable infrastructure that supports data-intensive analyses. This Web service model is modeled after the successful SkyServer approach [19] in which multi-terabyte astronomy data archives have been available to researchers [12] for some time.

The data have been generated by a large-scale simulation performed using a HPC facility. Within the database cluster, data are partitioned across several nodes. The Web server issues queries to the database cluster asynchronously using multiple connections per node in order to leverage the multicore architecture of each node. Data from a simulation are stored in the database as binary large objects (BLOBs) indexed by the (Morton) Z-curve [20, 21]. The Z-curve maps the 3D data to the one dimensional index space. It passes through each location in space once and only once (i.e., space filling). It is easy to compute and provides good spatial locality, which is important in order to support contiguous data access for typical usage patterns of the databases.

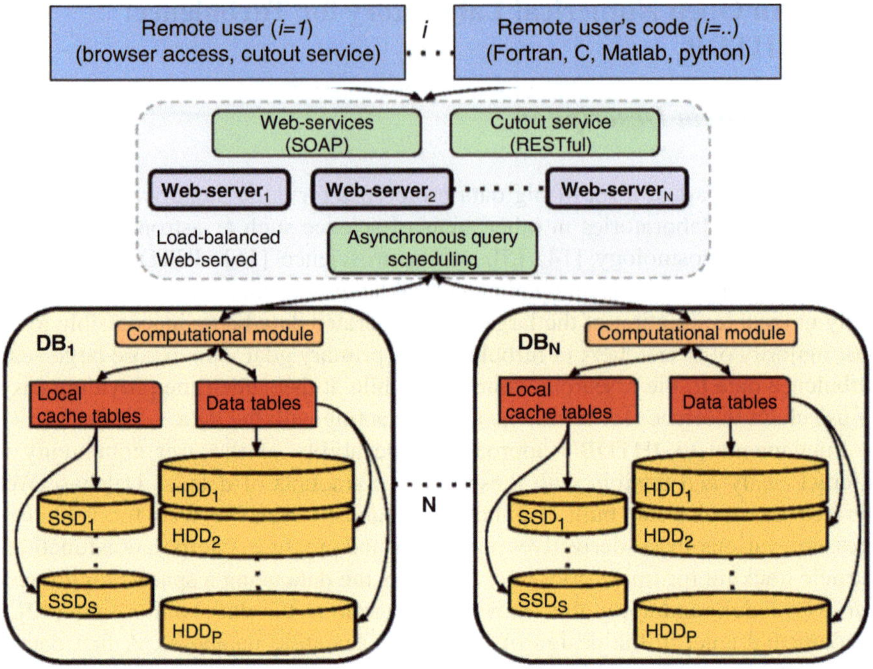

Fig. 27.1 Schematic of the existing architecture of the JHTDB Open Numerical Turbulence Laboratory, showing logical layout of the remote users (clients), Web server, and the database cluster (sketch adapted from Kanov et al. [18])

There are four types of flows presently available (over 200 TBytes) in the open turbulence laboratory: (1) a forced isotropic turbulence case at a Taylor-scale Reynolds number of $R_\lambda = 433$ from a pseudo-spectral simulation on 1024^4 space-time data points [17]. (2) A 56 TB database of steady-state magnetohydrodynamic (MHD) turbulence, obtained from a 1024^3 pseudo-spectral simulation of unit magnetic-Prandtl-number, incompressible MHD, forced with a Taylor–Green flow. Taylor-scale Reynolds numbers are $Re_{\lambda,u} = 186$ and $R_{\lambda,b} = 144$ for the velocity and magnetic fields, respectively. The velocity, pressure, magnetic field, and Coulomb-gauge vector potential are stored on 1024^4 space-time grid points, spanning about one large-scale eddy turnover time. (3) A channel flow dataset at $Re_\tau = 1000$ on $2048\times1536\times512$ spatial data points and 4000 time steps, simulated in collaboration with researchers at Univ. Texas at Austin [22–24], yielding over 110 TBytes, and (4) a variable-density mixing flow on 1024^3 spatial and 1015 temporal data points simulated by Los Alamos (LANL) researchers [25, 26].

27.2.2 Data Access

A hallmark of the system is the data access philosophy, predicated on the notion of *virtual sensors*. We base our virtual sensor data access philosophy on an experimental analogy: an experimenter would like to place sensors at specified positions and times (x, y, z, t) and "measure" velocities, pressure, and other derived quantities there. This can be a one-time measurement, or as function of time at a physical (Eulerian) location or the points can move with the flow as fluid particle trackers. Derived quantities can be based on a variety of operators, like the gradient or Laplacian of a field, or applying various filters [27] and thresholds [28]. Such data access patterns also enable the users to run time backwards [29], impossible in a direct simulation of a dissipative system. Snapshots are saved frequently enough so that one can interpolate field values smoothly enough. Sensors can back-track their original trajectory and one can see where they came from, all the way back to the initial conditions. Furthermore, virtual sensors placed anywhere in the 4D flow domain typically require interpolations. For spatial interpolation, Lagrange polynomial [17] or spline interpolation [24] methods are used. For Lagrange polynomials the client has the option between no interpolation, 4th order, 6th order, or 8th order interpolation. For splines, the client can select between 3rd and 5th order methods. In the case of the isotropic, MHD and variable-density mixing databases, interpolation with uniform grid spacing is employed. For cases with non-equal grids, a generalized barycentric Lagrange interpolation method is used (e.g., non-uniform grid spacing in the wall normal direction for the channel flow database). Spatial differentiation is based on various methods such as finite difference or differentiation of the interpolation splines/polynomials. Temporal interpolation is performed using piecewise cubic Hermite interpolating polynomial.

The existing Web services provide a convenient mechanism for remote clients to interact with the databases via the immersive approach. The Web services use Simple Object Access Protocol which provides a standard protocol for sending and receiving messages over the internet. Client functions are provided by the Web services for spatial and temporal interpolation, differentiation, fluid particle tracking, and other secondary calculations performed within the database. These calculations are in addition to client functions to request primary fields. Packaged libraries which allow easy use of the Web service functions are provided for C, Fortran, Matlab, and Python. Figure 27.2 shows a snippet of Matlab code to read a subset of the channel flow velocity and generate a contour plot.

At this stage, what is noteworthy about the availability of a 4D dataset "at your fingertips" and the ability to make "casual" queries from anywhere at any time, is that it is beginning to change how we think about the data. Researchers can come back to the same place in space and time and be sure to encounter the same values. They can follow and observe phenomena forward and backwards in time. The reader is invited to use any web-browser and visit http://turbulence.pha.jhu.edu/webquery/query.aspx, select (e.g.,) the "channel" dataset and the "*GetPressureGradient*" function, and enter a particular time between 0 and 25 along with x, y, z positions

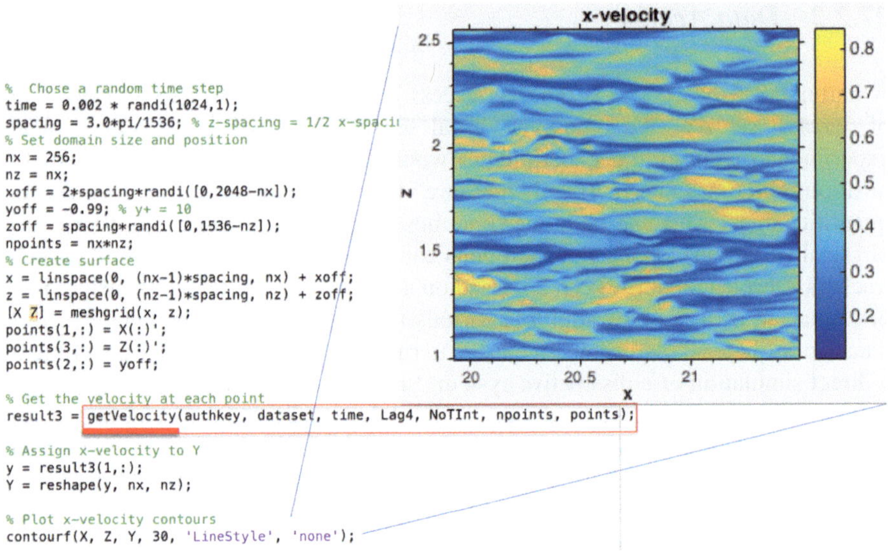

Fig. 27.2 Snippet of Matlab code for extraction of velocity data (using *getVelocity*) from JHTDB channel flow database at $Re_\tau = 1000$ and visualization of streamwise velocity in plane near the wall (at $y^+ = 10$)

in the range $(0 : 8\pi) \times (-1 : 1) \times (0 : 3\pi)$. The database will return the three components of pressure gradient at that position in space-time. Velocity, pressure, the velocity gradient tensor, and other quantities of interest are available by similar simple queries. There are tremendous benefits to being able to re-visit repeatedly the complex flow phenomena at leisure, as physical intuition about the complicated dynamics begins to develop.

27.2.3 Community Response and Impact

While researchers who themselves perform very large simulations could often (but not always) analyze their own data more effectively within original HPC environments, the rest of the community could either not do so effectively or not at all. Thus, many researchers started to access the JHTDB data and carry out their research in the open Numerical Laboratory. A recent analysis of JHTDB's usage patterns has been presented in Kanov et al. [18], as part of a special journal issue on open numerical laboratories [30]. Figure 2 in Kanov et al. [18] shows the worldwide distribution of requests to the turbulence database, measured by total number of points queried. There is heavy usage and it has become a global resource. Figure 27.3 displays the cumulative number of points queried from the system [31]. In 2015, the number of points has exceeded 12 trillion.

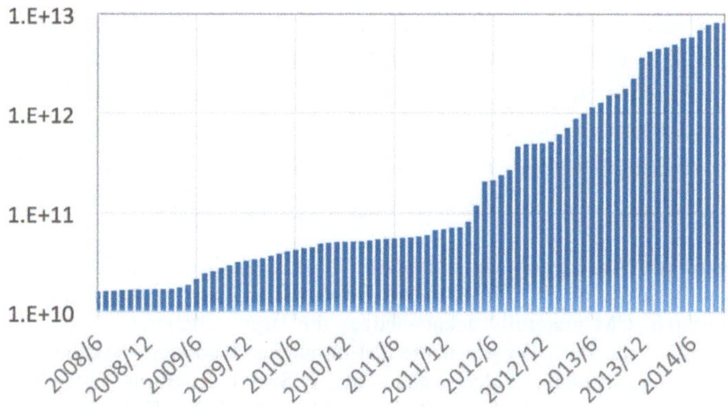

Fig. 27.3 Cumulative number of points queried as a function of time. As of July 2016 the number of points has exceeded 21 trillion [31]

The availability of massive, high-resolution turbulence datasets in web-accessible databases has already proven to have major impact on turbulence research and education. The system has been used to address many research questions in turbulence, both from the internal group that has generated and is curating the DNS data and from external users. From the internal set of users, a number of publications can be mentioned, such as Refs. [29, 32–47]. External users have used the databases in studies of extreme events in turbulence [48], shape evolution [49], characterization of particles–turbulence interactions [50], calibration of experimental measurement tools [51–54], the study of velocity gradient tensor properties [55–57], subgrid-scale model assessments [58, 59], and many others [49, 60–72]. Typically there are now about 1–2 individualized tokens requested and assigned to new outside users every week (a token is needed if a user wishes to download more than 4096 points in a single call).

In terms of education, JHDTB has been used in various classes around the world, as well as in student workshops (e.g., the "Tutorial School on Fluid Dynamics: Topics in Turbulence" held at the UMD College Park - see http://www2.cscamm.umd.edu/programs/trb10/, that has been repeated in 2015 http://burgers.umd.edu/school/). Especially the Matlab interface has been very popular to facilitate student access to turbulence data.

27.3 Conclusions

Turbulence research has arguably been at the forefront of "big data" for a long time, as it has been generating very large datasets both through numerical simulations and experiments. It is thus fitting for turbulence researchers to seek out and establish interdisciplinary collaborations especially with computer data scientists, and with them to propose and adapt new tools that enable wider access to turbulence data sets. Here we have presented details of a unique "open laboratory for turbulence" at the Johns Hopkins University as a example of the required paradigm shift where one has to move much of the analysis to the data rather than the traditional approach, which moved the data to our computers. It is hoped that broader access to data, from simulations and experiments, will further accelerate turbulence research in coming years.

Acknowledgements The authors congratulate Prof. William George, for whom this Festschrift is intended, on occasion of his 70th birthday and salute him for his many accomplishments in turbulence research. CM gratefully acknowledges the large collaborative and interdisciplinary effort that has made the JHTDB open numerical turbulence laboratory possible, specifically the following colleagues, scholars, students, and staff: A. Szalay, R. Burns, G. Eyink, S. Chen, E. Vishniac, T. Zaki, E. Perlman, Y. Li, T. Budavari, M. Wan, H. Yu, E. Frederix, K. Buerger, H. Aluie, J. Graham, C. Lalescu, K. Kanov, S. Hamilton, P. Johnson, D. Livescu, R.D. Moser, M. Lee and N. Malaya, K. Yang, J. Lee, J. Vandenberg, S. Werner, V. Paul, and G. Lemson. The funding from the National Science Foundation (grants CMMI-0941530 and CBET-1507469) is gratefully acknowledged. IM thanks the support of the Australian Research Council.

References

1. A. Hey, S. Tansley, K. Tolle, et al., The Fourth Paradigm: Data-intensive scientific discovery, Microsoft Research Redmond, WA, 2009
2. D. Donzis, P. Yeung, K. Sreenivasan, Dissipation and enstrophy in isotropic turbulence: resolution effects and scaling in direct numerical simulations. Phys. Fluids **20**, 045108 (2008)
3. T. Ishihara, T. Gotoh, Y. Kaneda, Study of high-Reynolds number isotropic turbulence by direct numerical simulation. Annu. Rev. Fluid Mech. **41**, 165–180 (2009)
4. X. Wu, P. Moin, Direct numerical simulation of turbulence in a nominally zero-pressure-gradient flat-plate boundary layer. J. Fluid Mech. **630**, 5–41 (2009)
5. Y. Khalighi, J. Nichols, S. Lele, F. Ham, P. Moin, Unstructured large eddy simulation for prediction of noise issued from turbulent jets in various configurations, AIAA paper 2886, 2011
6. M. Lee, R. Moser, Direct numerical simulation of turbulent channel flow up to $Re_\tau = 5200$. J. Fluid Mech. **774**, 395–415 (2015)
7. A. Smits, I. Marusic, Wall-bounded turbulence. Phys. Today **66**, 25–30 (2013)
8. I. Marusic, G.V. Candler, V. Interrante, P.K. Subbareddy, A. Moss, Real time feature extraction for the analysis of turbulent flows, in *Data Mining for Scientific and Engineering Applications*, ed. by R. Grossman, C. Kamath, P. Kegelmeyer, V. Kumar, R. Namburu (Springer, US, 2001), pp. 223–238
9. R.J. Adrian, J. Westerweel, *Particle Image Velocimetry* (Cambridge University Press, Cambridge, 2011)
10. A. Schröder, R. Geisler, K. Staack, G. Elsinga, F. Scarano, B. Wieneke, A. Henning, C. Poelma, J. Westerweel, Eulerian and Lagrangian views of a turbulent boundary layer flow using time-resolved tomographic PIV. Exp. Fluids **50**, 1071–1091 (2011)
11. C.M. de Silva, E. Gnanamanickam, C. Atkinson, N.A. Buchmann, N. Hutchins, J. Soria, I. Marusic, High spatial range velocity measurements in a high Reynolds number turbulent boundary layer. Phys. Fluids **26**, 025117 (2014)
12. A.S. Szalay, P. Kunszt, A. Thakar, J. Gray, D. Slutz, R.J. Brunner, Designing and mining multi-terabyte astronomy archives: the Sloan digital sky survey, in *Proceedings of the 2000 ACM SIGMOD International Conference on Management of Data* (2000), pp. 451–462

13. W. O'Mullane, J. Gray, N. Li, T. Budavari, M. Nieto-Santisteban, A. Szalay, Batch query system with interactive local storage for SDSS and the VO. in *Proceedings of the ADASS XIII, ASP Conference Series*, ed. by F. Ochsenbein, M. Allen, D. Egret (2004), pp. 314–372
14. G. Lemson, the Virgo Consortium, Halo and galaxy formation histories from the millennium simulation: public release of a VO-oriented and SQL-queryable database for studying the evolution of galaxies in the lambdaCDM cosmogony. arXiv:astro-ph/0608019 (2006)
15. R. Overzier, G. Lemson, R. Angulo, et al., The millennium run observatory: first light. MNRAS **428**, 778 (2013)
16. R. Burns, J. Vogelstein, A. Szalay, From cosmos to connectomes: the evolution of data-intensive science. Neuron **83**, 1249–1252 (2014)
17. Y. Li, E. Perlman, M. Wan, Y. Yang, C. Meneveau, R. Burns, S. Chen, A. Szalay, G. Eyink, A public turbulence database cluster and applications to study Lagrangian evolution of velocity increments in turbulence. J. Turbul. **9**, N31 (2008)
18. K. Kanov, C. Burns, R. Lalescu, G. Eyink, The Johns Hopkins turbulence databases: an open simulation laboratory for turbulence research. Comput. Sci. Eng. **17**(5), 10–17 (2015)
19. Y. Gu, R. Grossman, A. Szalay, A. Thakar, Distributing the Sloan Digital Sky Survey using UDT and sector, in *e-Science* (2006), p. 56
20. H. Samet, *Foundations of Multidimensional and Metric Data Structures* (Morgan Kaufmann Publishers, Los Altos, CA, 2006)
21. E. Perlman, R. Burns, Y. Li, C. Meneveau, Data exploration of turbulence simulations using a database cluster, in *SC07* 07 (2007)
22. J. Graham, K. Kanov, E. Givelberg, R. Burns, G. Eyink, A. Szalay, C. Meneveau, M. Lee, N. Malaya, R. Moser, A web-services accessible database for channel flow turbulence at Re-tau = 1000. Bull. Am. Phys. Soc. **58**, 236 (2013)
23. M. Lee, N. Malaya, R. Moser, Petascale direct numerical simulation of turbulent channel flow on up to 786k cores, in *Proceedings of SC13: International Conference for High Performance Computing, Networking, Storage and Analysis* (Association for Computing Machinery, New York, 2013), p. 61
24. J. Graham, K. Kanov, X. Yang, M. Lee, N. Malaya, C. Lalescu, R. Burns, G. Eyink, A. Szalay, R. Moser, C. Meneveau, A web services-accessible database of turbulent channel flow and its use for testing a new integral wall model for LES. J. Turbul. **17**(2), 181–215 (2016)
25. D. Livescu, J. Ristorcelli, Variable-density mixing in buoyancy-driven turbulence. J. Fluid Mech. **605**, 145–180 (2008)
26. D. Livescu, Numerical simulations of two-fluid turbulent mixing at large density ratios and applications to the Rayleigh–Taylor instability. Philos. Trans. R. Soc. A: Math. Phys. Eng. Sci. **371**, 20120185 (2013)
27. K. Kanov, R. Burns, G. Eyink, C. Meneveau, A. Szalay, Data-intensive spatial filtering in large numerical simulation datasets, in *International Conference for High Performance Computing, Networking, Storage and Analysis (SC)* (IEEE Computer Society Press Los Alamitos, CA, 2012)
28. K. Kanov, R. Burns, C. Lalescu, Efficient evaluation of threshold queries of derived fields in a numerical simulation database, in *Proceedings of the 18th International Conference on Extending Database Technology (EDBT)*, 2015
29. G. Eyink, E. Vishniac, C. Lalescu, H. Aluie, K. Kanov, K. Bürger, R. Burns, C. Meneveau, A. Szalay, Flux-freezing breakdown in high-conductivity magnetohydrodynamic turbulence. Nature **497**, 466–469 (2013)
30. F. Alexander, C. Meneveau, Foreword to the "open simulation laboratories" special issue. Comput. Sci. Eng. **17**(5), 7–9 (2015)
31. K. Kanov, Efficient evaluation of data-intensive batch-queries in open simulation laboratories, Ph.D. thesis, Johns Hopkins University, 2015
32. P. Johnson, C. Meneveau, Large-deviation joint statistics of the finite-time Lyapunov spectrum in isotropic turbulence. Phys. Fluids **27**(8), 085110 (2015)
33. G. Eyink, Stochastic flux freezing and magnetic dynamo. Phys. Rev. E **83**, 056405 (2011)

34. H. Yu, C. Meneveau, Lagrangian refined Kolmogorov similarity hypothesis for gradient time evolution and correlation in turbulent flows. Phys. Rev. Lett. **104**, 084502 (2010)
35. H. Yu, C. Meneveau, Scaling of conditional Lagrangian time correlation functions of velocity and pressure gradient magnitudes in isotropic turbulence. Flow Turbul. Combust. **85**, 457–472 (2010)
36. Y. Li, L. Chevillard, C. Meneveau, G. Eyink, Matrix exponential-based closures for the turbulent subgrid-scale stress tensor. Phys. Rev. E **79**, 016305 (2009)
37. K. Kanov, E. Perlman, R. Burns, Y. Ahmad, A. Szalay, I/O streaming evaluation of batch queries for data-intensive computational turbulence, in *Proceedings of 2011 International Conference for High Performance Computing, Networking, Storage and Analysis* (Association for Computing Machinery, New York, 2011), p. 29
38. X. Wang, E. Perlman, R. Burns, T. Malik, T. Budavári, C. Meneveau, A. Szalay, Jaws: job-aware workload scheduling for the exploration of turbulence simulations, in *Proceedings of the 2010 ACM/IEEE International Conference for High Performance Computing, Networking, Storage and Analysis* (IEEE Computer Society, Washington DC, 2010), pp. 1–11
39. H. Yu, K. Kanov, E. Perlman, J. Graham, E. Frederix, R. Burns, A. Szalay, G. Eyink, C. Meneveau, Studying Lagrangian dynamics of turbulence using on-demand fluid particle tracking in a public turbulence database. J. Turbul. **13**, N12 (2012)
40. M. Treib, K. Burger, F. Reichl, C. Meneveau, A. Szalay, R. Westermann, Turbulence visualization at the Terascale on desktop PCs. IEEE Trans. Vis. Comput. Graph. **18**, 2169–2177 (2012)
41. G. Eyink, D. Benveniste, Diffusion approximation in turbulent two-particle dispersion. Phys. Rev. E **88**, 041001 (2013)
42. J. Graham, E. Givelberg, K. Kanov, Run-time creation of the turbulent channel flow database by an HPC simulation using MPI-DB, in *Proceedings of the 20th European MPI Users' Group Meeting* (Association for Computing Machinery, New York, 2013), pp. 151–156
43. D. Benveniste, T. Drivas, Asymptotic results for backwards two-particle dispersion in a turbulent flow. Phys. Rev. E **89**, 041003 (2014)
44. C. Meneveau, Lagrangian dynamics and models of the velocity gradient tensor in turbulent flows. Annu. Rev. Fluid Mech. **43**, 219–245 (2011)
45. C. Meneveau, Germano identity-based subgrid-scale modeling: a brief survey of variations on a fertile theme. Phys. Fluids **24**, 121301 (2012)
46. J. Boschung, P. Schaefer, N. Peters, C. Meneveau, The local topology of stream-and vortex lines in turbulent flows. Phys. Fluids **26**, 045107 (2014)
47. Y. Ahmad, R. Burns, M. Kazhdan, C. Meneveau, A. Szalay, A. Terzis, Scientific data management at the Johns Hopkins Institute for Data Intensive Engineering and Science. ACM SIGMOD Rec. **39**, 18–23 (2011)
48. H. Xu, A. Pumir, G. Falkovich, E. Bodenschatz, M. Shats, H. Xia, N. Francois, G. Boffetta, Flight–crash events in turbulence. Proc. Natl. Acad. Sci. **111**, 7558–7563 (2014)
49. J. Jucha, H. Xu, A. Pumir, E. Bodenschatz, Time-reversal-symmetry breaking in turbulence. Phys. Rev. Lett. **113**, 054501 (2014)
50. K. Gustavsson, J. Einarsson, B. Mehlig, Tumbling of small axisymmetric particles in random and turbulent flows. Phys. Rev. Lett. **112**, 014501 (2014)
51. N. Buchmann, C. Willert, J. Soria, Pulsed, high-power LED illumination for tomographic particle image velocimetry. Exp. Fluids **53**, 1545–1560 (2012)
52. X. Liu, J. Katz, Vortex-corner interactions in a cavity shear layer elucidated by time-resolved measurements of the pressure field. J. Fluid Mech. **728**, 417–457 (2013)
53. D. Xu, J. Chen, Accurate estimate of turbulent dissipation rate using PIV data. Exp. Therm. Fluid Sci. **44**, 662–672 (2013)
54. D. Fiscaletti, J. Westerweel, G. Elsinga, Long-range μPIV to resolve the small scales in a jet at high Reynolds number. Exp. Fluids **55**, 1–15 (2014)
55. J.M. Lawson, J.R. Dawson, On velocity gradient dynamics and turbulent structure. J. Fluid Mech. **780**, 60–98 (2015)

56. B. Luethi, M. Holzer, A. Tsinober, Expanding the Q–R space to three dimensions. J. Fluid Mech. **641**, 497–507 (2010)
57. J. Cardesa, D. Mistry, L. Gan, J. Dawson, Invariants of the reduced velocity gradient tensor in turbulent flows. J. Fluid Mech. **716**, 597–615 (2013)
58. H. Lu, Assessment of the modulated gradient model in decaying isotropic turbulence. Theor. Appl. Mech. Lett. **1**, 041004 (2011)
59. A.G. Gungor, S. Menon, A new two-scale model for large eddy simulation of wall-bounded flows. Prog. Aerospace Sci. **46**, 28–45 (2010)
60. F. Grinstein, A. Gowardhan, A. Wachtor, Simulations of Richtmyer–Meshkov instabilities in planar shock-tube experiments. Phys. Fluids **23**, 034106 (2011)
61. W. Liu, E. Ribeiro, Scale and rotation invariant detection of singular patterns in vector flow fields, in *Structural, Syntactic, and Statistical Pattern Recognition* (Springer, Berlin, Heidelberg, 2010), pp. 522–531
62. P. Bhat, K. Subramanian, Fluctuation dynamos and their Faraday rotation signatures. Month. Not. R. Astron. Soc. **429**, 2469–2481 (2013)
63. C. Keylock, T. Tokyay, G. Constantinescu, A method for characterising the sensitivity of turbulent flow fields to the structure of inlet turbulence. J. Turbul. **12**, N45 (2011)
64. M. Holzner, M. Guala, B. Lüthi, A. Liberzon, N. Nikitin, W. Kinzelbach, A. Tsinober, Viscous tilting and production of vorticity in homogeneous turbulence. Phys. Fluids **22**, 061701 (2010)
65. C. Wu, T. Chang, Rank-ordered multifractal analysis (ROMA) of probability distributions in fluid turbulence. Nonlinear Processes Geophys. **18**, 261–268 (2011)
66. C. Keylock, K. Nishimura, J. Peinke, A classification scheme for turbulence based on the velocity-intermittency structure with an application to near-wall flow and with implications for bed load transport. J. Geophys. Res.: Earth Surf. (2003–2012) **117**, F1 (2012)
67. W. Liu, E. Ribeiro, Detecting singular patterns in 2D vector fields using weighted Laurent polynomial. Pattern Recognit. **45**, 3912–3925 (2012)
68. M. Mishra, X. Liu, M. Skote, C.-W. Fu, Kolmogorov spectrum consistent optimization for multi-scale flow decomposition. Phys. Fluids **26**, 055106 (2014)
69. L. Moriconi, R. Pereira, Vorticity statistics and the time scales of turbulent strain. Phys. Rev. E **88**, 013005 (2013)
70. F. Grinstein, A. Gowardhan, J. Ristorcelli, A. Wachtor, On coarse-grained simulations of turbulent material mixing. Phys. Scripta **86**, 058203 (2012)
71. T. Chang, C. Wu, M. Echim, H. Lamy, M. Vogelsberger, L. Hernquist, D. Sijacki, Complexity phenomena and ROMA of the earth's magnetospheric cusp, hydrodynamic turbulence, and the cosmic web. Pure Appl. Geophys. **172**(7), 2025–2043 (2015)
72. A. Pumir, H. Xu, G. Boffetta, G. Falkovich, E. Bodenschatz, Redistribution of kinetic energy in turbulent flows. Phys. Rev. X **4**, 041006 (2014)

Chapter 28
Public Dissemination of Raw Turbulence Data

Juan A. Sillero and Javier Jiménez

28.1 Introduction

Perhaps because of the natural tendency of engineering research to proprietary information, the traditional working model of fluid mechanics has been that data are generated and analysed within the same group. Databases have always existed, and professional societies have specially contributed to generate and maintain them, but they usually contain processed results from which it is difficult to extract anything beyond what had been anticipated by the data compiler. The difference between raw data and what is eventually made public has worsened with the appearance of increasingly large simulation and experimental data sets. These are usually archived, but because they are bulky and difficult to read, they are often lost without further use. In some cases in which several interesting computations or experiments have been performed within the same institution, workshops and personal contacts have provided a mechanism for the use of unprocessed data by other groups, but the democratization of computer power, with data being generated at many places around the World, is making that model increasingly cumbersome and hard to justify.

Other communities have used a different working method for some time. High-energy physics, meteorology and molecular biology come to mind as examples in which there are specialists who generate data and others who exploit them, collaborating by means of more or less public databases to which the former contribute and from where the latter draw their raw material. The process of collaboration is not trivial, and each community has developed its own protocols for data interchange, archiving, quarantine periods and attribution. The loss is a certain 'personal' way of doing science, which should probably not be completely

J.A. Sillero (✉) • J. Jiménez
School of Aeronautics, Universidad Politécnica de Madrid, 28040 Madrid, Spain
e-mail: sillero@torroja.dmt.upm.es; jimenez@torroja.dmt.upm.es

A. Pollard et al. (eds.), *Whither Turbulence and Big Data in the 21st Century?*,
DOI 10.1007/978-3-319-41217-7_28

abandoned, but the gain is a greater freedom in what can be observed, and the greater use to which the data are put. An instructive case is astrophysics, where the division of labour dates at least to the analysis and publication by Kepler of Tycho Brahé's data in 1627. This eventually led to his planetary laws, and probably to Newton's dynamics, but it was not free of controversy; Kepler had to fight Tycho's heirs for ownership of the data, and eventually paid for the publication himself [6]. Astrophysical data are today mostly in the public domain under well-defined rules and formats.

Whether fluid mechanics should follow this model is open to discussion, and we believe that it should be discussed. Many groups make information public in their private websites, but there is no standard way of disseminating the availability of new data, no generally agreed way of storing or reading them and, in truth, little incentive to prepare the data for general use. An important problem is long-term curation. In our experience, large data sets have a useful lifetime of around 10 years, after which they are easier to recompute or remeasure than to retrieve from a database. Even if limited, this time is longer than most research contracts, and it is unclear how the cost of maintaining and preparing data is to be supported after the contract expires. Moreover, large simulation or experimental data have peculiarities that are in many cases only understood by their originator (typically a student or postdoc), and these constraints may be forgotten once the originator leaves the research group. For example, if the results of a spectral code are differentiated using low-order finite differences the resulting errors are often unacceptable. Again, the useful lifetime of a data set typically exceeds the span of a thesis or a postdoctoral stay.

Broadly speaking, there are currently two models. The first one is a service centre offering data and on-site postprocessing facilities for a particular kind of problems [7]. The second is a more unstructured facility where data are stored and indexed with minimal on-site processing capabilities [2, 3, 5]. The former are very useful for the purpose for which they are developed but, in our personal experience, tend to be cumbersome when they are used for a very different purpose. They are also critically subject to the long-term funding issues mentioned above, particularly if restricted to a particular problem. The unstructured services typically require more computer power on the side of the user. They also require continuing funding, but they are broader and easier to justify. The curation problem is also different in both cases. While the first one typically requires the continued attention of somebody who is familiar with the fluid mechanics of the problem, the second, if properly documented, only requires computer expertise. Thus, the first model tends to be permanently associated with the group that generated the data while, in principle, the second could be outsourced to a more generic institution.

There is a certain urgency to the discussion of whether and how our data should be shared. An increasing number of funding agencies and scientific journals are beginning to insist on open public access of research results (while others specifically forbid it, and the effect of intellectual property laws is uncertain). There are increasingly common warnings from the agencies that, if we do not decide on a viable dissemination policy and format, they will do it for us without regard to any issues that may differentiate our application from others. On the other hand, several supranational initiatives are starting to form for the long-term storage and standardized access of large scientific data [1, 4], and the time to start exerting our influence over them as a community is probably now.

28.2 The Database of the Universidad Politécnica de Madrid

To illustrate these concepts, we will briefly describe our experience with the public databases hosted by our research group, which mostly contain results of large fluid-mechanical simulations. There are two types of data sets. The first ones are large collections ($\sim 100\,\mathrm{TB}$) of raw flow snapshots. They may include as many as $\sim 10^5$ moderately large binary files ($\sim 1\,\mathrm{GB}$), fewer ($\sim 10^3$) larger ones ($\sim 100\,\mathrm{GB}$) or a mixture of both. The files can often only be used with the original numerical code that generated them, which is often unique to each set. These data are routinely mined by local users in a 'write once, read many' fashion, over a characteristic period of about ten years. Once checked, cleaned and after initial publication from the data originator, most of them are made public without any particular access control [5]. The necessary codes are also made public under an open-source 'as is' license.

The second type of files are obtained by postprocessing the original ones. They include flow fields expressed in primitive variables in physical space, rather than in the typically reduced set of variables and spectral representations of the original files. Another common type of postprocessed results is the temporal evolution of 'objects' defined in particular ways, some of them obtained on-the-fly during the simulations. The accumulated size of these data is typically several times larger than the original ones, and they are not intended for unrestricted public use. Although they are occasionally provided to outside researchers on the basis of personal collaborations, they are not carefully curated, and often not even well documented.

Some statistics of the file size are shown in Fig. 28.1. Figure 28.1a presents a compilation of the sorted file size versus the number of files (circles) together with their cumulative size (squares). The figure shows that there are $O(10^5)$ files larger than 1 GB, accounting for 65 % of the total used storage ($\approx 550\,\mathrm{TB}$). Figure 28.1b is a histogram of the size distribution of our current files, which we believe to be typical of those used in turbulence research. Traditional databases are of limited use for this application. The size of a typical transaction is larger than those in most commercial services, such as financial ones and personal records, whose characteristic entry size is a few KB, or even than those in scientific databases, such as particle physics or genomics, where the characteristic entry size is a few MB.

Our postprocessing infrastructure is optimized not only for performance and price, but also, most importantly, for easy of maintenance. Besides larger clusters dedicated to simulations, the core of the system is a collection of Linux servers with relatively large memory ($\geq 128\,\mathrm{GB}$). They are connected to a dedicated network file system (NFS) server attached to a centralized network attached storage that

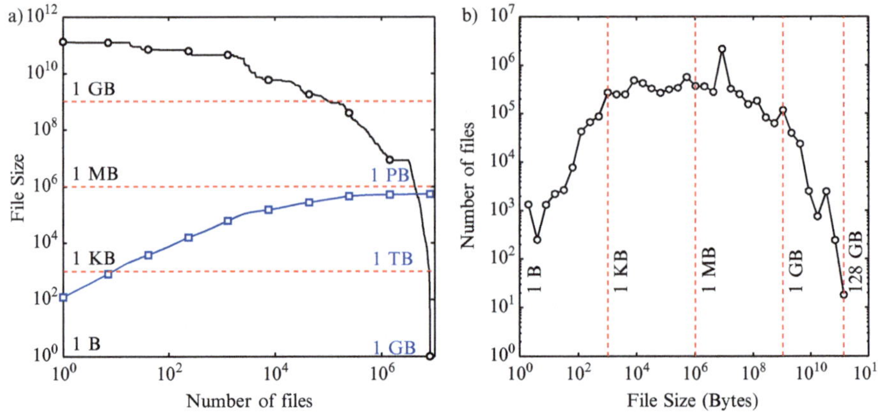

Fig. 28.1 File size distribution in our current data centre storage. (**a**) Sorted file size (Bytes) as a function of: *open circle*, the total number of files; and *blue open triangle*, its cumulative size (GBytes). (**b**) Histogram of the file size distribution

requires a relatively powerful front-end with substantial processing capabilities. The hardware reserved for the open-access database is a high-density commodity JBOD disk enclosure with SAS interface, accepting sixty 3.5" drives in a 4U rack space. Because of the relatively large size of the files, they are mostly accessed sequentially rather than randomly, and our solution uses exclusively traditional mechanical 4 TB hard drives. Moreover, to test the long-term behaviour of a 'low-cost' storage system with decent performance, we have limited ourselves to SATA drives instead of the more expensive SAS ones.

This choice implies losing some of the benefits of the SAS interface, such as better command queuing and error checking algorithms, full duplex communications allowing multipath, higher mean time between failure or lower bit error rate. To compensate for this possible degradation of data integrity, we chose a ZFS file system [9] that guarantees integrity at disk-block level. This file system differs from others in being specifically designed to protect the data from silent data corruption [8] caused by multiple factors such as bit rot, current spikes and firmware bugs. ZFS was conceived to control all the layers in a modern file system, from logical volumes down to block devices. For this reason, expensive hardware RAID cards are no longer needed to control which information goes into each block device. ZFS offers its own software RAID solution using copy-on-write semantics (COW), avoiding problems such as the "write hole" phenomenon [9] in traditional storage appliances. To protect the integrity of the data in the event of failure of one or more disks, we have configured ZFS in Raidz2 mode, allowing the loss of up to two disks for every group of ten. On top of that, ZFS allows the use of different lossless compression algorithms to perform on-the-fly compression of the data at block level. Depending on the installed capacity, the current price of such an storage system varies from 90 to 110 €/TB, which is within the funding range of an individual research group. It delivers I/O throughputs in excess of 2.5 GB/s between the front-end node and the JBOD, and about 700 MB/s to the NFS clients.

Although their large size makes the use of full data set by external researchers difficult, our experience is that downloads of individual files are reasonably common. Practical data transfer rates from our university to other academic institutions are at most 1–2 TB per day, although it may be hard to sustain those throughputs in the case of several concurrent requests. After about a year of continued used, we have experienced no critical loss of data.

28.2.1 User Interface

Although data dissemination through personal contacts and visits remains important and will probably always remain so, there is an increased demand from external groups for a more impersonal interaction, which could not be satisfied until the system just described was implemented.

At present, the data open to unrestricted external use is about 150 TB, and provides the scientific community with an efficient way to download about 15 different DNS databases of incompressible channel and boundary layer flows at Reynolds numbers $Re_\tau \approx 180$–4200. Information about specific databases can be found at http://torroja.dmt.upm.es/turbdata/index. The database also provides utility tools to convert the data files between formats (e.g. from binary to HDF5), or to manipulate the original data.

Even if the requirements are simple in principle, their execution is challenging for several reasons. Firstly, there has to be a physical infrastructure able to host hundreds of terabytes, although not necessarily as fast as for a local service. This was already discussed with some detail in the previous section. Secondly, any centralized architecture in which one server interacts with many clients can strongly degrade the availability of the files to the local group, requires large bandwidth and is expensive to run. Finally, it is important to retain the ability to resume partial file downloads when transferring large datasets, saving time and bandwidth resources.

Our first attempt was to adapt a peer-to-peer protocol such as BitTorrent to distribute large scientific files. This protocol benefits from several factors such as integrity, robustness, simplicity and aggregate bandwidth capabilities, making it a reasonable choice to transfer large amounts of data from a local digital repository to the scientific community via the Internet network. We implemented a Torrent site for part of our databases without finding any unsurmountable technical difficulties. However, we encountered a number of practical ones that discouraged us from continuing its use: many home institutions block P2P networks, the BitTorrent protocol listens to a large number of ports in the local machines (associated with the trackers), and it is not straightforward to automate the indexation of hundreds of thousands of files. Most importantly, the essence of the P2P networks is to share a file with many peers simultaneously, and to complete a download in at most few hours. Neither has proved to be the case with our databases. The observed access pattern is most often a single user downloading a few Terabytes for a couple of days, and rarely there are several users simultaneously accessing the data. This motivated us to consider alternative solutions.

Our presently preferred solution has been to implement a simple but practical file repository using an Apache HTTP server. The files are served to the clients through a dedicated 1 Gbps Ethernet network without impacting on the availability of the files to the local group, which is served by a private high-speed QDR Infiniband network (40 Gbps). The HTTP protocol allows resumption of partial downloads in case of failure, although some local web servers are configured to avoid header range requests, preventing resumes. Our Apache web server is configured to accept such requests, and users should be able to continue unfinished downloads at any time, either using terminal commands like wget, or through a generic web browser.

28.2.2 Statistics of Outside Use of the Data

Besides the local data use by members of the group, data analytics collected in the open part of our website shows the increasing interest of researches in raw and postprocessed turbulence databases, which is also reflected in the increasing number of citations to the associated papers. Figure 28.2 shows a summary of the audience for the part of our website listing one- and two-point statistics for different turbulent boundary layers, based on geographical location. Since the open repository described above was brought online in February 2015 we have received about 600 visits per month: 48 % correspond to direct searches, 40 % to referrals and the remainder to organic searches and social networks. Around 20 % of the total hits are return visitors who check our databases often to either download new data or to check for content updates. For instance, 18 % of the sessions come from a total of 251 universities from around the world.

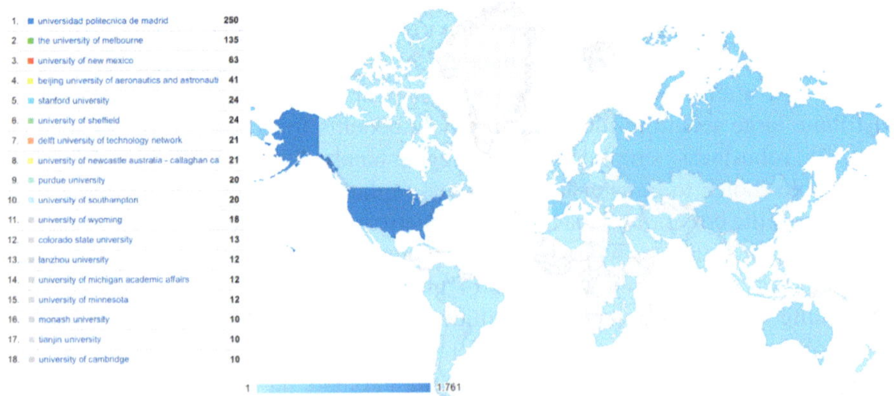

Fig. 28.2 Statistics for the *postprocessed* Boundary Layer databases based on the geographical location. Data provided by Google Analytics between October 2013 and November 2015. *Source*: http://torroja.dmt.upm.es/turbdata/blayers/

During the same period, the raw-data part of the web page averaged 100–150 sessions per month. About 37 % of them correspond to higher education institutions, including more than 40 universities.

These statistics lead us to believe in the potential impact of a web-service that allows full access by the scientific community to raw databases, probably including applications beyond turbulence research.

Acknowledgements This work was funded by the Multiflow programme of the European Research Council.

References

1. http://eudat.eu/
2. http://euhit.icm.edu.pl/services/turbase
3. https://www.opensciencedatacloud.org/
4. https://rd-alliance.org/
5. http://torroja.dmt.upm.es/turbdata/
6. J. Hannan, *God's Philosophers: How the Medieval World Laid the Foundations of Modern Science* (Icon Books, London, 2009). ISBN:9781848310704
7. Y. Li, E. Perlman, M. Wan, Y. Yang, C. Meneveau, R. Burns, S. Chen, A. Szalay, G. Eyink, A public turbulence database cluster and applications to study Lagrangian evolution of velocity increments in turbulence. J. Turbul. **9**, N31 (2008)
8. B. Panzer-Steindel, Data integrity. Technical report, CERN/IT, April 2007
9. Wikipedia, ZFS (2013), http://en.wikipedia.org/wiki/ZFS

Chapter 29
Reacting LES@2030: Near Diskless and Near Real-Time Computing for Design?

S. Menon, R. Ranjan, and J.C. Oefelein

29.1 Challenges for Future Computing

The combustion research and design community is diverse and geographically distributed. It aims to provide a predictive understanding of the complex multiphysics, and multiscale processes that are present in a variety of systems such as transportation, propulsion, and power systems. Challenges involve treatment of turbulence, advanced fuels, multiphase flows, and catalytic systems, to name a few. A key objective is the construction of predictive models that can ultimately be assembled into engineering design tools for development and optimization of device-scale combustion systems.

Given the many challenges, the combustion community relies on interdisciplinary communication and integration of research and design tools through a variety of experimental and simulation approaches. Establishing synergy across the full hierarchy of computational resources (e.g., desktop computers, mid-scale computer clusters, and leadership-class computing facilities) is an integral part of the workflow. The scale at which both computational and experimental data acquisition techniques are producing information is now occurring at unprecedented rates. The growth in data output, coupled with the inherent need for collaborative exchange of information, is now challenging many assumptions about how the scientific process can efficiently operate.

While petascale computing has enabled the application of direct numerical simulation (DNS) for the treatment of three-dimensional reacting flows with detailed chemistry, the largest DNS runs to date are off by more than an order of

S. Menon (✉) • R. Ranjan
Georgia Institute of Technology, Atlanta, GA, USA
e-mail: suresh.menon@ae.gatech.edu

J.C. Oefelein
Sandia National Laboratories, Livermore, CA, USA

A. Pollard et al. (eds.), *Whither Turbulence and Big Data in the 21st Century?*,
DOI 10.1007/978-3-319-41217-7_29

magnitude in Reynolds number compared to practical devices. This presents a gap on the order of 1000 in computational requirements. Likewise, domains associated with practical devices are typically on the order of 1000 times larger and require 1000 times the amount of run-time compared to what is possible with state of the art DNS. Thus, results from DNS are potentially useful to better select underlying physical assumptions used in modeling and to understand fundamentals related to small-scale phenomena. However, it is unlikely that DNS will ever be feasible for design, even in the long term. Thus, some form of modeling will always be required.

Opposite to DNS, the Reynolds-averaged Navier–Stokes (RANS) approximation employs filtering in time to derive the governing conservation equations for the mean state. For this approach, all dynamic degrees of freedom smaller than the largest energy containing eddies in a flow are modeled to minimize the cost of the calculations. For these reasons, RANS is currently the primary method used for industry-relevant engineering calculations since it is the least costly. However, since the models must represent dynamic interactions over the full range of spatial and temporal scales in a flow, they are the least universal in character and provide the lowest level of fidelity. These models inherently involve many tuning constants that must be calibrated on a case-by-case basis. Thus, RANS is by definition limited in terms of true predictive capabilities.

The large eddy simulation (LES) technique falls between the two limits of DNS and RANS. The large energetic-scales are resolved directly and the subgrid-scales are modeled. Just like one chooses the resolution at which a photographic image is resolved, one can conceptually choose the resolution at which pertinent unsteady broadband structures of a flow are resolved with LES if validated subgrid-scale (SGS) models are available. However, as the spatial and temporal resolution is increased, the cost associated with a calculation increases, both with respect to CPU time and the volume of data generated. But the range of scales over which the system of SGS models must work becomes proportionately less and they tend to be universal in character. Given these attributes, LES can represent a range of scales from the DNS limit to RANS and is now widely used as both a tool for scientific discovery (using high-fidelity albeit more expensive first principles models) and for engineering design (using less expensive albeit less universal engineering-based models).

With the advent of massively parallel computers and the rapid increase in the number of cores in a configuration, it has been suggested that the next generation exascale computer will involve millions of processors and simulations on such machines will require specialized fault-crash tolerant algorithms to adjust for local core failures. The scalable exaFLOP algorithms being designed from the bottom up [8, 23] are still in their infancy, and most likely will change rapidly as architectures and compilers evolve. However, for the majority of researchers, such large systems and their associated supporting infrastructure (storage, staff, etc.) will never be available. Therefore, alternate strategies are needed and need to be developed concurrently, especially for engineering design. Of particular interest to aircraft and power gas turbine engine manufacturers (often called original equipment manufacturer, OEM) is an ability to predict and eliminate many unwanted features in their next generation designs. For example, combustion instability, lean blow out, altitude relight, cold start, etc., are all key design concerns. Since these are highly transient phenomena, LES has been proposed as a viable strategy and although there still remains a lot of unresolved issues. Nevertheless, this chapter focuses on the potential of LES methodology to impact engineering design studies in the future and what is any, are the new challenges to consider.

There are many variants of LES and their associated SGS closures and no one specific approach is highlighted here (citations are provided just as example reference since there are many more recent publications). From a practical point of view, LES still has not replaced RANS in the design cycle used by the OEMs. There are many reasons of which some are: (a) RANS, especially steady state variants use commercial codes that are user-friendly and provide rapid turnaround (e.g., 1–2 weeks on 64–128 cores) of very complex and realistic designs. Users who have very little understanding of the underlying mathematics and/or physics can do this rapidly, (b) LES is an unsteady time-accurate approach and statistical convergence requires significantly more time, typically 2–4 weeks for large problems (and typically requiring more than 100 million cells on more than $\mathcal{O}(1000)$ cores). It has been limited to laboratory scale problems in most studies although some recent studies are starting to simulate realistic design rigs [6, 10], (c) LES accuracy is often questioned since grid convergence studies are very difficult to do, and most OEMs have limited access to large systems. This makes it very difficult for engineering teams to use LES as a predictive tool, (d) the very large datasets generated from LES is very difficult to post-process and interrogate for its rich transient signature, often requiring new parallel and in-situ processing tools that are still being developed, and finally, (e) most OEM users understand the limitations of RANS and although they may not believe in all the predicted results they are satisfied if RANS can provide some systematic trends that can help in the design. Since RANS type studies (even unsteady RANS, called URANS) can be systematically conducted regularly by engineering teams, the current trend in OEMs is to use LES only in a limited manner to focus on one or two key areas of interest. Given all the cost overhead (computational and user), such studies require very experienced researchers.

While RANS has served the community well in the past, many of the current design challenges involve phenomenological processes that it cannot represent. For example, for gas turbine aircraft engines, lean-blowout (LBO), altitude relight, cold start, and combustion instability are key areas where designers are interested in understanding and resolving before a hardware is built and tested [16, 21]. Rocket engines have other challenges including supercritical and trans-critical mixing and combustion instability, as well as multi-injector effects [15, 17]. Many if not all of these physics are unsteady and cannot be captured using steady state models and URANS strategy has not provided any reliable insight. Apart from accuracy and robustness of the SGS closures used by researchers, LES requires scalable parallel computing algorithms and tools for rapid and efficient data analysis of the 4D-datasets (three-dimensional in space and in time). Although these tools are also needed by DNS solvers, the range of problems of interest for LES typically involve more complex geometries and complex multiphase physics in a highly turbulent reacting environment.

Parallel computing and processing tools now have to consider the evolving trend in the next generation HPC system that is heading towards a hybrid architecture, which comprise of both shared and distributed computing and processing nodes. The programming languages and supporting libraries are still being developed to facilitate efficient simulations on hybrid systems. Currently, a majority of legacy codes have some form of code acceleration based on hybrid approaches, such as pthreads, multi-cores, graphics processing units (GPUs), and co-processing. This trend would soon become a standard for the solvers suited for simulation on next generation HPC systems. However, users from OEMs are not interested nor are they going to be qualified to write such codes or let alone debug them for problems. New exaFLOP algorithms for engineering use are still a long way off and their use in engineering design although possible, is not going to be the norm.

The alternate is to optimize the existing capability to the best possible level using available resources and then determine how to rapidly transition the results towards design goals as the hybrid architectures evolve over the next 15 years or so. It is noted that most OEMs can access perhaps 1000+ cores over a limited time but have access to 100–200 cores on a regular basis. However, next generation systems that employ the Intel's Many Integrated Cores [11] or GPUs [22] in hybrid systems can already provide 100–1000 cores on a desk-side system. The use of such systems on a regular basis by design engineers will soon be practical. Therefore, LES codes that can exploit systems are being developed by many researchers (including in OEMs). However, even if there are codes that can run optimally on 1000+ core desk-side systems there remains another major roadblock for its use on a regular basis. This restriction is related to storage requirements and the ability to process and post-process the data generated from these simulations.

For example, we are currently conducting LES to analyze the effects of fuel sensitivity on various aspects of combustion dynamics within gas turbines [18]. LES with 29 species (with finite-rate kinetics), polydisperse spray (Lagrangian droplet particles $\sim$ typically $\mathcal{O}(100\text{K})$ parcels with multicomponent fuel properties) and using approximately 10 million finite-volume cells are being run on around 1500 cores. Figure 29.1a shows a schematic of the model swirl spray combustor, which is

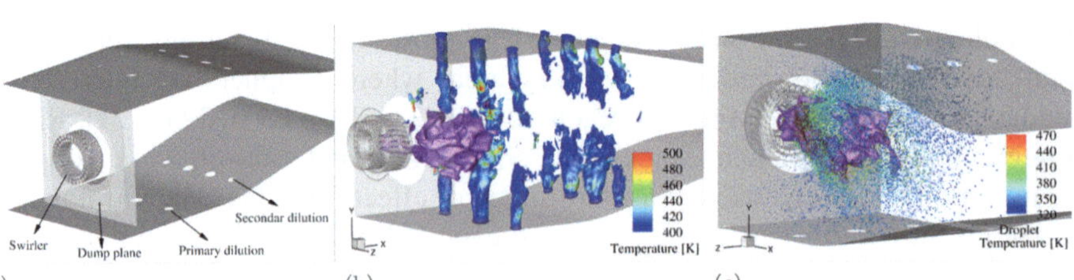

Fig. 29.1 Schematic of the UDRI rig [4] highlighting its various components (**a**), flame surface shown in *purple color* (identified through iso-surface of $T = 1800\,\text{K}$) interacting with the primary and secondary dilution jets (identified through iso-surface of $|v| = 40\,\text{m/s}$) (**b**), and flame surface overlaid with the spray droplets (sized by their radius and *colored* by temperature) (**c**)

being studied experimentally as well [4]. The geometrical complexity is apparent as such combustors include many complex components such as swirlers, combustion chamber, cold plenum, fuel injector, effusion, and dilution holes. Figure 29.1b shows the interaction of a typical flame surface (identified by an iso-contour of temperature of $T = 1800\,\mathrm{K}$) with the cold dilution jets entering the combustion chamber through the primary and secondary dilution holes. The swirling flow acts like an aerodynamics flame holder to anchor the flame. Figure 29.1c shows the flame surface overlaid with the spray droplets and shows how the swirl disperses the spray in the transverse direction but since the droplets continuously evaporate, they mix and burn in the central core and recirculation bubble sustains the auto-ignition process and keeps the flame stable. Analysis of such long-term simulations requires many types of data output and in-situ processing to reduce the demand on the user. For such simulations, typical data storage requirement exceeds over 20 GB for a single instantaneous field of all variables and particle fields (such files although stored infrequently are required for restarts!). If a grid sensitivity analysis is needed and the resolution is doubled along each direction, then the size of the data can quickly reach over $\mathcal{O}(100)$ GB for one instant. To analyze features of combustion dynamics, such as thermo-acoustic instabilities, burning mode behavior, ignition transients, and LBO, we need multiple instantaneous snapshots (perhaps not for all variables but, for example, the three velocity field, pressure, temperature, key species mass fractions, etc.). Typical snapshots (used, for example, to analyze results show in Fig. 29.1) can be as large as 5 GB per instant. Additional data such as vorticity field, strain rate, scalar dissipation, unsteady heat release, and key reaction rates are also needed to understand the flame-turbulence-acoustic interactions that are associated with combustion dynamics. Clearly, storing such huge datasets, which can reach up to an order of 10 TB for a single simulation, is still possible and will be possible in the future but no commodity cluster or workstation even with 1000+ cores can be expected to have a large high-speed data archival system with 100+TBs or PBs of accessible storage. Another challenge is how to access and analyze such big size data, that will be discussed below.

As another example, Fig. 29.2 shows a recent LES aimed at understanding the dynamics of high-pressure fuel injection processes [12]. Liquid n-dodecane at 363 K is injected into a quiescent gaseous mixture at 900 K and 60 bar, which are typical conditions in advanced Diesel engines. Results from a high-resolution case (2 μm spacing across a 0.09 mm injector nozzle) are shown. A three-dimensional rendering of the injected fuel jet is in the center. The red iso-surface marks the central liquid core where large density gradients exist. These gradients induce high shear forces through strongly coupled interactions between turbulence and non-ideal multicomponent thermodynamics and transport processes. Instabilities form in these regions creating flow structures that entrain air and generate intense turbulence. These structures are highlighted by the yellow (500 m/s) to blue (50 m/s) iso-surfaces. Molecular diffusion completes the mixing, as shown by the more diffused nature of the temperature field in the lower panel (363 K in dark blue to 900 K in red). The level of detail captured by simulations such as this require grids that are on the order of 100-million cells to provide the level of spatial resolution required.

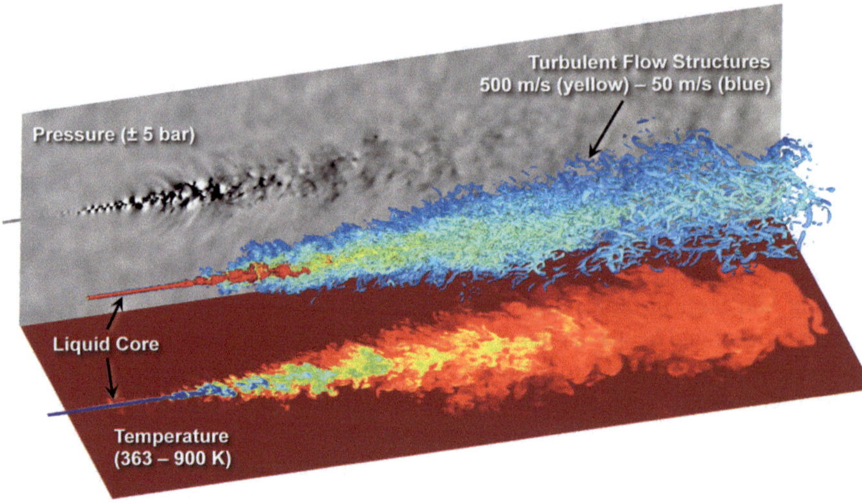

Fig. 29.2 Three-dimensional rendering liquid n-dodecane injected into nitrogen at $t = 200\,\mu$s. The *red* iso-surface marks where density is 200 kg/m^3, the *blue* iso-surface marks a Q-criterion threshold that localizes coherent turbulent structures, the *bottom panel* shows the temperature and rear the pressure

Comparable reductions in the integration time step are also required. Making full use of all the available data (e.g., statistical, transient, rare event, etc.) requires a significant ensemble of runs over relevant design spaces. This requires sets of LES cases that typically take or the order of 10 days per run on mid-scale clusters rated at 50 teraflops, which leads to storage needs on the order of 100 terabytes per dataset. Extensible and scalable I/O using NetCDF or HDF5 is typically required, and the data that is output must be formatted in a manner that is easily assessable for post-processing, which itself must be automated where possible to reduce the labor required for detailed analysis.

Although the above examples are for cases that are currently being pursued they are not something a typical engineer would tackle or even be willing to analyze. Future simulations aimed at design must involve more complex computational domains, longer run-times, and increasingly sophisticated systems of SGS models to represent the multitude of multiphysics processes associated with advanced combustion systems. A full engine simulation (e.g., gas turbine or reciprocating) will require grid sizes on the order of 1-billion cells, and potentially datasets that approach 1 PB. Interactive and in-situ data processing and visualization will become a common necessity to offset the overall size and cost of data that is stored. A typical LES simulation 10 years from now may nominally involve computer systems that range from $\mathcal{O}(10)$-petaflops (for local "mid-scale" clusters) to $\mathcal{O}(1)$-exaflops (for leadership-class supercomputer centers). Datasets that are on the order of petabytes will potentially be generated. However, the amount of bulk data generated per case will be offset by in-situ "on-the-fly" data processing. Data transfer rates will also need to be optimized. For example, if 200 files are written out at approximately 1 file per hour in a 7–10 day run, the data generation rate will approach rates on the order of 20-GB/s. If the data is moved elsewhere at the rate it is generated, then networks will need an even higher bandwidth to account for overheads due to protocol, metadata, contention, etc. Anticipated goals are to stream data to an analysis and rendering machine as it is produced rather than waiting until the run is complete. Then, known analysis tools can be applied to the data as it is generated to get an initial understanding of the underlying physics. Subsequent iterative analysis can then be performed off-line.

The paradigm shift outlined above will require automated workflow scripts that facilitate data streaming, morphing, archival, and analysis. Data will need to be moved to platforms and archival storage within larger supercomputing centers. Reduced data will need to be transferred to local facilities for refined analysis and rendering. To strike a balance between the size of data to be moved and speed at which it can be generated, typical simulations in the future will be instrumented with in-situ data processing capabilities such as feature detection, segmentation, and tracking to enable data reduction and querying on-the-fly. This will effectively reduce the volume of data required for further analysis and enabling steering of adaptive I/O. In addition, web-based portals developed for sharing data will need to be developed. The goal is to work towards a scalable, extensible framework for analyzing large data. The framework will ideally adopt standardized formats, translators, graphics, combustion analysis software, parallel feature detection and tracking libraries, and query tools that can operate on portions of the data where it resides. Reduced data and remote visualization results will then most likely be transferred via high-speed networks.

To make LES useful in the design cycle some additional new thinking is needed even given all the potential of in-situ processing discussed above. We assume for the sake of argument that in the near future (e.g., 2030) desktop (desk-side) workstation/server will easily contain 1000–4000 cores—perhaps with a hybrid architecture using CPU-GPU or CPU-MIC or some other combination. We could also assume that 100 TFLOP is feasible on these systems. For example, Intel Knights Landing (KNL) and the future Knights Hill (KNH) system using 64-core dual-chips on a node (for 128 cores) can achieve about 6/16 TFLOPS double-/single-precision performance [19]. Typical turnaround for some problems could be overnight (e.g., 12 h between 5 pm and 5 am) with automated script running post-processing so that the engineer can review the results with his morning coffee! or it can be longer say 2–3 days (e.g., weekend) or even more specialized (1–2) weeks.

Assuming such simulations become routine, the challenge remains on how users can process the 4D datasets. In a traditional approach of data analysis, various types of files are stored in LES: (a) Restart files (infrequent but must contain all the key variables to restart the simulation), (b) Statistics files (these contain time-averaged statistics and may contain many variables, including many in-situ processed variables, e.g., Reynolds stresses, Rayleigh criterion, etc.), (c) Trace files (typically single point time evolution data of field variables at various locations to extract spectra, two-point correlations, etc.), and (d) time-sequence of the entire

3D flow for some specific features that one wishes to process and analyze for visualization and/or extract new physics (these instantaneous snapshots can rapidly cause increase in the storage demand and efficient data transfer over network requirements).

The above approach implicitly implies that (a) the user knows what he is looking for and (b) the chosen fields are sufficient for make conclusions from the simulations. Processing the data from even the current LES takes $\mathcal{O}(10)$ or more CPU hours (including accessing the data from the archival system). The advantage of this approach is that the stored data can be interrogated multiple times as needed since it forms a stored database. Stored DNS fields are used in this manner [3]. Therefore, the data analysis approach can evolve as we analyze the data. However, this approach is not feasible when it comes to huge datasets generated from LES of realistic combustors, especially when both the time-averaged and transient features have to be analyzed. It is noted that currently DNS fields are only stored at a few random instants [9] and can easily run into 100s of TB at this time. Future DNS will require O(100) Pbs of storage [2].

An alternative approach is to perform online data analysis (ODA), which comprises of approaches such as UDF-based analysis and in-situ visualization/co-processing [1, 5, 14]. Here data is post-processed while the simulation is ongoing and only the data we hope will reveal the physics of our interest is stored. There are several advantages and disadvantages with this approach. The ODA approach has low storage requirement and this allows for data extraction at a high frequency that may be essential to investigate multi-dimensional high-frequency combustion dynamics. Since the stored data is limited, efficient data access and efficient processing can help the user to rapidly investigate the physics and determine his/her next course of action. On the other hand, a major issue is that often we are not aware of what to look for other than the usual suspects (velocity, temperature, pressure, heat release rate, flame surface, etc.). With the ODA approach, if we have not saved such data, then the only option is to run the simulation again to extract the new features. Therefore, ODA approach requires a significant thought-process during the planning stage of the simulation—even experienced LES researchers often overlook some of the key features. For engineering professionals more interested in getting a quick insight into small changes in the design this approach can still be feasible provided proper post-processing tools are integrated within the simulation software.

Another major issue with the ODA approach, which needs to be addressed in coming years, is the parallel scalability of the data extraction techniques and file I/O. Recently, strong scalability was achieved for in situ post-processing up to 60K cores using *libsim/FieldView* integration with an LES solver [5]. Further improvement in the scaling up to $\mathcal{O}(100)$ K cores are currently underway. Some other approaches for efficient ODA are (a) data staging to secondary memory through ADIOS interface to the solver [1, 13], (b) Paraview co-processing [7], and (c) MPI-I/O virtual file driver of HDF5 redirecting the data in parallel to a distributed shared memory buffer over multiple TCP connections [20]—all of which are currently being developed but still remains to be optimized. Such strategies or perhaps even more optimal tools need to become easily accessible to typical future engineer to allow for the ODA approach to be fully exploited in the design cycle.

29.2 So Recompute Rather than Store?

The increases in the speed and size of computers coupled with advances in flow solvers continue to facilitate more comprehensive and quantitative science and engineering across wider and more relevant parameter spaces. There are currently two orders of magnitude difference in speed/size from the fastest computer in the world to the 500th fastest. Similarly, there has been a four order of magnitude change in the speed/size of all classes of machines over the last 15 years, and this trend is expected to continue, albeit while requiring much more code optimization to truly make use of the potential peak system speeds. This has already led to a shifting paradigm in how simulations are used. In the near past, typical cases involved detailed analysis of simple flames and comparisons with experiments at only a few target conditions. These conditions are typically at low-pressure, with idealized thermodynamics, transport, and chemistry. They involved 1–10-million CPU h/case, using 1–10-million cells/grid. Current and future cases involve detailed parametric analysis of fully coupled multiscale physics, with complex fuel-oxidizer combinations and reactions, high-pressure thermodynamics, transport, chemical kinetics, and heat transfer. In addition, advanced uncertainty quantification techniques are now being applied to wide parameter spaces, using 10–100-million CPU h/case, and 100–1000-million cells/grid.

Future LES code capabilities will continue to be refactored to efficiently scale on the newest architectures with improvements in attaining a greater fraction of theoretical peak architecture speeds. Continued development of data analysis tools and methods in-situ data reduction done in parallel on mid-scale clusters and parallel storage systems will also occur. This will include interactive remote data processing and visualization. Precise data storage needs will have to be defined based on a balance between the volume of data stored, and the cost of doing analysis on-the-fly. We envision that @2030 desktop workstations will be able to process large-scale realistic LES setups using high-fidelity codes. Reliable data will be available overnight, but other than token data being stored for checkpointing, statistics, and backup, most of the analysis will be facilitated in-situ and in parallel to the simulation. Thus, data storage will become minimal, with limited backup, and erased and replaced for parametric studies. This in term will begin to minimize the complexity of current storage systems, such as Lustre. How the future evolves will depend significantly on how multi-disciplinary studies are focused to achieve this goal rather than an ad hoc top-down approach where each area evolves and adapts to what is available at a given time. Further developments in many fronts are critically needed since after all, 2030 is less than 15 years away!

References

1. J.C. Bennett, H. Abbasi, P.-T. Bremer, R. Grout, A. Gyulassy, T. Jin, S. Klasky, H. Kolla, M. Parashar, V. Pascucci, et al., Combining in-situ and in-transit processing to enable extreme-scale scientific analysis, in *2012 International Conference for High Performance Computing, Networking, Storage and Analysis (SC)* (Institute of Electrical and Electronics Engineers (IEEE), New Jersey 2012), pp. 1–9
2. J.H. Chen, Combustion co-design center: exascale simulation of combustion in turbulence application/proxy deep dive. Tech. Rep., The Technical Report Number SAND2012-8619C, Sandia National Laboratories, 2012
3. J.H. Chen, A. Choudhary, B. De Supinski, M. DeVries, E. Hawkes, S. Klasky, W. Liao, K. Ma, J. Mellor-Crummey, N. Podhorszki, et al., Terascale direct numerical simulations of turbulent combustion using S3D. Comput. Sci. Discov. **2**(1), 015001 (2009)
4. M. Colket, J. Heyne, M. Rumizen, J.T. Edwards, M. Gupta, W.M. Roquemore, J.P. Moder, J.M. Tishkoff, C. Li, An overview of the national jet fuels combustion program. AIAA-2016-0177, 1–24 (2016)
5. E.P.N. Duque, B.J. Whitlock, C.P. Stone, R. Ranjan, S. Menon, The impact of in situ data processing and analytics upon weak scaling of CFD solvers and workflows, in *27th International Conference on Parallel Computational Fluid Dynamics* (2015), pp. 1–8
6. T. Echekki, E. Mastorakos, *Turbulent Combustion Modeling: Advances, New Trends and Perspectives*, vol. 95 (Springer Science & Business Media, New York, 2010)
7. N. Fabian, K. Moreland, D. Thompson, A.C. Bauer, P. Marion, B. Geveci, M. Rasquin, K.E. Jan, The paraview coprocessing library: a scalable, general purpose in situ visualization library, in *2011 IEEE Symposium on Large Data Analysis and Visualization (LDAV)* (Institute of Electrical and Electronics Engineers (IEEE), New Jersey, 2011), pp. 89–96
8. X. Guo, M. Lange, G. Gorman, L. Mitchell, M. Weiland, Developing a scalable hybrid MPI/OpenMP unstructured finite element model. Comput. Fluids **110**, 227–234 (2015)
9. E.R. Hawkes, O. Chatakonda, H. Kolla, A.R. Kerstein, J.H. Chen, A petascale direct numerical simulation study of the modelling of flame wrinkling for large-eddy simulations in intense turbulence. Combust. Flame **159**(8), 2690–2703 (2012)
10. J. Janicka, A. Sadiki, M. Schäfer, C. Heeger, *Flow and Combustion in Advanced Gas Turbine Combustors* (Springer Science & Business Media, New York 2012)
11. J. Jeffers, J. Reinders, *Intel Xeon Phi Coprocessor High-Performance Programming* (Newnes, Elsevier Inc., New York, 2013)
12. G. Lacaze, A. Misdariis, A. Ruiz, J.C. Oefelein, Analysis of high-pressure diesel fuel injection processes using LES with real-fluid thermodynamics and transport. Proc. Combust. Inst. **35**, 1603–1611 (2015)
13. J.F. Lofstead, S. Klasky, K. Schwan, N. Podhorszki, C. Jin, Flexible IO and integration for scientific codes through the adaptable IO system (ADIOS), in *Proceedings of the 6th International Workshop on Challenges of Large Applications in Distributed Environments* (Association for Computing Machinery, New York, 2008), pp. 15–24
14. K.-L. Ma, C. Wang, H. Yu, A. Tikhonova, In-situ processing and visualization for ultrascale simulations. J. Phys.: Conf. Ser. **78**, 012043 (2007). IOP Publishing
15. M. Masquelet, S. Menon, Y. Jin, R. Friedrich, Simulation of unsteady combustion in a LOX-GH_2 fueled rocket engine. Aerosp. Sci. Technol. **13**(8), 466–474 (2009)
16. S. Menon, Co emission and combustion dynamics near lean-blowout in gas turbine engines, in *ASME Turbo Expo 2004: Power for Land, Sea, and Air* (American Society of Mechanical Engineers, New York, 2004), pp. 153–160
17. J.C. Oefelein, V. Yang, Modeling high-pressure mixing and combustion processes in liquid rocket engines. J. Propul. Power **14**(5), 843–857 (1998)
18. R. Ranjan, A. Panchal, G. Hannebique, S. Menon, Towards numerical prediction of jet fuels sensitivity of flame dynamics in a swirl spray combustion system, in *52nd AIAA Joint Propulsion Conference* (Salt Lake City, 2016)

19. S. Saini, H. Jin, D. Jespersen, S. Cheung, J. Djomehri, J. Chang, R. Hood, Early multi-node performance evaluation of a knights corner (KNC) based NASA supercomputer, in *Parallel and Distributed Processing Symposium Workshop (IPDPSW), 2015 IEEE International* (Institute of Electrical and Electronics Engineers (IEEE), New Jersey 2015), pp. 57–67
20. J. Soumagne, J. Biddiscombe, J. Clarke, An HDF5 MPI virtual file driver for parallel in-situ post-processing, in *Recent Advances in the Message Passing Interface* (Springer-Verlag, Berlin Heidelberg 2010), pp. 62–71
21. S. Srinivasan, R. Ranjan, S. Menon, Flame dynamics during combustion instability in a high-pressure, shear-coaxial injector combustor. Flow Turbul. Combust. **94**(1), 237–262 (2015)
22. J. Tölke, M. Krafczyk, TeraFLOP computing on a desktop PC with GPUs for 3D CFD. Int. J. Comput. Fluid Dyn. **22**(7), 443–456 (2008)
23. R. Yokota, L.A. Barba, A tuned and scalable fast multipole method as a preeminent algorithm for exascale systems. Int. J. High Perform. Comput. Appl. **26**(4), 337–346 (2012)

Part VIII
Discussion 1: Challenges in Turbulence in the Twenty-First Century—What Problems We Should Focus On in the Next 20 Years?

Chapter 30
Whither Turbulence and Big Data for the Twenty-First Century

Discussion 1: Challenges in Turbulence in the Twenty-First Century—What Problems Should We Focus on in the Next 20 Years?

Andrew Pollard

Fazle Hussein (Chair for the Discussion): Okay, we had two exciting days of new research, new ideas, new results. Now it is time for us to share with each other what are we going to do in coming years for the rest of the century. Since many of you have ideas we also have, I suggest that both you and I say nothing, but let you come and say. I had approached personally Christos and Jean-Paul, so maybe some of you can say a few words and then more or less everybody has to say something. Christos where are you?

Christos Vassilicos: This is Christos. Fazle's question is about the future and I'm guessing about 20 years would be about right, which is a long time. So the first problem I think we should look at is of course what has been discussed in the last talk, non-equilibrium turbulence. This type of turbulence is a kind of turbulence that appears now to be the norm. Kolmogorov-type equilibrium appears to be the exception, as it happens only whenever you balance energy injection rate with dissipation instantaneously or nearly so. Most turbulent flows are unsteady in the sense that their turbulent kinetic energy varies in time or in space for spatially developing flows. In such cases, the turbulence is out of two-point Kolmogorov equilibrium and you have non-equilibrium turbulence. The good news is that there is a dissipation law, which appears to be common in many such flows, including different types of grid turbulence, whether the grid is fractal or regular, axisymmetric wake turbulence, and even DNS of periodic turbulence, unless you go very far downstream or in time where something else happens, which

Andrew Pollard is the reporter for the discussions and that most speakers have made minor edits to the transcription that appears here.

A. Pollard (✉)
Department of Mechanical & Materials Engineering, Queen's University, Kingston, ON, Canada
e-mail: andrew.pollard@queensu.ca

A. Pollard et al. (eds.), *Whither Turbulence and Big Data in the 21st Century?*,
DOI 10.1007/978-3-319-41217-7_30

again not Kolmogorov equilibrium but nevertheless exhibits the classical turbulence dissipation scaling. The problem now is to understand the inter-scale physics of non-equilibrium turbulence and its various laws and scaling and determine ranges of validity across an increasing number of turbulent flows. The accompanying problem is to devise turbulence one-point and/or two-point turbulence models, which take non-equilibrium physics and scaling into account. I expect a first theory of non-equilibrium turbulence to surface within a year or so, but the hard work of investigating it in many different cases of unsteady turbulent flows and testing new turbulence models will and should take years, perhaps a decade or two depending on how many colleagues decide to take up the challenge.

Second, we have been discussing self-preservation/self-similarity and many of us here just assume it in our thinking and procedures or just test for it even though self-preserving profiles exist hand in hand with dependencies on inlet/initial conditions, even though some profiles are self-similar and others are not. It is about time we work towards some deeper understanding of the causes of self-preservation and understand why we can assume it, when and under what conditions. The Navier–Stokes equations admit not only such a statistically self-preserving solutions but also many more solutions and it is not clear why one is chosen against another.

The third problem is wall turbulence given the large amounts of new data produced over the past about 20 years or so on canonical wall flows such as turbulent boundary layers and turbulent pipe flow by the Princeton, Lille, and Melbourne groups. Clearly there is now a need to start theorizing about it as this may be the moment in time where new theory will be catalytic to make the most of these new data and future experimental campaigns. New theories should include basics such as the mean flow profile, turbulence profiles and the second turbulence peak, its relation to underlying flow structure, fluctuating pressure, the role of dissipation and cascades in such flows, etc. This is something which should be and I believe is coming, sooner because so much new data is available than ever before.

Luciano Castillo: So your point on that one is to look at the second peak in the turbulent boundary layer?

Christos Vassilicos: No, not only. I am talking about boundary layer structure, various flows profiles, new theories based on new assumptions, integral scale profiles, etc. There is enough data now that perhaps we can do some new theory, in fact we should do new theory to get the most of the data and decide on new measurements.

The fourth problem is the one introduced by Corrsin and Kistler in 1955 and concerns interfaces, which appear in various ways and circumstances in many turbulent flows. Turbulence models (RANS or sub-grid models for LES) based on eddy viscosities are not in any obvious way in agreement with interfaces and their physics. In this man-made sense, turbulence is not supposed to have interfaces. This is another phenomenon that we have to understand better in relation to turbulence and turbulence modeling.

These are the four problems I would work on in the next 20 years.

Fazle Hussein: I recommend we discuss each persons comments as we go along.

Ivan Marusic: I agree with Christos, I think the four problems that he listed are all important. I'll comment on the third one, since I've been involved with a lot of these experimental data that have been generated. What is new? Well if I look back over the past 20 years, our field has arguably been at the forefront of generating big data, just by the nature of the problem, and this comes both from numerical simulations and experiments. Now we are starting to look at things in four dimensions (space and time) and with high resolution, where 20 years ago we were still looking at single point measurements and then trying to uncover the three dimensional structure and so forth. Perhaps 20 years ago DNS and experiment were thought of by some as being in competition, but now it is recognized that the techniques are complementary and most beneficial when used synergistically. Large-eddy simulation is a good example of a technique that greatly benefits from both experiment and DNS, and most likely within the next 20 years it will become the industrial standard for computing. With these new simulations and experiments comes a great richness and abundance of data. What is needed, however, are new theories that use these data and takes our understanding to the next level.

One of the things I didn't get a chance to present during my talk, but one of the things that becomes very obvious when you look across the logarithmic layer or the outer layer of the boundary layer in instantaneous PIV results with a very large dynamic range, and in fact Michel (Stanislas) showed this in his talk is this, you see very well organized and distinct variations of momentum across the layer. So when I look at the logarithmic layer, I don't see a logarithmic profile with perturbations about the mean profile, as would be typical in a Reynolds decomposition, rather I see a staircase with discrete steps. Through ensemble averaging these staircases, which appear jittered in space from frame to frame, produce on average the logarithmic law. Now that's an observation or hypothesis, which requires further testing. However, if it is true then this highlights the need for new theories that take into account those discrete steps. In future efforts we need new theories that go well beyond just POD, and we need to move to dynamically model that use the observations that the new data present to us. The search for new theories will undoubtedly benefit from engagement with people in other communities in what is a very rich area.

As a further comment I think the work that Will Anderson and others showed on sparse roughness, including Vincenzo's (Armenio) work in geophysical flows, nicely highlights that significant advances are likely to come in the coming years on the long-standing of rough wall flows. Roughness has always been the can of worms, but I think we're ready to tackle it now in a clear way. This includes looking at practical types of rough flows, well beyond homogenous rough surfaces. This and other applications of wall turbulence lead to working and contributing to problems of great societal importance, and really I think it is an exciting time to be in the field. When I started working in this area, many outside the field regarded turbulence as a dead area, but the reality is just the opposite. There are so many important applications that we can contribute to. So, I see it as a good time to be in the field.

Fazle Hussein: Ivan was already on the list, later. Try to preserve your comments. Let's stick to the comments of Christos addressed. And we'll come back to anyone who wants to start a new topic, that's fine. Anyone else want to respond to what Christos said?

Jean-Paul Bonnet: I'm interested to say something, not about turbulent boundary layers, but I have this impression that there is two communities in fact, one which is here, which is very deeply engaged in turbulence, knows about spectrum, knows about equilibrium, non-equilibrium, and a lot of younger people that just consider us here that turbulence is bad or wrong. We have to be careful to maintain this culture of real turbulence from the beginning. Because due to measuring techniques and CFD and so on that do not require any models. And this is particularly important for flow control strategies because there is some success in flow control strategies where you can disregard the physics or just concentrate on stability analysis and different stages of open-loop control I think its dangerous to leave this culture in the background to a smaller community. And difficult problem that I think we are facing in the future at least for flow control is could we analyze for different scales as they appear. For example, dissipation and production, how is it possible to maybe revisit the concept of cascade? In time or in space? When you are acting and not letting the flow evolve in equilibrium but by putting it out of equilibrium.

Christos Vassilicos: you have to work to keep it there.

Fazle Hussein: Jean-Paul I guess, maybe we will have a combination of both new comments and response; I was trying to withhold my own comments. I would like to inject myself now. And also I would to see that each of us try to limit our comments to about 2 min each time, you'll have another round. So my one comment, my two pennies worth to add at this time is I think we have had a lot of noise about structure of turbulence, including myself but we have not brought it to an analytical or mathematical form for it to be used in modeling or understanding turbulence. And one example I will give which I have been a proponent of in the early 1980s that reconnection which is never complete. The succession reconnection, so I propose that the reconnection cascade is the perhaps tractable and viable turbulence cascade mechanism. Reconnection I think is also very important in mixing as well as also in noise generations. So here is one example this structured aspect may give some insight and some link to analytical framework in turbulence, which has been lacking. So this is one contribution, somebody else, who wants to talk?

Bernd Noack: I would like to emphasize Ivan's statement that we need to reach out to other communities. Our progress in turbulence control would have been limited if it were done only by fluid dynamicists and if it were not significantly shaped by control theorists. Let us take turbulence theory as another example; the turbulence cascade has been discovered by Kolmogorov, a mathematician. We can also take the work of Kraichnan, a physicist and Einstein's last assistant. Currently big progresses are introduced by machine learning. Turbulence can get addressed from many interdisciplinary aspects. For instance, there is no dynamical systems guy in this room, no statistical physicist, no Bayesian theorist, and no maximum entropy expert.

Fazle Hussein: No, Mark Glauser is a dynamical systems guy. There are people who do dynamical systems.

Bernd Noack: Dynamical systems from the physics community. I think what we need to do is reach other to other communities. If I look here it's a closed club. When we discuss the stress–strain relationship, always in terms of eddy viscosity between stress and strain. We can ask for the most likely relationship of what was entropy and so on. So the questions to ask here in this room in terms of turbulence modeling, it could be much richer if we had physicist, mathematicians, computer science people, and a number of other audiences in this room.

Fazle Hussein: I think you have a very good point; I don't think any of us keep anybody out and lock the door. I mean they're welcome, but I think we should seek them out also. I totally agree with you. I personally have an interest in coherent structures from a completely different point of view, its maybe somewhat of a viewed as a joke but it is not. Try to keep comments for 1 min, we can go around again so that everybody gets a chance.

Vincenzo Armenio: So let's go really quickly on an example. Today, someone asks you for a solution to a complex industrial problem, for instance. You can use your RANS and your answer got very bad results. You try to use your LES for practical problem. Basically there is a lack of tools to solve practical problems. So, this is a very important issue and now LES is useless for these kinds of problems because the time needed for pre-processing, running, and post-processing is sum of the magnitude much larger the time we get for RANS and the results are comparable. So why to use LES? So in the next 20 years maybe LES will become a practical tool, but if there is a focus on this. The second point is if all people working in turbulence, we can imagine some problem but if we don't have feedback by other communities of their problems from simulation like roughness, like problems of control, and other problems in different field how can we interact with them? We need interaction with other communities which use the results of our activity for some practical problems. Otherwise, we will not be able to attract funding for our fundamental studies.

Andrew Pollard: I think one of the things I want to try to encourage is a cross collaboration idea not only among different communities but also within our own community. We all have our own experimental facilities, for example, and some of us have reasonable computing resources but we need to cross boundaries to enable linkage to bring these two together. So, one thing is to identify a number of what we would call canonical-type problems that are of broad interest; for example, Luminita has some experimental data on her variable viscosity jets. Can we find somebody to do a DNS and link them together and then you start asking some more interesting questions because you have access to data coming from the same conditions whether that be the inlet conditions, the boundary conditions, and so on. And there is a cross communication between individuals and groups. I think as we move foreword with funding issues nationally then its going to become more important to do it on an international basis. So EuHIT is a prime example and we're going to see something tomorrow about the WINDEEE Dome at a Canadian university (Western). These are unique facilities that we always need to be aware of what their capabilities are and learn how to communicate and share.

Fazle Hussein: Thank you Andrew. What I recommend as you make comments try to suggest actionable items. Something we can do and what we can do about it.

Mark Glauser: I would like to add to Bernd's comments. I strongly feel that we need to get much more connected to computer science and computer engineering communities. Perhaps that's Bernd and I both have worked closely with flow control pipes and embedded systems and so on. So, I think this is a critical thing for us to do. I have been fortunate enough to work with some of my colleges in this space and I think that it has really paid off in some interesting things that could have an impact on the field. Another thing I think we need to think about how we even talk about our field a little bit differently. Let me give you an example. Last year I was a member of a panel that evaluated NASA's fundamental aeronautics program and of course as many of you know in the states NASA has the aeronautics piece cut, cut, cut, cut, and so they during our briefings they showed us a wonderful example of our colleague Israel Wygnanski who wanted to be here. He was involved with a team that got open-loop flow control on a tail of a 757 and demonstrated just how much you could do with un steady flow control. He probably will be able to reduce the size of that tail by 50 %. And it was in todays AIAA announcement, the Boeing demonstrated this in flight test. Now let me tell you the problem of how this was presented to us. I was part of this panel and we're at NASA Langley, and they brief us on this and they show us all this wonderful experimental data that they had from the big tunnel at NASA Ames and then we ask where's the corresponding CFD? Uhm, blank looks from the group in the room. Then we probed a little bit deeper and guess what, the CFD didn't work. Why? Because all the CFD data they were focusing on was for steady problems. Steady and the mean problems. They weren't focused on the unsteadiness. And we basically said to them, you're complaining because they're shutting your facilities down because you can do everything with simulation and I said you need to stop BS-ing the management. You need to be honest and say our CFD sucks and we need to be able to develop CFD that's calibrated and developed with careful experimentation. And that leads me to one other thing then I'll shut up. Basically the idea that fluid information triad, this is an idea that we had a working group within AIAA for about 6 years, the idea that we bring theory, experiment, and simulation together to solve these hard problems. We need all of these techniques. And we need to know how to fuse these and bring these ideas together at different levels of fidelity. So I'll hand it off but focus on the unsteadiness.

Paolo Orlandi: As I pointed out in my lecture dealing with turbulent flows, we are lucky with respect to other research fields because we have the model, that is, the system of Navier–Stokes equations. The most difficult part is to impose the boundary conditions of the problem of interest, ranging from applications occurring in our everyday life to those necessary for the industries to produce more efficient goods. The common view is that in real problems turbulent models are requested. For several years in the past we have relied on RANS closures, giving satisfactory results only in flows deeply investigated by laboratory experiments, with the result to assert that the universality of the model is impossible to be achieved. Subsequently there was the hope that the LES could lead to the desired universality, but also with this model based on physical grounds, it was observed that ad-hoc tuning is needed. The result achieved after more than 30 years is that a valuable tool is the DNS largely improved by efficient discretization techniques and by the possibility to use a huge number of processors in parallel. The use of MPI routines allowed to perform simulations at Reynolds number comparable to those in laboratory, and the possibility to evaluate any quantity of interest has led to a better understanding of the complex physics of turbulent flows. Therefore, to have a large impact on the society our community should demonstrate that the output of our research is the basis to build a better world. This can be obtained by building a strong link between those involved in numerics and experiments, with the DNS focused on the preliminary approach to understand the effects of changes in the boundary conditions at low or intermediate Reynolds numbers, and only after in laboratory experiments the variations with the Reynolds number could be understood. In this meeting, including my presentation, we observed a large amount of work devoted to understand the occurrence of strange behaviors in turbulent flows, e.g., the second bump in the spectra in wall-turbulent flows or the non-equilibrium Kolmogorov behavior in grid turbulence, of interest only to stimulate discussions in our small community, but not useful to demonstrate that we could have a large impact in our everyday life.

Carsten Westegaard: I just want to pick up on two things and Mark's just said one of them. It boils down to controls, which is a really important topic for the future. Coupling fluid mechanical solutions with controls. I mean it's true for aerospace, it's true for wind energy, it's true for the wind turbines, it's true for the wind farm, it's true for all of the industrial processes. And controls are really an area where there is a lot of sophistication but it's been very difficult to couple with a mechanical solution, which has a practical implementation. Simply, because it takes an enormous time to solve the problem in fluid mechanic solution. So that's one area, another area I want to reiterate is large scales, so don't sweat the small stuff. Large scales are super important, wind energy resolves landscape presentation, all environmental we human beings are polluting the earth and big scale flows are super important in this context. So anything that's big and large scales and large convection or large mixing problems, those are super important. So don't sweat the small stuff, they aren't going to be that important to us in 25 years.

Roger Arndt: I would like to underscore some things that Mark said. It's bouncing around here a lot of things about theory, about simulation, about experiments. And people are doing a lot of experiments. I think a very important point is to try and have a research program that combines all three in the same program. And I've been doing this for over 50 years and I can tell you there is just an incredible amount of good things that come out of having these, say a numerical student and say an experimentalist working together. You would be amazed by how many things about numerical simulations that are really screwed up. I'm not trying to be facetious but I've had cases where they've been using boundary conditions at the aft end of an experiment but the experiment has been doing it on the front end. There is just so much that comes out of having a good interaction. And I know it's hard to do in this funding environment, but I think it's extremely important to take that into account.

Julio Soria: This comes from Tony Perry—Theory is the brain, numerics is one muscle, and experiments are the other muscle but really the theory has to be in both.

Because, otherwise both muscles cannot talk to each other. So, I think underpinning thing is the theoretical framework that helps to build the body of understanding. You can't do one without the theory.

Peter Johansson: GKN Aerospace produces parts for aero engines to major manufacturers such as Pratt and Whitney, GE, and Rolls Royce. I would like to add a couple of comments from the industry perspective. Basically, we have 2 problems. The first is to predict pressure differences to compute drag or performance. The second is to predict separation margin. Our current design systems are heavily based on RANS with experimental validation support. Whenever we want to go outside the currently known design space, we do not know what margins we have and have to validate the computations. We have some research programs focused on LES and combinations of LES and RANS and such. Also here, validation is a key and access to reliable experimental data is of chief importance also for the industry.

Horia Hangan: We have a long tradition in wind engineering. It's a group that does work on wind energy and with the construction industry mainly, and now with other industries, for 60 years while keeping the academic hat on at the same time. So I am glad to hear, frankly that our thinking has moved towards un-stationary because all the wind phenomena we study now don't, we study tornados we study things that have nothing to do with stationary flows. They are un-stationary in nature and I think there is a huge opportunity for turbulence people to step in there and actually lead the game because if they don't do that its going to be lead in the wrong direction either by very practical engineers that are interested in modifying codes in a wrong way or by industry that is going to do CFD without understanding the differences in peak pressure that happens in un-stationary phenomena as apposed to a peak pressure that happens in a stationary phenomena. Which are basic things, so I think it's a huge opportunity for this community but it has to take the opportunity and become meaningful to what the industry is and needs. So I'm very happy to hear that that's leading up from there. And of course I think it's time for new experiments, and I hope I will try to exemplify that because we are doing wind tunnels and wind tunnels and flowing one direction and stationary for I don't know how many years now. So if we follow this direction we have to follow a new thinking in terms of new experiments.

Bernd Noack: As said before, we need to combine experiments and simulations and see where it ends up. If I look at what we discuss here, it is quite different from what we discuss in interdisciplinary fora which is often nourished by big dreams. In nuclear fusion one can get billions of Euros for something that might not even materialize in the foreseeable future and this nuclear fusion dream is present in many discussions. They do interdisciplinary research and convince politicians and other deciders that they are doing a highly relevant research. If one goes to CERN or the machine learning community you will find similar outcomes. We need to offer our benefits to society which could be drag reduction, noise reduction, and so on to make our area attractive and sexy for the smartest brains that the scientific community has to offer. We could, for instance, articulate well-defined benchmark problems like drag reduction of a D-shaped body and ask other people to integrate their control ideas. We could invite other audiences and create some sort of benchmarking for this problem explaining the relevance it has for greening traffic. So I would like to have a view on our problems from somebody else, from another community, from a young student, from physicists and mathematicians. An important question is how we can attract those people to our research field.

Bill George: I think I have been working in this field at least as long as anybody here, technically 50 years I think. And I think we are our own worst enemy. I remember Rodger (Arndt, ed.) making a comment one time when he was at NSF; it was that they were with-holding 20 % of their budget for extraordinary proposals. So each division had 20 % of its budget held back, if I remember the number, and if you had really good proposals you could go forward as a program manager and compete for that money. And Roger's comment was "Every proposal I send out I get one excellent, two goods and three poors." And it doesn't matter if it's John Lumley or Fazle Hussain or someone I've never heard of (as the P.I.), I can guarantee there will be three poors and two goods. And I've got to compete with these physicists who will have five excellents. I really think we are our own worst enemy. We are automatically hostile to new ideas and papers. It shouldn't take 12 years to produce a paper. We seem to operate under the presumption that if it's not our idea it's not good. And if it violates anything in the past it's not good. And as long as we approach our field this way we don't have a prayer of competing for reasonable resources.

The other comment I would make is as a field we've actually had the opposite problem. Instead of buckling down to the difficult task of actually solving the Navier–Stokes equations and understanding the solutions to the Navier–Stokes equations, using the tools that we have like POD and wavelets, we've instead opted for every new idea that came along. In my career first it was intermittency. And then it was chaos, then log normality. Everybody had a new idea and any new idea anywhere in the world, the turbulence community would jump on it and eventually it would turn out to be nothing. And so decades later, we're back to the same old two things, the only two things that actually decomposed the Navier–Stokes equations that I know of are Lumley's original POD and wavelets. We have to decompose the flow in some way in order to reduce its dimensions. If we don't do that with something that is analytical and mathematical, we are never going to make progress.

Alfredo Pinelli: I agree with George, we've seen that there's a large vast investigation on large structures in wall-bounded flows. It's not clear to me, maybe I'm ignorant, what is the effect of those large structures with the correlations representing the decrease in pressure at the wall on all those techniques that were represented in the 1990s for controlling skin fraction drag reduction. For example, we know that oscillating wall does not work very well when you increase the Reynolds number. So I think we should try to focus and try to understand what is the impact on those large structures on affecting the possibilities of controlling skin friction drag. And another thing I would like to say is that concerning POD also, I remember the work of Wallace in the 1990s again, trying to simplify dynamic system for close to the wall region, trying to control it. And all of a sudden it seems to completely disappear from the scene, what happened to that, why is there no more interest?

Fazle Hussein: Thank you. Let us focus on actionable items things we can do something about. You're right but . . .

Martin Wosnik: I want to comment on a couple of things, one is new areas, one is about how to collaborate but as Bernd and Bill pointed out we have a funding problem. Most of the people here report work being funded by DFG or aerospace or something, and that's really a small part of the pie. And Bernd hinted at drag reduction, which ultimately energy conservation problem, but also energy conversion is a big area we can really move in to. I tried to calculate this 15 years ago, what the cost is of not understanding turbulence and I couldn't quite do it, but it was a large number. It's hard calculating it. We have to communicate with society the cost of not funding us. That's the point and there's a lot we can do in energy. The wind industry is already doing it, in some way without any standards. One example is if you observe turbines from a second law (of thermodynamics, ed.) point of view, there is a certain amount of energy in front of the turbine and a certain amount downstream. The difference is going somewhere; it's not all going to shaft work. It's dissipating through the wake. People are actually working with us and decoding turbines within the atmospheric boundary layer, etc., without really understanding how these things work, and if we can help them figure these things out there's actually a big monetary award for that. Lastly, in terms of collaboration, not everybody can do high quality experiments and simulations at the same time, so I advocate strongly to sharing data and making things open. We started to get interested in cross flow axis turbine some years ago, very simple but highly interesting devices from a fluid mechanics point of view, a very unsteady similar to helicopter aerodynamics. We started putting data sets out in public, via big share, for example, the whole CAD file, its processing code, data everything. You never know who is going to actually pick it up and work with it.

Fazle Hussein: There is lots of funding, if one is doing experiments and one is doing corresponding simulation. I think there is a substantial amount of funding of that and we already have benefits from such efforts. So I think it's a good solution.

Martin Wosnik: But you don't have to wait for funding for both, just do what you can and put it out there. Somebody will pick it up and work with it.

Andrew Pollard: Coming back to this, basically two fields of computation and experiment. One of the comments that came back from a funding review was that they were really happy to support the research when its linked, in my particular case, to the medical community. But they did not want to fund anything associated with fundamental turbulence. So the real issue that we're facing, I think, is the business of trying to link what we're doing to what are benefits to society as perceived by society.

Sedat Tardu: Just a few comments, the first one is about the large scales, and drags reduction. Large scales transport large amounts of UV from the outer layer, but in order to control the drag we need to focus on the wall. And, don't believe that controlling the large-scale structure will affect at least in some way change the drag reduction. Second thing is about the order. So most of the time you have huge order in the wall turbulence and you have to look at some specific methods and gradients. But even if you do that you have to determine some quantities which is very well ordered. I got an example which is complicated; you have a sort of synchronization near the wall; you have coherent vortices that in some way contribute to the wall turbulence. But it is here that you really need to go deep to understand it. More and more now if you want to find the funding they ask you what is the important part to the social point of view and it is a very hard question because most of the time they want it to be understood by normal people what is the energy benefit. Most of us say that because we have to say something. But most of the time it makes no sense. So the biology component is a very good example because it's limited in its applications. Of course oscillatory flows are important; so what we need to do is find a list.

Luciano Castillo: Is this very limited?

Sedat Tardu: I think this is limited, so we're a bit behind society and I don't know how to answer the question what is the social impact that we having.

Fazle Hussein: Thanks Sedat, I actually beg to differ with you on one thing. With DNS, which does not use any assumption, we showed a 55 % drag reduction. Unlike LEBU's, which was oversold, they showed drag reduction as if you are cutting the structures. They forgot to recognize the LEBUs add drag also.

John Naughton: We've talked about combining experiment and computation, but I think even within these fields we have to as a community we have to do more projects that combine expertise, within the experimental community and within the computational community and in experiments where multiple techniques are brought to bearer the same facility, you're doing surface measurements as well as flow field measurements; you're using multiple techniques with flow field measurements. You're combining structural types of measurements with statistical types of measurements, and then taking analysis techniques that are spanning, and I'm thinking of the work of Jean-Paul and Mark did in the shear layer, demonstrating everything wavelengths to the complementary POD. We as a community need to do more because I think we leverage what we learn in any given experiment or computation when we do those things. So breaking it down beyond just a trial, even within those things doing a better job of taking our expertise and rather trying to do it all by ourselves, which the single PI model did for years, we do a better job at combining because that way we're going to advance. We're not going to do it individually.

Ivan Marusic: I wanted to bring up a point, and several of the speakers have touched on it, it has to do with the way we approach science and the way that we organize ourselves. Mark mentioned that there is a funding problem, and I agree, but I think even more important, it's not the funding problem, it's a confidence problem. There is plenty of money. 20 years ago our main meeting was the APS-DFD meeting with only modest attendance and membership numbers. Now the DFD is the third largest division in membership in the APS. However, we don't act as the third largest group, and hence the senior executives and administration of APS still don't appreciate the quality and importance of the work that is performed in our community. This extends to the broader community not having an appreciation of the important societal impact that can make. The problem partly is that we all operate in the way that we did 20 years ago, and we haven't changed our mode of operations. Large funding opportunities will come once we make our case properly, and this requires effective lobbying. We should be looking at other communities to see how they operate. A good example is the field of Astronomy; try justifying the societal impact of what happened 13.6 billion years ago. However, they are able to do this in a compelling way that catches the public's imagination. The problems we work on are equally rich in physics and in depth. We need to present the case and make clear the grand challenge and importance of turbulence, beyond just an engineering challenge. Perhaps, the reason we've had so little progress over a century, and we have had progress, is that the problem is so difficult. This has to work to our advantage. We must and should consider the practical aspects of the problem relevant for industry, but at the same time we should not shy away from the fundamental challenge and, respectively, ask for the levels of funding required to tackle the problem. A standard practice in many communities is to have a decadal plan where as a community they collectively agree what the major funding support should go for the next 10 years. This level of organization and coordination is what you need when you're talking about a multi-billion dollar investment. We don't look at anything like that, so I think collectively we need to be better organized, and we need to decide what our grand challenge problems are as a community. For that you have to go through a certain process and we right now don't have any such mechanisms that allow that. It also requires a degree of cooperation that we presently do not have. I think we need to start thinking bigger picture and considering broader funding opportunities including philanthropy. Here, one needs to present a vision, and it is one that we as a community need to present. We have a beautiful problem to work on. We need to concentrate on establishing plans and working towards them collectively where the sum is so much more than the parts. This will likely mean collaboration at a large scale as is typically seem in the physics community.

Fazle Hussein: I have had direct interaction with some of the physics leaders in the USA and I know how they work. And I think that's the only way we can get ahead. But I don't want to inject myself. I know Paolo and Mark are also getting restless but I would like to see some of you who have not yet spoken, speak first.

Will Anderson: I am speaking a bit from the junior side having only just getting started in my career. I am really convinced for the need for these collaborations and not just these collaborations within the community, but these are really diverse collaborations. I think that is where the biggest intellectual arbitrage is, to put it one way, that's where the big opportunity is. So, we have to talk about sometime, some of our frontier challenges being in energy and in medicine, but in truth we're going to consume the energy that we need, even though we know that there is carbon emissions. I think an added frontier problem would be things like anthropogenic effects, associated with our activities. The one I'm really interested in is dust. Dust is a fluid mechanics problem, but there are all kinds of evidence that dust and other sediments that have been at rest for along time and haven't been mobilized in a while because of climate change and things like this. And these are all fluid mechanics problems. And the way we can access them, in my opinion, is through collaborating with other scientist. These people have great problems not to mention they're delightful to work with, they are really pleasant people. So that would be my thing. The other comment I would make, I think its already been said, but a little bit on outreach. When we do these NSF courses we have to talk about outreach, but actually its very important and thinking about how we can impact the community. We had this great experience last night with Prof. Larsen, I mean people were just enthralled by the penguin and it was really interesting. I mean last week I went to a community college in Dallas and spoke to people about fluid mechanics and different scales, and this is the stuff that's really going to get people excited. It's going to get us on the front page of University newsletters and stuff.

Fazle Hussein: Before we go to Mark and Paolo. I would like to encourage any of you that have not spoken and wish to, to participate, and raise your hand.

Charles Tinney: One of the things I've noticed, being a new entry to the field relative to most of you in the audience, I do a lot of reviewing; perhaps others are pushing papers and reviews off to me. What stands out to me is that we have enormous databases out there; this ties in with the theme of the meeting: Big Data. It seems to me that we're not spending enough time properly documenting boundary conditions for these data sets. Perhaps it is a consequence of laziness, or from the distractions associated with having too many research initiatives to monitor. I see this more and more. Likewise, there is a lot of really rich information in past data sets that can be used and exploited in new ways to answer and address current problems before going to the lab and taking more data. And it just seems that some of these things we could do a better job documenting. I see this time and time again when reviewing outside research, and other items that come across my desk. This is an item that I have become adamant about with my own students while preparing manuscripts for archival publication. Any time you go into the lab there's got to be better documentation of boundary conditions and so fourth so that we can make better use of these data sets in the future.

Julio Soria: Sometimes there's data out there, so you don't have do the experiment. You just have to analyze the data!

Fazle Hussein: In this line there's a lot of papers written in bio-physics and bio-medical engineering only using data from the NIH website. They don't take any data, they are already there.

Mark Glauser: I would like to build on Ivan's comments, for the last 2.5 years I've been on the US army science board and I'm really the only fluid mechanics type on this board. I had to think really hard about how to explain what we do and the importance of what we do to the cyber warriors and all these other types of folks that we're competing within some sense. When I mention fluid dynamics, aerodynamics, and turbulence they just kind of have this perception that it's a mature field and that it's ho-hum. But I think if we talk about it a little bit differently there are opportunities. Working with these types I've explained it to them as a high dimensional system governed by nonlinear partial differential equations, with stochastic solutions and highly unsteady. Nonlinear, unsteady, three dimensional, then they get it. Its very interesting and then the start, of course then you have to go further with it. I think taking the time to really explain more about the importance of this field and build on what Ivan said, the richness of this field and so on. But this has been a very interesting experience for me to have to try to explain why is it relevant? Why does it matter? Especially when they're dealing with cyber security issues, and sensing and fusion and all the kinds of things they think about. I just wanted to build on that again. How we talk about our field is really important.

Paolo Orlandi: In order to get more research money from politicians we have to act as the theoretical physics community, which was able to obtain a large amount of resources to capture the neutrino or to prove the existence of the dark matter. My view is that if the answers to these questions are delayed by 100 years our everyday life will not suffer too much. On the other hand, with more research money directed to turbulence research, larges changes could be obtained to go towards a better world. We know that sometimes politicians are narrow-minded and that they are attracted by fundamental questions as that from were we come from. Therefore, we should do a large effort to show that turbulence, let us say the $k^{5/3}$ laws, is the law governing the formation of the universe, and that, at a smaller scale, it might help to predict the fluctuations of the world economy.

Fazle Hussein: Paolo, thank you. There is cardio-vascular problems, there is energy. These are not minor problems and I think educated people in the society are aware and we have to harp on them. Let's get anybody who has not spoken speak.

Andrew Pollard: I want to come back to what Ivan was saying because I think this is a challenge that he's laid down, but as one who has participated in two of these exercises, one for high performance computing, and one on how to handle scientific data on a blue ribbon panel and how absolutely difficult it is to coordinate and get people to sit on the same page. So now we have people in civil engineering, physics, etc., and we have people in a variety of different areas where fluid mechanics plays an important role. I think what you are suggesting is admirable; however, the question is how do you do it? I want to come back to what Paolo just said about our issue is the communication piece. One of the biggest articles, I think its been quoted in a popular press, is the one that was put out by Parviz Moin and John Kim talking turbulence with super computers, it was in Scientific American I think. What I believe will be an obvious opportunity for us is that you want to move down the line of creating an international panel that would somehow consider these interesting problems that need to be tackled to benefit society, you need to find 3, 4, or 5 of these ideas and start writing these type of popular articles to get it into the public space. And then you build from that. And you create that broad community. So this means that if you take ERCOFTAC, if you take APS, if you take Japanese science community, etc., you need to find people that can come together and perhaps start asking some of the broader questions so they can get the attention of the politicians and others who are in the funding loop, so we can move foreword.

Luciano Castillo: Actually the Horia Hangan had a very good point earlier, and he was talking about the wind community and I mention this because the expectation is that most of the population will be living in mega cities. And as the weather is changing, that means we have to do better predictions to save lives. I working a big project with a group that means that we are working on low dimensional approaches. And even with POD to get it low order resolution to do weather predictions. And these are already problems and we're talking about saving lives. These are issues that are very very important, how we do save life and how we use better imagery resources as we grow. So I think that could perhaps be a very good idea. And I would suggest, Fazle, that maybe a summary of this discussion, we could put an article in the NAE (US National Academy of Engineering, ed.) as a way to disseminate this discussion and how turbulence can play a big role in society.

Michel Stanislas: I would like to contribute on two points. The first one is the cooperation aspect. I would like to say that in Europe we've had the chance to have the possibility of being funded for cooperative projects. On my side I have the chance to participate in two communities, from some years it was a PIV community, which was good funded and we had several projects of covering all of Europe and about 15–20 teams working together. And I have to say that the result was very positive because the progress which was done in Europe, on PIV and on the new ideas which popped up was very significant. On top of that, all the post docs that participate on this program, nowadays are all leaders of the field. So it also contributed to renew the people and to form very well-trained guys who are now contributing to the field. I have to say that I coordinated one project in turbulence and I'm sorry to say the turbulence community is not the same. The ability of the turbulence community to cooperate is much less. It was much easier to establish exchanges of knowledge and know how along the PIV then it was in turbulence when you put 3 or sometimes 2 turbulence theoreticians in the same room, very quickly the entropy is growing. I think the community really needs to think about how to put some things on the table like exchange and can share knowledge. I remember when we started on the PIV each of us were developing out own algorithm and people were very frightened to compare because, "ah, my algorithm might not be as good as yours." In fact, we enforced the comparison and everyone benefited because all the algorithms weren't perfect. We got better resolution because you were trying to understand their reasons. In the end everybody benefited. And now you can see there are these PIV challenges where people compare and exchange. I would like to make another contribution, which is on the next 20 years. I think we cannot expect to predict flows in the next 20 years without a turbulence model. I don't see how it could be possible to predict turbulence in the next 20 years without a turbulence model. They are based on RANS and LES, Kolmogorov's theory and log-law, and on zero pressure gradient. All the turbulence models are based on that. And when you look at the last 10 years, what are they doing on the wall? they use the log-law model to jump the near-wall region. We have to find new ideas and invent something different. When people talk about dynamical systems this is probably the right way to go, you may have to make a jump in the turbulence modeling.

Gina Lee-Glauser (VP Research, Syracuse University at time of the meeting, now retired): As a non-turbulence person and being a senior administrator, I wanted to make a point. I think some of the points that I'm resonating with that your community is a very exclusive community and it's very hard to collaborate because you think your problem is hardest and you dismiss everybody else. And I think this goes into, Bill, your comment related to a few comments, because if it is you are not coming from this lineage, everyone else is just wrong. And therefore you are attacking your own community from getting funded. And that's part of the issue and I remember when Mark and John Lumley and other folks were trying to get an NSF Science and Technology Center, one of the premier centers NSF has and they went all the way up to the final competition it was a reviewers comment in a summary said that the problem was too hard. And I thought the Science and Technology Center grant which is about $20 Million that they were supposed to work on hard problems and it was rejected. One on the comments was that is was prominently stated that it was too hard of a problem. I remember my husband Mark Glauser making fun of me as I worked in systems theory for control systems of a structure and now you guys have time to do the controls and the control theorems that you guys are working on are very fundamental. I know people are talking about you need to collaborate, but you cannot continue to knock down another collaborators and expect me to collaborate with you.

Fazle Hussein: I think Mark stands corrected and he is very sorry . . . .(laughter!)

Christos Vassilicos: A quick comment about the history of science, there seems to be two different types of scientific revolution. There are moments where I feel like a shelf and there are moments when I have ideas and there are no new discoveries and then a field becomes very political. The moment where a discovery is made, like DNA, we are asking why would your thesis get so much money. Actually they made some discoveries in LES, they have some ideas and it worked and then the money flowed. I think the politicians are aware and if they're happening right now (the new idea, ed.) then all the rest will follow.

Luminita Danaila: I agree with most of the things that were mentioned. I will emphasize something to me seems very important. We have to look at real flows which signifies the Reynolds numbers are moderate, and not infinite. Moreover, real flows involve real fluids, with variable physical and chemical properties. This morning, my talk was on variable viscosity and density. Both of them involve other communities, e.g., the two-phase flows. This is an example of a very different community, but often they have questions similar to ours. There are bridges to be put and we might use common investigation methods. Now, regarding the tools to be used, theory is very important and the analytical approach is part of that. We use experiments and simulations and we have to put all of them to understand, model, and ultimately predict real flows. But theory is important, we should not only be happy to write equations and to close a budget. We need to go further and use matters of predictive tools. People in combustion, for instance, need to determine the shape of the spectrum (or, of the second-order structure function), because one has to predict the time necessary to mixing to be performed, or the spatial location where the smallest scales are created. High-order structure functions, for instance, are used to investigate internal intermittency, and the latter is not only a fundamental issue, but it is also equally important for practical applications. It is crucial to know if mixing is homogeneous, then you have a flatness factor very close to 3. If this factor is very different from 3, then mixing is highly intermittent. I would finally put emphasis on the prediction of real flows, involving real (i.e., heterogeneous) fluids.

Vincenzo Armenio: Lets start from Paolo's point. Let's go to the fundamental problems. Fluid mechanics is now important for the safety of the planet. Someone, before us realized this, for example, and created Intergovernmental Panel on Climate Change. They have collected a lot of money from the world for what is a fluid dynamic problem. There are fluid dynamic problems with energy, there are fluid dynamic problems with our body, for bio-medical applications. There is someone here with that idea, but not starting from the fundamentals, but from the very practical end associated with the climate. We are missing a lot of occasions acting as a closed community because we were not able to discuss with these people work in practical fluid mechanics for the next century. According to Luminita's points I strongly suggest us to get interdisciplinary family where turbulence can be combined with other fundamental problems within fluid mechanics because they don't care about turbulence, they don't know this stuff. And so they are able to get money, stay in their world and influence from agencies and governments, and so on.

Luminita Danaila: Thank you all... we have to close and we can continue this discussion on Thursday afternoon.

Part IX
Discussion 2: Large Data: Opportunities for Collaborations

Chapter 31
Whither Turbulence and Big Data for the Twenty-First Century

Discussion 2: Large Data: Opportunities for Collaborations

Andrew Pollard, Thomas J. Hacker, and Shirley Dyke

Professors Tom Hacker and Shirley Dyke joined the meeting by video link from Purdue University.

Luciano Castillo: This session is the beginning of our broader discussion on Big Data. I invited Tom Hacker, who is an Associate Professor of Computer and Information Technology at Purdue and Shirley Dyke, who is a professor in both Mechanical Engineering and Civil Engineering also at Purdue to tell us about a project they are part of. I've been interfacing with them for about a year now. We've been talking about how can we bring our large turbulence data problems to existing infrastructure and they have already a big center actually which is inside infrastructure. And this project they already have at Purdue includes a large group, the USA, in Europe, and everywhere. We have a project that hopefully will do all of these, but I will let them talk and the idea is to ask questions, so that way you have a better sense of what they are trying to do. So if you look at this screen here, that's some questions that they pose. Basically, they want to understand what do they need, what do you need actually in order to be able use this large database. And I will try to explain to you the perks, from my perspective, can we have a large experiment and simulations that will be included in their database in such a way that the turbulence community could download the data. At least that's what I see. So these are some of the questions, please think about some of the questions you would like to ask. And then when Bernd and Mark come in to do the discussion, Shirley and Tom will stay in. So that way they can try to be part of the discussion group.

This discussion is an edited transcription of a presentation and discussion period, with the majority of the editing done by AP, TJH and SD.

A. Pollard (✉) • T.J. Hacker • S. Dyke
Department of Mechanical & Materials Engineering, Queen's University, Kingston, ON, Canada
e-mail: pollard@me.queensu.ca

A. Pollard et al. (eds.), *Whither Turbulence and Big Data in the 21st Century*,
DOI 10.1007/978-3-319-41217-7_31

Tom Hacker: Okay great, thanks a lot for the introduction and good afternoon everybody. What we're going to talk about today is some of the work we've been doing at Purdue to develop and operate the cyberinfrastructure for scientific research, for the Network for Earthquake Engineering Simulation. So there has been a real transition in the last 20 years or so, research in the past was powered and based on theory, experiment, and computational simulation. Computing was used to support research through the simulation of physical systems and the analysis of data collected from physical experiments. Over time the advances in computational abilities have been powered by Moore's Law. In the past, many researchers could increase the performance of their application by exploiting the increase in single core performance made possible by Moore's Law by moving their software to new hardware every few years. This worked well for applications in the past, since most systems contained only a single processor core. Fields that have greatly benefited from this environment include areas such as computational chemistry and automobile design. There is a great website created by the Council on Competitiveness [http://www.compete.org/initiatives/compete-innovation/23-hpcac] that talks about some of the applications and use of high performance computing in science and industry. So over the last decade or so there has been a great shift in the flow of changes in computing technology that have really changed the landscape. So what you see today is that the wide availability of multicore CPUs and also a technology called simultaneous multithreading, which allows you to more efficiently use those cores. Also the growing use of hardware such as graphics processing units within systems and the Intel Phi processor, and this is really changing the way that computing can be done through the use of this technology that is readily available for the community. In addition to these new capabilities on a system level, there are other new technologies that are now also available. One example is the availability of a high-speed Internet connection at many institutions. At Purdue we have a 100 gigabit connection to campus, which we can use for projects such as the Compact Muon Solenoid project. Also the pervasive availability of wireless Internet technologies such as 4G LTE and traditional Wi-Fi with really good bandwidth is an example of another new technology. Another example is small scale computing systems that are available now, such as the Raspberry Pi or an Intel Edison board that are becoming important in the development and deployment of the Internet of Things to connect small scale systems and sensors to the Internet. So a couple of good examples are a distributed network of radiation sensors that can be connected to the Internet to share observations on a website [http://www.uradmonitor.com/]. There's a small kit available that can be used to set up your own small scale radiation sensors and to distribute the information collected from the sensors to the web. Also a project at Purdue, called the CAM2 project that's able to collect video streaming information from tens of thousands of video cameras from across the world along with a facility to analyze images collected from these cameras. Another technology we're seeing right now is consumer grade low cost video conferencing systems such as Skype and WebEx. So these have really been changing the landscape for science and engineering. Also, software used today needs to be able to use multiple cores simultaneously, there's also this huge avalanche of data available from physical sensors everywhere. And these data can be collected from a swarm of sensors in the field, and this data goes beyond simple numerical data and includes streaming video, images, and audio. What's happening now because of all this new landscape is that distance is becoming less of a variable in collaboration across the country and across the world. So these capabilities are adding new tools that change the ways in which research is conducted. The traditional triad of theory, experimentation, and simulation has added big data now; it's called the fourth paradigm that is based on the availability of the large amount of data that can be processed and analyzed. Also the ability to have widely distributed research collaboration groups and also the sharing of large scale research facilities. So some examples of this kind of work today in Europe is the large hadron collider at CERN, where there are several detectors that include the Compact Muon Solenoid and the Atlas detector. We are helping the CMS group here at Purdue with a tier two computing center with thousands of computing cores and four petabytes of storage that is used to process data from the CMS at the large hadron collider. Another example is a large scale synoptic survey telescope project, which is going to be built in Chile on a mountaintop. The data from this telescope will be processed in a base site in Chile and will be moved to a central archive site in the USA over a high-speed network. Another example of a large facility is the Network for Earthquake Engineering Simulation. So in NEES there were 14 earthquake engineering laboratories, and experimental data generated at the sites were brought together and archived in a central data repository that gave researchers access to raw and processed data. So these are three examples of three large scale shared experimental science facilities for which cyberinfrastructure and these new computing capabilities are a significant component of how these research communities can do their work. So the NSF funded NEES project and the NEES earthquake engineering laboratories are transitioning into a new NSF program called the Natural Hazards Engineering Research infrastructure that will add wind engineering to earthquake engineering. At Purdue we're continuing to operate the cyberinfrastructure as a bridge to whoever the new program will be awarded. So the NEES research facility included a wide range of capabilities such as shake tables, a tsunami wave basin, centrifuges, a large scale displacement facility as well as the cyberinfrastructure. The characteristics of the experiments conducted at these laboratories really require a substantial expense for building specimens. The specimens are often tested to failure. So because of this it's very important to capture and preserve the data from experiments. And the types of data that are collected depend on the type of facility. For example, the type of sensors and the type of experiments you conduct with a wave basin versus a centrifuge are very different. So the specific characteristics of the data depend on experimental design of the specimen. So here's an example of some of the information collected from one of the experiments at the Minnesota site. Carol French collected this information on a multi-story concrete wall. This is really interesting, what you can see here is the structure is covered with the sensors. Whenever they stress the wall new cracks appear in the wall. The lab facility people mark the cracks with a marker and number them and then take an image of this wall. So the data collected during this experiment over the course of potentially weeks or days or months sometimes, is that to actually collect the sensor data, you have to know where the sensors are located, collect the image data of where the damage was inflicted and also have all these data be properly organized, sorted, and arranged so that others can go back and look at the data later and make sense out of it. So this is an example of this information. So the data we get from this would be numeric data from sensors, video data, if you have a video camera, or image data as well as field notes. So we operate the NEES cyberinfrastructure to help this community. There are factors that affect the sharing and the use of this type of data. One of the factors is high complexity, for example, time series sensor data. We have basically many sensors all delivering data, they use sensor metadata to allow the community to make sense of what data is from each sensor. Also another factor is time and distance, because experiments can often take weeks, days, or months and researchers are not usually on site for the entire experiment. Cyberinfrastructure capabilities allow people to remotely participate in their experiments. Also the quantity and size of the file, for example, videos are very large, or a very large number of JPEG files that are made from a video. Context for understanding is another factor. The researcher needs to provide context that is critical for others to understand or review data. Another factor is communication; there is a need for effective communication throughout collaboration facilities and also for sharing documents and data. So for the NEES cyberinfrastructure the very important principle we have is that communities drive the requirements and services provided by the cyberinfrastructure.

Luciano Castillo**:** So this is what you have, a network group of similar facilities and all of these facilities are part of this network and they're loading data and then all the users beyond that network could then access the raw data and process it into whatever they want.

Tom Hacker: That's correct, yeah. So the experiments are conducted at the NEES facilities and laboratories and after a certain amount of time when the project is complete the researchers are required to publicly release their data as a project and then all that data is available to anyone around the world. So if anyone wants to look at this, you just go to the website NEES.org and you can register for an account and you can download and use tools on the public data. As I said the communities drive the cyberinfrastructure. The cyberinfrastructure needs fall into different categories. An important one is data and there is a need for central curated repository for experimental data that can be cited, that's very important. For long-term longevity and usability, the data needs to be something that can be cited. Another aspect is computational. And there are widely distributed computing capabilities for simulation that we provide that can be used to simulate physical systems and analyze experimental data. They also need to build and run software as a service on a central server, or be able to download and install software on their own computers. So the NEES-hub provides for both of these. Another area is collaboration within the community. To provide support for collaborative facilities we provide within the NEES-hub the group space and the wikis and also a central WebEx facility. There are some similarities between NEES and that CMS project; one is a large scale shared physical experimental infrastructure that requires considerable effort to build and to conduct experiments. And both of these experiments generally involve the collection of data over the course of the experiment. So what we see with the large hadron collider and the LSST is that there are other projects that are examples of emerging large scale shared facilities for which cyber infrastructure and data are an integral part of how they do their work. So the cyberinfrastructure we develop is driven by the highest priority means identified by civil engineering researchers. We provide the NEES-hub science gateway which leverages HUB-zero that is also called a science gateway that is built on leveraging open source and commercial software. The Project Warehouse is the curated authoritative data repository for NEES experimental simulation data. And also have another facility called DataStore for non-curated imported community-contributed data sets that are meant for sharing. So the NEES software is really the science gateway, the nexus for the NEES community that provides an integrated environment for the NEES users. It supports the use of tools, computation, data, and collaboration all within the hub environment. And it facilitates access for users and allows them to navigate resources for high performance computing systems, curated data, community data as well as education and outreach resources. And NEES helps researchers connect with tools, data, projects, and with each other. Some I'm just going to a show a demonstration of the NEES-hub, this is just an overview of it. So what you can do is go to NEES and register, what I'm going to show you, you can do as well. I'm going to log in and on the opening page you can see information about the calendar, what's happening in the community, we can then get an overview of the NEES projects tools and resources available for the community. Learning and outreach tools, there's some course materials and training materials. The Project Warehouse is the central repository; we have some featured projects here. Then we have some simulation capabilities that can be accessed using the NEES cyberinfrastructure. We have some demos and some examples. Information about NEES experimental sites is available. The collaboration tools, and we can also look then at tools and we have over 65 tools installed right now. All these can be run within the hub environment. I'll show you that later. We also have community databases that I will describe a bit more about this, we have about 15 databases right now. These are community-contributed data sets that aren't necessarily created by NEES. So the NEES cyberinfrastructure provides an application execution environment where we extensively use virtualization that supports Linux or Windows tools with direct access to data and the central repository. And a collaboration space provides access to WebEx, groups, wikis, and site scheduling. So the NEES data repository that contains the simulation and experimental data from these researchers is really in an important group asset. The critical data needs to be stored and needs to be accessible for many years because much of the testing is destructive; the data is consequently expensive and difficult to reproduce. So we really focus a lot on the long-term accessibility of the data and as many different types or formats of data collected. There's data from physical experiments, computational simulation, data from sensors, videos, images as well as sensor calibration information and documents. So the NEES data repository is contained in the project warehouse. The project warehouse has over 1.1 million files that are stored in a wide array of file formats. We have 315 public and private project that are stored in the NEES-hub. So we provide a structure in the Project Warehouse to allow users to easily find and navigate information, which I'll show you really quickly. We have a project overview page that shows a project description and the PIs, navigation tabs to allow users to easily find experiments, team members, files, and users can read files directly using the tools in the NEES-hub without the need to download and install the software of the data. So I'm going to show you a demonstration of the Project Warehouse now. So you can do this as well if you wish. Go in to Project Warehouse, I've already logged in, and we're going to scroll down, and select one of the projects. We'll pick project number 33. And what you can see here first is a project overview page, with a brief description of the project. We're going to look at the experiments conducted as part of this project. Click on the experiments tab. And there are several experiments that were conducted for this project, then we're going to go look at one experiment in more detail. And we can see a description of the experiment, some of the parameters attached, there's the date and location of the experiment. We can also quickly view some of the images collected from the experiments and browse through them to find something interesting. What we see on the right is a pie chart showing us some of the file formats of the data files. We can then go back and view the data directly using the indeed browsing tool. So we're launching in DEED in the NEES-hub, and then what we are going to do it look at the engineering drawings of a structure, click on the sensor based on its placement, and then directly plot the data within the tool. This allows us to quickly go through the sensor data collected, and the image data collected from the experiments and determine whether or not we're interested in this data. You can actually run those tools on the data directly within the hub. We can then find more information about the team members on the project, and if we want to go directly to the file we can actually go to the file browser and you're actually looking at the structure of where the data is stored. We can then go to one of the experiments and get a look at the data files and you can select one of the Excel spreadsheets containing some of the sensor data collected for the experiment and then put it into Excel. So what we show here is basically, you can both quickly navigate the data within the NEES-hub environment and run tools on it. Or you could download the data if you wish and use your own software on it. So we provide both options for the NEES community. So the other aspect of our cyberinfrastructure is the community databases, this is a non-curated facility within the NEES-hub that allows members of the community to host and create their own data sets. We're currently hosting 21 data sets today, a partial list is right here, some of the interesting ones are the ACI circular column data base, the publications data base for ACI we have some data for some of the published papers from the Journal of Earthquake Engineering and also the shear wave velocity database. And I'll show you a quick demonstration of that here. So if you log into NEES-hub you can go into the databases section, we can scroll down and select one of the databases, we'll pick the Haiti Earthquake database, well go to the data view. So during events like this, engineers will go out and collect information like field notes on the structures you saw that were damaged. So we can upload the field notes, we also collect images as well. And these databases have allowed these researchers to quickly share their information through the NEES-hub databases. Here are some of the images collected and we can quickly browse through those. And so part of the metadata that's collected is information on the building and some of the information about the location. And this is all then disseminated through the public database. We also provide tools on the NEES-hub. We can go and select a Linux tool here, let's say GOYA-C analysis. Launch the tool directly within the hub, now this is running all in the hub. And now users can go and directly interact with the tool directly in the hub environment, without the need to download and install the software. We can also go ahead and find another tool for Windows, we can launch that tool in the hub. And we can look at the data within the hub. Let me just skip a head here a bit so we can actually view the information directly. So in terms of users we have 143,000 users over the last 12 months now, and it's been growing over time. As I said we have 315 private and public projects, 2 million files and over the last year over 317,000 files downloaded from the project warehouse. So we made good progress in NEES in both developing the CI software approach and the cyberinfrastructure to address the needs of the community. So what I think we're going to see for the next generation of cyberinfrastructure is some new items, such as a scalable internet library of curated scientific data. We're going to see fully integrated communication and collaboration systems linked with experimental sites, a greater sharing of limited physical experimental infrastructure. Some examples we talked about are the CMS detector and the LSST telescope. And these lab locations could be at difficult to access locations such as the Antarctic, which allows a community to share a large scale facility. Also I think we are likely going to see the ability for cyberinfrastructure to allow the scientific community to create and manage your own designed infrastructure. For example, a small cheap radiation sensors or the CAM2 system we have here at Purdue. So in conclusion we developed for the NEES community a distributed cyber infrastructure that provides curated data repository capabilities, the ability for users to upload and run tools on these data and the ability for users to collaborate and share information across their scientific communities. It's been an exciting time to build this and I think you can see from our usage numbers that we're getting some interest in it and the feedback we've been getting from the community has been really positive. That's all I have, I really appreciate your time and I think Luciano has some questions for discussion for you as well.

Phil Lavoie: Thanks for the talk Tom. So a lot of the data you are talking about is for special equipment so presumably there is some form of optimum in a sense because there is a cost associated with their upkeep and sharing this data. Do you get a sense of what the break-even point is to get a new data set by running the experiments again versus keeping the data set available for the community?

Tom Hacker: So I think I heard the question. So the question is what's the cost of acquiring data versus curating it for long-term use, is that right?

Phil Lavoie: Well, there is going to be some sort of balance where the two are going to be equal so presumably this is a small experiment, for example, it might be cheaper to redo it rather than keep the data.

Tom Hacker: So there are a few phases the first is where experimental data is generated at experimental laboratory sites and in this phase the data are collected and then the site personnel and the laboratory manager and the PI can first see if the data is clean, that it's accurate. Once they're satisfied with the data the next phase is uploading it to the NEES cyberinfrastructure, which requires a number of tools to do that. After that the data librarian or data curator goes through the data sets with the PI and makes sure all the necessary metadata is there, all the information that is needed, correct file formats, an then the data becomes a curated entity. So in terms of costs and labor, I think in the front end on the first phase of this a lot of effort is spent by the researcher in the laboratory facility, and on the back end a lot of the effort is spent by the data librarian. It's a pretty manual process on the data librarian side. We've been developing tools to help create a pipeline to automate some of that process. But once the data is created and stored it's basically an oracle database and a file system. So the cost of maintenance over time goes down. So the front end is where most of the work happens, but it's really an effort between the PI at the laboratory and the data curator to identify those critical data sets that need to be curated.

Shirley Dyke: Let me add just one thing. I've been both a NEES researcher and I've been part of the NEES cyber team. Performing each of these experiments within NEES is going to cost on the order of $1 million USD, for each of the projects I should say. It's a 4-year effort and these require building a large concrete specimen and then testing it to failure is a process that takes a couple of years. So this is the type of data that is much more valuable to store then to reproduce as a new experiment over and over again.

Mark Glauser: I wanted to ask you about your efforts in bringing in, for example, third world scientists; it seems to me there should be some really opportunities to bring in some folks who maybe don't have the infrastructure. Can you talk to us and little bit about that?

Tom Hacker: Yeah, that's a great question. So what we found in the NEES user profile is that we have visits and use of the NEES-hub from all around the world, 211 countries. I have a slide I could show you that shows you the distribution. So what we're finding is that in terms of the data that's in the NEES cyberinfrastructure right now, it's getting worldwide views, which is fantastic. Now, I think what's going to be happening like you mentioned is that over time, as other countries have more experimental capabilities and facilities, they may try to integrate them into cyberinfrastructures and other types of facilities. I'm hoping that over time that the whole cyberinfrastructure community will be able to provide facilities that a lot of people start to integrate those types of data. So the CAM2 system is one that I mentioned, is one example. There are other emerging examples out there that are kind of bottom up facilities becoming more available over the internet. Did that answer your question?

Mark: Yeah, Thanks.

Andrew Pollard: Setting up this capability is not cheap; it takes dedication by an awful lot of people and a lot of coordination. Could you address those three points, and I know we all want to try and do things for our general community, but at the end of the day who gets the glory out of this? In other words, why would somebody want to do this? I would like to know why you would want to do it. Again, it's to help us here in Corsica to frame our broader discussion after we are finished speaking with you.

Tom Hacker: Okay, thanks. I'll give what I think is the answer. I think what's happened in science is that in the past we had more single investigators and small group laboratories. So it might be a nexus at a university and you had to go to that place to use those facilities. So what's happening now is that some of the large projects like the large hadron collider, like the LSST, like NEES, or the ANL advanced photon source. Because science communities are finding that it's more effective to pool their efforts into a larger shared facility and that requires that the use of computing technology and cyberinfrastructure to bring those communities together virtually. So you don't have to go to that physical place to be a part of the community. So I think that's an emerging thing we will see. At least in high-energy physics, they got to the point where they needed to do that in order to advance the field into the next level. In CMS and actually in the physics community, the communities have worked out some of the issues with credit, from what I understand, determining who gets credit for what in terms of the community. I think on the cyberinfrastructure side my community is really happy to help these communities succeed, which has been especially true for physics and for NEES. For me as a computer scientist, my kind of credit comes from the fact that we can build these really large scale systems that are really effective and meet the needs of the community. That's really important to me because that's what I wanted to get out of working on them. So I think what you're going to see is that there is going to be a necessity of building these kinds of cyberinfrastructure systems that would exploit all of these computing technologies to help bring people together virtually. This is because the nature of the science is changing over time. I think that with cyberinfrastructure, running it and developing it is more like maintaining a garden rather than building a building. In a garden you have to worry about planting seeds, watering them, maintaining things, pulling out the weeds, which is like cyber security. And helping that community, that garden will evolve over time as its needs of its users change. And that's something that we're really aware of over time. So I think that's what I see.

Shirley Dyke: I think Tom did a very nice job of explaining this transformation that is occurring right now. It's becoming much more common in many disciplines to share data. Maybe not all, but in many disciplines to take advantage of all the sharing of data and use it to its full potential. In earthquake engineering for decades, we've been trying to reuse data. And one just hopes, as certain experiments go back to the 1950s, that someone is able to give you a complete set of their data, maybe from an experiment they did in 1965. Otherwise researchers try and scan figures and then guess what the data was and what the missing points are. But if you can get their original data, you can use it for many more purposes than what the original researchers expected to see it used for. I have been a developer of benchmark problems, where we were trying to share within the community some common problems and how to try different strategies to tackle those common problems and see what we learn as a community. And I would like to see that happen in the earthquake engineering field, so that's I guess why I've been part of this project for the last 5 years. Actually I've been part of the project in many ways for a much longer time, but for this particular effort, for NEES-hub, I've been a part of it for about 5 years.

Tom Hacker: I just want to mention to you that if you go to the project warehouse you'll actually find experiments that were conducted in 1970s and their data is in the Project Warehouse. <https://nees.org/warehouse/welcome>.

Ivan Marusic: Thank you for your talk and participation. My question relates to the size of data that can be shared. In our community we already share fully processed data; for example, time series statistics and the sort of data that would appear in plots in a paper, say. However, a real potential utility is if we could actually share the raw data that has been acquired in space and/or time. But now we're talking about multiple terabytes of data, so could you please comment on what the limits are in terms of the size that you can reasonably shared.

Tom Hacker: That's a great question. For the NEES projects we tend to see lots of smaller files, so I think the total repository is about 32 terabytes right now. It's not large in volume but it's large in complexity and number of files. For the CMS tier two center here in Purdue, the data that we have are in large files that are more homogenous compared with files from NEES. So we have 4 petabytes of data that we run at Purdue. So in terms of the limits, I think from the computing side we can build our systems out of modern components to get petabytes pretty quickly, it's not technologically that difficult. The complexity comes in, or the difficulties come in both in having a decent network speed and being able to handle the data transfer and the human time needed to actually add the annotation metadata for the duration of the data sets so that others coming along later will actually have enough information to actually use the information. There are two extremes and one extreme is the CMS tier two center where we have 4 petabytes of storage, you know we have 1000s of jobs running at a time. On the other end is the NEES in terms of the complexity and the number of files and also the amount of annotations needed for all the data. So I think in technological capabilities we handle both those situations.

Martin Wosnik: So we're just in the process of connecting a number of our facilities to an optical kind of parallel internet, it's called the DMZ, <https://en.wikipedia.org/wiki/DMZ_%28computing%29> that has to be connected outside of firewalls to ensure the really high data transfer speed. So you can transfer really large data sets at very high speeds. For example, we can connect with 200–300 gigabit per second. Are you connected to the internet or do you foresee being connected to non-optical networks so we can actually transfer large data sets like the type just mentioned? Because we actually transfer terabytes worth of files to look at the raw data.

Tom Hacker: Could you repeat the question because I couldn't understand it.

Martin Wosnik: So this goes back to the large data sets that Ivan Marusic just mentioned. We literally have to transfer terabytes to be able to draw data that is full of space and time information. Several facilities around the USA are connecting to the secondary Internet, the DMZ optical Internet, to be able to do that. Is your subject connected to it? Do you have plans to connect to it? And what is the cost of doing that on a large scale? I don't have a good handle on the cost of that.

Tom Hacker: That's a great question. So for the CMS, tier two center here, we sort of have a science DMZ at Purdue and we actually have our systems connected to 100-gigabyte network that goes up to Chicago. And we just joined the LHCOne network <https://twiki.cern.ch/twiki/bin/view/LHCONE/WebHome> to try and help that. So in terms of the large amounts of data, like for CMS, the whole community working on the setting up science DMZs, let me explain what that is. A science DMZ is an area of the network at an institution that's outside the usual intrusion protection system in terms of cybersecurity. They're designed to be very high-speed systems with very high-speed connections. So for CMS we absolutely require that because we have petabytes of data. Beyond this kind of science DMZ at Purdue, the community is building these virtual networks. Now on the NEES side, because we don't have petabytes of data, we have more users uploading data on their own computers in their own laboratories, we have more emphasis on software tools the users can use to upload their data and do tractions and things like that. So I think the issues always been about transfers and capability. I think the way that you would approach the architecture of the cyberinfrastructure would depend on the needs of the community, the size of the data sets you're moving, and the institutional capabilities. Here at Purdue, on the CMS project we pay a yearly fee, which is not extraordinary high, it's well within the budget for access to this high-speed network. For the NEES network we don't need to have a science DMZ so we don't need to pay that. Does that answer your question? I think it really depends on scale on the data transfer.

Luciano Castillo: Anybody else has a question for Tom before we move to the discussion. Tom one of the things that we see here is the amazing facilities around the world that we have, and from that perspective that could be very very good to have access to that data. But the challenge I see here is how you would be able to handle those big data that were started from those experiments. Even though experiments of simulations, maybe we could discuss that later. What we are going to do now is transition into the discussion. We're going to leave you in the sky, Tom and Shirley, if you want to stay on. That way if you have anything to ask you can add to our discussion, but let the speakers first. Martin Oberlack and Bernd Noack have agreed to lead this broader discussion.

Martin Oberlack: Thanks a lot. I was asked by the organizers to very briefly present something I just got this morning from Andrew Pollard, and that is the John Hopkins Turbulence Database, initiated by Charles Meneveau. I have never used it but I can comment something about it. The basic idea is that Charles actually stores data, and invites others to use it. Some of you may have actually used it before and I understand it's a type of MATlab tool you can use to access the data and essentially do anything you want with the data. According to that discussion, which I listened to, the astronomy and other physics communities have huge databases. From what I understand it's easy to use, it's similar to web-type of use, of course the problem is cost. It costs money as you just heard before. So there is a long-term problem to keep this web page running. There was also the criticism of whether it contains real physics. I don't share that criticism. Maybe at some point it should be moved out of universities, these things become too big, too expensive but I understand from the discussion that Andrew and Charles had separately before this meeting think that the turbulence community may be too small for this. The question is what can be done on an international level. Charles Meneveau gave at some point the advice because before he started that database there were a lot of meetings on various things, like data in the form of metadata and whatnot. You can see how to get this thing started. But he did say "well if we just start doing it and then we can probably sort out questions later or otherwise well never get started." In this year, there are about 30 papers this year coming out as a result of accessing of the database. I think about 30 % is probably from his group. I put down a couple of questions: let's say we collect turbulence data, the question is what equations do we agree on for compressible or in-compressible flow? there was an earlier discussion on numerical schemes, which one is accurate enough to trust the data. Well we understand that Paolo is a fan of second order schemes, other people use spectral schemes, discontinuous Galerkin is becoming extremely popular. Spatial and temporal resolution is a key question so it should be above or below Kolmogorov scales. Some people want data just significantly lower than Kolmogorov. If you're just interested in mean velocity, then maybe shear stresses and Reynolds stresses, then of course you can think of many geometries. You can think of isotropic flows and they also include shear flows and others. My personal opinion, I have essentially no experience in terms of machine learning and related ideas, but was it because of small data. In turbulence we need to have working hypothesis for modeling or something from first principles, otherwise to me it doesn't make sense. I'll stop here and over to Bernd.

Bernd Noack: For the purpose of discussion, I want to counterbalance Martin's position. I am a reduced order modeler, so I count my degrees of freedom in my answer, preferring 5 or less. I want to raise a couple of issues, annoy a bit and then try to stimulate with questions. I will make a bridge to the excellent discussion we had on Tuesday, to the current discussion on the data. If you subject a lot of data to stupid questions you will get stupid answers. If you have some right data and ask smart questions you may also get smart answers. Hence, the question: is it the data or the question that play a big role and maybe we should start with questions before we go to the data. An example of a less relevant question is: should we be concerned about 17th order structure function, if already the first moment is 30 % wrong? And if the 17th order structure function has no relation to the first moment. A smart question is what kind of approach for turbulence control yields greater profit or greener energy? There are some communities who do an excellent job in marketing their research questions. For instance, other communities make a good living from looking for life in the universe. We could ask should we look for life in the universe, if human life on planet earth can be expected to become increasingly miserable. If you look on fluid mechanics, they're a real contribution to our future living quality. What will the future hold for fluid mechanics? Some people say we have to increase our understanding. What is understanding? It's a sequence of ASCII characters in text and there is no element of cross-validation.

There is no rocket that will blow up, and no PDF file that will destroy itself. If you can do prediction, you should be able to do control. If the control does not work, the prediction capability is invalidated. This is a much harder problem. We have heard here that control takes bigger and bigger shares just for these tasks. Is knowing the Navier–Stokes equation a competitive advantage for solving these crucial problems or will this task be solved in the future by other audiences? Let me be a bit provocative. The machine learning methods become better and better, the brains become poorer and poorer. Many fluid mechanics problems will increasingly be solved by machine learning researchers. One possibility to keep a prominent role of fluid mechanics is formulating clearly defined attractive benchmark problems posing challenges also for other communities. Fluid dynamicists are necessary for experiments and simulation. This is something that should always stay in our hands. We should think about big questions, the survival of mankind may be affected by greener traffic, renewable energy, and by a number of problems to which we can really make a contribution.

Luciano Castillo: How can we solve the big data problem we have right now or do we need to solve it?

Bernd Noack: Well the question is what are the big questions. The next important step is big plans. We want to have an element of cross-validation. If somebody says I have looked at one terabyte of data and we found the theory validated, are we going to believe him/her or not? There is a cross-validation if you say you have a dedicated experiment which can test any controller. It is difficult to do good control design based on poorly performing model. These problems will attract a much larger community as opposed to handling over some data. What are data? Data are dead corpses of life dynamics plants. It is easy to make a good funeral talk about a dead person because it can rarely be disproved. This is an example of over-fitting. We would like to have an element of cross-validation. I'm a bit provocative for the reason of stimulating the discussion.

Paolo Orlandi: The topic is big data, therefore to initiate we have to define what is big, because few years ago the data from the channel flow at $Re = 180$ were big data, in fact several scholars were going to the CTR (Stanford) in order to use the data to evaluate new statistics and improvements for Reynolds Averaged closures (Durbin) or large eddy simulations (dynamic subgrid model) were achieved. Now, the situation is different, perhaps the big data are for Re_tau of 5000 or 4000. From our experience when you are beyond 4 terabytes for one field, you cannot save anymore this amount of data. This brings to a new way that instead of the data we have to share the codes for the numerical simulations as is done with the experimental facility, for instance, what Saddoughi did on the big low speed NASA wind tunnel. If I want to use a big wind tunnel I would go there hoping that someone let me use the facilities to look at what I want. For instance, I'm going to change the boundary conditions if I wish to look at the effects of rough surfaces. We have to act as the physicists which created the CERN to design different experiments. So we have to build experimental facilities devoted to understand the complex physics of turbulent flows. Experimental facilities for practical applications, as those for racing cars or aeroplanes cannot be used in the same way. From the numerical side the direction is similar; engineering codes are different and they cannot be shared, instead codes for fundamental questions should be shared among different groups. These validated codes (without saying which are best) should be located into some big computational centers without saving anymore an enormous amount of data. The post-processing is done during the simulation. To conclude not any more archives of big data, everyone is going to produce his own data and to do what he likes with a facility that can be either numerical or experimental.

Bernd Noack: What you mean is that you have to have a reproducible plan? Which means producing the dynamics, which you can integrate yourself?

Paolo Orlandi: It is not important whether numerical or experimental techniques are used each one has advantages and disadvantages. However, the best results are obtained by coordinating the two techniques; soon this effort will be realized in Italy within the CICLOPE project designed to understand in more detail wall bounded flows. At the moment a very long pipe has been accurately designed, built, and tested, which will attract scholars from several countries to measure the statistics important to characterize wall bounded flows. There is also the hope that other scholars will produce efficient numerical methods to perform simulations, by using the CINECA machines, in the same range of Reynolds numbers of the experiments.

Michel Stanislas: It is not a question, but a comment. There is an ongoing initiative at the European level, which is doing exactly that which is called EUHIT. Which is a European network of large facilities for turbulence and it is managed by Eberhard Bodenschatz. Inside this initiative there is also a database, which is built, called TURBASE. Where large data sets of turbulence are supposed to be stored.

Martin Oberlack: Is there already data up there?

Michel Stanislas: They are starting to load data up from the consortium. The database is also open to everybody. The objective is to make it live after the end of the project and to collect large data sets from turbulence.

Martin Oberlack: One question to that. The consortium is that it is basically experimental, right? I mean the EUIT is a consortium of people doing large scale experiments and then invite other people? Isn't it also numerical?

Michel Stanislas: It is not a consortium of experiments. It's a consortium of large facilities, which are made available to everybody. So any of you can apply to come and use any of the facilities and the turbase is open to larger data than on the experiment and data of numerical or experimental facility can reside there.

Ivan Marusic: My impression, however, is that it is restricted to the EU. And again, the data we are referring to is processed, small scale, data. Even so, I very much like this idea and really we are about 30 years behind big physics, the other areas, in this regard. If we are going to be talking about being competitive and having a presence in broader scientific landscape we have to start acting at that same level of capability and co-operation as our colleagues in Astronomy and other prominent areas. I applaud the EU program but I'm also skeptical about the fact that it doesn't really have the funding to enable people to actually come and work there in a meaningful way. However, it is worth persisting. Collaboration is certainly needed and if enough people all come together in an organized campaign, we can do something none of us can do individually. A challenge that remains is how to effectively share and get the most of the measurements. As an example, in our large wind tunnel at Melbourne we have collaborated with Julio Soria's group and pooled our resources, with multiple cameras and lasers, to acquire extremely large PIV data sets with unprecedented dynamic range. However, beyond the usual processed results we have not even scratched the surface in terms of what information could be extracted from the full raw data set. This is because we are set up to handle or share such large data sets.

Similarly, some years back in 2005, about six groups collaborated and went out to the salt flats of Utah, where we pooled all our individual anemometers and set up an unprecedented measurement array. Again those extensive raw data sets are effectively idle. We've been through maybe 2 % of it, but again the problem remaining of how to effectively manage, share, and use those data.

Steve Brunton: So first Bernd, I like your distinction that we need to be asking the right questions with our big data. And I think that goes to the point that you also need to have the right data. Big data is often confused with correct data or the right data to answer scientific questions. And so I just want to re-emphasize this point that you still need human intelligence to figure out what features of the data are likely relevant to answer the big questions, so that you can collect the right data. It doesn't matter how big the data is if it is not the right data. I think having these organizations and these reproducible data generation schemes is important to test and validate hypotheses. Also, good data should provide answers and insights that motivate deeper questions and motivate how to collect more data in the future. So I don't think that the current path of simply collecting a huge amount of data will necessarily lead to us having all of our questions answered. There is a feedback for how we can intelligently collect data with the right questions in mind, and based on the information we learn from the data, we can design better future data collection.

Martin Wosnik: I just want to point out that there are already a number of startup or commercial platforms that provide similar services as the one discussed within the John Hopkins and the EU efforts. For example, QUICKSHARE and GITHUB, these are things I learned from my grad students. We've started to publish data sets on those platforms. The limitation right now is, I don't know the exact number, but both of them are limited to single digit gigabytes. You can upload a lot of data, but its not terabytes. And we try to upload full experimental protocols, and full data for people to access and for people to assess whether this is an experiment you want to compare simulations to. But this is mostly in the solar energy area. So almost all big data that exists commercially can be used only to an extent, but the really good data, we need optical networks to move terabytes around without having to travel around with hard drives in a suitcase. That doesn't really exist yet, right? That is the next step we would actually have to move to, to join forces and actually get this in place, because it doesn't exist yet.

Andrew Pollard: I would like to draw in Tom Hacker and Shirley Dyke from Purdue. There are two issues that I think we need to face. Number one is clearly what the under current is here: I've got data, I've got these results, etc., and I want to be able to share them. In my opinion, what we need is to either create a broader platform on which to work, starting with whatever NEES is doing, whatever EUHIT is attempting on the large scale experimental infrastructure front, plus whatever may be happening in other parts of the world. From this we need to figure out, both from and for the turbulence community, how they got to where they are today. I also posed the questioned earlier, who takes the lead to create the framework under which those databases are now being submitted and curated, etc. Otherwise, as a community, we'll end up being fragmented and we will have something in Europe that may or may not be open to or conform to what another group is implementing. We need both the data and the framework. I am not sure where we are as a community at this point since any effort seems to be local (Hopkins database or EUHIT, etc.).

Tom Hacker: So your question was, what framework do we use to develop this curation approach in databases?

Shirley Dyke: Are you talking about the facilities?

Andrew Pollard: Not the facilities, because that's the side of infrastructure that is acquired through various other channels and funding agencies. It's being able to address what you want to create within that infrastructure, how did you create the strategic and operational framework for NEES as it currently stands? How did you bring the community together? Who took the leadership? And so on.

Shirley: So this approach was all conceptualized around 1995, and what happened essentially is that the National Science Foundation of the USA offered a series of workshops and discussed many of these questions. How do we ensure the quality of the data? How do we make sure the researchers are going to put the data into the repository, and the facilities were created for NEES in parallel with the cyberinfrastructure. So it was all done in parallel while developing what we call the headquarters organization. There was a headquarters office in charge and the funding was provided to the facilities as well as to the cyberinfrastructure organization. So there were a number of years where it was in a development phase, and there was a lot of discussion about what do we need, what functionality do we need? How do we build these capabilities? How do we reuse? And since then a major culture shift that has taken place. It took several years for people to understand and accept this new mode, because people wanted to retain their own data for as long as they wanted, as long as they could write papers. And so sharing the data with other researchers became a requirement of using these facilities and getting those types of projects and funding. After about 10–15 years it kind of started coming together, the community had accepted and adopted this concept that sharing data would be a great strategy to working towards these problems together and the cyberinfrastructure has always been supported by the National Science Foundation. It is also viewed as the most essential element, if they were to lose everything, they still do not want to lose the cyberinfrastructure. So if you saw the cyberinfrastructure in the past, it has really come together in these last 5 years. It's a foundation, it's a clear asset that they're not willing to lose. So, the facilities can go on competing for other resources from other sources, but in terms of the cyberinfrastructure it's always been supported by NSF and there is not really an expectation that would be possible to change in the near future. Did that help answer your question?

Andrew Pollard: I think so. I was hoping you would have highlighted that in our community we all use cyberinfrastructure for our calculations, experimental data storage, etc. Even so, irrespective of who or what arm of a government agency, say, provides either the infrastructure or the money to purchase and support it, I think only NSF has funded the serving up of these data (the Hopkins example) and even there the turbulence community piggy-backed on the physicist, etc. So, more broadly speaking it is formulating how to create a strategic and operational framework whereby our turbulence community can come together in the same way that these other communities have. But we are 15 years behind some of these communities at this point. So, that just puts into perspective that the younger generation is going to have to try and take the lead on some of this: if you drive this, you will be the ultimate beneficiaries.

Mark Glauser: I would like to shift gears a little bit here and think about just not how we handle these large data sets but from the point of view of sharing and so on. But thinking about how we bring new analysis tools to these large data sets. I think of Steve's talk and I think of Bernd's talk, and I think as a community we need to really start to think about how we bring the modern tools that are coming out of computer science and computer engineering into our field. I think this is a really critical thing. These are new tools that can help us solve this, or at least make progress on this very very difficult problem. And so I just throw that out there, I think we really need to be looking at these tools. And you saw a wonderful talk from Steve, and he talked about these ideas of compressive sensing and how they are opening new pads for us and solving some issues we've had for a long time now, the sensor placement problem for flow control and so on that we've got new machinery that we can use. We still need the Navier–Stokes equations; we still need the other machinery that we have. But we need to bring these new tools into our field.

Michel Stanislas: I have two comments. The first one, as a civil servant: we need to at least try to keep the data of the experiment we are doing and to put them in a format that would allow other people to use them afterwards. And we are to put that into some form of database. We see the significant difficulty for university people because this requires money and man power. And also it requires the involvement of the scientist to give a correct description of the data because if you just have the data, but you don't have the correct description by the one who did generate it then it is sometimes very difficult to use them. My second comment is that, about the clever ideas, I have an example, you never know what someone else can do with your data because we are generating our spatially distributed, time resolved data. Sometime you design your experiment for one purpose but afterwards it can be used very well for something else. The example I could give is what Jean-Marc presented. The experiment was set up to build the full correlation tensor in order to design low order dynamical system representative and consistent with near-wall turbulence. And finally it was also used to look at the dissipation, so that idea came after the original experiment. We say "oh this data set, we could also use it for that." That way the two data sets describe our available, it's like when you have an idea the first thing you do go into the literature and see if something available, now we could go into data sets on the databases and this saves you a lot of money because then you don't need to set the experiment by yourself, you can just start working on these data.

Martin Oberlack: I have a question also to the people at Purdue and also maybe someone has used the machinery that Charles Meneveau set up. Does anybody know anyone who has used that?

So what my question is, from what I understand from looking at part of the discussion, I understand that most of the people with experiments are also the people who have DNS. The amount of data is so huge that there is no way to ship them. I mean over FTP or something of that sort. So the point is we have to keep the data they would produce, more or less. The question is now, there is so few software machinery we have, either from Charles Meneveau or from the people at Purdue, can be sort of say take this software machinery and can we transfer it to the people who have the data so that they can make use of the data, well also let's say using a sort of local machines otherwise it won't work. That is sort to say that my question may be first to you, Ivan, and also to the people at Purdue. Can you comment on that Ivan?

Ivan Marusic: I can.

Martin Oberlack: Is that doable.

Ivan Marusic: I've used the Johns Hopkins database.

Tom Hacker: I think the question you had was if you have a lot of data you are generating at a facility you need kind of process it first, at least to get winnowed down to some insights instead a lot of raw data. So what's the machinery to do that if you can't transport it. I think the answer is probably that you can certainly use local computing systems to do your first stage analysis, to winnow that data down and get some insights out of it and then from that, these reduced data sets can be transported to a central repository or to a central computing center for further insight, for further analysis or for saving the data. And I notice that the CMS project at LHC, they do a lot of filtering first at CERN before the data sets are moved to the tier one centers at the country level. And then at Purdue, jobs are run for the physics codes that request the data sets to be transferred offline between Fermi and Purdue for later processing. So there have been a lot of approaches developed to do that and I don't think you need to have either extreme of purely centralized computing or purely local computing. You can look at it as the tiered approach. You can do some local computing first to kind of winnow down the data and get some actually insights but also couple that with centralized computing that can look more broadly across the data.

Shirley Dyke: One more comment to add to that, is that if researchers have the data using this type of central repository, the researchers are trying to figure out which data sets do I need to download locally so I can do additional processing of those data sets, what we do at NEES is that the centralized repository is integrated with tools, so the tools can act upon that data, we can plot it, we can analyze it, and we can do whatever we wish to do basically. And then the researcher can decide if that data is something that they really want, then they download the data they want, and not everything at once. So the best procedure is to try to see if you really want it before you download it. And if possible do the analysis in a centralized repository that's possible with the tools available.

Ivan Marusic: And that's actually my experience with the John Hopkins site. It depends on the question you're asking. I have a student that is looking at developments in the stereo PIV, and different algorithms for PIV. Here, we only need a very small subset in physical space but with full time resolution. The JHU database works extremely well for that and avoids the major problem for us of having to deal with the full data set. I went to the other extreme 15 years ago with Bob Moser who sent me his entire channel flow data set. This was done by sending him a RAID system, which at that time cost about $10,000, and for his team to tediously copy over all the files, which took considerable time and effort, made worse by the later coding issues that needed to be resolved. So that's not an efficient way to do things.

Luciano Castillo: In the sense of Charles Meneveau, he would tell you that the database is a massive amount of work, just to handle that database. So this is a challenge that we have.

Fazle Hussain: Perhaps what I want to say is that is very obvious and well known, the most massive use of such database and resulting the massive number of papers is NIH, I'm not involved in this but all of you who are, you may want to look into this.

Julio Soria: Just a few comments. It is my experience that in fact bringing data from the our DNS of an APG-TBL that we did on the superMUC [https://www.lrz.de/services/compute/supermuc/] in Germany is possible, because they don't have storage. We transported the data via the Internet to Australia. And in some sense that's an impossible task, but it's not completely impossible. You just set it up the copying with "rsync" and it does it by itself. It just keeps on restarting, and we're talking terabytes, in fact tens of terabytes for one DNS. So it's not impossible, it's actually quite possible to get data around to different places. Comment number 2, is that if I do an experiment, I have a particular aim, a particular idea and as long as it's well documented and I think that the data out are OK, then the data may be useful for others with other aims. This means that colleagues do not have to go build a rig, develop instrumentation, and do the experiment again. If the experiment is well documented and everything is there, this is a very useful path. Doing experiments is a very time-consuming task, as is DNS, but with DNS there is another aspect that with time it will actually becoming more feasible to re-compute the DNS rather than store it, for example, say you want to do Re_tau = 180, you don't store that data anymore, you just compute it because you can do it on our laptop. So as time evolves, some of the simulation are not required to be stored because everybody can simulate the problem and I think that's part of our thinking: what you need to really do is store well-documented codes. If the codes are well documented, and the user knows how to use them properly, then you do the simulation and you can test your hypothesis on the fly post-processing as you do the simulation and get the data. So those are a couple of points nothing is steady, these things evolve. And you need to adapt to it as you go along.

Martin Oberlack: I have a quick question directed directly to you. How did you transfer this large amount of data, did you chop it?

Julio Soria: You just use "rsync-a" and copy.

Martin Oberlack: How long did it take for say a terabyte?

Julio Soria: Each restart file is 300 GB. and we can transport a couple of those overnight or per day. It's possible.

Jean-Paul Bonnet: I just want to stress one point, experiments are very important to give you the flow physics behind the information, the vorticity, the pressure, etc. The example of the water was a very obvious in that you can have facilities that you can share, and then when you want to store the data, the way you store it can be very different. I wanted to once again remind you that Joel Delville spent a lot of time, providing some very huge data bases and we have this European Flow Control consortium and I think that sharing of experiments data is critically important.

Andrew Pollard: I'd like to go back to the Purdue folks. They are involved in computing and it was mentioned earlier about data transfer speeds and how it continues to improve with each passing year. There are multiple light-paths available through optical fiber, which means that you can get very large data sets transferred through national backbones, but not necessarily through the "last mile" connected to your local computer. Can you speculate on where the bandwidth throughput is heading with time?

Shirley Dyke: I think this might be a question for Tom, and he just stepped out for a minute.

Andrew Pollard: I know, for example, in Canada, we just invested another $100 million to update our research-based fiber optic system across the country. So we know that these systems are being put in place. So if we start talking in terms of terabyte data sets, then what we are concerned about as a community is about whether we can share them in a reasonable time (and by reasonable time, I mean time measured in hours not days/weeks). We heard just a moment ago that if you wanted to do a low Reynolds number simulation, you can now do it on your own and you don't have to transfer data beyond your own lab. But as the complexity of the problems we want to tackle increase, then we know that the bandwidths are also going to increase. So the question is can you speculate where we might be in 5 years, for example, in terms of data transfer rates?

Shirley Dyke: That's a really good question, I'm sure that Tom could give you a really good answer. In terms of speculating on technology, he is very good at that. We'll ask him when he gets back.

Stephano Leonardi: I just want to confirm what we are all saying, I was downloading data from Texas (Moser's group), you can download a couple of terabyte per night. The problem is actually the space to store the data, the space in your cluster to process it, which means you have the processors to do the job. That means transfer speed if you are talking about 20 terabytes of data set, would take 1 week, 2 weeks or so?

Julio Soria: Just to add a quick one onto that, I don't know if some of you have downloaded data from the grid, but in fact some researchers have now set it up as a torrent. So you can use parallel paths and it's actually the quickest way to download the data, such as big huge DNS data. So there are many paths.

Michel Stanislas: There is one question in which I would like to raise, in relation to this database citing. We are all scientists, and we are for the moment we are evaluating essentially on our presentation. And I think we want to set a dynamic, and push people to make available and acknowledge them for providing the data. We should find a way to be rewarded for doing that. This comes into the recognition of the work people are providing to the community because otherwise it's much better to make a paper, put the data aside, do another experiment to make another paper. So this is a real question because I think all the technical problems will be solved, if everybody thinks that placing data somewhere and making it available and open, such a system is going to provide more recognition and more good evaluations.

Steve Brunton: One comment on that. If you look at some of the most highly cited publications in our field they are often cited because they provide data for other people to use, so that is a motivation in terms of citation. There should be a direct benefit to sharing good data.

Andrew Pollard: So Tom can you speak about an earlier question?

Phil Lavoie: So I just wanted to address your point actually. The robotics community has actually plans to do this, to publish a data set and that becomes a publication. We don't have that in fluid mechanics yet, and maybe that's a nice item for all the associated editors and editors of various journals. That maybe something like this should be thought about and potentially implemented. I don't know how that would work if it was a terabyte data set. But at least for some basic experimental data sets, if only for validation purposes; it might be a good way to start on that part. And from there we can start to use the data set, and then there is a paper to cite and so on.

Martin Oberlack: Maybe now we can get the answer from the Purdue people here. Can you're hear us?

Tom Hacker: I think the question is can you predict Internet speeds? Is there a predictable trend that you can use to design your experiments? Is that right?

The growth in network capabilities is actually fairly predictable over time. It used to be, I don't remember the exact numbers, there is a graph I can show you here that on the screen. In computing you can look at trends of memory size CPU capabilities, network speed, disc size, and it's a fairly predictable trend. So it is possible to predict these trends and then ask the question in 5 years from now, how fast will the network speeds be, how much storage power cost, how fast will the processor be? It is definitely predictable.

Unknown: So I just have a question Tom, maybe not a question but a comment is that I know it's coming from a lot of places in the USA from the funding agency, there is now a data management plan as part of proposals and so I think rather than us all responding to this individually, it's going to help if there are some collective ways to do this so we are not all trying to re-invent the wheel and doing things that we're actually really bad at. And I actually think we should turn around and ask the agencies that we get our funding from, how are you going to provide the infrastructure to actually do what you're asking us to do. Because these are kind of irritating things. I respect the idea but one of these unfunded things where we keep talking about it. So I think as a community we can be proactive and go back to some of the agencies and say hey we're willing to do this but we need some help. It might be good timing to say it's a lot of work in the next 5 years.

Bill George: I would add to these comments. My biggest problem is always paying the student to finish their thesis, much less having enough time to archive the data. To make things even worse, most universities I have been associated with can hardly wait to take away your email address and for sure not provide storage to archive data.

Julio Soria: We have to do that. And they (administration, etc.) have actually provided the facilities, we have actually had that for 2 or 3 years. We have an entire data management system and you can ask what we have, what we have is access to terabytes of original data and the aim is make the data available publicly.

Andrew Pollard: I just wanted to say there is a copy of the dedication volume of Tennekes and Lumley (presented earlier to Bill G.) floating around here somewhere and if you have not signed it, please do so at the restaurant tonight.

Martin Oberlack: Well I'm not actually an expert on big data, so I'm not sure why it was asked to chair this session, but I learned a lot. But I think there are a couple of nice things underway that we can transfer big data. Which is good, and it's good to be reminded about EUHIT and there is other things still going on.

Michel Stanislas: As we talk about EUHIT there is a work shop in May at our place (Lille, ed.), you can follow this link. This is going to generate big data, because it is a big experiment, which is done jointly with Australia and Germany. And if you want to come it's a 2-day workshop on the 18 and 19 (May, 2016, ed.) and you can see the experiment running.

Luciano Castillo: I just want to add that we don't need to duplicate effort, which is a major outcome here. Now, I want to thank everybody, thank you so much Bernd and Martin for leading this and thank you. Last Tuesday we had the discussion on the big questions in turbulence and perhaps I would like Julio and Fazle to maybe finish up this discussion. So maybe we could take another 20 or 30 min or so and maybe that way we will be ready to get our cocktails at 7:30.

Fazle Hussain: Hello, please sit down so we can finish quickly. Last time we were supposed to talk about big questions and big ideas and we ended up complaining about how difficult it is and funding issues. So today we will limit our conversation to big ideas, big concepts, and I would request that when you speak, do so within 20 s, you can have your turn again. I think George Batchelor said clear writing is a reflection of clear thinking, and sloppy writing is a reflection of sloppy thinking, or words to that effect. This comes from his article in JFM, "Preoccupations of a Journal Editor" (Volume 106, pp. 1–25, 1981, ed.). I would recommend you read it, it's quite a masterpiece. Now, of course, I think that sloppy speaking reflects sloppy thinking. So let's get to the point.

Martin Wosnik: In terms of big ideas that I see being asked in the next 10 years, is what happens to turbulence when things are unsteady or when the boundary conditions are still important. A lot of what we know is when we are very very far from boundaries and from an engineering perspective that is certainly not very useful. Fortunately there are some hints, the dynamical system that is turbulence has some dynamics that isn't necessarily "universal," like it's not going to change for every boundary condition. Finding out why that's the case and what other physics we can plug back into our models I think will be an interesting question that will come up in the next 10 years.

Martin Oberlack: Well it's probably to be expected what I am going to say, is that if you ask physicists what they think, they will tell you something about symmetries. And I think that is also true for turbulence.

Fazle Hussain: Paolo has a lot of things to say. Come on this is your chance.

Paolo Orlandi: Perhaps we have to direct our efforts on the energy containing eddies rather than on the energy cascade. The energy containing eddies are linked to the boundary conditions, by changing these we are acting directly on the big eddies. The energy cascade is the same for any flow and it is difficult to act on this process. If we wish to control the flow we have to act on the energy containing eddies. For instance, by rotating a pipe we are creating helical vortices and we can have drag reduction. The reason is that helical vortices are more ordered structures one way to reduce the drag. A similar phenomenon occurs in presence of riblets were the lateral motion of the near-wall structures is reduced. A further example is the effect of the stable stratification were pancake eddies are created which reduce the vertical mixing. To finish, rough walls increase the flow isotropy near the walls, the structures have a greater disorder and the drag increases. We have to direct the numerical and experimental tools to understand better the complex physics of vortex dynamics to find the strategies to act on the energy containing eddies to get useful results to improve ours everyday day life strongly affected by fluid dynamics.

Andrew Pollard: In terms of big questions, I think it's not just turbulence per se that we need to consider. I think it's where turbulence occurs in various other systems that are accompanied by other physic-chemical processes; combustion obviously comes to mind. Some of my research focuses on flows in the human body in a variety of different ways, whether it be fluid structure interaction, how can we mimic speech and so on and pre-surgery planning for obstructed airways with all the attendant fluid–structure interaction and possible impacts on speech patterns. Understanding the fundamentals of anisotropic, non-homogeneous, and unsteady elements of turbulence all play significant roles in bio-fluid flow that can and will have tremendous societal impacts.

Jean-Paul Bonnet: I'm not sure if we can disconnect the turbulence from cascade. I think from an experimental point of view it is difficult to promote, for example, small scale, or dissipative scale without altering the rest of the flow, this is difficult. But I think that from CFD this can be possible if we add some flow forcing in such a way that you can interact with the cascade to give some new ideas.

Bernd Noack: Let me emphasize Andrew's point. We are an engineering community and write our proposals to make a contribution to greener traffic, greener energy, and so on. And we can make a big contribution. I would say a big question is how can we reduce the cost of traffic, how can we green the energy, etc. And we can make a big contribution to our own future by focusing on these types of engineering benefits of turbulence.

Bill George: I certainly wouldn't disagree with that. One of John Lumley's strengths was that he worked simultaneously with applications. Many of his fundamental contributions really came out of applications and problems that he encountered with the Navy in particular. I think that has been true in my experience as well. My lab has always been devoted in some way to real problems.

Let me just make another comment now. I think it is easy to be trapped in the language and paradigms of the past and I would suggest this whole cascade idea is one of them. If there is anything that is coming out of the new theoretical work it's that there really isn't a cascade. All the scales are affected simultaneously. Now this doesn't mean that vortices don't break down into smaller and smaller vortices. It doesn't have to be a spectral cascade in the traditional way that we thought. So it's really important to not get trapped in this bad thinking and old language.

Fazle Hussain: I want to say something; let me repeat what I said Tuesday. I think I was a major contributor to lifting turbulence from statistical physics to structural turbulence, except we fail because there is no conceptual or theoretical foundation how can structure ideas be used? So this is one missing link. Are you personally interested in reconnection as being the inherent component of cascade and noise production. These are some of the things I would like to see studied. One example in geophysical flow. A recent discovery of life lengthening protein found in jellyfish, before which jellyfish was a nuisance. But it turns out it is the primary contributor to balancing temperature and mixing in the ocean. So there are many interesting things we may not have thought of before and I think they can be found that they have appeal.

Julio Soria: I have to say something. One thing that I actually found quite interesting, and that is probably going into a little bit towards what Bernd Noack was saying. He was putting it that the concept of using these genetic algorithms and machine learning flow control may only be the way to look at it. It may also be a way to effectively search for other possible mechanisms to explain turbulence based on the data we have—mechanisms that we haven't even thought about. Because these algorithms go outside the box of the way we necessarily think. It might actually be quite interesting to let them loose to develop ideas (mechanisms) rather than the current use that has been talked about, namely how to control the flow, coming up with optimal control strategies. You could ask the question what's the optimum physical mechanism that explains the data. Because we all have preconceived ideas, from our background from what we know, and sometimes that constrains the way we see the world. Sometimes we got to be outside that part that we view that we have in our brain and this might be away of in fact of removing the shackles and look at in from a different way. This might just be the method that in fact allows us to go around from outside the box and look for other mechanisms. After that I think we can all go have some cocktails!!

MIX
Papier aus verantwortungsvollen Quellen
Paper from responsible sources
FSC® C105338

If you have any concerns about our products,
you can contact us on
ProductSafety@springernature.com

In case Publisher is established outside the EU,
the EU authorized representative is:
Springer Nature Customer Service Center GmbH
Europaplatz 3, 69115 Heidelberg, Germany

Printed by Libri Plureos GmbH
in Hamburg, Germany